3rd International Conference on Terahertz and Microwave Radiation: Generation, Detection and Applications (TERA-2018)

EPJ Web of Conferences Volume 195 (2018)

Nizhny Novgorod, Russia
22 – 25 October 2018

Editor:

A.A. Silaev

ISBN: 978-1-5108-7560-9

Printed from e-media with permission by:

Curran Associates, Inc.
57 Morehouse Lane
Red Hook, NY 12571

CURRAN ASSOCIATES INC. proceedings.com

Some format issues inherent in the e-media version may also appear in this print version.

For additional information, please contact EDP Sciences – Web of Conferences at the address below.

EDP Sciences – Web of Conferences
17, Avenue du Hoggar
Parc d'Activité de Courtabœuf
BP 112
F-91944 Les Ulis Cedex A
France

Phone: +33 (0) 1 69 18 75 75
Fax: +33 (0) 1 69 28 84 91

contact-edps@webofconferences.org

Additional copies of this publication are available from:

Curran Associates, Inc.
57 Morehouse Lane
Red Hook, NY 12571 USA
Phone: 845-758-0400
Fax: 845-758-2633
Email: curran@proceedings.com
Web: www.proceedings.com

TABLE OF CONTENTS

Conference Chairman – **Mikhail Glyavin,** Institute of Applied Physics RAS
Conference Co-chair – **Boris Knyazev,** Budker Institute of Nuclear Physics SB RAS
Conference Co-chair – **Alexander Shkurinov,** Lomonosov Moscow State University

Organized by
Institute of Applied Physics, Russian Academy of Sciences. http://iapras.ru/
Budker Institute of Nuclear Physics SB RAS, www.inp.nsk.su/
M.V. Lomonosov Moscow State University, https://www.msu.ru/
GYCOM Ltd. http://www.gycom.ru/

Sponsored by
Russian Foundation for Basic Research
(grants №18-02-20141, №18-02-20142, №18-02-20143)
http://www.rfbr.ru/rffi/ru/
TYDEX www.tydex.com
Keysight Technologies, Inc. www.keysight.com
OES Specpostavka www.oessp.ru
Rohde&Schwarz www.rohde-schwarz.com
TeraSense www.terasense.com

International Advisory Committee

S. Alberti	Ecole Polytechnique Fédérale de Lausanne, Switzerland
Michael I. Bakunov	Lobachevsky State University of Nizhny Novgorod,Russia
J.-L. Coutaz	University of Savoie, France
G.G. Denisov	Institute of Applied Physics RAS, Nizhny Novgorod, Russia
Gian Piero Gallerano	ENEA, Italy
T. Idehara	University of Fukui, Japan
Young Uk Jeong	Korea Atomic Energy Research Institute, Daejeon, South Korea
Kodo Kawase	Nagoya University, Japan
R.A. Khabibullin	Institute of Ultra High Frequency Semiconductor Electronics RAS, Moscow, Russia
D.R. Khokhlov	Lomonosov Moscow State University, Moscow, Russia
S.A. Kozlov	ITMO University, Saint-Petersburg, Russia
G.N. Kulipanov	Budker Institute of Nuclear Physics, Novosibirsk, Russia
Jean Leotin	High Magnetic Field National Laboratory of Toulouse, France
A.G. Litvak	Institute of Applied Physics RAS, Nizhny Novgorod, Russia
Paolo De Natale	Istituto Nazionale di Ottica, Italy
Hiroyuki Nojiri	Institute for Materials Research, Tohoku University, Japan
Gun-Sik Park	Seoul National University, Seoul, South Korea
N.A. Vinokurov	Budker Institute of Nuclear Physics, Novosibirsk, Russia
V.L. Vaks	Institute for Physics of Microstructures RAS, Nizhny Novgorod, Russia
H. Schneider	Helmholtz Zentrum Dresden-Rossendorf, Dresden, Germany

edp sciences

Web of Conferences

Statement of Peer review

In submitting conference proceedings to *Web of Conferences*, the editors of the proceedings certify to the Publisher that

1. They adhere to its **Policy on Publishing Integrity** in order to safeguard good scientific practice in publishing.
2. All articles have been subjected to peer review administered by the proceedings editors.
3. Reviews have been conducted by expert referees, who have been requested to provide unbiased and constructive comments aimed, whenever possible, at improving the work.
4. Proceedings editors have taken all reasonable steps to ensure the quality of the materials they publish and their decision to accept or reject a paper for publication has been based only on the merits of the work and the relevance to the journal.

Title, date and place of the conference

3rd International Conference "Terahertz and Microwave Radiation: Generation, Detection and Applications" (TERA – 2018), October 22-25, 2018
Nizhny Novgorod, Russia

Proceedings editor(s):

Dr. A.A. Silaev

Date and editor's signature

October 25, 2018

17, avenue du Hoggar - PA de Courtabœuf – BP 112 - 91944 Les Ulis Cedex A (France) Tel. : 33 (0)1 69 18 75 75 - Fax : 33(0)1 69 07 45 17 – www.edpsciences.org

EPJ Web of Conferences **195**, 00001 (2018) https://doi.org/10.1051/epjconf/201819500001
TERA-2018

From millimeter to microns – IAP RAS powerful sources for various applications

M. Yu. Glyavin, G. G. Denisov, E.A. Khazanov

Institute of Applied Physics RAS, Nizhny Novgorod, Russia, glyavin@appl.sci-nnov.ru

The last decade has contributed to the rapid progress in the development of THz sources, in particular gyrotrons. Although in comparison with the classical microwave tubes the gyrotrons are characterized by greater volume and weight due to the presence of bulky parts (such as superconducting magnets and massive collectors where the energy of the spent electron beam is dissipated) they are much more compact and can easily be embedded in a sophisticated laboratory equipment (e.g. spectrometers, technological systems, etc.) than other devices with a comparable value of Pf^2 such as free-electron lasers (FELs) and radiation sources based on electron accelerators. All these advantageous features have opened the road to many novel and prospective applications of gyrotrons as radiation sources in a great number of high power THz technologies, advanced spectroscopic techniques, plasma science and materials processing, fusion research as well as in many other scientific and technological fields.

The gyrotron developed at IAP RAS jointly with GYCOM Ltd (Nizhny Novgorod) provides CW radiation with a power from 1W up to 300 kW in the frequency range 0.26-0.52 THz [1,2] and up to 1 MW at frequencies 0.14-0.17 THz [3]. It is important, that power up to 10W needed for spectroscopy applications can be obtained at low voltage about 1.5 kV or low beam current – tens mA.

The experiment on frequency stabilization was carried out using a continuous-wave (CW) gyrotron operating at a frequency of 263 GHz and an output power of up to 1 kW with an electron beam formed by a triode-type magnetron injection gun. The phase lock loop control of anode voltage has been used and the width of the frequency spectrum was decreased to 1 Hz [4], which corresponds $\Delta f\ /\ f\ =\ 3*10^{-12}$ with measurement time of a few seconds. The long-term stability was defined by reference clock ($\delta f / f \sim 10^{-9}$ for quartz clock and up to $\delta f / f \sim 10^{-12}$ for rubidium clock).

It is interesting, that the same scheme can be used for quick power modulation and data transmitting. Experiment was made based on 1kW/0.26THz tube and linear dependence between control (anode) voltage and output frequency was observed. As a result, high quality transmission of sound and pseudo random bit sequence (with a speed up to 1.5Mbit/s) has been obtained [3,5].

The prototype of CW 250GHz/200 kW gyrotron has been successfully developed. The power up to 300 kW in 40 microsecond pulses with efficiency more than 30% was demonstrated [6] with a full accordance with theoretical estimation. The gas discharge experiments, initiated by mentioned microwaves, demonstrate unique plasma parameters. The installation diamond window, needed for CW tests, is in progress.

The gyrotrons with pulsed magnetic fields operating at the fundamental harmonic produce up to 100 kW power level at 0.7 THz [7] and 1 kW at 1 THz frequency [8]. The future rising of operating frequency (especially excitation of high harmonics due to limit of reasonable magnetic field) needs improved methods of mode selection, which can be divided to electron-optical and electro-dynamical ones. The most successful realization of electron-optical methods is a gyrotron with axis-encircling electron beam (see, for example [9]). The paper [9] presents the pulsed third harmonic 1THz tube with a power level several hundred Watts. The CW version of LOG based on the cryomagnet is under tests at the moment.

The second harmonic CW gyrotron with improved mode selection based on double electron beam has been tested [10] jointly with FIR UF (Fukui, Japan). The wide step tuning of frequency by excitation of various modes was demonstrated in the range 0.4-0.75 THz. The stable operation at the frequency 0.76 THz on the second harmonic with power level about few Watts has been obtained, which useful for modern NMR/DNP spectroscopy applications.

The feasibility of a high-power sub-THz gyrotron with smooth wideband frequency tuning suitable for direct measurement of the positronium hyperfine structure is demonstrated numerically using both averaged equations and PIC-code simulations. Analytical estimates show that the frequency tuning through the excitation of axial modes can be achieved in a gyrotron with a short cavity driven by an electron beam with a high current. Simulations [11] and preliminary experiments demonstrates possibility of wide (about 10%) fine frequency tuning. An output power of 0.5-1 kW can be obtained at a frequency of about 0.2 THz within a 10-GHz band, which are the parameters needed for testing of quantum electrodynamics predictions through the spectroscopy of positronium.

The pulsed gyrotrons have been used successfully for initiation of localized gas discharges. Such a plasma is promising for development of both a point source of multi-charged ions and a source of high-energy ultraviolet (extreme ultraviolet EUV or XUV) [12,13]. The gas discharge has been successfully obtained also with 0.26 THz/CW IAP RAS gyrotron mentioned above.

The gyrotron looks as a most promising source for the high-resolution molecular spectroscopy in a gas mixture. A significant improvement of spectrum quality due to power growth in contrast with traditional BWO has allowed observing the theoretically predicted earlier transitions in SO_2 molecule [14]. Sensitivity of a radio-acoustic detector was increased about three orders due to high power. It is important to mention, that during last experimental company we use not only power on fundamental harmonic, but also gyrotron radiation produced due to nonlinear effects at high harmonic simultaneously with the fundamental. Though the generated power at the second and third harmonics is two orders and three orders, respectively, lower than at the first one, such a power level (1-10 W) is significantly higher than in other sources. In this experiment, the SO_2 spectrum line was measured at the frequency range 0.26-0.8 THz.

One more interesting feature of powerful sub-THz radiation is high speed production of pure nanopowders by material evaporation and condensation. The use of a subterahertz gyrotron setup with output frequency of 263 GHz and a nominal power of 1 kW as a radiation source to obtain nanoscale particles of metal oxides by the evaporation-condensation technique is demonstrated. An increase in the substance evaporation rate by more than three times compared with similar experiments using a technological gyrotron setup with a frequency of 24 GHz and a nominal power of 7 kW (the power density on the sample surface was about 13 kW/cm^2) as a radiation source was demonstrated. An increase in the radiation frequency leads to the improvement of the heating efficiency due to an increase of the absorption coefficient and provides a greater power density due to a better focusing of the radiation [15].

Despite the requirement for strong operating magnetic fields, mode competition, high ohmic losses, etc., the THz frequency range has been already achieved both by the pulse and CW gyrodevices. A number of applications rapidly increase with the development of radiation sources.

The basic distinguishing feature of the IAP RAS approach to creating powerful femtosecond laser complexes is the use of optical parametric chirped pulse amplification (OPCPA) instead of conventional laser amplification. The femtosecond laser complex PEARL created at IAP RAS is based on these principles. This complex generate pulses with power up to 0.56 PW. The experience gained in creating PEARL and its principles are currently used for constructing at IAP RAS of the laser complex PEARL-10 with a power over 5 PW. The concept of an exawatt laser system - the XCELS project [16] - has been developed. Appearance of such sources opens up absolutely new horizons for basic research and unique applications. Generation of high-energy, highly collimated bunches of charged particles and creation of secondary sources of powerful X-ray and gamma radiation during the interaction of superintense laser radiation with matter are the most topical problems of modern science. Present-day high-power laser complexes can lay the basis of a new compact technology of particle acceleration and generation of coherent X-ray and gamma radiation. Such technologies have great promise for use in diverse fields of science and engineering. In medicine, high-energy bunches of protons and ions are used for therapy of a variety of cancers. Electron bunches can be used for injection into free-electron lasers.

Acknowledgements. The gyrotron development program supported by the RSF grant 14-12-00887.

References

1. M.Glyavin et al. High-power terahertz sources for spectroscopy and material diagnostics // *Physics-Uspekhi*, 59, 6, 595-604 (2016)
2. M.Yu. Glyavin et al. Experimental tests of 263 GHz gyrotron for spectroscopy applications and diagnostic of various media // *Rev. Sci. Instr.*, 86(5), 054705 (2015)
3. G.Denisov. New development of Megawatt power gyrotrons in Russia // *20th Joint Workshop on Electron Cyclotron Emission and ECRH*, Greifswald, Germany. May 14-17, 2018.
4. A.Fokin et al. High power sub-terahertz microwave source with record frequency stability up to 1 Hz // *Scientific Reports* 8, 4317 (2018)
5. A.Fokin, A.Tsvetkov, private communication
6. G.G.Denisov et al. First experimental tests of powerful 250GHz gyrotron for future fusion research and CTS diagnostics // *Rev.Sci.Instr.* 89(8), 084702 (2018). DOI:10.1063/1.5040242
7. V.L. Bratman et al. Gyrotron Development for High Power THz Technologies at IAP RAS // *J Infrared Millimeter and Terahertz Waves,* 33(7), 715-723 (2012). DOI: 10.1007/s10762-012-9898-6
8. M.Yu.Glyavin et al. Experimental Study of the Pulsed Terahertz Gyrotron with Record-Breaking Power and Efficiency Parameters // *Radiophysics and Quantum Electronics*, 56, 8-9, 497-507 (2014)
9. V.L.Bratman et al. Large-Orbit Gyrotron Operation in the Terahertz Frequency Range // *Phys. Rev. Lett.* 102, 245101 (2009).
10. T.Idehara et al. A Novel THz-Band Double-Beam Gyrotron for High-Field DNP-NMR Spectroscopy // *Rev. Sci. Instr.*, 88, 094708 (2017).
11. A.E.Fedotov et al. Frequency Tunable sub-THz Gyrotron for Direct Measurements of Positronium Hyperfine Structure // *IEEE Trans. On Electr. Devices*, 46, 7, 2465-2469 (2018).
12. M.Yu.Glyavin et al. A point-like source of extreme ultraviolet radiation based on a discharge in a non-uniform gas flow, sustained by powerful gyrotron radiation of terahertz frequency band // *Applied Physics Letters* 105, 174101 (2014).
13. A.Sidorov et al. Gas discharge powered by the focused beam of the high-intensive electromagnetic waves of the terahertz frequency band // *Journal of Physics D: Applied Physics* (2018).
14. M.A.Koshelev et al. Molecular gas spectroscopy using radioacoustic detection and high-power coherent subterahertz radiation sources // *Journal of Molecular Spectroscopy*, 331, 9–16 (2017).
15. A.V.Vodopyanov et al. Application of the 263GHz/1kW gyrotron setup to produce a metal oxide nanopowder by the evaporation-condensation technique // *Vacuum*, 145, 340-346, (2017).
16. www.iapras.ru/english/science/las_phys/gen_lasf4.html (Online available 01/10/2018)

EPJ Web of Conferences **195**, 00002 (2018) https://doi.org/10.1051/epjconf/201819500002
TERA-2018

Recent experiments at NovoFEL user stations

B. A. Knyazev[1,2], I. A. Azarov[3], E. N. Chesnokov[4], Yu. Yu. Choporova[1,2], V. V. Gerasimov[1,2], Ya. I. Gorbachev[1], Ya. V. Getmanov[1,2], B. G. Goldenberg[1], O. E. Kameshkov[1,2], P. V. Koshlyakov[4], I. A. Kotelnikov[1,2], A. S. Kozlov[4], V. V. Kubarev[1,2], G. N. Kulipanov[1], S. B. Malyshkin[4], A. K. Nikitin[2,5], P. A. Nikitin[6], N. D. Osintseva[1], V. S. Pavelyev[1,7], S. E. Peltek[8], A. K. Petrov[4], V. M. Popik[1], T. V. Salikova[1], M. A. Scheglov[1], S. S. Seredniakov[1], V. N. Shastin[9], O. A. Shevchenko[1], V. A. Shvets[1,2], D. A. Skorokhod[1], A. N. Skrinsky[1], S. L. Veber[10], N. A. Vinokurov[1,2], V. B. Voloshinov[6], R. Kh. Zhukavin[9]

[1]Budker Institute of Nuclear Physics, Siberian Branch of Russian Academy of Sciences, Novosibirsk, Russia, b.a.knyazev@inp.nsk.su
[2]Novosibirsk State University, Novosibirsk, Russia
[3]Rzhanov Institute of Semiconductor Physics, Siberian Branch of Russian Academy of Sciences, Novosibirsk, Russia
[4]Voevodsky Institute of Chemical Kinetics and Combustion, Siberian Branch of Russian Academy of Sciences, Novosibirsk, Russia
[5]Scientific and Technological Center of Unique Instrumentation, Russian Academy of Sciences, Moscow, Russia
[6]Lomonosov Moscow State University, Moscow, Russia
[7]Samara State University, Samara, Russia
[8]Institute of Cytology and Genetics, Siberian Branch of Russian Academy of Sciences, Novosibirsk, Russia
[9]Institute for Physics of Microstructures, Russian Academy of Sciences, Nizhny Novgorod, Russia
[10]International Tomography Center, Siberian Branch of Russian Academy of Sciences, Novosibirsk, Russia

The Novosibirsk free-electron laser (NovoFEL) is a source of monochromatic frequency-tunable radiation [1]. This facility belongs to the Siberian Synchrotron and Terahertz Radiation Center, which is open for Russian and foreign users. The NovoFEL consists of three laser oscillators, the supposed generation regions of which cover three regions of the spectrum, shown in Fig. 1 (a) with light rectangles. The spectral ranges of generation currently available to users are marked with darker rectangles. The first laser system of the facility, generating terahertz radiation in the range of 90 – 240 μm, was commissioned in 2003, and most studies performed by users to date relates to this range. Experiments using radiation of 40-50 μm have been started recently. The studies that had been performed at the first six workstations by 2015 are described in the review [2]. Since then, the facility was significantly upgraded, and now 11 workstations, the location of which is shown in Fig. 2, are in operation. Several more stations are under construction. The directions of research conducted at the stations are understandable from their names, but are not limited by them.

The laser beams at the inputs to the user stations are Gaussian beams (Fig. 1(b)). The radiation is an infinite sequence of 30-100 ps pulses with a standard repetition rate of 5.6 MHz and average power of up to 100-200 W. Regimes in which the radiation is coherent are described in [3]. Such parameters enabled the development of several superfast and high-resolution techniques of molecular spectroscopy [4,5], and the high pulse power allowed ignition of a continuous optical discharge in gases at the atmospheric pressure [6].

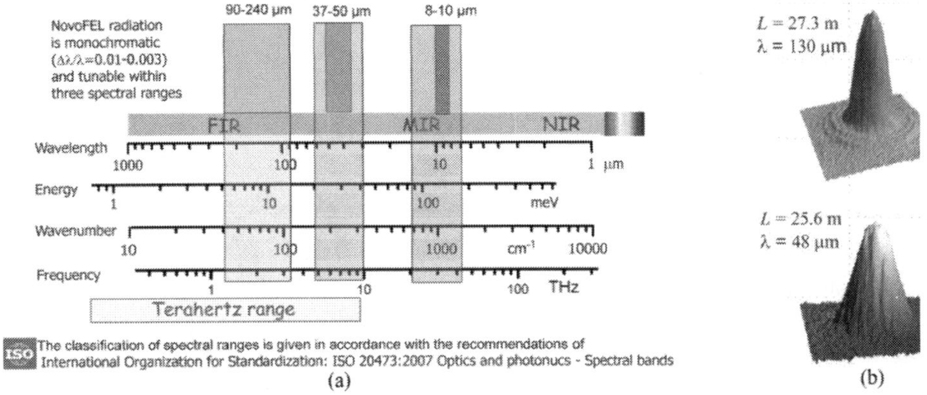

Fig. 1. (a) Generation ranges of the Novosibirsk free electron laser; (b) beam shapes at the inputs to workstations (*L* is the distance from the laser oscillator).

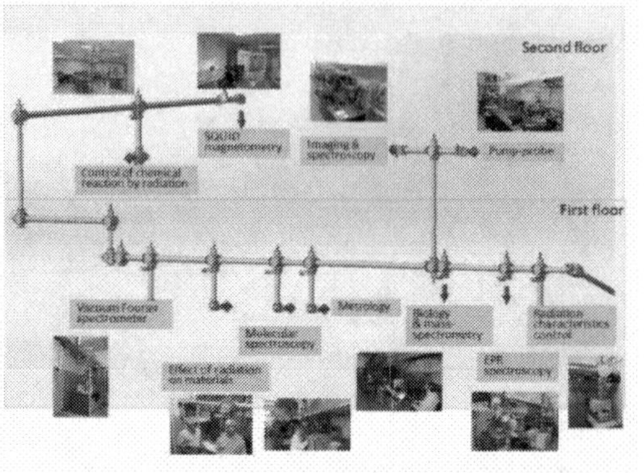

Fig. 2. Beamline (filled with dry nitrogen) and workstations at the Novosibirsk free electron laser facility.

Many applications require transformation of a Gaussian laser beam into beams of different mode structure or concentration of radiation in a predetermined volume or on an area. A number of diffractive optical elements made of diamond or high-resistivity silicon have been designed and fabricated for this purpose [7-8]. In particular, using diffractive optics made it possible to transform the laser radiation into Bessel beams with an orbital angular momentum ("vortex beams") [9-10]. The latter were converted into arrays of vortex beams via their diffraction on a 2D amplitude array of circular openings [11]. Acousto-optical deflection of a terahertz vortex beam was demonstrated in [12], and production of nanosized metal particles by acoustic waves generated in liquids by the NovoFEL pulsed-intermittent radiation was described in [13].

Classical holography with a monochromatic terahertz radiation source was first demonstrated in [14]. The NovoFEL radiation enabled achievement of record resolution for the terahertz range in the internal reflection ellipsometry [15]. Studies of terahertz surface plasmon polaritons on metal-dielectric-air interfaces revealed their particularities in comparison with the visible range plasmons [18-20].

The effect of THz radiation on biological objects, from cells to organisms, is under investigation at the biological station of NovoFEL (see, e. g., [21]). Two new workstations have been commissioned at the facility. A one-color pump-probe setup was applied to research on the relaxation time in semiconductors with shallow donors at cryogenic temperatures [22]. The electron paramagnetic resonance station [23] enables exploration of the influence of high-power THz or mid-IR radiation on spin systems.

The study was supported by the Russian Science Foundation (grant 14-50-00080). The experiments were carried out at the collective research center supported by the Ministry of Education and Science of the Russian Federation (project RFMEFI62117X0012).

References

1. *Shevchenko O. A. et al.* Novosibirsk free electron laser facility // These Proceedings, 2018.

2. *Kulipanov G. N et al.* Novosibirsk free electron laser as a user facility // IEEE Trans. on THz Sci. Techn. 2015. V. 5, No. 5. P. 798-809.

3. *Kubarev, V. V.* Instabilities, coherency, and spectra of the NovoFEL radiation // EPJ Web of Conf. 2017. V. 149. 05007.

4. *Chesnokov, E. N., et al.* The influence of magnetic field on the echo-like free induction decay in NO_2 // Chem. Phys. Lett. 2016. V. 662. P. 62–66.

5. *Chesnokov, et al.* Heterodyne Method of Detection of Molecular Gas in the Terahertz Region Using the Beats Between Free Induction Decay Signals // IEEE Trans. on THz Sci. Techn. 2017. V. 7, No. 2. P. 144–150.

6. *Kubarev, V. V. et al.* Threshold conditions for terahertz laser discharge in atmospheric gases // J. Infrared, Millimeter, and THz Waves. 2017. V 38, No. 6. P. 787-798.

7. *Pavelyev, V. S. et al.* Fabrication of high-effective silicon diffractive optics for the terahertz range by femtosecond laser ablation // Phys. Procedia. 2016. V. 84, P. 177-174.

8. *Agafonov, A.N. et al.* Focusing of Novosibirsk free electron laser (NovoFEL) radiation into paraxial segment // J. Modern Opt. 2016. V. 63. No. 11. P. 1051-1054.

9. *Choporova, Y. Y. et al.*, High-power Bessel beams with orbital angular momentum in the terahertz range // Phys. Rev. A. 2017. Vol. 96. No. 2. 023846.

10. *Knyazev, B., Serbo, V.*, Beams of photons with a nonzero projection of the orbital angular momentum - new results // Phys. Uspekhi, 2018. V. 61. No. 5. P. 449-479.

11. *Knyazev B.A. et al.* Quasi-Talbot effect with vortex beams and formation of vortex beamlet arrays // Opt. Expr. 2018. V. 26. No. 11. 14174.

12. *Nikitin, P. A. et al.* Deflection of terahertz vortex beam in nonpolar liquids by means of acousto-optics // Phys. Procedia (2016)

13. *Kozlov, A.S. et al.* Formation of Nanosized Metal Hydrosols under the Influence of Novosibirsk Terahertz free Electron Laser Radiation // Phys. Procedia. 2016. V. 84. P. 131–134.

14. *Choporova, Yu. Yu. et al.* Classical Holography in the Terahertz Range: Recording and Reconstruction Techniques // IEEE Trans. on THz Sci. Techn. 2015. V. 5, No. 5. P. 836–844.

15. *Azarov, I. A. et al.* Measurement of optical constants of liquids in the terahertz range using internal reflection ellipsometry // J. Infrared Millimeter THz Waves, 2018, submitted.

18. *Gerasimov, V. V. et al.* Growth of terahertz surface plasmon propagation length due to thin-layer dielectric coating // JOSA B. 2016. V. 33, No. 11. P. 2196-2203.

19. *Gerasimov, V. V. et al.* Wave-vector spectrum of monochromatic terahertz surface plasmon polaritons on real surfaces // Phys. Procedia. 2016. .V. 84. P. 157-164.

20. *Knyazev, B. A. et al.* Generation of terahertz surface plasmon polaritons using nondiffractive Bessel beams with orbital angular momentum // Phys. Rev. Lett. 2015. V. 115. No. 16. 163901.

21. *Goryachkovskaya, T. N. et al.* The impact of terahertz radiation on an extremophilic archaean Halorubrum saccharovorum proteome // Vavilov Journal of Genetics and Breeding. 2016. V. 20. No. 6. P. 869-875.

22. *R. Kh. Zhukavin et al.* Low-temperature intraceter relaxation times of shallow donors in germanium // JETP Lett. 2017. V. 106, No. 9, P. 571-575.

23. *Veber, S. L. et al.* X-band EPR setup with THz light excitation of Novosibirsk free electron laser: goals, means, useful extras // J. Magn. Res. 2018. V. 288. P. 11-22.

Biomedical applications of microwave radiation: innovative approaches

Andrew Martusevich[1,3], A. V. Kostrov[2]

[1]Privolzhsky Research Medical University, Nizhny Novgorod, Russia, cryst-mart@yandex.ru
[2]Institute of Applied Physics, Nizhny Novgorod, Russia
[3]Kirov State Medical University, Kirov, Russia

Now it is widely known postulate that all the major discoveries in recent years occur at the junction of different sciences. In this regard, one of the most important innovative areas uniting physics, biology and medicine is the use of microwave technologies. In this regard, the purpose of this work was to clarify the practical possibilities of diagnostic and therapeutic microwave technologies.

Microwave technology in medical imaging

Medical visualization is currently one of the most dynamically developing areas of medical science. Analysis of the literature on the methodology of microwave imaging, implemented in Biomedicine, allows us to identify two main sensing techniques that differ in the estimated parameters: microwave thermometry (passive) and near-field resonance tomography.

In the framework of microwave thermometry (radiometry), the volume dynamics of the subsurface temperature distribution is recorded by the trapping device. At the same time, it is known that in a number of pathological conditions (for example, in oncomammology) there is a significant local hyperthermia, which will be detected by a microwave sensor.

On the contrary, in recent decades actively developed by resonant near-field microwave imaging, allowing you to visualize the structure of biological tissues on the basis of spatial distribution their electrodynamic characteristics - dielectric constant (ε) and conductivity (σ) [Fig. 1], characterizing the type and structural features of the biological object. The possibility of using this type of tomography for diagnostic purposes is due to the fact that the formation of pathological changes in biological tissues is a distinct change in their electrodynamic characteristics.

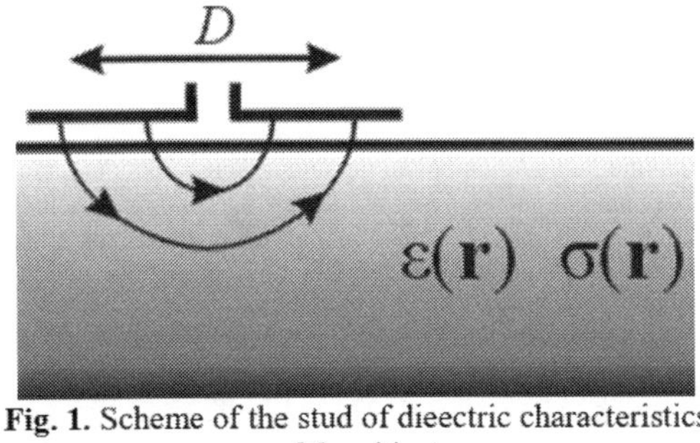

Fig. 1. Scheme of the stud of dieectric characteristics of the object

Near-field resonance microwave diagnostics is based on the measurement of electrodynamic characteristics of biological tissues. Work in this direction with biomedical positions are relatively recent, but the physical principles underlying near-field microwave tomography, studied much more fully. Thus, the fundamental basis of the method is the measurement of impedances located on the surface of the bio-object of a system of electrically small antennas with different scales of localization of the probing electric field. The spatial distribution of dielectric permittivity and electrical conductivity of tissues is restored by the values of impedances (active and reactive resistance) of antennas on the basis of a special mathematical apparatus.

Measuring the impedance of antennas is implemented in a special resonant sensor. The sensor is a microwave resonator in the form of a transmission line segment (coaxial, two-wire or strip) [1]. At one end of this segment there is a probing near-field antenna (measuring capacitance), at the opposite end there is a magnetic frame. Excitation of the resonator and reception of its response is carried out by means of magnetic coupling loops located near the magnetic frame of the resonator. All elements of the resonance system, except for the measuring capacitance, are located in a metal cylindrical body. Own resonant frequency of the sensor, as a rule, is in the range of 600-800 MHz, the characteristic quality factor – 150.

The scheme illustrating deep sounding of biological tissue within the near-field microwave tomography is presented in Fig. 2. Near-field antenna is located on the surface of the object under study, its quasi-static electric field penetrates into the medium to a depth determined by the design of the antenna and the size of its aperture (D). The depth of penetration of the electric field into the medium will increase with increasing aperture D.

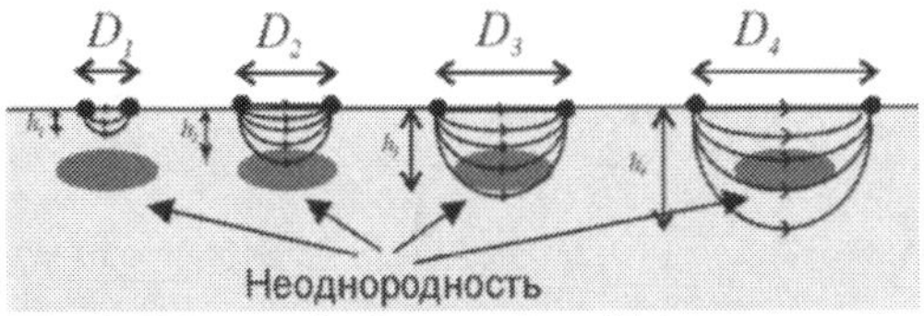

Fig. 2. Scheme of a deep sensing of inhomogeneous media by the method of resonance near-field microwave tomography

The study of heterogeneous biological media is carried out as follows. First, the measurements are carried out by the sensor with the lowest D_1 value and, accordingly, with the lowest sounding depth h_1, the measurement results reflect the integral properties of the medium in the surface layer. For sensor with sensing h_2 depth I integral properties of the medium in the near-surface layer of greater thickness, etc. Knowing the responses of measuring systems with different sensing depths, it is possible to restore the deep structure of the inhomogeneous medium.

Have presently developed near-field diagnostic systems key value "the size of the aperture of the probe (D) / wavelength of the probing microwave signal (λ)" comes to the level of 10^{-5}-10^{-6} that in comparison with the wave techniques gives the opportunity to examine the condition and structural features of the tissue on their electro-dynamic properties with subwavelength spatial resolution (much smaller than the wavelength λ).

Thus, near-field resonance microwave sensing, thanks to its physical principles, makes it possible to vary the depth and surface area of the study of biological objects, using electrodynamic characteristics of the medium - its dielectric permeability and conductivity – as diagnostic criteria.

Microwave technologies for treatment

Plasma medicine is one of the most modern synthetic scientific directions, born at the junction of plasma and Biomedicine physics, dealing with fundamental and applied issues of interaction between plasma and living matter. In this case, the greatest attention of researchers attracts the field of plasma medicine associated with the disclosure of biological and sanogenetic effects of cold plasma. Currently, the cold plasma is usually understood as ionized gas of different composition, cooled to a temperature comparable to the physiological (30-40 C). In recent decades for this factor a number of effects is potentially useful for medical purposes have been demonstrated experimentally and clinically. Among them, the most well studied and described antibacterial activity associated with the direct damaging effect of cold plasma on the cell wall of microorganisms. On the other hand, with a more thorough analysis of even this aspect of the problem, it reveals numerous poorly studied issues, including the effectiveness of the action, the absence of damage to the own tissues of the macroorganism, etc.

Special interest, in our opinion, have a bioregulatory properties of cold plasma in relation to functional and metabolic parameters of living systems. Thus, the pro-regenerative activity of the factor [6, 14], its ability to inhibit tumor growth [4, 10], etc. that was demonstrated earlier. Undoubtedly, the given empirical facts require multistage verification, but the very presence of such data testifies to the presence of secondary, indirect biological effects in cold plasma, which makes it possible to consider it as a potential bioregulator.

In our previous studies, is performing both in vitro and in vivo, the existence of secondary bioregulatory properties in cold plasma was shown and verified. In particular, even a short-term treatment (1-3 min.) of blood samples by the flow of helium cold plasma led to the formation of the biosystem response (by metabolic and physico-chemical criteria), and it is important to emphasize the dose dependence of the detected shifts [2, 3]. It should also be noted that the nature of the changes significantly differed from the "pattern" of the response to a similar effect of non-ionized helium flow from the same source [3]. Interestingly, the reaction of blood isolated from the body as a simple model biosystem as a whole was co-directed to shifts of the same blood parameters of rats exposed to short-term (1-2 min.) exposure to cold plasma on the pre-epilated areas of the back. As in the case of in vitro experiments, in this case shown moderate antioxidant activity of the factor, its positive effect on the intermediate link in the energy exchange etc. in addition, it was demonstrated modulating effect on the parameters of systemic and local hemodynamics.

Thus, plasma biomedicine is a promising, dynamically developing direction, integrating biophysical and biomedical approaches and capable of forming fundamentally new medical technologies, potentially useful for the correction of various diseases, pathological conditions, injuries and burns.

References

1. *Martusevich A.K., Peretyagin S.P., Vanin A.F.* The study of some porducts is generating with medical device for the synthesis of NO-containing cold plasma // Medical physics. 2012. No 4. P. 80-86.

2. *Martusevich A.K., Soloveva A.G., Yanin D.V., Galka A.G., Krasnova S.Yu.* The influence of gelium cold plasma on the parameters of blood oxidative metabolism in vitro // Gerald of New Medical Technologies. 2017. V. 24, No 3. P. 163-166.

3. *Martusevich A.K., Yanin D.V., Bogomolova E.B., Galka A.G., Klemenova I.A., Kostrov A.V.* Possibilities and perspectives of the use of microwave tomography in sin testing // Biomedical Radioelectronics. 2017. No 12. P. 3-12.

Generating High Power Terahertz and Far Infrared Electromagnetic Radiation with Relativistic Electrons

N. A. Vinokurov[1, 2]

[1]Budker Institute of Nuclear Physics SB RAS, Novosibirsk, Russia, vinokurov@inp.nsk.su
[2]Novosibirsk State University, Novosibirsk, Russia

Relativistic electron beams are used successfully to generate electromagnetic waves. Both narrow-band and short-pulse THz generators of such type are described with examples of the free electron lasers (FELs) and prebunched-beam devices.

FELs (see, e.g., [1]) convert the power of relativistic electron beams to narrow-band radiation. The main component of an FEL is a special spatially periodic magnetic system, called undulator (see, e.g., [2]). The undulator magnetic field curves electron trajectories; therefore, electrons emit narrow-band electromagnetic radiation in free space. The scheme of the FEL amplifier is shown in Fig. 1.

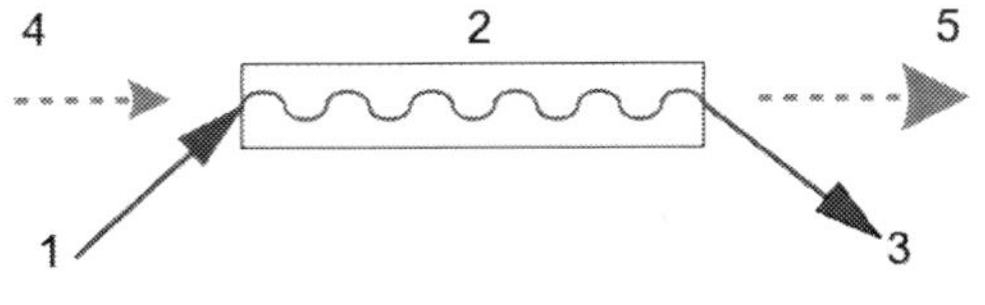

Fig. 1. The scheme of FEL amplifier: 1 – incoming electron beam, 2 – undulator, 3 – used electron beam, 4 – incoming electromagnetic wave, 5 – amplified electromagnetic wave

FELs use a phenomenon of the stimulated undulator radiation. i. e. bunching of electrons by an external electromagnetic wave and an increase of the wave field amplitude by addition of the field emitted by the bunched electron beam. On the other hand, FEL can be considered as travelling wave tube. Then undulator is necessary to provide (i) interaction of electrons with transverse electromagnetic wave and (ii) fast space harmonic of transverse electron current.

The particular problem of terahertz FELs is large diffraction divergence of the amplified wave. It require either large aperture of undulator, or the use of waveguide inside undulator. The first solution leads to the increase of the undulator period and corresponding increase of the electron energy. Moreover, big diameter of radiation beam leads to lower amplitude of electromagnetic field (at fixed power) and lower amplification. The second option is suffers from significant power losses. Therefore, it requires high enough FEL gain. For high average power FEL, the sufficient waveguide cooling has to be provided.

Another issue of FELs is the way to vary the radiation wavelength. The wavelength of the undulator radiation

$$\lambda = \frac{\lambda_u}{2\gamma^2}\left(1+\frac{K^2}{2}\right) \quad (1)$$

depends on the undulator period λ_u, the electron energy γmc^2 and the undulator parameter $K = eB\lambda_u/(2\pi mc^2)$ (c is the velocity of light, e and m are the charge and the mass of electron, correspondingly), which is proportional to the amplitude B of magnetic field. The most straightforward way is to vary the electron energy. However, in fact, it is rather complicated, as require changing many parameters of an electron accelerator, which is typically optimized for some particular energy. The variation of the magnetic field amplitude is the most common now. It is achieved by variation of current in the coils of electromagnetic undulators or by variation of the gap between upper and lower parts of permanent magnet undulators. The drawback of the field amplitude variation is that at shortest wavelength of the tunability range the undulator performance is far from the optimal one. Indeed, for such a low field amplitude, one can build an undulator with shorter period and larger pole numbers at the same length. Last decade we developed the third option – the variation of the undulator period [3, 4]. Variable period permanent magnet undulators were built and tested successfully [5, 6]. Next year we plan to install the variable period undulator to Novosibirsk FEL facility (see [7, 8]). It will replace one of the old electromagnetic undulators there.

For some applications, short high-field pulses are required. Contemporary accelerator technology provides very short electron bunches. The common way to do it is the use of radiofrequency (RF) electron guns with photocathodes (see, e.g., [9]). Using commercial lasers one can obtain few-picosecond electron bunches with electric charge Q of about 1 nC and normalized emittance of about 1 micron. The bunch charge is limited by Coulomb repulsion of electrons inside the RF gun. After acceleration, the bunch can be compressed to sub-picosecond duration T. Then the energy of the field around the bunch, which can be converted to the electromagnetic pulse energy, $W \sim Q^2/(cT)$ may reach 1 mJ. One of the possible radiators in this case is transition radiation, when the bunch passes through a conducting foil. The efficiency of such radiator may be enhanced using many foils and combining signals from them. The example of such a device, the multifoil radiator [10], is shown in Fig. 2. Short electron bunch passes through the conical foil stack along the cone axis z. After the bunch has passed through the gap between two of the foil disks, the induced electromagnetic wave propagates radially outward. As the pulses reach the foil boundaries, they merge and form the conical radially polarized wave. One can also say that this radiator is the Cherenkov one with strongly anisotropic media. The device was tested [11] with the short bunch from laser plasma wakefield accelerator.

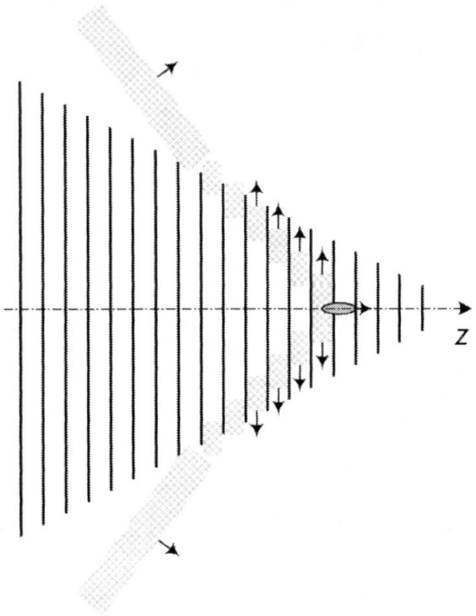

Fig. 2. The scheme of the conical multifoil radiator

Acknowledgment

This work was supported by the Russian Science Foundation (project No. 14-50-00080).

References

1. *Vinokurov, N. A., Shevchenko, O. A.* Free electron lasers and their development at Budker Institute of Nuclear Physics SB RAS // Phys. Usp. 2018. V. 61, No. 5. P. 435–448.

2. *Vinokurov, N. A., Levichev, E. B.* Undulators and wigglers for the production of radiation and other applications // Phys. Usp. 2015. V. 58, No. 9. P. 850–871.

3. *Shenoy, G. K., Lewellen, J.W., Shu, D., Vinokurov, N.A.* Variable-period undulators as synchrotron radiation sources // J. Synchrotron Radiat. 2003. V. 10. P. 205–213.

4. *Vinokurov, N. A., Shevchenko, O. A., Tcheskidov, V. G.* Variable-period permanent magnet undulators // Phys. Rev. Spec. Top. - Accel. Beams. 2011. V. 14, P. 040701–1 - 040701–7.

5. *J. Mun, et al.* Variable-period permanent-magnet helical undulator // Phys. Rev. Spec. Top. - Accel. Beams. 2014. V. 17, P. 080701–1 - 080701–6.

6. *Davidyuk, I.V., Shevchenko, O. A., Tcheskidov, V. G., Vinokurov, N. A.* Results of test of prototype of variable period undulator // Nucl. Instr. & Meth. A. 2017. V. 871, P. 77–82.

7. *Shevchenko, O. A. et al.* The Novosibirsk Free Electron Laser – unique source of terahertz and infrared coherent radiation // Phys. Procedia 2016. V. 84. P. 13–18.

8. *Knyazev, B. A. et al.* Novosibirsk free electron laser as a user facility // Phys. Procedia 2016. V. 84. P. 27–34.

9. *Vinokurov, N. A. et al.* Electron guns at the Budker Institute of Nuclear Physics SB RAS: prospects for the use of photocathodes with nanosecond and subpicosecond laser drivers // Phys. Usp. 2017. V. 60, No. 10. P. 1034–1038.

10. *Vinokurov, N. A., Jeong, Y. U.* Generating High-Power Short Terahertz Electromagnetic Pulses with a Multifoil Radiator // Phys. Rev. Lett. 2013. V. 110, No. 6. P. 064805-1 – 064805-5.

11. *Jo, J. et al.* High-power ultrashort terahertz pulses generated by a multifoil radiator with laser-accelerated electron pulses // Proc. of FEL2015, Daejeon, Korea. 2015. P. 739-740. http://accelconf.web.cern.ch/AccelConf/FEL2015/papers/wep082.pdf .

Femtosecond Acoustics and Terahertz Ultrasonics

Chi-Kuang Sun

Graduate Institute of Photonics and Optoelectronics, National Taiwan University, Taipei, Taiwan, sun@ntu.edu.tw

Acoustic wave is one of the most natural tools for communication and imaging. It is used for clinical imaging, to noninvasively probe physical structures under the surface as well as thermal dynamics of objects. However, considering the bandwidth (GHz) of current ultrasonic technologies, the electrically generated acoustic wave has a pulse duration much longer than 1 ns and a wavelength longer than 10 μm, which is far from adequate for molecular-level imaging, not to mention the femtosecond dynamics of many interacted molecules, including liquid water. Besides, it is believed that a wave-particle transition of THz acoustic waves must come into play when they are interacting with thermal phonons in all kinds of materials, including thin films, bulk materials, and nanostructures. To resolve the mysterious thermal dynamics in amorphous materials as well as the energy (heat) transport and dissipation from solids into liquids, an ultra-broadband THz acoustic source is thus urgently needed.

In this presentation, I would like to present the result of a new research direction: femtosecond acoustics and terahertz ultrasonics. The physical principle to convert a femtosecond optical pulse into a femtosecond acoustic pulse will be introduced. With a short pulse width, the spectral bandwidth can thus exceed THz. Combined with the ultraslow sound velocity, the application of this new tool is two folds. One is for THz phonon spectroscopy, which is critical to resolve the heat transport mysteries. The other is for in situ noninvasive sub-surface atomic level resolution imaging, which was not possible before. Examples will be given in this presentation.

Since it's proposed by H. Maris 30 years ago, "picosecond ultrasonic" has been developed as a complete technique able to study various acoustic phenomena and parameters from thickness measurements [1] to hypersound attenuation [2,3]. Based on the use of an ultrafast laser, this technique allows the generation and detection of coherent acoustic phonons (CAP) in the sub-THz range. Even though picosecond ultrasonic has been developed for more than 30 years, its frequency range can only reach 400GHz, which is still far from THz. In 1999, by femtosecond laser excitation of piezoelectric semiconductor quantum wells, we discovered a simple way to generate giant coherent acoustic phonons (acoustic waves) with a frequency over 2 THz [4]. This piezoelectric nano-layer based technique was later-on termed "nano-ultrasonics," with a focus to emphasize its capability to provide nano-meter spatial resolution ultrasonic images [5]. After over one decade's study, we have made this technique a powerful tool to image and characterize nanoscale materials and interfaces. In recent years, this method has not only been successfully applied to characterize the ångström-level surface and interface roughness [6], but also the acoustic attenuation up to 650 GHz in amorphous SiO_2 and more recently we showed the possibility to measure the acoustic attenuation spectrum in ice up to 1 THz, thus providing critical thermal dynamics parameters [7].

During my presentation, I will show the effort of the current program to not only continuously apply the THz acoustic spectroscopy 1). for the investigation of the Boson peak issue in amorphous solids and 2). of the collisionless decay of the coherent phonons in crystals, but also 3). to work on the atomic resolution imaging in solid-liquid water interface.

Amorphous material systems have impacted human lives in numerous aspects. Glasses, polymers, and even bio-tissues are significant matters made with amorphous structures. Despite the long appearance, the physical natures of the amorphous material system are still far from well understood. Unlike crystalline solids, amorphous materials exhibit a number of anomalous thermal properties. These anomalies were first discovered half a century ago by Berman, who noticed the unexpected instance of the thermal conductivity $\kappa(T)$ anomaly for vitreous silica (v-SiO_2). He measured the thermal conductivities of several v-SiO_2 samples in the temperature range between 2.2 and 90K and found the plateau behavior of thermal conductivities at around 10K, corresponding to a phonon peak frequency of ~1THz, with a magnitude several orders smaller. Since then, several other techniques, such as Raman scattering, inelastic neutron scattering, and X-ray scattering, had been carried out to explore this anomalous thermal phenomenon. Although many had witnessed the anomaly, none of them were able to provide a comprehensive understanding of its origin.

In crystals, the low-frequency density of states (DOS), $g(\omega)$, follows the so-called Debye model well; that is, by simply considering possible modes of sound waves which match the boundary conditions, one obtains the square frequency dependence of the DOS, $g(\omega)\propto\omega^2$. In contrast, amorphous material systems show an excess contribution to the DOS at the THz frequency regime (or at ~5meV energy level appeared in inelastic scattering spectra), which in turn results in a decreased thermal conductivity or an increased specific heat, as evidenced by previous vitreous silica experimental results. The excess in DOS is usually identified as a peak in a reduced DOS representation, $g(\omega)/\omega^2$ vs ω. At the time of its discovery, little was known about this feature except for its bosonic character. Therefore, it became known as the "boson peak" (BP). The physical origin as well as dynamics of the Boson Peak remained puzzling and controversial due to the fact that no acoustic attenuation spectral measurement in this frequency range can be directly performed in amorphous materials to verify the correctness of different theoretical models.

In contrast to amorphous solids, in past decades, hypersound wave propagation in crystalline solids have drawn much attention in a wide variety of scientific studies focusing on topics such as the spontaneous decay of longitudinal acoustic phonons, picosecond ultrasonics, nanoultrasonics, and the development of acoustic phonon lasers. In the frequency range of 0.4-3THz, coherent acoustic phonons (or hypersound waves) in crystal are physically modeled to be attenuated due to collisions with thermal phonons. However, different theories were debated with no experimental proof to distinguish their correctness and appropriate frequency ranges.

In crystalline dielectrics, this hypersound propagation loss can arise from both extrinsic and intrinsic mechanisms. The extrinsic mechanisms are commonly dominated by phonon-scattering processes at lattice imperfections. When these extrinsic mechanisms are absent, the acoustic loss will ultimately be determined by intrinsic anharmonic decay. Based on low frequency data, the mostly accepted picture is as follow. In most solids, thermal phonons have a lifetime, τ, on the order of picosecond to femtosecond. When the angular frequency, ω, of the acoustic waves is small compared to the thermal phonon lifetime with $\omega\tau \ll 1$ (usually in the sub-THz range), the propagation of acoustic waves will drive the thermal phonons out of equilibrium, and causing the interaction and resulting in the acoustic wave attenuation. Due to the fact that the acoustic waves are with a wavelength much longer than most room-temperature thermal phonons, the propagating coherent phonons will be scattered by all different thermal phonons and the momentum conservation is not a concern, due to the large uncertainty. This model well describes all lower-frequency (below 400GHz) measurement on different solids.

However it has been long debated and postulated that, when $\omega\tau \gg 1$ (usually above THz at room temperature), acoustic waves have to be considered as coherent phonons and the attenuation of coherent phonons due to their interaction with thermal phonons will be limited due to specific selection rules for energy and momentum conservation. That is to say that, only selected thermal phonons with specific momentums and energies involve in this "3-phonon interaction" process, thus making THz acoustic wave attenuation "saturated." In the transition range between these two regimes, $\omega\tau \gg 1$, the acoustic frequency will be around THz for room temperature crystals.

In this frequency range, the coherent acoustic phonon attenuation processes have never been explored and have been under debate for almost four decades due to the lack of reliable THz acoustic transducers. Applying our developed THz nano-ultrasound technique and studying the frequency dependent behavior of coherent acoustic phonons in this ambiguous transition range is therefore crucial since it will bring critical understanding for thermal properties in solids.

For nanoimaging with a femtosecond acoustic pulse, one can take advantage of the ultraslow sound velocity, which is usually on the order of 3000 m/s and the ultrahigh temporal resolution which is usually on the order of 100 fs. It can be noted that within 100 fs, a femtosecond acoustic pulse travels only a distance of 3 angstroms, assuming a 3000 m/s velocity. Femtosecond time-resolved sound propagation imaging can thus provide a spatial resolution down to 1.5 angstroms assuming an acoustic round trip path. To study the performance of femtosecond acoustics applied on *in situ* monitoring a chemical reaction at a solid/liquid interface, the well-known anode-oxidation process occurring during the GaN-based photo-electro-chemical (PEC) water splitting was taken as our model. As a result, sub-atomic resolution imaging was observed in situ noninvasively [8].

This project is sponsored by Ministry of Science and Technology of Taiwan under 106-2112-M-002 -004 -MY3 and National Taiwan University under 107L880404.

References

1. *Thomsen, C., Grahn, H. T., Maris H. J., Tauc, J.* Surface generation and detection of phonons by picosecond light pulses // Phys. Rev. B, 1986. V. 34, No. 6, P. 4129-4138.

2. *Lin, H. N., Stoner, R. J., Maris, H. J., Tauc, J.* Phonon attenuation and velocity measurements in transparent materials by picosecond acoustic interferometry // J. Appl. Phys., 1991. V. 69, No. 7. P. 3816-3822.

3. *Duquesne, J.-Y., Perrin, B.* Ultrasonic attenuation in a quasicrystal studied by picosecond acoustics as a function of temperature and frequency // *Phys. Rev. B,* 2003. V. 68, No. 13. 134205.

4. *Sun, C. K., Liang, J. C., Yu, X Y.* Coherent acoustic phonon oscillations in semiconductor multiple-quantum-wells with piezoelectric fields // *Physical Review Letters* 2000. V. 84, No. 1, P. 179-182.

5. *Lin, K.-H., Lai, C.-M., Pan, C.-C., Chyi, J.-I., Shi, J.-W., Sun, S.-Z., Chang, C.-F., Sun, C.-K.* Spatial manipulation of nanoacoustic waves with a nanoscale spot size // Nature Nanotechnology 2007. V. **2**, No. 11, P. 704-708.

6. *Wen, Y.-C., Hsieh, C.-L., Lin, K.-H., Chen, H.-P., Chin, S.-C., Hsiao, C.-L., Lin, Y.-T., Chang, C.-S., Chang, Y.-C., Tu, L.-W., Sun, C.-K.* Specular scattering probability of acoustic phonons in atomically flat interfaces // Physical Review Letters 2009. V. 103, No. 26. 264301.

7. *Mante, P.-A., Chen, C.-C., Wen, Y.-C., Sheu, J.-K., Sun, C.-K.* Thermal boundary resistance between GaN and cubic ice and THz acoustic attenuation of cubic ice from complex acoustic impedance measurements // Physical Review Letters 2013. V. 111, No. 22. 225901.

8. *Shen, C.-C., Weng, M.-Y., Sheu, J.-K., Yao, Y.-T., Sun, C.-K.* In situ monitoring of chemical reactions at a solid-water interface by femtosecond acoustics // Journal of Physical Chemistry Letters 2017. V. 8, No. 21. P. 5430-5437.

Terahertz and mm-wave applications at ENEA-Frascati

G. P. Gallerano, S. Ceccuzzi, A. Doria, E. Giovenale, G. L. Ravera, M. Zerbini, G. Galatola-Teka[2]

[1]ENEA – Fusion Physics Division – Radiation Sources, Antennas and Diagnostics Laboratory
Via E. Fermi 45, 00044 Frascati – Italy, gianpiero.gallerano@enea.it
[2]ENEA Phd Student

Frequencies in the THz region are usually considered as too high for electronic devices, due to the technological limits in reducing the size of components as frequency increases, and too low for photonic devices, since the corresponding photon energy is comparable to those of thermal excitations at room temperature. A variety of applications of THz radiation are being developed worldwide. The majority of applications exploits the spectral fingerprints of the chemical species contained in the material under investigation and takes advantage from the fact that THz radiation can penetrate through plastic, cardboard and clothes, without causing the hazards of ionizing radiation (X-ray) and suffering at the same time less scattering than near-infrared radiation.

This wide variety of applications requires an equally wide range of sources and radiation characteristics. This is the reason why two Free Electron Lasers (FELs) have been designed, developed and realised at the ENEA Frascati Research Centre [1]. Thanks to their intrinsic generation mechanism they have the peculiar advantage of being tunable sources.

The ENEA Compact FEL is based on a microtron electron accelerator capable of generating a relativistic beam at an energy of 2.3 MeV with a current of 300 mA, in 4 μs pulses. A magnetic undulator, designed and realised by ENEA, made of 8 permanent magnets (SmCo) periods, provides the generation of radiation in the range 100-150 GHz with a peak power of 1.5 kW over the macropulse. Due to the RF nature of the accelerator, a sub-structure inside the pulse generates a train of micro-pulses of about 60 ps duration, equally spaced by 330 ps, with a peak radiation power of about 10 kW. The average CW power can be varied according to the repetition rate of the RF system driving the accelerator providing about 15 mW at 2.5 Hz (Fig. 1).

Fig. 1. The ENEA Compact FEL

The second ENEA FEL source is named FEL-CATS (Compact Advanced THz Source) and it is a unique device in the FEL panorama. It is a compact source based on a Linac type accelerator (2.3 MeV of energy and 250 mA of average current over 5 μs). It includes an additional RF device (called PMD), placed downstream the Linac, capable of manipulating the electrons in the longitudinal phase-space in such a way to create an adequate level of order in the bunch, to enhance the electromagnetic generation mechanism. This unique source, due to the presence of the PMD, does not make use of an optical resonator, thus extending its tuning capability over almost one octave. In fact, it generates radiation from 400 μm (0.7 THz) to 700 μm (0.4 THz) with a power level of several kW over the microsecond pulse duration (see Fig. 2).

Fig. 2. The ENEA FEL-CATS source

Beside FEL sources, our laboratory has been equipped with an extensive assortment of other coherent sources, both lasers and electron devices as a part of the Center of Excellence for Terahertz Radiation and Applications (CETRA). A Far InfraRed (FIR) gas laser based on a CO_2 laser, pumping a gas cell, is available as well as a Ti:Sa laser oscillator that generates a train ultra-short pulses (about 130 fs) in the Near InfraRed (NIR) with an average CW power in excess of 1W. Other compact, portable, solid state sources are available in the THz laboratory to cover the low frequency spectrum: two YIG oscillators that emit in the ranges 8-20 and 20-40 GHz respectively, with an average CW power of 5-10 mW; an IMPATT diode at 97 GHz with about 80 mW of output power and a second IMPATT diode emitting at 140 GHz with about 20 mW of power. Finally a frequency multiplier in the range 220-330 GHz is also available to test high frequency components.

The CETRA Laboratory has realised an extensive variety of applications in different scientific and technological fields, like living plants biology, human biology, solid state physics and cultural heritage protection. The most notable activity was the pioneering THz BRIDGE project (2001 – 2004), a European project that opened the way to the study of the interaction of THz radiation with biological systems following a streamline of increasing complexity, from bio-molecules to living cells and tissues. At present the utilization capability of the laboratory resources takes place mainly in the context of scientific collaborations funded through national or international projects.

Recent applications in the biomedical field, are carried out in the frame of the GREAM project [2-3], financed by the Italian Ministry of Defence, and was realised in collaboration with the Scientific Department of Army Medical Center of Rome, and the Universities of Rome "La Sapienza", "Tor Vergata" and "Roma Tre". It focused on the study of potential genotoxic effects induced on different kind of human fibroblasts, after irradiation with the Compact FEL source, in the 100-150 GHz frequency range, and with the YIG source at 25 GHz. In order to correctly characterize the sample exposition, a preliminary study on the Specific Absorption Rate (SAR) has been performed, also by means of simulations, as reported in Fig. 3.

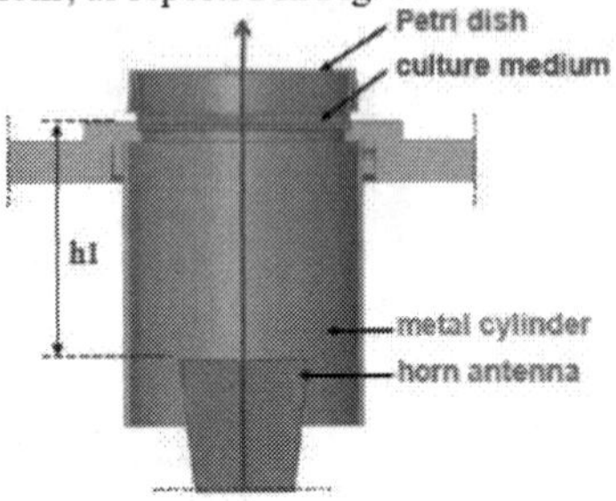

a)

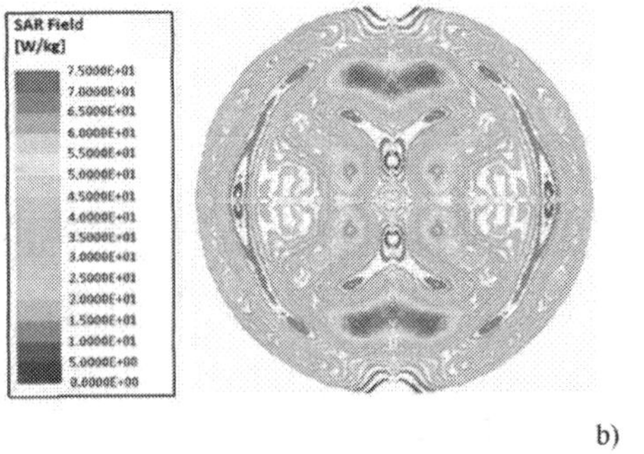

b)

Fig. 3. (a) Detail of the setup showing the position of the petri dish; (b) SAR distribution on the cell layer at the bottom of the petri dish

In the field of Cultural Heritage preservation, a specific project, THz-ARTE, was carried out in the frame of a bilateral collaboration between Italy and Japan. For this project the CETRA Laboratory has realized an innovative imaging system [4] for the analysis of artwork samples, which were prepared by the NICT Institute in Tokio and by the CNR-IFAC Institute in Florence (see Fig. 4).

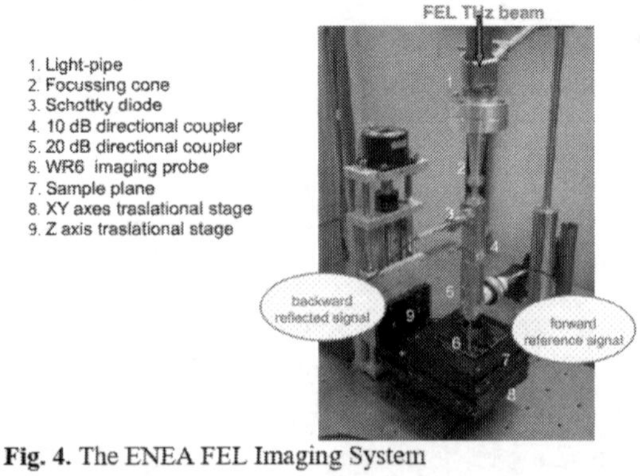

Fig. 4. The ENEA FEL Imaging System

A second reflective imaging system, employing the 97 GHz IMPATT source, has been realised to be suitable for "on field" operation. It exploits the penetrating features of the MM wave radiation together with a phase sensitive detection technique. This imaging system has been successfully operated on real ancient paintings (see Fig. 5).

Fig. 5. Portable ENEA Imaging system applied to a "Fresco" painting on the "tavella"—St. John Baptist—by Alessandro Gherardini (1655-1726)

References

1. *Biedron, S., Gallerano, G. P., et al.* Compact, High Power Electron Beam based Terahertz Sources // Proc. Of the IEEE 2007. V. 95, No. 8 P. 1666-1678.

2. *Franchini, V., et al.* Study of the Effects of 0.15 Terahertz Radiation on Genome Integrity of Adult Fibroblasts // Environmental and Molecular Mutagenesis 2018 V. 59, P. 476-487.

3. *Franchini, V., et al.* Biological Effects of 25 to 150 GHz Radiation after in Vitro Exposure of Human Fibroblasts: a Comparison of Experimental Results // J Infrared Milli Terahz Waves 2018, V. 39 P. 899–911.

4. *Gallerano, G. P., Doria, A., Germini, M., Giovenale, E., Messina, G., Spassovsky, I.* Phase-Sensitive Reflective Imaging Device in the mm-wave and Terahertz Regions // J Infrared Milli Terahz Waves 2009. V. 30 P. 1351–1361.

5. *Doria, A. Gallerano, G. P., Giovenale, E., Casini, A., Cucci, C., Picollo, M., Poggesi, M., Stefani, L., Fukunaga, K., Tamassia, M.*, Vis-NIR Hyper-spectral and Terahertz Imaging Investigations on a Fresco Painting on "Tavella" by Alessandro Gherardini // J Infrared Milli Terahz Waves 2017, V. 38, P. 390–402

 https://doi.org/10.1051/epjconf/201819501001

Study of mutual phase locking of two gyrotrons coupled with delay

A.B. Adilova[1], N.M. Ryskin[1,2]

[1]Saratov State University, Saratov, Russia, adilovaab@gmail.com
[2]Saratov Branch, Institute of Radio Engineering and Electronics, Saratov, Russia

Introduction

The most important application of high-power continuous-wave (CW) gyrotrons is electron cyclotron plasma heating for nuclear fusion [1]. For such application, a large number of gyrotrons should be used to obtain a very high power. For example, in ITER it is planned to use 26 gyrotrons with 1 MW power level. Providing coherence of radiation of large number of high-power microwave sources is a challenging task. One of the possible solutions is either injection locking by external master oscillator with stabilized frequency [2],[3] or mutual phase locking [4].

In this paper, we present the results of theoretical analysis and numerical simulation of mutual phase locking of two coupled gyrotrons.

Modified quasilinear model

In [5], we developed a basic theory of mutual phase locking of two limit-cycle oscillators coupled with delay. The delay in coupling provides essential features of the pattern of synchronization. In particular, it is very sensitive to the phase shift of the coupling signal propagating between two counterparts. However, the analysis presented in [5] is valid only in the case of weak coupling and small frequency mismatch. In [6], we considered a more general model of two coupled gyrotron oscillators, which are assumed identical except a small frequency mismatch. The model is described by two delay-differential equations (DDE)

$$\frac{dA_1}{dt} + i\frac{\Delta}{2}A_1 + A_1 = I_s\Phi(A_1)\cdot A_1 + \rho e^{-i\psi}A_2(t-\tau),$$
$$\frac{dA_2}{dt} - i\frac{\Delta}{2}A_2 + A_2 = I_s\Phi(A_2)\cdot A_2 + \rho e^{-i\psi}A_1(t-\tau). \quad (1)$$

Here, $A_{1,2}$ are slow amplitudes, Δ is normalized frequency mismatch, Φ is the complex gain factor, I_s is the normalized beam current parameter, ρ is the coupling strength, $\tau = \omega_0 t_d/2Q$ is the normalized delay time t_d, $\psi = \omega_0 t_d$. In the quasi-linear theory [7], the complex gain factor is obtained as a polynomial expansion $\Phi = \alpha - \beta|A|^2 - \gamma|A|^4 - \ldots$ as a result of approximate solution of the electron motion equations. However, this approximation is valid only at a slight excess of the start-oscillation current and, consequently at low efficiency. Therefore, to extend the range of validity of the model, we obtain $\Phi(|A|^2)$ by fitting the numerical solution of the equations of time-domain gyrotron theory with fixed Gaussian profile of the RF field [7].

Numerical simulation in a wide range of parameters confirms the basic pattern of synchronization revealed in [5]. The results of simulation for the DDE model (1) and time-domain gyrotron theory with fixed Gaussian profile of the RF field [7] are in good agreement with each other.

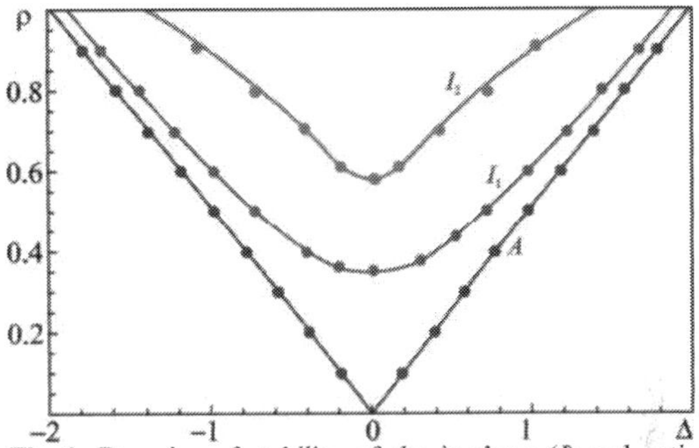

Fig. 1. Domains of stability of the in-phase (I) and anti-phase (A) mode of synchronization on the Δ, ρ plane for $I_s = 0.05$, $\psi = 0.4\pi$, and $\tau = 3.0$.

The study of mutual phase locking of two gyrotrons in the soft-excitation mode ($\Delta_H = 0.4$) [6] revealed that the delay in coupling strongly affects the pattern of phase-locking regimes. In particular, it is very sensitive to the phase shift of the coupling signal propagating between two counterparts.

Numerical simulation in a wide range of parameters confirms the basic pattern of synchronization revealed in [5]. With the increase of the delay, there appear new modes of synchronization, i.e. the multistability becomes more pronounced. Fig. 1 illustrates the situation when there exist two in-phase and one anti-phase modes.

In addition, in Fig. 1 the stability domains calculated according the time-domain gyrotron theory with fixed Gaussian profile of the RF field [7] are shown with circles. They are in good agreement with the results of simulation for the DDE model (1).

Synchronization of two gyrotrons in the hard excitation mode

Maximum efficiency in a gyrotron is usually attained in a hard excitation mode [7]. Thus, we studied synchronization of two coupled gyrotrons with Gaussian profile of the RF field with the parameters μ=15, $I_s = 0.06$, $\Delta_H = 0.53$ when the orbital efficiency is nearly 70%.

The system of two mutually coupled gyrotrons demonstrates a behavior which has much common with dynamics of a single gyrotron oscillator driven by an external signal [2]. In [2], we found a number of differences from the well-studied pattern of synchronization of the generator with soft excitation.

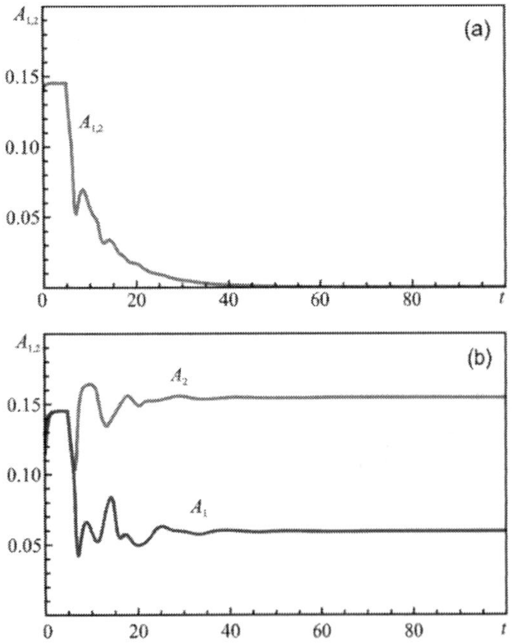

Fig. 2. Time histories of the amplitudes at μ=15, $\Delta_H = 0.53$, $I_s = 0.06$, ρ = 0.5, ψ = 0.15π, τ = 5.

In particular, since bistability presents in the autonomous system, it also appears in the non-coupled system. The synchronization mode with high efficiency establishes only if the initial perturbation is large enough that is typical for systems with hard self-excitation. Moreover, for certain values of the parameters, the synchronization mode turns out to be extremely sensitive to the initial phase of the oscillations and an external signal may cause a breakdown into the low-amplitude mode instead of synchronization. In Fig. 2, examples of the time histories for a sufficiently large coupling are presented. When both gyrotrons start from the same initial amplitudes (Fig. 2(a)), the oscillations are excited in both gyrotrons and the efficiency reaches its maximum value. However, at $t = 5$ the signal from one gyrotron starts to affect the other one and vice versa. As a result, the oscillations of both gyrotrons are completely suppressed. Under somewhat different initial condition, only one of the gyrotrons suppresses the other, as is shown in Fig. 2(b).

On the contrary, coupling of the gyrotrons may lead to the excitation of one gyrotron by another. Fig. 3 illustrates the case when the initial amplitude of the second gyrotron is zero. In the hard excitation regime, this is a stable state. However, when the signal from the first gyrotron enters the resonator of the second one, the oscillation of the second gyrotron starts to grow and finally the synchronization mode is established.

Synchronization of coupled multimode gyrotrons

Finally, we considered mutual synchronization of two multimode gyrotrons. The simulation shows that in this case instead of oscillation quench there occurs excitation of a spurious mode.

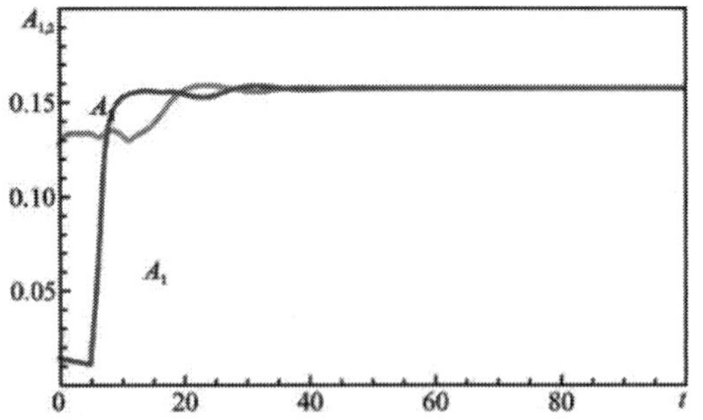

Fig. 3. Time histories of amplitudes of gyrotrons at ρ = 0.3, ψ = 0.4π, τ = 5.

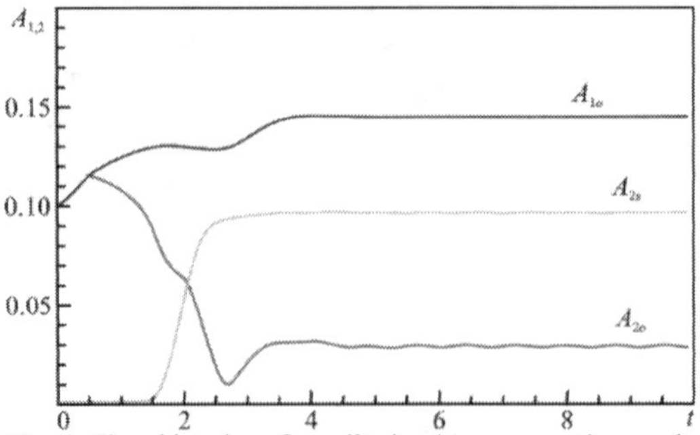

Fig. 4. Time histories of amplitudes (A_{1o} - operation mode in the first gyrotron, A_{2o} and A_{2s} - operation and spurious modes in the second gyrotron) at ρ = 0.5, ψ = 0.4π, τ = 0.5.

This is illustrated by Fig. 4 where time histories of the operating modes in two gyrotrons are plotted (cf. Fig. 2(b)). Or, on the contrary, at different initial conditions build-up of the synchronization mode may occur while the spurious modes in both gyrotrons are completely suppressed.

Acknowledgment

This work is partly supported by the Russian Foundation for Basic Research grant No. 18-02-00839.

References

1. *Thumm M.* Recent advances in the worldwide fusion gyrotron development // IEEE Trans. Plasma Sci., 2014. V. 42, no. 3, pp. 590-599.
2. *Yakunina K.A., Kuznetsov A.P., Ryskin N.M.* Injection locking of an electronic maser in the hard excitation mode // Phys. Plasmas. 2015. V. 22, No 11. 113107.
3. *Bakunin V.L., Denisov G.G., Novozhilova Y.V.* Zones of frequency locking by an external signal in a multimode gyrotron of a megawatt power level // Radiophys. Quantum Electron., 2016. V. 58, No. 12. P. 893-904.
4. *Rozental R.M., Ginzburg N.S., Glyavin M.Yu., Sergeev A.S., Zotova I.V.* Mutual synchronization of weakly coupled gyrotrons // Phys. Plasmas. 2015. V. 22, No. 9. 093118.
5. *Usacheva S.A., Ryskin N.M.*, Phase locking of two limit cycle oscillators with delay coupling // Chaos. 2014. V. 24, No. 2. 023123.
6. *Adilova A.B., Gerasimova S.A., Ryskin N.M.*, Synchronization of delay-coupled gyrotron oscillators // EPJ Web of Conferences. 2017. V. 149. P. 04029.
7. *Nusinovich G.S.* Introduction to the Physics of Gyrotrons. Johns Hopkins University Press, Baltimore, 2004.

https://doi.org/10.1051/epjconf/201819501002

High power THz-range Wave generation based on Transformation of Plasma Waves Pumped by High-current Relativistic Electron Beam

A.V. Arzhannikov[1,2], V.V. Annenkov[1,2], V.S. Burmasov[1,2], I.A. Ivanov[1,2], A.A. Kasatov[1,2], S.A. Kuznetsov[1,2], M.A. Makarov[1], K.I. Mekler[1], S.V. Polosatkin[1,2], V.V. Postupaev[1,2], A.F. Rovenskikh[1], D.A. Samtsov[1,2], S.L. Sinitsky[1,2], V.F. Sklyarov[1,2], V.D. Stepanov[1,2], I.V. Timofeev[1,2], and E.P. Volchok[1,2]

[1]Budker Institute of Nuclear Physics Russian Academy of Science, Novosibirsk, Russia, arzhan1@ngs.ru
[2]Novosibirsk State University, Novosibirsk, 630090, Russia

Introduction

The beam-plasma interaction system allows one to generate high-power sub-mm waves (frequency interval 0.1÷1 THz) by usage of mechanism of plasma wave transformation [1]. This way gives possibility to achieve multi megawatt power with the promptly varying frequency that can be necessary for various practical applications. The original project of a sub-mm wave based on transformation of plasma waves pumped by a kA-current relativistic electron beam is developed at the GOL-PET facility. We present novel results on the study of mechanisms of sub-mm wave emission by the processes of plasma wave transformation in electromagnetic one in case of the strong beam-plasma interaction.

Facility and results of experiments

To study the conditions associated with the generation of electromagnetic radiation, the specialized facility GOL-PET was constructed. The GOL-PET facility consists of an open magnetic trap with a multiple-mirror or uniform magnetic field of mean value B = 4.2 T and length L = 2.4 m between the end mirrors, see Fig.1. The end mirrors have a strong field up to B = 8 T. A plasma column with the density $n_e \approx (0.2 \div 5) \times 10^{15}$ cm^{-3} and a diameter of 7 cm is created by a longitudinal high-current discharge. U-2 accelerator producing a high current relativistic electron beam (REB) is mounted at one end of the trap. It produces the beam with the current I ~ 10 kA, the pulse duration 6 μs at the electron kinetic energy $E_e \approx 0.8$ MeV. The REB is injected into the end of the plasma column and where the beam diameter is 4 cm at the magnetic field 4 T. The radial profile of the plasma density is measured in 9-dots over the column diameter at a distance z = 0.83 m from the entrance mirror by diagnostics based on Thomson scattering. Dynamics of the average plasma density is measured by Michelson interferometer at z = 1.16 m. Pulsed current transformers measure the electron beam current at different axial points. Electron energy of REB is determined according with accelerator voltage in the U-2 diode. The transfer of energy from REB to plasma is calculated from results of measurements of diamagnetism of the plasma column. The 8-channel sub-mm polychromator, as well as a number of single detectors with bandpass filters at their entrance are utilized to study properties of radiation emitted by plasma [2].

Previous experiments showed that the changes of the power value of the emitted electromagnetic waves correlated with the variation in the efficiency of plasma heating by the electron beam. [1-3]. In a previous series of experiments on the simultaneous registration of radiation power emitted along the axis of the plasma column and in its perpendicular direction conducted at varying the plasma density from 10^{14} up to 10^{15} cm^{-3}, we observed that the emission in the transverse direction was only at relatively low plasma densities (ne < 5×10^{14} cm^{-3}). For higher value of plasma density, the emission in the band above 300 GHz was observed along the axis of the device only. Just as in the earlier experiments on the GOL-3, the EM-wave emission mainly exists simultaneously with the increasing of the plasma diamagnetism.

A series of experiments for measuring the temporal dynamics of radiation spectra emitted along the axis carried out for the plasma density higher than 5×10^{14} cm^{-3}. It was done by the 8-channel polychromator with semiconductor diodes for the frequency interval from 0.1 up to 0.5 THz. For the interval 0.5 ÷ 0.9 THz the measurements were done by single plasmonic detectors and an additional 2-channel system of cryogenic sensors. Experiments showed that at plasma density $5\times10^{14} \div 10^{15}$ cm^{-3} the spectral composition of the emission along the axis of the plasma column depended on radial gradient of the plasma density [2, 3]. The increase of the gradient caused to strongly increasing the spectral power density in the frequency band 150 ÷ 300 GHz. An example of signals from various sensors for one of the pulses of the GOL-PET facility (#1945) in case of the strong plasma density gradient (greater than 10^{15} cm^{-4}) is presented in Fig. 2. The signal of the spectral power density from the diodes in the interval 150 - 300 GHz is dumped in 50 times for presenting here. Taking into account, this dumping factor we can say that the power in the spectral interval 150 ÷ 300 GHz is 25 kW/(sr·cm^2) and the spectral power density is greater than its value in the interval 400 ÷ 800 GHz approximately in two orders of the value.

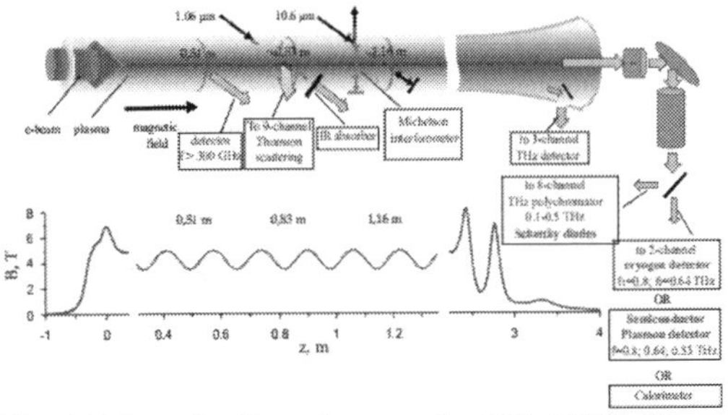

Fig. 1. Schematic of experiments at the GOL-PET facility

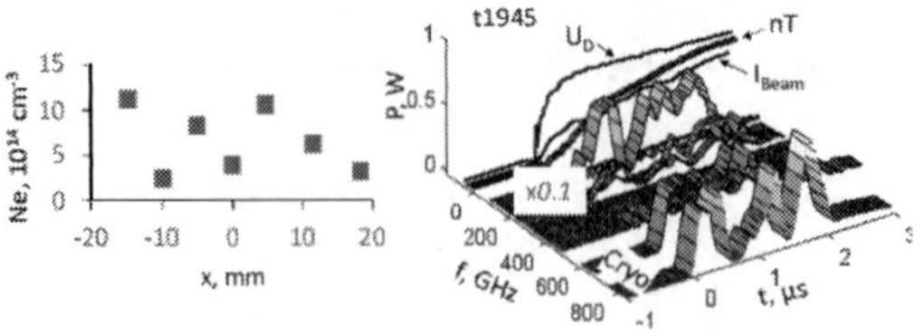

Fig. 2. Results of measurements in pulse #1945 at the GOL-PET facility. Left – plasma density distribution from 8-channel Tomson scattering system; right – signals from various sensors. U_D – diode voltage of the accelerator (MeV), I_{Beam} – beam current (10 kA), nT – plasma pressure from diamagnetic loop, other signals – radiation power in frequency selective detectors in Watts.

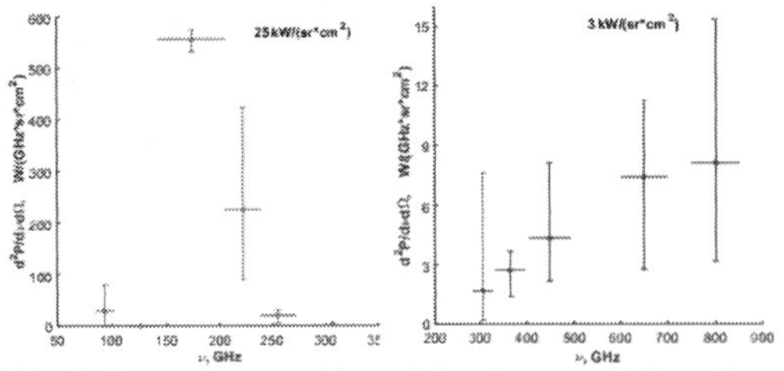

Fig. 3. Spectral composition of the plasma emission along the axis in case of strong plasma density gradient in transverse direction

Short analysis of measurement results The signals in the frequency band 125 ÷ 250 GHz are interpreted as radiation near the plasma frequency, Fig.3 (spectral band in left side of the picture). Such radiation can be produced through the linear mode conversion of upper-hybrid branch of plasma oscillations in region with plasma density gradients [4]. The observed width of the spectral line corresponds to the limited range of plasma density (3-6) $\times 10^{14}$ cm^{-3}. It means that intense beam-driven oscillations are localized near the bottom of density wells visible in Fig. 2 (left). If the plasma density in such a narrow well is modulated in the longitudinal direction due to, for example, the modulation instability, sub-luminal beam-driven modes can resonate with slow waveguide electromagnetic modes [5]. Then, these waveguide modes escape from the local density wells in the predominantly longitudinal direction via regular longitudinal inhomogeneity of plasma density on the scale of the magnetic field corrugation (10 cm). As to the high-frequency band 400 ÷ 800 GHz, Fig.3 (spectral band in right side), on our opinion, is a result of the coalescence of two upper hybrid plasma waves into electromagnetic ones [6-8]. Since the sub-mm wave emission at the plasma density higher than 5×10^{14} cm^{-3} is realized along the axis, we suppose that additional feature exists for longitudinal redirection of electromagnetic waves in the GOL-PET device. Note that the tendency to this redirection is qualitatively reproduced in theory [7].

Conclusion

It is found for the plasma density (0.5 ÷ 1.0) $\times10^{15}$ см$^{-3}$ that the radiation spectrum of the flux propagating along the axis of the beam-plasma system is mainly concentrated in two clearly distinct regions with the high level of spectral power density. The first region is located in the frequency interval f_1=0.15 ÷ 0.25 THz, the second one is in the interval f_2=0.4 ÷ 0.8 THz. The emission observed in the frequency interval f_1 has the power in the flux on a megawatt level and is interpreted as result of the linear conversion of the upper-hybrid branch of plasma oscillations to the electromagnetic radiation in regions of strong plasma density gradients. We interpret the emission in the interval f_2 as result of merging of these two plasma oscillations into the electromagnetic wave at a high level of plasma turbulence.

This work is supported by the Russian Foundation for Basic Research (grant 18-02-00232).

References

1. *A. V. Arzhannikov, et al.,* // Transactions of Fusion Science and Technology, 2013, vol. 63, pp. 82-87.
2. *I. A. Ivanov, et al.,* // AIP Conf. Proc., 2016, vol.1771, p.070009, doi.org/10.1063/1.4964233.
3. *A.V. Arzhannikov, et al.,* // AIP Conf. Proc., 2016, v.1771, p.070004, doi.org/10.1063/1.4964228.
4. *I.V. Timofeev, V.V. Annenkov and A.V. Arzhannikov* // Phys. Plasmas 2015, vol.22, p.113109.
5. *C.Miao, J.P.Palastro, and T.M. Antonsen* // Physics of Plasmas, 2017, vol.24, p.043109, doi.org/10.1063/1.4981218.
6. *I.V.Timofeev* // Phys. Plasmas 2012, vol.19, p.044501.
7. *A.V.Arzhannikov and I.V.Timofeev* // Plasma Phys. Control. Fusion, 2012, vol.54, p.105004.
8. *A. V. Arzhannikov et al.,* // IEEE Trans. Terahertz Sci. Technol., 2016, vol. 6, no 2, pp.245–252.

Prospective THz Gyrotrons for High-Field Magneto-Resonance Spectroscopy

V.L. Bratman[1,2], A.E. Fedotov[2], Yu.K. Kalynov[2], and V.N. Manuilov[2,3]

[1]Ariel University, Ariel, Israel, v_bratman@yahoo.com
[2]Institute of Applied Physics of Russian Academy of Sciences, Nizhny Novgorod, Russia
[3]Nizhny Novgorod State University, Nizhny Novgorod, Russia

Abstract

A high-harmonic Large Orbit Gyrotron and a low-voltage gyrotrino placed inside a spectrometer cryomagnet enable greatly simplify terahertz systems for magneto-resonance spectrometers. Large Orbit Gyrotrons provide a powerful third-harmonic generation at frequencies of 1 THz and 0.394 THz in pulsed and CW regimes, respectively, at significantly lower magnetic fields than conventional gyrotrons. According to simulations the gyrotrino with the voltage of 1.5 kV and frequency of 0.264 THz can generate a power of tens of watts; a possibility to operate at such a low voltage is demonstrated in the existing gyrotron with three-electrode magnetron-injection gun.

Introduction

THz medium power gyrotrons are in high demand for high-field DNP-NMR spectroscopy. However, only few installations are now equipped with such fairly expensive generators. Large Orbit Gyrotrons (LOGs) and gyrotrinos may make this method accessible for many laboratories due to the considerable simplification of the gyrotron magnetic system.

Conventional gyrotrons commonly operate at the fundamental or second cyclotron harmonics. The latter allows using simpler cryomagnets with a half-value of the magnetic field. A greater effect can be achieved in LOGs, capable of selective operation even at the third and fourth harmonics due to applying an axis-encircling electron beam [1-5], in contrast to a hollow poly-axis beam in a conventional gyrotron.

Another option is integration of a compact gyrotrino in single cryomagnet with an NMR spectrometer [6, 7]. This is based on the proximity of the electron cyclotron frequency and the frequency of the paramagnetic resonance in a sample. The integration eliminates the need for an additional cryomagnet and a long THz transmission line with high losses. An exact frequency matching can be achieved at a very low gyrotron voltage, namely 1-2 kV [6], or at special profiling the cryomagnet field [7].

High-Harmonic Large Orbit Gyrotrons

In a LOG cavity (Fig. 1), an axis-encircling electron beam can excite only co-rotating modes with azimuthal indices equal to the number of the resonant cyclotron harmonic [1-5]. This strong selection rule prevents the excitation of most of parasitic modes and makes it possible to operate at higher harmonics.

At the Institute of Applied Physics, LOGs, are being studied in parallel with conventional gyrotrons, for more than 20 years to obtain higher frequencies at lower magnetic fields. First experiments were carried out at high-current electron accelerators with moderately relativistic electron energy of (250 – 400) keV. These LOGs selectively generated high output powers at harmonics s=2-5 and frequencies from 20 GHz up to 0.4 THz [5]. At relativistic energy, electron-wave coupling at high harmonics can be stronger than at the fundamental resonance, but a problem of neighboring harmonics discrimination is fairly complex. In addition, use of relativistic energy decreases the cyclotron electron frequency and reduces the frequency gain.

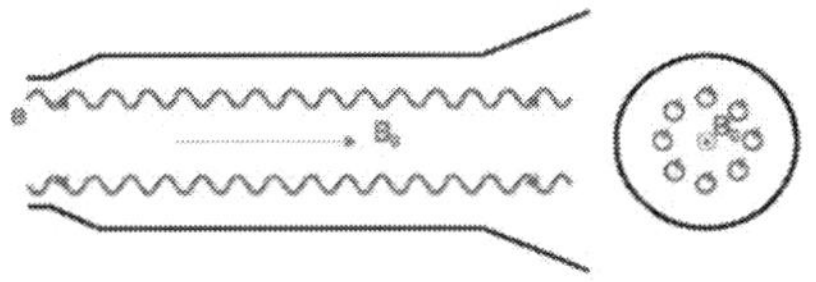

a)

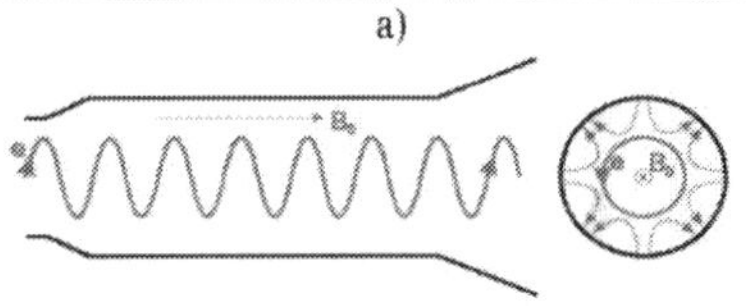

b)

Fig. 1. Electron beams in conventional gyrotron (a) and high-harmonic Large Orbit Gyrotron (b)

The further progress in development of LOGs was associated with reducing the operation voltage that enables replacing the accelerators by simpler high-voltage modulators, as well as with increasing the radiation frequencies. In the 80-kV LOG (Fig. 2a) [4], a stable single-mode second- and third-harmonic generation with a power of 0.3–1.8 kW was obtained in 10-μs pulses at four frequencies in the range of 0.55–1.00 THz at magnetic fields 10.5–13.6 T. A relatively low efficiency of this generator ~1% was caused by great Ohmic losses in a long gyrotron cavity at THz frequencies. This LOG was used for scientific applications and for testing novel versions of THz cavities [5] aimed to enhance efficiency by decreasing Ohmic losses, as well as to make possible the fourth-harmonic operation.

An important step in LOG progress is a development of a 30 kV/0.7 A gyrotron [5] with a 5 T cryomagnet designed for a CW operation (Fig. 2b). In this oscillator, an electron beam with a large pitch-factor of 1.5 is formed in a cusp gun and then the transverse electron velocity increases in an increasing magnetic field. The main scope of this setup is operation at the second, third and fourth harmonics at the

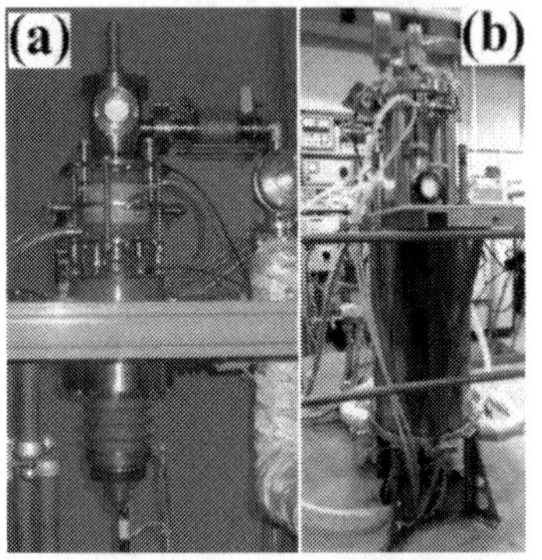

Fig. 2. LOGs with parameters 1 THz/80 kV (a) and 0.394 THz /30 kV (b).

typical DNP frequencies of 0.26 THz, 0.39 THz, and 0.52 THz with the power level of hundreds of watts.

In a good accordance with simulations, experimental zones of excitation of the second-harmonic $TE_{2,5}$ and third-harmonic $TE_{3,7}$ modes are well separated by magnetic field. Generation with parameters 800 W/ 0.267 THz and 300 W/ 0.394 THz is obtained at the fields B=5.02 T and B=4.93 T, respectively. Competition of these two modes is observed at intermediate magnetic fields. An increase in operating magnetic field up to 6.3 T and in the electron voltage up to 45 kV should allow achieving frequencies up to 0.65 THz at the fourth cyclotron harmonic [5].

Gyrotrino

Integration of a THz generator and an NMR spectrometer in single cryomagnet requires the exact matching of the DNP and gyrotron frequencies. Because of relativistic dependence of the cyclotron frequency on electron energy this is possible at an extremely low voltage of 1-2 kV only [6]. The very possibility of operating at a such low voltage was verified in an existing CW gyrotron [8]. A three-electrode magnetron injection gun of the gyrotron was initially designed for a higher voltage; to obtain an acceptable electron pitch factor for the low voltage, a high positive potential was applied to a modulating anode while keeping a negative cathode potential in order to increase the transverse particle velocity in the emitter region and decrease the longitudinal velocity in a region between the anode and the cavity. As a result, a stable generation with the frequency of 0.25 THz was observed at whole range of electron energies in the gyrotron cavity from 15 keV down to 1.5 keV.

According to simulations [10], a gyrotrino with an operating $TE_{6,2}$ mode, a voltage of 1.5-1.8 kV, a current of 200 mA, a pitch factor of 1.2 and a small cavity length can provide a power up to 15 W at the frequency of 0.264 THz with a required magnetic field about of 9.42 T (Fig. 3b). Since the gyrotron cavity is very close to the sample, a short transmission line is needed, that reduces the radiation losses and decreases a required THz power. A magnetic field disturbance induced by the low-voltage electron beam at the sam-

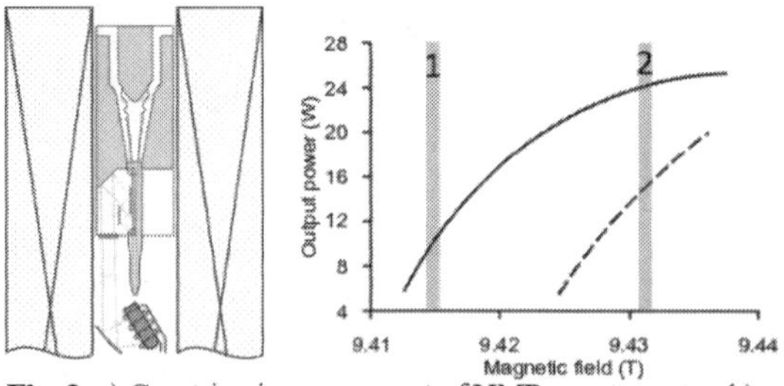

Fig. 3. a) Gyrotrino in cryomagnet of NMR spectrometer; b) calculated power vs. magnetic field for 1.5 kV / 200 mA (solid line) and 2 kV / 150 mA (dotted). Lines 1 and 2 show two regimes of gyrotron-DNP frequency matching.

ple is about of 10^{-8} and can be neglected. To save a limited space in the uniform magnetic field region, the radiation output from the gyrotron cavity can be directed toward the electron gun (Fig. 3a). Due to lack of space, the electron beam is to be collected in a cavity cut-off narrowing in a strong field [10].

The gyrotrino project was supported by Russian Science Foundation under grant No. 16-12-10445. Authors are grateful to M.Yu. Glyavin, A.P. Fokin, I.V. Osharin, and N.A. Zavolsky for collaboration.

References

1. *Jory, H.* Investigation of electronic interaction with optical resonators for microwave generation and amplification // Varian Associates, Palo Alto, CA, USA, R&D Tech. Rep. ECOM-01873-F. 1968.

2. *McDermott, D.B., Luhmann, Jr, N.C., Kupiszewski, A., and Jory, H.R.* Small-signal theory of a large-orbit electron-cyclotron harmonic maser // Phys. Fluids. 1983. V. 26. No. 7. P. 1936–1941.

3. *Lawson, W., Destler, W.W., and Striffler, C.D.* High-power microwave generation from a large-orbit gyrotron in vane and hole-and-slot conducting wall geometries // IEEE Trans. Plasma Sci. 1985. V. 13. No. 6. P. 444–453.

4. *Bratman, V.L., Kalynov, Yu.K., and Manuilov, V.N.* Large-orbit gyrotron operation in the terahertz frequency range // Phys. Rev. Lett. 2009. V. 102. No. 24. P. 245101.

5. *Bandurkin, I.V., Bratman, V.L., Kalynov, et al.* Terahertz Large-Orbit High-Harmonic Gyrotrons at IAP RAS: Recent Experiments and New Designs // IEEE Trans. Electron Devices. 2018. V. 65. No. 6. P. 2287-2293.

6. *Bratman, V.L., Fedotov, A.E., Kalynov, Yu.K., and Samoson, A.* THz Gyrotron and BWO Designed for Operation in DNP-NMR Spectrometer Magnet // J. Infrared, Millimeter, and THz Waves. 2013. V. 34. P. 837.

7. *Sirigiri, J.R., Maly, T.* Integrated high-frequency generator system utilizing the magnetic field of the target application // U.S. Patent No. 8,786,284, 22. Jul. 2014.

8. *Glyavin, M.Yu., Denisov, G.G., Zapevalov, et al.* High power terahertz sources for spectroscopy and material diagnostics // Phys.-Usp. 2016. V. 59. P. 595.

9. *Bratman, V.L., Fedotov, A.E., Fokin, A.P., et al.* Operation of a sub-terahertz CW gyrotron with an extremely low voltage // Physics of Plasmas. 2017. V. 24. P. 113105.

10. *Bratman, V.L., Fedotov, A.E., Kalynov, et al.* Numerical study of a low-voltage gyrotron (“gyrotrino”) for DNP/NMR spectroscopy. // IEEE Transactions on Plasma Science. 2017. V. 45. No. 4. P. 644.

Fast Magnetic Measurements Of 8.6 m Undulator

I. V. Davidyuk[1,2], Ya. I. Gorbachev[1], O. A. Shevchenko[1]

[1] Budker INP SB RAS, Novosibirsk, Russia, daveduke@outlook.com
[2] Novosibirsk State University, Novosibirsk, Russia

Abstract

The design of the large-aperture variable-period undulator (VPU) developed for the Novosibirsk free electron laser (FEL) is shortly described. High amplitude of on-axis field is achieved due to the arc shape of the magnet blocks and poles of the undulator. To conduct magnetic measurements and fine tuning of the undulator as well as to perform real time position tracking of magnet blocks, pulsed wire technique was adopted.

Introduction

A permanent magnet variable-period undulator with large (1.4) ratio of the aperture to the minimum period and possibility to change the number of poles was designed to upgrade the first stage FEL of the Novosibirsk FEL facility. The Novosibirsk FEL is a source of radiation in the THz range. The new undulator will replace the existing insertion device section which consists of two 4-meter-long undulators and the three-pole buncher. That will shift long-wave border of the tuning range from 200 μm to 450 μm. A comparison of parameters of former and new undulators is given in the following table:

Parameter	EM undulators	VPU
Period, cm	12	10 – 16
Gap (inner diameter), cm	8	14
Number of periods	2×32	50 – 80
Radiation wavelength, μm	90 – 240	82 – 450
Deflection parameter	0 – 1.1	0.45 – 1.9
Full length, cm	2x400	860

The idea of a VPU with free moving poles was proposed in [1]. At the minimum period lu_{min}, the developed undulator is a conventional hybrid device with neodymium magnets and poles made of soft steel with large permeability. All poles in the undulator are split into halves, and so the undulator is two arrays (upper and lower) of separate removable blocks. Each block is $lu_{min}/2$ long and consists of a magnet embraced by two pole halves and the whole structure is set in an aluminum frame. The pairs of blocks in upper and lower arrays are connected and placed on a carriage that can move freely along the undulator, see Fig. 1. Adjacent magnets in an array always have antiparallel longitudinal magnetizations, so they experience strong repulsion from each other. Due to repulsion force and the low-friction between guide rail and carriages, the undulator period can be changed via shifting the terminal blocks. This procedure, which changes the overall length of the undulator, is performed by pushers that can be remotely controlled. More details about similar undulator can be found in [2].

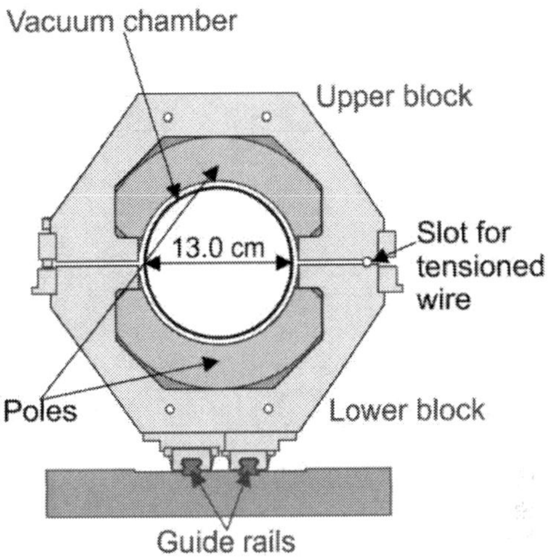

Fig. 1. Front view of the undulator

Obviously blocks in the undulator can create inhomogeneous distribution which may lead to phase errors between the radiation field and the electron transverse motion. The construction of the undulator provides possibility to make in situ measurements of magnetic field distribution in real time to monitor blocks' position errors and calculate the radiation spectrum via pulsed wire method. Slot for the stretched wire in undulator block is shown in Fig. 1.

The pulsed wire technique for magnetic measurements was proposed in [3] and since then it has been used for measurements of field when undulator gap cannot fit the support of the Hall probe or to speed-up the tuning of the undulator [4]. This method allows several measurements per second, thus it would significantly reduce the time of magnetic investigation of the developed 8.6 meters long VPU for the first FEL of Novosibirsk FEL facility.

Measurement system

The basic components of the measurement setup are a wire stretched along the axis of investigated magnetic field, a current pulse generator and a detector of the wire displacement. When current is passing through the wire it interacts with the magnetic field. Flexural waves that travel through the wire towards detector contain information about transverse field profile in the region of interest.

Theoretical analysis of this technique is presented in [5] where the stretched wire is treated as a thin rod in order to take dispersion into account. The expression for wire displacement at the detector position $z = 0$

reads:

$$x(0,t) = \frac{1}{2\sqrt{\mu T}} *$$
$$* \int_{t-\tau}^{t} I(t-t') \int_0^{v_0 t'} B_{eff}(z) dz\, dt', \quad (1)$$

where μ is the mass per unit length of the wire, T is the tension in the wire, τ is current pulse duration, v_0 is $\sqrt{T/\mu}$ and B_{eff} is defined by the following expression:

$$B_{eff}(z) = \int_{-\infty}^{\infty} A(k) e^{-i\frac{\omega(k)}{v_0}z} dk, \quad (2)$$

where $A(k)$ is the Fourier transform of the magnetic field in the path of the current and $\omega(k)$ is the dispersion relations for the flexural waves in the wire [5]. Using above expressions one can correct effects of dispersion, current pulse form averaging and restore the magnetic field distribution from the data obtained by the detector.

An experimental setup was constructed in order to adopt the technique. A beryllium copper wire with a diameter of 200 μm and length of 10 m was stretched by applying tension close to yield point. A capacitor based current pulse generator can generate pulses with duration of 50 μs and amplitude of 10 A. A homemade optocoupler is used as the wire position detector, it is based on fd-24k photodiode and provide 0.14 mV/μm gradient. Since the size of the undulator do not allow avoiding intersection of the main signal and reflections by elongation of the wire and the period of the undulator can be changed during operation of the FEL it was decided to use dampers based on oil drops, see [6], placed on the ridges of an automated pantograph.

Several trial launches of the system were performed, see example in Fig. 2.

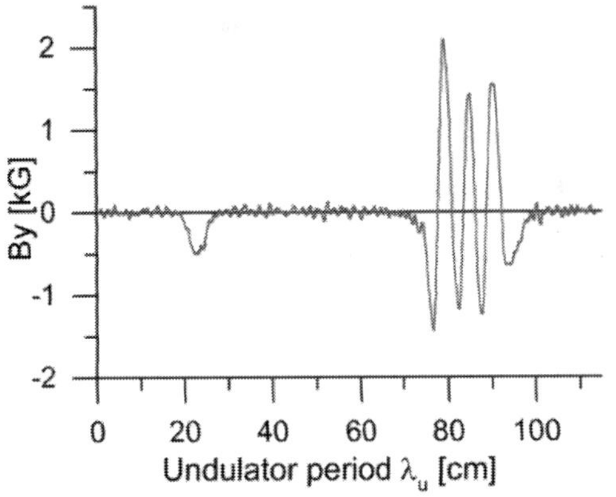

Fig. 2. Test measurement

The field distribution was reconstructed from the measured first integral data by differentiating, no filter was applied.

Discussion and conclusion

The described experimental setup is under active development now. Test measurements of reference magnet and VPU short prototype were conducted, see Fig. 2, and obtained results indicate that detector sensitivity should be improved in order to obtain larger signal to noise ratio.

The task to commission a pulsed wire measurements setup over 10 m long has not been challenged yet. It is yet to be determined whether complete compensation of signal distortion effects e.g. wire sagging influence, attenuation and dispersion is possible during postprocessing. Alternatively, complementary components, like wire supports and additional detector, could be applied to minimize the influence of distortion effects.

Currently authors attempt to increase the sensitivity of the detector and suppress impact from external noises in order to acquire clean data.

Acknowledgements

The work was supported by the Russian Science Foundation (project No. 14-50-00080).

References

1. *Vinokurov N.A., Shevchenko O.A., Tcheskidov V.G.* Variable-period permanent magnet undulators // Phys. Rev. Spec. Top. - Accel. Beams. 2011. V. 14. No. 4. P. 040701.
2. *Davidyuk I.V., Shevchenko O.A., Tcheskidov V.G., Vinokurov N.A.* Modeling and designing of variable-period and variable-pole-number undulator // Phys. Rev. Accel. Beams. 2016. V. 19. No. 2. P. 020701.
3. *Warren R.W.* Limitations on the use of the pulsed-wire field measuring technique // Nucl. Instruments Methods Phys. Res. Sect. A Accel. Spectrometers, Detect. Assoc. Equip. 1988. V. 272. No. 1–2. P. 257–263.
4. *Fan T.C., Lin F.Y., Hwang C.S., Hsu I.C.* Pulsed wire magnetic field measurements on undulator U10P // PACS2001. Proceedings of the 2001 Particle Accelerator Conference (Cat. No.01CH37268). Chicago: IEEE, 2001. P. 2775–2777.
5. *Kumar V., Mishra G.* Analysis of pulsed wire method for field integral measurements in undulators // Pramana. 2010. V. 74. No. 5. P. 743–753.
6. *Varfolomeev A.A. et al.* Wire method for magnetic field measurements in long undulators // Nucl. Instruments Methods Phys. Res. Sect. A Accel. Spectrometers, Detect. Assoc. Equip. 1995. V. 359. No. 1–2. P. 93–96.

Frequency control in subterahertz gyrotrons

A. Bogdashov, A. Fokin, A. Fedotov, M. Glyavin, M. Morozkin, Yu. Novozhilova, M. Proyavin, R. Rozental, A. Sedov, A. Tsvetkov, I. Zotova and G. Denisov

Institute of Applied Physics of the Russian Academy of Sciences, Nizhny Novgorod, Russia, ap.fokin@mail.ru

Subterahertz gyrotrons are promising for application in different fields of science and technology, like spectroscopy [1], media diagnostics [2] and materials processing [3]. Many of the application require the possibility of precise frequency tuning or stabilization. In this paper we present the experimental investigation of different ways of frequency control in medium-power continuous-wave gyrotrons carried out in the Institute of Applied Physics RAS. The capabilities and limitations of different methods are presented.

Magnetic field variation

The magnetic field variation provides the biggest range of frequency tuning. Excitation of different transverse modes with close caustic radius enables step frequency tuning in range up to 200 GHz [4]. The fine frequency tuning with band of up to 6 % of carrier frequency can be obtained by consequent excitation of high-order longitudinal modes. It is possible due to the use of a short cavity with low Q-factor and operation at low transverse modes in order to increase the electron-wave coupling. The proof of concept experiment on low frequency (using the 12 GHz gyrotron setup) demonstrated the 4% tuning band without any special optimization of the system (see Fig. 1). Based on this concept the project of the 200 GHz gyrotron with 1 kW output power and 10 GHz band was developed [5].

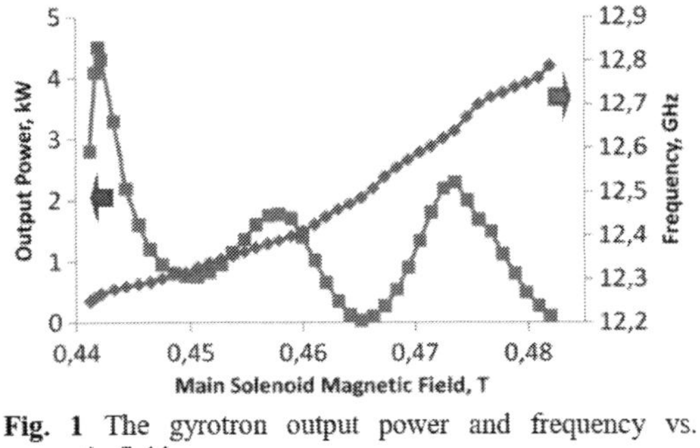

Fig. 1 The gyrotron output power and frequency vs. magnetic field

The frequency tuning speed of such method is limited by the parameters of the cryomagnets, mainly the maximum current variation rates, which are less than 0.1% of maximum field per minute. The possible solution is to apply the additional coil either in the cathode region or at the cavity. The latter approach was tested and demonstrated the possibility of operating mode switching with sweeping frequency up to 2 kHz. The same method can be applied for fast power modulation with frequency up to 10 kHz (Fig. 2), which is limited by auxiliary coil inductance and alternating filed screening [6].

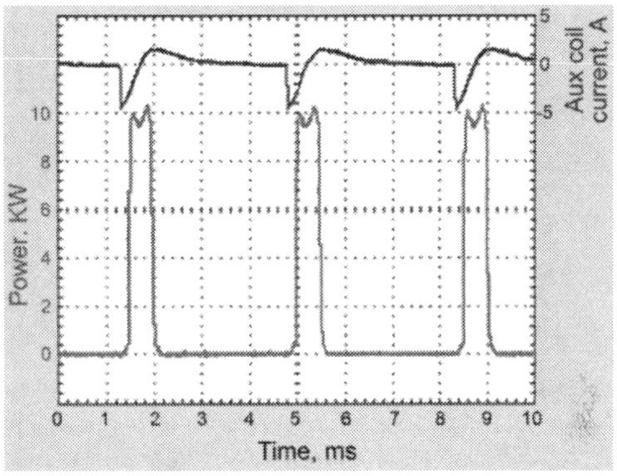

Fig. 2 Oscillograms of auxiliary coil current (top) and gyrotron output power (bottom)

Cavity temperature variation

In the recent experiments, the possibility of tuning the frequency of the 0.26 THz gyrotron [7] by more than 1 GHz was demonstrated by the simultaneous changing of the magnetic field of the main solenoid and the temperature of the cavity coolant. The measured frequency sensitivity of 4 MHz/°C allowed to tune the frequency in more than 200 MHz range and close the power gap between different longitudinal modes (Fig. 3).

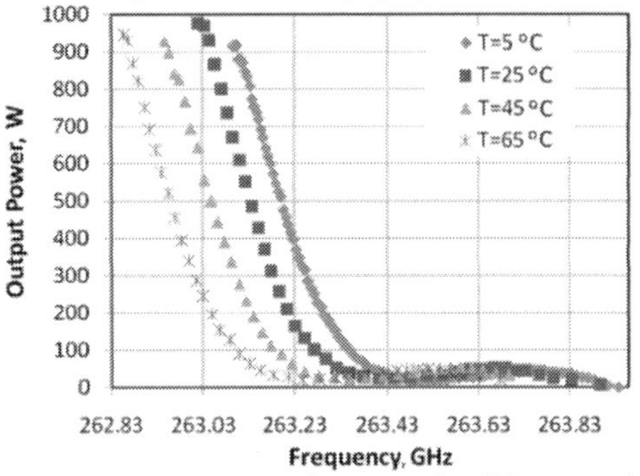

Fig. 3. Output power vs. frequency for different cavity coolant temperatures

Voltage variation

The fastest way of frequency control is the variation of the voltage at one of the electrodes of the magnetron-injection gun or the potential of the gyrotron cavity [8, 9]. The modulation of cathode voltage requires complex and expensive power supply; the speed of cavity voltage variation is limited by big capacitance, so the most effective way of frequency control is the anode voltage variation.

For this purpose, the special control unit was developed, that allowed voltage variation in range of 1 kV with speed more than 1 kV/μs. The control system based on this unit opened up the possibility of frequency tuning in 20 to 50 MHz range with up to 200 kHz modulation frequency. Application of the system for frequency stabilization by phase-lock loop against reference oscillator allowed to achieve the spectrum width of the gyrotron radiation of 1 Hz [10].

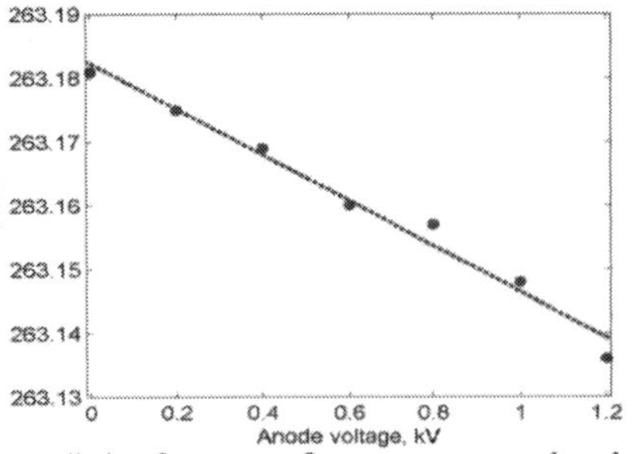

Fig. 4. Radiation frequency of a gyrotron vs. anode voltage

External signal and reflections

The presence of external signal or the power reflected from the load can be used for power and frequency control [11]. Experiments on influence of the signal, reflected from distant non-resonant load [12] show the possibility of frequency stabilization, while proof of concept experiments with mobile reflector demonstrate the means of slow power and frequency control (Fig. 5)

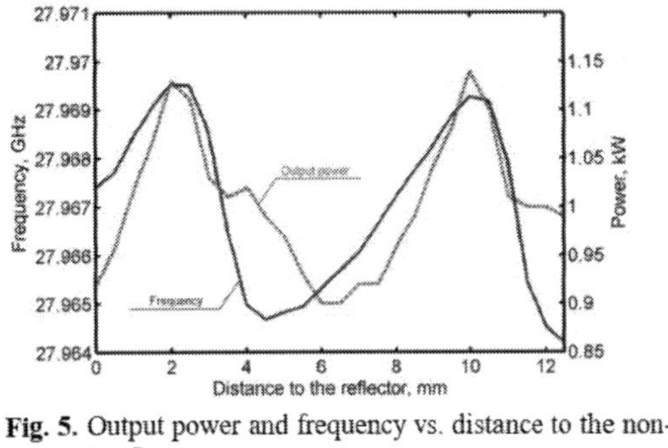

Fig. 5. Output power and frequency vs. distance to the non-resonant reflector

Conclusion

The IAP RAS gyrotron research team developed and tested a number of approaches to frequency control in sub-THz gyrotrons, including excitation of high-order longitudinal modes, cavity temperature sweeping, anode voltage variation and use of reflections. Beyond the scope of this work are mechanical means of frequency tuning, for example [13], which are surely promising, but their use in sub-THz frequency range is limited by the precision requirements.

Investigated approaches open up new prospects for the successful development of new medium power THz band gyrotrons with unique capabilities for modern applications

The work was supported by Russian Foundation for Basic Research grant № 17-02-00183

References

1. *Blank M, Rosay M, Engelke F* Instrumentation for solid-state dynamic nuclear polarization with magic angle spinning NMR // J. Magn. Reson., 2016, V. 264 P. 88-98

2. *Yamaguchi, Y. et al.*, High-power pulsed gyrotron for 300 GHz-band collective Thomson scattering diagnostics in the Large Helical Device // Nuclear Fusion, 2015, V. 55 013002

3. Vodopyanov, A. et al., Application of the 263GHz/1 kW gyrotron setup to produce a metal oxide nanopowder by the evaporation-condensation technique // Vacuum, 2017, V. 145, P. 340-346

4. Idehara, T. et al., A novel THz-band double-beam gyrotron for high-field DNP-NMR spectroscopy // Review of Scientific Instruments, 2017, V. 88, 094708

5. Fedotov, A.E. et al., Frequency Tunable sub-THz Gyrotron for Direct Measurements of Positronium Hyperfine Structure // Journal of infrared, millimeter and terahertz waves, 2018, V. 39, No. 10, P. 975–983

6. Glyavin, M.Yu., Luchinin, A.G., Morozkin, M.V., Ka-band 10 kW CW gyrotron with the wide-band fast frequency sweep // Rev. Sci. Instrum., 2012, V. 83, 074706

7. Glyavin M.Y. et al. Experimental tests of 263 GHz gyrotron for spectroscopy applications and diagnostic of various media // Review of Scientific Instruments, 2015, V. 86, No. 5, P. 054705

8. Idehara, T., Mitsudo, S., Ogawa, I. Development of High-Frequency, Highly Stable Gyrotrons as Millimeter to Submillimeter Wave Radiation Sources // IEEE Trans. Plasma Sci., 2004, V. 32, P. 910–916

9. Golubiatnikov, G.Y. et al. Gyrotron frequency control by a phase lock system // Tech. Phys. Lett., 2006, V. 32, P. 650

10. Fokin, A. et al., High-power sub-terahertz source with a record frequency stability at up to 1 Hz // Scientific Reports, 2018, V. 8, 4317

11. Novozhilova, Yu. et al., Gyrotron frequency stabilization under the influence of external monochromatic signal or wave reflected from the load: Review // Izvestiya VUZ, Applied Nonlinear Dynamics, 2017, V. 25, No. 1, P. 4-34

12. Glyavin, M.Yu. et al., Frequency Stabilization in a Sub-Terahertz Gyrotron With Delayed Reflections of Output Radiation // IEEE Transactions on Plasma Science, 2018, V. 46, No. 7, P. 2465 – 2469

13. Sabchevski, S, Idehara, T. Resonant Cavities for Frequency Tunable Gyrotrons // Int J Infrared Milli Waves, 2008, V. 29, P. 1–22

Problems of Amplifier Klystron Advancing into the Terahertz Band

A.D. Grigoriev[1,2]

[1]Saint Petersburg State Electrotechnical University "LETI", St. Petersburg, Russia, adgrigoriev@mail.ru,
[2]"Svetlana-Electronpribor" JSC

Electromagnetic oscillations and waves of the terahertz band have some important preferences compared with their neighbors. But lack of power coherent sources impedes wide application of these waves in radar and telecommunication systems.

Amplifier klystrons due to their high efficiency and relatively small dimensions are considered as one of the most perspective power sources of submillimeter and terahertz radiation. For example, the CPI Inc. (Canada) produces a number of Extented Interaction Klystrons (EIK) with working frequencies from 94 to 280 GHz and output pulse power from 3 to 0.03 kW [1]. In Russia, the "Svetlana" JSC produces the 8-mm wavelength amplifier klystron with output power 6 kW.

In spite of these advances, design and manufacturing of terahertz amplifier klystrons encounter a number of serious problems, namely:

- Very high current density of the electron beam;
- High cathode current density;
- Low cavities impedance;
- Small gap length and consequently little threshold voltage.
- Small collector area and dissipated heat.

To overcome these problems some techniques were proposed [2]. Among them are using of extended interaction (multigap) cavities, sheet electron beams and cathodes with high current density. But to obtain desirable effect the designer have to choose thoroughly the main device parameters starting from project input data.

The main input data for amplifier klystron design are its working frequency f_0, output power P_{out}, bandwidth Δf and amplification factor G. Basing on these data, the designer have to estimate device efficiency η, which can change from 10 to 80 %. The more is working frequency and bandwidth, the less is efficiency.

When efficiency is estimated, one can calculate electron beam current $I_0=P_{out}/(\eta U_0)$, where U_0 – accelerating voltage. Increasing of the voltage leads to decreasing beam current density, increasing gap length and optimal value of the cavity R/Q ratio. As a rule, such values cannot be achieved in a single gap cavity, so multigap cavities have to be used. Besides, one has to use complicated power sources.

Decreasing accelerating voltage, vice verse, decrease demanding R/Q ratio, but increases beam current. Large magnetic field is needed to transfer such beam throw the interaction region. And cathodes with high current density have to be used.

The efficient way to overcome these difficulties is using a sheet electron beam. With a given beam current a sheet beam with cross-section $w{\times}d$ has current density approximately w/d times less than a pencil beam with diameter d. But, cavities designed for interaction with a sheet beam have R/Q ratio much less than cavities for a pencil beam.

The mentioned considerations were taken into account in the project of 95 Ghz high power klystron proposed by the "Svetlana-Electronpribor" JSC in collaboration with LETI. The pulsed output power of the device is no less than 1 kW, relative bandwidth ≥1% and amplification factor ≥45 dB. After examining parameters of the existing devices, efficiency of the klystron was set at 10 % level. It is rather low value, but taking in account possible manufacturing difficulties, this value seems reasonable.

The accelerating voltage was set to 10 kV, which is rather low compared with, for example, CPI EIKs with U_0=20…25 kV. Chosen value helps us to design klystron cavities and obtain demanded bandwidth. The beam current in our case have to be 1 A. A pencil beam with such current has current density more than 5 Ka/cm^2, which is unreal, so we chose sheet current with 1×0.1 mm^2 cross-section.

Three-gaps cavities composed from a section of wide comb slow wave system were chosen as an input and idle resonators and four-gaps cavity of the same type – as an output resonator. The working mode of the all cavities is 2π because this mode well separated from the neighboring ones and has positive electron load, preventing self-oscillations.

To find optimal cavities configuration we used computer simulation code RFS [3], based on vector finite element method. Calculated cavity parameters are given in the table, where eigenfrequency f_0, Q-factor, R/Q ratio, interaction factor M and relative electron load Ge/G_0 are listed.

Cavity type	f_0, GHz	Q_0	R/Q, Ohm	M	G_{er}
Input & idle	94.3667	1120	83.1	0.654	0.3
Output	95.0617	1310	112	0.65	0.2

Fig. 1 shows the computer cavity models with main dimensions. Input and output cavities are coupled with waveguides by slots in their sidewall. Ceramic rods are used for cavity tuning (not shown on Fig. 1).

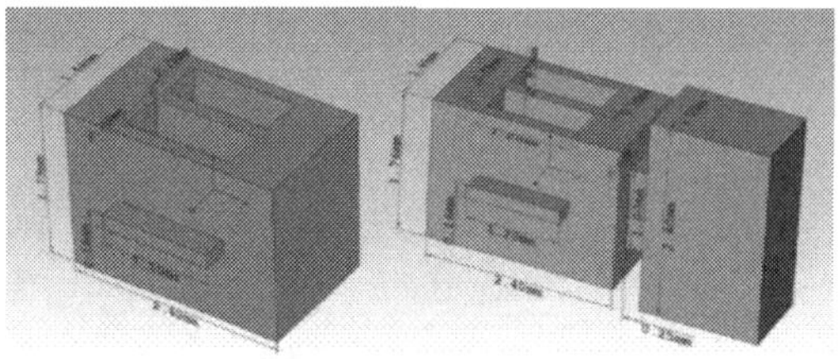

Fig. 1. Computer models of the input cavity (left) and output cavity with waveguide (right).

Input and output cavities frequency charts are shown on Fig. 2 for beam current 0.8 A and accelerating voltage 10 kV. As can be seen, the output cavity coupling slot provides bandwidth about 1 GHz, i. e. almost 1 %.

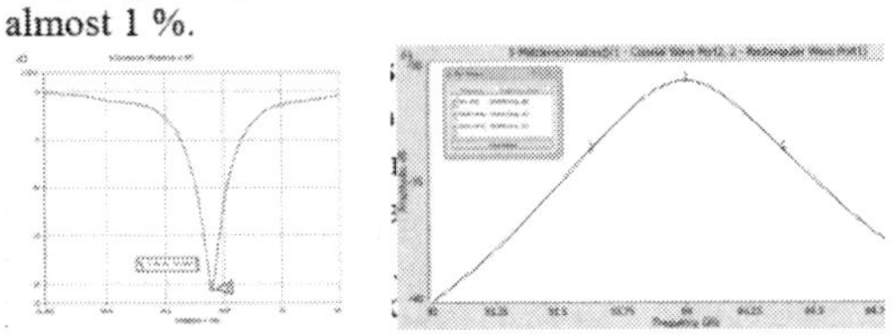

Fig. 2. Input (a) and output (b) cavities frequency charts.

The model of the klystron interaction region is shown on Fig. 3. The model contains 6 cavities, their eigen frequencies were tuned according to the skirtron scheme, where eigen frequencies of all intermediate (idle) cavities are higher than f_0. Input cavity eigen frequency is lower than f_0, and output cavity frequency is equal to f_0. Electron gun, collector and magnetic focusing system are not shown.

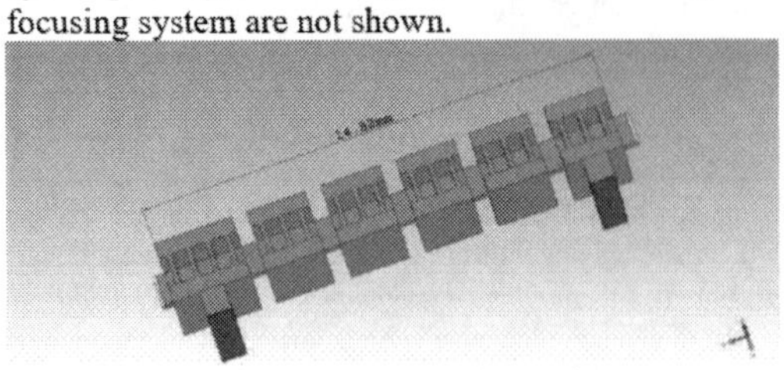

Fig. 3. Model of the klystron interaction region.

Field-particle interaction was modeled by Particle-in-Cell (PIC) method. Total number of particles in the interaction region was 26500, simulation time 15 ns, which was sufficient to reach stationary regime. Fig. 4. Shows calculated amplitude chart of the klystron at the central frequency. Klystron amplification factor changes from 60 dB for small signal regime to nearly 50 dB for the saturation point. These values are greater than targeted one.

Output power as a function of frequency at the input power level 8 mW is depicted on Fig. 5. The -3 dB bandwidth is 380 MHz, which is somewhat less than needed, but we hope that implementation of output filter circuit would give 500 MHz, or 0.5 % from carrying frequency.

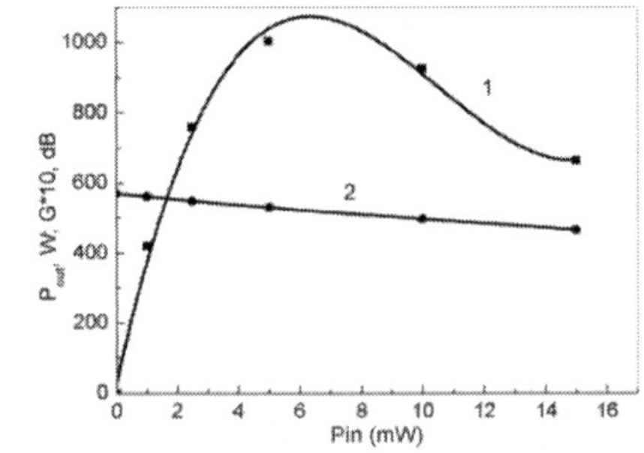

Fig. 4. Output power (1) and amplification factor (2) vs input power

The project confirmed the possibility of design and manufacturing 3-millimeter wavelength power klystron, using existing technological equipment.

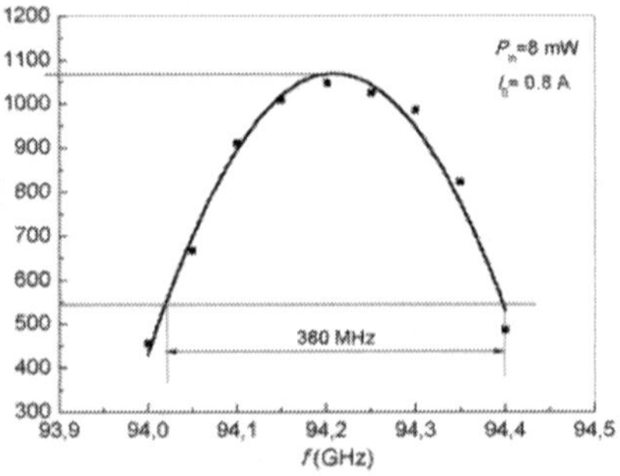

Fig. 5. Output power vs frequency

References

1. http://citeseerx.ist.psu.edu/viewdoc/download?doi=10.1.1.734.2121&rep=rep1&type=pdf .

2. *Cariotakis, G.* High power klystrons: theory and practice at the Stanford Linear Accelerator Center // SLAC-PUB 10620. 2005. 138 pp.

3. *Grigoriev A. D., Salimov R.V., Tikhonov R.I.* Cellular handsets antenna modeling by vector finite element method // Journal of communication technology and electronics. 2012, V.57, No 3, pp 261-270.

A

Electron-optics systems with decreased life-time of trapped electrons for terahertz gyrotrons

V.N.Manuilov[1,2], A.L.Goldenberg[1], M.Yu.Glyavin[1], K.A.Leshcheva[1,2]

[1]Institute of Applied Physics, Russian Academy of Sciences, N. Novgorod, Russia
[2]Lobachevsky State University of Nizhni Novgorod, N. Novgorod, Russia, e-mail: manuilov@rf.unn.ru

One of the most promising applications of novel gyrotrons is a high resolution spectroscopy [1]. For this goal, scientific society needs sources with rather low microwave power (typically, from 10W up to 1 KW) and frequencies 0.3-1.0 THz or even more. For the consumer, low operating voltage is preferable. Such frequency range can be obtained mainly at the second or even third cyclotron harmonic due to present cryomagnets limitations. Same time with frequency value, frequency tunability is very important for spectroscopy applications. It is a well known, that in a "short" cavity with a reduced length of homogeneous section (and, correspondently, lower Q-factor) tuning range becomes bigger. Unfortunately, decreasing of cavity length proportionally reduce the normalized length of the interaction space and, finally, push down gyrotron efficiency. Efficiency can be restored or even increase by rise of the electron pitch-factor g (ratio of gyration and longitudinal velocities) from 1.1-1.3 (at present tube) to 1.7-2.0. But in such case increase a number of electrons, reflected from the magnetic mirror (fig.1), that causes the beam instabilities. Suppression of this effect is possible by decreasing of the life-time of trapped electrons [2]. An effective way to eject trapped electrons is to install on the second anode in the triode gun the longitudinal ribs (fig.2).

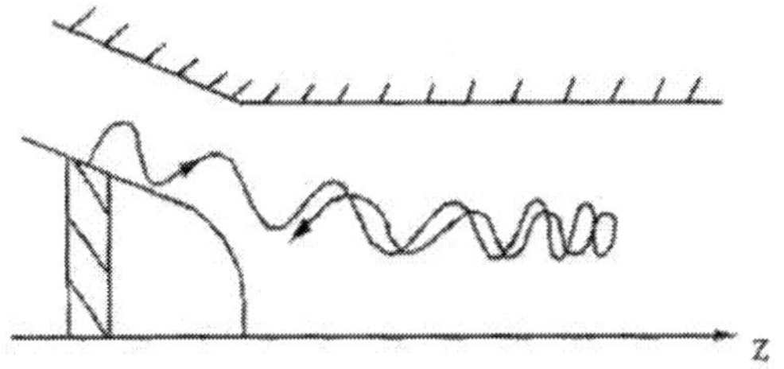

Fig. 1. The reflection of electrons from a magnetic mirror

The principle of extraction in a such gun is based on the particle drift in the crossed electric and magnetic fields. In traditional configuration of the above mentioned electron-optic system, such ribs intercept not only reflected particles, but also essential part of the primary electron beam, passed from the cathode to the cavity. To combine the small intersection of the primary beam with quick (during 1-2 longitudinal oscillations in the trap) catch of trapped electrons it is necessary to use the sectioned cathode, where emission takes place only from some sections spaced apart from azimuth angle, and/or special complex geometry of the ribs.

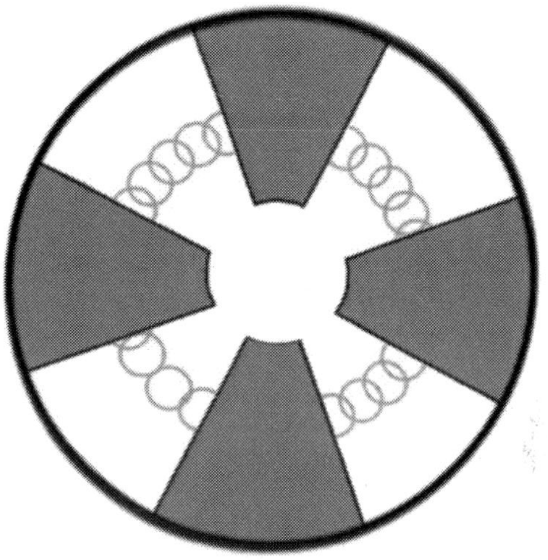

Fig. 2. The longitudinal ribs in the triode gun

In the report the preliminary geometry estimations of the sectioned electron-optic system are performed, including the profile and position of the ribs. Two versions of electron-optic system for low-voltage (16-20 kV) gyrotrons operating at the frequencies 10 GHz and 300 GHz are considered. The numerical simulation of both systems is performed. The possibility to catch the trapped electrons during first 1-2 longitudinal oscillations in the trap is shown. Same time reasonable thermal load on the longitudinal ribs is obtained, good enough for CW operation. So, proposed system looks promising for the formation of the stable helical electron beam and looks reasonable for future gyrotrons for spectroscopy applications.

The reported study was funded by RFBR according to the research project № 18-32-00142

References

1. *J.H.Booske, R.J.Dobbs, C.D.Joye, C.L.Kory, G.R.Neil, G.Park, J.Park, R.J.Temkin*. Vacuum electronic high power terahertz sources. // IEEE Trans. on Terahertz Science and Technology, 2011, V.1, No.1. pp.54-75

2. *E. V. Ilyakov, I. S. Kulagin, V. N. Manuilov, A. S. Shevchenko* Experiments on the formation of an intense helical electron beam under conditions of picking-up of the electrons reflected from the magnetic mirror // *Radiophysics and Quantum Electronics*, 2007, Volume 50, Issue 9, pp. 713–719.

Development of the Prototype of High Power Sub-THz Gyrotron for Advanced Fusion Power Plant (DEMO)

M.Morozkin[1], G.Denisov[1], E.Tai[2], E.Soluyanova[2], A.Sedov[1], A.Fokin[1], A.Kuftin[1], A.Tsvetkov[1], M.Bakulin[2], E.Sokolov[2], V.Malygin[1], M.Proyavin[1], V.Zapevalov[1], O. Mocheneva, and M.Glyavin[1]

[1]Institute of applied physics RAS, Nizhny Novgorod, Russia, morozkin@ipfran.ru
[2]Gycom Ltd., Nizhny Novgorod, Russia,

Next stage of the development of controlled nuclear fusion systems after the ITER project in accordance with the European Fusion Roadmap [1] is a demonstrational reactor DEMO, which should be the first thermonuclear reactor generating electrical energy. The parameters of the gyrotrons in the electron cyclotron heating system of the plasma of this reactor are not yet definitively specified, but frequencies of about 230-250 GHz and power up to 2 MW are discussed with an efficiency of more than 60% (full efficiency with energy recovery).

At present, most of the participants of the ITER project, which are involved in creating electron-cyclotron plasma heating systems, have already begun developing prototypes of radiation sources that meet the requirements of the DEMO project [2]. In the Institute of Applied Physics and Gycom Ltd. for these purposes, a prototype gyrotron with a frequency of 250 GHz and a radiation power of up to 200 kW in continuous mode was designed and tested (see fig. 1), intended for use in advanced fusion installations.

Fig. 1. Photo of the gyrotron in the cryomagnet.

The main factor limiting the output power in gyrotrons for plasma heating is the limitation of thermal loads on the cavity. Therefore, to increase the output power, it is necessary to increase the dimensions of the interaction space and use high-order modes. However, for the prototype gyrotron with an operating frequency of 250 GHz, the maximum size of the resonator and the electron-optical system was limited by the magnetic system. To generate the cyclotron frequency at the first harmonic, a magnetic field of more than 9.5 T is required, which is available in the cryomagnet at the IAP RAS, which has a warm bore of 100 mm in diameter. The warm bore diameter determined the dimensions of the electronic optics and cavity taking into account the technical requirements for cooling systems.

As a result, to ensure stable generation at the selected frequency, the $TE_{19.8}$ mode was selected as the working mode with a cavity radius of 9.34 mm.

Investigation of the coupling factors of the electron beam with the modes of the cylindrical cavity made it possible to determine the optimum radius of the electron beam in the cavity, which simultaneously provides both a large coupling coefficient with the working mode and the smallest coupling with parasitic modes. Due to the close coupling factors of the working mode $TE_{19.8}$ and the parasitic modes $TE_{18.8}$ and $TE_{20.8}$, the optimal radius of the electron beam is R_{beam} = 3.93 mm, which differs slightly from the radius of maximum coupling coefficient with the working mode R_{opt} = 3.85 mm.

To ensure the effective interaction of the electron beam with the high-frequency field, the cavity length was optimized. The choice of length is determined by the compromise between the increase in the efficiency of the interaction of electrons with the RF field and the limitation of ohmic losses in the walls of the cavity (see Fig. 2). As a result of modeling, the length of the homogeneous section was chosen to be L = 10 mm.

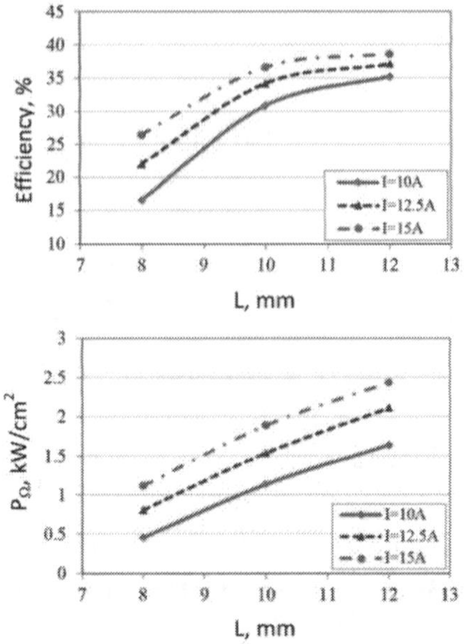

Fig. 2. Dependence of the efficiency and the density of ohmic losses on the length of a homogeneous part of the cavity

The calculated parameters of the developed gyrotron are given in the following table.

Main parameters of the gyrotron

Operating frequency, GHz	250
Operating mode	$TE_{19,8}$
Cyclotron harmonic	1
Accelerating voltage U_0	55 kV
Depressed collector potential	30 kV
Magnetic field in the cavity	9.6-9.7 T
Beam current (nominal in CW mode)	12 A
Beam current (pulsed mode)	20 A
Cavity radius	9.34 mm
Cavity length (homogeneous part)	10 mm
Beam radius inside cavity	3.93 mm
Pitch factor, not less than	1.1

The tube is equipped with an internal quasi-optical wave beam converter which provides a transformation of the operating mode into a Gaussian TEM_{00} beam and couples it radially to the vacuum window. The converter consists of a shaped waveguide, a quadratic mirror, four flat mirrors, and a synthesized mirror, which allows directing the wave beam in the output window.

The first experimental tests of the developed gyrotron were carried out at the IAP RAS on a high-frequency gyrotron setup equipped with a dry cryomagnet Jastec JMTD10T100 with a warm bore of 100 mm in diameter and a magnetic field of up to 10 T. Due to the limitations of the power sources existing in the IAP RAS, the first experiments were carried out in a pulsed mode. For this tests a removable BN output window with a diameter of 66 mm was manufactured and installed. The thickness of the window was chosen to be about 3.1 mm in order to minimize the reflections at the operating frequency of 250 GHz. The pulse duration in the experiment was 20-40 μs with a repetition frequency of 10 Hz. The power was measured with a water calorimetric dummy load equipped with thermal sensors in the inlet and outlet nozzles, the calibration was carried out with a heater in the circuit of a calorimeter with a known power of 100 W. The output power of the gyrotron was calculated on the basis of the measurement of the steady-state average power in the pulsed mode, taking into account the duty cycle of 2500.

The frequency of the output radiation was measured using a resonant-cavity wave meter. The registered value of 249.74 GHz has been obtained at the following operating parameters: accelerating (cathode) voltage U_0 = 55 kV, beam current I_b = 12.5 A, and magnetic field B = 9.625 T. These results are in agreement with the simulation data.

Reduction of the thermal load on the resonator during operation in the pulsed mode made it possible to test gyrotron with parameters exceeding the nominal design values. Thus, with an increase in the electron beam current, it was possible to reach a power of 330 kW at a beam current of 20 A and an accelerating voltage of 55 kV. The obtained experimental results are shown in Figures 3 and 4.

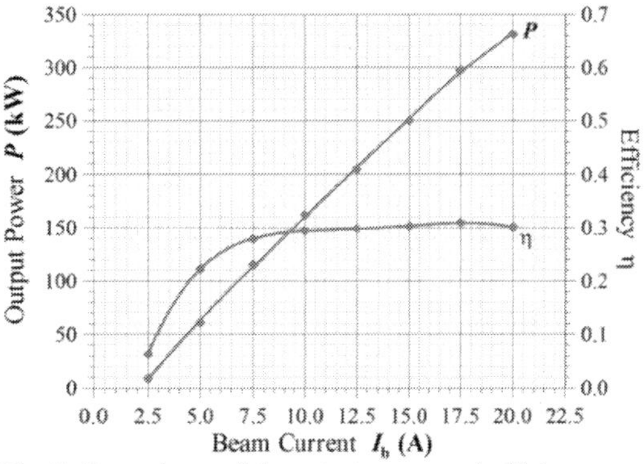

Fig. 3. Dependence of the output power and efficiency on the beam current, U_0 = 55 kV.

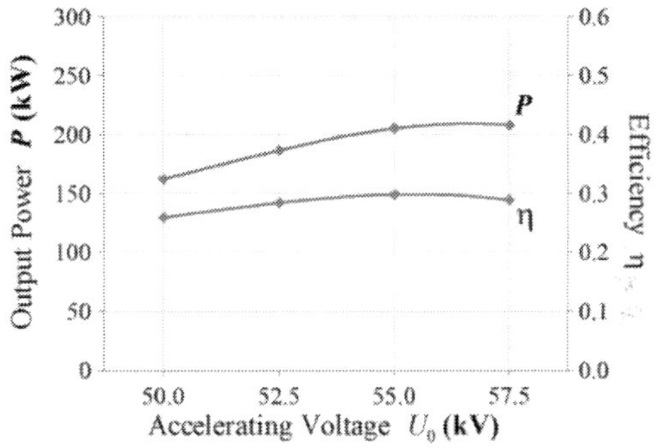

Fig. 4. Dependence of the output power and efficiency on the accelerating voltage, I_b = 12.5 A.

Measurements of the transverse distribution of the microwave beam formed by the built-in quasi-optical gyrotron converter were made, and the content of the TEM_{00} wave (Gaussian beam) was estimated. The measurement was performed according to the thermal imaging technique [3] in several cross sections during the propagation of a wave beam. As the analysis showed, the content of a Gaussian wave beam in the reconstructed wave beam is 98.6%.

Thus, the tested 250-GHz gyrotron showed output parameters close to the calculated ones: the power of 220 kW at a beam current of 12.5 A (corresponds to the beam current in a continuous mode) and 330 kW at a beam current of 20 A. The generation efficiency was about 30% (without energy recovery), the content of the TEM_{00} wave in the output beam is 98.6%.

This work was supported by Russian scientific fund, grant 14-12-00887.

References

1. *https://www.euro-fusion.org/eurofusion/roadmap/*
2. *J. Jelonnek et al. Design considerations for future DEMO gyrotrons:* A review on related gyrotron activities within EUROfusion // Fusion Eng. Des. 2017. V. 123, P. 241–246.
3. *S. O. Kuznezov, V. I. Malygin* Determination of gyrotron wave beam parameters // Int. J. Infrared Millimeter Waves 1991. V. 12, No. 11, P. 1241–1252.

Thermal regimes and THz generation from BSCCO mesas

L.S. Revin[1], E.A. Vopilkin[1], A.L. Pankratov[1], S.A. Kraev[1], A.A. Yablokov[1], S.A. Churin[1], A.B. Kulakov[2]

[1]Institute for Physics of Microstructures of Russian Academy of Sciences, Nizhny Novgorod, Russia, alp@ipmras.ru
[2] Institute of Solid State Physics of RAS

At present, the question of creating generators of subTHz and THz ranges based on $Bi_2Sr_2CaCu_2O_{8+x}$ (BSCCO) mesastructures is of great interest [1]. BSCCO is a high-temperature superconductor with strong anisotropy, which leads to the appearance of an internal Josephson effect [2]. In other words, CuO layers represent a series of Josephson chains formed on an atomic scale. The first success in observing of electromagnetic radiation from BSCCO structures was achieved in 2007 [3]. The observed radiation had a power of the order of 0.5 μW and a frequency up to 0.85 THz. Since that time the radiation power has been increased to a value of hundreds of microwatts [4]. These achievements make it possible to use BSCCO mesostructures as sources for noise spectroscopy [5, 6].

In our group the technology of electrolytic build-up of a BSCCO/Cu heterostructures is worked out, representing copper covering over a thin layer of gold, that allows to get rigid single BSCCO mesastructures with a good heatsink. To obtain the described samples, we used a BSCCO single crystal, fabricated at ISSP RAS, of several microns thick with lateral dimensions of several millimeters. On one of its surfaces, a layer of 50 nm thick gold was deposited by thermal deposition. After this a layer of copper with thickness of about 30 microns was electrolytically built up on top of the gold layer. Then a layer of 50 nm thick gold was deposited on BSSCO second surface by thermal deposition. After that, using the original method of fast chemical etching [7], developed in our team, the cylindrical single mesastructures were fabricated. The obtained sample was glued to the copper holder, Fig. 1, after which it is taped to the contact pads and installed in a cryostat.

Fig. 1. Photo of the mesas connected by wires. Inset: Schematic view of the Si substrate with BSCCO sample, attached to the copper holder

To measure the radiated power, the room temperature detector (Golay cell) together with the horn and infrared filters were used, Fig. 2. Although the obtained samples were not intended for effective power radiation, since antennas and lenses were not supposed, the level of the expected signal was sufficient. The calculated diagram had maxima at 45° from the plane and only a small amount of radiation was detected outside the cryostat window with the holder parallel to the cryostat plate.

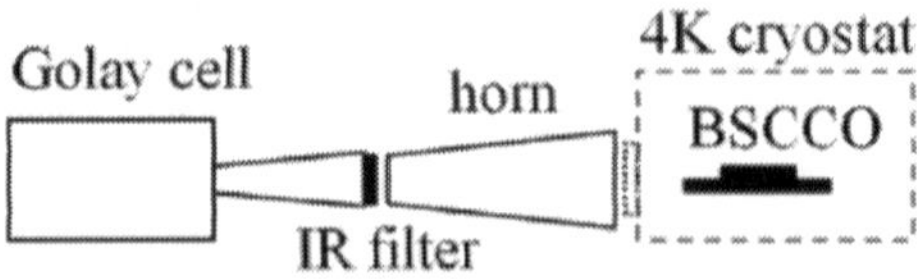

Fig. 2. Schematic of the experimental setup

In Fig. 3 the current-voltage characteristic of BSCCO mesa is presented, forward and backward branches are measured. The shape of the curve indicates the presence of Josephson generation in the structure. On the forward branch there are transitions from the superconducting state to the resistive one both of the individual junctions of the layered superconductor and of the entire structure. On the backward branch there is an area of negative differential resistance, which indicates the presence of a "hot spot" inside the mesastructure, the heating region of the structure, which is the channel for synchronization of electromagnetic waves in the superconductor-insulator-superconductor layers.

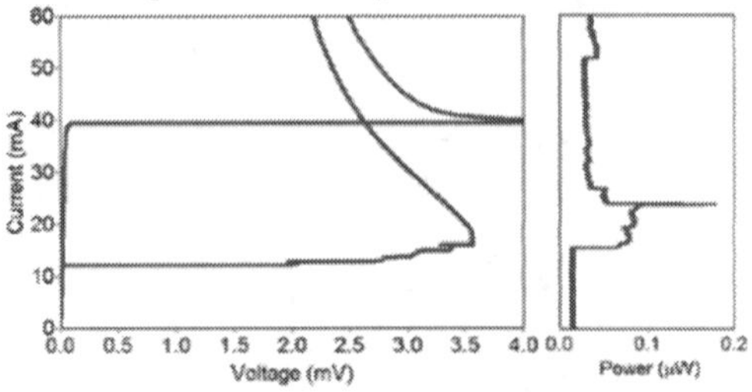

Fig. 3. The voltage and received power as a function of bias current

As a result, the statistics of the obtained structures were processed, and the characteristics of the samples were compared with the world level. Thus, the critical current density of the obtained samples J_c = 30 - 70 A/cm^2 is close to the literature data J_c = 80-250 A/cm^2. The gap voltage in terms of one junction V_g =

0.6 - 0.7 mV is also close to the literature data V_g = 1 - 1.4 mV.

Experiments on detecting the radiation from BSCCO sample allow to obtain a signal of the order 0.05–0.08 μW. Radiation was observed at the plate temperature of 18 K when the substrate was attached to a copper holder by a rubber glue. Conversely, when the substrate was attached by indium, no radiation was observed, presumably due to a too-low holder temperature at a better heat sink or not enough overheating due to a too-efficient heat sink.

Fig. 4. Photo of BSCCO single crystal fabricated by Bridgman technique

Recently, first BSCCO single crystal samples have been fabricated at IPM RAS by Bridgman technique, having a plain surface of order 3 by 10 mm, see Fig. 4. The samples have been measured in a cryostat and critical temperature of 84K has been achieved for oxigen annealed samples, see Fig. 5. Currently, new BSCCO single mesas with Cu sublayer from these single crystals are under fabrication.

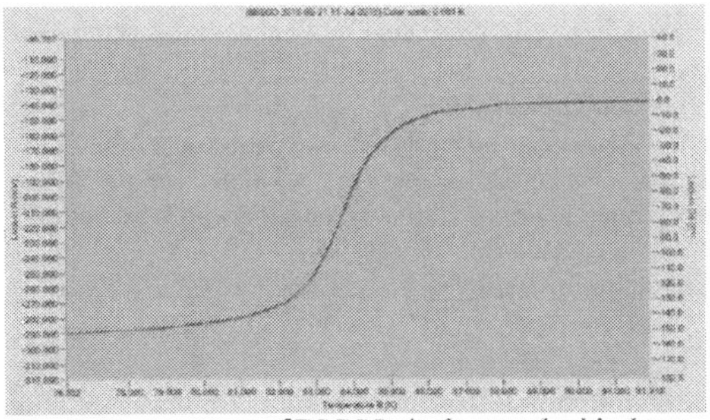

Fig. 5. Measurement of BSCCO single crystal critical temperature

In conclusion, we report on the progress of the wet etching method for the fabrication of BSCCO mesastructures, making it possible to obtain samples with a large thickness on the order of 1-15 μm. The current-voltage curves of BSCCO mesas are measured at different plate temperatures and there is a significant dependence on the thermal interface. It is possible to determine the power (of the order of 0.1 μW) outside the cryostat by a room temperature detector. BSCCO single crystals, having critical temperature of 84K were fabricated by Bridgman technique.

The work is supported by RSF (project 16-19-10478).

References

1. *Welp, U., Kadowaki, K., Kleiner, R.,* Superconducting emitters of THz radiation // Nat. Photonics. 2013, Vol. 7, pp. 702

2. *Kleiner R.,, Steinmeyer, F., Kunkel G., Mueller P.,* Intrinsic Josephson effects in Bi2Sr2CaCu2O8 single crystals // Phys. Rev. Lett. 1992, Vol. 68, p. 2394.

3. *Ozyuzer, L., Koshelev, A.E., Kurter, C., Gopalsami, N., et al* Emission of Coherent THz Radiation from Superconductors // Science 318, 2007, p. 1291.

4. *Yamaki, K., Tsujimoto, M., Yamamoto, T., Furukawa, A., Kashiwagi, T., Minami, H., Kadowaki, K.,* High-power terahertz electromagnetic wave emission from high-T c superconducting Bi2Sr2CaCu2O8+δ mesa structures // Opt. Express. 2011. Vol. 19, P. 3193.

5. *Revin, L.S., Vaks, V.L., Koshelets, V.P., Wang, H., //* EPJ Web of Conferences. 2017, Vol. 132, P. 03042.

6. *Sobakinskàya, E.A., Vax, V.L., Kinev, N.V., Koshelets, V.P., Wang, H., //* Proc. 1st Russian Microwave Conf. 2013, p. 1015.

7. *Vopilkin E.A., Chiginev A.V., Revin L.S., Tropanova A.N., Shuleshova I.Yu., Okhapkin A.I., Shovkun A.D., Kulakov A.B., Pankratov A.L.,* Quick and reliable technology for fabrication of stand-alone BSCCO mesas, // Supercond. Sci. Technol. 2015, Vol. 28, P. 045006.

Development of powerful long-pulse Bragg FELs operating from sub-THz to THz bands based on linear induction accelerators: recent results and projects

N.Yu. Peskov[1], N.S. Ginzburg[1], A.M. Malkin[1], A.S. Sergeev[1], V.Yu. Zaslavsky[1], A.K. Kaminsky[2], S.N. Sedykh[2], I.I. Golubev[2], S.M. Golubykh[2], A.P. Kozlov[2], A.I. Sidorov[2], A.V. Arzhannikov[3], D.A. Nikiforov[3], S.L. Sinitsky[3], D.I. Skovorodin[3] and A.A. Starostenko[3]

[1]Institute of Applied Physics RAS, Nizhny Novgorod, Russia, peskov@appl.sci-nnov.ru
[2]Joint Institute for Nuclear Research, Dubna, Russia,
[3]Budker Institute of Nuclear Physics RAS, Novosibirsk, Russia

Introduction

To date, a series of successful experiments on the realization of powerful high-efficiency free-electron masers (FEMs) has been carried out in collaboration between JINR (Dubna) and IAP RAS (N.Novgorod) based on linac LIU-3000 0.8 MeV / 200 A / 250 ns. The use of high-selective Bragg resonators of new types allowed stable regime of narrow-band generation to be obtained in these experiments up to W-band at multi-MW powers [1, 2]. The radiation parameters achieved allows JINR-IAP FEM to be used in several applications including testing of components for high-gradient accelerators, biology-medical studies, physics of nanoparticles, etc [3].

The aim of present work is advance of powerful long-pulse FELs into the THz frequencies. Such project is initiated based on the linear induction accelerators, which are developing in the BINP RAS (Novosibirsk) 5 - 20 MeV / 2 kA / 200 ns. In this paper, we discuss the basic parameters of such FELs and the results of their computer simulations. As a microwave system capable to provide stable narrow-band operation in strongly oversize interaction space in these FELs, we consider so-called advanced Bragg structures with the feedback loop involving propagating and quasi-cutoff waves (Fig. 1) [4, 5].

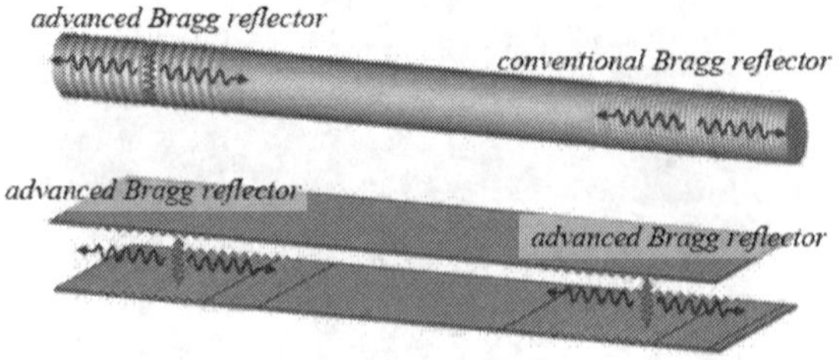

Fig. 1. Different schemes of Bragg resonators: **(a)** combined two-mirror resonator based on advanced and conventional Bragg reflectors of cylindrical geometry and **(b)** two-mirror resonator consisted from two advanced Bragg reflectors of planar geometry. The partial wave-fluxes in both reflectors are shown by the arrows.

JINR-IAP FEM operating from Ka- to W-band

The JINR-IAP FEM is driven by induction linac LIU-3000 (JINR). A reversed guide magnetic field configuration is used in the FEM, which possesses low sensitivity to the initial beam spread and, as a result, provides high-efficiency interaction. For operation from Ka- to W-bands, helical wigglers having periods from 6 cm to 3 cm correspondingly were constructed. Enhance in amplitude of the transverse magnetic field in short-period wigglers alongside with refining its transverse homogeneity was achieved by optimization of the currents distribution in the wiggler winding. To improve the quality of helical electron beam formation at short wavelengths, the slowly up-tapered wiggler entrance was optimized as well [2].

For operation in different parts of the millimeter wavelength band a series of combined two-mirror Bragg resonators, consisted from the up-stream high-selective advanced Bragg reflector and down-stream weakly-reflecting conventional Bragg reflector (Fig.1a), was elaborated. Resonators of such type allow decrease in Ohmic losses associated with excitation of the cut-off mode. For operation at 30 GHz, combined Bragg resonator was constructed with the oversized parameter $Ø/\lambda \sim 2$ and included advanced Bragg reflector with a feedback loop formed by two counter-propagating $TE_{1,1}$ waves and cutoff $TE_{1,2}$ wave. Resonator for V-band (60 GHz) was composed with advanced up-stream Bragg reflector having feedback loop $TE_{1,1} \leftrightarrow TM_{1,2}$ - cutoff $\leftrightarrow TE_{1,1}$ at the oversize parameter $Ø/\lambda \sim 3$. For W-band (80 GHz) advanced Bragg reflectors based on excitation of the feedback cutoff wave of $TE_{1,5}$ - type were designed to provide an effective FEM operation at oversize parameter $Ø/\lambda \sim 5$ (Fig. 2). In accordance with the 3D simulations, effective narrow-band reflections were demonstrated in "cold" test of novel Bragg structures. In all resonators described above the reflection band in advanced Bragg structures was measured 0.5 - 0.7 GHz with maximum power reflection up to 80 - 90%, while the conventional structures had ~ 60% reflection in much broader band of 2.5 - 3 GHz.

In the proof-of-principle experiments at LIU-3000 a narrow-band operation of novel scheme of FEM-oscillators was obtained under design parameters. At Ka-band, stable single-mode operation at the frequency of 30.2 GHz was observed in accordance with simulations. The output power amounted up to 20 MW (efficiency about 15 - 20%) with the spectrum width of 6 - 7 MHz (measured by heterodyne technique) close to the theoretical limit (Fig. 3).

In the experiments at V- and W-band, the radiation spectrum was measured by means of cut-off filters set with the accuracy of about 1 GHz. Both FEM demonstrated the oscillation frequency belonging to the designed feedback loop of the hybrid resonator in

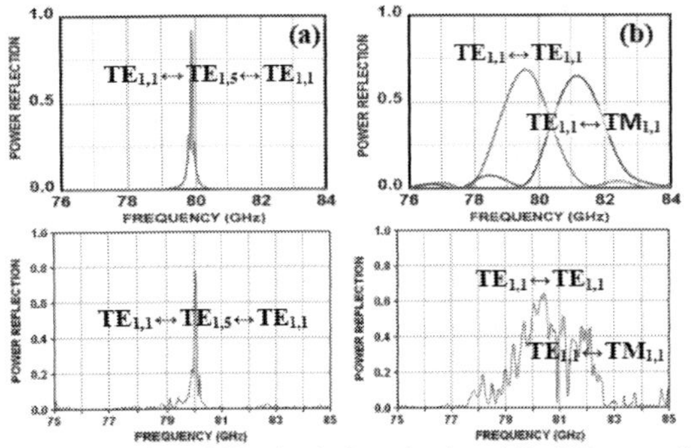

Fig. 2. Results of 3D simulations (top) and "cold" tests of Bragg structures of different types at W-band. Frequency dependence of reflection of **(a)** advanced (l_{adv} = 12 cm) and **(b)** conventional (l_{con} = 8 cm) reflectors.

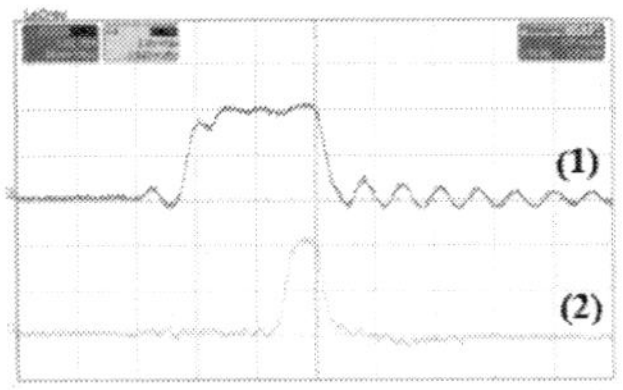

Fig. 3. Results of the JINR-IAP FEM experiments in W-band. Typical oscilloscope traces of (1) beam current and (2) output RF-pulse (100 ns / div.).

the vicinity of 59 GHz and 80 GHz correspondingly. In the resonators of optimal geometry the output power (measured by calorimeter) in both frequency regions amounted to 5 - 7 MW when the beam current was about 70 - 100 A (Fig. 3), which corresponded to the electron efficiency of up to 10 - 12%.

Project of powerful Bragg FEL for THz-band

Further increase of the radiation frequency in the JINR-IAP FEM is restricted by the wiggler period and electron beam energy. To advance FEL of this type into THz frequencies, it is attractive to exploit more powerful induction linac, which was elaborated currently at BINP RAS (the LIU-2 accelerator) [6]. This accelerator generates electron beam with the current of kA-level and energy of 2 - 5 MeV (with the possibility to increase electrons energy up to 20 MeV).

Based on this beam, we initiated new project of multi-MW long-pulse THz FEL. In fact, the use of the beam with specified particles energy allows realization of the FEL in the range 1 - 10 THz using the wiggler of 3 - 6 cm period. For the realized beam current of ~ 1 - 2 kA, the radiation power is estimated on the level of 10 - 100 MW even under the electron efficiency of 1 - 0.1% (which, obviously, would decrease with increase of the radiation frequency).

Results of simulation of the FEL driven by 5 MeV / 1 kA electron beam generated by the LIU - 2 are presented in Fig. 4. This beam is focused by the guide magnetic field of ~ 0.15 T (in reversed configuration) and operating transverse electron velocities in the beam are pumped by the wiggler having 4 cm spatial period. For this FEL-oscillator we consider two-mirror resonator consisting of advanced Bragg reflectors of planar geometry (Fig.1b). Simulations demonstrated that Bragg structures of such type

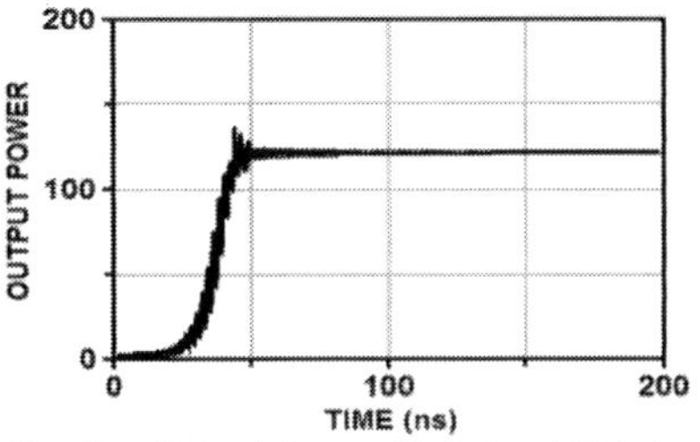

Fig. 4. Results of simulations of THz-band FEL based on linac LIU-2 (BINP RAS).

possess high reflectivity and ensure selectivity of the resonator for the transverse size (gap) up to 20 wavelengths. This size is sufficient for usage with the intense relativistic electron beams. For operation at 1 THz, we designed the resonator with 6 mm gap and the Bragg structures of 20 cm (up-stream) and 10 cm (down-stream) long having corrugation of the 0.3 mm period and 10 - 15 μm depth, a regular section of 50 - 70 cm long between them. According to the simulations, under design parameters the electron efficiency could achieve 2 - 3% and the output power reaches up to 50 - 100 MW. The Ohmic losses in this case do not exceed 25 - 30% from the radiated power.

Summarizing, novel conception of powerful THz-band Bragg FEL is developed. Original feature of this concept (in comparison with the others for now operating THz FEL) is the possibility to use electron beams with high-current (kA-level) and long-pulse (hundreds ns) duration. Such beams would allow achieving record levels of the radiation power of 10 - 100 MW and pulse energy up to 10 J for the THz-band. Operability of key components of electrodynamic system for such FEL - advanced Bragg resonators - was demonstrated in the proof-of-principle experiments up to W-band under oversized parameter Ø/λ ~ 5.

This work is partially supported by the Russian Foundation for Basic Research (grant # 18-02-40009).

References

1. *A.K. Kaminsky, E.A. Perelstein, S.N. Sedykh, e.a.* Demonstration of powerful 30-GHz FEM operation with resonant load // Tech. Phys. Lett. 2010. V.36 No.5 P.37.
2. *N.Yu. Peskov, N.S. Ginzburg, A.K. Kaminsky, e.a.* Powerful FEM-oscillators with advanced Bragg resonators operating in a single mode regime from Ka- to W-band // Proc. of the 41th Int. Conf. on IR, MM and THz Waves, Copenhagen, Denmark, 2016, p.F2E.01.
3. *N.S. Ginzburg, I.I. Golubev, A.K. Kaminsky, e.a.* Experiment on pulse heating and surface degradation of a copper cavity powered by powerful 30 GHz free electron maser // Phys. Rev. Accel. Beams 2011. V.14. No.4. P.041002.
4. *N.S. Ginzburg, A.M. Malkin, N.Yu. Peskov, e.a.* Improving selectivity of free electron maser with 1D Bragg resonator using coupling of propagating and trapped waves // Phys. Rev. Accel. Beams 2005. V.8. P.040705.
5. *N.S. Ginzburg, V.Y. Zaslavsky, I.V. Zotova, e.a.* Terahertz Free-Electron Lasers with Bragg structures based on coupling between travelling and quasi-cutoff wases // JETP Lett. 2010. V.91. No.6. P.266.
6. *P.V.Logachev, G.I.Kuznetsov, A.A Korepanov, e.a.* LIU-2 linear induction accelerator // Instruments and Experimental Techniques. 2013. V.56. No.6. P.672.

Generation of Ultra-Short Microwave Pulses in a Tunable Gyrotron with Subsequent Compression

R.M. Rozental[1], I.V. Zotova[1], N.S. Ginzburg[1], V.P. Tarakanov[2,3]

[1] Institute of Applied Physics RAS, Nizhny Novgorod, Russia, rrz@appl.sci-nnov.ru
[2] Moscow Engineering Physics Institute, Moscow, Russia
[3] Joint Institute for High Temperatures, RAS, Moscow, Russia

Generation of ultrashort electromagnetic pulses is of interest for many applications. A well-known method for obtaining such pulses is based on the compression of a frequency-modulated signal in a dispersive medium [1]. In principle, as such media, one can use regular waveguides. However, in this the frequency range optimal for compression is very close to the critical frequency, where the dispersion of the wave is most pronounced. This creates problems for ensuring matching in the operating band of the input chirped source. In [2,3] for solution of this problem, the compression in the helical waveguides is proposed. But in high-frequencies band such systems are rather complicated for manufacturing.

In this paper we suggest the two-stage scheme of generation of multi kilowatts subnanosecond microwave pulses based on a gyrotron matching directly with a compression section in the form of a regular waveguide. It is well-known, that gyrotrons typically have a rather narrow band of frequency tuning (~0.1-0.3%). However, as it was shown in [4], significant widening of the generation band in gyrotrons can be achieved with using of a shortened interaction space due to the weaker sensitivity to the velocity spread in the electron beam. The required frequency tuning of 5% can be provided due to excitation of several axial modes.

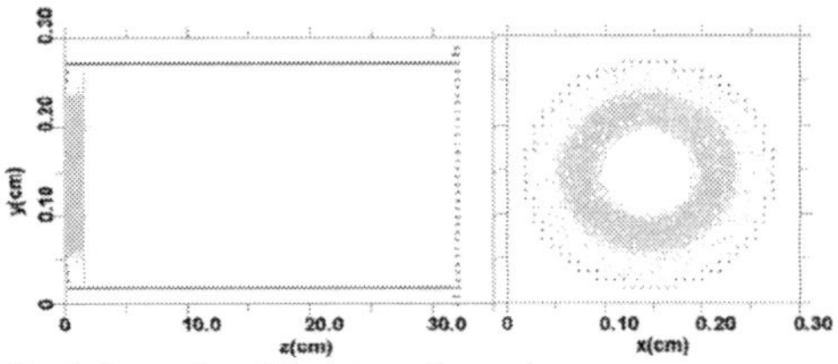

Fig. 1 Geometry of the interaction region.

Further, results of particle-in-cell simulations of such a system are presented based on the KARAT code [5]. In Fig.1 the geometry of the interaction region is shown, where the first section with an electron beam corresponds to a gyrotron with rather short resonator, while the second one is the section of compression. The Ohmoc losses were taken into account. The electron beam with the current of 2 A, excites the TE_{21} mode at the frequency of about 200 GHz. the operating Due to variation of an accelerating voltage from 90 to 15 kV, the operating frequency of a gyrotron varies from 200 to 210 GHz in the pulse with a duration of 12 ns (Fig.2a). The radiation power at the output of a gyrotron doesn't exceed 8 kW. In the second stage, the radiation is compressed in a regular waveguide connected directly to the gyrotron resonator. As a result, the output radiation represents the subnanosecond pulse with power of 80 kW.

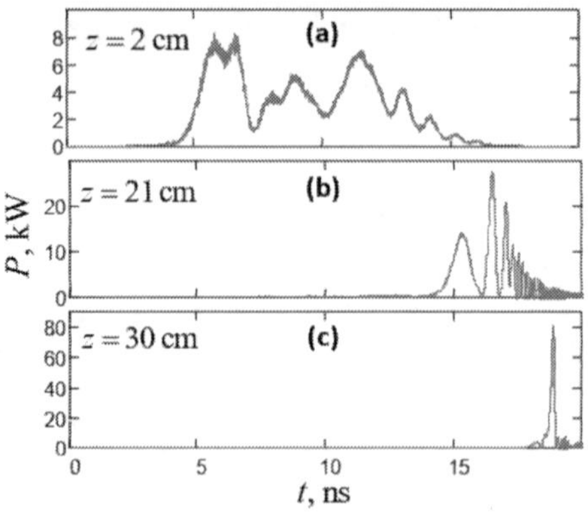

Fig. 2. Evolution of the radiated pulse in the section of compression.

This research was performed within the framework of the RFBR project No. 18-08-00717.

References

1. *Petelin, M.* Microwave pulse compressors // Proc. Int. Workshop on Strong Microwaves in Plasmas. 1997. V. 2. P. 903.

2. *Samsonov, S.V., Phelps, A.D.R., Bratman, V.L., Burt, G., Denisov, G.G., Cross, A.W., Ronald, K., He, W., Yin, H.* Compression of Frequency-Modulated Pulses using Helically Corrugated Waveguides and Its Potential for Generating Multigigawatt rf Radiation // Phys. Rev. Lett. 2004. V. 92. No.11. Art.no. 118301.

3. *Bratman, V.L., Denisov, G.G., Kolganov, N.G., Mishakin, S.V., Samsonov, S.V., Cross, A.W., He, W., Zhang, L., McStravick, M.C., Whyte, G., Young, A.R., Ronald, K., Robertson, C.W., Phelps, A.D.R.* Generation of 3 GW microwave pulses in X-band from a combination of a relativistic backward-wave oscillator and a helical-waveguide compressor // Phys. Plasmas. 2010. V. 17. No. 11. Art.no. 110703.

4. *Fedotov, A.E., Rozental, R.M., Zotova, I.V., Ginzburg, N.S., Sergeev, A.S., Tarakanov, V.P., Glyavin, M.Yu., Idehara T.* Frequency Tunable sub-THz Gyrotron for Direct Measurements of Positronium Hyperfine Structure // J Infrared Milli Terahz Waves. 2018. V. 39. No. 10. P. 975.

5. *Tarakanov, V.P.* Code KARAT in simulations of power microwave sources including Cherenkov plasma devices, vircators, orotron, E-field sensor, calorimeter etc. // EPJ Web of Conferences. 2017. V. 331. Art.no. 04024.

Gyrotrons with Shortened Cavities as Tunable Sources of Powerful Sub-Terahertz Radiation for Spectroscopic Applications

M.Yu. Glyavin, A.E. Fedotov, R.M. Rozental, I.V. Zotova, A.S. Zuev, V.N. Manuilov, M.Yu. Tretyakov, D.S. Makarov

Institute of Applied Physics RAS, Nizhny Novgorod, Russia, zotova@appl.sci-nnov.ru

Studies of forbidden resonance transitions are among the most complicated and most interesting problems in molecular gas spectroscopy. Typical transitions of this type are quadrupole transitions. Such transitions should be found in the spectra of all molecules, but currently only single vibrational-rotational lines of nonpolar molecules such as H_2, N_2, CO_2, etc. can be detected. The intensity of quadrupole transitions is very small. Therefore, an extraordinary sensitivity of the spectrometers is required for their observation and analysis. Most part of the spectrometers detecting radiation transmitted through a gas or re-emitted by a gas, have reached a sensitivity close to their theoretical limit. Only spectrometers using the method of radioacoustic detection of radiation absorption (RAD) by gases can provide a reserve of increasing sensitivity due to an increase in the power of the probing radiation [1]. Implementation of this idea requires the development of a CW coherent frequency-tunable radiation source.

In this paper, we discuss the possibility of developing powerful (>100 W) subterahertz gyrotron with frequency tuning band of about 3-5%, which is aimed to be used for RAD spectroscopy of quadrupole transitions of CO_2 at the frequency of 163 GHz. As it was demonstrated in [2], significant widening of the generation band in gyrotrons can be achieved with using of a shortened interaction space due to the weaker sensitivity to the velocity spread in the electron beam. At the same time, the shortening of the resonator will require increasing the current of the gyrotron electron beam. However, the required current value can be substantially reduced by going to operation at low transverse modes due to the growth of the electron-wave coupling coefficient. With a certain optimization, such approaches can provide a sufficiently wide band of gyrotron generation.

Further, results of PIC simulations are presented with using the KARAT code [3]. For 15 keV/0.4 A electron beam with pitch factor of 1 the TE_{13} mode, counter-rotating with electron motion in the guiding magnetic field, was chosen as an operating one. In Fig.1 the geometry of magnetron-injection gun (MIG) with above indicated parameters are presented. The gyrotron resonator was formed by a section of a circular waveguide bounded by a cutoff narrowing at the cathode end and a smooth widening at the collector end. The radius and the length of the regular part of resonator was 0.25 cm and 2.5 cm, correspondingly.

The simulations evidence that for the velocity spread of 20% the radiation power exceeds 100 W in the entire frequency tuning band of 4.1 GHz (2.5%) (Fig.2). For magnetic fields from 5.87 to 6.35 T, stationary single-mode excitation of the operating TE_{13} mode takes place. For magnetic field values outside this region, the excitation of the parasitic transverse TE_{42} and TE_{32} modes occurs, which limits the further increase in the bandwidth.

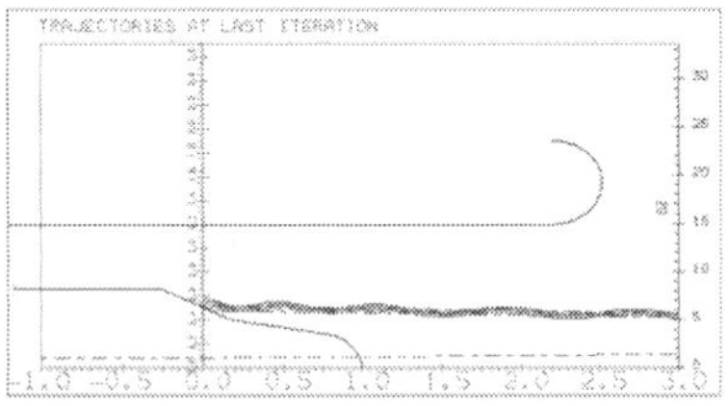

Fig. 1 Geometry of the MIG cathode part for a 163 GHz frequency-tunable gyrotron.

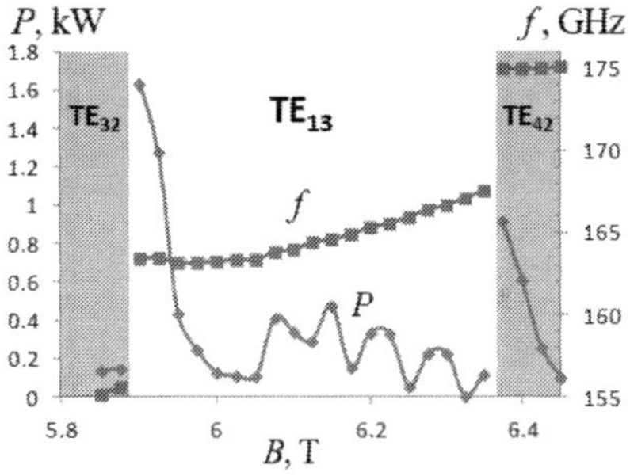

Fig. 2 Results of PIC simulations. Dependence of output power and operating frequency on the guiding magnetic field.

The work is supported by the Russian Science Foundation under grant No. 18-12-00394.

References

1. *Koshelev, M.A., Tsvetkov, A.I,. Morozkin, M.V., Glyavin, M.Yu., Tretyakov, M.Yu.* Molecular gas spectroscopy using radioacoustic detection and high-power coherent subterahertz radiation sources // J. Mol. Spectrosc. 2017. V.331. P. 9–16.

2. *Fedotov, A.E., Rozental, R.M., Zotova, I.V., Ginzburg, N.S., Sergeev, A.S., Tarakanov, V.P., Glyavin, M.Yu., Idehara T.* Frequency Tunable sub-THz Gyrotron for Direct Measurements of Positronium Hyperfine Structure // J Infrared Milli Terahz Waves. 2018. V. 39.

3. *Tarakanov, V.P.* Code KARAT in simulations of power microwave sources including Cherenkov plasma devices, vircators, orotron, E-field sensor, calorimeter etc. // EPJ Web of Conferences. 2017. V. 331. Art.no. 04024.

Rogue-waves generation in the terahertz region

R.M. Rozental, N.S.Ginzburg, A.M. Malkin, A.S. Sergeev, V.Yu. Zaslavsky, I.V. Zotova
Institute of Applied Physics RAS, Nizhny Novgorod, Russia, rrz@appl.sci-nnov.ru

One of the promising schemes of high-power terahertz radiation generators are gyrotrons with planar configuration of interaction space [1]. In such a configuration, by reducing the distance between the plates of the cavity, it is possible to obtain a large current parameter sufficient for occurrence of rogue waves – ultrashort pulses with a peak power exceeding the power of the unperturbed electron beam [2].

For description of rogue waves in gyrotron the self-consistent time-domain model was used:

$$i\frac{\partial^2 a}{\partial Z^2}+\frac{\partial a}{\partial \tau}=\frac{iI_0}{2\pi}\int_0^{2\pi}\frac{\hat{p}_\perp}{\hat{p}_\parallel}d\theta_0,$$

$$\left[\frac{\partial}{\partial Z}+\frac{g^2}{4}\frac{\partial}{\partial \tau}\right]p_\perp+i\frac{p_\perp}{p_\parallel}\left(\Delta-1+|p_\perp|^2+\frac{p_\parallel{}^2-1}{g^2}\right)=i\frac{a}{\hat{p}_\parallel}+\frac{\beta_{\perp 0}^2}{2}\frac{\partial a}{\partial Z}$$

$$\left[\frac{\partial}{\partial Z}+\frac{g^2}{4}\frac{\partial}{\partial \tau}\right]p_\parallel=-g^2\frac{\beta_{\perp 0}^2}{2}\mathrm{Re}\left(\frac{\partial a}{\partial Z}\frac{p_\perp{}^*}{p_\parallel}\right)$$

where $a(Z,\tau)$ – normalized amplitude of the operating wave; Z,τ – normalized coordinate and time; $p_{\perp,\parallel}$ – normalized particle moments, I_0 – current parameter, Δ – initial mismatch between the wave cutoff frequency, g – initial pitch-ratio.

In simulations we assumed that the 300 GHz gyrotron excited by the planar electron beam with energy of about 200 keV, current density of about 2 kA/cm and pitch-factor of about 1.0 at the lowest TE mode of the planar waveguide. With this parameters the value of the dimensionless current parameter $I_0 \approx 1$, which is which is enough for the radiation of rogue waves.

All processes were analyzed over a time interval of several tens of nanoseconds, which corresponds to the typical pulse duration in electron guns based on explosive emission cathodes. Simulation showed that in a sufficiently wide range of magnetic fields in the system, it is possible to generate rogue waves with a peak power of several hundred megawatts (Fig.1).

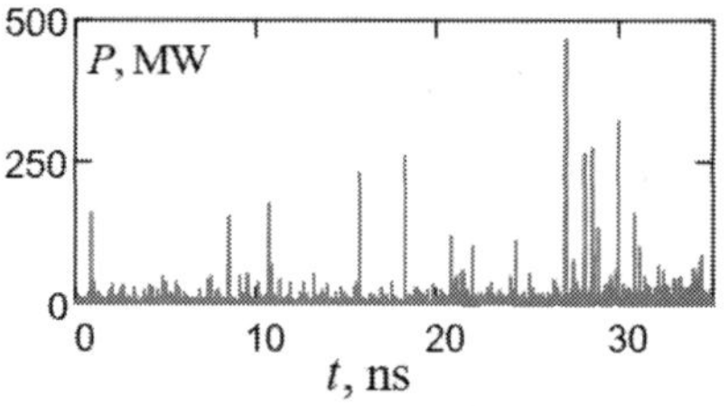

Fig. 1. The time dependence of the output power for simulation of planar gyrotron based on the averaged equations.

The obtained results have been confirmed by direct PIC simulations at KARAT code [3]. In the simulation, pulses of a duration of the order of 10 ps were obtained with a peak power reaching hundreds of megawatts (Fig.2a). The total width of the emission spectrum reached 150 GHz (Fig. 2b).

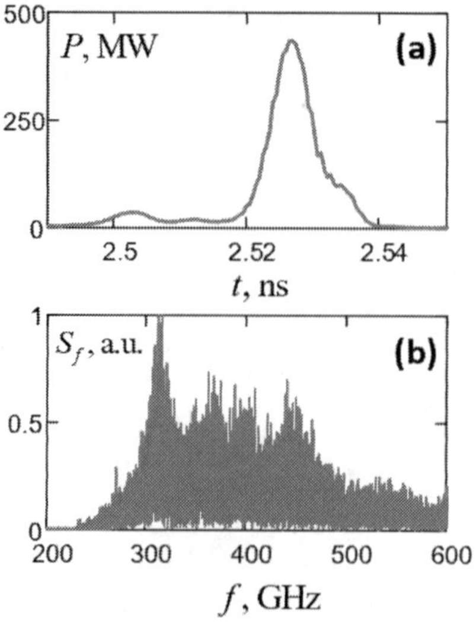

Fig. 2. Results of PIC simulation: (a) - a typical form of an rogue wave; (b) is the total spectrum of the output radiation.

This research was performed within the framework of the RFBR project No. 17-08-01077.

References

1. *Zaslavsky, V.Yu., Ginzburg, N.S., Glyavin, M.Yu., Zheleznov, I.V., Zotova, I.V.* Three-dimensional particle-in-cell modeling of terahertz gyrotrons with cylindrical and planar configurations of the interaction space // Phys. Plasmas. 2013. V. 20. Art.no. 043103

2. *Ginzburg, N.S., Rozental, R.M., Sergeev, A.S., Fedotov, A.E., Zotova, I.V., Tarakanov, V.P.* Generation of Rogue Waves in Gyrotrons Operating in the Regime of Developed Turbulence // Phys. Rev. Lett. 2017. V.119. Art. no. 034801.

3. *Tarakanov, V.P.* Code KARAT in simulations of power microwave sources including Cherenkov plasma devices, vircators, orotron, E-field sensor, calorimeter etc. // EPJ Web of Conferences. 2017. V. 331. Art.no. 04024.

Development and Modeling of Miniaturized Traveling-Wave Tubes in Millimeter and Sub-THz Bands

Nikita M. Ryskin[1,2]

[1]Saratov Branch, V.A. Kotel'nikov Institute of Radio Engineering and Electronics RAS, Saratov, Russia
[1]Saratov State University, Saratov, Russia, RyskinNM@info.sgu.ru

1. Introduction

Development of compact vacuum-tube millimeter- and THz-band sources with 10-100 W output power would have a great potential for numerous applications such as high-data-rate wireless communications, security, spectroscopy, biomedical applications, etc. Among the vacuum tubes, microfabricated analogs of traveling-wave tube (TWT) amplifiers are the most promising. Over 50-W output power in G-band (0.22 THz) has recently been demonstrated [1,2].

Slow-wave structure (SWS) is a core part of a TWT serving to slow the electromagnetic wave down to the electron beam velocity. In this paper, we present the results of recent studies aimed at modeling and development of medium power TWTs with various SWS suitable for operation in millimeter and sub-THz frequency bands. Different types of TWTs are considered, such as folded-waveguide TWT, TWTs with sheet electron beam and dual-grating SWS, and TWTs with planar meander-line SWS on a dielectric substrate.

2. Folded-waveguide TWT

TWT with folded-waveguide (FW) SWS has numerous advantages due to its high coupling impedance, reasonably wide (~30%) bandwidth, relatively simple structure compatible with existing microfabrication technologies, thermal and mechanical robustness, and simple input/output coupling.

We designed and simulated FW TWTs operating at Q-band (38-54 GHz), V-band (48-90 GHz), and G-band (175-275 GHz). Electromagnetic parameters of the SWSs were calculated by using ANSYS HFSS simlator. FW SWS circuits at Q- and V-band were fabricated using CNC-milling machine. In Fig. 1, a picture of the 50-pitch Q-band SWS with input/output couplers is shown. Cold-test electromagnetic measurements of *S*-parameters of the circuits were performed using Keysight Technologies PNA N5227A vector network analyzer. Good transmission characteristics were measured. In particular, return loss (S11) is less than −10 dB and transmission loss (S21) is about −2 dB in the most part of the passband. Experimental measurements are in good agreement with numerical simulations using ANSYS HFSS.

In addition, small-signal and large-signal gain regimes of Q- and V-band FW TWT amplifiers driven by a 50-mA, 15-kV electron beam were simulated using the 1-D parametric code [3]. Parameters of the tubes were obtained, for which 30-40 dB small-signal gain and nearly 100 W saturated power are attained.

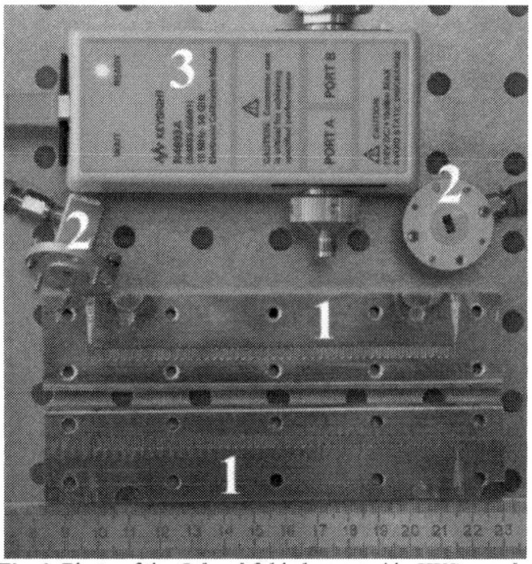

Fig. 1. Photo of the Q-band folded-waveguide SWS sample. 1 – folded waveguide; 2 – coax to waveguide adapters; 3 – electronic calibration module (ECal module)

3. Sheet-beam TWT dual-grating SWS

Using high-aspect-ratio sheet electron beams allows substantial reduce of current density and facilitates beam focusing and transportation in a narrow beam tunnel. In [3,4], we studied feasibility of a 0.2-THz sheet-beam TWT amplifier with a staggered grating SWS. An electron gun with a 0.7×0.1 mm^2 thermionic cathode producing intensive sheet electron beam with over 120 A/cm^2 current density has been developed [3,4]. However, such a gun can operate only in a short-pulse mode with 10 us pulse duration and 2000 off-duty cycle. At the same time, the required focusing magnetic field is as high as 1.12 T.

In order to overcome the problems mentioned above, we designed the electron-optic system (EOS) with a converging sheet electron beam having 10-times compression in vertical direction using the method of synthesis described in [5]. This allows a significant reduce of the beam thickness as well as of the cathode load and focusing magnetic field (0.8 T). Schematic of the electron gun is presented in Fig. 2. The design was verified by 3-D simulation using LORENTZ-3EM simulator [6]. The results are in a fairly good agreement, except that the beam thickness obtained in 3-D analysis is somewhat higher than expected (~75 um) that can be attributed to thermal and angle spread of the emitted electrons.

The gun was fabricated and up to 140-mA beam current was measured.

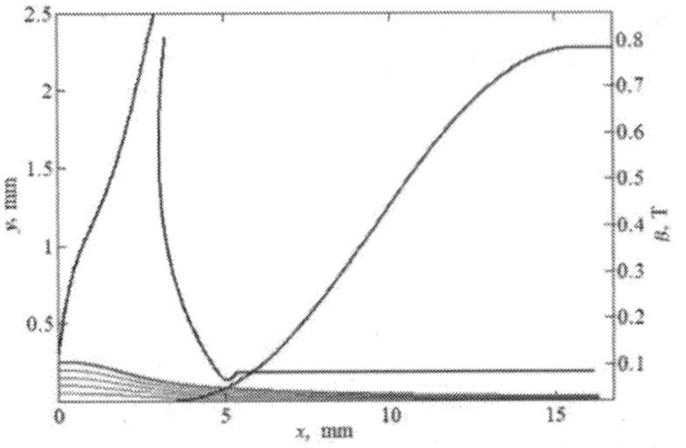

Fig. 2. Configuration of the gun electrodes, particle trajectories, and the magnetic field profile

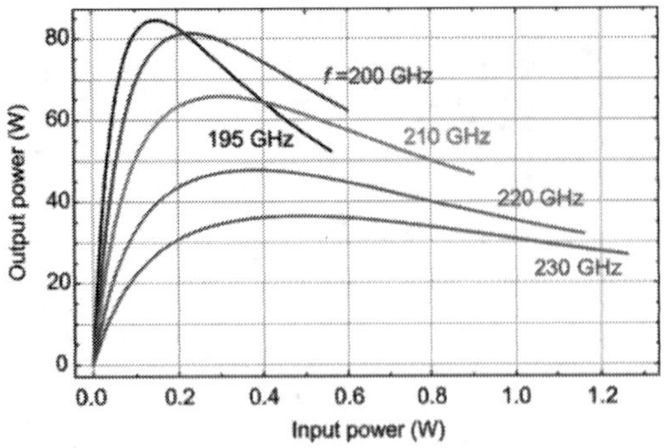

Fig. 3. Output power versus input power of the sheet-beam TWT at different frequencies

Performance of the TWT with 0.1-A, 20-kV sheet beam and dual-grating SWS was simulated. Parameters of the SWS were the same as in [3,4] except the beam tunnel height, which was reduced from 200 um to 150 um, owing to reduced beam thickness. This results in significant increase of the coupling impedance and the gain factor. For the 40-mm-length TWT circuit, over 30-dB small-signal gain was observed in the simulations (instead of ~20 dB in [3]). In Fig. 3, output power versus input power at different frequencies is plotted. Saturated power over 80 W is observed, which is close to [3]. However, the saturation is attained at the input power about 0.1 W, which is nearly 10 times less than in [3].

4. TWT with planar microstrip meander-line SWS

In order to reduce the tube size and dimensions, as well as the dc voltage, planar microstrip meander-line SWS on dielectric substrates have been proposed. We fabricated SWS circuits in V- and W-band using the novel technology utilizing magnetron sputtering and laser ablation methods [7]. Several samples of V-band and W-band meander SWS circuits on quartz substrates were fabricated. The strip thickness was increased up to 10 um that should increase robustness of the circuits. In Fig. 4, photo of the 50-pitch SWSs with matching elements at both ends is shown.

Cold-test measurement of the electromagnetic parameters shows good transmission characteristics. Measured transmission loss of the structure did not exceed –2.5 dB.

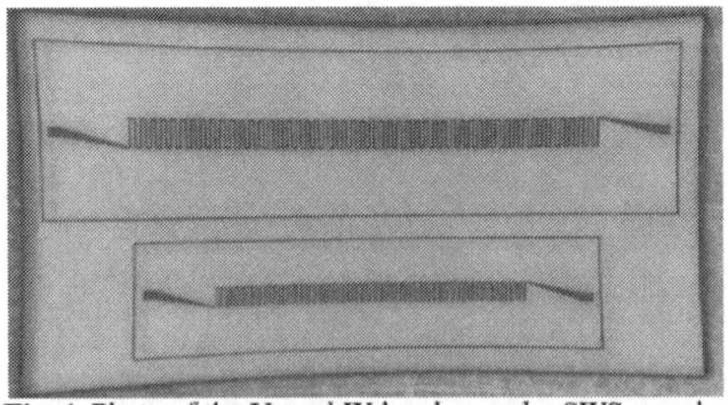

Fig. 4. Photo of the V- and W-band meander SWS samples fabricated by laser ablation

The developed SWSs provide high slow-down factor ($c/v_{ph} \sim 7 \div 9$) and thus are suitable for operation with a low-voltage (3÷5 kV) electron beam. Small-signal and large-signal gain characteristics of the TWT driven by a 100-mA sheet electron beam were simulated. For a 1-cm-length SWS, the peak gain is about 15-20 dB. Owing to strong dispersion of the meander-line SWS, the bandwidth of the amplifier is rather narrow (2–3 GHz). However, the central frequency may be easily tuned in a wide range by varying the beam voltage. The simulations predict over 60 W output power at saturation.

Acknowledgment

This research is supported by Russian Science Foundation under Grant # 17-12-01160.

References

1. *Joye C.D., Cook A.M., Calame J.P., Abe D.K.,et al.* Demonstration of a high power, wideband 220-GHz traveling wave amplifier fabricated by UV-LIGA // IEEE Trans. Electron Devices. 2014. V. 61, No. 6, P. 1672–1678.

2. *Baig A., Gamzina D., Kimura T., Atkinson J. et al.,* Performance of a nano-CNC machined 220-GHz traveling wave tube amplifier // IEEE Trans. Electron Devices. 2017. V. 64, No. 6. P. 2390-2397.

3. *Karetnikova T.A., Rozhnev A.G., Ryskin N.M., Fedotov A.E., Mishakin S.V., Ginzburg N.S.* Gain analysis of a 0.2-THz traveling-wave tube with sheet electron beam and staggered grating slow wave structure / // IEEE Trans. Electron Devices. 2018. V. 65, No. 6. P. 2129-2134.

4. *Karetnikova T.A., Rozhnev A.G., Ryskin N.M., Torgashov G.V. et al.* Modeling a subterahertz traveling-wave tube with a slow-wave structure of the double grating type and a sheet electron beam // J. Commun. Technol. Electron. 2016. V. 61, No. 1. P. 50–55.

5. *Yu.G. Gamayunov, E.V. Patrusheva.* Synthesis of systems formation of converging sheet electron beams at partial magnetic shielding of the cathode // J. Commun. Technol. Electronics. 2017. Vol. 62, No. 11, 1291-1297.

6. *Karetnikova T.A., Rozhnev A.G., Ryskin N.M., Burtsev A.A., Navrotsky I.A., Danilushkin A.V.* Design and simulation of a 0.2-THz traveling-wave tube with a converging sheet electron beam // Proc. 19th International Vacuum Electronics Conference (IVEC 2018). 24-26 Apr. 2018, Monterey, CA, USA. P. 161-162.

7. *Ryskin N.M., Rozhnev A.G., Starodubov A.V., Serdobintsev A.A., et al.* Planar microstrip slow-wave structure for a low-voltage V-band traveling-wave tube with a sheet electron beam // IEEE Electron Device Lett. 2018. V. 39, No. 5. P. 757-760.

EPJ Web of Conferences **195**, 01015 (2018) https://doi.org/10.1051/epjconf/201819501015
TERA-2018

High-harmonic gyrotrons with irregular microwave systems

A.V. Savilov, I.V. Bandurkin, M.Yu. Glyavin, Yu.K. Kalynov, Yu.S. Oparina, I.V. Osharin, N.A. Zavolsky

Institute of Applied Physics RAS, N. Novgorod, Russia, savilov@appl.sci-nnov.ru

A natural way to increase the gyrotron frequency is the use of operation at higher cyclotron harmonics. However, at moderate voltages, the intensity of the electron-wave coupling decreases rapidly with the increasing harmonic number. Due to this fact, the danger of excitation of parasitic close-to-cutoff waves at lower harmonics becomes a serious problem. This is true especially in low-relativistic gyrotrons with low output powers (and, therefore, low operating currents) at least due to two reasons. First, low-harmonic waves possess bigger electron-wave coupling factors. Second, longer operating cavities are required, in general, to excite high-harmonic operating waves. The latter results also in extremely high diffraction Q-factors for the operating near-cutoff waves and, therefore, a great share of ohmic losses. In this work, we describe several schemes of cavities with short irregularities designed to improve operation of low-relativistic high-harmonic gyrotrons.

Cavities with decreased ohmic losses

A method to combine a long-length interaction region of a high-harmonic gyrotron and a relatively low diffraction Q factor was proposed in Refs. [1,2]. This method is based on the use of a sectioned system (Fig. 1 a) with a klystron-like character of the electron-wave interaction: particles interact with a near-cutoff wave only in the input and output cavities (sections 1 and 3), whereas the drift region (section 2) provides coupling between these cavities as well as bunching of electrons (Fig. 1). The diffraction Q factor in such a scheme is determined mainly by the sum of lengths of the two cavities, $Q_{\rm dif} \propto (L_1 + L_3)^2$. At the same time, the starting current is determined by the total length of the system, $(L_1 + L_2 + L_3)$. This scheme was realized in the pulsed large-orbit 80 kV gyrotron [3]. Selective excitation at the third harmonic was achieved at a magnetic field close to 10.2 T. The output rf signal with a power of 100–250 W at a frequency of 0.74 THz corresponded to the transverse mode $TE_{3,5}$. According to simulations, the share of the ohmic losses in this experiment was relatively low (20%–25% of the rf wave power emitted from the electron beam) as compared to the first regular-cavity experiment under this third-harmonic gyrotron [4].

A different method for decreasing the diffraction Q-factor of the operating gyrotron mode is proposed in works [5,6]. This is based on exciting a higher axial mode (possessing a relatively low diffractive Q-factor) in the gyrotron-type regime (i.e. at the electron-wave resonance condition $\omega \approx N\Omega$ without Doppler terms). This method is based on the use of a periodically sectioned cavity; every two sections are separated by a short phase corrector providing the π-shift of the wave phase (Fig. 2). This approach is used to provide operation at the fourth cyclotron harmonic in the 30 keV/0.7A CW large-orbit gyrotron [7] based on a 5 T cryomagnet. The main scope of this setup is to provide operation at the second, third, and fourth cyclotron harmonics at the frequencies 0.26 THz, 0.39 THz, and 0.52 THz, respectively, with the output power level of hundreds of Watts. In first experiments, the use of a regular 19 mm cavity provided selective excitation of the mode $TE_{2,5}$ at the second harmonic (0.267 THz) and of the mode $TE_{3,7}$ at the third harmonic (0.394 THz) at slightly different magnetic fields. However, the 0.52 THz fourth-harmonic gyrotron requires a long (50-60 wavelengths) operating cavity, so that the diffraction Q-factor of the lowest axial mode (over 100,000) is much greater than the ohmic Q-factor (~10,000), and the share of ohmic losses exceeds 90%. This leads to as low output efficiency of this gyrotron as 0.2% We designed a cavity consisting of 5 sections and based on the excitation of the mode $TE_{4,5}$ with 5 axial variations (Fig. 2). Although the cavity is relatively long (~ 90 wavelength), sectioning makes possible to reduce the diffraction Q-factor of the operating wave down to ~30,000, and to reduce the share of ohmic losses down to ~60%. This provides an increase in the output wave efficiency up to ~1.% (the output power is ~ 100 W).

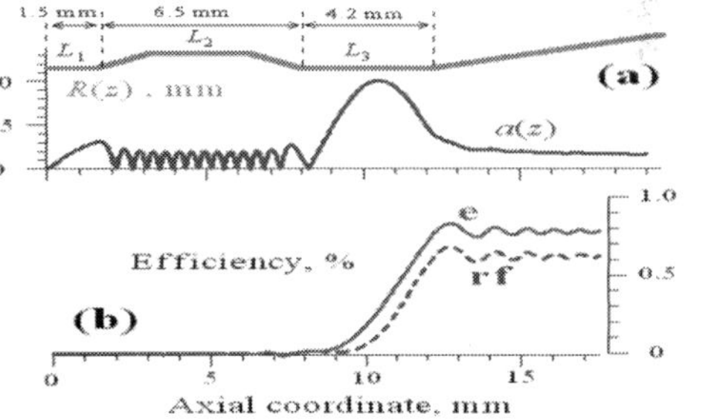

Fig. 1. (a): Sectioned cavity of the third-harmonic large-orbit gyrotron. (b): Results of simulation of the operation at a frequency of 0.74 THz; electron efficiency (e) and rf wave efficiency (rf) versus the axial coordinate.

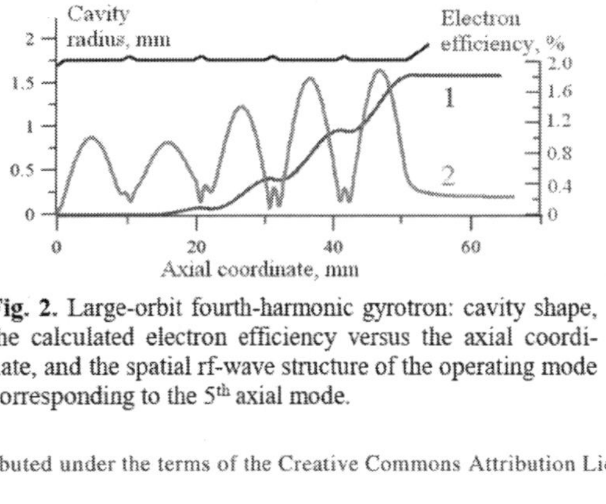

Fig. 2. Large-orbit fourth-harmonic gyrotron: cavity shape, the calculated electron efficiency versus the axial coordinate, and the spatial rf-wave structure of the operating mode corresponding to the 5th axial mode.

Selective irregularities

A method for suppression of the fundamental-harmonic parasitic waves in second-harmonic gyrotrons based on a simple principle [8] can be used in low-power short-wavelength gyrotrons with a weak electron-wave interaction and, therefore, with long (tens wavelengths) operating cavities. This method is based on the use a quasi-regular cavity with either one or several short irregularities (phase correctors). Evidently, the phase incursion of a near-cutoff wave inside the irregularity is proportional to the frequency, $\phi(\omega) \sim \omega$. For the operating second-harmonic wave, $\phi(\omega) = 2\pi$ and, therefore, such an irregularity does not effect on the excitation of this wave. As for fundamental-harmonic near-cutoff parasitic waves, their phase shift on the same corrector is twice smaller, π (Fig. 3 a). The π-shift of the phase changes the "sign" of the electron-wave coupling (shifts decelerated electron to the accelerating phase of the wave); this lead to a significant degradation of the interaction of electrons with the parasitic waves. This method is valid only for low-mode resonators, when the corrector does not provide scattering of the waves into lower transverse modes. In the case of operation at high radial modes, this approach can be developed by the use of the mode transformation effect (Fig. 3 b) [9]. If the operating high-harmonic mode is $TE_{m,n}$ then the radius of the widening corresponds to the cutoff $TE_{m,n+1}$ wave at the same frequency. Thus, the operating $TE_{m,n}$ wave is converted into the $TE_{m,n+1}$ wave inside the widening with no transformation losses. However, the irregularity is not resonant for the parasitic near-cutoff fundamental-harmonic waves, and a significant reduction of the diffraction Q-factors of these waves is provided due to their scattering. Using this approach, we designed a selective cavity for a pulsed 60 keV gyrotron to achieve second-harmonic excitation of a very developed mode ($TE_{63,15}$) at a frequency of ~1.2 THz.

One more approach is based on reducing the cavity length due to the use of a special output irregularity [10]. The irregularity increases the Q-factor of the operating high-harmonic wave due to the reflection, but it almost does not reflect parasitic fundamental-harmonic waves (Fig. 3 c). In addition, the irregularity spoils the axial structure of the parasitic wave. This approach can provide a significant increase in the starting current of parasitic close-to-cutoff waves with no considerable effect on the generation of the operating high-harmonic wave.

The work is supported by the Russian Science Foundation, project # 17-19-01605.

References

1. *Savilov, A. V.* High-harmonic gyrotron with sectioned cavity // Appl. Phys. Lett. 2009. V. 95, P. 073503.

2. *Bandurkin, I. V., Kalynov, Y. K., Savilov, A. V.* High-harmonic gyrotron with sectioned cavity // Phys. Plasmas. 2010. V. 17. No. 8. P. 073101.

3. *Bandurkin, I. V., Kalynov, Y. K., Savilov, A. V.* Experimental realization of the high-harmonic gyrotron oscillator with a klystron-like sectioned cavity // IEEE Trans. Electron Devices. 2015. V. 62. No. 7. P. 2356-2359.

4. *Bratman, V. L., Kalynov, Y. K., Manuilov, V. N.* Large-orbit gyrotron operation in the terahertz frequency range // Phys. Rev. Lett. 2009. V. 102, No. 24. P. 245101.

5. *Bandurkin, I. V., Kalynov, Y. K., Osharin, I. V., Savilov, A. V.* Gyrotron with a sectioned cavity based on excitation of a far-from-cutoff operating mode // Phys. Plasmas. 2016. V. 23. N. 1. P. 013113.

6. *Bandurkin, I.V., Kalynov, Y.K., Makhalov, P.B., Osharin, I.V., Savilov, A.V., Zheleznov, I.V.* Simulations of sectioned cavity for high-harmonic gyrotron // IEEE Trans. Electron Devices. 2017. V. 64 (1). P. 300.

7. Bandurkin, I.V., Bratman, V.L., Kalynov, Y.K., Osharin, I.V., Savilov, A.V. Terahertz large-orbit high-harmonic gyrotrons at IAP RAS // IEEE Trans. Electron Devices. 2018. V. 65. P. 2287.

8. *Bandurkin, I. V., Kalynov, Y. K., Osharin, I. V., Savilov, A. V.* A method for suppression of spurious fundamental-harmonic waves in gyrotrons operating at the second cyclotron harmonic // Phys. Plasmas. 2016. V. 23. No. 5. P. 053116.

9. *Bandurkin, I. V., Glyavin, M.Y., Kuzikov, S.V., Makhalov, P.B., Osharin, I. V., Savilov, A. V.* Method of providing the high cyclotron harmonic operation selectivity in a gyrotron with a spatially developed operating mode // IEEE Trans. Electron Devices. 2017. V. 64. P. 3893.

11. *Oparina, Yu.S., Savilov, A.V.* Improvement of mode selectivity of high-harmonic gyrotrons by using opeating cavities with short output reflector // J. IR MM THz Waves 2018. V. 39. P. 595.

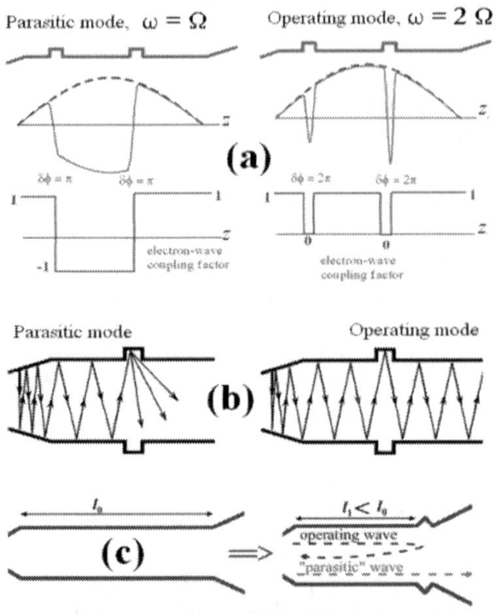

Fig. 3. Gyrotron cavities with selective irregularities. (a): Phase corrector, which provides the 2π phase incursion for the operating second-harmonic wave, and the π phase incursion for the parasitic fundamental-harmonic wave. (b): An irregularity providing scattering of the parasitic near-cutoff wave. (c): Shortening of the gyrotron operating cavity by the use of the reflecting output irregularity.

THz radiation of stabilized dense electron bunches

A.V. Savilov[1], I.V. Bandurkin[1], I.S. Kurakin[1], Yu.S. Oparina[1], V.L. Bratman[1,2], N. Balal[2], Yu. Lurie[2]

[1] Institute of Applied Physics RAS, N. Novgorod, Russia, savilov@appl.sci-nnov.ru
[2] Ariel University, Ariel, Israel

Laser-driven photo-injectors are capable to form very compact and dense electron bunches with a particles energy of 3-6 MeV, picosecond and subpicosecond duration, and charge of the order of 1 nC. When moving in the periodic field of the undulator with a period of a few centimeters, such bunches can generate coherent radiation in the near-terahertz range. The power and duration of such a generation is limited due to an increase in the axial length of the electron bunch under the action of the Coulomb repulsion, which under normal conditions does not allow for the effective implementation of such a scheme for dense nC bunches [1]. Therefore, a special methods of stabilization of the axial length of the operating bunch during its motion over a long enough electron-wave interaction region should be used. The report describes two such methods, namely, the so-called "negative-mass" stabilization and the radiative compression of electron bunches.

Negative-mass stabilization and compression

The negative-mass regime of the electron motion is realized in a combination of periodic undulator field and relatively strong homogeneous axial magnetic field (Fig. 1 a). The cyclotron frequency corresponding to the axial field should be slightly higher than the undulator bounce-frequency of the particle. In this case, the Coulomb field inside the bunch leads not to repulsion of electrons but to their mutual attraction [1]. This effect is a result of an abnormal dependence of the velocity of undulator oscillations of electrons on the cyclotron frequency (Fig. 1 b). Let us consider a particle moving in the front of the bunch (Fig. 1 c). The Coulomb field accelerates this electron, and its relativistic cyclotron frequency decreases. The electron approaches the undulator-cyclotron resonance, which is accompanied by a resonant increase in its transverse velocity; this leads to decrease in the axial electron velocity.

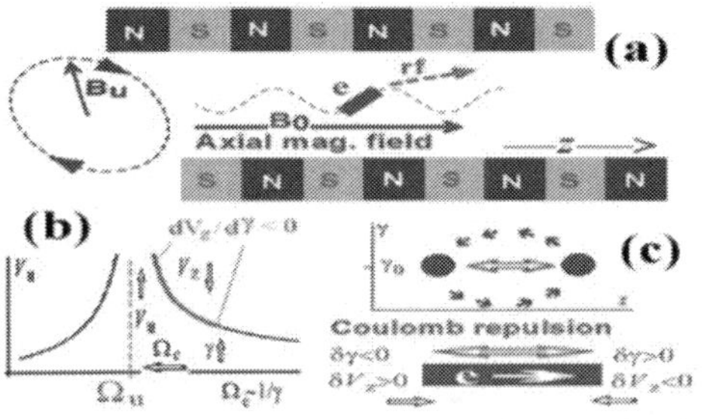

Fig. 1. (a): Electron motion in the combined helical undulator and uniformaxial fields. (b) Characteristic dependence of the transverse electron velocity on the cyclotron frequency. (c) Coulomb attraction and oscillations of electrons in the "negative-mass" regime.

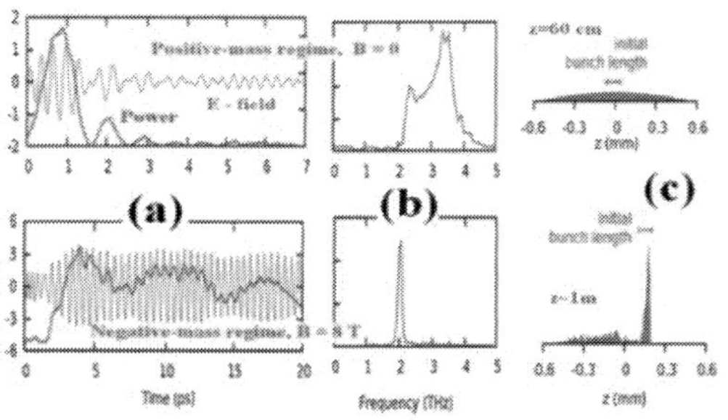

Fig. 2. (a): Radiated wave field and power (red) in the "positive-mass" and "negative-mass" regimes. (b): Spectra of the radiated rf signal. (c): Axial distributions of the charge in the bunch after the 60 cm trip in the positive-mass regime and after the 1m trip in the negative-mass regime.

Thus, increasing the energy of the particle causes it to slow down in the longitudinal direction. The opposite situation takes place for the particle moving in the tail of the bunch; this particle is decelerated by the coulomb filed, and this leads to an increase in the axial velocity. Overall, the Coulomb interaction of particles leads to compression of the bunch in the longitudinal direction. Such a behavior of electrons can be regarded as a consequence of its effective mass being negative [2,3], and this is very similar to the negative-mass effect in cyclotron masers [4-7].

Numerical simulations show [1,8] that in a certain range of parameters, significant improve of the radiation characteristics is possible when using the described scheme. Figure 2 illustrates an example, where an electron bunch with the particles energy of 5.5 MeV scatters the undulator field having a period of 2.5 cm into the wave with a frequency of about 2 THz. In this case, the resonant cyclotron frequency corresponds to the axial field of about 5 T, and effective negative mass regime is realized at magnetic fields near 8 T and the undulator field amplitude of about 0.2 T (Fig. 2). Simulations predict generation of a 20 ps rf pulse with a power of ~ 10 MW. According to simulations [8], the negative-mass regime can provide axial stabilization of extremely dense (with charges up to several nC) picosecond bunches. A prototype of the negative-mass undulator with a 8 T axial magnetic field is designed and tested in the experiment [9]. There is also an interesting option to use this regime for the axial compression of dense bunches by their own Coulomb fields, when the undulator is used not for the radiation but for providing the negative-mass electron motion only [10]. Since in this situation there is no goal to achieve THz radiation, undulators with longer periods can be used, and a lower magnitudes of the axial magnetic field are required to provide the negative-mass electron motion.

Radiative compression

An alternative compression method can be the undulator super-radiative radiation of a long-wavelength wave in an auxiliary long-period undulator [11]. A relatively long electron bunch propagates in a simple (with no axial magnetic field) helical undulator inside a waveguide (Fig. 3 a) and radiates a short wave packet propagating with a group velocity equal to the electron axial velocity (Fig. 3 b). Since the wavelength of the radiated wave is longer than the bunch length, the emission has the spontaneous coherent character. Such emission results in axial compression of the bunch (Fig. 3 c). This is due to the fact that in the situation described above the phase of the radiated wave is correlated in a certain way with the electron bunch phase, namely, there is a $\pi/4$ shift between the bunch center and maximum of the decelerating wave phase (Fig. 3 d). In the case of a bunch being four times shorter than the wavelength, the bunch is compressed in the radiated field, because the front of the bunch is placed in the maximum of the decelerating wave phase whereas the tail is placed close to the "zero" wave field.

This effect can be used for creation of a "bicolor" THz source based on the spontaneous emission from a short bunch, so that the super-radiation of the auxiliary long-wavelength wave is used to compress the bunch down to a size shorter than the wavelength of the short-wavelength wave. Figure 4 illustrates an example, where a 0.1 nC 3 MeV 0.6 mm bunch moves in two undulators ($\lambda_{u,1} = 2.3$ cm and $\lambda_{u,2} = 1.1$ cm). At the first stage, electron oscillations in the first undulator lead to the spontaneous coherent radiation of the long-wavelength ($\lambda_1 = 1.2$ mm) compressing wave. During this process, the bunch is compressed several times. This makes possible the spontaneous coherent radiation of the short-wavelength ($\lambda_2 = 0.3$ mm) wave in the second undulator with a relatively high (~10%) electron efficiency.

The work is supported by IAP RAS Project 0035-2014-0012, Russian Foundation for Basic Research (Project 18-32-00351 and 16-02-00794), and the Israeli Ministry of Science, Technology and Space.

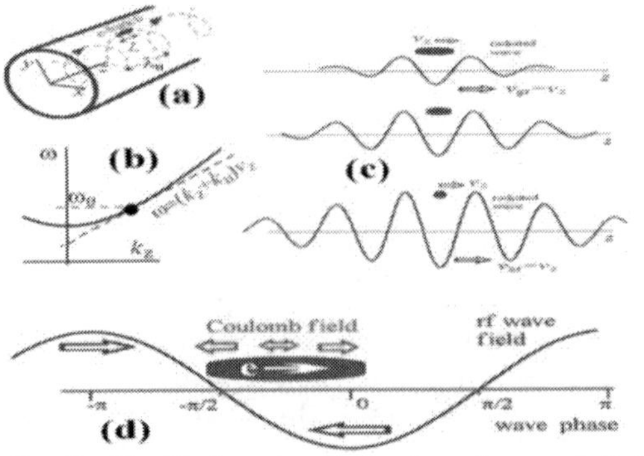

Fig. 3. (a): Electron bunch moving along a helical undulator. (b): Dispersion diagram of the operating waveguide mode. (c) Super-radiative radiation and the bunch compression. (d): Bunch phase with respect to the wave.

References

1. *Balal, N., Bandurkin, I. V., Bratman, V. L., Magory, E., Savilov, A. V.* Negative-mass mitigation of Coulomb repulsion for terahertz undulator radiation of electron bunches // Appl. Phys. Lett. 2015. V. 107. P. 163505.

2. *Freund, H. P., Sprangle, P.* Unstable electrostatic beam modes in free electron-laser systems // Phys. Rev. A. 1983. V. 28. P. 1835.

3. *Ginzburg, N. S., Peskov, N. Yu.* Non-linear theory of relativistic ubitrons with electron beams formed in an adiabatically increasing undulator field and a uniform longitudinal magnetic field // Tech. Phys. 1988. V. 58. P. 859.

4. *Bratman, V. L.* Instability of orbital motion in a layer of electrons rotating in a uniform magnetic field // Tech. Phys. 1976. V. 46. P. 2030.

5. *Bondeson, A., Antonsen, T.* Space-charge instabilities in gyrotron beams // Int. J. Electron. 1986. V. 61. P. 855.

6. *Bratman, V. L., Savilov, A. V.* Phase mixing of bunches and decrease of negative-mass instability increments in cyclotron resonance masers // Phys. Plasmas. 1995. V. 2. P. 557.

7. *Savilov, A. V.* Negative-mass instability in magnetron-injection guns // Phys. Plasmas. 1997. V. 4. P. 2276.

8. *Lurie, Yu., Bratman, V. L., Savilov, A. V.* Energy enhancement and spectrum narrowing in terahertz electron sources due to negative mass instability // Phys. Rev. AB. 2016. V. 19. P. 050704.

9. *Balal, N., Bandurkin, I.V., Bratman, V.L., Fedotov, A.E.* Helical undulator based on redistribution of uniform magnetic field // Phys. Rev. AB. 2017. V. 20. P. 122401.

10. *Bandurkin, I.V., Kurakin, I.S., Savilov, A. V.* Compression of a photoinjector electron bunch in the negative-mass undulator // Phys. Rev. AB. 2017. V. 20. P. 020704.

11. *Bandurkin, I.V., Oparina, Y.S., Savilov, A.V.* Super-radiative self-compression of photo-injector electron bunches // Appl. Phys. Lett. 2017. V. 110. P. 263508.

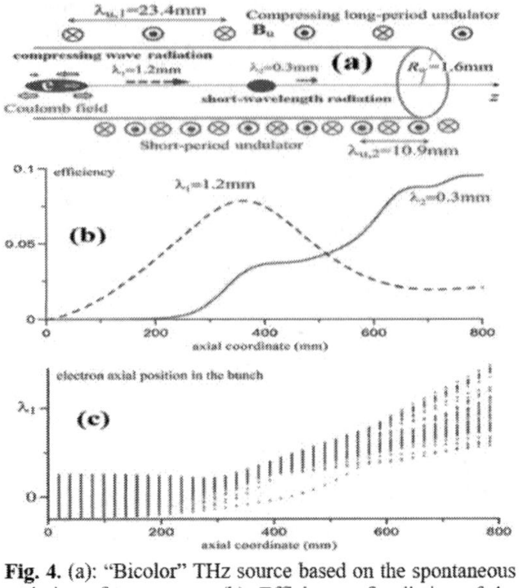

Fig. 4. (a): "Bicolor" THz source based on the spontaneous emission of two waves. (b): Efficiency of radiation of the two waves versus the axial coordinate of the bunch. (c): Evolution of axial distribution of electrons inside the bunch.

3D Printed Periodic Structures for Subterahertz Sources

D. I. Sobolev, M.D. Proyavin, N.Yu. Peskov, V.Yu. Zaslavsky, V.V. Parshin

Institute of Applied Physics, Nizhny Novgorod, Russia, nrtnm@mail.ru

Recent advances in 3D printing technology made it a favorable choice in many scientific applications, where large-scale production is not required. There are several different materials and printing methods available, which are more or less suitable for subterahertz devices. Selective laser sintering (SLS) printers can reach tolerances down to several microns, although these are on the expensive side, and in the case of metallic printing the surface roughness is unsatisfactory for radiation with millimeter and submillimeter wavelengths without additional finishing. Moreover, SLS printing of copper is very difficult due to its high reflection coefficient at commonly used laser wavelengths. Stereolithography (SLA) photopolymer printers have very high accuracy, but these plastics usually have high absorption in microwave and terahertz region. Fused-deposition modelling (FDM) printers are cheap, widely available and can use several well-known plastics with low absorption, such as nylon or HDPE, but the precision is less than for the former types.

To study the best possible solutions, we analyzed the dielectric permittivity and loss tangent of plastics used in different 3D printer types. Sample discs were printed and then placed inside the two-mirror resonator with high Q factor [1], and measurements were made for frequencies in 94-150 GHz range. Additional measurements were made for plastic waveguide inserts in 26-40 GHz range. Dielectric properties in different ranges have only a slight difference and can therefore be extrapolated to higher frequencies. Measurement results are collected in the table.

Plastic	Printer	Re ε	tan δ
Polyethylene terephthalate glycol (PETG) 1	FDM	2.31	$1.5\cdot10^{-3}$
PETG 2	FDM	2.47	$1.6\cdot10^{-3}$
Polylactic Acid (PLA)	FDM	2.27	$6.2\cdot10^{-3}$
Sterol-butadiene-sterol (SBS) 1	FDM	2.22	$1.6\cdot10^{-3}$
SBS 2	FDM	2.40	$1.3\cdot10^{-3}$
Visijet SL Clear	SLA	1.8	$3.5\cdot10^{-2}$
Visijet SL Flex	SLA	1.8	$3.5\cdot10^{-2}$
Visijet SL Hi-Temp	SLA	1.8	$3.5\cdot10^{-2}$
Polyamid	SLS	1.7	$3.5\cdot10^{-3}$

Sterol-Butadiene-Sterol has the best dielectric properties among the measured samples, and it was chosen to test the concept of two-dimensional planar Bragg resonator with chessboard-like grating, which can be used in distributed feedback lasers [2]. The parameters of Bragg structure were chosen in accordance with dielectric properties of the media and precision of FDM printers, which is acceptable for frequencies up to 100 GHz. Size of the square resonator is 180 mm, and average thickness is 2 mm. Grating periods in orthogonal directions are both equal to 4 mm, and grating depth is 0.4 mm. The 3D-printed structure is shown on Fig. 1.

The properties of this resonator were calculated in CST MWS for Re ε=2.4 and tan $\delta=2\cdot10^{-3}$, and the zone of effective scattering was near 60 GHz. To conduct the low-power measurement, the resonator had to be excited by a wave beam with a plane phase front, which was be made by a planar quasioptical transmission line. The transmission line consists of two parallel metallic planes, parabolic mirror and single-mode waveguide ending in the focus of the mirror. After the reflection, the planar wave beam is very close to the TEM-mode of planar waveguide. The wave beam is transmitted through the resonator and then received into single-mode waveguide by the symmetrical quasioptical transmission line.

Fig. 1. Dielectric two-dimensional plane Bragg resonator printed from SBS plastic

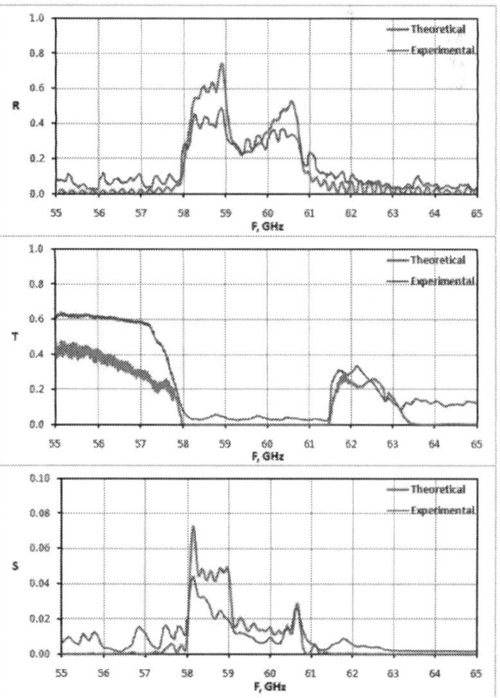

Fig. 2. Reflection (R), transmission (T) and orthogonal scattering (S) for dielectric two-dimensional plane Bragg resonator

Measurement results for reflection, transmission and orthogonal scattering are shown on the Fig. 2. There is a good agreement between calculations and experiment for both frequencies of the effective Bragg scattering and amplitudes of the scattering coefficients. The use of higher precision printers will allow manufacture of resonators suitable for submillimeter wavelengths.

Another promising field for printed dielectric structures is low-loss windows. Calculations basing on thermal and dielectric properties of printable plastics give the power of several tens of watts, which is sufficient for many applications, e.g. subterahertz gyrotron systems used for spectroscopy [3]. Window reflection can be minimized by the use of gratings of different shapes, like two-dimensional pyramidal or one-dimensional close to sinusoidal [4]. The shape of the grating requires fine details to be printed with high precision, and single-nozzle FDM printers flatten the top parts of the shape, which leads to the increase of the reflection. SLS plastic printers have significantly better precision. To compare different gratings, three discs were printed of polyamid: the first one has no gratings on the sides, the second one has pyramids with height 2 mm and base width 1 mm on both sides, and the third one has special one-dimensional periodic grating with height 3 mm and period 2 mm on both sides. The discs are shown on Fig. 3.

The low-power tests were conducted on the same resonator setup, which was used for the tests of dielectric properties. The measurements performed at frequencies 70-200 GHz show the significant reduction of reflection: both types of gratings effectively suppressed the reflection by more than 10 dB. This is sufficient for many subterahertz gyrotrons, and additional tests of mechanical and vacuum properties will be performed later.

This work was supported by RSCF grant 18-48-520022. Authors are grateful to DS Group for their help with manufacture of samples.

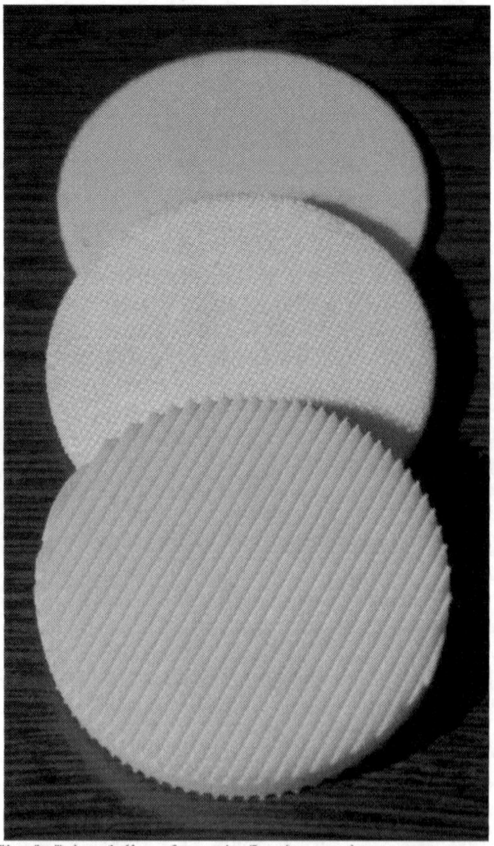

Fig. 3. Printed discs for antireflection grating tests: top – no gratings, middle – pyramidal gratings, bottom – one-dimensional gratings.

References

1. *Dryagin, Y.A. & Parshin, V.V.* A method to measure dielectric parameters in 5–0.5 millimeter wavelength band // Int. J. Infrared Milli. Waves. 1987. V. 13, No. 7. P. 1023-1032.

2. *Ginzburg, N.S., Peskov, N.Yu., Sergeev, A.S.* Dynamics of free-electron lasers with two-dimension distributed feedback // Optics Commun. 1994. V.112. P.151-156.

3. *Glyavin, M.Yu., Idehara, T., Sabchevski, S.P.* Development of THz gyrotrons at IAP RAS and FIR UF and their applications in physical research and high-power THz technologies // IEEE Trans. Terahertz Sci. Techn. 2015. V. 5, No. 5, P. 788-797.

4. *Vlasov, S. N., Koposova, E. V.* Suppressing the reflection from an interface between dielectrics by corrugating their surfaces // Techn. Phys. 1993. V. 38, No. 7. P. 592-595.

Recent progress in development and application of sub-THz gyrotrons in University of Fukui

Y. Tatematsu

Research Center for Development of Far-infrared Region, University of Fukui, Fukui, Japan, tatema@fir.u-fukui.ac.jp

I. Introduction

Till 2005, Research Center for Development of Far-Infrared Region, University of Fukui (herein referred to as FIR-UF) had developed gyrotrons to aim at increasing their frequency. For this purpose, the Gyrotron FU series were developed[1-3]. Fundamental, second and third harmonic oscillations were observed, and the breakthrough of 1 THz oscillation in gyrotron was achieved for the first time in the world[4].

Since 2006, gyrotrons have been developed in FIR-UF for the purpose of applications such as DNP-NMR spectroscopy[5,6], sintering and direct measurement of energy level of positronium hyperfine structure[7]. Gyrotrons developed for this purpose are Gyrotron FU CW series[8-13], Gyrotron FU CW G series[14-19] and pulsed gyrotrons[20-28].

II. Development of Gyrotron

Gyrotron FU CW I was developed in the collaboration between IAP, RAS and FIR-UF[10]. It was fabricated by Gycom. It was used for material processing and sintering. Some gyrotrons applying for DNP-NMR spectroscopies were developed, such as Gyrotrons FU CW II, IV, VI and VII[9,11-13]. For this application, the frequency tuning is necessary to adjust the sensitivity to maximum. With Gyrotron FU CW IV, a continuous frequency variation width of 6 GHz was observed[11]. The oscillation mode was fundamental TE_{12} mode. The mechanism of continuous frequency variation was explained as the interaction with gyro-backward waves.

Recently Gyrotron FU CW XA was developed. It is installed on an 8 T superconducting magnet. It is a linear type gyrotron without a built-in mode converter. The cavity is a normal hollow one with up and down tapers at the both ends of the straight section. The radius and length of the straight section of the cavity are 2.997 mm and 20 mm, respectively. Although this cavity was not specially designed for the purpose of frequency tuneability, frequency variations more than 1 GHz were measured with four modes, TE_{12}, TE_{13}, TE_{33} and TE_{14}, in this gyrotron. The frequency bands are 90, 140, 180, and 190 GHz, respectively[29].

For applications, conversion to Gaussian beam is necessary. About Gaussian beam output gyrotrons in FIR-UF, the Gyrotron FU CW I is the first, which was fabricated in Russia[10]. Gaussian beam output gyrotrons developed in FIR-UF are called as Gyrotron FU CW G series[14-19]. Gyrotron FU CW GI oscillates in 203 GHz frequency[14] and was used for direct measurement of the energy level of positoronium hyperfine structure[7]. Gyrotrons FU CW GII and GIII oscillate in second harmonic[15-17]. Gyrotron FU CW GIV showed frequency variations in two frequency ranges. The frequency varies with the widths of 5 GHz in 190 GHz band for the fundamental mode and of 3 GHz in 390 GHz band for the second harmonic mode[30].

Gyrotron FU CW GV is a multi-frequency gyrotron[18]. It is of sealed-off type and was constructed by New Japan Radio Company. It was installed on a 10 T magnet. The mode converter consists of a conventional Vlasov launcher and four mirrors. Although it was designed for the $TE_{10,6}$ mode, Gaussian beams were radiated in oscillations of ten modes. Their oscillation frequencies are from 162 GHz to 265 GHz about 10 GHz step. The transmittance of a window disk depends on the wave frequency. Some of them have low transmittance. To prevent the drop in transmittance due to the frequency-dependence, a distance variable double-disk window was adopted. With variation of the distance between the two disks, the transmittance of the electromagnetic wave periodically changed because of the wave interference. The transmittance can be almost one by setting with appropriate distances for every frequency.

In Gyrotrons FU CW VI and VIA, frequency variation around 460 GHz was achieved in a second harmonic oscillation using a 10 T magnet[19].

We also developed pulsed high power gyrotron series[20-28]. They are developed in the purpose of a light source for the Collective Thomson Scattering (CTS) measurement in LHD, NIFS. First, we developed 400 GHz frequency, 2nd harmonic oscillation gyrotrons using 8 T superconductive magnet. However, for the required target power of 100 kW injection into the plasma, 2nd harmonic oscillation did not reach the goal power. So, we changed to 300 GHz gyrotron development using 12 T magnet in fundamental oscillation. With the beam voltage of 65 kV and current of 15 A, power of 320 kW and efficiency of 33 percent are achieved[28]. This gyrotron was moved to LHD of NIFS in July, 2018 and will be applied to CTS experiment.

III. Applications of Gyrotron

As examples of applications of sub-THz gyrotrons, there are sintering, DNP-NMR spectroscopy, positronium energy level measurement, CTS, pulsed ESR, millimeter wave discharge, light emission from crystals and study for safety data base of sub-THz radiation.

Gyrotron FU CW GI was applied for direct measurement of hyperfine structure of positronium[7]. This is a collaboration with Tokyo University. Positoronium is a bound state of a positron and an electron. Ground-state positronium has two spin eigenstates, ortho-state and para-state. The energy difference of the two states correspond to about 203.4 GHz. The measurements of the energy difference were done in the past, however, they were measurement used the Zeeman effect in a static magnetic field. It may be affected by the magnetic field, and the result has a slightly difference from the theoretically calculated values. So, we did directly measurement of the energy level using electromagnetic wave. The expected spectrum of the gamma ray annihilation probability is a function of frequency. So, for this measurement, a light source with frequency tunable between 201 to 206 GHz and high power more than 100 W with Gaussian beam output is necessary. However, frequency of a gyrotron is not widely continuously changed to keep high power. Therefore, to change the frequency, we changed the cavity itself. We prepared a lot of cavities whose diameters were slightly different each other. Changing the cavities, the intensity of gamma-ray produced by pair annihilation was measured. The energy difference of the hyperfine structure of positronium was measured with the central energy of 203.39 ± 0.11 GHz.

Another application of gyrotron is light emission from ZnO crystal[31]. This is a collaboration with Osaka University. Spectra of irradiation of a ZnO crystal by gyrotron exhibit a broad and asymmetric peak around 600 nm with a longer-wavelength tail. When the duty ratio of gyrotron irradiation increased, the peak emission intensity increases rapidly. Yellow light emitted from the ZnO crystal. The temperature was more than 1000 degree by the IR camera measurement. When heated up to 1250K using an electric heater without gyrotron irradiation, the crystal does not exhibit any visible emission. Therefore, this result indicates that the yellow emission of the gyrotron irradiated crystal is not a simple blackbody radiation but is caused by a combination of thermal effects and the gyrotron's sub-THz electric field.

When sub-THz wave from a gyrotron is radiated to nano-scale cages of $Ca_{12}Al_{14}O_{33}$ crystal, visible light was emitted from the crystal[32]. It was explained the mechanism of lightening that oxygen ions inside the cage rattle due to sub-THz wave and emit visible light. Conversion from THz wave to visible light is useful for THz range research to specify the wave position. This is a collaboration with Tokyo Institute of Technology and Hirosaki University.

Gyrotron is also applied for the study of safety data base of sub-THz radiation for eyes. This is a collaboration with Tokyo Metropolitan University and Kanazawa Medical University. They had the experiments at frequencies of 40, 75, 90 GHz irradiation so far. In the present experiment, we use electromagnetic wave with the frequency of 162 GHz, one of oscillation of Gyrotron FU CW GV. Sub-THz wave is irradiated into rabbit eyes, and the threshold of power or power density to cause injury on the eyes is being investigated.

A pulsed high power gyrotron of 303 GHz will be applied for CTS measurement in LHD, NIFS. The electromagnetic wave generated by a gyrotron is injected into plasma. From the observation of the scattering wave, we can obtain the information of ion temperture. Due to the magnetic field distribution of LHD, 300 or 400 GHz is the best frequencies for CTS measurement in LHD.

IV. Summary

In FIR-UF, Gyrotron FU CW series, Gyrotron FU CW G series and pulsed gyrotron series have been developed for the purpose of various applications for the current decade and they have been and will be applied for various fields.

References

1. *Idehara, T. et al.*, Phys. Fluids **B 5**, 1377 (1993).
2. *Hong, K. D. et al.*, J. Appl. Phys. **74**, 5250 (1993).
3. *Shimizu, Y. et al.*, Phys. Plasmas **2**, 2110 (1995).
4. *Idehara, T. et al.*, Int. J. Infrared. Milli. Waves. **27**, 319 (2006).
5. *Matsuki, Y. et al.*, J. Mag. Res. **225**, 1 (2012).
6. *Matsuki, Y. et al.*, J. Mag. Res. **264**, 107 (2016).
7. *Miyazaki, A. et al*, Prog. Theo. Exp. Phys. **2015**, 011C01 (2015).
8. La Agusu et al., Int. J. Infrared. Milli. Waves. **28**, 315 (2007).
9. *Idehara, T. et al.*, Int. J. Infrared. Milli. Waves. **28**, 433 (2006).
10. *Zapevalov, V. E. et al.*, Radiophys. Quant. Electron. **50**, 420 (2007)
11. *Chang T.-H. et al.*, j Appi. Phys. **105**, 063304 (2009).
12. *Idehara, T. et al.*, J. Infrared Milli. Terahertz Waves **31**, 763 (2010).
13. *Idehara, T. et al.*, J. Infrared Milli. Terahertz Waves **31**, 775 (2010).
14. *Tatematsu et al.*, J. Infrared Milli. Terahertz Waves **33**, 292 (2012).
15. *Tatematsu et al.*, J. Infrared Milli. Terahertz Waves **35**, 169 (2014).
16. *Tatematsu et al.*, J. Infrared Milli. Terahertz Waves **35**, 517 (2014).
17. *Tatematsu et al.*, Phys Plasmas **21**, 083113 (2014).
18. *Tatematsu et al.*, J. Infrared Milli. Terahertz Waves **36**, 697 (2015).
19. *Idehara, T. et al.*, J. Infrared Milli. Terahertz Waves **36**, 819 (2015).
20. *Notake, T. et al.*, Rev. Sci. Instrum. **79**, 732 (2008).
21 *Notake, T. et al.*, Plasma Fusion Res. **4**, 011 (2009).
22. *Notake, T. et al.*, Phys. Rev. Lett **103**, 225002 (2009).
23. *Saito, T. et al.*, Plasma Fusion Res. **7**, 1206003 (2012).
24. *Saito, T. et al.*, Phys. Plasmas **19**, 063106 (2012).
25. *Yamaguchi, Y. et al.*, Plasma Fus Res. **8**, 1205165 (2013).
26. *Yamaguchi, Y. et al.*, Nuclear Fusion **55**, 13002 (2015).
27. *Yamaguchi, Y. et al.*, J. Instrumentation **10**, C10002, (2015).
28. *Saito, T. et al.*, Plasma Fusion Res. **12**, 1206013-1-1206013-2, 2017.
29. *Tatematsu et al.*, 42nd IRMMW-THz, RA2.2, 2017.
30. *Tatematsu et al.*, 41st IRMMW-THz, H5P.21.13, 2016.
31. *Kato, K. et al.*, Appl. Phys. Lett. **111**, 031108 (2017).
32. *Toda, Y. et al.*, ACS Nano **11**, 12358 (2017).

Quasi-fractal PBG structures for Multi-Beam Devices

V. A. Tsarev[1], D. A. Nesterov[2], P. D. Shalaev[3]

[1]Yuri Gagarin Saratov State Technical University, Saratov, Russia, tsarev_va@mail.ru
[2]Yuri Gagarin Saratov State Technical University, Saratov, Russia, enlucioro@gmail.com
[3]AO "NPP "Almaz", Saratov, Russia, p.d.shalaev@yandex.ru

Over the last years, the great attention of microwave vacuum tube and accelerator researchers therefore drawn to so-called photonic crystal resonator cavities, formed by a defect lattice (photonic band gap structure-PBG)[1-4].

Very strict requirements applied to the resonator systems used in such devices. They must have a high characteristic impedance $\rho = R/Q_0$ and increased (compared to the operating wavelength) dimensions. In addition, they must have a rigid and technological structure having a high unloaded Q-factor and a good heat-dissipating ability.

In the defect area it is appeared local electromagnetic waves (defect modes) similar to electromagnetic waves of ordinary cavity resonators.

The advantages of PBG are increased overall dimensions, high unloaded Q-factor, comparative simplicity of engineering implementation, as well as the possibility to suppress undesirable (parasitic oscillation).

Due to these advantages, the metal photonic band gap structure has potentialities in the areas of high-energy accelerators, multi-beams klystrons.

Up to the present moment, it was studied only one gap single defects and multiple defects of PBG for single-beam and multi-beam klystrons [5, 6]. This did not allow us to realize a large gain factor, a high efficiency and a wide band in such devices.

This paper presents the results of 3D computer simulations of a double-gap PBG resonator with enhanced impedance of interaction.

The transformation of the defect zone from two-dimensional to three-dimensional was accomplished by inserting an additional X-shaped quasi-fractal resonance element in the cross section in the central part of the interaction region (fig. 1). This transformation was performed by inserting an additional X-shaped quasi-fractal resonance element in transverse section in central part of the interaction area.

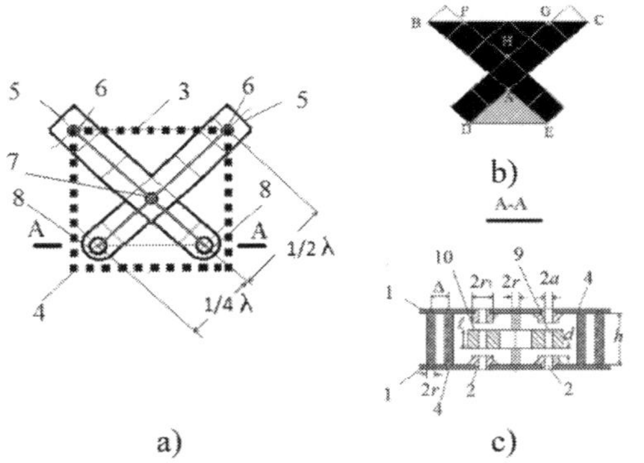

Fig. 1. The structure of the resonator: a) fractal scheme, b) top view, c) transverse section

The interaction area is confined by side covers 1 with circular holes 2 for transition of two electron beams in longitudinal direction and in transverse direction by a conditional square 3 that defines boundary of the longitudinal rods location 4, which restrict two-dimensional defect area. The resonance rods 5, forming the additional X-shaped resonance element, are oriented on the diagonal of the conditional square in the two-dimensional photonic crystal. This element consisted of two similar triangles (ADF and ABC) with cutout in them (FGH and ADE), which are also scale lines of the initial triangles. The drawn figure is similar to Serpinski triangle fractal [7]. In the center of diagonal intersection at the distance $l_1 = \lambda/2$ from electric contact place between resonance rods with longitudinal rods 6 is located a central longitudinal metal rod 7, having electric contact with side covers. At the free ends of the resonance rods it is made circular holes 8 for electron beams transition that are coaxial with circular holes on side covers. The full length of both resonance rods is equal $l_2 = 3/4\lambda$, where λ is TEM – mode operating wavelength. As a result of these transformations it is formed two double high-frequency gaps. The lowest frequency modes in three-dimensional photonic band gap are quasi-TEM modes, which electric и magnetic fields have opposite directions in each gap (π-mode). However, for operation in millimeter and sub-millimeter wavelength ranges in-phase modes similar to TM_{110} and TM_{210} in ordinary prismatic resonant cavities is of the greatest interest.

Simulation results

We have designed a quasi-fractal double-gap cavity based on PBG structure. 3D-models of studied resonance system are illustrated in table.

Table. Computer models of studied PBG structure

Model 1	Model 2	Model 3	Model 4
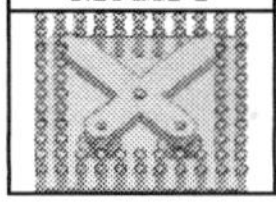	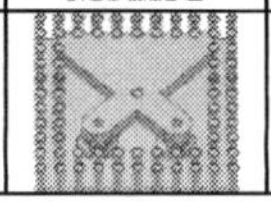	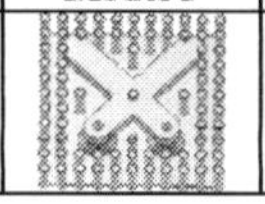	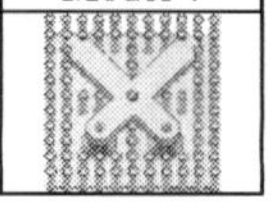

It had the following dimensions: lattice depth 10 mm, metal rods diameter 2 mm, spacing between the rods, 3.7 mm, drift tube radius 1.0 mm; gap length 1.5 mm; resonance rods thickness in along-track beam direction 3 mm. It was studied the influence of number n additional longitudinal rods that are inserted into 3D defect area, at a frequency of indicated modes. During the simulation, it is also determined characteristic impedance (shunt impedance, fig.2) and unloaded Q-factor cavity (fig.3).

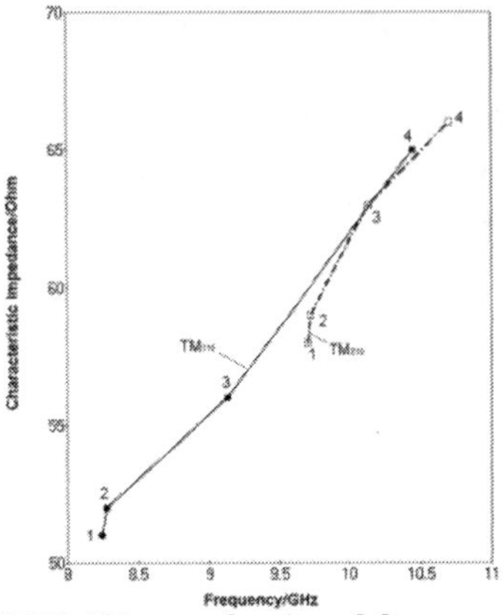

Fig. 2. HOM R/Q_0 as a function of frequency. Numbers indicate the model numbers indicated in table.

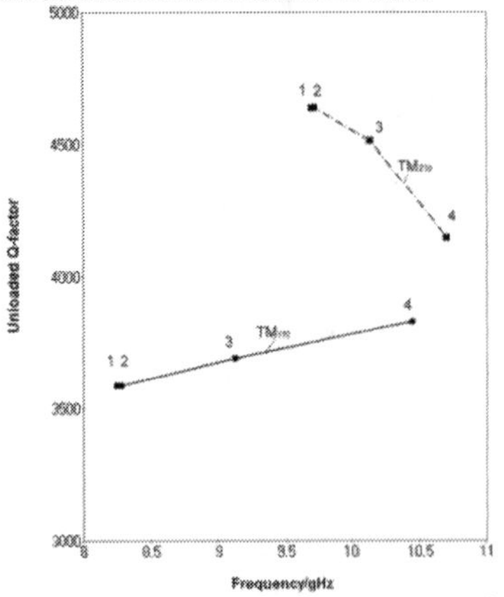

Fig. 3. HOM Q_0 as a function of frequency. Numbers indicate the model numbers indicated in table.

From fig. 2 it is clear that increasing numbers of rods in the cavity volume leads to characteristic impedance increasing of both TM_{110} (1.3 times) and TM_{210} modes (1.2 times). The dependence of unloaded Q-factor on rod numbers (fig. 3) for TM_{110} mode has linear increasing form while for TM_{210} mode Q-factor it falls. Considering both dependence we can conclude: stunt impedance linearly grows with the rods number for TM_{110} and has a maximum at a number of rods $n = 2$ (fig. 4). The dependence of stunt impedance for TM_{210} on frequency has the parabola form, a maximum is at $n = 1$. Having said so this mode has higher stunt impedance values then at TM_{110} mode for any n.

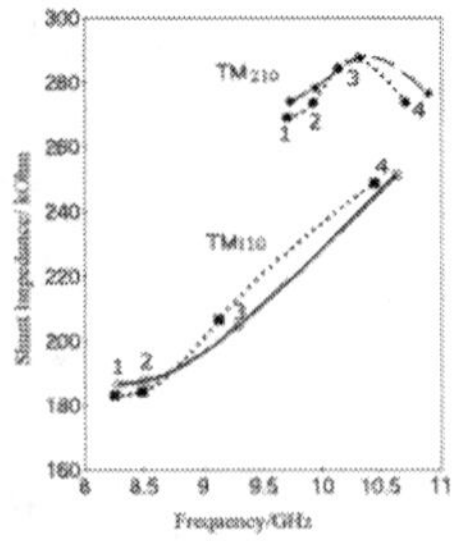

Fig. 4. Shunt impedance for HOM as a function of frequency. Numbers indicate the model numbers indicated in table. Dotted lines represent cavities with the rods, dashed and solid lines represent cavities with solid walls

The presence of the rods in quasi-fractal cavity models leads to frequency decreasing as compared with cavity models with solid walls.

The coupled quasi-fractal double-gap cavity array as a fractal electrodynamic system that capable to slow down electromagnetic wave considerably can be used for TWTs.

Due to increased stunt impedance, quasi-fractal PBG structures for Multi-Beam Devices has potentialities in the areas of high-energy accelerators, multi-beams klystrons, terahertz radiation sources, etc.

References

1. *Ozbay E., Temelkuran B., and Bayindir M.* Microwave applications of photonic crystals // Progress In Electromagnetics Research, 2003. Vol. 41, pp. 185–209.

2. *Ashutosh, Jain P.K.* Design and analysis of metallic photonic band gap cavity for a gyrotron // Journal of Microwaves, Optoelectronics and Electromagnetic Applications. 2012. Vol. 11, № 2. P. 242.

3. *Han S.-T., Jeon S.-G., Shin Y.-M., Jang K.-H., So J.-K., Kim J.-H., Chang S.-S., Park G.-S.* Experimental investigations on miniaturized high-frequency vacuum electron devices // IEEE Trans. Plasma Sci. 2005. Vol. 33, № 2. P. 679.

4. *Jeon S.-G., Shin Y.-M., Jang K.-H., Han S.-T., So J.-K., Joo Y.-D., Park G.-S.* High order mode formation of externally coupled hybrid photonic-band-gap cavity // Appl. Phys. Lett. 2007. Vol. 90, № 2. 021112.

5. *Smirnov A. V. et. al.* PBG Cavities for Single-Beam and Multi-Beam Electron Devices // Proceedings of Particle Accelerator Conference, 2003, pp. 1153-1155.

6. *Xu, Y. and Seviour, Rebecca.* Design of Photonic Crystal Klystrons. In: Proceedings of the 1st International Particle Accelerator Conference IPAC 2010. JACoW, Kyoto, Japan, pp. 4002-4004. ISBN 978-92-9083-352-9.

7. Sierpinski triangle // Wikipedia, the free encyclopedia. – URL: https://en.wikipedia.org/wiki/Sierpinski_triangle. – 06.09.18.

Generation of Ultrashort Microwave Pulses in Passive Mode-Locked Electron Oscillators with Homogeneous and Inhomogeneous Line Broadening

M. N. Vilkov, N. S. Ginzburg, I. V. Zotova, A. S. Sergeev

Institute of Applied Physics RAS, Nizhny Novgorod, Russia, vilkovmn@ipfran.ru

Introduction

In the laser physics the generation of ultrashort pulses (USP) via passive mode-locking [1] is broadly used. Gain media with homogeneous and inhomogeneous line broadening are applied [2]. The analog of homogeneous and inhomogeneous line broadening in microwave electronics is electron-wave interaction with beams having small or large initial energy spread. USP electron oscillators with passive mode locking with homogeneous line broadening were investigated in [3-4]. In this paper we will study peculiarities of the USP microwave oscillators with both homogeneous and inhomogeneous line broadening.

Model and basic equations

The schema of USP generators is presented in Fig. 1. It consists of an electronic amplifier, a nonlinear (saturable) absorber in the feedback circuit, and a partially transparent mirror for output of the signal.

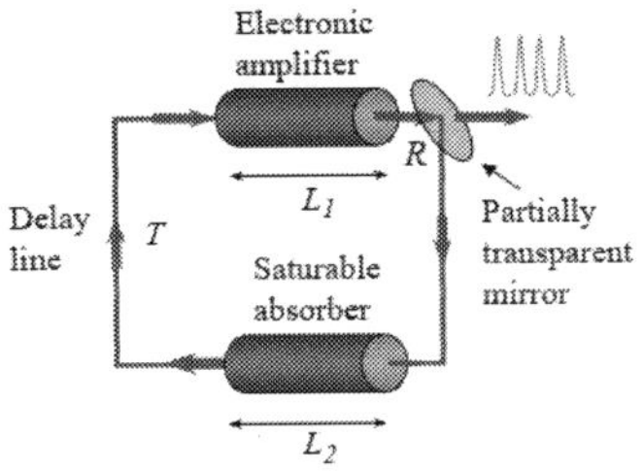

Fig. 1. Principal scheme of a USP oscillator with a saturable absorber in the feedback loop.

We will use a rather general model of the electronic amplifier with prevailing inertial particle bunching [5]. Under conditions of a relatively small energy change of particles $|1-\varepsilon/\bar{\varepsilon}|\ll 1$ (where $\varepsilon = mc^2\gamma$, γ is the relativistic factor, $\bar{\varepsilon}$ is an average value of electron energy in the initial distribution), the process of amplification can be described by the universal system of equations [5,6]

$$\frac{\partial a_n}{\partial Z}+\frac{\partial a_n}{\partial \tau}=J,\qquad \frac{\partial u}{\partial Z}=\mathrm{Re}\left(a_n e^{i\theta}\right),\quad \frac{\partial \theta}{\partial Z}=u. \tag{1}$$

The boundary condition for electrons is

$$u\big|_{Z=0}=u_0,\quad \theta\big|_{Z=0}=\theta_0\in[0;2\pi). \tag{2}$$

We will suppose initial energy spread of electrons of the beam is normal

$$f_0(u_0)=\frac{1}{\sqrt{2\pi}\sigma}e^{-\frac{u_0^2}{2\sigma^2}} \tag{3}$$

where $\sigma=\mu C^{-1}\Delta\varepsilon/\bar{\varepsilon}$ is a parameter characterizing the electron energy spread, $J=1/\pi\times \int_{-\infty}^{+\infty}\int_0^{2\pi} f_0(u_0)e^{-i\theta}d\theta_0 du_0$ is an amplitude of the high-frequency electron current, θ is an electron phase relative to a synchronous running wave, $u=\mu C^{-1}(1-\varepsilon/\bar{\varepsilon})$ is the normalized electron energy variations, $Z=C\omega z/c$, $\tau=C\omega(t-z/V_{\|0})\times (c/V_{gr}-c/V_{\|0})^{-1}$, $a_n=\chi\mu C^{-2}eA/mc$ is the normalized wave amplitude, n is the number of the field passages through the feedback loop, $L=C\omega l/c$ is the length of amplification region, $C=(eI_0\chi^2\mu/mc^3\bar{\gamma}N)^{1/3}$ is the parameter of amplification (Pierce' parameter), I_0 is the beam current, χ is the electron–wave coupling coefficient (see [6]), N is the operating mode norm, and $V_{gr}=d\omega/dh$ is the wave group velocity.

The signal passage in a feedback circuit with absorber is described by the equation

$$\frac{\partial a_n}{\partial Z}+\frac{\partial a_n}{\partial \tau}+\nu(a_n)a_n=0. \tag{4}$$

where the coefficient of attenuation ν for a saturable absorber can be specified as

$$\nu(a_n)=\frac{\nu_0}{1+\upsilon|a_n|^2}. \tag{5}$$

In accordance with the scheme shown in Fig. 1 boundary conditions can be presented in the form

$$a_{n,in}^{(2)}(\tau)=Ra_{n,out}^{(1)}(\tau),\qquad a_{n+1,in}^{(1)}(\tau)=b_{n,out}^{(2)}(\tau-T) \tag{6}$$

where index "1" and "2" corresponds to section of amplification and absorbing, respectively, R < 1 is the reflection coefficient, T is a delay time.

Electron efficiency of the oscillator is determined by the relationships

$$\eta=\frac{C}{\mu\left(1-\bar{\gamma}^{-1}\right)}\hat{\eta},\qquad \hat{\eta}=\frac{1}{2\pi}\int_0^{2\pi} u\left(Z=L_1\right)f_0(u_0)d\theta_0 du_0. \tag{7}$$

Results of simulations

Results of simulations of USP generation via passive mode-locking with homogeneous (the initial electron energy spread is absent) and inhomogeneous line broadening (the initial electron energy spread $\sigma=2$) are presented in Fig. 2. For both cases one can

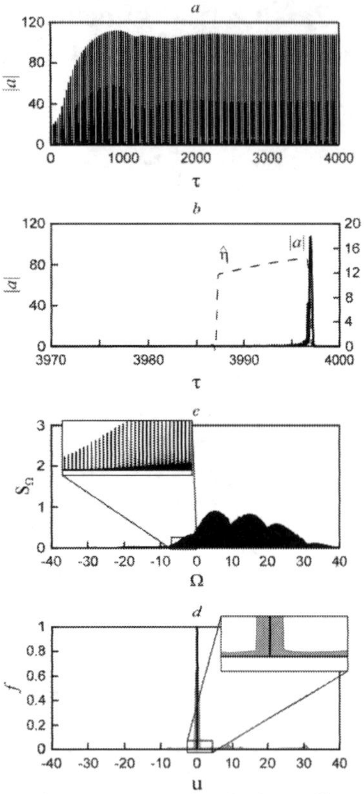

Fig. 2. Setting on a USP regime in the oscillator with homogeneous line broadening (L_1=10, R=0.9, L_2=5, ν_0=1.5, υ=0.6, T=20, $\sigma\to0$): (a) time dependence of the field amplitude, (b) detailed profile of pulses and the instant electron efficiency in the extended time scale, (c) radiation spectrum, and (d) electron energy distribution function at the initial (black line) and final (red line) stages.

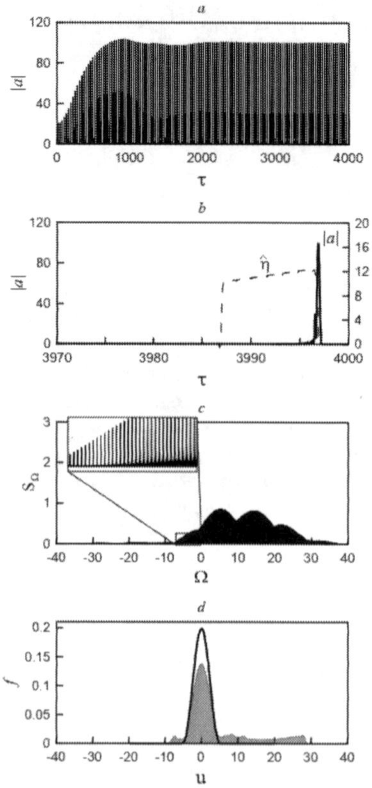

Fig. 3. Setting on a USP regime in the oscillator with inhomogeneous line broadening (σ = 2, other parameters are same as in Fig. 2): (a) time dependence of the field amplitude, (b) detailed profile of pulses and the instant electron efficiency in the extended time scale, (c) radiation spectrum, and (d) electron energy distribution function at the initial (black line) and final (red line) stages.

see that there is possibility of arrangement of mode locking regime accompanied by USP pulse production. For the same beam current, gain parameter and the length of amplification section the characteristic of generated pulses in regime with homogeneous and inhomogeneous line broadening is very closed. Nevertheless the peak amplitude of USP pulses is larger for the homogeneous line broadening.

Thus based on our analysis we can conclude that passive mode locking is an effective way to exploiting of powerful electron beams possessed by substantial energy spread (like high current relativistic electron beams) for generation of coherent radiation.

This work was supported by Russian Scientific Foundation (RSCF), grant No. 16-42-01078.

References

1. *Haus H.A.* Mode-Locking of Lasers // IEEE J. of Select. Topics in Quant. Electron. 2000. V. 6, No. 6. P. 1173-1185.

2. *Yariv A.* Quantum electronics // J. Wiley and Sons. 1975

3. *Ginzburg N.S., Denisov G.G., Vilkov M.N., Zotova I.V. Sergeev A.S.* Generation of a Periodic Sequence of Powerful Ultrashort Pulses in a Traveling Wave Tube with Bleachable Absorber in the Feedback Loop //Technical Physics Letters, 2015, V. 41, No. 9. P. 836–839

4. *Ginzburg N.S., Denisov G.G., Vilkov M.N. et al.* Generation of "gigantic" ultra-short microwave pulses based on passive mode-locking effect in electron oscillators with saturable absorber in the feedback loop // Phys. Plasmas. 2016. V. 23. P. 050702-1-4.

5. *Bratman V.L., Ginzburg N.S., Kovalev N.F.*, et al., in Relativistic High Frequency Electronics, Ed. by A. V. Gaponov-Grekhov (IPF RAS, Gorky 1979), p. 249

6. *Ginzburg N. S. Sergeev A. S.*, Sov. Phys. Tech. Phys. 1991. V. 36, P. 665.

Transformation of High-Power Gyrotron Output Radiation Frequency under Conditions of Raman Scattering on Additional Electron Beam

L.A. Yurovskiy, N. S. Ginzburg, A.M. Malkin, A.S. Sergeev, I.V. Zotova

Institute of Applied Physic RAS, Nizhny Novgorod, Russia, ginzburd@appl.sci-nnov.ru

Development of microwave oscillators and of gyrotrons in particular [1] increases the field of potential applications, for which the radiation spectrum control is essential. For example, in gyrotron applications as a source for plasma heating, set frequencies are chosen based on a compromise between the absorption in central and peripheral areas of plasma. Along with this, the investigations show [2] that plasma heating can be much more efficient if the frequency tuning of the generated signal is possible in the gyrotrons involved. In this paper, we propose using the Stimulated Raman Backscattering (SRB) regime on the additional weakly relativistic electron beam. Currently, this regime is under investigations for compression and amplification of optical frequency pulses scattered on the plasma with neglection of frequency transformation [3-5]. However, in microwave frequency range, if the electron beam is used, Doppler shifts values become of the same order as the pump and the scattered signal frequencies. Moreover, the feature of this regime is energy transfer from higher-frequency (HF) pump wave concurrent with the electrons to the lower-frequency (LF) backward propagating scattered wave. Energy transfer, accompanied by the development of the absolute instability, leads to the system self-excitation in the absence of external resonators. This allows for a smooth frequency tuning by varying energy of the electron beam.

The model, where SRB regime take place, is presented in Fig.1. Here the pump wave concurrent with the electrons is scattered on the magnetized cylindrical electron beam in the waveguide, into the backward signal with a frequency downshift. The strong guiding field can increase the transverse oscillations of the electrons in the pump field. However, the magnetic field should be far from the values corresponding to the cyclotron resonance with the pump wave. Otherwise, excitation of parasitic cyclotron oscillations leads to the violation of combination synchronism conditions.

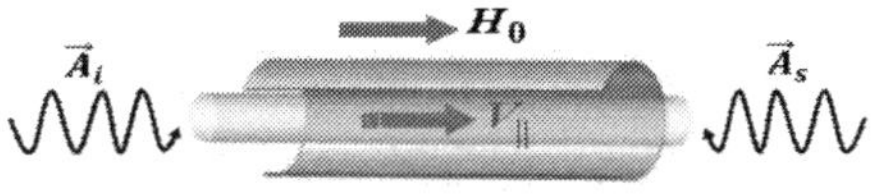

Fig. 1. The model of scattering of the concurrent pump wave into the LF backward wave

In case of sufficiently large density of electron beam, it can be considered as a material medium supporting the space-charge waves. Pump wave leads to the excitation of the Fast Space-Charge Wave (FSCW) possessing positive energy in the absorbing beam. As a result, selecting appropriate combination synchronism detuning, leads to emerging of SRB on space-charge waves with next synchronism conditions:

$$\omega_i - h_i V_{||} - \omega_s - h_s V_{||} = \omega_{P_{||}} \qquad (1)$$

where $\omega_{P_{||}}$ is the longitudinal plasma frequency, i,s – indexes of pump and scattered waves correspondingly.

This process described by a three-wave interaction equations, which include the equations for a HF pump wave, an LF scattered wave and a FSCW:

$$\frac{dA_b}{dz} = A_i A_s \qquad \frac{dA_i}{dz} = -A_b A_s \qquad \frac{dA_s}{dz} = -A_b A_i \qquad (2)$$

with boundary conditions

$$A_b\big|_{z=0} = 0 \qquad A_s\big|_{z=L} = 0 \qquad A_i\big|_{z=0} = A_{i0}$$

where A_b, A_i, A_s - normalized amplitudes of space-charge wave, pump wave and scattered wave correspondingly, L - normalized length of scattering region.

Equations (2) possess the following integrals:

$$A_i^2 - A_s^2 = A_{i0}^2 - A_{s0}^2 \qquad A_s^2 + A_b^2 = A_{s0}^2 \qquad (3)$$

where $A_{s0} = A_s\big|_{\zeta=0}$ is the signal wave amplitude at the output of the electrodynamic system. Taking into account (3), the equation for the signal wave amplitude reduced to the form:

$$\frac{dA_s}{dz} = -\sqrt{A_{i0}^2 - A_{s0}^2 + A_s^2}\sqrt{A_{s0}^2 - A_s^2} \qquad (4)$$

In the steady state oscillations regime, the longitudinal distributions of the wave amplitudes are represented by Jacobi elliptic functions (Fig.2).

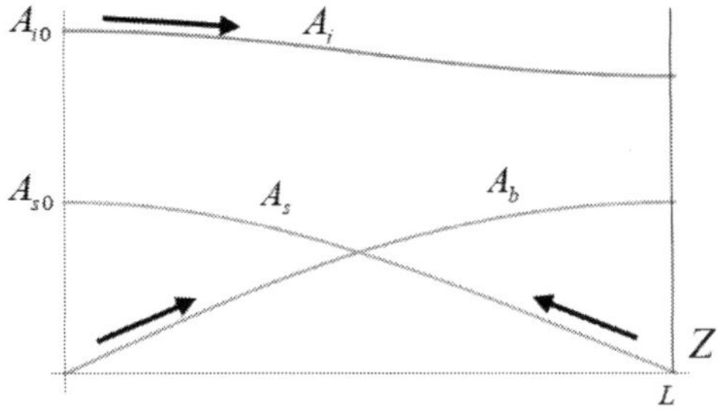

Fig. 2. Longitudinal amplitude distributions in the steady-state oscillations regime

For instance, integrating (4) yields:

$$A_s = A_{s0} cn[A_{i0} z, \sqrt{\Re}] \qquad (5)$$

Here $\Re = A_{s0}^2 / A_{i0}^2$ is the quantum yield (the number of pump quants transformed into the signal quants), to be found from equation:

$$A_{i0} L = K(\sqrt{\Re}) \qquad (6)$$

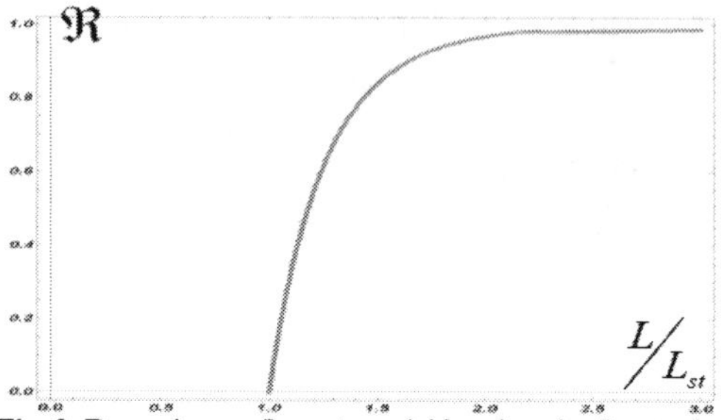

Fig. 3. Dependence of quantum yield on length of scattering region

where $K(x)$ is the complete elliptic integral of the first kind. Fig. 3 shows that quantum yield tends asymptotically to unity as the length of scattering region L increases and when L is two times greater than starting value L_{st} almost all pump power transferred to the scattered signal. Starting value L_{st} can be found from the self-excitation condition, derived from (6) at $\Re \to 0$:

$$A_{i0} L_{st} = \pi/2 \quad (7)$$

In physical variables, equation (7) can be rewritten for the oscillator starting current:

$$I_{st} = 2\pi^4 \cdot I_0 \frac{P_0^2}{P^2} \cdot \frac{\omega_i^2 h_s^2 h_i^2}{T^4 \omega_c^4 l^4} \cdot \beta_{\parallel 0}^7 \cdot \frac{r_w^8 D_1 \left(N_s^{m_s,p_s}\right)^2 \left(N_i^{m_i,p_i}\right)^2}{r_b^2}$$

Here l - the length of scattering region, T - coupling coefficient, $I_0 = 17kA$, $P_0 = 8.7GW$, D_1 - space-charge forces depression coefficient, $N_j^{m_j,p_j}$ - waveguide mode norm and P is the input pump power.

Estimations were carried out for the input pump power $P = 1MW$ at wavelength $\lambda_i = 2mm$. Radius and length of the scattering section were chosen to be $r_w = 6mm$ and $l = 30cm$, correspondingly. Magnetic field H_0 was taken 3% off cyclotron resonance value. Both pump and scattered waves were assumed to be $TE_{1,1}$ modes of the cylindrical waveguide. Calculations show (Fig.4) that there exists an optimal radius of the electron beam $r_{b_{min}}$ providing the minimum value of the starting current. For $TE_{1,1}$ mode $r_{b_{min}} = 1.6mm$.

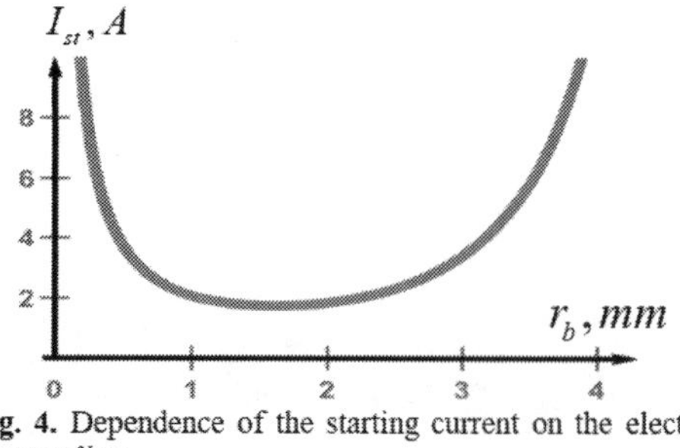

Fig. 4. Dependence of the starting current on the electron beam radius

Further calculations were carried out for $r_b = r_{b_{min}}$. As can be seen from (Fig.5-a), in the case when electron energy is in the range of values $U = 2-20keV$ starting current does not exceed $6A$ and remains much less than the limiting current of the cylindrical waveguide I_{max}. The magnitude of frequency tuning (Fig.1.4-b), is found from the synchronism condition (1) and is 20-40% of the pumping frequency, when electron energy is varied in the same interval of electron energies as in (Fig.1.4-a). The change in the frequency of the scattered wave in the range of 5-7% is achieved by varying the electron energy within the limits of $\pm 2.5keV$.

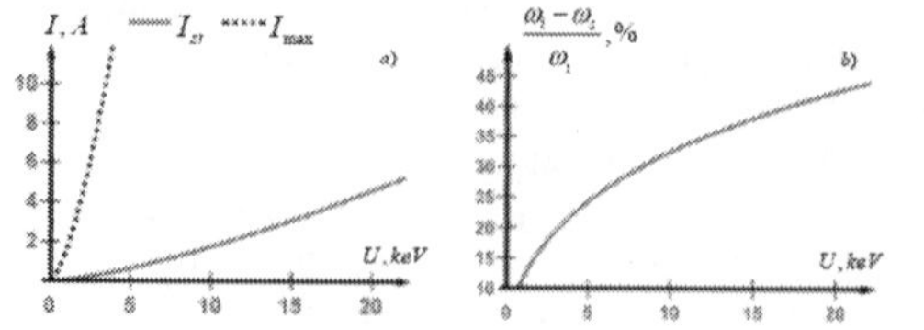

Fig. 5. a) Dependence of limiting (dotted curve) and starting (solid curve) current on electron beam energy, b) Dependence of relative magnitude of frequency tuning on electron beam energy

Magnitude estimations of the electric field on the waveguide surface shows that its values does not exceed $17kV/cm$ and therefore breakdown will not develop. Also an important parameter for estimation is the typical scale of the fields insertion L_{sw}. To obtain radiation on the combination synchronism, without oscillations at cyclotron resonance $\omega_s - h_s V_{\parallel} \approx \omega_H$, an adiabatically smooth insertion of the fields is necessary $\left(\Omega - \omega_{H_0}\right) L_{sw} V_{\parallel 0}^{-1} \sim 2\pi$, when only stimulated oscillations of the electrons are excited. The estimation of the length L_{sw} for the electron energy range $U = 2-20keV$ shows that L_{sw} it is within the limits of $2.5-3cm$.

This work was supported by the Russian Scientific Foundation (RSCF), Grant No. 16-42-01078.

References

1. *A.G. Litvak, G.G. Denisov, et al.* Development in Russia of 170 GHz/ 1MW/ CW Gyrotron for ITER // 2011 International Conference on Infrared, Millimeter and Terahertz Waves. 2-7 Oct. 2011. Houston, USA.
2. *H. Zohm, M. Thumm.* On the use of step-tunable gyrotrons in ITER. // Journal of Physics: Conference Series. Vol. 25. 2005. pp. 274-282.
3. *V. M. Malkin, G. Shvets, N.J. Fisch.* Fast Compression of Laser Beams to Highly Overcritical Powers. // Physics Review Letters. Vol. 82. No. 22. 1999. pp. 4448-4451.
4. *I.Y. Dodin, G.M. Fraiman, V.M. Malkin, N.J. Fisch.* Amplification of Short Laser Pulses by Raman Backscattering in Capillary Plasmas. // Journal of Experimental and Theoretical Physics. Vol. 95. No. 4. 2002. pp. 625-638.
5. *A.A. Balakin, N.J. Fisch, G.M. Fraiman, V.M. Malkin, Z. Toroker.* Numerical modeling of quasitransient backward Raman amplification of laser pulses in moderately undercritical plasmas with multicharged ions. // Physics of Plasmas. Vol. 18. 2011. pp. 102311/1-7.

Various types of echelette resonators for gyrotrons

V.E.Zapevalov, S.N.Vlasov, E.V.Koposova, A.N.Kuftin, A.B.Paveliev, N.A.Zavolsky

Federal Research Center Institute of Applied Physics, RAS, Nizhny Novgorod, Russia, *zapev@appl.sci-nnov.ru*,

In recent years, significant progress has been made in the development of high-performance gyrotrons and various applications based on them [1, 2]. In the gyrotrons of the millimeter wavelength range the resonators (cavities) in the form of waveguides of variable cross section are used. In the operating regime, they radiate into the output waveguide one rotating operating TE_{mp} mode with azimuthal m and radial p indices. These resonators have the so-called minimum diffraction Q-factor, determined by the ratio of the resonator length to the wavelength of the radiation. The existence of this quantity is due to the operation of the resonator at the quasicritical frequency of the working mode and the presence of a "diffractive" single-mode output from the resonator. The existence of a lower limit of the diffraction Q-factor (at a finite value of the ohmic Q-factor) limits the power that can be obtained from a gyrotron with a given mode.The increase in power is achieved by increasing the surface area of the resonator, for which it is necessary to increase its cross section. However, this is possible up to a certain limit because of the thickening of the spectrum of eigen frequencies of the resonator and the deterioration of the mode selection. Thus, the possibilities of increasing the output power and operating frequency are limited, especially in gyrotrons working on gyrofrequency harmonics. Overcoming this limitation is possible with the use of resonators in which the minimum value of the diffraction Q-factor is not limited so rigidly by their length. In this paper the results of a study of several variants of gyrotrons with echelette-type resonators are presented.

Echelette resonator is one of the possible resonators with the required properties. In comparison with conventional resonators, the power flux density to the wall here can be reduced severalfold. The principles of constructing echelette resonators are described in [3-6]. In such resonators the operating modes are formed by reflection of waves from corrugated surfaces in an autocollimation regime. Three types of echelette resonators are discussed: a) two-mirror resonator, in which the field is represented as waves propagating towards each other, b) three-mirror resonator, in which only one mirror is corrugated, and the field represents a wave traveling along the azimuth, c) an axially symmetric resonator with a corrugated surface. Resonators a) and c) are investigated in this paper in more detail. The electron beam is assumed to be axially symmetric, which has by now been well developed. However, in some cases, in particular for a two-mirror circuit, it is advisable to use ribbon beams to increase the efficiency and additional selection.

In Fig. 1 the profile of the axially symmetric echelette resonator, RF field distribution (a), and the photo of this resonator (b) are presented. The resonator has sinusoidal corrugation and was designed to operate at the 1st harmonic of the gyrofrequency and a symmetrical operating mode.

The operating frequency of the gyrotron is 150 GHz and the accelerating voltage - 80 kV. With the chosen parameters of the resonator and the electron magnetron-injector gun described in [7], the theoretical value of Q ≈ 4200 provides the achievement of the maximum value of electron efficiency.

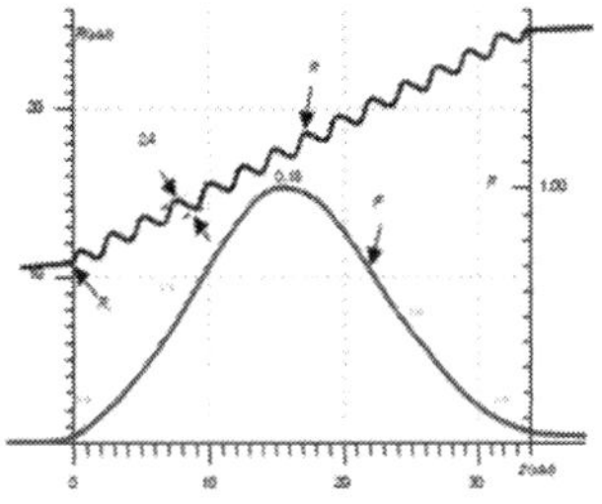

(a)

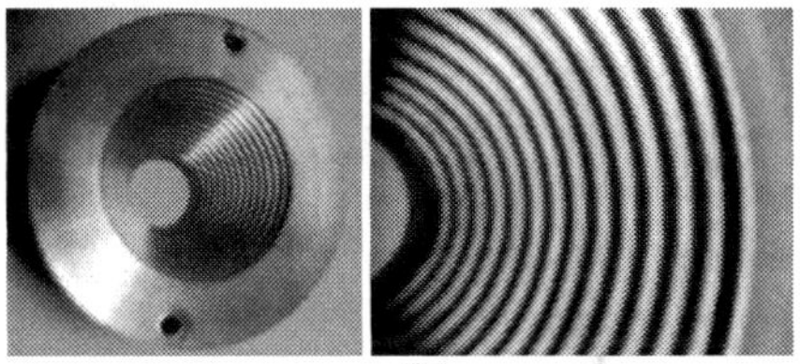

(b)

Fig. 1. The profile of the echeltte resonator and the distribution of the high-frequency field (a) and the photos of the resonator and its part (b)

An experimental study of the gyrotron was carried out according to the routine scheme using the apparatus described in [8]. The dependences of the pulse output power P (pulse duration $\tau = 50\ \mu s$ at repetition frequency $f_0 = 5\ Hz$) on the magnetic induction in cryomagnet B_0 (current in the solenoid) were obtained at a voltage $U = 80\ kV$ for different values of the electron beam current I_b taken from cathode (see Fig. 2).

The experimental data demonstrates only one oscillation in the magnetic field tuning band. A high value of the output efficiency $\eta \approx 32$ % took place at the current of the electron beam $I_b = 35\ A$. Thus, the axially symmetric echelette gyrotron has a much more sparse spectrum of eigenfrequencies and a noticeably lower specific thermal load on the resonator walls compared with the canonical gyrotron (1.8 times) with a close efficiency. The decrease in the thermal load is due to both a decrease in the quality factor in comparison with a cylindrical resonator of similar length and a certain increase in the surface.

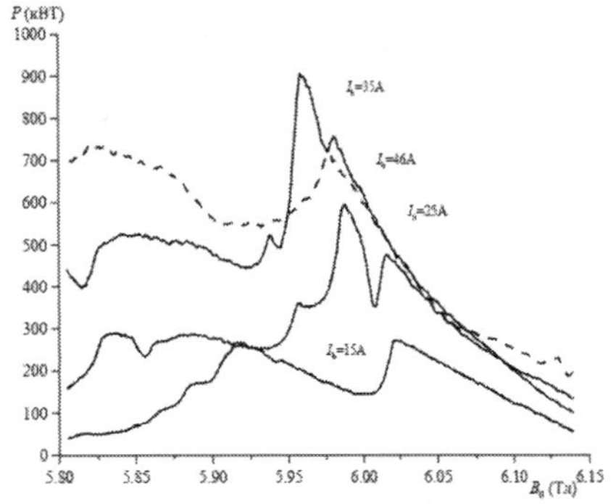

Fig. 2. Dependence of the output power of the gyrotron on the magnetic field in the resonator for different values of the electron beam current

For a number of applications a two-mirror version of the echelette gyrotron is interested. In this variant, there are some possibilities for frequency tuning and the radiation output system is simplified but with the use of a cylindrical beam an efficiency decrease of 1.5-2 times is inevitable. This scheme is the development of gyrotrons with a quasioptical resonator with additional selection due to the echelette mirror. Fig. 3 shows a gyrotron in which a two-mirror echelette resonator is used. Here the RF field can be represented in the form of waves propagating towards each other.

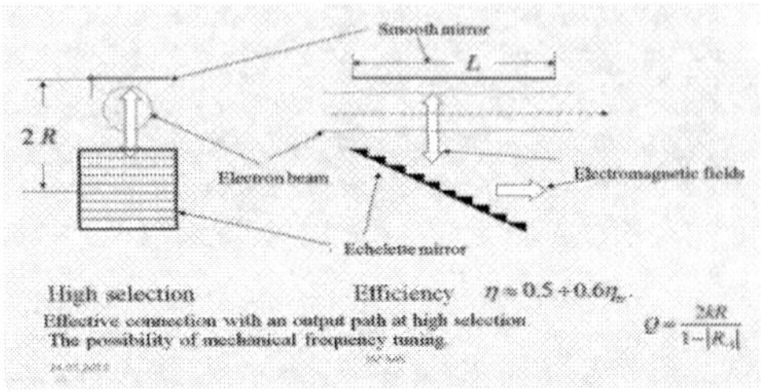

Fig. 3 Scheme of the gyrotron with a two-mirror echelette resonator

Calculations carried out for gyrotrons with two-mirror echelette resonators operating at the third harmonic of the gyrofrequency demonstrate the promises of these systems. Estimations of the *Q*-factor of resonators and start-up regimes for these gyrotrons were performed. The electron beam was assumed to be axially symmetric and had the same design parameters as indicated above. The possibilities of smooth tuning of the frequency in such systems were investigated. Calculations show that the use of two-mirror echelette resonators in gyrotrons operating at the 1st to 3rd harmonics of the gyrofrequency makes it possible to realize a smooth tuning of their operating frequency in the band of about 1% by varying the distance between the mirrors and corresponding correction of the magnetic field.

To increase the efficiency and additional selection in a two-mirror scheme it is expedient to use ribbon beams. For the same purposes, it is also possible to develop cylindrical beams with axially inhomogeneous emission. The development of electron-optical systems for the formation of such beams has its own specifics and is in the initial stage.

A three-mirror echelette resonator consisting of one echelette (corrugated) and two smooth mirrors possesses all the advantages of a two-mirror resonator exceeding its efficiency in the use of an electron beam of a cylindrical configuration.

Theoretical and experimental studies have shown that the axially symmetric echelette gyrotron has a significantly more rare spectrum of eigenfrequencies and a lower thermal load (by a factor of 1.8) on the resonator walls compared to the canonical gyrotron with close efficiencies. This allows us to hope for obtaining in the echelette gyrotrons more power at long-pulsed and continuous regimes. The output radiation in the form of a conical wavebeam can be converted, for example, into the TE_{01} wave of a waveguide of circular cross section.

The performed analytical estimations and results of numerical simulation show the possibility to realize the smooth frequency tuning within the band of about 1 % in the gyrotrons with two-mirror echelette resonators operating on the 1-3th harmonics of the gyrofrequency.

Echelette gyrotrons have great prospects, but in their calculation and manufacturing technology serious problems are encountered. A complex structure of the output radiation requires the development of new variants of the converter.

Acknowledgements

The work was supported by the Russian Foundation for Basic Research (Grant 18-02-00832 A)..

References

1. *Nusinovich G.S., Thumm M., Petelin M.I.* The Gyrotron at 50: Historical Overview. // J. Infrared Mm THz Waves. 2014. V. 35, No. 4, P. 325–381.

2. *Zapevalov V.E.* Evolution of gyrotrons // Radiophysics and Quantum electronics. 2011. V. LIV, No. 8-9, P. 559-572.

3. *Kosarev E.L.* Resolution of the Spectrum of an Open Resonator with the Aid of an Echelette Grating // JETP Letters. 1966. V. 3, No. 7, P. 190-192.

4. *Vlasov S.N., Koposova E.V., Pavel'ev A.B., Khizhnyak V.I.* Gyrotrons with echelette resonators // Radiophysics and Quantum Electronics. 1996. V. 39, No. 6, P. 458–462.

5. *Vlasov S.N., Zavolsky N.A., Zapevalov V.E., Koposova E.V., Moiseev M.A.* Axisymmetric multistage cavity resonators // Radiophysics and Quantum electronics. 2009. V. 52, No. 9, P. 642–654.

6. *Vlasov S.N., Koposova E.V.* Traveling-wave echelette cavity // Radiophysics and Quantum electronics. 2014. V. 56, No. 10, P. 658-663.

7. *Zapevalov V.E, Kuftin AN, Lygin V.K.* Numerical simulation and experimental study of an electron-optical system of a megawatt gyrotron with step frequency tuning in the range 100–170 GHz. 2007. V. 50, No. 9, P. 702–712.

8. Zapevalov V.E., Bogdashov A.A., G.G. Denisov, Kuftin A.N., *Lygin V.K., Moiseev M.A., Chirkov A.V.* Development of a prototype of a 1-MW 105-156-GHz multi-frequency gyrotrons gyrotrons // Radiophysics and Quantum electronics. 2004. V. 47, No. 5-6, P. 396–404.

Generation of Powerful Subterahertz Superradiance Pulses for High-Gradient Acceleration of Charged Particles

A. Vikharev[1], N. Ginzburg[1,2], S. Kuzikov[1], I. Zotova[1], M.Yalandin[3]

[1]Institute of Applied Physics RAS, Nizhny Novgorod, Russia, alvikharev@appl.sci-nnov.ru
[2]N. I. Lobachevsky State University, Nizhny Novgorod, Russia
[3]Institute of Electrophysics, UB RAS, Ekaterinburg, Russia

In currently existing conventional linear accelerators fed by klystrons operating in S-band and X-band, acceleration gradients typically not exceeded 100 MV/m [1]. This is the reason why for electrons acceleration up to high energies long and, correspondingly, cost-demanding setups are required. Substantial increase in acceleration gradient would allow for practical implementation of colliders with TeV power level of colliding particles; for substantial improvement of parameters and/or reducing the cost of X-ray free electron lasers currently being designed; for development of compact accelerators for radiography with sub-nanosecond synchronization and definition, for medical, industrial and other applications, etc. Correspondingly, this problem became a subject of rather intense investigation including the development of brand new concepts such as laser-plasma and beam-plasma acceleration, and also acceleration in optical nanostructures. However, investigations on improving the conventional schemes of accelerators based on electrons acceleration in hollow metallic non-superconducting ("warm") structures fed by RF radiation remain relevant. Increasing the acceleration gradient in so-called "warm" accelerators operating at room temperature is limited by effects of near-wall breakdown and pulsed thermal heating of the walls which limit the accelerating field values and deterioration of the structures surface under cyclic action of powerful microwave radiation. Despite the fact that investigations of breakdown phenomena and of walls heating have been conducted for several decades already, the only efficient method of suppressing these negative effects is considered to be the use of systems operating in the short pulse regimes when those effects have no time for settling. At that, shortening of RF pulse with simultaneous increase of its frequency, which would increase the shunt resistance considerably, would allow for relatively easy way of reaching accelerating fields of about 200 MV/m and more without exceeding threshold heat loads and surface breakdown fields. Whereas at higher frequencies the output power of RF klystrons conventionally used in accelerators decreases, the problem of design and development of high-power high-efficiency pulse sources operating in millimeter wavelength range.

In this paper we propose to use for the purpose of high-gradient particles acceleration the new class of pulsed relativistic RF generators based on the Cherenkov superradiance (SR) of high-current extended electron bunches. Currently, SR generators provide record-breaking power level in mm wavelength range up to 2 GW in ultrashort pulses with a duration of 250-300 ps [2]. Investigations conducted lately show viability both of further increase of SR pulses peak power at 8-mm wavelength band (due to coherent summation of such pulses [3,4]) and for advancement of SR generators into the short-wavelength part of the millimeter band with a power level of 50-70 MW due to using oversized electrodynamic structures [5].

In experiments on particles acceleration the experimental setup [4] consisting of two synchronized high-voltage RADAN modulator could be used. Based on the electron beam in the first section, pulsed high-frequency radiation would be generated. The accelerating structure would be positioned on the course of the electron beam in the second section. Thus, a nanosecond or subnanosecond RF radiation pulse after transmission through the quasi-optical line and reflection from the special reflector at the cathode side would feed the positioned structure in the regime of wave co-moving with the accelerated beam. Further results of PIC simulations are presented for two experimental configurations.

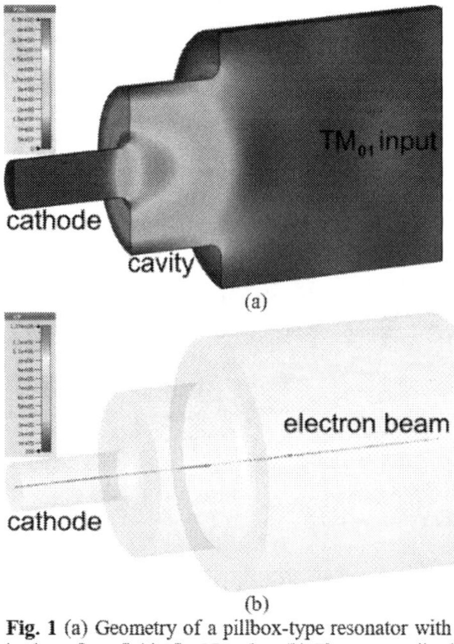

Fig. 1 (a) Geometry of a pillbox-type resonator with distribution of em field of a SR pulse. (b) The energy distribution of an electron beam with an initial energy of 300 keV (the maximum energy of the accelerated fractions is of 1.3 MeV, the acceleration gain is of 550 MV/m).

In the first experiments, SR pulses would be used at 38 GHz frequency with a duration of up to 250 ps and power of ~500 MW. In this combination of parameters, acceleration in short half-wave resonator is of interest, which allows to obtain rather high acceleration gradient. Pillbox-type resonator at TM010 mode (see Fig. 1) with loaded quality factor of Q = 30 at the resonance frequency is capable of providing maximum accelerating field on the axis when fed by a pulse with duration of 250 ps. According to simulations, maximum acceleration gradient reaches 550 MV/m with surface waves of up to 650 MV/m. Maximum energy gain at the parameters stated would be more than 1 MeV.

It should be noted, that implementation of the described scheme at higher frequency, 140 GHz, is of unquestionable interest. At the pulse parameters (200 ps, 50 MW) obtained in experiments on pulse generation [5], acceleration gradient twice as high could be obtained as the parameters of the experiment to be conducted at 38 GHz frequency. Accelerated electrons energy would be less (up to 700 keV), but due to smaller resonator dimensions, accelerating gradient and surface fields appear to be larger.

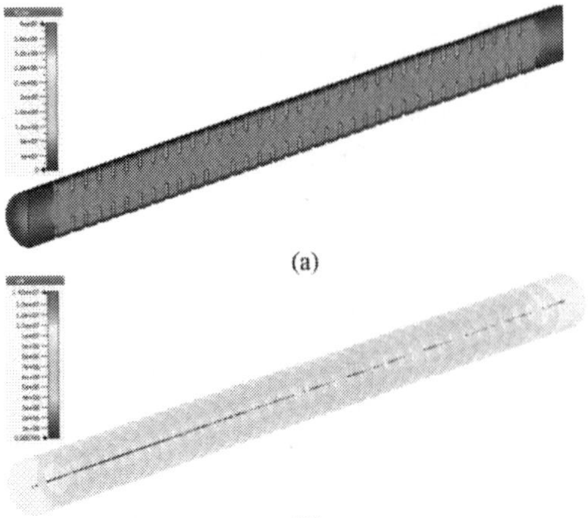

(a)

(b)

Fig. 2 (a) Geometry of an accelerating structure with distribution of em field. (b) The energy distribution of an electron beam with an initial energy of 300 keV (the maximum energy of the accelerated fractions is of 14 MeV).

In the second experimental configuration, a relatively long (2.5 ns) electromagnetic pulse with a power of up to 100 MW at the 38 GHz frequency generated by a relativistic BWO would be used for feeding of periodic hollow metallic structure with a length of 10 cm (see Fig. 2). Such $2\pi/3$ structure consists of coupled accelerating cells with a length of 3 mm. Group velocity in the structure should be 11% of the velocity of light according to the considerations of its full feeding by a 2.5 cm pulse, shunt resistance of about 90 MOhm/m and the cells Q-factor ~3500. These parameters are close to those of the CLIC structure at a frequency of 30 GHz (Rsh = 97.5 Mohm/m, Q = 3900, Vgr/c = 7%). According to simulations, averaged over the length acceleration gradient would be 220 MV/m with gained accelerated particles energy over 20 MeV. Such an experiment is of considerable interest because it simulates a so-called RF gun, which represents a source of electron beams of accelerators with higher energies (in the presence of synchronized laser providing the photoemission of the electron bunch). Implementation of such gun layout would allow for decrease of the dark current, breakdown frequency and potentially higher brightness of accelerated electron bunches comparing to conventional design.

This research was performed within the framework of the RAS Presidium project No. 10

References

1. *A. Degiovanni, S. Doebert, W. Farabolini, A. Grudiev, J. Kovermann, E. Montesinos, G. Riddone, I. Syratchev, R. Wegner, W. Wuensch, A. Solodko, B. Woolley,* High-gradient test results from a CLIC prototype accelerating structure: TD26CC, Proceedings of IPAC2014, Dresden, Germany, WEPME015. P. 22885-2287.

2. *Rostov, V.V., Romanchenko, I.V., Pedos, M.S., Rukin, S.N., Sharypov, K.A., Shpak, V.G., Shunailov, S.A., Ul'masculov, M.R., Yalandin, M.I.* Superradiant Ka-band Cherenkov oscillator with 2-GW peak power // Physics of Plasmas 2016. V. 23, No. 9. P. 093103 (1-4).

3. *Ginzburg, N.S., Cross, A.W., Golovanov, A.A., Mesyats, G.A., Pedos, M.S., Phelps, A.D.R., Romanchenko, I.V. Rostov, V. V., Rukin, S.N., Sharypov, K.A., Shpak, V.G., Shunailov, S.A., Ulmaskulov, M.R., Yalandin, M.I., Zotova, I.V.* Generation of electromagnetic fields of extremely high intensity by coherent summation of Cherenkov superradiance pulses// Phys. Rev. Lett. 2015. V. 115. P.114802 (1-4).

4. *Mesyats, G. A., Ginzburg, N. S., Golovanov, A. A., Denisov, G. G., Romanchenko, I. V., Rostov, V. V., Sharypov, K. A., Shpak, V. G., Shunailov, S. A., Ulmaskulov, M. R., Yalandin, M. I., Zotova, I. V.* //Phase-imposing initiation of Cherenkov superradiance emission by an ultrashort-seed microwave pulse // Phys.Rev.Lett. 2017. V.118. P. 264801.

5. *Ginzburg, N.S., Malkin, A.M., Sergeev, A.S., Zheleznov, I.V., Zotova, I V., Zaslavsky, V.Yu., Boltachev, G.Sh., Sharypov, K.A., Shunailov, S.A., Ul'masculov, M.R., Yalandin, M.I.* Generation of sub-terahertz superradiance pulses based on excitation of a surface wave by relativistic electron bunches moving in oversized corrugated waveguides // Phys. Rev. Lett. 2016. V. 117. P. 204801.

The project of third harmonic medium power W-band gyrotron

A.S. Zuev[1], A.P. Fokin[1], M.Y. Glyavin[1], R.M. Rozental[1], A.S. Sedov[1], E.S. Semenov[1]

[1]IAP RAS, Nizhniy Novgorod, Russia, alan.zuev@yandex.ru

One of the most interesting ranges is the W-band due to the presence of a transmission line in this range in the atmosphere, which is used for communication, radiolocation, security and for other applications [1-4]. These applications require continuous and pulsed W-band sources. One of the powerful devices in this band is the gyrotron [5]. The gyrotron is a vacuum electronics device based on the interaction between an electron beam and microwave fields, which coupling is achieved by the cyclotron resonance condition. High power subterahertz and terahertz radiation gyrotrons require strong external static magnetic field typically provided by a superconducting magnet. Such magnets are quite expensive; moreover they require special operating conditions.

One way to reduce the cost of this device is to develop gyrotrons operating at higher harmonics on the cyclotron frequency, in which the required external magnetic field in the interaction space is significantly less and allows the use of low-cost versions of magnetic systems. At present, there are projects of "warm" (i. e., with liquid or oil cooling) magnets with a cavity diameter sufficient for the gyrotron and the magnetic field level of more than 1 T [6]. The operation at high harmonics on the cyclotron frequency in gyrotrons encounters the problem of competition of modes, synchronous with lower harmonics of the electron beam. In addition, in order to reach the effective electron-wave interaction for operation at high harmonics, on needs to use resonators with higher quality factor. It leads to an increase in ohmic losses and, consequently, a drop in the output power for the same parameters of the electron beam. In this report, we consider the project of a gyrotron at frequency 95 GHz of medium power operating at the third harmonic. To solve the problem of mode competition, it is planned to use a relatively unusual method - the addition of an annular diaphragm in the output part of the resonator (fig 1). By introducing the annular diaphragm, it is possible to reduce the diffraction quality factor and, respectively, to increase the starting current of the competing mode.

For the operation of the gyrotron at the third harmonic, mode $TE_{1,3}$ was chosen. This mode is isolated from modes operating at the fundamental harmonic, and is almost isolated from modes synchronous with the second harmonic. The nearest competitor for it is the counterrunning wave $TE_{1,2}$ on the second harmonic. The gyrotron was designed for operation at an accelerating voltage 30 kV. We used a model with a non-fixed structure of the RF field taking into account the velocity spread of the electron beam. The dependences of the starting currents on the magnetic field for the working and competing modes were calculated and shown in figure 2. The presence of a parasitic mode in the excitation region of the operating mode limits the operating current to 3 A and ultimately determines the maximum possible output power of the gyrotron. In this case, at the operating current 3 A the optimum length of resonator is 62 mm with the output efficiency 5.8%, while the share of ohmic losses is 64%. On the other hand, if one can effectively suppress the parasitic mode, it would be possible to increase the operating current, at which the optimum length of the resonator becomes less. In particular, at operating current 10 A the maximum efficiency 11.4% is achieved with the optimum resonator length 49 mm. Figure 3 shows the dependence of the share of ohmic losses and the output efficiency on the resonator length at operating currents 3 A and 10 A. In addition, the length of the resonator is limited by the characteristic size of the magnetic systems. For further calculations, the resonator length 50 mm with the output efficiency 4.7% at current 3 A and the output efficiency 11.3% at current 10 A was chosen. For a longer resonator length it becomes difficult to realize a homogeneous section of the magnetic field in the interaction space.

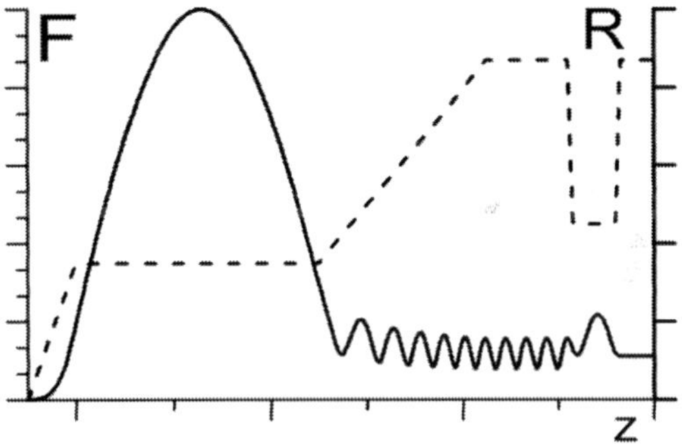

Fig. 1. General view of the cavity profile with an annular diaphragm (dashed line) and longitudinal structure of RF field (solid line).

One of the electrodynamic methods of suppressing the parasitic mode in the gyrotron is the addition of an annular diaphragm in the output part of the resonator. This method of solving the problems of mode competition was proposed earlier, for example, in [7], where the influence of the "external reflector" on the quality factors of the neighboring longitudinal modes was considered. By solution of the cold problem of the longitudinal field structure in the entire volume of the gyrotron electrodynamic system, the parameters of the annular diaphragm for optimal mode selection were chosen. The system with the annular diaphragm drops the diffractive quality factor of the competing $TE_{1,2}$ mode by more than four times, while the diffraction quality factor of the operating mode remains at the same level. Figure 2 shows that the addition of the annular diaphragm results in a significant increase of the starting current of the parasitic mode $TE_{1,2}$ in the field of excitation of the operating mode. Also the starting current of the $TE_{1,3}$ mode nearly unchanged. The new electrodynamic structure allows to increase the operating current up to 10A and thus to increase the output power by 7 times.

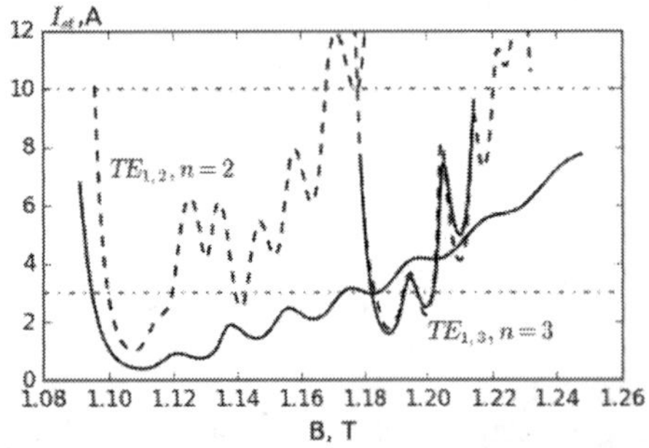

Fig. 2. Dependence of the starting currents on the magnetic field for the operating mode and the main competing mode in a gyrotron at a frequency of 95 GHz with a traditional electrodynamic structure (solid lines) and with diaphragm (dashed lines).

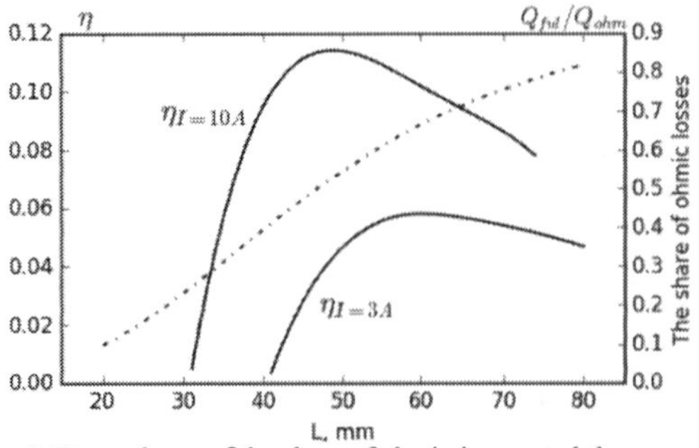

Fig. 3. Dependence of the share of ohmic losses and the output efficiency in the gyrotron on the resonator length at operating currents of 3 A and 10 A.

The estimation of the stability of this method of mode selection for errors in the manufacture of a gyrotron allows us to hope for scaling this method to the region of higher frequencies. Further development of the system consists of optimization of the resonator profile, which will reduce the share of ohmic losses while maintaining high efficiency, and also exclude the possibility of transforming the operating mode into modes with other azimuthal variations. In addition, an experimental verification of this selection method is planned.

The paper is based on the research results, which was conducted in the framework of projects of the Russian Science Foundation No17-79-10422.

References

1. Bratman, V. L., Litvak, A. G., Suvorov, E. V. Mastering the terahertz domain: sources and applications // Physics-Uspekhi. 2011. V. 54, No. 8, P. 837-870.
2. Kumar, N., Singh, U., Singh, T. P., Sinha, A. K. A Review on the Applications of High Power, High Frequency Microwave Source: Gyrotron // Journal of Fusion Energy. 2011. V. 30, No. 4. P. 257–276.
3. Manheimer, W. M. On the possibility of high power gyrotrons for super range resolution radar and atmospheric sensing // International Journal of Electronics. 1992. V. 72, No. 5-6. P. 1165–1189.
4. Pilossof, M., Einat, M. A 95 GHz mid-power gyrotron for medical applications measurements // Review of Scientific Instruments. 2015. V. 86, No. 1. P. 016113.
5. Nusinovich, G. S., Thumm, M. K. A., Petelin, M. I. The Gyrotron at 50: Historical Overview // Journal of Infrared, Millimeter, and Terahertz Waves. 2014. V. 35, No. 4. P. 325–381.
6. Proyavin, M. D., Glyavin, M. Yu., Manuilov, V. N. Development of high-efficient gyrotron based complex for industrial applications. // EPJ Web Conf. 2017. No. 149, P. 04034.
7. Rozental, P. M., Ginzburg, H. C., Zaitsev, N. I., Ilyakov E.V., Kulagin I.S. Controllable spectrum of an axial-mode gyrotron with external reflections // Tecnical Physics. the Russian Journal of Appl. Phys. 2006. V. 76, No. 1. P. 78–81.

Terahertz-Range Gyrodevices of Planar Geometry

V.Yu. Zaslavsky, N.S. Ginzburg, I.V. Zotova, M.Yu. Glyavin,
A.S. Sergeev, R.M. Rozental, A.M. Malkin, I.V. Zheleznov

Institute of Applied Physics of the Russian Academy of Sciences, Niznhny Novgorod, Russia, zas-vladislav@ipfran.ru

For drastic increase in the output power of short-wavelength gyrotrons, we suggest to use the planar scheme with a sheet electron beam and transverse (with respect to the electrons translation velocity) electromagnetic energy extraction. The main advantage of this scheme comparing to the conventional cylindrical geometry is the possibility of effective mode selection over the open transverse coordinate in combination with radiation out-coupling, which leads to substantial reduction of Ohmic losses [1, 2]. The formation of ribbon helical electron beam (HEB) in planar magnetron-injection gun (MIG) was demonstrated in [3]. In this paper we consider gyrotrons and gyroklystrons of planar geometry driven by sheet electron beams. Possibility of power increasing and selectivity improvement in the terahertz band is shown.

High-Order Cyclotron Harmonic Excitation in a THz-Range Planar Gyrotron

An additional advantage of the planar scheme is the peculiarity associated with excitation of odd ($s = 1,3...$) and even ($s = 2,4...$) cyclotron harmonics. Under assumption that the sheet electron beam is injected in the middle of the cavity between the plates, the interaction at odd cyclotron harmonics occurs only for resonator modes with odd transverse indexes $n = 1,3...$, while interaction at even harmonics occurs only for modes with even transverse indexes $n = 2,4...$. Moreover, for example, for operation at the second harmonic it is beneficial to use an even resonator mode with indexes n equal to doubled even number. In this case interaction at second harmonic will be not accompanied by simultaneous excitation of a lower order mode at the first cyclotron harmonic due to the coupling factor for the 1st harmonic with an even mode is equal to zero. From the other hand, for excitation at the odd cyclotron harmonic the number of the resonator mode should not be dividable by s. For example, for $s = 3$, the resonator mode number may be $n = 5,7,11...$. In this case the parasitic mode at the 2nd cyclotron harmonic is not excited. At the same time, excitation at the 1st cyclotron harmonic can be suppressed due to sensitivity to the spread of electrons velocity and due to additional diffraction losses caused by some penetration of low-frequency running wave through collector narrowing.

Further results of simulations of 1.2 THz planar gyrotron operating at the 3d cyclotron harmonic are presented. Simulations were carried out in the frame of averaged multi-modes approaches:

$$s_n \frac{\partial a_n}{\partial \tau} + i\frac{\partial^2 a_n}{\partial X^2} + i\frac{\partial^2 a_n}{\partial Z^2} + (i\Delta_n + \sigma_n)a_n = \frac{I_n}{2\pi}\int_0^{2\pi} p^s d\vartheta_0,$$

$$\frac{\partial p}{\partial \tau} + \frac{g^2}{4}\frac{\partial p}{\partial Z} + ip\left(|p|^2 - 1\right) = -\sum_n a_n\left(p^*\right)^{s-1}.$$

Here a_n is the amplitude of the TE_n mode excited at the s_n harmonic of cyclotron gyrofrequency, p is the transverse electron momentum, Δ_n is the cyclotron resonance detuning, I_n is the parameter proportional to electron current and the coupling factor $G_n(y)$, σ_n is the parameter of Ohmic losses proportional to the skin depth. Due to the diffraction extraction of radiation is realized in transverse x direction (Fig.1a), at the edges of interaction space the nonreflecting boundary conditions may be applied. In the direction of translational motion of electrons the waveguide is considered to be closed by cut-off necks. Thus we can apply zero boundary condition along z coordinate.

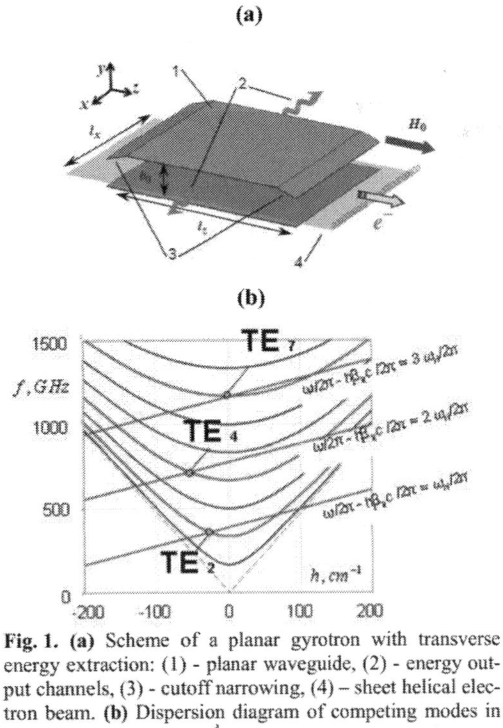

Fig. 1. (a) Scheme of a planar gyrotron with transverse energy extraction: (1) - planar waveguide, (2) - energy output channels, (3) - cutoff narrowing, (4) – sheet helical electron beam. **(b)** Dispersion diagram of competing modes in the case of exciting the 3d cyclotron harmonic.

Simulations were carried out for following parameters: energy of electrons of 30 keV, electron current of 6 A, pitch-factor of 1, the distance between plates of 0.9 mm, resonator length and width of

20 mm both. Three modes exciting at the 1st, 2nd and 3d cyclotron harmonics were included in considera-tion. Results of simulations (see Fig.2) confirm the possibility of selective excitation of the operating TE_7 mode at the frequency of 1.2 THz.

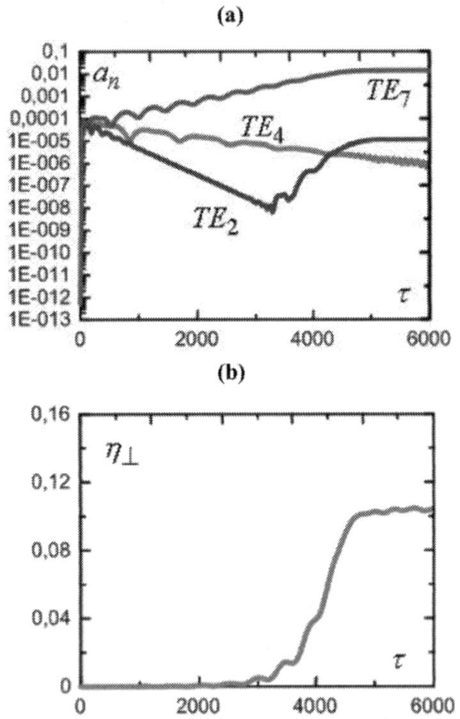

Fig. 2. Temporal dependencies of amplitudes of competing modes **(a)** and the transverse electron efficiency **(b)**.

Planar Gyro-Klystrons

Using of planar configuration in gyro-klystrons is of interest as a method of increasing the power of output radiation. The considered scheme is presented in Fig. 3a. Analysis was carried out based on 3D PIC (Particle-in-cell) simulations using the code KARAT. Parameters of electron beam were following: energy of 20 kV, total current of 2 A, pitch-factor of 1.3. For distance between plates of 2.2 mm and excitation of the TE_3 mode the operating frequency was of 200 GHz. According to simulations for input power of 10 W output power achieves 4.5 kW. It corresponds to the amplification depth of 26 dB and efficiency of about 10%.

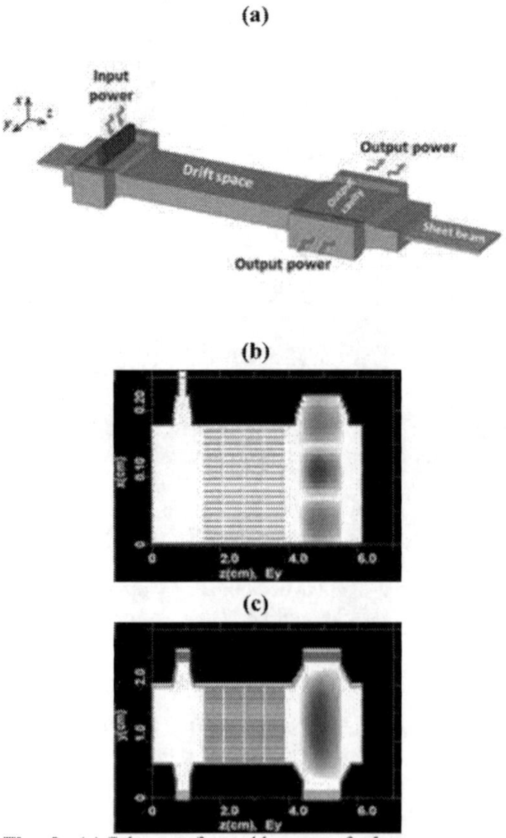

Fig. 3. **(a)** Scheme of gyroklystron of planar geometry. **(b), (c)** Field distributions in the input and output resonator over x and y coordinates.

The work was supported by the Russian Science Foundation under grant No. 18-19-00704.

References

1. *Ginzburg N.S., Zotova I.V., Sergeev A.S., Zaslavsky V.Yu., Zheleznov I.V.*. High-power terahertz-range planar gyrotrons with transverse energy extraction// Phys. Rev. Lett. 2012. V. 108. Art.no. 105101.

2. *Zaslavsky V.Yu., Ginzburg N.S., Glyavin M.Yu., et al.* Three-dimensional particle-in-cell modeling of terahertz gyrotrons with cylindrical and planar configurations of the interaction space. // Physics of Plasmas. 2013. V. 20. P. 043103.

3. *Manuilov V.N., Zaslavsky V.Yu., Ginzburg N.S., et al.* Electron-optical systems for planar gyrotrons. // Physics of Plasmas. 2014. V. 21. P. 023106.

EPJ Web of Conferences **195**, 01026 (2018) https://doi.org/10.1051/epjconf/201819501026
TERA-2018

Design of mm-wave slow-wave devices with sheet and hollow electron beams

A. E. Fedotov[1], V. L. Bratman[1,2], P. B. Makhalov[1], and V. N. Manuilov[1,3]

[1] Institute of Applied Physics, Nizhny Novgorod, Russia, fedotov@appl.sci-nnov.ru
[2]Ariel University, Ariel, Israel
[3]Nizhny Novgorod State University, Nizhny Novgorod, Russia

Abstract

The frequency increase of the slow-wave electron devices is accompanied by inevitable increase of the electron current density and ohmic losses which strongly restricts the attainable power. In recent years, the use of the spatially-developed sheet electron beams is considered as the major way to develop the medium-power slow-wave devices at mm and sub-THz waves. Hollow electron beams is another configuration which could be used for this purpose. The designs of the oscillators and amplifiers in Ka-band and W-band with both sheet and hollow electron beams are considered and compared.

Introduction

The development of the powerful amplifiers in the millimeter and sub-terahertz wavelength ranges is one of the most topical problem in vacuum microwave electronics. In particular, millimeter-wave amplifiers with an output power of several hundred watts are of interest for spectroscopy, telecommunications, remote sensing, and also as preamplifiers for even more powerful gyro-amplifiers. The difficulty of development of traveling wave tubes (TWT) and extended-interaction klystrons (EIK) with the required parameters is caused largely by a very high required beam current density (up to 1 kA / cm^2) if conventional electron beam with pencil-like configuration is used. Due to evanescent nature of the synchronous space harmonic of the operating mode, this beam with high energy density should be guided in close proximity to the surface of the slow-wave structure in order to provide efficient electron-wave interaction. The non-perfection of the beam guiding system and high microwave fields inevitably results in some current deposition and, hence, a high heat load on the small elements of the slow-wave structure, in addition to ohmic losses. In order to alleviate the thermal regime of the devices, a use of planar sheet electron beams for millimeter-wave TWTs and EIKs are extensively studied in recent years [1-3]. The large width of the planar beam allows decrease of the required current density but increase of the total electron current. Therefore, the CW or average output power for the planar-beam slow-wave amplifier can be significantly higher than for the pencil-beam device. An alternative configuration of the spatially-developed electron beam is a cylindrical hollow thin-wall beam [4,5]. In this paper we consider both these options for development of the slow-wave device with medium power at the millimeter wavelengths.

Design of sheet-beam slow-wave TWT and BWO

Two key points in the development of the sheet-beam slow-wave devices are slow-wave structure and electron optical system. Presented design is based on a slow-wave structure consisting of two gratings shifted by half period relative to each other (Fig. 1). This structure is widely explored in recent years since it is simple, wideband, easy to manufacture and compatible with a planar electron beam [6]. The sheet beam can be provided by quasi-Pierce electron gun and then solenoidally focused and guided. The beam guiding in the solenoidal magnetic field seems to be simpler than periodic-permanent-magnet focusing system which is traditional for pencil-beam devices at longer waves. To reduce the magnetic compression ratio, a thermionic cathode with high current density of 30 A/cm^2 can be used, similar to cathodes of low-power submillimeter backward-wave oscillators, clinotrons and orotrons. Simulations predict that the beam which is 0.2 mm thick and 1.8 mm wide with a current of 300 mA (current density 80 A/cm^2) can compressed and guided between gratings in 0.65 T magnetic field with less than 5 % of the total current being lost. Simulation made for W-beam TWT shows the possibility of obtaining an output power of 250 W at a voltage of 20 kV. The calculated gain is 26 dB in the linear regime and about of 21 dB in the saturation. For initial experimental verification of the electron-optical system, a BWO with similar periodic structure was designed with the power of 200-300 W and 7 % tuning band (Fig. 2). The same cathode makes it possible to design the Ka-band BWO and TWT with 500-W output power using beam guiding in the uniform magnetic field, without beam compression.

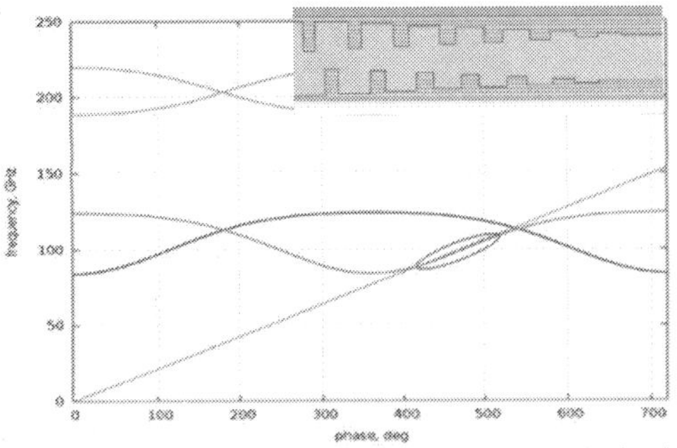

Fig. 1. Calculated dispersion of the operating mode for designed W-band TWT and line of synchronism with 20-kV electron beam. Inset: the staggered-gratings slow-wave structure with the smooth taper of the gratings near the collector end of the interaction region.

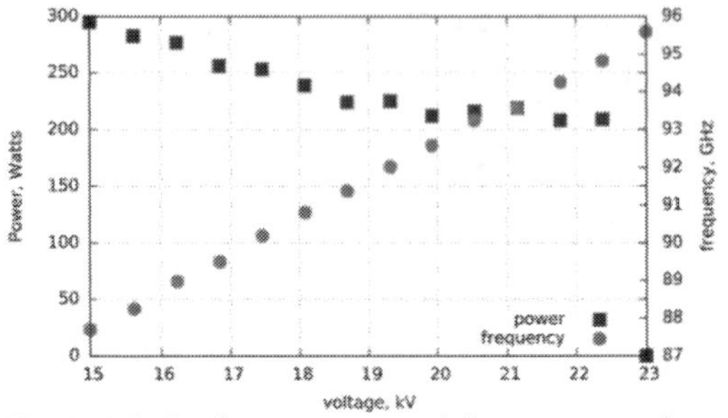

Fig. 2. Calculated output power and frequency vs. voltage for sheet-beam W-band BWO with the staggered-gratings slow-wave structure.

Design of hollow-beam BWOs

The use of hollow electron beams permits a significant increase in the diameter of the beam tunnel in comparison with the conventional pencil-beam slow-wave electron devices. Hollow beam can efficiently interact with a lowest azimuthally-symmetric TM mode of the azimuthally-symmetric slow-wave structure. This configuration is similar to the relativistic high-current electron devices. However, the slow-wave structures with deep corrugation should be used in non-relativistic devices due to weak electron-wave coupling in the millimeter wavelength range. W-band and Q-band axisymmetric BWOs has been designed on the base of 30 kV/1 A hollow thin-wall electron beam with an outer diameters of 1.6 mm and 3.4 mm, respectively. This beam is produced in a Pierce-like electron gun with the hundredfold magnetic compression (Fig. 3). Simulations based on the averaged equations and 3D PIC-code predict an output power of 300-500 W (depending on the beam wall thickness) for W-band BWO and about of 800 W for Q-band BWO. 5% frequency tuning is simulated for the W-band device. To reduce ohmic losses, the wave reflections at the collector end of slow-wave structures are minimized by using specially designed last matching tooth (Fig. 4).

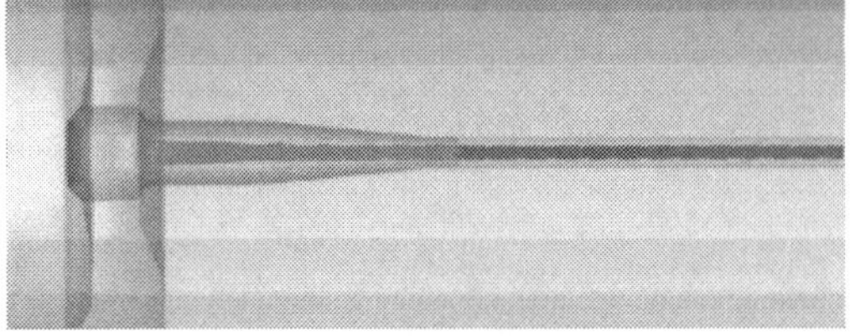

Fig. 3. Designed electron gun producing 30-kV, 1 A hollow electron beam with simulated electron trajectories.

In conclusion, let us compare both configurations of the spatially-developed electron beams. The slow-wave structures for sheet beam devices are easier for manufacturing, and they could provide wider frequency band. The electron-wave coupling for sheet-beam devices could be stronger since electrons travel between two gratings, while for hollow beam there is single grating at the outer side of the beam. On the other hand, the formation of the sheet electron beam is difficult due to diocotron instability, but this difficulty can be avoided if cathodes with high emission density are used.

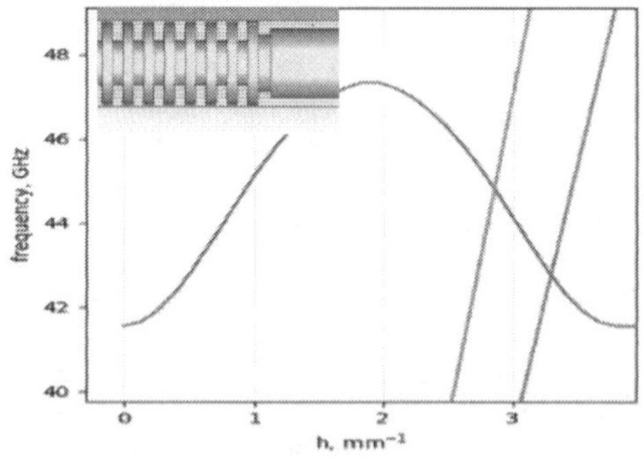

Fig. 4. Calculated dispersion of the operating mode for designed Q-band BWO and line of synchronism with 20- and 30-kV hollow beams. Inset: the axisymmetric slow-wave structure with the matching tooth at the collector end.

The development of the planar-beam devices is supported by Russian Science Foundation under grant No. 18-79-10252. The development of the hollow-beam devices is supported by RFBR under grant No. 16-08-01174.

References

1. *Pershing, D. E., Nguyen, K. T., Abe, D. K., Wright, E., Larsen, P. B., Pasour, J., Cooke, S. J., Balkcum, A., Wood, F. N., Myers, R. E., and Levush, B.* Demonstration of a wideband 10-kW Ka-band sheet beam TWT amplifier. // IEEE Transactions on Electron Devices. 2014. V. 61, No. 6. P. 1637-1642.

2. *Pasour, J., Wright, E., Nguyen, K. T., Balkcum, A., Wood, F. N., Myers, R. E., and Levush, B.* Demonstration of a multikilowatt, solenoidally focused sheet beam amplifier at 94 GHz. // IEEE Transactions on Electron Devices. 2014. V. 61, No. 6. P. 1630-1636.

3. *Karetnikova, T. A., Rozhnev, A. G., Ryskin, N. M., Fedotov, A. E., Mishakin, S. V., and Ginzburg, N. S.* Gain analysis of a 0.2-THz traveling-wave tube with sheet electron beam and staggered grating slow wave structure. // IEEE Transactions on Electron Devices. 2018. V. 65, No. 6. P. 2129-2134.

4. *Bratman, V. L., Fedotov, A.E., and Makhalov, P. B.* High-frequency devices with weakly-relativistic hollow thin-wall electron beams. // Physics of Plasmas. 2012. V. 19, No. 2. Art. 020704.

5. *Bratman, V. L., Fedotov, A. E., Makhalov, P. B., and Manuilov, V. N.* Design and numerical analysis of W-band oscillators with hollow electron beam. // IEEE Transactions on Electron Devices. 2014. Vol. 61, No. 6. P. 1795-1799.

6. *Shin, Y. M., Barnett, L. R., and Luhmann, N. C.* Phase-shifted traveling-wave-tube circuit for ultrawideband high-power submillimeter-wave generation. // IEEE Transactions on Electron Devices. 2009. V. 56, No. 5. P. 706-712.

EPJ Web of Conferences **195**, 01027 (2018)
TERA-2018 https://doi.org/10.1051/epjconf/201819501027

Generation of Sub-Terahertz Surface Waves by Relativistic Electron Beams: Quasioptical Theory, Simulations and Experiments

Naum S. Ginzburg[1], A.M. Malkin[1], I.V. Zheleznov[1], V.Yu. Zaslavsky[1], A.S. Sergeev[1], I.V. Zotova[1], and M.I. Yalandin[2]

[1]Institute of Applied Physics RAS, Nizhny Novgorod, Russia, ginzburg@appl.sci-nnov.ru
[2]Institute of Electrophysics RAS, Ekaterinburg, Russia

Advancement of vacuum electronic devices into sub-THz and THz frequency ranges calls for oversized beam-wave interaction space due to the fact that that the dimensions of the beam guiding systems can not be reduced lower than the millimeter scale. Thus, in order to provide coherent THz radiation from the spatially extended beams, excitation of surface modes existing in 1D and 2D corrugated systems appears to be attractive [1,2,5,6].

In this paper, we present recent results of theoretical and experimental studies of sub-terahertz generation based on excitation of surface waves by electron beams and extended bunches. Using oversized slow-wave structures allows for a significant increase of total current and, correspondingly, radiation power. Based on superradiance of electron bunches, 150 ps superradiant pulses with a central frequency of 0.14 THz, and an extremely high peak power of 50-70 MW were obtained in the joint effort by the Institute of Electrophysics RAS and IAP RAS. We also report of the first successful experiments on the cylindrical 0.03 THz Cherenkov oscillator with a 2D corrugation conducted at IAP RAS with an output power of 1.5 - 2 MW.

Generation of Sub-THz SR Pulses Based on Excitation of Surface Waves in Oversized Waveguides

Cherenkov SR of electron bunch exciting the surface wave in an oversized 1D corrugated cylindrical waveguide (Fig. 1a) can be considered within quasioptical approach [1]. In this case the radiation field near a shallow corrugation is presented as a sum of two counter-propagating *TM* polarized wave-beams:

$$H_{\varphi} = \mathrm{Re}(A_{+}(z,r,t)e^{i\bar{\omega}t-ikz} + A_{-}(z,r,t)e^{i\bar{\omega}t+ikz}), \quad (1)$$

propagation and mutual coupling of which is described by two non-uniform parabolic equations. The synchronous interaction of electrons with a forward partial waves leads to development a self-bunching and formation of powerful SR pulse.

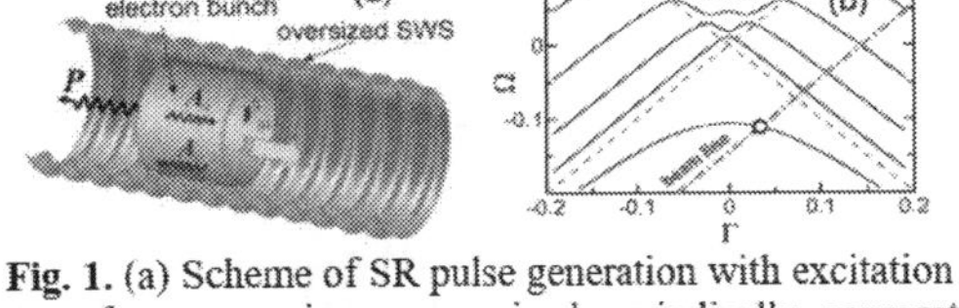

Fig. 1. (a) Scheme of SR pulse generation with excitation of a surface wave in an oversized periodically corrugated waveguide. (b) Dispersion characteristics of a corrugated waveguide and an electron beam.

Simulations show that the most optimal conditions for SR emission correspond to excitation of the backward surface wave near the Bragg frequency (π-regime, Fig.1b). For parameters of an electron bunch formed by an accelerator RADAN (electron energy of 300 keV, a total current of 2 kA, a bunch duration of 500 ps) and a corrugated waveguide with the mean radius of 3.75 mm, corrugation period of 0.825 mm, and corrugation depth of 0.36 mm the operating frequency in the resonant point is of 0.14 THz ($2r_0/\lambda \approx 3.5$). In this case the power of generated SR pulse emitted in $-z$ direction achieves ~200 MW for pulse duration of ~200 ps (Fig. 2a). As it is seen in Fig. 2b the instant spatial structure of the partial wave corresponds to formation of the evanescent surface wave with the field amplitude exponentially decaying from the corrugation.

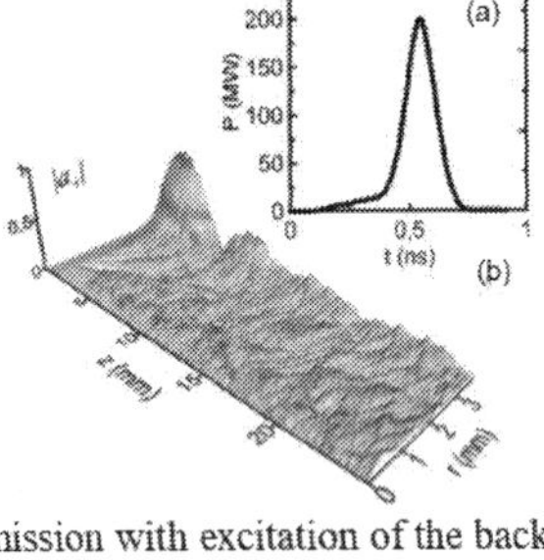

Fig. 2. SR emission with excitation of the backward surface wave: (a) generated SR pulse, (b) the structure of the forward partial wave.

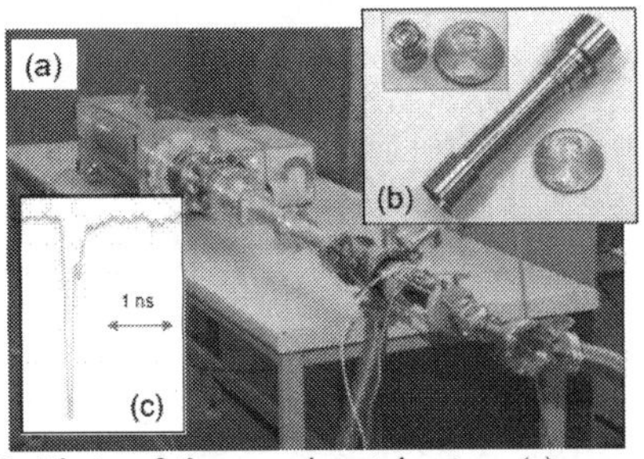

Fig. 3. Photo of the experimental set-up (a), corrugated waveguide and coaxial reflector (b) used for observation of superradiance with excitation of a surface wave. (c) Oscilloscope trace of the 0.14 THz SR pulse with duration of 150 ps and peak power up to 70 MW.

Based on a theoretical analysis, experiments on observation of the sub-terahertz SR pulse generation were carried out in IEP RAS (Ekaterinburg). Photo of the experimental set-up is shown in Fig. 3. A typical oscilloscope trace of generated SR pulses with a duration of about 150 ps and a rise time of 100 ps reconstructed in the "power-time" coordinates is presented in Fig. 3c. Frequency measurements using a set of cut-off waveguide filters show that the pulse spectrum has a central frequency in the interval 0.13-0.15 THz. The peak power of generated SR pulses was estimated by integrating the detector signal over the directional pattern and achieved of 50-70 MW, that strongly exceeds the value obtained in the previous sub-terahertz experiments [2] with single-mode waveguides.

Ka-band surface-wave oscillator based on 2D periodical corrugated structure

For spatially extended relativistic electron beams, the use of two-dimensional (2D) distributed feedback is beneficial for providing spatial coherence of radiation and can be exploited in order to increase the total radiation power in the microwave generators [3]. Such 2D feedback can be realized in planar or coaxial 2D Bragg structures (resonators) having double-periodic corrugation (Fig.4a)

$$r = \frac{\tilde{r}}{4}\left[\cos(\bar{M}\varphi - \bar{h}_z z) + \cos(\bar{M}\varphi + \bar{h}_z z)\right], \quad (2)$$

which provides coupling and mutual scattering of the four wavebeams (Fig.4b),

$$\vec{H} = \mathrm{Re}\left[\left(\vec{x}_0(C_z^+ e^{-ihz} + C_z^- e^{ihz}) + \vec{z}_0(C_x^+ e^{-ihx} + C_x^- e^{ihx})\right)e^{i\omega t}\right]$$

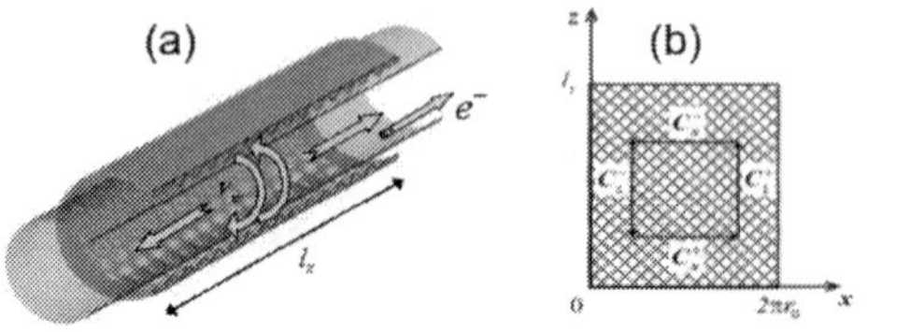

Fig. 4. (a) Scheme of an oversized SWO with 2D corrugated structure. Directions of propagation of the partial wave fluxes and tubular electron beam are shown. (b) Diagram illustrating coupling of partial waves at the 2D corrugation.

Experimental studies of free electron masers (FEMs) based on the novel feedback mechanism have been performed in Ka-band and in W-band in collaboration with the Institute of Applied Physics RAS [4]. As a result, narrow-band generation with an output power of 50 - 100 MW, which is a record for millimeter wavelength FEMs, was obtained.

At present, theoretical and experimental studies of Cherenkov masers with 2D distributed feedback are in progress [5,6]. Among relativistic masers of such type, surface wave oscillators (SWO) appear to be preferable due to the larger values of the electron-wave coupling impedance. Besides, formation of a surface mode ensures the regular field distribution along the coordinate directed perpendicularly to the corrugated surface and, thus, can solve the problem of mode selection over this coordinate. In SWOs, a 2D periodic structure can be exploited both as a slow-wave system and as a highly selective Bragg resonator simultaneously. It provides effective mode control over azimuthal coordinate.

Numerical simulations within the quasi-optical model and using 3D numerical codes show that the resulting mode to be excited in such system depends on the accelerating voltage rise time. In order to excite an azimuthally symmetric mode, this value should be small in the scale of the field excitation increment.

Experimental investigations of the SWO with 2D slow-wave structure based on the 300 keV / 100 A / 4 μs SATURN thermionic accelerator were conducted at IAP RAS [6]. The results are presented in Fig.5. Narrow-spectrum excitation of the 3rd azimuthal mode was observed.

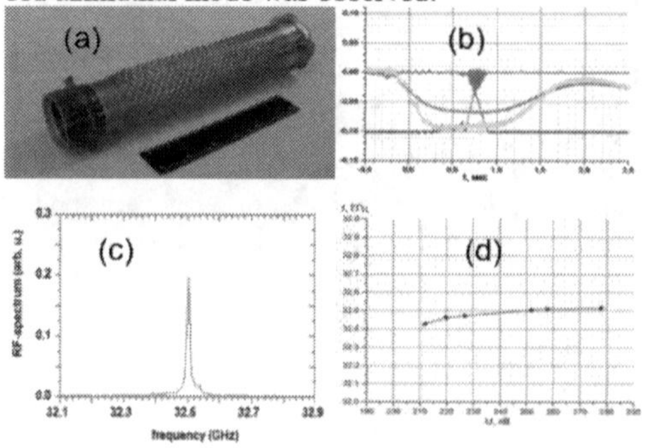

Fig. 5. Results of experimental studies of oversized Ka-band SWO based on the SATURN accelerator: (a) photograph of double periodic slow-wave structure; (b) typical oscilloscope traces of the accelerating voltage (green curve), beam current (red curve), signal from a heterodyne mixer (brown curve) and output RF-pulse (blue curve); (c) spectrum of the output radiation and (d) dependence of the radiation frequency on the accelerating voltage.

This research was performed within the framework of the state task (projects No. 0035-2014-012), and, in part, by RFBR Grant No. 17-08-01072

References

1. *Ginzburg, N.S., Malkin, A.M., Sergeev, A.S. et al.* Generation of Sub-Terahertz Superradiance Pulses Based on Excitation of a Surface Wave by Relativistic Electron Bunches Moving in Oversized Corrugated Waveguides // Phys. Rev. Lett. 2016. V.117 P.204801
2. *Yalandin, M.I., Shpak, V.G., Shunailov, S.A., et al* Generation of powerful subnanosecond microwave pulses in the range of 38–150 GHz// IEEE Trans. on Plasma Sci. 2000. V.28, Iss.5 P.1615
3. *N.S.Ginzburg, N.Yu.Peskov, A.S.Sergeev,* Dynamics of free-electron lasers with two-dimensional distributed feedback // Optics Commun., **112,** 151 (1994).
4. *A.V.Arzhannikov, N.S.Ginzburg, P.V.Kalinin, et al.,* Using Two-Dimensional Distributed Feedback for Synchronization of Radiation from Two Parallel-Sheet Electron Beams in a Free-Electron Maser // Phys. Rev. Lett. V. **117**, 114801 (2016).
5. *N.S.Ginzburg, A.M.Malkin, A.S.Sergeev, and V.Yu.Zaslavsky,* Powerful surface-wave oscillators with two-dimensional periodic structures // Appl. Phys. Lett. 100, 143510 (2012)
6. *N.S.Ginzburg, E.V.Ilyakov, I.S.Kulagin, et al.*, Theoretical and experimental studies of relativistic oversized Ka-band surface-wave oscillator based on 2D periodical corrugated structure // Phys. Rev. AB. V. **21**, 080701 (2018).

Powerful surface-wave oscillators with one-dimensional and two-dimensional periodic planar structures

V.Yu. Zaslavsky, N.S. Ginzburg, A.M. Malkin, A.S. Sergeev

Institute of Applied Physics of the Russian Academy of Sciences, Niznhny Novgorod, Russia, zas-vladislav@ipfran.ru

Within the framework of a quasi-optical approach, we develop self-consistent theory of relativistic surface-wave oscillators [1]. Presenting the radiation field as a sum of two counter-propagating wave-beams coupled on a shallow corrugated surface, we describe formation of an evanescent slow wave. Dispersion characteristics of the evanescent wave following from this method are in good compliance with those found from the direct CST simulations. Considering excitation of the slow wave by a sheet electron beam, we simulate linear and nonlinear stages of interaction, which allows us to determine oscillation threshold conditions, electron efficiency and output coupling. The transition from the model of surface-wave oscillator operating in the π - mode regime to the canonical model of relativistic BWO is considered. We also described a modified scheme of planar relativistic surface-wave oscillators exploiting two-dimensional periodic gratings [2]. Additional transverse propagating waves emerging on these gratings synchronize the emission from a wide sheet rectilinear electron beam allowing realization of a Cherenkov millimeter-wave oscillators with subgigawatt output power level.

Simulations of surface-wave oscillators with conventional single periodic slow-wave structures in the framework of a 3D model

We start with consideration of a surface-wave oscillator driven by a sheet electron beam with a finite width of l_x^e over the x axis (Fig. 1a). We should assume that the field structure is non-fixed over all three spatial coordinates. Thus, 3D quasi-optical model was used in simulations [3].

Results of simulations are depicted in Fig. 1 for the electron beam width of $l_x^e = 10\ cm$. At a relatively small width of the beam $l_x^e = 5\ cm$ the transverse (over x) field structure possesses a regular symmetric distribution (an exponential decay in the y-direction perpendicularly to the corrugated structure takes place in all variants). For wider beams, for example, for the electron beam width of $l_x^e = 10\ cm$, multistability regimes take place. For various initial conditions, we observed the excitation of both symmetrical and anti-symmetrical modes, as shown in Fig. 1. For explanation we analyzed the dependence of temporal gain on beam width for symmetrical and anti-symmetrical modes. For rather small widths, gain for the symmetrical mode strongly exceeds gain for anti-symmetrical one. Under such conditions, a regular symmetric distribution sets on in the steady-state regime. With increasing beam width, gains for symmetrical and anti-symmetrical modes (and other modes) become very close, which leads to multistability.

Thus, diffraction of radiation in the x-direction is sufficient for formation of regular field structure over this coordinate while the beam size is restricted by Fresnel condition, $N_F = l_x^{e2} / 4 l_z \lambda \le 1$. For practical estimations of the conventional scheme of a surface-wave oscillator we can take the beam width of 4-5 wavelengths, which corresponds to the regular field distribution. For a relatively high electron efficiency of about 10 %, the total output power amounts about 40 MW. For a metallic (brass) surface, Ohmic losses did not exceed several per cent of radiated power.

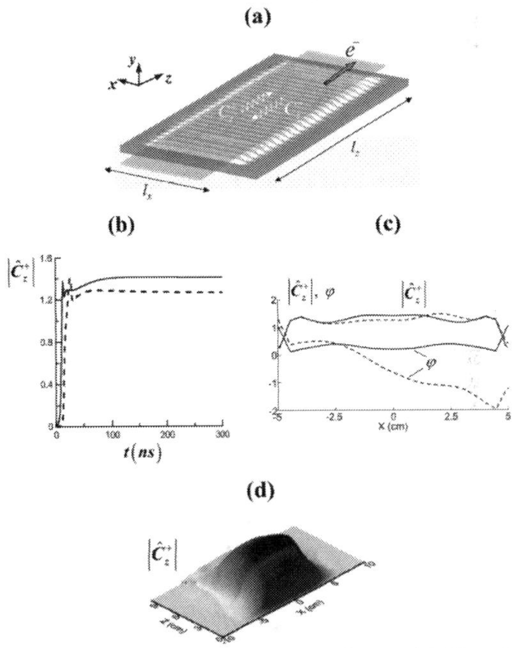

Fig. 1. **(a)** Scheme of surface-wave oscillator with 1D periodical gratings driven by a rectilinear sheet electron beam. Simulations of surface-wave oscillator based on 3D quasi-optical model for electron beam width $l_x^e = 10cm$: **(b)** Temporal dependence of amplitude $\left|\hat{C}_z^{\,+}\right|$, **(c)** distributions of amplitude $\left|\hat{C}_z^{\,+}\right|$ and phase $\varphi = \arg\left(\hat{C}_z^{\,+}\right)$ at the output cross-section $z = l_z$ ($I_0 = 0.3$ kA/cm, $\gamma_0 = 3$, $l_z = 10\ cm$), **(d)** spatial distribution of this amplitude in the steady-state regime in the cross-section y = const. Solid line corresponds to excitation of symmetrical mode, dashed line - to anti-symmetrical one.

It should be noted that in 3D quasi-optical analysis we considered an idealized model of grating infinite in the transverse x direction. On the basis of *CST STUDIO SUITE* software, we simulated a more realistic model of the grating with the same transverse width as the electron beam. In this case we found more severe restrictions on stability of single-frequency oscillation regime. In fact, for electron beam and corrugation parameters indicated above, we observed single-frequency oscillation at the system width of up to 4-5 wavelengths. For larger width values, multifrequency generation regimes with simultaneous excitation of several modes with different numbers of variations over the x coordinate are more typical.

Surface-Wave Oscillators with Two-Dimensional Periodic Structures

In the case when the beam width substantially exceeds the wavelength, we suggest a more effective mechanism for provision of transverse radiation coherence based on the improved version of the slow-wave structure, namely a double periodic grating [2, 3]. Similarly to free electron masers (FEMs) with 2D distributed feedback, these gratings allow organizing additional transverse (with respect to the translational velocity of the electrons) electromagnetic energy fluxes that synchronize the radiation from wide sheet electron beams (Fig. 2).

Based on quasi-optical approach [2, 3], we simulated the operation of a powerful W-band relativistic surface-wave oscillator driven by a sheet beam with parameters close to those of the beam realized at a high-current ELMI accelerator (Budker Institute, Novosibirsk). We took the electron energy of 1 MeV, the beam width of l_x=27 cm, and the linear current density of I_0=0.3 kA/cm.

Onset of steady-state oscillations is shown in Fig. 2b. Temporal dependencies of energy fluxes radiated from different edges of the interaction space demonstrate that in the variant under consideration the most part of the power is radiated in the backward and forward directions (Fig. 2c). The power associated with transverse fluxes is relatively small. In accordance with the negative frequency, the shift from the Bragg frequency in the steady-state regime the partial wave fields are evanescent. In the (z, x) plane the partial wave amplitudes possess bell-shaped profiles which are also similar to those found for the "cold" mode. Electron efficiency of about 10% at the chosen parameters corresponds to the integral radiation power of ~0.9 GW.

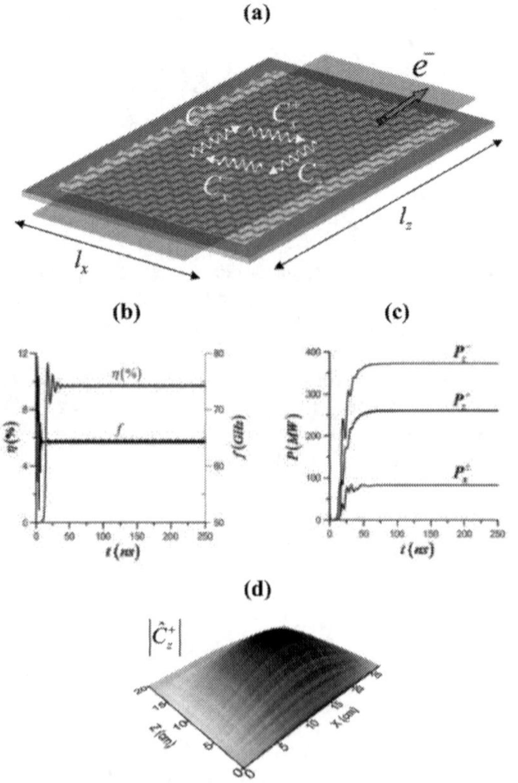

Fig. 2. (**a**) Scheme of surface-wave oscillator with 2D periodical grating. Temporal dependencies of (**b**) the efficiency and (**c**) the radiation power emitted in the different directions during the onset of a steady-state regime. (**d**) Spatial distribution of partial wave-beam $|\hat{C}_z^+|$ in the steady-state regime ($I_0 \approx 0.3$ kA/cm, $\gamma_0 = 3$, l_z=19.6 cm, l_x=27 cm).

The work was supported by the Russian Science Foundation under grant No. 18-79-10252.

References

1. *N.S. Ginzburg, A.M. Malkin, A.S. Sergeev, V.Yu. Zaslavsky*. Quasi-optical theory of relativistic submillimeter surface-wave oscillators. Applied Physics Letters. 2011. V. 99. P. 121505.

2. *Ginzburg N.S., Malkin A.M., Sergeev A.S., and Zaslavsky V.Yu.* Powerful surface-wave oscillators with two-dimensional periodic structures. // Applied Physics Letters. 2012. V.100. P. 143510.

3. *N.S. Ginzburg, A.M. Malkin, A.S. Sergeev, V.Yu. Zaslavsky*. Quasi-Optical Theory of Relativistic Surface-Wave Oscillators with 1D and 2D Periodic Planar Structures. // Physics of Plasmas. 2013. V. 20, P. 113104.

Novel schemes for Compact FELs in the THz region: ENEA Experience and Perspectives

A. Doria, G. P. Gallerano

ENEA- Fusion Physics Division – Radiation Sources, Antennas and Diagnostics Laboratory
Via E. Fermi 45, 00044 Frascati – Italy, andrea.doria@enea.it

The rapid advance of Terahertz technologies, in terms of radiation generators, systems and scientific or industrial applications, have put a particular focus on compact sources with challenging performances in terms of generated power (peak and/or average), radiation time structure and frequency tunability.

Free Electrons based sources are probably the best candidates to express such a versatility; there are a number of schemes that have been investigated over the years to generate coherent radiation from free electrons in the mm-wave and Terahertz regions of the spectrum, covering a wide frequency range from approximately 100 GHz to 10 THz. At such long wavelengths, good performance in terms of output power and gain can be achieved with a short length of the interaction region (< 50 cm) and with different mechanisms of energy transfer from the electron beam to the radiation field. These include the magnetic undulator, dielectric loaded waveguides for Cerenkov emission, and metal grating devices, based on the Smith-Purcell effect. This study will proposes novel schemes for exploring the limits in performance of radio-frequency driven free electron devices in terms of ultra-short pulse duration, wide bandwidth operation and energy recovery for near CW operation.

Auston Switch and Optical rectification

Conventional THz emitters have gained popularity due to their characteristics and peculiarities. The so called Auston Switch, named after its inventor, and the Optical Rectifiers are the most widely used devices [1]. The Switch is made up with a coplanar strip antenna deposited on a non absorbing substrate. Along the strips there is a point where the gap is very small, of the order of 10 μm. If a powerful short laser pulse is focalised on this gap, the photons generate free carriers that drift under the action of the electric field generated by this bias voltage. This transient current gives rise to a time dependent dipole moment that acts as a source of a time dependent electric field according to expression. The result is the generation of an electric pulse which duration, and thus spectrum are related to the duration of the laser pulse. A picosecond or a sub-picosecond laser pulse generate a radiation burst in the THz range that is propagated and irradiated by the antenna.

The above illustrated approach suffers of some drawbacks. The most important is the limited bandwidth of the conventional micro-strip antennas that introduce distortion and frequency dispersion on the THz pulse ending with a rapid lengthening of the pulse.

To overcame these limitations electro-optic devices, realised with a non-absorbing material, are used. The interaction mechanism is now a second order process: the rapidly oscillating electric field of the laser pulse excite in the medium a polarisation vector which is proportional to the modulus of the electric field itself. This polarisation vector is the source for an Hertzian vector potential that generates a radiation field. This is a rectification process because the rapid oscillation of the electric field of the laser pulse are compensated in the frequency domain and only the envelope of this pulse remains. The important feature of this device is that, being the medium non absorbing, the polarisation is very fast in following the laser envelope; this means that there is no physical limit in the bandwidth of the THz pulse that can be generated with this device. The only drawback is that being this a second order process, the intensities of the THz radiation are smaller respect to the previous cases, but it is now well compensated by the more powerful short-pulse laser source now available also commercially.

A Free Electron Device as THz Radiator

"Flexibility" is the most relevant feature a Free Electron Device (FED) because any component of such a source can, in principle, be designed in order to emphasise a specific characteristics of the generated radiation [2]. Broadband emission is, in fact, ensured by a short interaction region and by avoiding optical resonators, that usually filter frequency components; these precautions, together with the use of a low energy electron beam, allow an easy and compact design of the source. The most significant peculiarity of a FED is the coherence of an electron beam generated by a Radio-Frequency (RF) accelerator. The coherent emission dominates when the electron bunch length is comparable to wavelength of the radiation to be emitted. As a result coherent emission dominates in the THz range or at longer wavelengths. Another characteristics of RF based FED is the coherence among the bunches. An RF accelerator generates a train of bunches, and if the correlation among bunches is good, the radiation will be emitted at discrete frequencies which are harmonics of the RF. This is exactly what we have measured experimentally in our compact FED in ENEA [3-4]. A further degree of coherence can be exploited when a proper energy-phase correlation is introduced in the electron beam, before entering the interaction region, in order to minimise the negative interferences among the single electrons' emissions. Several techniques can be applied to get such a correlation, but the final result is a single-pass emission of a broad-band short pulse emission as reported in Fig. 1(result obtained for an electron beam with $\gamma=15$, $I_{AV}=20\ mA$ and an undulator of $\lambda_u=2.5\ cm$ period and $K=1.45$ as magnetic parameter).

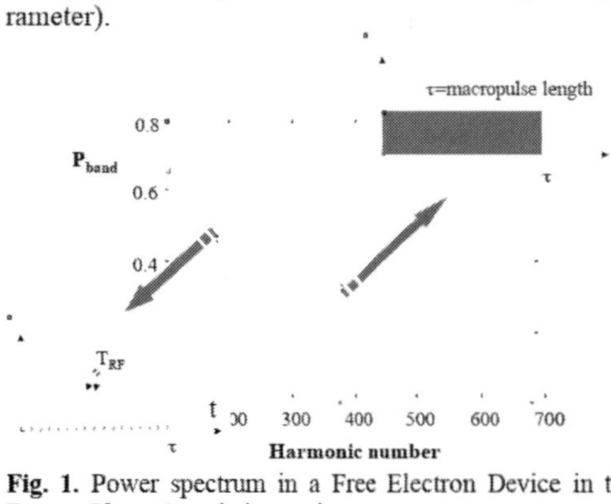

Fig. 1. Power spectrum in a Free Electron Device in the Energy-Phase Correlation regime

Analysing Fig. 1 it is evident that in the Energy-Phase correlation regime it is possible to isolate the single harmonic, with an interferometer, still having an average power for the single frequency around 0.5 W. This is not possible with the conventional THz sources. Moreover another interesting result is that the single frequency, being an harmonic of the RF has a temporal structure equal to that of the RF macropulse. If we look at the whole bandwidth, the temporal structure is the well known train of microbunches separated by the RF period. In conclusion FED can be considered a convenient flexible and powerful source for the generation of coherent radiation in the THz spectral region.

A way to get a correlated bunching in the THz region is the use of a two frequency RF device. Such cavities, in fact, offer appealing possibilities to control the bunch length of an electron beam generated by a RF accelerator. The use of a double frequency cavity requires that the second frequency be an harmonic of the fundamental one. The electrons passing through the cavity will see a field that is the result of the sum of the two fields of the fundamental and its harmonic according to:

$$V(\phi) = V_0\left[\sin(\phi+\phi_S) + k\sin(n\phi+\phi_n)\right] \qquad (1)$$

There are some design parameters that can be set like the harmonic number n, the relative amplitude ratio k that contribute to the bunch length, but the most relevant is the relative phase (ϕ_s-ϕ_n) between the fundamental and the harmonic. These parameters establish the slope of the total field in the cavity. Two different regimes can be in principle realised: one in which one can realise a bunch-lengthening, which spread can help in damping coherent instabilities that often cut down the bunch lifetime in re-circulated accelerators; and a second in which the particles bunch reduces with respect to the reference electron. In Fig. 2 we report the behaviour of the electron bunch intensity profile as a function of the phase ϕ, for different harmonics n in the RF cavity ($k=-0.6$, $\phi_n=\pi$, $\phi_s=0$).

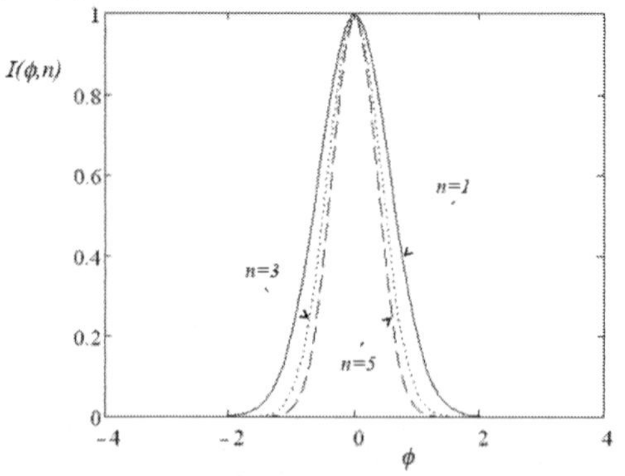

Fig. 2. Bunch shortening in a double frequency RF cavity for different harmonics.

One of the problems suffered by RF accelerator based FED, both normal-conducting and superconducting, is the relatively low efficiency. Considering the basic generation mechanism the power extraction from the electron beam in favour of the radiation never exceed few per cent, both in high gain and low gain regimes. In addition to the use od auto-resonant mechanisms (CARM), a possibility to increase the efficiency of an FED is to design a scheme for recovering the electron beam kinetic energy remained after the interaction. The energy recovered can be used to accelerate a new bunch of electrons thus lowering the accelerator radio-frequency power requirements. The possibility to manage a lower RF power and, at the same time, an exhausted electron beam of low energy, is very useful when designing the device. Generally speaking there are two possibilities for the energy recovering: one is to recycle the electron beam after the interaction region. The second possibility is to recover only the kinetic energy of the electrons before the beam dumps.

Energy recovery devices are particularly useful when designing a radiation source based on superconducting RF accelerators, or on normal conducting cavities working at very low temperatures. These accelerators, in fact, can guarantee a long bunch operation or a quasi-CW operation.

References

1. *Sakai, K.* Terahertz Optoelectronics // Berlin: Springer-Verlag. 2005.

2. *Dattoli, G., Doria, A., Sabia, E., Artioli, M,* Charged Beam Dynamics, Particle Accelerators and Free Electron Lasers // Bristol: IOP Publishing. 2017.

3. *Biedron, S., Gallerano, G. P., et al.* Compact, High Power Electron Beam based Terahertz Sources // Proc. Of the IEEE 2007. V. 95, No. 8, P. 1666-1678.

4. Doria, A. Relativistic Electron based THz Sources: Principles of Operation and the ENEA Experience // Physics and application of Terahertz Radiation, edited by M. Perenzoni and D. J. Paul. Dordrecht: Springer. 2014. P. 123-148.

Towards Fully Automated Systems for the Generation of Very High Order Modes in Oversized Waveguides

T. Ruess[1], K. A. Avramidis[1], M. Fuchs, G. Gantenbein[1], S. Illy[1], F.-C. Lutz[1], A. Marek[1], S. Ruess[1,2], T. Rzesnicki[1], M. Thumm[1,2], D. Wagner[3], J. Weggen[1], and J. Jelonnek[1,2]

[1]IHM, [2]IHE, Karlsruhe Institute of Technology (KIT), Kaiserstr. 12, D-76131 Karlsruhe, Germany
[3]Max Planck Institute for Plasma Physics, Boltzmannstr. 2, D-85748 Garching, Germany

Introduction

Fusion plasma experiments require very powerful heating systems to achieve the needed high plasma temperatures. For Electron Cyclotron Resonance Heating and Current Drive (ECRH&CD), the gyrotron oscillator is the only known source which is capable to produce a megawatt-level RF output power [1] in the range from 30 GHz up to 300 GHz for a possible future DEMO EC heating system. The excitation of the main operating mode in the gyrotron cavity and the separation of this mode from the different possible competing modes is a very challenging task due to the very high order of that modes. A similar problem exists for the excitation of the wanted high-order mode in low-power test systems to validate the quasi-optical output system of that kind of gyrotrons. But, the availability of such a test system is vital during the gyrotron design and validation phases. Particularly, the launcher of the quasi-optical output coupler, which converts the rotating high-order operating cavity mode into a linearly polarized Gaussian output beam, is the critical component here. The existing test set-up at Karlsruhe Institute of Technology (KIT) which basic idea bases on [2], includes a quasi-optical mode generator using fully manually operated mechanical sliding units [3]. Obviously, those manual sliding units suffer from a large hysteresis, hence, require a very time-consuming adjustment. Here the first steps towards a fully automated test system are presented.

First steps towards a fully automated mode generator setup

Moving towards a fully automated test system at KIT required three major steps: the introduction of a fully automated mechanical adjustment of the mode generator, the replacement of the frequency multiplier and harmonic mixer [4] by a new vector network analyzer which is even capable to cover the complete frequency range from 140 GHz to 330 GHz using different frequency extension modules and an intelligent algorithm which finds autonomous a setup configuration to excite the correct mode.

The sliding units were to replace by two precision linear stages, as depicted in Fig. 1, which can be handled by a computer. It allows to run experiments including an adjustment of the quasi-parabolic (q.-p.) mirror system without any human attendance.

The proposed system has been validated by the excitation of the $TE_{28,8}$-mode at 140 GHz used for the 1 MW, CW gyrotrons for the stellerator W7-X at IPP Greifswald, Germany [5]. Recently, an upgraded 1.5 MW version of this gyrotron, with the possibility of operation also at 175 GHz, is under discussion. Therefore, a mode generator cavity for the $TE_{28,10}$/$TE_{36,12}$-cavity modes operating at 140/175 GHz, respectively, has been designed, fabricated and successfully tested [6]. The cavity of the generator contains an inner conductor and is especially designed for the low power measurement setup. Its design has been studied and optimized employing a scattering matrix code [7]. A frequency shift occurs due to the perforated midsection area. First results are shown in [6], where the operation frequency are determined at 140.155 GHz and 175.992 GHz. The remaining frequency shift of around 155 MHz can be eliminated by increasing the radius of the inner rod which leads to a lower eigenvalue and thus a lower frequency. The measured $TE_{28,10}$ mode pattern at 140.01 GHz is shown in Fig. 2 and 3. The highest mode content is determined by 92.2 % and the counter-rotating amount is calculated to be 0.57 %. The designed quality factor is 2506 and is in very good agreement with the measured 2522. The adjustment of this mode was done using the old setup, as well. The comparison show, that a time saving of more than 94 % is achieved using the new setup with the advanced algorithm.

Recently, programs for the determination of the frequency, acquisition of the mode pattern and the mode pattern evaluation are running separately. In the future, those programs will run sequential and an already written intelligent algorithm makes the decision if the acquired mode pattern is matching to the theoretical one or not. The algorithm has two main functions, namely (i) finding the ideal position for the q.-p. mirror and (ii) the determination of the mode and its quality. Regarding point (ii), five techniques are

Fig. 1. Depiction of the mode generator setup.

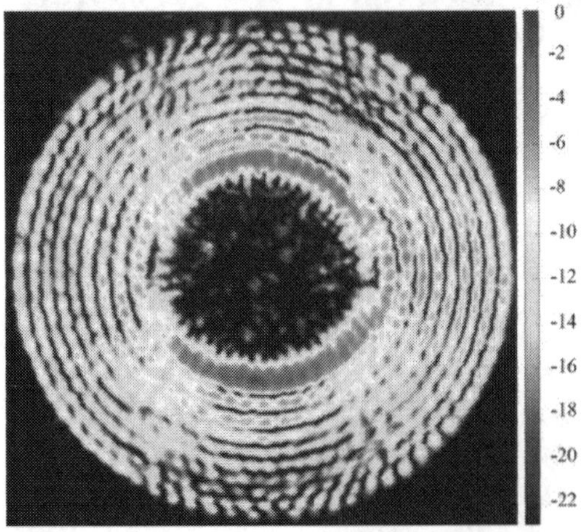

Fig. 2. Amplitude pattern of the $TE_{28,10}$-mode at 140.01 GHz with a resolution of the measurement arm of 0.2x0.2 mm.

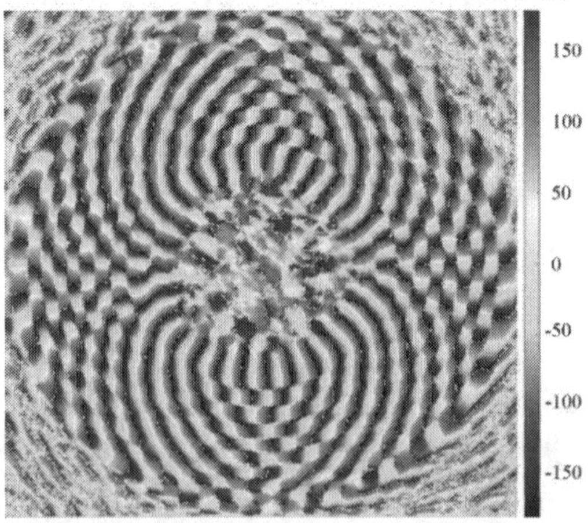

Fig. 3. Phase diagram of the $TE_{28,10}$ mode pattern at 140.01 GHz with a resolution of the measurement arm of 0.2x0.2 mm.

implemented. It can determine the radial index by counting the number of rings and the azimuthal mode index by evaluating the phase jumps. Further, the scalar mode content, the quality factor and the amount of the counter-rotating mode are calculated.

Tolerance study of the quasi-parabolic mirror

The adjustment of the q.-p. mirror is the most important parameter for a correct excitation of the mode. A tolerance study has been started proving the required adjustment accuracy. The precision linear tables under use have a position error of < 25 µm/100 mm and a repetition failure (bidirectional) of < 15 µm. For this study, different q.-p. mirror positions are tested and the scalar mode content and the amount of the counter-rotating mode is calculated at each position and summarized in Tab. 1. Any deviation in x- and y-direction can be assumed and measured using the two precision linear tables. The values Δx and Δy are the differences between the actual and the ideal position. A deviation

Tab. 1. Summary of the tolerance study regarding different positions of the q.-p. mirror in respect to the ideal one.

ID	Δx [mm]	Δy [mm]	Scalar mode content [%]	counter-rotating mode [%]
1	0.0	0.0	92.2	0.57
2	-0.8	0.0	88.6	4.15
3	-0.8	+0.7	90.2	1.43
4	-0.4	0.0	91.0	1.25
5	-0.4	+0.7	92.0	1.13
6	0	+0.7	92.0	1.05
7	-0.2	-0.2	91.5	1.51
8	-0.2	0.0	91.1	1.35
9	-0.8	-0.2	92.1	1.00
10	-0.8	0.0	92.1	1.05

of Δx = 3 mm leads to a mode pattern where several modes are superimposed and no dominant mode could be identified. A deviation of Δx = 1.5 mm leads to a mode pattern where several modes are superimposed as well, but the $TE_{28,10}$-mode could be identified as dominant. Nevertheless, the quality is not sufficient. In general, a deviation of the ideal mirror position reduces the scalar mode content and increases the amount of the counter-rotating mode. But, the impact of a deviation of roughly 0.8 mm is not incredible high for this specific cavity, as can be seen in Tab. 1. In the next steps, further tolerance analysis will be made.

References

1. *Rzesnicki, T.* et al., 2.2-MW record power of the 170-GHz european pre-prototype coaxial-cavity gyrotron for ITER // *IEEE Trans, on Plasma Science,* Vol. 38, No. 6, pp. 1141-1149, 2010.

2. *Alexandrov, N. L.* et al., Low-power excitation of gyrotron-type modes in a cylindrical waveguide using quasi-optical techniques // *Int. Journal of Electronics,* 79, pp. 215-216, 1995.

3. *Ruess, T.* et al, 2018 status on the measurement capabilities for fusion gyrotrons at KIT/IHM // *30th Joint Russian-German Meeting on ECRH and Gyrotrons,* EPJ Web of Conferences (2018).

4. *Losert, M., Jin, J. and Rzesnicki, T.* RF beam parameter measurements of quasi-optical mode converters in the mw range // *IEEE Trans. On Plasma Science*, vol. 41, no. 3, 2013.

5. *Braune, H.* et al., Gyrotron operation during the first W7-X campaign – handling and reliability // in *Proc. IRMMW*, Copenhagen, Denmark, pp. 1-2, 2013.

6. *Avramidis, K. A.* et al., Studies towards an upgraded 1.5 MW gyrotron for W7-X // *20th Joint Workshop on Electron Cyclotron Emission (ECE) and Electron Cyclotron Resonance Heating (ECRH),* Greifswald, Germany, May 2018.

7. *Wagner, D., Pretterebner, J., Thumm, M.*, Design of coaxial gyrotron cavities using a scattering matrix code // *5th Joint Russian-German Meeting on ECRH and Gyrotrons*, pp. 555-565, 1993.

EPJ Web of Conferences **195**, 01031 (2018) https://doi.org/10.1051/epjconf/201819501031
TERA-2018

YBaCuO Josephson generators fabricated by preliminary topology masks

L.S. Revin[1,2], A.L. Pankratov[1,2,3], D.V. Masterov[1], A.E. Parafin[1], S.A. Pavlov[1], A.V. Chiginev[1,2], I.V. Rakut'[2,3], A.V. Gordeeva[1,2], V.O. Zbrozhek[2], A.V. Blagodatkin[1,2], L.S. Kuzmin[2,4]

[1]Institute for Physics of Microstructures of Russian Academy of Sciences, Nizhny Novgorod, Russia, rls@ipmras.ru
[2]Nizhny Novgorod State Technical University n.a. R.E. Alexeev, Nizhny Novgorod, Russia
[3]Lobachevsky Nizhny Novgorod State University, Nizhny Novgorod, Russia
[4]Chalmers University of Technology, Gothenburg, Sweden

The investigation of flux-flow regimes in the Josephson junction is of practical interest. In a long junction the mode of fluxon motion may occur under the action of external magnetic field in which solitons are created at one edge of the junction, move along the junction, and are radiated at the other edge. Such regimes can be useful in creation of THz band oscillators [1], heterodyne and Hilbert spectrometers.

It is well known that the properties of thin YBCO films are very sensitive to the processes of formation of planar structures. New technology of preliminary mask (PM) with CeO_2 buffer layer, which determines the necessary topology of the structure directly during the growth of the YBCO film has been developed [2-4]. The exclusion of the etching operation of the YBCO film from the technological process is important for structures with weak links, where the film is most sensitive to possible destructive factors. The PM method has been successfully used to create Josephson junctions on a bicrystal YSZ substrate with a sublayer of epitaxial cerium dioxide CeO_2 [2,4].

As the results of preliminary study, the samples of Josephson junctions with 6-350 μm length and 0.3 μm thickness have been fabricated on symmetric 24° [001]-tilt bicrystal YSZ substrate.

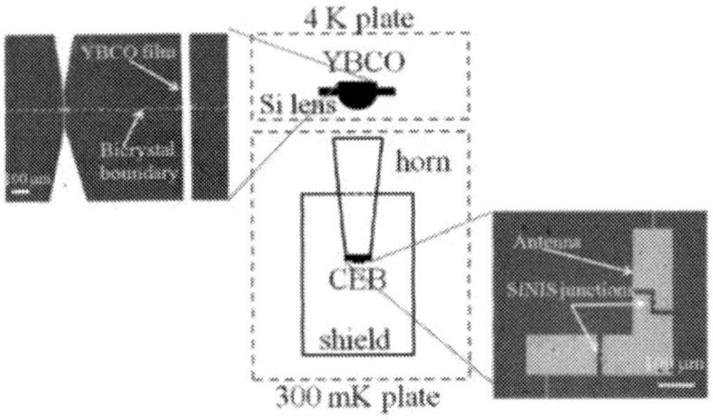

Fig. 1. Schematic of the experiment and samples design.

To register the radiation from long bicrystal junction the sample of Cold Electron Bolometer (CEB) [5-8] was used. CEBs represent SINIS junctions with nanoabsorber made of aluminum with suppressed superconductivity. One CEB can receive up to 0.5 pW without going into saturation. Based on the requirements of the power load the 2D array of CEB realized as meander-type structure. CEBs are connected in series and parallel at DC for optimal matching with the amplifier. A pair of half lambda dipoles with wide electrodes is connected to another pair by high-inductive 1 μm wide lines. The design is optimized for the frequency band of 240-280 GHz.

In figure 1 the scheme of experimental setup as well as samples design are presented. The Josephson oscillator chip consists of several junctions integrated with planar dipole antennas. The substrate with the sample was attached to Si hyperhemisphere lens at the 4 K plate. The bolometer with horn on the back side of silicon substrate was placed at the 300 mK plate. The absorbing shield was used to avoid reflections. The magnetic field perpendicular to the grain boundary of HTSC sample was produced by a current through the copper wire coil, having the inner diameter more than order of magnitude larger than the junction length.

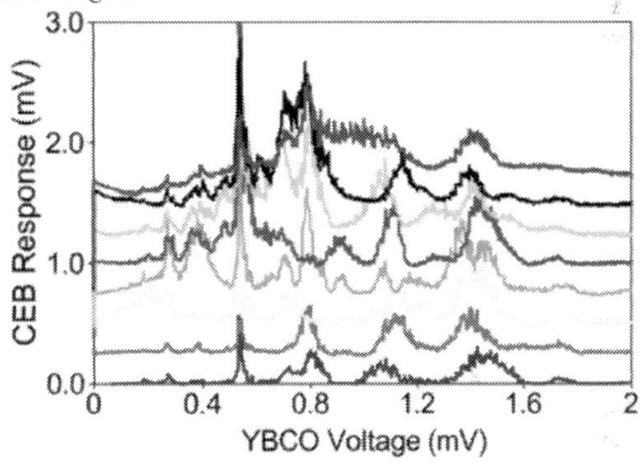

Fig. 2. CEB response for a different magnetic field through the YBCO generator. For clarity, each curve from the second is shifted relative to the other by 0.25 mV.

Changing the current through the Josephson junction the bolometer response was measured as well as the voltage on YBCO oscillator (Josephson frequency $f = 2eV_{YBCO}/\hbar$). Figure 2 shows the bolometer response (received power) for different values of the external magnetic field. For clarity each curve from the second one is shifted relative to the other by 0.25 mV. It can be seen that the response represents several distinct peaks and a pedestal. The maximum of this pedestal shifts towards higher voltage with increasing of magnetic field.

Finding the maximum response for each frequency for all values of the magnetic field, we obtain the frequency response of the radiating system together with the receiving system, figure 3. Comparing the result with the CEB amplitude-frequency characteristic based on the calibrated backward-wave source (enlarged part of Fig. 3), it can be concluded that the signal peaks are determined by the characteristic of the receiving system rather than the oscillating one.

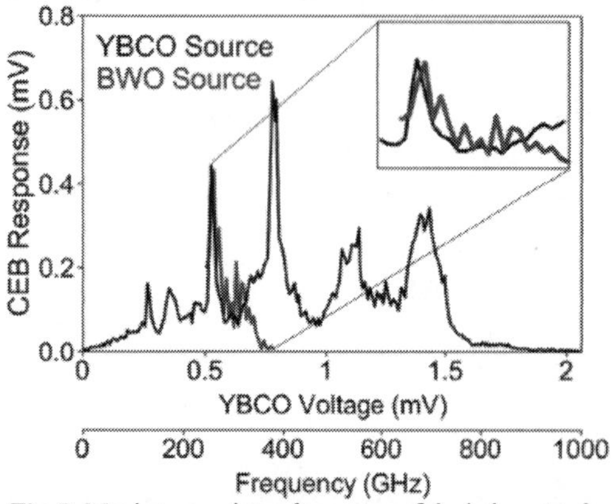

Fig. 3. Maximum registered response of the bolometer depending on the frequency of radiation. Comparison with experiments using BWO.

In conclusion, the samples of YBCO long Josephson junctions on YSZ bicrystal substrate have been fabricated using original technology of preliminary topology mask and good characteristics of the samples have been achieved. The observation of flux-flow regime in long grain-boundary junctions is evidenced. Using the cold-electron bolometer, the subTHz emission up to 900 GHz was registered. Magnetic field controlled regime make long junctions promising candidates for THz applications. It can also be used as a THz band cryogenic network analyzer in combination with a cryogenic bolometer.

The work is supported by RSF (projects 16-19-10478 and 16-19-10468).

References

1. *Stepantsov, E., Tarasov, M., Kalabukhov, A., Kuzmin, L., Claeson T.,* THz Josephson properties of grain boundary YBaCuO junctions on symmetric, tilted bicrystal sapphire substrates // Journal of Appl. Phys. 2004. V. 96, No. 6. P. 3357.

2. *Masterov, D.V., Parafin, A.E., Revin, L.S., Chiginev, A.V., Skorokhodov, E.V., Yunin, P.A., Pankratov, A.L.,* YBCO long Josephson junctions on bicrystal $Zr_{1-x}Y_xO_2$ substrates fabricated by preliminary topology masks // Supercon. Sci. Technol., vol. 30, no. 2, pp. 025007, 2017.

3. *Masterov, D.V., Pavlov, S.A., Parafin, A.E.,* A new approach to formation of the topology of planar structures on the basis of a YBCO high-temperature superconductor // Physics of the Solid State. 2017. vol. 59, pp. 2133-2136.

4. *Revin, L.S., Pankratov, A.L., Masterov, D.V., Parafin, A.E., Pavlov, S.A., Chiginev, A.V., Skorokhodov, E.V.,* Features of Long YBCO Josephson Junctions Fabricated by Preliminary Topology Mask // IEEE Trans. on Appl. Supercon. vol. 28, no. 7, pp. 1100505, 2018.

5. *Kuzmin, L., Hoste, S., Ausloos, M.,* An Array of Cold-Electron Bolometers with SIN Tunnel Junctions and JFET readout for Cosmology Instruments // 8th European Conference on Applied Superconductivity. 2008. vol. 97, pp. 17.

6. *Kuzmin, L.S., Devyatov, I.A., Golubev, D., et al.* "Cold-electron" bolometer with electronic microrefrigeration and the general noise analysis // Millimeter and Submillimeter Waves Iv. 1998. vol. 34 pp. 193-199.

7. *A.V. Gordeeva, V.O. Zbrozhek, A.L. Pankratov, L.S. Revin, V.A. Shamporov, A.A. Gunbina, L.S. Kuzmin,* Appl. Phys. Lett., 110, 162603 (2017).

8. *L.S. Kuzmin, A.L. Pankratov, A.V. Gordeeva, V.O. Zbrozhek, L.S. Revin, V.A. Shamporov, A.A. Gunbina, S. Masi, P. de Bernardis,* "Realization of Cold-Electron Bolometers with Ultimate Sensitivity due to Strong Electron Self-Cooling", Proceedings of 16th International Superconductive Electronics Conference (ISEC'2017), DOI: 10.1109/ISEC.2017.8314194

EPJ Web of Conferences **195**, 02001 (2018) https://doi.org/10.1051/epjconf/201819502001
TERA-2018

THz stimulated emission at interband transitions in HgTe/CdHgTe quantum wells

V.I.Gavrilenko[1], V.V.Rumyantsev[1], A.A.Dubinov[1], S.V.Morozov[1], N.N.Mikhailov[2], S.A.Dvoretsky[2], F.Teppe[3], C.Sirtori[4]

[1]Institute for Physics of Microstructures, Nizhny Novgorod, Russia, gavr@ipmras.ru
[2]A.V.Rzhanov Institute of Semiconductor Physics, Siberian Branch of Russian Academy of Sciences, Novosibirsk, Russia
[3]Charles Coulomb Laboratory, University of Montpellier & CNRS, Montpellier, France
[4]Laboratory of Materials and Quantum Phenomena, University Paris-Diderot, Paris, France

Terahertz (THz) spectral range still lacks compact and efficient radiation sources. Quantum cascade lasers (QCLs) based on GaAs and InP demonstrate remarkable performance in the spectral range from 1 THz to 5THz and above 15 THz[1] the "gap" resulting from strong two-phonon absorption. The interband lasers are a straight-forward alternative, but in narrow-gap materials the non-radiative Auger recombination is expected to be very effective. Lead salt diode lasers are known to operate up to 50 μm[2]. (though their figures of merit are very limited). The factor that mitigates the Auger recombination in PbSnSe is the symmetry between electron and hole dispersion laws. For energy and momentum conservation laws to be fulfilled, the summarized kinetic energy of the three particles involved in the Auger process has to be over a certain threshold energy E_{th}[3] that decreases if there are heavy mass carries, e.g. heavy holes.

Symmetric energy-momentum laws are realized in HgTe/CdTe based quantum wells (QWs). In HgCdTe QW structures with bandgap 60 - 80 meV we observed the increase of carrier lifetimes with the pumping intensity up to several microseconds due to saturation of Shockley-Read-Hall (SRH) recombination centers, inferring that optical gain can be obtained under reasonable pumping intensity[4]. Earlier the longest wavelength of 5.3 μm was obtained in HgCdTe-based lasers that did not exploit QWs in its design[5]. In this work, we report stimulated emission (SE) from HgCdTe structures at wavelengths up to λ ~ 20 μm, and demonstrate possibilities for HgCdTe based lasers to enter 5 – 15 THz frequency domain.

Structures under study were grown by molecular beam epitaxy on semi-insulating GaAs(013)[6]. Energy spectra were calculated in the framework of Kane 8x8 model. Structure #1 contains ten 12 nm thick $Hg_{0.87}Cd_{0.13}Te/Cd_{0.65}Hg_{0.35}Te$ QWs and structure #2 - five 5.4 nm thick $Hg_{0.91}Cd_{0.09}Te/Cd_{0.6}Hg_{0.4}Te$ QWs. The structures were designed so as to effectively confine light to in-plane direction, therefore the "active" region (5 - 12 QWs) was placed at the antinode position of TE_0 mode in a thick (several micrometers) dielectric waveguide. Due to a quite specific growth direction (013), naturally cleaved facets do not form the Fabri-Perot resonator. Thus, the stimulated emission (SE) studied in this work results from single-pass amplification.

SE spectra for two samples under study at 8K and a "critical" temperature T_{max}, above which no SE is observed, are given in Fig. 1. The emission intensity demonstrates clear threshold dependence on pumping power, typical for the onset of SE. Another telltale sign of the SE is photoluminescence (PL) line narrowing.

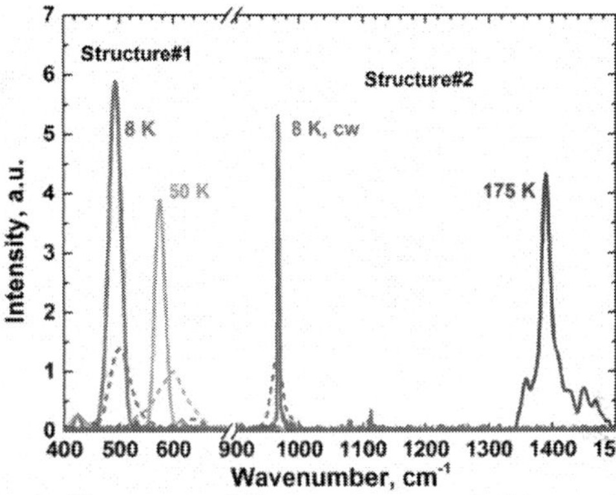

Fig. 1. SE spectra at different temperatures (solid curves) obtained under pulsed pumping with 2.3 μm wavelength and 10 kW/cm² intensity for Structure #1 (at both temperatures) and 65 kW/cm² for Structure #2 at 175K. SE for Structure #2 at $T = 8K$ was obtained at cw excitation with 0.9 μm wavelength 7 W/cm² intensity. Dash curves show PL spectra obtained with the same cw pumping source at 5 W/cm² intensity for Structure #1 and 1 W/cm² for Structure #2 at 8 K.

The corresponding thresholds are 5 kW/cm² and 120 W/cm² for SE wavelength ~20 μm (15 THz) and ~10 μm (30 THz) at low temperatures. However, these values are obtained for "below barrier" excitation ($\hbar\omega < E_g$ in barriers), i.e. when non-equilibrium carriers are generated in QWs only. Taking for estimation the QW absorption as 1% one can find the corresponding carrier density in each QW $n_{th} = 1.4\times10^{11}$ cm^{-2} for $\lambda_p = 2.3$ μm and $I_{th} = 0.12$ kW/cm² pumping intensity. This n_{th} value agrees fairly well with our previous calculations[7]. It also allows estimating the equivalent threshold current density for 5 QWs placed into a *p-n* junction as $j_{th} = 5en_{th}/\tau_{pulse}$ (e is the elementary charge), that for $\tau_{pulse} = 10$ ns gives $j_{th} = 11$ A/cm². This threshold is low enough to obtain SE under cw excitation: Fig 1. shows the corresponding spectrum, measured with cw pumping at 0.9 μm wavelength.

Obviously, the threshold grows with the SE wavelength and the maximum "operating" temperature T_{max} gets lower. To understand the energetic scale that determines T_{max} let us consider the energy spectrum of Structure #1, presented in Fig.2. One can see that the hole effective mass increases dramatically with k and the dispersion laws in the valence band and the conduction band are no longer quasi-symmetrical. When holes reach larger-mass region the Auger recombination becomes efficient. We have calculated the threshold energies E_{th} for CCHC Auger process (the energetic threshold for CHHH process is high compared to CCHC process, therefore the latter is less important) given in Fig. 2. For both structures under study we get $k_B T_{max} \approx E_{th}/2$. In Fig. 2 one can see also that the detrimental impact of side maxima in the valence band can be reduced in QW of pure HgTe with the same bandgap as in Structure #1. As easy to see in this case the threshold energy increases significantly.

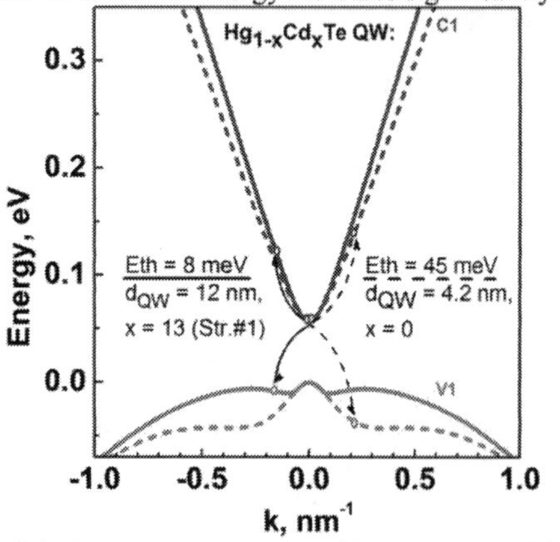

Fig. 2. Calculated energy spectrum of Structure #1 (solid lines) and HgTe/CdHgTe QW with the same bandgap (dash lines), $T = 0$. Threshold energies and carrier configuration are shown for Auger CCHC process.

Effective carrier temperature is affected by their heating by the pumping radiation. At the excitation with radiation wavelength 2.3 μm the carrier heating effect is noticeable already in QW structures emitting at wavelengths μm about 14 μm. For the Structure #1 emitting at $\lambda \sim 20$ μm the SE quenching with the increase of pumping power takes place just after the SE arising. We have demonstrated that the pumping with CO_2-laser ($\lambda = 10.6$ μm) drastically decrease the carrier heating and the SE intensity monotonously grows with pump power. Thus, the effect of SE quenching with the pump power, observed in Ref.8 is not fundamental but results from using the short wavelength excitation.

Carrier lifetimes are expected to decrease for QW structures with narrower bandgap. The sub-nanosecond carrier lifetimes have been explored via the pump-probe measurements of a sample's transmission using THz free electron laser FELBE at Helmholtz-Zentrum Dresden-Rossendorf[9]. For HgCdTe QW with $E_g = 20$ meV ($f \sim 4.8$ THz) the carrier lifetime proved to be about 100 ps for carrier density 10^{11} cm^{-2} that is sufficient to achieve the population inversion. One can estimate a threshold pumping intensity of 10 kW/cm^2 for an optically pumped laser exploiting such QWs as an active media. Thus, HgCdTe QWs should be able to provide amplification of radiation at interband transitions down to 5 THz ($\lambda = 60$ μm).

In conclusion, we demonstrated stimulated emission at wavelengths up to 20.3 μm (14.7 THz) from HgCdTe based waveguide quantum well heterostructures. Nonradiative Auger recombination is shown to be suppressed compared to bulk HgCdTe solid solutions due to the "symmetry" of electron and hole energy-momentum laws. Results of time-resolved pave the way to obtain stimulated emission in wide THz range down to 5 THz.

This work was financially supported by the Russian Foundation for Basic Research (Grants #16-32-60172, #18-52- 16013) and Ministry of Education and Science of the Russian Federation (Grant MK-4399.2018.2). The work was done using equipment of Common Research Center "Physics and technology of micro- and nanostructures" at IPM RAS.

References

1. *Vitiello, M. S., Scalari, G., Williams, B., De Natale, P.* Quantum cascade lasers: 20 years of challenges // Opt. Express. 2015. V.23, No. 4. P. 5167-5182.

2. *Maremyanin, K. V., Ikonnikov, A. V., Bovkun, L. S., Rumyantsev, V. V., Chizhevskii, E. G., Zasavitskii, I.,I., Gavrilenko, V. I.* Terahertz injection lasers based on the PbSnSe solid solution with emission wavelength up to 50 μm and their use for magnetospectroscopy of semiconductors // Semiconductors. 2018. V. 52, No 11 (in press).

3. *Abakumov, V. N., Perel, V. I., Yassievich, I. N.* Nonradiative Recombination in Semiconductors. North-Holland: Elsevier Science Publishers. 1991.

4. *Morozov, S. V., Rumyantsev, V. V., Antonov, A. V., Kadykov, A. M., Maremyanin, K. V., Kudryavtsev, K. E., Mikhailov, N. N., Dvoretskii, S. A., Gavrilenko, V. I.* Time resolved photoluminescence spectroscopy of narrow gap $Hg_{1-x}Cd_xTe/Cd_yHg_{1-y}Te$ quantum well heterostructures // Appl. Phys. Lett. 2014. V. 105, No 2. P. 022102.

5. *Arias, J. M., Zandian, M., Zucca, R., Singh, J.* HgCdTe infrared diode lasers grown by MBE // Semicond. Sci. Technol. 1993. V. 8, No 1S. P. S255-S260.

6. *Dvoretsky, S., Mikhailov, N., Sidorov, Yu. Shvets, V., Danilov, S., Wittman, B., Ganichev, S.* Growth of HgTe Quantum Wells for IR to THz Detectors // J. Electronic Mater. 2010. V. 39, No. 7. P. 918-923.

7. *Morozov, S. V., Joludev, M. S., Antonov, A. V., Rumyantsev V. V., Gavrilenko, V. I., Aleshkin, V. Ya., Dubinov, A. A., Mikhailov, N. N., Dvoretskiy, S. A., Drachenko, O., Winnerl, S., Schneider, H., Helm, M.* Study of lifetimes and photoconductivity relaxation in heterostructures with $Hg_xCd_{1-x}Te/Cd_yHg_{1-y}Te$ quantum wells // Semiconductors. 2012. V. 46, No 11. P. 1362-1366.

8. *Morozov, S. V., Rumyantsev, V. V., Fadeev, M. A., Zholudev, M. S., Kudryavtsev, K. E., Antonov, A. V., Kadykov, A. M., Dubinov, A. A., Mikhailov, N. N., Dvoretsky, S. A., Gavrilenko, V. I.* Stimulated emission from HgCdTe quantum well heterostructures at wavelengths up to 19.5 μm // Appl. Phys. Lett. 2017. V. 111, No 19. P. 192101.

9. *Ruffenach, S., Kadykov, A., Rumyantsev, V. V., Torres, J., Coquillat, D., But, D., Krishtopenko, S.S., Consejo, C., Knap, W., Winnerl, S.. Helm, M., Fadeev, M. A., Mikhailov, N. N., Dvoretskii, S. A., Gavrilenko, V. I., Morozov, S. V., Teppe, F.* HgCdTe-based heterostructures for terahertz photonics // APL Materials. 2017. V. 5, No. 3. P. 035503.

EPJ Web of Conferences **195**, 02002 (2018) https://doi.org/10.1051/epjconf/201819502002
TERA-2018

Electronic Tunneling and Electric Domains in GaAs/AlAs Superlattices at Room Temperature

I. V. Altukhov[1], M. S. Kagan[1], S. K. Paprotskiy[1], N. A. Khvalkovskiy[1], I. S. Vasil'evskii[2], A. N. Vinichenko[2]

[1]V.A. Kotel'nikov Institute of Radio Engineering and Electronics, Russian Ac. Sci., Moscow, Russia, kagan@cplire.ru
[2]National Research Nuclear University MEPhI, Moscow, Russia

One of the main problems in the development of semiconductor sources of the THz radiation is their operation at room temperature. Operation frequencies of room-temperature hot-electron bulk semiconductors such as Gunn diodes are limited by the time of development of the negative differential conductivity and the dielectric relaxation time. The former time in most semiconductors is usually in the range of picoseconds. The Maxwell relaxation time at the conventional doping level on the order of 10^{15}cm^{-3} is also $\sim 10^{-12}$ s. Therefore, the operation frequency band of hot-electron devices does not exceed 100–200 GHz. This limit is eliminated by using the laser scheme, when the energy of a photon is determined by the difference between energy levels with the inverse population and does not depend on the inertia of the conductivity relaxation. However, all THz semiconductor lasers operate at cryogenic temperatures. It is hardly possible to implement the population inversion at room temperature, since $kT = 25$ meV and the energy of THz photons is ~10 meV. Therefore, it is reasonable to search for systems with a negative high-frequency differential conductivity, which can excite oscillations in the corresponding cavity. Good examples are resonant-tunneling diodes, in which the generation frequency has been reached 1.92 THz [1]. Another suitable semiconductor systems are superlattices (SLs).

The electron transport in semiconductor SLs has been studied in detail in recent decades [2], mainly in connection with the predicted amplification of Bloch waves, which can make it possible to create tunable sources of THz radiation. The main hindrance to the Bloch wave amplification is the formation of electrical domains owing to the static negative differential conductivity (NDC) arising in superlattices at resonant tunneling between confined states in neighboring quantum wells [3]. The domain formation in superlattices eliminates the Bloch gain. On the other hand, the samples with the electrical domains can exhibit the dynamic NDC (the real part of electrical impedance is negative) in some frequency range (similarly to bulk materials [4-8]) and can be used for high-frequency generation in a suitable resonant cavity.

Superlattices for the THz range should satisfy several conditions. To get the short conductivity relaxation time, which is the time of tunneling in SLs, one has to use SLs with thin barriers. To prevent a thermal exchange with carriers between succeeding confined states in SLs, we need to have narrow quantum wells and the large band offset. For fast space charge redistribution in SLs with electric domains, the high doping level is needed to obtain the short dielectric relaxation time. Here we present studies of tunneling electronic transport in short-period GaAs/AlAs superlattices (SLs) with electric domains.

The MBE grown GaAs/AlAs SLs consisted of 100 periods of 4 nm GaAs/2 nm AlAs between heavily doped (~10^{19}cm^{-3}) n$^+$ cap layer and n$^+$ substrate. GaAs quantum wells (QWs) were Si doped with concentration of $2*10^{17}$cm^{-3}. The ring-shaped mesa structures with the diameters from 10 to 15 μm and 0.8 to 1.5 μm widths (Fig. 1) formed the distributed THz cavity with resonant frequencies corresponding to the free-space wavelengths from ~110 to ~170 μm. Triangular voltage pulses with a sweep-up times of 0.5 to 10 μs and sweep-down times up to 100 μs applied to the sample were used to record I-V curves. Measurements were performed at room temperature.

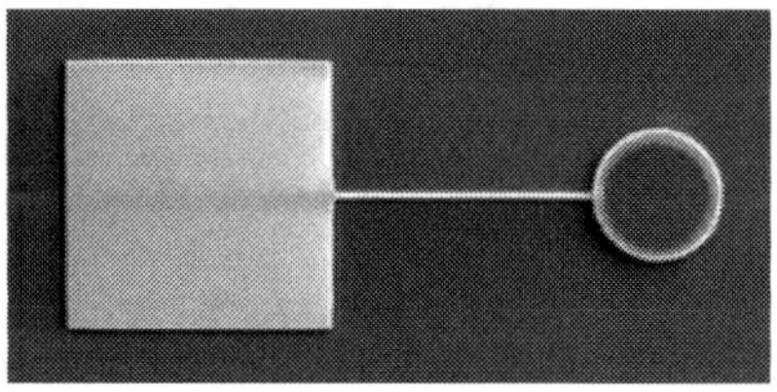

Fig. 1. View of the sample with the contact and current lead

The moving domains are formed at some threshold voltage and became apparent in a sharp decrease of current (up to 50% - Fig. 2). It was found that this threshold voltage changes considerably under the change of the cavity [9]. The cavity was changed by deposition of a droplet of conductive silver epoxy paste covering the entire top part of the mesa structure (see insets in Fig. 2). Fig. 2 shows the current–voltage

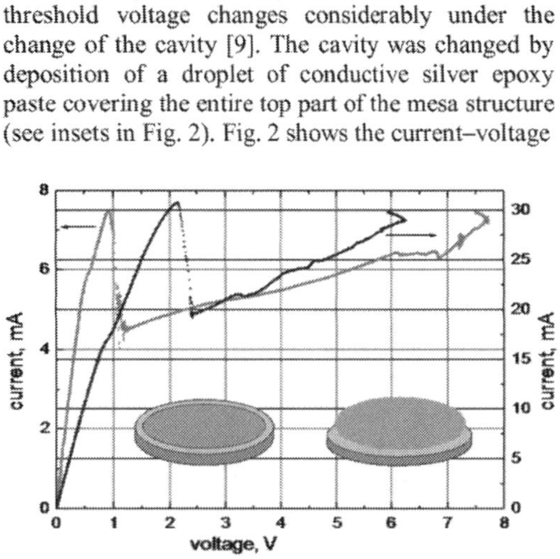

Fig. 2. Current–voltage characteristics of the GaAs/AlAs SLs under a change of the cavity: (lower curve) before and (upper curve) after the deposition of a droplet.

characteristics of the structure before and after the deposition of this droplet. The current through the sample after the deposition increased, what is natural because the contact area increased. In fact, the increase in the current approximately corresponds to the increase in the area. In addition, the threshold field of the domain formation changed, which was generally surprising because the thickness of the structure (the distance between current contacts) remained the same.

The possible origin of the change in the threshold field at changing the contact area is the appearance of an alternating field with sufficiently high amplitude in the cavity. Indeed, if an ac voltage of comparable amplitude is applied to the sample with a nonlinear current–voltage characteristic in addition to the dc voltage, an additional dc voltage appears because of the detection and shifts the operating point. Thus, this experiment indicates the excitation of THz oscillations in the cavity owing to the negative resistance of the superlattice with domains. We note that an analogous effect of an alternating electric field on the shape of direct current–voltage characteristics was observed in the bulk GaAs with the Gunn effect [10].

For not resonated structures, several features appeared in I-V curves (Fig. 3). First one is the hysteresis at forward and backward bias sweep. The size and shape of hysteresis loop depends on the peak voltage of the triangular pulse. The maximal loop observed when the current grows up to its peak value in the I-V curve. In this case the current saturation appears in some voltage range (Fig. 3), being essentially larger at bias sweep-down. The current saturation is due to static domain formation. So, the current hysteresis can be explained by a transition from moving to static field domain at sweeping the bias up and the return at its sweeping down. The large asymmetry of the transition times was found: the transition from travelling to static domain regime passes in a time shorter 0.1 microsecond, while the time of the reverse transition - from static to travelling domain regime – is ~100 microseconds. It is seen from I-V curves in Fig. 3 measured at different sweep-down duration. The long time is referred to deep impurities at the SL-buffer layer boundary.

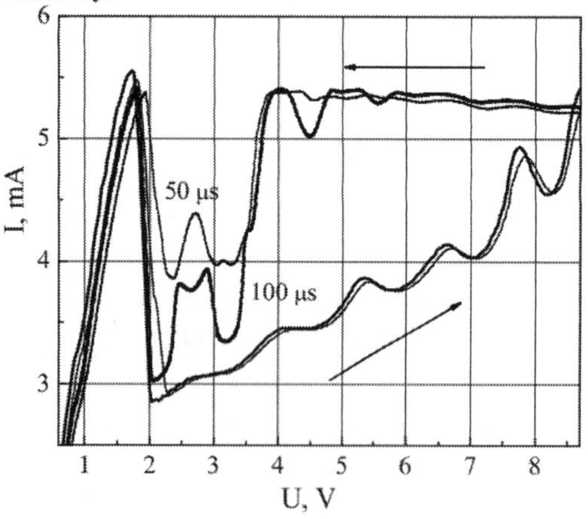

Fig. 3. I-V curves at different voltage sweep-down times

The second feature is the series of almost voltage-periodic maxima in current-voltage characteristics. The maxima are on a background of current rise in the mean. The current rise is evidence of triangular form of the domain: the negative charge at one of domain boundaries (in one quantum well) is due to free electrons with the concentration of more than one order larger than the donor concentration supporting the necessary field step, while the positive charge at opposite domain boundary is due to positively charged donors [11, 12]. The linear spatial grows of electric field inside the domain is just the result of donor depletion.

The positions of the maxima in the I-V curves did not depend on the frequency of THz cavity, they coincided at different ring cavity diameters, while essentially changed at cooling. The average period of the maxima at nitrogen temperature was by 1.5 times larger than that at room temperature. The origin of these maxima is attributed to the optical phonon assisted tunneling between quantum wells inside the triangular domain.

The work was supported by RFBR grant 16-29-03135

References

1. *Maekawa, T., Kanaya, H., Suzuki, S., Asada, M.* Oscillation up to 1.92 THz in resonant tunneling diode by reduced conduction loss // Applied Physics Express 2016. V. 9, No. 2. P. 024101
2. *Wacker, A.* Semiconductor superlattices: a model system for nonlinear transport // Physics Reports, 2002. V. 357, P. 1
3. *Klappenberger, F., Alekseev, K. N., Renk, K. F.*, et al. Ultrafast creation and annihilation of space-charge domains in a semiconductor superlattice observed by use of Terahertz fields // Eur. Phys. J. B 2004. V. 39, P. 483
4. *Thim, H. W.*, Linear microwave amplification with Gunn oscillators // IEEE Trans. Electron. Dev. 1967. V. 14, P. 517
5. *Hakki, B. W.* Amplification in Two-Valley Semiconductors // J. Appl. Phys. 1967. V. 38, P. 808
6. *Zhdanova, N. G., Kagan, M. S., Kalashnikov, S. G.* Impedance of semiconductor with static domain // Sov. Phys. Semicond. 1974. V. 8, P. 1121; 1126
7. *Altukhov, I. V., Kagan, M. S., Kalashnikov, S. G., et al.* Electromagnetic wave amplification by Gunn diodes with moving domains // Sov. Tech. Phys. Lett. 1980. V. 6, P. 237
8. *Kagan, M.S., Landsberg, E.G., Chernyshov, I.V.* Negative conductivity due to vibrations of the wall of a static domain, Sov. Phys. Semicond. 1984. V. 18, P. 615
9. *Altukhov, I. V. Dizhur, S. E. Kagan, M. S.*, et al. Effect of a Terahertz Cavity on the Conductivity of Short-Period GaAs/AlAs Superlattices // JETP Letters 2016. V. 103, No. 2, P. 122
10. *Altukhov, I. V. Kagan, M. S. Kalashnikov, S. G., et al.* Electrical instability of a semicondutor with a negative differential conductivity due to simultaneous heating of electrons by static and alternating electric fields // Sov. Phys. Semicond. 1978. V. 12, P. 172
11. *Suris R. A.* Inhomogeneous structures in semiconductors with superlattices // Sov. Phys. Semicond. 1973. V. 7, No. 8 P. 1035
12. *Bonilla, L. L., Grahn H. T.* Non-linear dynamics of semiconductor superlattices // Rep. Prog. Phys. 2005. V. 68, No. 3, P. 577

Wideband Josephson THz flux-flow oscillator integrated with the slot lens antenna and the harmonic mixer

N.V.Kinev[1], K.I. Rudakov[1,2,3], L.V. Filippenko[1], A.M. Baryshev[3], V.P. Koshelets[1]

[1]Kotel'nikov Institute of Radio Engineering and Electronics of RAS, 125009 Moscow, Russia, nickolay@hitech.cplire.ru
[2]Moscow Institute of Physics and Technology, 141701 Dolgoprudny, Russia
[3]University of Groningen, 9712 CP Groningen, Netherlands

Introduction

Terahertz (THz) wideband sources are required nowdays in many field, such as spectroscopy, astronomy and radio physics. We present a new implementation for a flux-flow oscillator (FFO) based on a long Josephson junction providing the THz emission in the open space. The FFO was used earlier in our team merely as an on-chip heterodyne for the SIS receiver [1-2], it has a really wide operating range up to 100% of central frequency and a power of about 1 μW. Based on the Nb/AlN/NbN trilayer the FFO with dimensions 400x16 μm² provides the output radiation from 200 GHz up to 750 GHz with the spectral linewidth of about 1 MHz. The upper operating frequency can reach 1 THz. The phase-lock loop (PLL) is used for the phase locking of the emission collecting up to 97% of the output THz power in the peak with a width of about 40 kHz. Such oscillator implemented as an external source could be useful for many tasks – gas spectroscopy, heterodyne receivers, etc.

Concept of the THz oscillator & numerical simulations

The principle of the FFO operation is discussed elsewhere [3-4]. The operating frequency is defined strictly by the Josephson equation

$$hf = 2eV, \quad (1)$$

where V is the DC voltage on the junction. The original idea of this work is the integration of the FFO with a harmonic mixer (HM) for the feedback locking loop and a transmitting slot antenna on a single chip (fig. 1a) placed on the back surface of the elliptical lens (fig.1b). The chip substrate and the lens are both made of silicon. Thus, the main task is the coupling of the oscillator having low output impedance (less than 1 Ω) to the lens antenna having high impedance (tens of Ω) and forming a beam pattern required for applications, and simultaneous coupling to the harmonic mixer based on SIS junction having the area about 1 μm² and the impedance of about several Ω. Coupling to the antenna should be as high as possible, while coupling to the HM should be just enough for properly PLL operation and not take away much power (commonly 10-20 %). Both couplings with the antenna and the HM should be in the same frequency range, which is required to be as wide as possible. The slot antenna is fabricated of superconducting Nb thin film, as well as the transmitting microstrip lines between the FFO, antenna and HM (the thickness is about 200-450 nm). The chip with the planar integrated circuit mounted on the lens is placed in the liquid helium cryogenic system with the temperature 4.2 K.

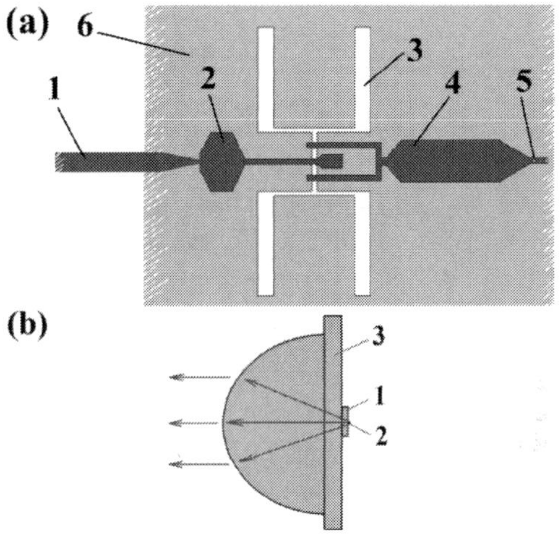

Fig. 1. (a) layout of the planar structure of THz oscillator (1) based on a long SIS junction coupled by microstrip line (2) to slot antenna (3) and by microstrip line (4) to harmonic SIS mixer (5), metallization layer (6) of antenna is also the bottom electrode of the microstrip and the SIS junctions;
(b) scheme of the chip (1) with oscillator and antenna shown on (a) placed at the far focus (2) of the silicon lens (3). The scheme (b) is not to scale.

The main calculations for the designs of the oscillator coupled to the antenna and HM are made by using the specialized software for microwave 3D modeling. Three designs of the "oscillator & antenna structure" were developed for the central frequencies 350 GHz, 450 GHz and 600 GHz, hereinafter called A-350, A-450 and A-600 respectively. More than 70% of oscillator output power is radiated in the ranges of 250 – 410 GHz, 330 – 570 GHz and 420 – 700 GHz for three designs, the calculation results for the emitted power vs frequency will be shown together with the experimental results in the next section. Designs A-450 and A-600 contains all the elements shown in fig.1a, but A-350 doesn't contain the HM with a corresponding coupling line and has no ability to lock the oscillator. A-350 design will be upgraded in the next step of this work. In the fig. 2 the calculated beam patterns at the fixed frequency for 350 GHz, 450 GHz and 600 GHz antenna designs are shown, the main power is concentrated in the center lobe.

Experimental results: preliminary testing

The batch of experimental samples based on Nb/AlN/NbN was fabricated according to the developed designs. The current density of the SIS trilayes on the batch is about j_c = 10kA/cm², this corresponds to parameter R_nS of about 20 Ω·μm². The quality factor of the junctions defined as the ratio of "sub-gap" resistance to normal resistance R_j/R_n is about 30.

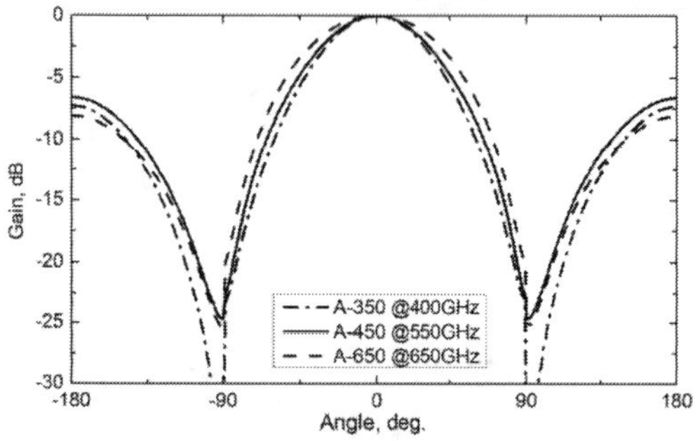

Fig. 2. Numerically simulated beam patterns of three antenna designs at the fixed frequency in the region of operation.

The pumping of the SIS harmonic mixer by FFO power leads to the appearance of current quasiparticle step on the IVC which is measured directly in the whole frequency range of the FFO operation. In the fig.3 the experimental results for frequency dependences of HM pumping are presented together with the calculation results of the absorbed power for the designs A-450 and A-600. There is a good agreement of the experiment with numerical simulations.

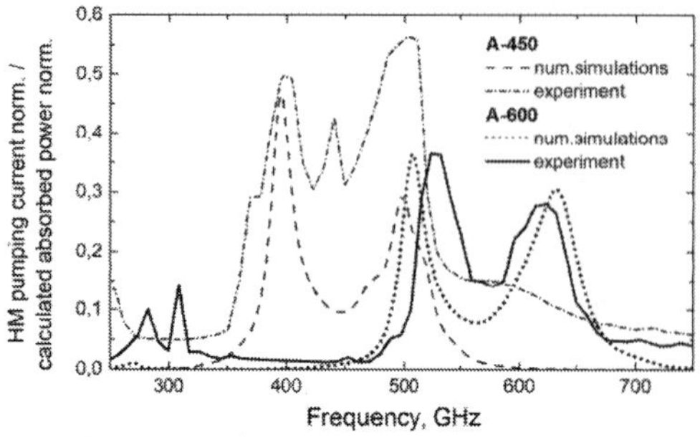

Fig. 3. Experimental results of HM pumping by FFO power and calculation results of the absorbed FFO power by HM. The HM pumping current is normalized to the "current jump" at the gap SIS voltage. The absorbed power is normalized to the total output FFO power.

For the measurements of the FFO emission in the open space by lens antenna the superconducting integrated spectrometer (SISP) is used [5]. Two liquid helium cryostats for the oscillator and the receiver were used and set opposite each other, so that the quasioptical windows of the cryostats were oriented toward one another. The local oscillator of 19-21 GHz is used for the mixing by the HM of the FFO signal and the *n*-th harmonics of the oscillator. Additionally, the PLL system is used for locking the signal. The radiated spectral line could be measured simultaneously by the HM in the intermediate frequency (IF) range 0 - 800 MHz and by the spectrometer in the IF range 4 - 8 GHz. The emission in the open space was successfully obtained in the wide region at some frequency points of operation and studied by the SISP as well as by the HM and the feedback loop. The results of the first measurements together with the transmitting antenna characteristics are shown in the fig.4. The spectral lines emitted by FFO have the Lorenzian shape with a linewidth of about 2-15 MHz; the ratio of signal (spectral line) power to noise power is shown on the right axis in fig.4. Numerical simulations for A-350 (250 - 410 GHz range) are not presented since an experiment still was not carried out at THz frequencies for this design. Such experiment requires another type of receiver (a Golay cell or a Si bolometer).

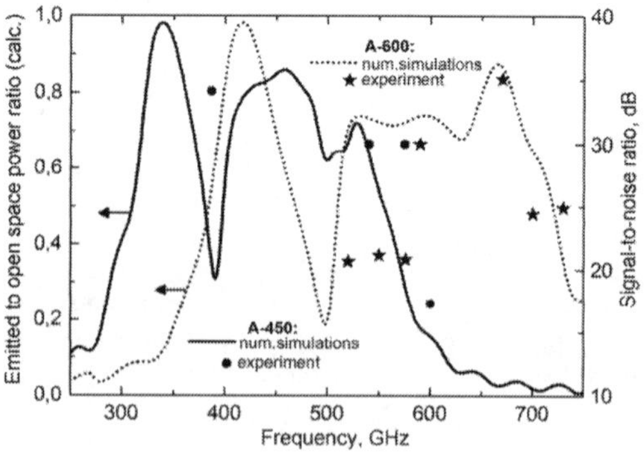

Fig. 4. Calculation results of power emitted by antenna to open space normalized to the total output FFO power (left axis); experimental points of output emission detected by SISP at some frequencies (right axis).

Conclusion

The superconducting THz oscillator based on the long Josephson junction with unidirectional flow of the fluxons (a flux-flow oscillator or a FFO) is an encouraging solution of the THz source for the tasks where wideband frequency tuning is required and the high power is not necessary. We proposed the idea and implementation for the external THz source based on the FFO integrated with the harmonic mixer and the transmitting slot antenna integrated on a single chip with the oscillator. The lens is used forming the narrow beam pattern. Three antenna designs that covers the 0.25 - 0.7 THz region are developed and numerically simulated, two designs for 0.33 - 0.57 THz and 0.42 – 0.7 THz are experimentally studied. The emission to open space is measured by external high resolution spectrometer.

This work is supported by Russian Science Foundation (project № 17-79-20343).

References

1. *Koshelets, V.P., Shitov, S.V.* Integrated Superconducting Receivers // Supercond. Sci. Technol. 2000. V. 13. P. R53-R69.

2. *Lange, G., Boersma, D., Dercksen, J. et al.* Development and Characterization of the Superconducting Integrated Receiver Channel of the TELIS Atmospheric Sounder // Supercond. Sci. Technol. 2010. V. 23. 045016 (8pp).

3. *Nagatsuma, T., Enpuku, K., Irie, F., Yoshida, K.* Flux-flow type Josephson oscillator for millimeter and submillimeter wave region // J. Appl. Phys. 1983. V. 54 P. 3302-3309.

4. *Zhang, Y.* Theoretical and experimental studies of the flux-flow type Josephson oscillator. 1991. Gothenburg: Chalmers University of Technology.

5. *Koshelets, V.P., Dmitriev, P.N. et al.* Superconducting Integrated Terahertz Spectrometers. // IEEE Trans. Terahertz Sci. Technol. 2015. Vol. 5, No. 4. P. 687-694.

Evidence of synchronization of large Josephson-junction arrays by traveling electromagnetic waves

M.A. Galin[1,2], E. A. Borodianskyi[2], V. V. Kurin[1], I. A. Shereshevskiy[1], N. K. Vdovicheva[1], V. M. Krasnov[2], and A.M. Klushin[1]

[1]Institute for Physics of Microstructures RAS, Nizhny Novgorod, Russia, a_klushin@ipmras.ru
[2]Department of Physics, Stockholm University, AlbaNova University Center, Stockholm, Sweden

The Josephson effect can be used for the generation of high-frequency electromagnetic radiation [1]. The frequency is limited only by the superconducting energy gap. For achieving a practically important mW level of emission power the arrays containing $N \sim 10^4$–10^5 JJs would be needed. Such arrays would have a typical size of $L \sim 1$ cm, which is much larger than the wavelength λ even at sub-THz frequencies. The substantial phase delays between JJs are the significant problem impeding the synchronization of large JJ arrays.

It was recently suggested [2, 3] that the synchronization of large JJ arrays can be mediated by a unidirectional traveling wave along the array. The main fingerprint of the traveling-wave regime in an oscillating system is a strong forward-backward asymmetry of the radiation pattern. This is qualitatively different from the resonant, standing-wave case, which per definition has a symmetric radiation pattern. Thus, the shape of the radiation pattern allows a clear distinction of the two scenarios.

In our work we have studied angular dependence of electromagnetic wave emission from two large Nb/NbSi/Nb JJ series arrays having different design. For both arrays we observe a significant forward-backward asymmetry of the measured radiation patterns which is interpreted by the involvement of the traveling-wave mechanism of synchronization. We have also performed numerical simulations of the Josephson traveling-wave antenna which have supported our conclusions.

The studied arrays contain 6972 and 9000 JJs which are located on silicon substrates, on the area of 5×5 mm. The first sample was named as linear array while the second was named as meander array. More details about characterization of the arrays can be found in Ref. [4]. Electromagnetic radiation was detected by a high-purity *n*-doped InSb bolometer which was preliminarily calibrated. Measurements are performed in a closed-cycle cryostat with a sample-in-gas cooling and the base temperature $T = 1.8$ K. Array can be rotated by special gear mechanism while the detector is permanently fixed. The detector is located at approximately 1 cm from the geometrical center of the array. Position of the detector face-to-face on top of the array corresponds to the angle $\alpha = 90°$. The relative accuracy of the rotator is 0.02°.

We performed measurements of current-voltage characteristics (IVCs) at different α simultaneously with measurements of the detected power. It was obtained that critical current of both JJ arrays is $I_c \sim 3$ mA and the steps in the IVCs appear in the range $U \sim 2$–2.5 V.

The angular measurements were performed in the stable steps of IVCs. An analysis of the angular dependencies of the emission power reveals a forward-backward asymmetry of the radiation patterns for both samples. In the Fig.1 the radiation pattern measured for the meander array was shown. It represents decreases of zero-bias resistance $-\Delta R_{det}$ relative to the case of unbiased array. Measurement was made at $U = 2.223$ V corresponding to oscillation frequency $f = 119.3$ GHz. As seen, the radiation pattern demonstrates a pronounced asymmetry in emission between forward ($\alpha < 90°$) and backward ($\alpha > 90°$) directions. Similar, but less prominent, asymmetry of radiation power is demonstrated by the linear array. Such asymmetry is inconsistent with the resonant mechanism of synchronization of the arrays by standing waves because standing waves per definition should have forward-backward symmetric radiation patterns.

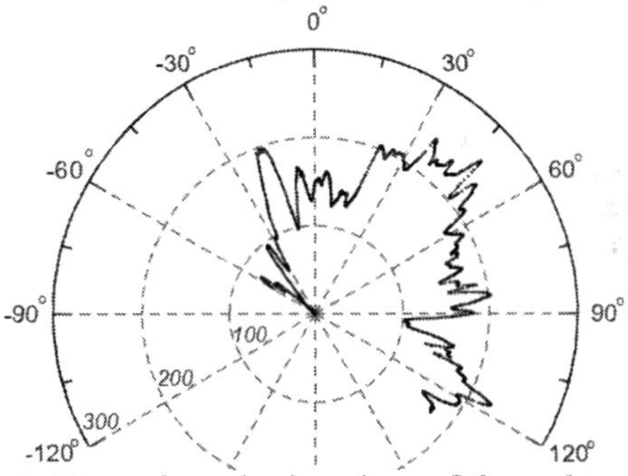

Fig. 1. Measured angular dependence of the ac-detected signal $-\Delta R_{det}$ (Ω) for meander array

The emission power reached the maximum values $P_{em} \sim 80 - 90$ μW for both arrays. For the linear array it occurs at the step in the resistive branch of IVC. However for the meander array it occurs outside the step in the reverse branch of IVC near $U = 1.8$ V (see Fig.2, inset). This suggests that traveling waves play a more prominent role in the emitted electromagnetic field for the meander, than for the linear array.

In Fig.2 the dependence of detected emission power of meander array along the reverse branch of the IVC is demonstrated. Here the dc-voltage response was registered $-\Delta U_{det}$ and the data was obtained at $\alpha = 40°$ when the maximal power was attained (Fig.2, inset). During the sweeping from a large bias current downward to zero the junctions switch in sequence from the resistive into the zero-voltage state. Thus, variation of the emitted power with bias voltage

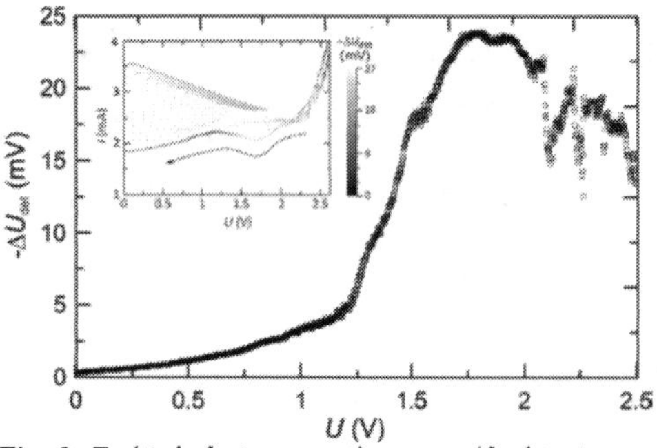

Fig. 2. Emitted electromagnetic power (dc-detector response, $-\Delta U_{det}$) along the reverse branch of the IVCs of meander arrays at $\alpha = 40°$. In the inset the corresponding IVCs obtained by repeated back-and-forth sweeping of bias current is shown. The gray bar indicates the detected power (dc-detector response)

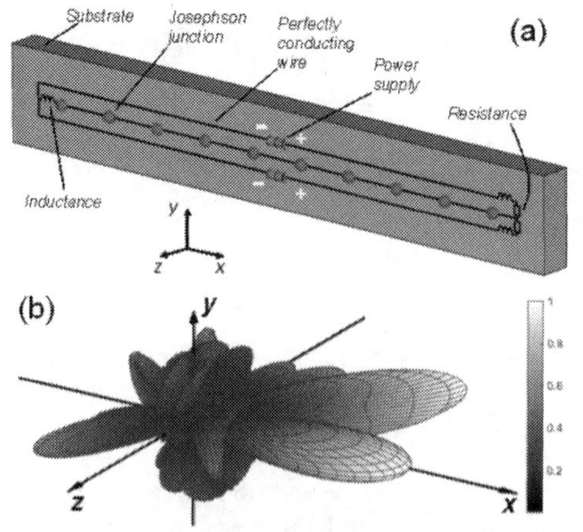

Fig. 3. The model of Josephson traveling-wave antenna (a) and simulated three-dimensional radiation pattern (b)

is primarily due to variation of the number of oscillating junctions N_a, approximately proportional to voltage U. It is seen that a rapid superlinear enhancement of the emission power occurs at 1.3 V. It is known that for the case, when all the oscillating junctions are in the exactly identical state, the emission power should increase as $P_{em} \sim N^2$. Thus the data in Fig.2 indicate the regime of superradiant emission in the meander array. It is suggested that a nonresonant traveling-wave mechanism of synchronization contributes to the observed superradiant emission in the meander array, along with persisting standing-wave resonances that is indicated by the presence of steps in the IVCs.

As suggested earlier [2, 3] large JJ arrays may act as Josephson traveling-wave antennas. Such antennas have asymmetric radiation patterns with a maximum in the direction of propagation of the wave at an angle $\alpha = \arccos h/k$, where h is the wave number of current oscillations in the antenna and k is the wave number of the emitted wave in vacuum. For analysis of nonlinear dynamics of the Josephson traveling-wave antennas, we develop a numerical code for solving Maxwell equations by the FDTD method in combination with self-consistent solution of junction dynamics within the RSJ model.

Our simulated array contains 10 JJs, linear lumped passive elements and two voltage sources (Fig.3a). Elements in the arrays are connected by perfectly conducting wires. The array is placed on dielectric substrate a dielectric permittivity 10. Junction parameters used in simulations are $I_c = 2.5$ mA, normal resistance 0.5 Ω, and the McCamber parameter 0.2. We inserted passive elements (resistances and inductances with the values of 25 Ω and 200 pH, respectively) which are different at the opposite sides of the circuits. These elements are needed for simulating the dynamical violation of the symmetrical state that has to take place in real JJ arrays.

Numerically simulated radiation pattern are shown in Fig. 3b. The chosen radiation frequency $f = 136.6$ GHz provides the wavelength of the current wave considerably smaller than the dimension of the circuits

It is seen that array exhibit a pronounced forward-backward asymmetry of radiation patterns with maximum in the direction of positive x-axis direction. The value of directivity in the maximum power radiation is 4.6 dBi. The radiation pattern has two large lobes with equal amplitudes. One of them is directed strictly along the x-axis that indicates a surface plasmon excited in the array. The relation $h/k > 1$ is realized in this case. For another large lobe $h/k < 1$, i.e. an ordinary traveling wave exists with the finite angular deviation α from the x-axis which is $\alpha = 42°$ in this case.

The aim of these simulations was rather to demonstrate that the main evidence for the traveling-wave regime is the forward-backward asymmetry of the radiation pattern with a maximum in the traveling-wave direction. This is qualitatively consistent with the observed forward-backward asymmetry of emission from the studied large JJ arrays.

The work was supported by the Russian Foundation for Basic Research (Grants No. 16-32-00686 and No. 18-02-00912), the Swedish Foundation for International Cooperation in Research and Higher Education (Grant No. IG2013-5453), and the Swedish Research Council (Grant No. 621-2014-4314). V. V. K. and A. M. K. would also like to acknowledge the partial support of this research by the Russian Science Foundation (Grant No. 15-12-10020).

References

1. *Yanson, I. K., Svistunov, V. M., Dmitrenko, I. M.* Experimental observation of the tunnel effect for cooper pairs with the emission of photons // Zh. Eksp. Teor. Fiz. 1965. V. 48, P. 976 [Sov. Phys. JETP 1965. V. 21, P. 650–652].

2. *Galin, M. A., Klushin, A. M., Kurin, V. V., Seliverstov, S. V., Finkel, M. I., Goltsman, G. N., Müeller, F., Scheller, T., Semenov, A. D.* Towards local oscillators based on arrays of niobium Josephson junctions // Supercond. Sci. Technol. V. 28, P. 055002.

3. *Kurin, V. V., Vdovicheva, N. K., Shereshevsky, I. A.* Josephson traveling-wave antennas // Radiophys. Quantum Electron. 2017. V. 59, P. 922–936.

4. *Galin, M. A., Borodianskyi, E. A., Kurin, V. V., Shereshevskiy, I. A., Vdovicheva, N. K., Krasnov, V. M., Klushin, A. M.*, Synchronization of large Josephson-junction arrays by traveling electromagnetic waves // Phys. Rev. Applied. 2018. V. 9, P. 054032.

Long-Pulsed Modulation Regimes of Subterahertz Nanosecond Waveguide Switches

Maxim L. Kulygin[1], G. G. Denisov[1,2], A. P. Fokin[1], E. A. Novikov[1], S. H. Salahetdinov[1], S. V. Shubin[1], I. A. Litovsky[2]

[1]Institute of Applied Physics RAS, Nizhny Novgorod, Russia, kmaxim@appl.sci-nnov.ru
[2]Lobachevski State University, Nizhny Novgorod, Russia

Introduction

Sub-terahertz resonator cavity-based switches [1, 2] with an active semiconductor element of Gallium Arsenide (GaAs) driven by laser pulses are used to cut microwave signal from a generator to series of wave packets. Intrinsic properties of Gallium Arsenide crystal help to achieve nanosecond level of switching performance with green driving laser emission [3]. The resonant waveguide cavity construction ensures low phase distortion to the packets' at the output [4]. The output phases are linked to each other once the input microwave signal is coherent, e.g. a gyrotron in a phase stabilization regime [4, 5]. The prospective application of the switches is the new generation of DNP-NMR (Dynamic Nuclear Polarization – Nuclear Magnetic Resonance) spectroscopy featuring low power on probing objects and expanded resolution.

Overview of the switch

A photo of the 260 GHz waveguide semiconductor switch used for the second-long pulse modes is shown at Fig. 1. Microwave input and output are symmetrical IEA WR3 standard rectangular waveguides – 0.864 x 0.432 mm – to the left and right flanges shaped according to UG 387 international standard. There is a rectangular input hole with the dimensions of about 0.55 mm in the center for driving laser pulses. A semiconductor plate of gallium arsenide is installed right under the laser input hole. The switch can be slightly fine-tuned by about a 10 GHz with a mechanical tuning screw (not shown on the picture).

Microwave power measurement problem

Testing the switch for the first time demonstrated safe operation with no heating issues with about 30 mW test power source at the input. The maximum possible microwave power the switch can commutate is limited by an IEA WR3 waveguide standard safety value for 260 GHz frequency band; it is about 20 W, about two times lower than the breakdown value. A trivial way to raise the power up to the maximum of 20 W is a challenge because of very limited measurement precision. Modern gyrotrons are usually designed to provide power up to kilo- and megawatts, and commonly used power meters (or wattmeters) are calorimeters rated to the corresponding maximum power levels. They produce errors of worse than 100% for 20 W and lower powers. The lower powers are possible to register with semiconductor detectors.

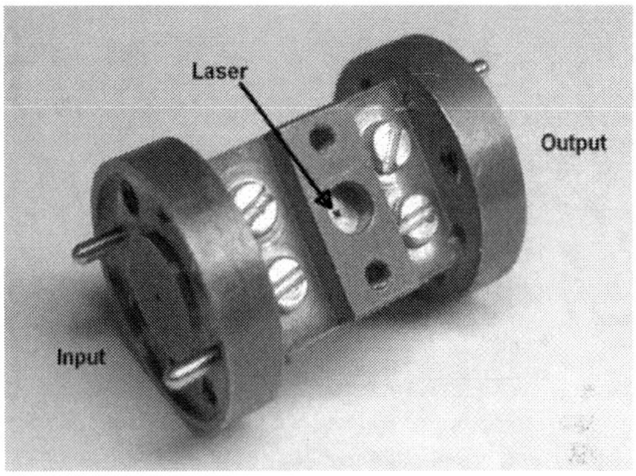

Fig. 1. Photo of the 260 GHz resonant cavity-based switch.

But there are still problems of matching the detectors preventing spurious reflection. Up to now there are no high-precision commercial wattmeters for low powers for frequency band of 260 GHz [6].

Changing the problem to laser measurements

Meanwhile, the most of lasers are known for more than 50 years, and their power is easy to measure with high precision in wide ranges of powers for common wavelengths. So substituting the Ohmic heating mechanism by commutated microwaves with an appropriate laser absorption in the semiconductor, we can recalculate the laser power value to the corresponding microwave one. The semiconductor plate heated by an infrared laser demonstrates the same temperature distribution as it is heated by a commutated microwaves. Both the functions decay smoothly into the depth, and the maximum values for the functions are achieved at the plate's surface. For a 1.06 μm laser it is known that about $R = 1/3$ of the power is reflected from gallium arsenide, and the rest is absorbed in the semiconductor. The microwave power insertion loss coefficient for 260 GHz switches is about -0.5 dB, i.e. about $T = 10\%$. Let us consider the switch is able to endure the maximum laser power of P_L. So, for the maximum endurable microwave power we can write:

$$P_M = \frac{(1-R)}{T} \cdot P_L \sim 7 \cdot P_L \qquad (1)$$

The experiments have shown that the 260 GHz waveguide semiconductor switches designed for nanoseconds are able to make pulses up to a second. Therefore the switches cover at least nine orders by duration, from 1 nanosecond to 1 second and even longer. The switching capabilities at longer durations depend on the pulse length and form, the power and energy of the laser pulse, and the microwave power the switch commutates. There is no case that should be called optimum for switching; it always depends on an application the switch is used within. The maximum commutated microwave power the switch should withstand can be estimated with a substitution of two heating mechanisms in gallium arsenide, the infrared laser one and the microwave Joule one. The switch has successfully passed a 1-second 1-Watt laser pulse, and 1 second is much more than all the relaxation times in the switch. So we can estimate the switch should withstand:

$$P_M = 7 \cdot P_L = 7W \tag{2}$$

The maximum safety transmittable power through an IEA standard WR3 waveguide, 0.864 x 0.432 mm, is about 20 W. While we have seen the semiconductor plate has damaged after the 20 W laser experiment, we have managed to capture the moment of blow and conclude the minimum operable pulse duration is 15 ms under that maximum laser power.

The switch has been repaired after the blow and further experiments have been performed with a 1 kHz pulsed laser. The laser has been shuttered on and off with an optical gate with operating durations from 100 µs to 1 ms to support different heating regimes of the semiconductor plate to make future blowouts virtually impossible.

THE VERY FIRST EXPERIMENT WITH GYROTRON

To confirm these simple theoretical estimations, (1) and (2), we performed a plain test with a gyrotron. The gyrotron [5] provides second-long microwave pulses at the frequency of about 258 GHz with the power rate of about 100 W and Gaussian beam aperture of about 35 mm. The switch has an input rectangular waveguide with dimensions 0.864 x 0.432 mm corresponding to IEA WR3 standard. So to obtain the input power of several Watts we have attached a 7 x 4 mm horn to the input of the switch. Therefore the estimated power in the real experiment is about 3.5 W. The maximum length of the gyrotron's pulse is 8 s; it was limited by the power transmission line and cooling. No laser has been used so the switch is always in the ground state.

Fig.2 displays the main characteristic of the switch in the ground state when the switch is closed and the microwave power does not pass through, the insertion loss before and after the gyrotron's microwave pulse. We can see that after the pulse the main resonance became deeper, about -5.5 dB compared to about -4.7 dB before the test. The resonance always vanishes if the semiconductor plate is burned out. But now we can see quite a contrary case. It can be explained the way the semiconductor plate has not been burnt out under the 3.5 W 8 s gyrotron pulse, it even got better since only the remained dirt (occasional fingertips, organics and dust from the air) has been burnt out.

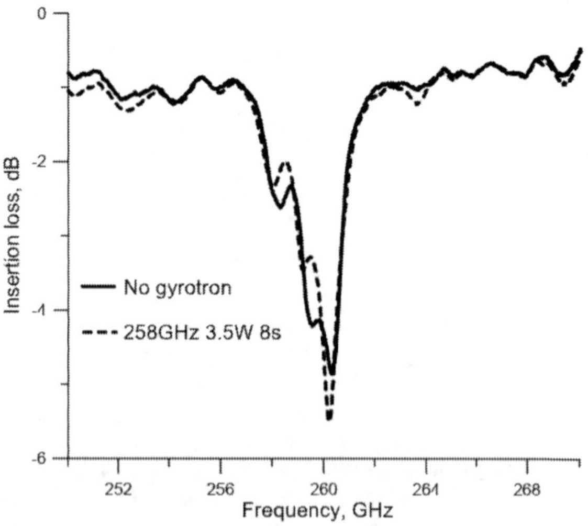

Fig. 2. Insertion loss of the switch in the ground state, solid line is before the gyrotron pulse, dashed line is after the pulse.

The switch has successfully passed the test and we can conclude that the gyrotron's radiation can be useful for final cleaning of the switch.

CONCLUSION

The second-long pulses have been produced with the switches intended for nanosecond-long pulses. So the switches have demonstrated the pulse range of about 10 orders, from 800 ps to 8 s at 260 GHz. Using the switches at the second-long pulse regimes is more important than just a scientific curiosity. It provides the ability to overcome the shortage of low-precision power measurement in sub-terahertz region by changing the problem to well-known low-cost commercial laser measurements that are easily available in every laser-equipped laboratory. We have predicted theoretically our switch is tough to withstand no less than 7 W of microwave commutated power. And we have proven it partially in the experiment with 3.5 W gyrotron. The switch has not been damaged; it's even got better quality. So the gyrotron test should be included in the switch production sequence further to improve quality.

The project is supported by the Russian Foundation for Basic Research, Grant No. 15-08-03158.

References

1. *Kulygin M., Denisov G.* Nanosecond Laser-Driven Semiconductor Switch for 70 GHz Microwave Radiation // J. Infrared mm THz w., V. 33, No. 6, p. 638-648.

2. *Kulygin M. L., et al.* Development of Waveguide Semicond. Switch. of Microw. Rad. in the 70- and 260-GHz Ranges // Radiophys. Quant. El., V. 57, No. 7, p. 509-518.

3. *Kulygin M., Denisov G., Vlasova K., et al.* Nanosecond Microwave Semiconductor Switches for 258…266 GHz // J. Infrared mm THz w., *V.* 36, No. 9, p. 845–855.

4. *Kulygin M. L., Shubin S. V., et al.* 260 GHz Laser-Driven Semiconductor Switches with Performance Rate up to Nanosecond. *Proc. IEEE COMCAS 2015 Int. Conf.*

5. *Glyavin M., et al,* Gyrotron Freq. Stab. by a Weak Ref. W. // *Radiophys. Quant. El.*, V. 59, No. 9, p. 673-683.

6. *Kulygin M., et al.* Subterahertz Nanosecond Switches Driven by Second-Long Laser Pulses // *IEEE Trans. THz Sci. Tech.*, V. 7, No. 2, p. 225-227.

Active Josephson traveling wave antennae as prospective terahertz oscillators

V. V. Kurin, N. K. Vdovicheva, I. A. Shereshevskii

Institute for Physics of Microstructures RAS, Nizhny Novgorod, Russia, kurin@ipmras.ru

A terahertz superconducting oscillators based on large amount of Josephson junction embedded in open system guiding traveling electromagnetic wave is theoretically considered and computer simulated. It is shown that such active Josephson antennae represent oscillator effectively radiated into open space with power scaled with the system size and number of junctions. Dynamics and directivity patterns of such Josephson antennae depending on bias current is investigated and it is shown that such oscillator with sufficiently large junction amount can be competitive with quantum cascade lasers.

It is well known that the Josephson Effect can be used for the generation of high-frequency electromagnetic radiation. The frequency is limited only by the superconducting energy gap. For low-Tc superconductors the frequency is in the sub-THz range and for high-Tc superconductors it can be up to approximately 20 THz. But the emission power from a single Josephson junction is quite small. It can be considerably enhanced provided coherent contribution of many, $N \gg 1$, Josephson junctions to radiated electromagnetic field.. For achieving a practically important mW level of emission power very large arrays containing $N \sim 10^4 - 10^5$ Josephson junctions would be needed. Taking into account the attainable integration density, such arrays would have a typical size of $L \sim 1$ cm, which is significantly larger than the wavelength λ even at sub-THz frequencies.

The larger the array is the more difficult is the synchronization. This is caused by large phase delays along the array. Recently, it was suggested [1] that the synchronization of extremely large arrays can be mediated by a unidirectional traveling wave along the array, qualitatively similar to the operation of a traveling wave Beverage-type antenna [2].

In this work, we consider such large-size Josephson systems and present their theoretical description and computer simulation. We demonstrate that in some classes of such systems, traveling-wave regimes are possible, under which all Josephson junctions are under identical electrodynamic conditions and make coherent contributions to the radiation field. Such systems, which will be called active Josephson traveling wave antennas, are scalable, i.e., in the case of optimal matching, the intensity of their radiation increases in proportion to the size of the system or the number of the Josephson junctions. An important component, which ensures coherence of the contributions made by a great number of junctions, is the lateral energy leakage from the transmission line (vertical radiation output), which prevents saturation of the nonlinearity of individual junctions with narrow dynamic ranges. Such systems can be used as a basis for Josephson oscillators producing power levels sufficient for practical applications. A simple qualitative theory of open Josephson traveling-wave lines was developed and conditions for existence and stability of travelling wave regimes were found [1].

If there exists the traveling wave of current along Josephson array with wave number h it will radiate electromagnetic wave to open space at an angle $\alpha = \sin^{-1} h/k$, where k is the wave number of the emitted wave in vacuum and α counted from normal to substrate. For $h/k < 1$ when current wave is faster than speed of light in vacuum it is an ordinary transverse electromagnetic wave, for homogeneous current with $h = 0$ the radiation will be directed strictly vertical. But for $h/k < 1$ when current wave is slow the angle α is imaginary and the traveling electromagnetic wave turns to a surface plasmon traveling along the interface of the wafer and radiated from the edge of structure.

It should be noted that due to not perfect matching conditions at edges of array the traveling wave would not be pure traveling and some mixture of opposite propagating wave will always be presented. But Standing Wave Ratio (SWR) should be final to provide the absence of AC field nodes so that all junctions will be pumped by common field. Since the AC current distribution and radiated field defined by cooperative dynamics of Josephson junctions should be found self consistently we undertook direct computer simulations of such active Josephson antennae.

A numerical simulation code based on the Finite Difference Time Domain (FDTD) method and a self-consistent solution of the nonlinear equations which describe the Josephson junctions has been developed. The results of simulation of some variants of Josephson traveling-wave antennas with different radiation patterns are presented.

An example chosen for the calculations is a Josephson system similar to the two folded wire traveling-wave antenna. The simulated Josephson system is a set of thin perfect conductors, lumped linear elements, such as capacitors, inductances, resistors, and sources of DC electromotive forces, and nonlinear active elements, Josephson junctions simulated by resistively and capacitive shunted model. These elements are located on the surface of a dielectric plate and interact via the electromagnetic field in the surrounding space. Calculations start from an initial distribution for Josephson phases and electric and magnetic fields on the computational grid and continue until a stationary state is reached. After that time average values of voltages and currents in Josephson junctions and Fourier amplitudes of the fields around the sample are

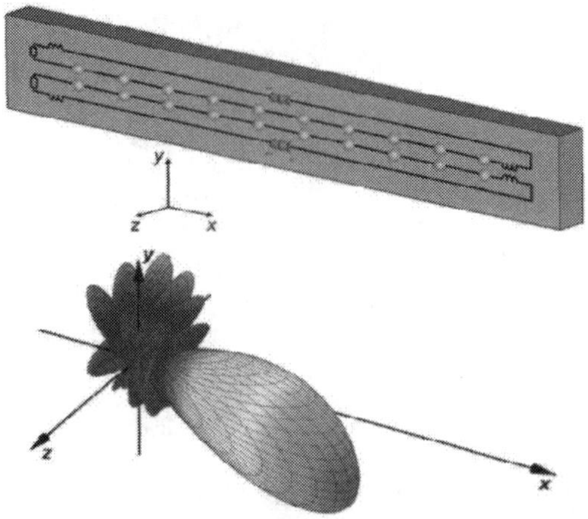

Fig. 1. Appearance and directivity pattern of two folded Josephson active antenna with separate DC bias demonstrating oblique angle radiation due to junction synchronization with fast traveling current wave. Yellow circles denote Josephson junctions, bias battery, resistor and inductances, serving for matching are denoted as usual.

obtained. The radiation pattern at a fixed frequency is calculated using a standard near-to-far field transformation [3]. Full power of emission was calculated by integration of the far field Pointing vector. Examples of such systems representing an antenna containing a set of straight and broken wires with different number of Josephson junctions imbedded into together with radiation patterns generated by them are shown on Fig. 1-3. The pictures shown demonstrate various regimes of radiation manifested itself by different radiation patterns which strongly depend on geometry of antennae and DC bias. When bias increases the oscillation frequency defined by Josephson relation $\hbar\omega = 2eV$ also increases, mode spectrum enriches and directivity patterns became more complicated. As it is common for oversized radiating system to control generation regimes and directivity pattern some complication of the system to provide working mode selection is needed. This work is now underway.

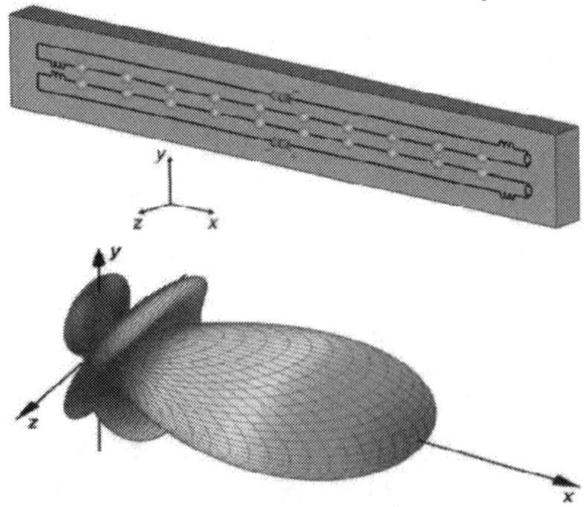

Fig. 2. The same Josephson active antenna as in Fig. 1 but radiated along the wafer due to synchronization with slow current wave.

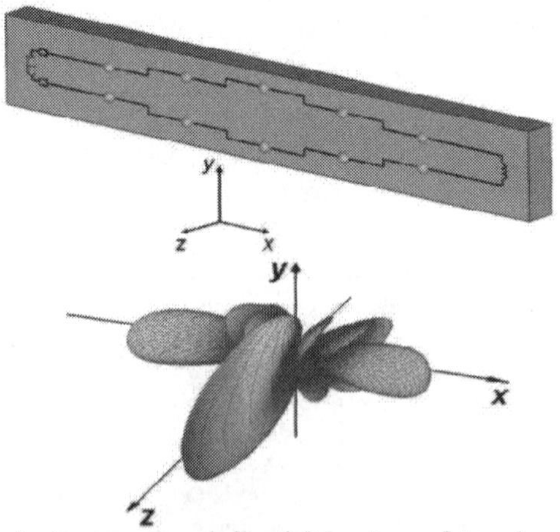

Fig. 3. Appearance and directivity pattern of Josephson active antenna consisting of two broken lines with common DC bias. It I seen pronounced lobe corresponding to vertical output of radiation due to junction synchronization with homogeneous mode of substrate.

Dependence of power of radiation and behavior of spectral properties of radiation on junction number, current and voltage spectra on individual junctions support suggested concept of array synchronization by traveling wave.

Recently [4] Josephson emission from such larger-than-the-wavelength Nb-NbSi-Nb junction arrays consisting of order 10^4 junctions was investigated experimentally and angular dependence of radiated energy was measured by small InSb semiconductor detector. The asymmetry of radiation pattern that was found support the suggested our concept of synchronization mediated by traveling wave. We argue that such a mechanism of synchronization opens a possibility for phase locking of very large arrays of oscillators what opens a way to develop effective Josephson oscillator of Terahertz range.

The work was supported by the Russian Foundation for Basic Research (Grants No. 18-02-00912), V. V. K. would also like to acknowledge the partial support of this research by the Russian Science Foundation (Grant No. 15-12-10020).

References

1. *V. V. Kurin, N.K.Vdovicheva, and I.A. Shereshevskii*, Josephson Traveling-Wave Antennas, Radiophysics and Quantum Electronics, **59**, 922, (2017)

2. *C. A. Balanis*, in Antenna Theory: Analysis and Design, 3rd ed. (Wiley Interscience, New York, 2005), Chap. 10.

3. A. *Taflove and S. C. Hagness*, in Computational Electrodynamics: The Finite-Difference Time-Domain Method, 3rd ed. (Artech House Inc., Boston and London, 2005), Chap. 8.

4. *M. A. Galin, E. A. Borodianskyi, V. V. Kurin, I. A. Shereshevskiy, N. K. Vdovicheva, V. M. Krasnov, A. M. Klushin,* Synchronization of Large Josephson-Junction Arrays by Traveling Electromagnetic Waves, Phys. Rev. Applied **9**, 054032 (2018)

Generation of THz radiation by photoconductive antennas on based thin films InGaAs and InGaAs/InAlAs.

A.A. Leontyev[1], K.A. Kuznetsov[1], G.B. Galiev[2], G.Kh. Kitaeva[1], V.V. Kornienko[1], E.A. Klimov[2], A.N. Klochkov[2], S.S. Pushkarev[2], and P.P. Maltsev[2]

[1] Lomonosov Moscow State University, Moscow 119991, Russia, aa.leontjev@physics.msu.ru
[2] Institute of Ultrahigh Frequency Semiconductor Electronics, RAS, Moscow 117105, Russia

Epitaxial low-temperature grown (LT) semiconductor arsenides (Al, Ga, In)As are widely used as materials for photoconductive antennas (PCA) generators and detectors of pulsed radiation in the terahertz (THz) frequency range [1–3]. It is the combination of subpicosecond carrier lifetime, relatively high mobility and high resistivity that makes LT-materials suitable for PCA applications. Lately, InGaAs has been investigated as a potential candidate for THz-PCA photoconductive material due to room-temperature band gap of 0.74 eV, which allows for 1.56 μm optical excitation with Er^{3+} fiber laser femtosecond pulses [4–6].

The low substrate temperatures result in a non-stoichiometric growth with the incorporation of excess arsenic in the crystal structure. The most common non-stoichiometry-related point defects in LT-arsenides are arsenic antisites with concentrations in the range 10^{17}–10^{19} cm^{-3} depending on the substrate temperature and arsenic overpressure [7–10]. Antisite-related defect band in the semiconductor energy bandgap play a significant role in carrier dynamics. Fast non-radiative recombination of photogenerated electrons and holes through antisite centers results in sub-picosecond carrier lifetimes in LT-materials at optimized growth and annealing conditions [11, 12]. It is generally agreed that main traps of photo-excited electrons are ionized antisite defects [13–15].

A possible approach to increase the resistivity of LT-InGaAs structures is to employ LT-InGaAs/InAlAs superlattices [6,13,16,17]. LT-InAlAs layers have a higher dark resistivity as compared to LT-InGaAs and exhibit deep trap states that are situated energetically below the antisite defect levels of adjacent InGaAs layers that results in a reduction of residual carrier concentration.

The Fig.1 shows the amplitude of THz radiation in time domain. It is seen that the signal from an InGaAs /InAlAs-based structure is 5-6 times higher due to a higher bias voltage, which is possible (without sample breakdown) due to higher sample resistance and lower dark current.

Fig.2 shows a comparison of the Fourier amplitude according to the materials of the antennas LT-InGaAs/InAlAs and LT-InGaAs. It is seen that the spectrum of the LT-InGaAs / InAlAs sample is slightly wider in the range from 0.1 THz to 0.6 THz than that of the LT-InGaAs sample. We explain this effect by the difference between the characteristic relaxation times of electrons in the transition from the conduction band to the antisites.

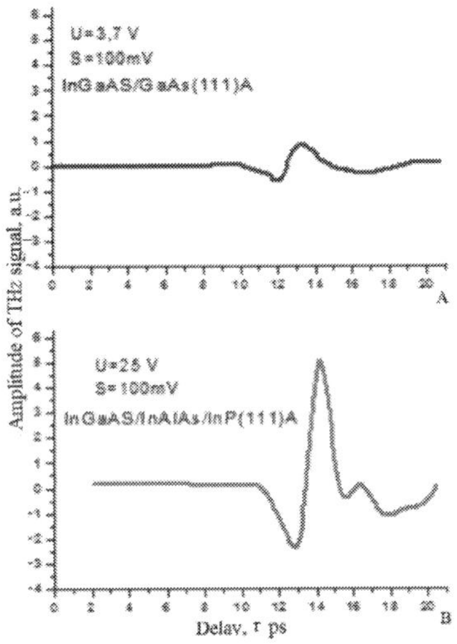

Fig.1. Time-domain waveforms detected for the following antenna materials: A) LT-InGaAs B) LT-InGaAs/InAlAs.

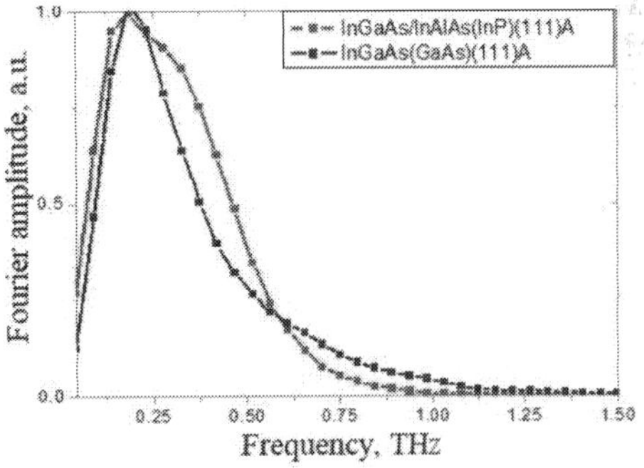

Fig. 2. Compare spectral amplitudes by antenna materials LT-InGaAs(Blue line), LT-InGaAs/InAlAs (Red line).

We determined the characteristic times of electron relaxation by the "pump-probe" spectroscopy method. Fig.3 shows the dependence of the normalized transmission in time domain for the samples of LT-InGaAs and LT-InGaAs / InAlAs.We used 2-exponentional model for description experimental curves. On figures τ_1 is an electron capture time (capture by charge As_{Ga} defects) [18,19], τ_2 is a recombination time of captured electrons and holes [17].

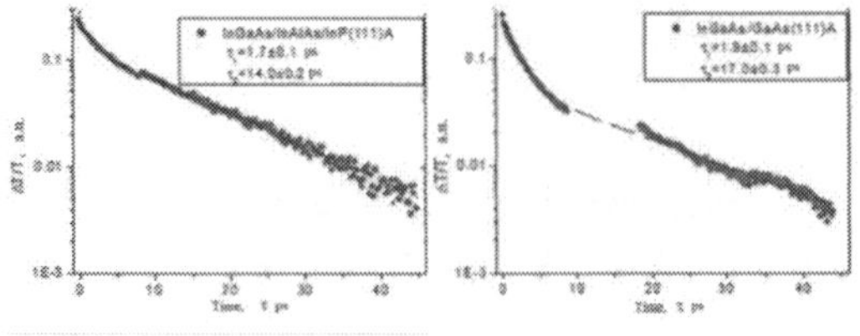

Fig. 3 Change the transmission coefficient in the time domain.

Due to the experimental fact that the characteristic relaxation times for the LT-InGaAs / InAlAs sample are less than for the LT-InGaAs, we observed the difference in the spectra for these samples.

Summing up, it was found that THz generation is about 5-6 times more efficient in the case of LT-InGaAs/InAlAs superlattice than LT-InGaAs generation. It is found that due to the shorter electron relaxation time in the superlattice, the spectrum of these samples is wider in the range of 0.1-0.6 THz.

This study was performed under the support of the RFBR grant № 16-02-258a.

References

1. *G.Kh. Kitaev.a* Terahertz generation by means of optical lasers // Laser Phys. Lett. V. 559, P. 20085.
2. *N.M. Burford and M.O. El-Shenawee.* Review of terahertz photoconductive antenna technology // Opt. Eng. 2017, V. 56, P. 10901.
3. *E. Castro-Camus and M. Alfaro.* Photoconductive devices for terahertz pulsed spectroscopy: a review // Photon. Res. 2016, V. 4, P. A35.
4. *A. Takazato M. Kamakura, T. Matsui,J. Kitagawa and Y. Kadoya.* Terahertz wave emission and detection using photoconductive antennas made on low-temperature-grown InGaAs with 1.56 μm pulse excitation // Appl. Phys.Lett. 2007, V. 91, P. 011102.
5. *I. Kostakis and M. Missous.* Optimization and temperature dependence characteristics of low temperature $In_{0.3}Ga_{0.7}As$ and $In_{0.53}Ga_{0.47}AsIn_{0.52}Al_{0.48}As$ semiconductor terahertz photoconductors // AIP Adv. 2013. V. 3, P. 092131.
6. *B. Sartorius, H. Roehle, H. Künzel, J. Böttcher, M. Schlak, D. Stanze, H. Venghaus and M. Schell.* All-fiber terahertz time-domain spectrometer operating at 1.5 μm telecom wavelengths Opt. Express 2008, 16 9565
7. *X. Liu, A. Prasad, J. Nishio, E.R. Weber, Z. Liliental-Weber and W. Walukiewicz.* Native point defects in low-temperature-grown GaAs // Appl. Phys. Lett. 1995. V. 67, P. 279.
8. *H. Künzel, J. Böttcher, R. Gibis and G. Urmann* Material properties of $Ga_{0.47}In_{0.53}As$ grown on InP by low-temperature molecular beam epitaxy // Appl. Phys. Lett. 1992. V. 61, P. 1347.
9. *R.A. Metzger, A.S. Brown, L.G. McCray and H.A. Henige.* Structural and electrical properties of low temperature GaInAs // J. Vac. Sci. Tech. B. 1993. V. 11, P. 798.
10. *M.R. Melloch, J.M. Woodall, E.S. Harmon, N. Otsuka, F.H. Pollak, D.D. Nolte, R.M. Feenstra and M.A. Lutz.* Low-temperature grown III-V materials // Annu. Rev. Mater. Sci. 1995. V. 25, P. 547.
11. *C. Baker, I.S. Gregory, W.R. Tribe, I.V. Bradley, M.J. Evans, E.H. Linfield and M. Missous.* Highly resistive annealed low-temperature-grown InGaAs with sub-500fs carrier lifetimes // Appl. Phys. Lett. 2004. V. 85, P. 4965.
12. *K.A. McIntosh, K.B. Nichols, S. Verghese and E.R. Brown.* Optical and terahertz power limits in the low-temperature-grown GaAs photomixers // Appl. Phys. Lett. 1997, V. 70, P. 354.
13. *B. Globisch, R.J.B. Dietz, D. Stanze, T. Göbel and M. Schell.* Carrier dynamics in beryllium doped low-temper-ature-grown InGaAs/InAlAs // Appl. Phys. Lett. 2014. V. 104, P. 172103.
14. *Z. Liliental-Weber, H.J. Cheng, S. Gupta, J. Whitaker, K. Nichols and F.W. Smith.* Structure and carrier lifetime in LT-GaAs // J. Electron. Mater. 1993. V. 22, P. 1465.
15. *U. Siegner, R. Fluck, G. Zhang and U. Keller.* Ultrafast high-intensity nonlinear absorption dynamics in low-temperature grown gallium arsenide // Appl. Phys. Lett. 1996. V. 69, P. 2566.
16. *H. Kuenzel, K. Biermann, D. Nickel and T. Elsaesser.* Low-temperature MBE growth and characteristics of InP-basedAlInAs/GaInAs MQW structures // J. Cryst. Growth. 2001. V. 284, P. 227
17. *R.J.B. Dietz, B. Globisch, H. Roehle, D. Stanze, T. Göbel and M. Schell.* Influence and adjustment of carrier lifetimesin InGaAs/InAlAs photoconductive pulsed terahertz detectors: 6 THz bandwidth and 90dB dynamic range // Opt.Express. 2014. V. 22, P. 19411.
18. *C. Baker, I.S. Gregory, W.R. Tribe, I.V. Bradley, M.J. Evans, E.H. Linfield and M. Missious.* Highly resistive annealed low-temperature-grown InGaAs with sub-500 fs carrier lifetimes // Appl. Phys. Lett. 2004, V. 85, P. 4965.
19. *B. Globisch, R. Dietz, R. Kohlhaas, T. Göbel, M. Schell, D. Alcer, M. Semtsiv, and W. Masselink.* Iron doped InGaAs: Competitive THz emitters and detectors fabricated from same photoconductor. // J. Appl. Phys. 2017. V. 121, P. 053102.

High Photoconductivity in Heavily Doped GaAs/AlAs Superlattices with Electric Domains

S. K. Paprotskiy[1], I. V. Altukhov[1], M. S. Kagan[1], N. A. Khvalkovskiy[1], I. S. Vasil'evskii[2], A. N. Vinichenko[2]

[1]V.A. Kotel'nikov Institute of Radio Engineering and Electronics, Russian Ac. Sci., Moscow, Russia, s.paprotskiy@gmail.com
[2]National Research Nuclear University MEPhI, Moscow, Russia

In last three decades, the electron transport in semiconductor superlattices (SLs) has been studied in detail [1, 2], mainly in connection with the predicted amplification of Bloch waves, which can make it possible to create tunable sources of THz radiation. The main hindrance to the Bloch wave amplification is the formation of electrical domains owing to the static negative differential conductivity (NDC) arising in superlattices at resonant tunneling between confined states in neighboring quantum wells. Many works were devoted also to studies of domain properties (see, e.g., [2] and references therein). Most of them were performed with weakly doped SLs at low temperatures under strong illumination [2-7]. The change in free carrier concentration under illumination gives rise to different domain regimes. It has been shown, in particular, that transition from travelling to static domain regime occurs at sufficiently high illumination intensity. In this work studies of effect of interband illumination on tunneling current in highly doped short-period superlattices at the domain formation are presented.

We used the MBE grown GaAs/AlAs SLs consisted of 100 periods of 4 nm GaAs/2 nm AlAs between heavily doped (~10^{19}cm^{-3}) n^+ cap layer and n^+ substrate. GaAs quantum wells (QWs) were Si doped with concentration of $2*10^{17}$cm^{-3}. A cross-section of the structure is shown in Fig. 1.

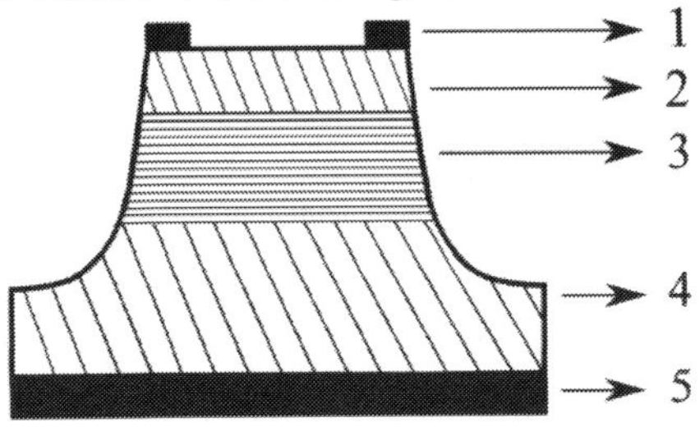

Fig. 1. Scheme of the structure: (*1*, *5*) top and bottom metal contacts, (*2*) top contact n^+ layer, (*3*) superlattice, and (*4*) heavily doped substrate.

Triangular voltage pulses with sweep-up times of 0.2 to 10 μs and sweep-down times up to 100 μs were used to record I-V curves. Measurements were performed at room temperature.

A sharp decrease of current at some threshold voltage (Fig. 2) indicates the moving domain formation. The current rise in the mean is an evidence of triangular form of the domain [2, 9]. The series of maxima in current-voltage characteristics is attributed to the optical phonon assisted tunneling between quantum wells inside the triangular domain.

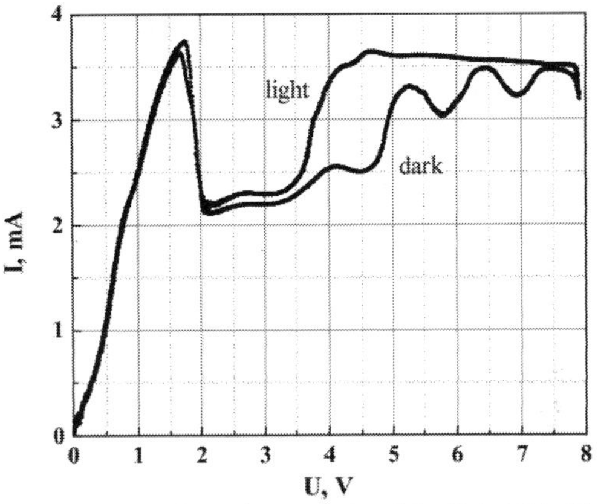

Fig. 2. Current–voltage characteristics of the GaAs/AlAs SL in the dark (lower curve) and under band-to-band illumination (upper curve).

A strong effect of rather weak illumination (with a LED or a filament bulb) on a tunneling current in GaAs/AlAs SLs with relatively high doping level was found under conditions of electric domains formation. Fig. 2 shows the current-voltage characteristics with and without illumination. The photocurrent at the domain regime could reach 50% of the dark current, while it was several orders of magnitude less at voltages below the threshold one (before the domain formation) as the free carrier concentration excited with light was considerably less than Si donor concentration. The measured photocurrent spectrum (Fig. 3) points to the band-to-band excitation. The spectral long-wavelength cutoff allows us to determine the GaAs QW gap and the energy of first size quantization level. These energies coincide with known values for GaAs QW.

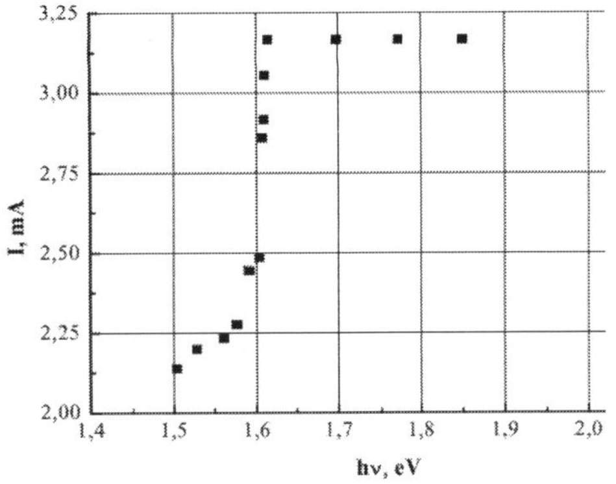

Fig. 3. Photocurrent spectrum

The large photocurrent at the domain formation is referred to the triangular form of high-field domain with wide region of donor depopulation. The free carrier concentration in this region is very small and additional concentration of electron-hole pairs created by interband illumination exceeds significantly a stationary concentration inside the domain, which results in a change of space charge at domain boundaries. This leads to a change of domain movement regime.

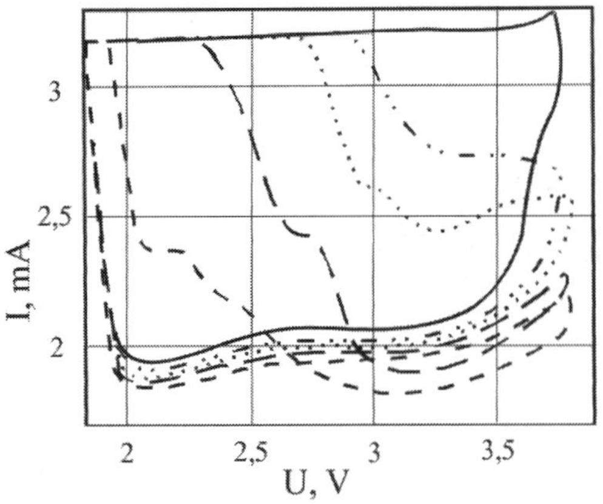

Fig. 4. Parts of current–voltage characteristics of the GaAs/AlAs SL in an intermediate range of domain existence at different light intensities.

The sufficiently intense illumination could initiate the travelling-to-static domain transition, which becomes apparent in current saturation regions in I-V curves (Fig. 2). Shown in Fig. 4 are parts of current-voltage characteristics corresponding to an intermediate voltage range of domain existence for different illumination intensities. I-V curves were measured at forward and backward bias sweep. The current hysteresis is connected with a large asymmetry of transition times from moving to static field domain and back at voltage sweep-up and sweep-down, respectively [10]. In this voltage range (compare with Fig. 2), the current saturation was observed only at backward voltage sweep.

The work was supported by RFBR grants 16-29-03135 and 16-29-09626

References

1. *Wacker, A.* Semiconductor superlattices: a model system for nonlinear transport // Physics Reports, 2002. V. 357, P. 1

2. *Bonilla, L. L., Grahn H. T.* Non-linear dynamics of semiconductor superlattices // Rep. Prog. Phys. 2005. V. 68, No. 3, P. 577

3. *Sun B. Q., Jiang D. S. and Wang X. J.* Current oscillations and stable electric field domains in doped GaAs/AlAs superlattices // *Semicond. Sci. Technol.* 1997, V. 12, No 4, P. 401

4. *Tomlinson A. M., Fox A. M., Cunningham J. E. and Jan W. Y.* Photocurrent self-oscillations in a spatially direct GaAs/AlGaAs superlattice // Appl. Phys. Lett. 1999, V. 75, P. 2067

5. *Kwok S. H., Grahn H. T. et al.* Nonresonant carrier transport through high-field domains in semiconductor superlattices // Phys. Rev. B. 1995. V. 51. P. 9943

6. *Kwok S. H., Norris T. B., Bonilla, L. L.,et al.* Domain-wall kinetics and tunneling-induced instabilities in superlattices // Phys. Rev. B 1995. V. 51. P. 10171

7. *Grahn H. T., Schneider H., v. Klitzing K.* Optical studies of electric field domains in GaAs/$Al_xGa_{1-x}As$ superlattices // Phys. Rev. B 1990. V. 41, P. 2890

8. *Ohtani N., Egami N., Grahn H. T., Ploog K. H., and Bonilla L. L.* Transition between static and dynamic electric-field domain formation in weakly coupled GaAs/AlAs superlattices // Phys. Rev. B 1998. V. 58, P. R7528(R)

9. *Suris R. A.* Inhomogeneous structures in semiconductors with superlattices // Sov. Phys. Semicond. 1973. V. 7, No. 8, P. 1035

10. *Altukhov I. V., Kagan M. S., Paprotskiy S. K., Khvalkovskiy N. A., Vasil'evskii I. S., Vinichenko A. N.* Electronic Tunneling and Electric Domains in GaAs/AlAs Superlattices at Room Temperature, this issue.

 https://doi.org/10.1051/epjconf/201819502009

Plasmonic terahertz antennas with high-aspect ratio metal gratings

A.E. Yachmenev[1,2], D.V. Lavrukhin[1,2], I.A. Glinskiy[1,2], R.A. Khabibullin[1,2,3], R.R. Galiev[1], A.Yu. Pavlov[1,2], Yu.G. Goncharov[2], I.E. Spektor[2], M. Ryzhii[5], T. Otsuji[6], K.I. Zaytsev[2,4] and D.S. Ponomarev[1,2,3]

[1]Institute of Ultra High Frequency Semiconductor Electronics of RAS, Moscow, Russia, ponomarev_dmitr@mail.ru
[2]Prokhorov General Physics Institute of RAS, Moscow, Russia
[3]Center for Photonics and 2D Materials, Moscow Institute of Physics and Technology, Dolgoprudny, Russia
[4]Bauman Moscow State Technical University, Moscow, Russia
[5]Department of Computer Science and Engineering, University of Aizu, Aizu-Wakamatsu, Japan
[6]Research Institute of Electrical Communication, Tohoku University, Japan

The state-of-art terahertz (THz) emitters and detectors are of great interest due to its wide applications in security systems, biology, medicine and material science [1–3]. Despite the variability of existing THz generation principles photoconductive antenna (PCA) is still promising THz emitter for broadband THz spectroscopy and imaging. Since its first demonstration by Auston's research on silicon substrate [4], numerous studies have been carried out. Despite the considerable progress in developing PCAs there are still challenges such as increasing an emitted THz power, an optical-to-THz conversion efficiency, a bandwidth of a PCA and its thermal stability.

In present article, we present a novel technology with a passivation approach for fabricating a dielectric-filled plasmonic PCA that allows us for reaching a 100 nm-height plasmonic gratings and strongly reduce leakage currents. In the proposed technology two different dielectrics are separately used: a Si_3N_4 for preliminary passivating a surface of a photoconductor to reduce a dark current flow and Al_2O_3 dielectric for filling and covering plasmonic gratings to increase an incident light coupling. This approach allows for sustaining a thermal stability of a PCA and is suitable for any photoconductive material, due to excluding a mesa etching, which is of particular importance for InGaAs-based PCAs.

It is well known that all types of PCAs, especially the log-periodic and log-spiral antennas, suffer from high leakage currents, because a dark current flows between the full length of the antenna bias lines which is hundred times longer than the optical excitation spot. In order to significantly decrease a dark current flow we have developed a novel technology for a PCA fabrication where an antenna metallization, including contact pads, is deposited on a photoconductor's surface preliminary covered (i.e. passivated) with a dielectric through etched windows. The schematic layout of the proposed PCA is depicted in Fig. 1.

Prior to fabrication process we carried out numerical calculations by means of a 2D finite element method-based solver for analyzing the interaction between a TM-polarized laser pump beam and plasmonic gratings. The calculation domain consisted of 4 periods of plasmonic gratings (4×200 nm) with width and distance between them equal to 100 nm, deposited on 3-μm GaAs substrate. First, we calculated an electric field near plasmonic gratings distribution versus a height (h) of metal gratings which showed that an increase of metal height leads to a continuous enhancement of an electric field and its maximum is shifted toward high values of h, which is difficult to being experimentally achieved according to chosen high aspect-ratio of gratings, which is defined as the ratio between a width of a grating to its period.

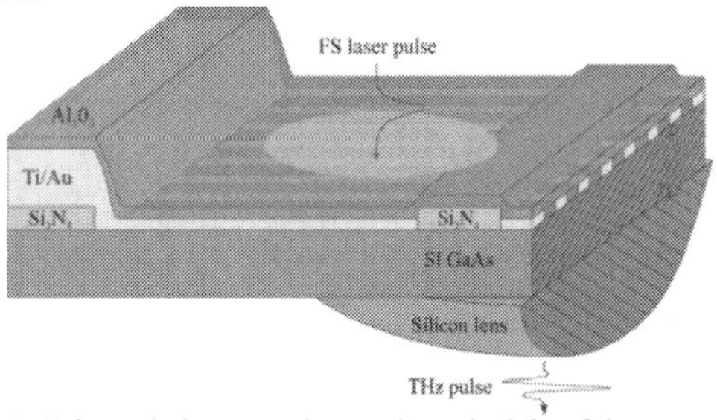

Fig. 1. Schematic layout and operation principle of the proposed dielectric-filled plasmonic THz emitter with high-aspect ratio metal gratings. A lower dielectric Si_3N_4 layer is used for reducing leakage currents (to sustain a thermal stability enhancement) while an upper Al_2O_3 dielectric serves for filling a distance between gratings and covering them to increase an incident light confinement. Also it acts as an anti-reflection coating layer

Thus we considered the case of h = 100 nm with aspect-ratio equal to 0.5, that to our knowledge has not been reached experimentally yet. Second, we estimated an optimum thickness of Al_2O_3 dielectric layer which entirely fills a distance between gratings to enhance a light coupling and simultaneously mitigate the Fresnel reflection losses for the case of h = 100 nm. We demonstrated that the dependency of an incident laser pump reflection coefficient on Al_2O_3 thickness (d) has a minimum which corresponds to 180 nm thickness of Al_2O_3. Afterwards, when fabricating the PCA, we used these parameters of the dielectric thickness and plasmonic gratings height, i.e. d = 180 nm and h = 100 nm, respectively.

Figure 2 shows scanning electron microscope (SEM) image of the fabricated dielectric-filled plasmonic PCA with log-spiral topology and a magnified image of plasmonic gratings. Initially we passivated GaAs photoconductor's surface by 230 nm Si_3N_4 dielectric, where the windows for 50/450 nm Ti/Au antenna electrodes were etched. Therefore antenna electrodes were deposited on a photoconductor's surface only through the etched windows in Si_3N_4 while a PCA's gap remained passivated. Plasmonic gratings were formed by electron-beam lithography with 18/75 nm Ti/Au metallization followed by lift-off. The PCA fabrication is partially described in Refs. [5,6].

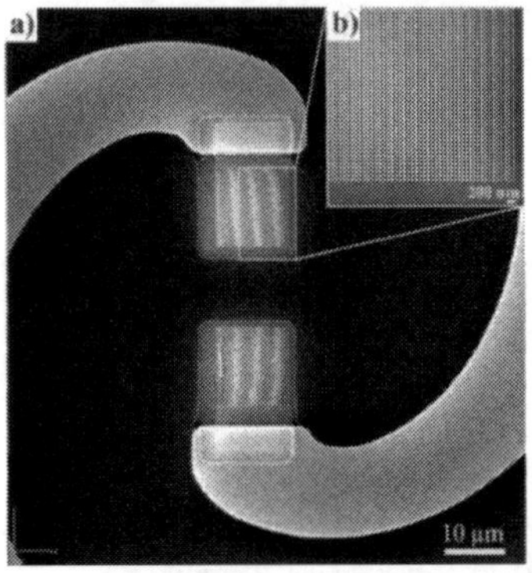

Fig. 2. SEM image of the fabricated plasmonic PCA (a) with a magnified image of plasmonic gratings (b)

The experimentally-measured current-voltage characteristics without laser illumination (see Fig. 3) demonstrated significant reduction of a dark current when a surface of GaAs is passivated by Si_3N_4 in comparison to a non-passivated one (more than three orders of a magnitude) demonstrating that the selected size of window in Si_3N_4 leads to strong decrease of a dark current, keeping a transient photocurrent remained.

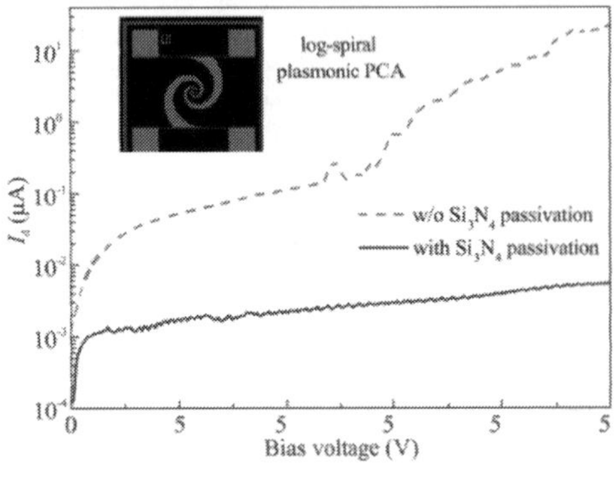

Fig. 3. Current-voltage characteristics without laser illumination measured for the proposed PCA when using Si_3N_4 dielectric (solid line) and without it (dashed line)

In order to demonstrate a performance of the proposed PCA we examined it by means of THz time-domain spectroscopy and compared to the traditional one. The both PCAs were used as a THz emitter, while compact all-fiber-based Toptica FemtoFErb780 laser was used as pump source with mean power up to 10 mW with a bias voltage equal to 30 V. As seen in Fig. 4 the spectra shape remain unchanged for the both PCAs, but the proposed PCA demonstrates 2000-100 times increase of the emitted THz power depending on laser pump (not attached in this article). Since PCAs are saturated at high laser pump due to electric field screening, we additionally investigated the emitted THz power versus laser pump in order to find out an optimum illumination regime. We showed that the transient photocurrent for the proposed and traditional PCAs is trending to converge at laser pump close 10 mW, though at low pump energies the transient photocurrent generated by the proposed PCA is up to 25 times higher than for the traditional one.

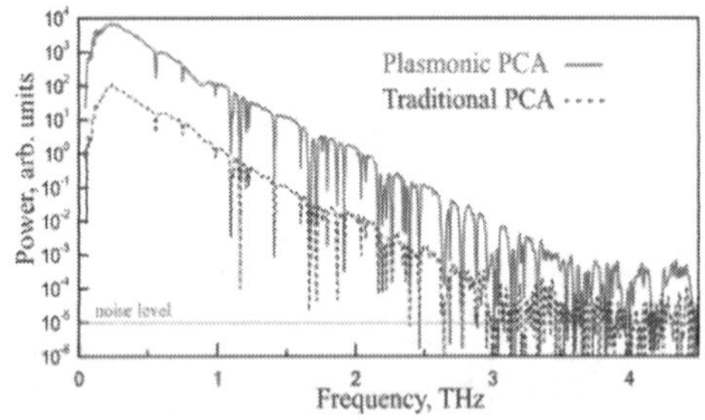

Fig. 4. Radiated THz power in the frequency-domain for the plasmonic and traditional PCAs based on SI-GaAs photoconductor

The work was supported by the Russian Scientific Foundation (Project No.18-79-10195) and RIEC ICRP (grant H30/A04). The numerical simulation of the PCA under consideration was supported by the President's grant of Russian Federation for young scholars (Project MK-5450.2018.2).

References

1. *Yardimci N., Cakmakyapan S., Hemmati S., Jarrahi M.* A high-power broadband terahertz source enabled by three-dimensional light confinement in a plasmonic nanocavity // Sci. Rep. 2017. V. 7. P. 4166.
2. *Ponomarev D. S., Lavrukhin D. V., Yachmenev A. E., Khabibullin R. A, Semenikhin I. E., Vyurkov V. V., Ryzhii M., Otsuji T., Ryzhii V.* Lateral terahertz hot-electron bolometer based on an array of Sn nanothreads in GaAs // J. Phys. D: Appl. Phys. 2018. V. 51. P. 135101.
3. *Ryzhii V., Ryzhii M., Svintsov D., Leiman V., Maltsev P., Ponomarev D., Mitin V., Shur M., Otsuji T.* Real-space-transfer mechanism of negative differential conductivity in gated graphene-phosphorene hybrid structures: Phenomenological heating model // arXiv: 2018. 1806.06227.
4. *Auston D.* Picosecond optoelectronic switching and gating in silicon // Appl. Phys. Lett. 1975. V. 26. P. 101.
5. *Ponomarev D.S., Khabibullin R.A., Yachmenev A.E., Pavlov A.Yu., Slapovskiy D.N., Glinskiy I.A., Lavrukhin D.V., Ruban O.A., Maltsev P.P.* Electrical and thermal properties of photoconductive antennas based on $In_xGa_{1-x}As$ (x>0.3) with a metamorphic buffer layer for the generation of terahertz radiation // Semicond. 2017. V. 51(9). P. 1218–1223.
6. *Lavrukhin D. V., Katyba G. M, Yachmenev A. E., Galiev R. R., Glinskiy I. A., Khabibullin R. A., Goncharov Yu. G., Spektor I. E., Khusyainov D. I., Buryakov A. M., Mishina E. D., Chernomyrdin N. V., Zaytsev K. I., and Ponomarev D. S.* Numerical simulations and experimental study of terahertz photoconductive antennas based on GaAs and its ternary compounds // Proc. SPIE 10680. 2018. V. 106801M.

Doubling of gyrotron radiation frequency due to nonlinear susceptibility in A3B5 semiconductors

Vladimir V. Rumyantsev[1], A. A. Dubinov[1], A.P. Fokin[2], V.V. Utochkin[1], M.Yu. Glyavin[2], S. V. Morozov[1]

[1]Institute for Physics of Microstructures of RAS, Nizhny Novgorod, Russian Federation, rumyantsev@ipmras.ru
[2]Institute of Applied Physics of RAS, Nizhny Novgorod, Russian Federation

At present, there is much activity aimed for development and application of different sources of radiation with terahertz (THz) frequency range, which are necessary for a variety of scientific and practical applications such as spectroscopy, radioastronomy, environmental monitoring, security and counter-terrorism [1]. Powerful sources of THz radiation like synchrotrons and free electron lasers (FELs) have high cost and large size, which limits their use even for scientific applications. Complicated solutions are required to increase the frequency of the vacuum sources such as a backward wave oscillator up to 1 THz and higher [2]. Semiconductor quantum cascade lasers in THz range operate only at cryogenic temperatures [3-5]. Molecular lasers, although operating at room temperature, are available only for a fixed set of frequencies [6]. THz radiation sources based on femtosecond lasers exhibit an extremely broad spectrum (~1 THz wide), which is not always acceptable for spectroscopy applications and have rather low output power [7]. Currently, the media of choice for optical rectification are LiNbO3 [8], ZnTe [9], GaP [10] crystals. They all have fairly high losses in the THz range and the conversion efficiency even for state-of-the-art experiments is approaching 10^{-3} [8] and in typical cases is an order of magnitude lower.

Alternative way to obtain intense THz radiation is to "multiply" the frequency of incident radiation using nonlinear susceptibility in semiconductors. This approach requires intense sources at the fundamental frequency in sub-THz range, among which there were mainly molecular lasers. However, recently a sufficient progress has been made in the development of sub-THz range gyrotrons. Continuous wave (cw) gyrotrons with frequencies of 460 GHz [11] and 527 GHz [12] have been demonstrated. IAP RAS recently developed unique pulse solenoids with magnetic fields up to 50 T that allowed implementing gyrotrons on the first cyclotron harmonic with the output frequency of up to 1.3 THz with pulse duration ~ 50 μs [13]. On the basis of "dry" cryomagnet with magnetic fields up to 10 T cw gyrotrons with frequencies 263 GHz [14] and 250 GHz were fabricated. A generator with 889 GHz frequency (in a synchronism with the third harmonic of the cyclotron frequency) was developed, as well as a pulsed generator with a frequency of ~1THz and power of ~10 W at the second harmonic of gyrofrequency [15]. These results were obtained by using the unique cryomagnet with the magnetic field of up to 20 T. In 2017, a double-beam gyrotron at the second harmonic with frequency of 0.8 THz was demonstrated [16].

For such sources, frequency doubling is enough to obtain frequencies above 1 THz, while previous works focused on generating the third harmonic of gyrotron radiation in n-type Ge and Si at much lower frequencies [17, 18]. Third harmonic generation was demonstrated in Ref. [17] with the effectiveness of 0.05% under gyrotron pumping with 70 GHz frequency, and in Ref. [18] under gyrotron pumping with 118 GHz frequency and 0.07% conversion efficiency. The frequency tripling is possible in semiconductor materials due to the third order nonlinearity, which is predominantly electronic (third order nonlinearity due to ionic motions is usually weaker) [19, 20]. The frequency doubling is possible in crystals lacking the inversion centre. In this work we consider the prospects of several A3B5 semiconductors for frequency doubling due to the second order susceptibility of the crystal lattice. It should be noted that nonlinear optical properties of A3B5 materials have been quite poorly studied in far infrared region where the frequency dispersion of second order nonlinear coefficient is substantial [21 - 24]. Second-order susceptibility and corresponding frequency doubling were investigated mainly in GaAs, using CH_3F laser [21] and free electron laser [22] as the radiation sources in the spectral ranges of 0.6 - 1.7 THz and 4 -- 6 THz, respectively. However, second order nonlinearity in GaAs is rather weak as can be seen from table:

Table. Second-order susceptibility of several A3B5 semiconductors

Semiconductor	Second-order susceptibility, χ [21, 25]
GaAs	$57 \cdot 10^{-10}$ cm/V
GaP	$18 \cdot 10^{-10}$ cm/V
GaSb	$150 \cdot 10^{-10}$ cm/V
InP	$345 \cdot 10^{-10}$ cm/V
InAs	$155 \cdot 10^{-10}$ cm/V
CdTe	$100 \cdot 10^{-10}$ cm/V [26]

The intensity of the second harmonic $I_{2\upsilon}$ can be calculated [27] as (c -- light velocity, n(υ) -- refractive index at the frequency υ, Δn = n(υ) - n(2υ))

$$I_{2v} = \frac{128\pi^5\chi^2 v^2 L^2}{n(v)^2 n(2v) c^3} I_v^2 \left(\frac{\sin 2\pi v \Delta n L / c}{2\pi v \Delta n L / c} \right)^2$$

Thus, the output intensity is proportional to the square of intensity at the fundamental frequency I_υ susceptibility χ, fundamental frequency υ and sample length L. As can been seen from the Table, the susceptibility in InP is 6 times higher than in GaAs, allowing sufficient enhancement of the second harmonic intensity. Note that maximum crystal length can be limited by refractive index dispersion and the optical losses. Typical frequencies of the lattice vibrations of optical transitions related to impurity absorption are considerably higher than 1 THz in semiconductors under consideration [25]. Therefore, the losses are mainly due to free carriers, density of which can be made negligibly low in state-of-the-art InP, reducing the absorption coefficient in the 200 -- 800 GHz region down to less than 1 cm^{-1} [28]. The refractive index change with frequency in the same range is extremely small $\Delta n \sim 0.002$ [28]. Thus, the crystal of several cm length can be used for frequency doubling. Fig. 1 gives the comparison of estimated output power of the second harmonic for GaAs and InP with L = 5 cm and fundamental radiation power of 50 kW/cm^2.

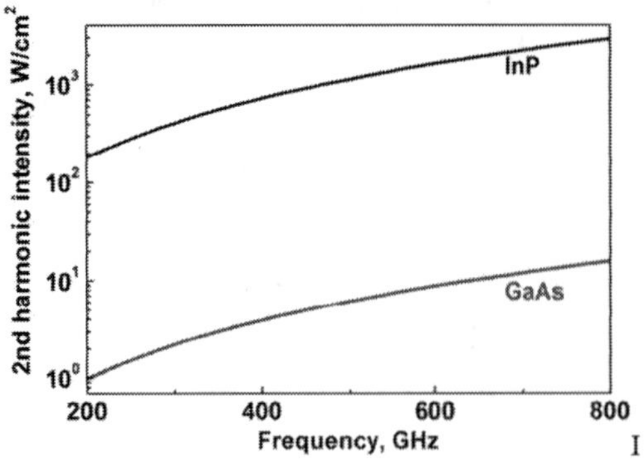

Fig. 1. The estimated second harmonic intensity vs. fundamental frequency for 5 cm long GaAs and InP crystals under fundamental radiation power of 50 kW/cm^2. Data on the refractive index for InP and GaAs is taken from Ref [28] and Ref [25], respectively

As can be seen, InP crystal is expected to provide about two-order enhancement over GaAs, studied in previous works [21-23]. We conclude that employing high-resistivity InP crystals is a promising route to obtain THz radiation by doubling the frequency of intense gyrotron radiation. The work was supported by Russian Science Foundation grant #18-79-10112.

References

1. *Van der Weide D.* Applications and Outlook for Electronic Terahertz Technology // Optics & Photonics News. 2003. V.14. No.4. P. 48-53

2. *Xiong Xu, Yanyu Wei, Fei Shen, Hairong Yin, Jin Xu et al.* A watt-class 1-THz backward-wave oscillator based on sine waveguide // Phys. Plasmas. 2012. 19, 013113.

3. *Williams B.S.* Terahertz quantum-cascade lasers // Nature Photonics. 2007. V.1. P. 517.

4. *Fathololoumi S., Dupont E., Chan C.W.I., Wasilewski Z.R., Laframboise S.R., Ban D., Matyas A., Jirauschek C., Hu Q., Liu H.C.* Terahertz quantum cascade lasers operating up to ~ 200 K with optimized oscillator strength and improved injection tunneling // Optics Express. 2012. 20 (4). 3866.

5. *Wang X., Shen C., Jiang T., Zhan Zh., Deng Q., Li W., Wu W., Yang N., Chu W., Duan S.*. High-power terahertz quantum cascade lasers with 0.23 W in continuous wave mode. // AIP Advances. 2016. 6. 075210.

6. *Dodel G.* On the history of far-infrared (FIR) gas lasers: Thirty-five years of research and application // Infrared Phys. Technol. 1999. 40 127–39

7. *Lewis R.A.* A review of terahertz sources // J. Phys. D: Appl. Phys. 2014. 47 374001

8. *Sergey B. Bodrov, Aleksey A. Murzanev, Yury A. Sergeev, Yury A. Malkov, and Andrey N. Stepanov* Terahertz generation by tilted-front laser pulses in weakly and strongly nonlinear regimes // Applied Physics Letters. 2013. 103. 251103

9. *Vidal S., Degert J., Tondusson M., Freysz E., and Oberlé J.*, Optimized terahertz generation via optical rectification in ZnTe crystals // J. Opt. Soc. Am. B. 2014. 31, 149-153.

10. *Taniuchi T., Nakanishi H.* Collinear phase-matched terahertz-wave generation in GaP crystal using a dual-wavelength optical parametric oscillator // Journ. Appl. Phys. 2004. v. 95, 7588.

11. Torrezan A.C., Han S.T., Mastovsky I., Shapiro M.A., Sirigiri J.R., Temkin R.J., Griffin R.G., Barnes A.B. Continuous-Wave Operation of a Frequency-Tunable 460-GHz Second-Harmonic Gyrotron for Enhanced Nuclear Magnetic Resonance. // IEEETrans. 2010. V.PS-38. №6.P.1150

12. Jawla S., Nanni E., Shapiro M. et al.// 2011 36th Int. Conf. on Infrared, Millimeter and Terahertz Waves (IRMMW-THz). Houston. 2-7 Oct. 2011. N.Y.IEEE, 2011. P. DOI 10.1109. IRMMWTHz.2011.61 05096

13. Glyavin M.Yu., Ginzburg N.S., Goldenberg A.L., Denisov G.G., Luchinin A.G., Manuilov V.N., Zapevalov V.E., Zotova I.V. THz Gyrotrons: Status and Possible Optimizations // Terahertz Science and Technology, V.5, No.2, 2012

14. Glyavin M. Yu., Chirkov A.V., Denisov G.G., et al. Experimental tests of a 263 GHz gyrotron for spectroscopic applications and diagnostics of various media. // Review of Scientific Instruments. 2015. 86. 054705.

15. Idehara T., Saito T., Ogawa I. The potential of the gyrotrons for development of the sub-terahertz and the terahertz frequency range — A review of novel and prospective applications, Proceedings of 2nd International Symposium on the Manipulation of Advanced Smart Materials // ThinSolidFilms. 2008. V. 517. No 4. P.1503, 2008

16. Idehara T., Glyavin M., Kuleshov A., Sabchevski S., Manuilov V., Zaslavsky V., Zotova I., Sedov A. A novel THz-band double-beam gyrotron for high-field DNP-NMR spectroscopy. // Rev.of Sci.Instrum. 2017. 88. 094708.

17 Keilmann F., Brazis R., Barkley H., Kasparek W., Thumm M. and Erckmann V. Millimeter-Wave Frequency Tripling in Bulk Semiconductors // EPL (Europhysics Letters), Volume 11, Number 4, p. 337

18 Narkowicz R., Siegrist M.R., Moreau Ph., Hogge J.P., Raguotis R. and Brazis R. Third-Order Susceptibility of Silicon Crystals Measured with Millimeter-Wave Gyrotron // Acta Physica Polonica A. 2011. V. 119, No. 4, p. 509

19. *Seeger K.*, Microwave frequency multiplication by hot electrons, // J. Appl. Phys. 1963. 34, 1608-1610.

20. *Nimtz G. and Seeger K.*, Microwave mixing by hot electrons in homogeneous semiconductors // J. Appl. Phys. 1968. 39, 2263-2266.

21. Mayer A. and Keilmann F., Far-infrared nonlinear optics. I. χ^2 near ionicresonance, // Phys. Rev. B. 1986. 33, 6954-6961.

22. Dekorsy T., Yakovlev V.A., Seidel W., Helm M., and Keimann F., Infrared phonon-polariton resonance of the nonlinear susceptibility in GaAs // Phys. Rev. Lett. 2003. 90, 055508-1/4.

23. McFee J.H., Boyd G.D., Schmidt P.H. Redetermination of the nonlinear optical coefficients of Te and GaAs by comparison with Ag3SbS3 // Appl. Phys. Lett. 1970. V.17 P.57.

24. Flytzanis C. Infrared Dispersion of Second-Order Electric Susceptibilities in Semiconducting Compounds // Phys.Rev. B. 1972. v.6, 1264.

25. *Madelung*, Semiconductors: Data Handbook (Springer-Verlag, New York, 2003).

26. *Akitt D., Johnson C., Coleman P.* Nonlinear susceptibility of CdTe // IEEE Journal of quantum electronics. 1970. V. QE - 6, 496.

27. Frits Zernike & John E. Midwinter Applied Nonlinear Optics John Wiley & Sons Inc. 1973

28. *Alyabyeva L.N., Zhukova E.S., Belkin M.A., and Gorshunov B.P.* Dielectric properties of semi-insulating Fe-doped InP in the terahertz spectral region // Scientific Reports. 2017. 7. 7360.

EPJ Web of Conferences **195**, 02011 (2018) https://doi.org/10.1051/epjconf/201819502011
TERA-2018

Molecular beam epitaxial growth of semiconductor heterostructures for THz electronics

Victor M. Ustinov[1,2]

[1]Ioffe institute, St. Petersburg, Russia, vmust@beam.ioffe.ru
[2]SHM R&E Center, RAS, St. Petersburg, Russia

In recent two decades semiconductor heterostructures emerged as a very promising basis for the development of various types of devices in a microwave frequency range. In the present work we discuss molecular beam epitaxial (MBE) growth of wide miniband superlattice (SL), heterostructure barrier varactor (HBV) and high electron mobility transistor (HEMT) structures, their characterization and application of these heterostructures for generators, detectors and frequency multipliers operating in the sub-THz and THz frequency range.

Superlattices.

The superlattice structures were grown by solid source molecular beam epitaxy using a Riber 32 machine. GaAs based structures were grown on GaAs substrates, consisted of several tens of periods of GaAs wells (3 - 4 nm thick) and AlAs barriers (about 1 nm thick) doped with silicon (10^{17} - 10^{18} cm^{-3}). The superlattice was embedded between graded layers of gradual composition and doping level to avoid abrupt heterojunctions. A n^+ GaAs or GaAs/InGaAs cap layer was provided for the formation of ohmic contacts. InP based SL structures were grown lattice matched to InP substrates and consisted of several tens of periods of $In_{0.53}Ga_{0.47}As$ wells (about 4 nm thick) and $In_{0.52}Ga_{0.48}As$ barriers (about 1.2 nm thick) doped with silicon (10^{17} - 10^{18} cm^{-3}). The superlattice was embedded between graded layers of gradual composition and doping level to avoid abrupt heterojunctions. InGaAs cap layer was provided for the formation of ohmic contacts. Mesa diodes (Fig. 1) with area of about 60 and 10000 μm^2 were formed using Au–Ge–Ni composition [1].

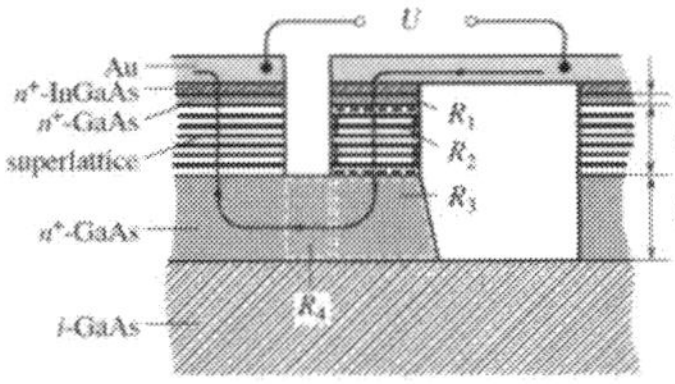

Fig. 1. Cross section of the GaAs based diode structure

Changing the widths of the well and barrier layers the lowest SL miniband width could be varied from about 20 up to 160 meV. These doped wide miniband superlattices exhibit negative differential conductance (NDC), Fig. 2, which is a consequence of Bragg reflection of miniband electrons accelerated by an electric field.

The negative differential mobility of a superlattice can give rise to travelling dipole domains and a self-sustained current oscillation at the transit frequency, which is the ratio of the domain velocity and the superlattice length. The fundamental oscillation frequency increased with the increase in the miniband width due to the increase of peak drift velocity [2]. The oscillation frequency can be controlled by changing the bias voltage applied to the diode. In comparison to a GaAs/AlAs superlattice, an InGaAs/InAlAs superlattice with the same quantum well and barrier thickness has the advantage of a larger miniband and, therefore, of a higher drift velocity due to the lower effective electron masses of the constituent materials and a lower barrier height. The highest fundamental frequency of an InGaAs/InAlAs superlattice oscillator is shown on Fig. 3. The power was ~80 mW corresponding to a conversion efficiency from DC to high-frequency power of 0.6% [3].

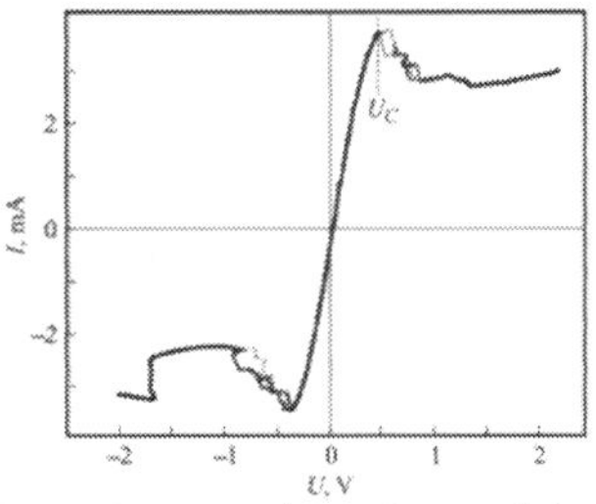

Fig. 2. Current-voltage characteristic of SL diode

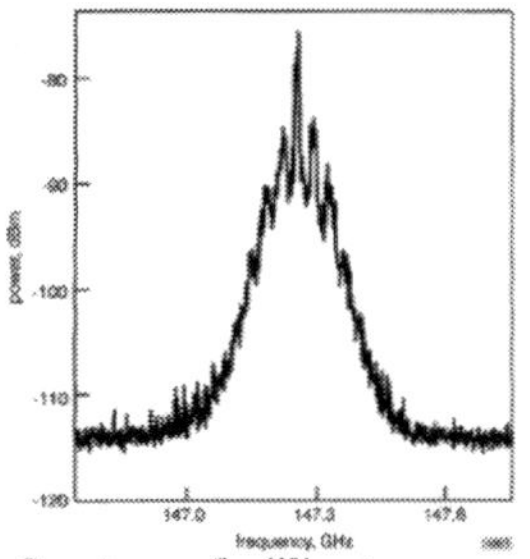

Fig. 3. Spectrum of millimetre wave emission from InGaAs/InAlAs superlattice device.

Strongly nonlinear U-V characteristic in the NDC region allows the use of the doped wide miniband superlattice device as a frequency multiplier in the microwave emission range, Fig. 4. The input frequency F = 150 GHz. Maximum harmonic number is n = 54, maximum harmonic frequency is 8.1 THz. Most part of radiation power of the frequency multiplier is concentrated in first 10 harmonics between 0.4 and 1.5 THz [4].

It was found that THz radiation affects Bloch oscillating miniband electrons. The direct current through a superlattice is reduced by a THz field and resonant structures in the current–voltage characteristic occur when the Bloch frequency is in resonance

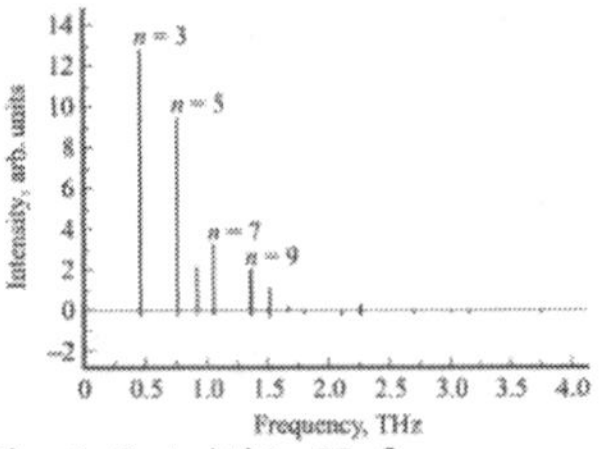

Fig. 4. GaAs/AlAs SL frequency multiplier harmonics intensity in the 0.4 – 4.1 THz range.

with the THz field or with its harmonics. In comparison to a GaAs/AlAs superlattice detector, the InGaAs/InAlAs superlattice detector has a responsivity, which is higher by an order of magnitude. The higher responsivity is caused by a higher current density due to a larger miniband of the InGaAs/InAlAs superlattice [5].

Heterostructure barrier varactors.

Heterostructure barrier varactor (HBV), Fig. 5, frequency multipliers demonstrate very promising results for applications in sub-THz and THz range. The best results for HBV multipliers have been achieved using MBE-grown heterostructures on InP substrates with several $In_{0.52}Al_{0.48}As/AlAs/In_{0.52}Al_{0.48}As$ barrier layers in the $In_{0.53}Ga_{0.47}As$ matrix. We have found that leakage current strongly decreases with the increase in the AlAs insertion thickness due to the reduction in the electron tunneling probability. Strain compensation InGaAs layers also decrease the leakage current at the same AlAs thickness. This may be attributed to the positive effect of the decrease in the overall amount of strain in the structure and better structure quality. Frequency tripler was fabricated using 6 mesa,18 barrier HBV diode, with a cross section area of 7x150 μm^2. Under CW operation at 94 GHz output frequency the maximal output power exceeds 100 mW at the conversion efficiency of about 14%. The estimated value of HBV cut-off frequency exceeds 1.2 THz [6].

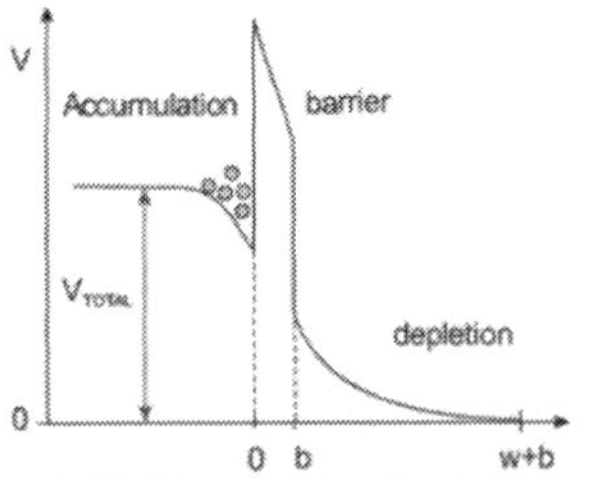

Fig. 5. HVB conduction band diagram at applied bias voltage.

HEMTs.

Nonlinear properties of the plasmonic excitations in a two-dimensional electron channel of HEMT can be used for detection of terahertz (THz) radiation. The plasmonic FET THz detectors are promising because they can operate in a broad THz frequency range at ambient temperature and exhibit a good balance between the responsivity and noise figure of merit. Along with the plasmonic nonlinearity, the asymmetry of the boundary conditions at the source and drain ends of the HEMT electron channel is mandatory for enabling a non-zero detection response. We fabricated a tight concatenation of four HEMT s connected in series on a single chip. It is shown that such HEMT chain exhibits strong THz photovoltaic response due to asymmetric form of the T-gate in each HEMT in the chain. We obtained the voltage responsivity above 1 kV/W in the unbiased mode of the detector operation [7].

References

1.*D. G. Pavel'ev, N. V. Demarina, Yu. I. Koshurinov, A. P. Vasil'ev, E. S. Semenova, A. E. Zhukov, V. M. Ustinov* Characteristics of planar diodes based on heavily doped GaAs/AlAs Superlattices in the terahertz frequency region // Semiconductors. 2004. V. 38, No. 9. P. 1105–1110.

2. *E. Schomburg, T. Blomeier, K. Hofbeck, J. Grenzer, S. Brandl, I. Lingott, A. A. Ignatov, K. F. Renk, D. G. Pavel'ev, Yu. Koschurinov, B. Ya. Melzer, V. M. Ustinov, S. V. Ivanov, A. Zhukov, P. S. Kop'ev* Current oscillation in superlattices with different miniband widths // PHYS. REV. B. 1998. V. 58, No. 7. P. 4035 – 4038.

3. *E. Schomburg, R. Scheuerer, S. Brandl, K.F. Renk, D.G. Pavel'ev, Yu. Koschurinov, V. Ustinov, A. Zhukov, A. Kovsh, P.S. Kop'ev* InGaAs/InAlAs superlattice oscillator at 147GHz // Electron. Lett. 1999. V. 35, No. 17. Online No: 19990973

4. *D. G. Paveliev, Y. I. Koshurinov, A. S. Ivanov, A. N. Panin, V. L. Vax, V. I. Gavrilenko, A. V. Antonov, V. M. Ustinov, A. E. Zhukov* Experimental study of frequency multipliers based on a GaAs/AlAs semiconductor superlattices in the terahertz frequency range // Semiconductors. 2012. V. 46, No. 1 P. 121–125.

5. *E. Schomburg, F. Klappenberger, M. Kratschmer, A. Vollnhals, R. Scheuerer, K.F. Renk, V. Ustinov, A. Zhukov, A. Kovsh* InGaAs/InAlAs superlattice detector for THz radiation // Physica E. 2002. V. 13, P. 912 – 915

6. *N. A. Maleev, V. A. Belyakov, A. P. Vasil'ev, M. A. Bobrov, S. A. Blokhin, M. M. Kulagina, A. G. Kuzmenkov, V. N. Nevedomskii, Yu. A. Guseva, S. N. Maleev, I. V. Ladenkov, E. L. Fefelova, A. G. Fefelov, V. M. Ustinov* Molecular-Beam Epitaxy of InGaAs/InAlAs/AlAs Structures for Heterobarrier Varactors // Semiconductors 2017. V. 51, No. 11. P. 1431–1434.

7. *V. V. Popov, D. M. Yermolaev, K. V. Maremyanin, V. E. Zemlyakov, N. A. Maleev, V. I. Gavrilenko, V. A. Bespalov, V. I. Yegorkin, V. M. Ustinov, S. Yu. Shapoval* Detection of terahertz radiation by tightly concatenated InGaAs field-effect transistors integrated on a single chip // Appl. Phys. Lett. 2014. V.104, 163508

EPJ Web of Conferences **195**, 02012 (2018)
TERA-2018

https://doi.org/10.1051/epjconf/201819502012

Theoretical and experimental studies of dielectric two-dimensional Bragg structures for development of spatially-extended heterolasers

N.Yu. Peskov, V.R. Baryshev, N.S. Ginzburg, E.R. Kocharovskaya, A.M. Malkin, D.M. Padozhnikov, M.D. Proyavin and V.Yu. Zaslavsky

Institute of Applied Physics RAS, Nizhny Novgorod, Russia, peskov@appl.sci-nnov.ru

Introduction

Two-dimensional distributed feedback (2D DF) was originally proposed [1] for obtaining super-power coherent radiation in relativistic masers based on spatially-extended electron beams. In this case, the 2D DF mechanism is realized in a 2D-periodic metallic Bragg structures with doubly periodic corrugation, which due to the arising in them of the transverse wave-flaxes allow synchronization of radiation of wide electron beams and the establishment of a single-mode oscillation regime when their transverse dimensions are in orders of magnitude greater than the radiation wavelength. To date, the operability of the new feedback mechanism has been experimentally demonstrated in the FEMs, which were elaborated in the millimeter wavelength range (from Ka- up to W-bands) under a record transverse size of the interaction space, reaching up to 50 wavelengths, and the output power level of ~ 50 - 100 MW [2, 3].

At the same time, high potential of the new feedback mechanism is not exhausted by microwave relativistic generators. In this aspect it should be noted that currently a 1D feedback mechanism, which is realized in "traditional" single-periodical Bragg structures based on the coupling and mutual scattering of two counter-propagating waves, is widely used in quantum DF lasers [4, 5]. However, the transverse sizes of such generators in the conditions of maintaining single-mode narrow-band oscillation are limited by several wavelengths, and further enhance of dimensions leads to a complication of the radiation spectrum and, thus, loss of its spatial coherence. One of the attractive ways to solve this problem in heterolasers is the use of novel 2D feedback mechanism (Fig. 1), which in this case can be realized in 2D Bragg structures of planar geometry with double-periodical modulation of the effective refractive index of the dielectric waveguide. According to the theoretical analysis carried out in [6], the use of these structures makes it possible to synchronize the radiation from the active medium characterized by large Fresnel parameters in both transverse directions.

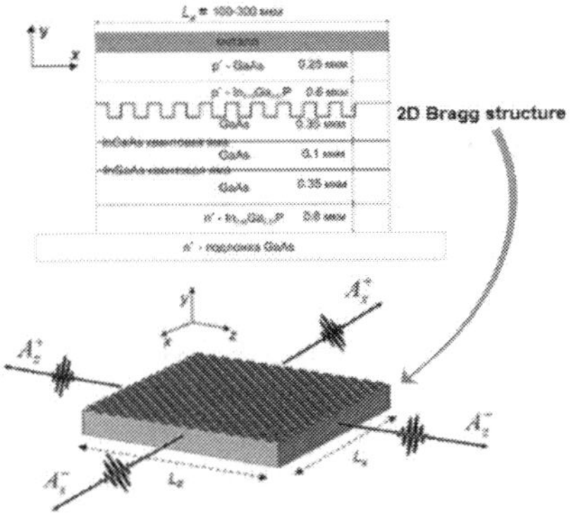

Fig. 1. Scheme of heterolaser based on 2D Bragg structure.

The paper presents results of theoretical analysis and computer simulations of the electrodynamic properties of dielectric 2D Bragg structures of planar geometry. "Cold" tests of prototypes of the structures were carried out in the W-band and demonstrated good coincidence with the design parameters. Possibility to advance structures of this type in the sub-mm ranges is discussed.

Results of simulations and "cold" tests in W-band

One of the main issues of the study was the analysis of acceptable technologies for the manufacture of planar dielectric Bragg structures. For operation at mm wavelengths, the technology based on the 3D computer printing, which characterized by a high speed of production and affordable cost of the models, was developed.

An important aspect for the discussed technology is the study of dielectric properties of plastics used for printing (which are practically not available in the standard descriptions and strongly depend on the quality and printing characteristics). Complex dielectric permittivities of different plastics were tested using a network analyzer in the Ka- and W-bands by measuring the reflection and transmission coefficients through a standard waveguide having the inserts from these dielectrics. The values of dielectric permittivity (complex and imaginary parts) were found by comparing the measured characteristics with the calculated ones (using iterative procedure). As a result of the experiments were selected plastics, the most suitable for the manufacture of the structures. Measured value of the dielectric constant was $\mathrm{Re}\,\varepsilon = 2.4$ and $\tan\delta = 0.0015$. At the same time, the experiments did not identify strong frequency dependences of these coefficients in tested bands.

Simulations of planar dielectric 2D Bragg structures were performed using the commercial code CST "Microwave Studio". Parameters for the simulations were chosen similar to the conditions of the experiments described below. Results of simulations of the frequency dependences of the integral scattering coefficients on the 2D structure (reflection, transmission, and transverse scattering coefficients) are presented in Fig. 2. It is showed good agreement with the results of previous theoretical analysis carried out in [6]. In the designed region of parameters the effective Bragg scattering zone corresponding to the coupling of the four partial wave-flaxes, which possess the structure of the TM-wave of the dielectric waveguide propagating over the corrugated surface, was verified.

Experimental study of 2D Bragg dielectric structures was conducted in the W-band. The thickness of the dielectric waveguide was 2 mm, the so-called "chessboard" corrugation was made on this waveguide with a period along each (x and z) directions $d_x = d_z = 4$ mm and a depth of 0.5 mm (Fig. 3). The transverse dimensions of the structure were 18 см × 18 см, the structure (dielectric waveguide) was placed on a substrate of duralumin.

A network analyzer P2-67 was used as the RF signal source. To form incident wave-beam (and receiving scattered radiation), the planar quasi-optical transmission line, which provides at its output a wide wave-beam with the structure of TEM-wave and plane phase, was exploited (Fig. 3). The results of "cold" measurements of the reflection and transmission coefficients on the structure described above are shown in Fig. 4. In accordance with the calculations, an effective Bragg scattering zone was observed in the vicinity of 60 GHz. Comparison Fig. 2 and 4 demonstrates good agreement between the simulations and "cold" tests results both on the position and width of the Bragg zone as well as on the amplitude of the integral scattering coefficients.

Summary and discussions

Thus, theoretical and experimental studies have confirmed the feasibility and operability of the 2D dielectric Bragg structures in the mm wavelength band. The available technology of manufacturing structures of this type using the 3D printing was developed. The good coincidence of the electrodynamic properties of the 2D structures measured in "cold" tests with the results of computer simulations based on commercial code CST "Microwave Studio" was demonstrated. The experimental studies have confirmed that the dielectric properties of plastics utilized in the proposed technology are adequate for manufacturing models of dielectric 2D structures with the accuracy acceptable for operation up to a short-wave part of the mm wavelength band. The further development of the 2D Bragg structures of the discussed type should be directed to advance into the sub-mm range when using the industrial 3D printers and the polymer photo printing technology, the claimed accuracy of which currently reaches up to 1 μm.

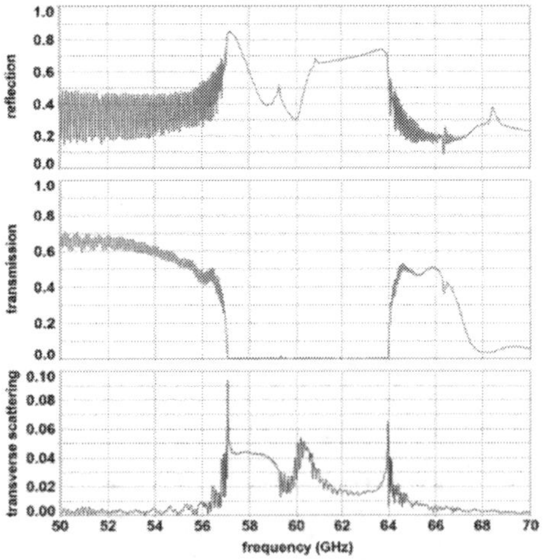

Fig. 2. Results of CST - simulations of dielectric 2D Bragg structure having parameters close to experimental prototype. Frequency dependencies of the reflection (top), transmission (middle) and transverse scattering (bottom) coefficients.

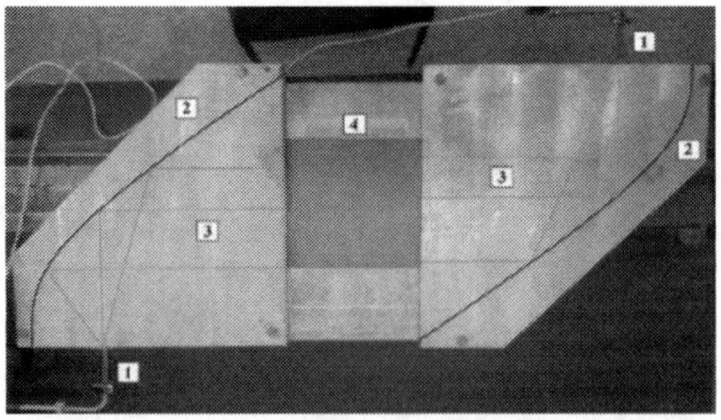

Fig. 3. Scheme of the "cold" tests: 1 - standard single-mode waveguides powered by the network analyzer, 2 - planar quasi-optical transmission lines, 3 - incident and transmitted wave-beams, 4 - dielectric 2D Bragg structure.

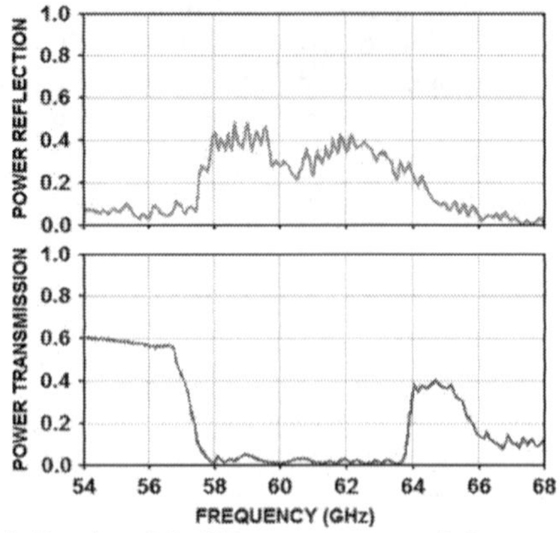

Fig. 4. Results of "cold" measurements of frequency dependencies of the reflection (top) and transmission (bottom) coefficients of dielectric 2D Bragg structure in the W-band.

This work is partially supported by the Russian Foundation for Basic Research (grant 18-48-520022).

References

1. *N.S.Ginzburg, N.Yu.Peskov, A.S.Sergeev*. Dynamics of free-electron lasers with two-dimension distributed feedback // Optics Commun. 1994. V.112. P.151.
2. *N.S.Ginzburg, N.Yu.Peskov, A.S.Sergeev, e.a.* Production of powerful spatially coherent radiation in planar and coaxial FEM exploiting two-dimensional distributed feedback // IEEE Trans. on Plasma Sci. 2009. V.37. No.9. P.1792.
3. *A.V.Arzhannikov, N.S.Ginzburg, P.V.Kalinin, e.a.* Using two-dimensional distributed feedback for synchronization of radiation from two parallel-sheet electron beams in a Free-Electron Maser // Phys. Rev. Lett. 2016. V.117. P.114801.
4. *H.Kogelnik, C.V.Shank*. Stimulated emission in a periodic structure // Appl. Phys. Lett. 1971. V.18, No.4. P.152.
5. *A.Yariv*. Quantum Electronics // John Wiley and Sons Inc., New York, 1975.
6. *N.S.Ginzburg, V.R.Baryshev, A.S.Sergeev, A.M.Malkin*. Dynamics of semiconductor lasers with two-dimensional distributed feedback // Phys. Rev. A. 2015. V.91. P.053806.

Terahertz lasers based on donor intracenter transitions in silicon

V.N. Shastin[1]

[1]Institute for Physics of Microstructures, Nizhniy Novgorod, Russia, shastin@ipmras

This talk presents a brief overview of terahertz (THz) lasing which is based on the intracenter transitions of impurity Coulomb centers in semiconductors. By now such kind of effects is obtained on group-V donor centers (phosphor P, antimony Sb, arsenic As, bismuth Bi). Energy diagram group-V donor states is shown on Fig.1.

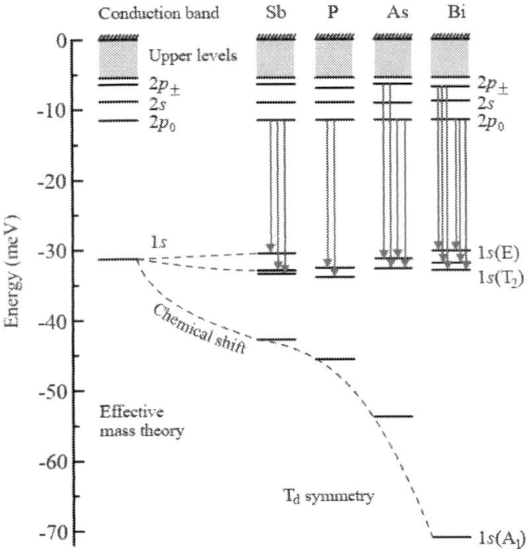

Fig. 1. Energy diagram group-V donor states in silicon. Arrows show possible "normal" laser transitions.

Depending on the pump parameters (frequency, intensity) one can reach either "normal" laser action which develops due to the population inversion between localized donor states or inversionless Raman lasing. Normal lasing occur under optical pumping by a midinfrared laser (for instance by a CO_2 laser) emitting at discrete lines originated from the $2p_0 \rightarrow 1s(T_2)$ and the $2p_\pm \rightarrow 1s(E,T_2)$ transitions subsequently from P, Sb and As, Bi centers in the wavelength range 4,5 – 6,4 THz [1]. Needed population of pointed states are controlled by phonon-assisted relaxation for cryogenic temperatures ($T < 20$ K). Under the same conditions and an intracenter optical excitation of donors one can obtain additionally Raman THz lasing on the transitions between so called "virtual" and 1s(E) final states. As shown the frequency of Raman lasing can be continuously changed between 4,5 – 6,4 THz varying the pump frequency. Normal and Raman lasing of donors were studied using free electron laser FELIX (Netherlands) facility.

Last years the experimental and theoretical activity was focused on the investigation of group-V donor lasing from axially strained silicon crystals. It has been shown that even moderate compressive stress applied along the [100] axis brings to a significant change of THz laser performance of group-V donors [1,2]. There is an optimal strain which may considerably reduce (up to two orders of magnitude) the laser threshold pump intensity increasing small signal gain and efficiency if it concerns normal lasing (see Table1).

Characteristics	Zero stress	Stress along [100] direction
Gain:		
Photo ionization (10,6 μm)	≤ 0,2 cm^{-1}	*1 cm^{-1}
Intracenter pumping (36 μm)	2–4 cm^{-1}	–
Quantum efficiency:		
Photo ionization (10,6 μm)	*≤1%	*≤10%
Photo ionization (17 μm)	–	*≤15%
Intracenter pumping (19–39 μm)	* 50%	–
Efficiency (10,6 μm):	*up to 10^{-3}	*up to 10^{-2}
Pump threshold (10,6 μm):	≥ 15 – 100 kW/cm^2	≥ 100 – 200 W/cm^2
Frequency tuning:	Absence	Δω/ω~1%
Operating temperature:	< 15 – 30 K	
Donor concentration:	(2 – 4)×10^{15} cm^{-3}	
Frequency range:	5,4 – 6,4 THz	

Tab. 1. Characteristics of "normal" THz silicon lasers; * – theoretical estimate.

It is clear that a value of the optimal strain depends on the doping agent. The influence of the axial strain on Raman lasing is still under continuation. The possible mechanisms behind the strain of the host crystal will be discussed.

In conclusion it is worth to touch the issue of another donor centers in the context of THz lasing from silicon.

References

1. *Pavlov, S.G., Zhukavin, R.Kh., Shastin, V.N., Hübers, H.-W.* The Physical principles of terahertz silicon lasers based on intracenter transitions // Physica Status Solidi B. 2013. V.250. P.9-36.

2. *Kovalevsky, K.A., Abrosimov, N.V., Zhukavin, R.Kh., Pavlov, S.G., Hübers, H.-W., Tsyplenkov, V.V., Shastin, V.N.* Terahertz lasers based on intracentre transitions of group V donors in uniaxially deformed silicon // Quantum Electronics. 2015. V.45 (2). P.113–120.

Frequency comb for THz metrology and spectroscopy

C. Bray[1], A. Cuisset[1], F. Hindle[1], JF. Lampin[2] and G. Mouret[1]

[1] Laboratoire de Physico-Chimie de l'Atmosphère, Université du Littoral Côte d'Opale, 189A Avenue Maurice Schumann, 59140 Dunkerque, France, mouret@univ-littoral.fr
[2] Institute of Electronics, Microelectronics and Nanotechnology, UMR 8520, University of Lille, Villeneuve d'Ascq, France

Frequency combs (FC) have radically changed the landscape of frequency metrology and high-resolution spectroscopy investigations extending tremendously the achievable resolution while increasing signal to noise ratio. By describing two experiments, we highlight that like other spectral ranges, THz domain can exploit a frequency comb.

First, the photomixing experiment we have developed at the LPCA during the last decade is presented on figure 1. This spectrometer is able to work between 300 and 3000 GHz with a frequency accuracy and a THz frequency linewidth suitable for high resolution THz spectroscopy [1-2]. The photomixing process is known from long time, and consist to convert an optical beat note into the THz domains. It is one of the best approach to cover a large part of the THz frequency gap. Thus, the frequency of the THz radiation produced is just the difference frequency of the two lasers working around 750 nm (and corresponding to a frequency around 400 THz) required to use of photomixer made in LTG GaAs. In principle, the frequency accuracy of the radiated THz radiation depends of the frequency accuracy of the pumps lasers. Clearly, there are no commercial solution to reach a precision better than 100 kHz often required for high resolution rotational spectroscopy. An alternative way consists to use a FC produce by a femtosecond laser as a ruler to control and to measure the difference frequency between the two pump lasers. In the present demonstration, the FC is produced by a frequency doubled erbium doped modelocked fiber laser with a stabilized repetition rate around 100 MHz. A beatnote between the laser and the nearest FC mode was isolated and phase locked to a local oscillator coherently locking the diode to the FC. The locking of the second laser to a different FC mode allows the difference frequency between the pump lasers to be synthesized. Feedback correction signal is applied to control the difference frequency of pump laser, and thus the THz frequency. The difference of frequency between the diode lasers can be expressed in Eq. (1) as:

$$f_{THz} = n \cdot f_{rr} \pm f_s \pm f_s \quad (1)$$

where f_{rr} is the repetition rate of the femtosecond laser, n is an integer, and f_s is the frequency of the synthesizer used to phase lock the beat signals against synthesizer to a precise frequency control. A classical commercial wavemeter is used to remove the ambiguity of the integer.

Various species have been study by use of this spectrometer from highly polar molecules (OCS, H_2CO) up to low and non-polar (CH_3D, CH_4) compounds and as well as radicals (OH, SO) to cite a fews [3-6].

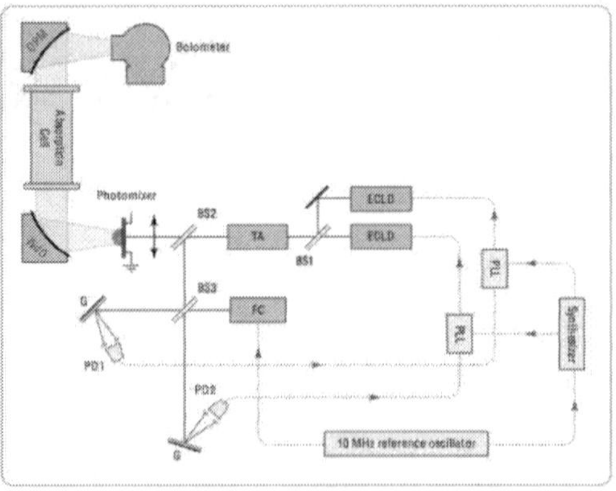

Fig. 1. Experimental setup of the THz photomixing spectrometer. BS1 is a beam splitter to overlap the two laser beams from the two External Cavity Diode Laser (ECDL) and seed Tapered Amplifier (TA). Then, the optical beat note is divided by a second beam splitter (BS2) to produce THz radiation and to manage the THz frequency. BS3 is used to overlap optical beat note and the frequency comb. Two grating (G) disperse the FC and the two lasers to ensure a heterodyne analysis between each ECDL and a part of FC thank to photodiodes (PD1 and PD2). The heterodyne signals are used to control the frequency of each laser.

Second, a very popular approach for THz investigations is the well-known THz Time Domain Spectrometer. Except the dual comb approach, the discrete spectral properties of the generated radiation which forms a THz offset free frequency comb is often ignored. However, the exploitation of the frequency comb structure of THz pulses should lead to an exceptional frequency resolution. To explore this approach, a dedicated heterodyne detection has been implemented around 200 GHz in order to analyse at very high resolution the THz FC (figure 2). Knowing exactly the repetition rate of the femtosecond laser used to produce the THz FC and intermediate frequency thanks to spectrum analyser, different components of the THz FC can be identified without ambiguity (figure 3). By tuning the repetition rate and/or the frequency of the local oscillator, high resolution spectroscopy can be performed [7].

Those two examples show the potential of the Frequency Comb to improve spectroscopic investigations in the THz domain.

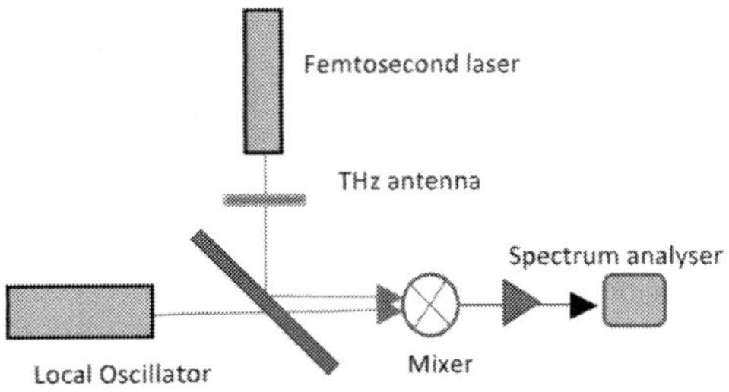

Fig. 2. Heterodyne spectrometer. The THz FC is produced by focusing optical fs pulses onto a a photoconductive dipole antenna. The THz FC and the local oscillator (formed by an amplifier multiplier chain driven by a synthesizer) is overlapped by a beam splitter to feed a subharmonic mixer. After amplification, the intermediate frequency (IF) signal is recorded by a classical spectrum analyzer.

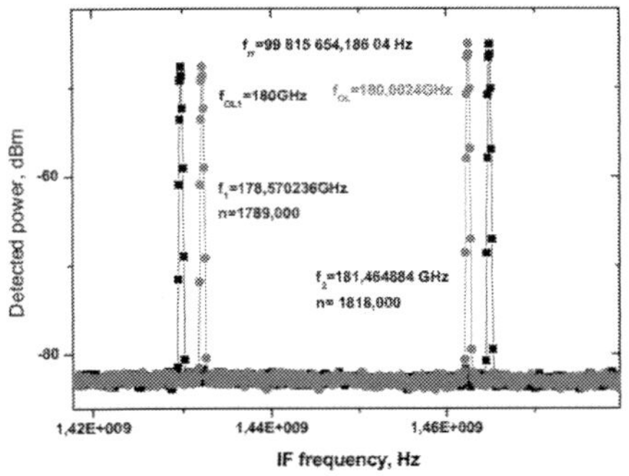

Fig. 3. High resolution spectra of the THZ FC obtained by use of heterodyne detection in which each component is clearly identified by an integer corresponding to an exact harmonic of the repetition rate.

Acknowledgements

This study was undertaken as part of the HEROES project which is financially supported by Agence Nationale pour la Recherche (project number ANR-16-CE30-0020).

References

1. *G. Mouret, et al.,* THz photomixing synthesizer based on a fiber frequency comb // Optics Express, Vol. 17(24), 22031-22040 (2009)

2. *F. Hindle, et al.,* Widely tunable THz synthesizer // Applied Physics B Vol. 104(4) 763-768 (2011)

3. *S. Eliet, et al.*, Rotational spectrum of formaldehyde reinvestigated using a photomixing THz synthesizer // Journal of Molecular Spectroscopy, Vol. 279, 12-15 (2012)

4. *M.A. Martin-Drumel, et al.,* New investigation on THz spectra of OH and SH radicals // Chemical Physics Letters, Vol. 550, 8 – 14 (2012)

5. *C. Bray, et al.,* CH_3D photomixing spectroscopy up to 2.5 THz: New set of rotational and dipole parameters, first THz self-broadening measurements // Journal of Quantitative Spectroscopy & Radiative Transfer, Vol. 189, 198-205 (2017)

6. *C. Bray, et al.* Spectral lines of methane measured up to 2.6 THz at sub-MHz accuracy with a CW-THz photomixing spectrometer: Line positions of rotational transitions induced by centrifugal distortion // *Journal of Quantitative Spectroscopy & Radiative Transfer, Vol. 203, 349-354 (2017).*

7. *S. Skryl et al.,* High-resolution terahertz spectroscopy with a single tunable frequency comb //Opt. Optics Express, Vol. 22 (26), 32276-32281 (2014).

EPJ Web of Conferences **195**, 03001 (2018) https://doi.org/10.1051/epjconf/201819503001
TERA-2018

A modified tilted-pulse-front excitation scheme for efficient terahertz generation in $LiNbO_3$

S.B. Bodrov,[1,2] A.N. Stepanov[2], E.A. Burova[1], and M.I. Bakunov[1]
[1]University of Nizhny Novgorod, Nizhny Novgorod 603950 Russia
[2]Institute of Applied Physics of the Russian Academy of Sciences, Nizhny Novgorod 603950 Russia

Pumping of $LiNbO_3$ by femtosecond optical pulses with tilted intensity front (pulse front) allows one to obtain high efficiency of optical-to-terahertz conversion (up to 1% at room temperature [1]), large energy of generated terahertz pulses (up to sub-mJ level [2]), and high strength of terahertz fields (exceeding 1 MV/cm [3]). Strong terahertz pulses are in demand for various applications, including terahertz-driven particle acceleration [4], terahertz nonlinear spectroscopy [5], and terahertz streaking techniques [6].

Tilted-pulse-front optical pulses are conventionally obtained by diffraction of a laser pulse off an optical grating. A lens (or two-lens telescope) is used to relay-image such pulses into the $LiNbO_3$ crystal [7,8]. In order to maximize the optical-to-terahertz conversion efficiency, the tilt angle α in the crystal should be close to a value defined by the equation $\cos\alpha = n_g / n_{\mathrm{THz}}$, where n_g is the optical group refractive index and n_{THz} is the terahertz phase refractive index. Additionally, it is postulated that a good quality of the terahertz beam can be achieved if the image of the grating is parallel both to the pulse front and to the exit surface of the crystal [7,8].

Here, we extend the analysis of the tilted-pulse-front excitation scheme [7,8] to a more general geometry, namely, with the pulse front not parallel to the exit surface of the crystal (the grating image is still assumed to be parallel to the surface).

In our theory, the generated terahertz radiation is found analytically from the Maxwell equations with a nonlinear polarization included as a source [9,10]. The nonlinear polarization is found as a result of optical rectification of the tilted-pulse-front optical pulse, which experiences effects of both material and angular dispersion.

At first, we consider the conventional geometry of the tilted-pulse-front excitation scheme, i.e., with the pulse front parallel to the exit surface of the crystal, but for an arbitrary orientation of the grating image plane, defined by the angle θ [Fig. 1(a)]. Figures 1(b)-(e) show the effects of the grating image orientation on the efficiency and spatial distribution of the generated spectrum along the exit surface of the crystal (y-axis). In the optimal case, when $\theta = \alpha$, the spectrum distribution is rather uniform [Fig. 1(d)]. For $\theta \neq \alpha$, the bandwidth and shape of the spectrum varies considerably with y [Figs. 1(c) and 1(e)]. The spectral non-uniformity across the terahertz beam can give strong divergence of the beam and prevent its further focusing. Interestingly, for $\theta < \alpha$, the spectral amplitude in the interval $-5 < y < -1$ mm is larger than for the optimal configuration $\theta \neq \alpha$ [Figs. 1(c) and 1(d)]. As a result, the energy of generated terahertz radiation does not decrease and even somewhat increases at a deviation of θ from the optimal value $\theta = \alpha$ to $\theta < \alpha$ [Fig. 1(b)]. In the case $\theta > \alpha$, the terahertz energy [Fig. 1(b)] and maximal spectral amplitude [Fig. 1(e)] decrease rapidly with deviation of θ from α.

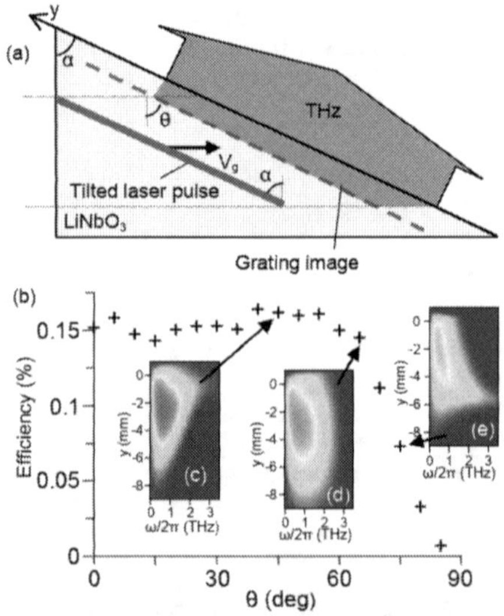

Fig. 1. (a) The conventional geometry of the tilted-pulse-front excitation technique with the pulse front parallel to the exit surface of the crystal. (b) Efficiency as a function of θ. (c-e) Distribution of the amplitude spectrum near the exit surface for (c) $\theta = 45°$, (d) $\theta = \alpha = 63.3°$, and (e) $\theta = 75°$. The optical pulse has a 3 mm transverse size and 100 fs duration.

Now let us consider a more general geometry, depicted in Fig. 2, with the pulse front not parallel to the exit surface of the crystal (the grating image plane is assumed to be parallel to the exit surface). In this geometry, the generated terahertz radiation propagates normally to the pump pulse front and, therefore, impinges obliquely on the exit surface of the crystal. To prevent total internal reflection of the radiation and minimize Fresnel losses, a Si prism should be attached to the exit surface of the $LiNbO_3$ crystal (Fig. 2).

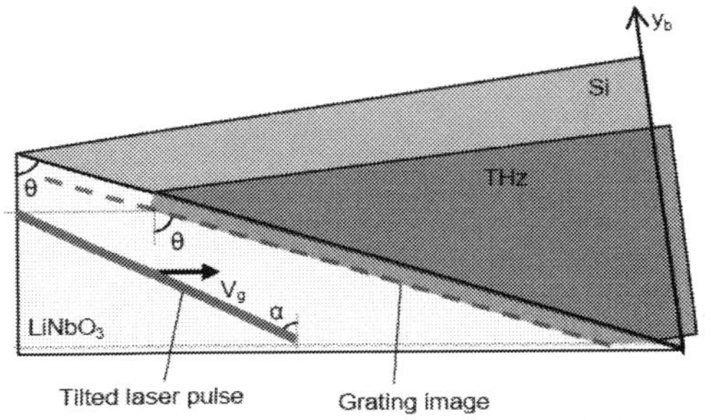

Fig. 2. The generalized tilted-pulse-front excitation geometry with the pulse front not parallel to the exit surface of the crystal.

Figures 3(a) and 3(b) show the distributions of the amplitude frequency spectrum along the exit boundary of the Si prism (y_b-axis) for two different angles $\theta = 45°$ and $\theta = 75°$, respectively. The distributions are rather uniform, similar to the optimal case in Fig. 1(d), and have practically the same spectral bandwidth for both angles. However, the maximal value of the spectral amplitude and size of its spatial localization differ substantially. The maximal value of the spectral density is two times higher for $\theta = 45°$ than for $\theta = 75°$, but the size of the terahertz beam is 2-3 times lager for $\theta = 75°$. The last fact is explained by a larger projection of the pump beam on the exit boundary of the $LiNbO_3$ crystal.

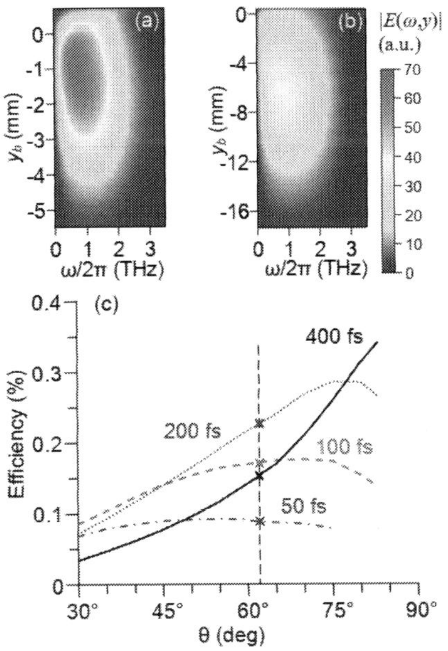

Fig. 3. Distribution of the amplitude spectrum along the exit boundary of the crystal for (a) $\theta = 45°$ and (b) $\theta = 75°$. (c) The conversion efficiency as a function of θ for different pulse durations. The optical pulse energy is fixed and corresponds to 10 GW/cm^2 at a pulse duration of 100 fs. The pump beam transverse size is 3 mm. Vertical line marks the optimal case $\theta = \alpha$.

Figure 3(c) shows the optical-to-terahertz conversion efficiency as a function of θ for different pump pulse durations. For a short pump pulse (50 fs), the efficiency is almost independent of θ in the interval from 30° to 75° and decreases rapidly for $\theta > 75°$. With increase of the pump pulse duration, the maximum becomes more pronounced and shifts to larger θ. The maximal efficiency also increases (for a fixed energy of the pump pulse) and for pulse duration >100 fs exceeds the efficiency in the optimal case. For the pump pulse of a 400 fs duration, the maximum efficiency at $\theta \approx 85°$ is about two times higher than in the optimal geometry, i.e., at $\theta = \alpha = 63.3°$.

To conclude, we have shown that a deviation of the grating image plane from the optimal orientation in the tilted-pulse-front excitation technique leads to a deterioration of the terahertz beam quality but does not affect significantly the terahertz energy (in the case $\theta < \alpha$). We have proposed a modified variant of the tilted-pulse-front excitation scheme with the pulse front not parallel to the exit surface of the crystal. We have shown that this geometry allows one to double the terahertz generation efficiency for long (>200 fs) laser pulses still preserving a high quality of the generated terahertz beam.

This work was supported by the Russian Science Foundation (RSF) (18-19-00486).

References

1. *S.-W. Huang, E. Granados, W. R. Huang, K.-H. Hong, L. E. Zapata, and F. X. Kärtner.* High conversion efficiency, high energy terahertz pulses by optical rectification in cryogenically cooled lithium niobate // Opt. Lett. 2013. V. 38. P. 796.

2. *J. A. Fülöp, Z. Ollmann, C. Lombosi, C. Skrobol, S. Klingebiel, L. Pálfalvi, F. Krausz, S. Karsch, and J. Hebling.* Efficient generation of THz pulses with 0.4 mJ energy // Opt. Express. 2014. V. 22. P. 20155.

3. *H. Hirori, A. Doi, F. Blanchard, and K. Tanaka.* Single-cycle terahertz pulses with amplitudes exceeding 1 MV/cm generated by optical rectification in $LiNbO_3$ // Appl. Phys. Lett. 2011. V. 98. 091106.

4. *E. A. Nanni, W. R. Huang, K.-H. Hong, K. Ravi, A. Fallahi, G. Moriena, R. J. Miller, and F. X. Kärtner.* Terahertz-driven linear electron acceleration // Nat. Commun. 2015. V. 6. P. 8486.

5. *S. Bodrov, Yu. Sergeev, A. Murzanev, and A. Stepanov* Terahertz induced optical birefringence in polar and nonpolar liquids // J. Chem. Phys. 2017. V. 147. P. 084507.

6. *U. Fruhling*, Light-field streaking for FELs, // J. Phys. B. 2011. V. 44. P. 243001.

7. *J. A. Fülöp, L. Pálfalvi, G. Almási, and J. Hebling.* Design of high-energy terahertz sources based on optical rectification // Opt. Express. 2010. V. 18. P. 12311.

8. *L. Tokodi, J. Hebling, and L. Pálfalvi.* Optimization of the tilted-pulse-front terahertz excitation setup containing telescope // J. Infrared Milli. Terahz. Waves. 2017. V. 38. P. 22.

9. *M. I. Bakunov, S. B. Bodrov, and E. A. Mashkovich.* Terahertz generation with tilted-front laser pulses: dynamic theory for low-absorbing crystals // J. Opt. Soc. Am. B 2011. V. 28. P. 1724.

10. *M. I. Bakunov and S. B. Bodrov.* Terahertz generation with tilted-front laser pulses in a contact-grating scheme // J. Opt. Soc. Am. 2014. V. 31. P. 2549.

Laser induced THz Sommerfeld waves along metal wire

A.V.Brantov[1,2], A.S.Kuratov[1], A. Maksimchuk[3], Yu. M. Aliev[2] and V.Yu.Bychenkov[1,2]

[1] Dukhov Research Institute of Automatics (VNIIA), Moscow, Russia, brantov@lebedev.ru
[2] P.N. Lebedev Physics Institute, Russian Academy of Sciences, Moscow, Russia
[3] Center for Ultrafast Optical Science, University of Michigan, Ann Arbor, USA

The thin metal wires can be utilized for guiding electron bunches [1, 2] as well as surface waves [3] during generation from irradiation of a wire target by short relativistically intense laser pulse. In such interaction a strong impulse type Thz surface wave can be excited at the metal surface due to a current of the high energy escaping electrons and electric sheath formation. These waves may propagate along wire without strong damping in the form of Sommerfeld wave. In this paper the electron bunch guiding along wire and supported by strong Sommerfeld surface wave is modeled and compared with experimental results.

To describe the initial stage of electron bunch formation and electromagnetic surface wave formation we have performed PIC simulation of laser-wire interaction considering a target as a plasma cylinder with given plasma density. The 3D simulations have been performed by using 3D PIC code Mandor and commercial code VSim (Vorpal). As example, we illustrate our findings with the following parameters. A linearly polarized laser pulse with maximum intensity of (1-3)×10^{19} W cm^{-2} has a Gaussian shape in both the temporal (FWHM duration τ_0 = 50 fs) and transverse dimensions (FWHM focal spot size d = 4 μm). The maximum intensity corresponds to dimensionless laser field amplitude a_0 = (2.7-4.7) for laser wavelength λ = 1μm. The laser is focused at the front surface of a cylindrical plasma target (with the axis along Y) with diameter of 8 μm consisting of electrons and heavy ions ($Z\, m_i/m_e$ = 4 × 1836). The plasma density is $5n_c$ (n_c is the electron critical density) inside interaction aria (cylinder piece of 10μm length from the focal spot center in both directions) and $2n_c$ outside. The simulation box has the size x×y×z = 20μm×200μm×20μm and the spatial resolution is Δ, x×Δ y×Δ z= 0.02μm ×0.1μm ×0.05μm.

The electron density after irradiation of a wire by laser pulse focused in the wire center is presented in Fig.1. Two symmetrical electron bunches propagated along wire from interaction area with velocity of the order of speed of light are clearly seen. Under the action of laser pulse, significant number of electrons inside a hot spot is heated to the high, multi-MeV, energies. The most of them form a sheath field electrostatic structure around wire as uncompensated positive target charge prevented them from escaping. At the same time, these electrons propagates along wire increasing a heated target surface, the length, L_0, of which is defined by the initial focal spot size, laser pulse duration and characteristic electron cooling time, i.e. $L_0 \sim d + c(\tau_0 + \tau_c)$ [4]. This heated area size roughly coincides with electron bunch longitudinal size, which is formed after a laser pulse termination. Because of the symmetry of electron bunch generation, several simulation runs have been performed with laser pulse, which was focused near the wire edge. It has been confirmed that the electron bunch length increases with laser pulse duration reaching the value $L_0 \sim c\tau_0$ for τ_0> 300 fs. The simulations show that all electrons forming a bunch originate from the laser-target interaction area and electrons moving along a wire are accompanied by the propagating surface electromagnetic wave (see Fig. 2). The amplitude of this electromagnetic wave is much high as compared to the intrinsic electron bunch filed.

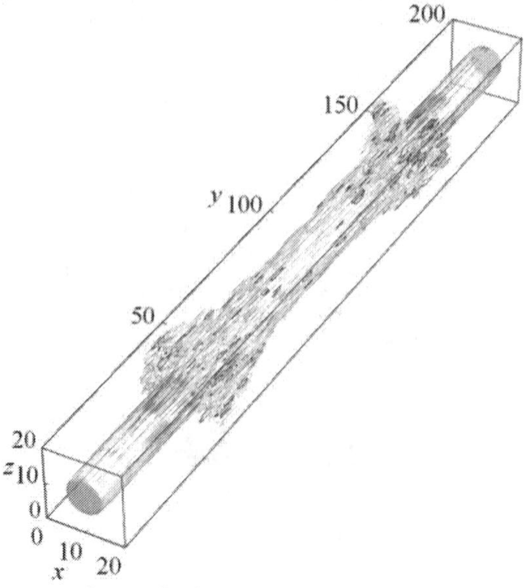

Fig. 1. Electron density contour plot at instant of 350 fs after wire irradiation by laser pulse with a_0 = 2.7.

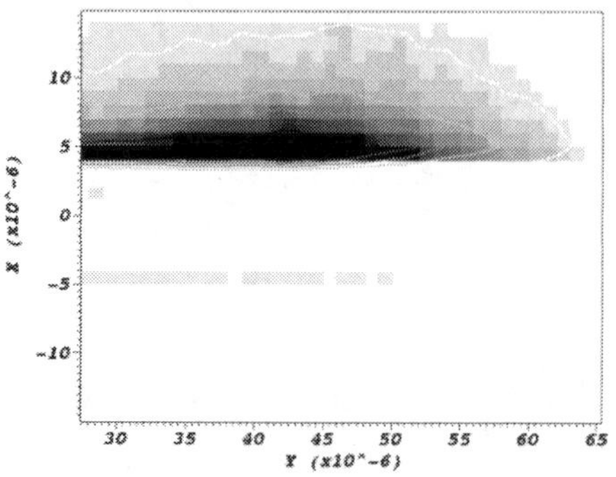

Fig. 2. Transversal (x>0) and longitudinal (x<0) electric field distributions in x-y plane at the instant of 350 fs after wire irradiation by laser pulse with a_0 = 4.7. Wire lies be-

tween x = -4 and x = 4 and a laser pulse is focused to y = 10μm.

The theory of Sommerfeld surface wave generation due to current of heated electrons has been developed. Based on solution of Maxwell equations the surface wave profile in a wave zone (far field) has been calculated. Expression for the magnetic field outside a wire with the radius R generated due to the radial electron current density, j_r, has the following form

$$B^s = \frac{2}{cR}\int \frac{d\omega\, k^s K_1(k_+^s r)}{k_+^{s2} d'(k^s,\omega)} e^{i(\omega t \pm k^s z)} \int_R^\infty dx \frac{x K_1(k_+^s x)}{K_0^2(k_+^s R)} j_r(x)$$

Here $k_+^2=k^2-\omega^2/c^2$, $k_-^2=k^2-\varepsilon\omega^2/c^2$, ε is wire dielectric permittivity, $I_{(0,1)}$, $K_{(0,1)}$ are the modified Bessel functions and k^s is a solution of the standard dispersion relation for Sommerfeld wave $d=0$, where

$$d(k,\omega) = \frac{K_1(k_+R)}{k_+K_0(k_+R)} + \frac{\varepsilon I_1(k_-R)}{k_-I_0(k_-R)}$$

As a radial current we have used both the current of the electrons forever leaving a target and the current of hot electrons forming a sheath structure. We assumed the Gaussian current profile along the wire with characteristic length of L_0, as well as the temporal Gaussian profile (with characteristic time scale of τ_0). We also assumed that heated electrons move in r-direction with speed of light and have exponential distribution with characteristic scales of their Debay length, r_{De}. Density of the hot electrons can be estimated from energy balance, assuming 30% laser energy deposition to the electrons, an average hot electron energy from simple ponderomotive scaling ~ $a_0\, m_ec^2$ and the current volume ~ $L_0 2\pi R r_{De}$. Some of the electrons may overcome potential barrier and escape from the target. Escaping electron density depends on magnitude of the electrostatic potential and can be estimated as ~ $a_0 m_e c^2/2\pi e^2 R^2$. Based on the above estimates, one can calculate a polar magnetic field of the surface wave (see. Fig. 3). We conclude, that electron current due to the escaping electrons and due to those, which form a sheath give comparable contribution to Sommerfeld wave generation.

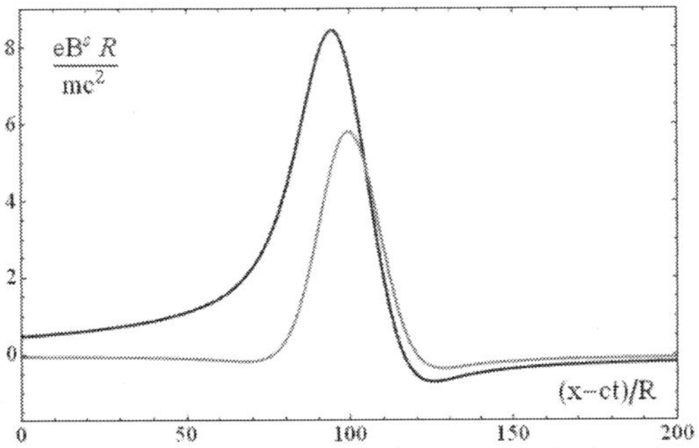

Fig. 3. Surface magnetic field at the tungsten wire boundary ($r=R=25$μm) for the laser pulse with $a_0 = 4.7$, d=4 μm and $\tau_0 = 400$ fs, generated due to the escaping electrons (black curve) and the sheath forming electrons (gray curve).

To conclude, the numerical simulations performed have shown a formation of relativistic electron bunches propagating along the wire together with electromagnetic surface waves in the process of wire target irradiation by short relativistically intense laser pulse.

This work was supported by the Russian Foundation for Basic Research (grants No 16-02-00088, 18-02-00452).

References

1. *Nakajima H., Tokita S., Inoue S., Hashida M., Sakabe S.* Divergence-free transport of laser-produced fast electrons along a meter-long wire target// Phys. Rev. Lett. 2013. V. 110. P. 155001

2. *A. Maksimchuk, P. Belancourt, M. J.-E. Manuel et al* Study of surface current confinement in high-intensity laser interactions with wire targets // Bull. Am. Phys. Soc. 2013. V. 58. P. 242; *A. Maksimchuk, P. Belancourt, P. Kordell et al.*, Guiding of high-energy electrons in high-intensity-laser interactions with wire targets through surface wave excitation // Bull. Am. Phys. Soc. 2014. V. 59, No. 15, NP8.00101

3. *Tokita S., Sakabe S., Nagashima T., Hashida M., Inou S.* Strong sub-terahertz surface waves generated on a metal wire by high-intensity laser pulses // Sci. Rep. 2015 V. 5. P. 8268.

4. *Kuratov A.S., Brantov A.V., Aliev Yu.M., Bychenkov V.Yu.* Laser-induced thermoelectric current as a source of generation of THz surface electromagnetic waves // Quantum Electronics 2018 V.48. No. 7. P. 653.

Modulation of two-color laser-induced filament terahertz emission by effective length variation

P. A. Chizhov[1], A. A. Ushakov[1,2], V. V. Bukin[1], S. V. Garnov[1]

[1]A. M. Prokhorov General Physics Institute RAS, Moscow, Russia, pvch@inbox.ru
[2]Lomonosov Moscow State University, Moscow, Russia

Terahertz (THz) radiation is of great scientific interest for its applications in spectroscopy and imaging [1]. Terahertz generation in femtosecond filament in gases has the specific features as compared with other THz sources [2, 3]. As the gaseous media are self-healing, one can use high-energy laser pulse focusing into small volumes without concern about material damage. Here one of the most effective ways to produce THz radiation is to use two-color (usually fundamental and second harmonics of fs-laser) laser pulse focusing for plasma creation [3]. However, in this case the efficiency of optical-to-terahertz conversion is highly dependent on mutual phase difference between harmonics. Moreover, for long filaments this difference varies along plasma channel. It has been demonstrated previously that this effect leads to off-axis peak in spatial THz distribution [4] and to possibility of waveform control by initial phase shift adjustment between harmonics [5]. In atmospheric air the distance for π phase walk-off between fundamental and second harmonic of Ti:Sapphire laser is about 25 mm.

In this paper, we show strong influence of phase matching between harmonics on overall THz power output and demonstrate that screening of a part of THz emission allows increasing the output power up to 20% of that for undisturbed emission. We also demonstrate the possibility of spectral modulation by means of π-retarder screens for a distinct THz frequency.

We use Ti:Sapphire laser system with following characteristics: 800 nm central wavelength, 40 fs pulse duration, 2.8 mJ pulse energy, ∅12 mm ($1/e^2$ level), 1 kHz repetition rate. A Glan-Taylor polarizer with a half-wave plate to control laser pulse energy is used.

A lens (∅25.4 mm, 1000 mm focal length) focuses laser radiation to create a filament in ambient air. A BBO crystal (SHG, 10x10x0.2 mm^3, I-type) converting part of radiation into second harmonic is placed on the optical pump path before the plasma. After the SHG a group delay compensator plate (TP) and a dual-wave plate ($\lambda/2$@800 nm+λ@400 nm) may be placed to provide more powerful THz generation due to use of collinear polarizations of harmonics in two-color pulse. The SHG and plates are mounted on a moving table to change air path of two-color pulse before the filament and so the initial phase difference between harmonics.

To collect THz radiation a system of two PTFE-lenses (first: ∅50 mm, 150 mm focal length; second: ∅50 mm, 100 mm focal length) is used. Laser pulse duration is lengthened up to ~120 fs by adding positive chirp to maximize THz output [6]. The length of visible plasma string is approximately 80 mm for maximal laser pulse energy of 2.3 mJ.

A metallic iris diaphragm with a ~1 mm diameter aperture centered at filament is mounted to screen THz radiation emitted before its position. Another moving table is utilized to control the position of the iris along the filament. Corresponding scheme is de-

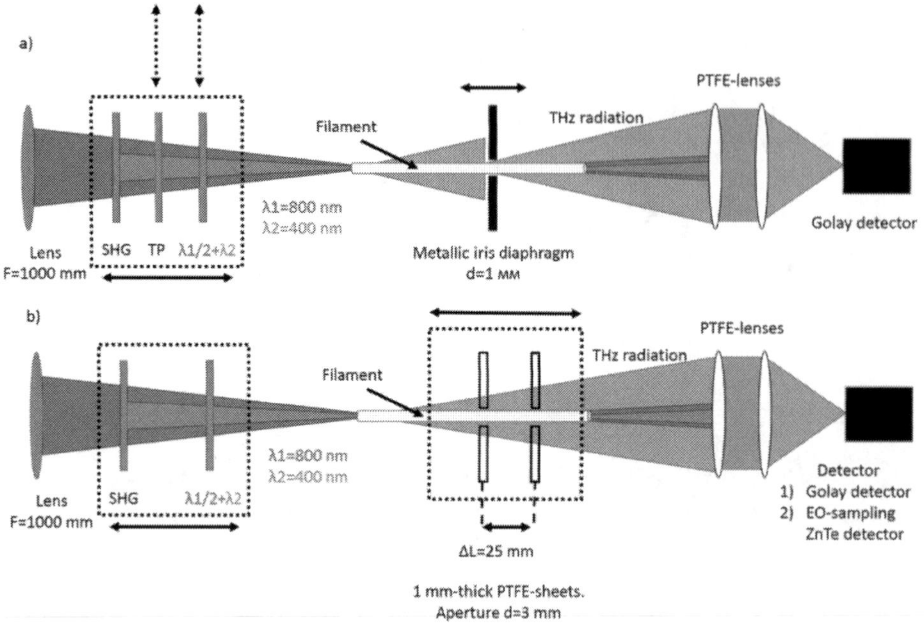

Fig. 1. Experimental schemes for: **(a)** screening of a part of the THz radiation with a metallic iris diaphragm; **(b)** spectral modulation by π-retarder PTFE-screens.

picted on Fig.1 (a).

The second thing we want to realize is spectral and spatial modulation of the emitted THz radiation. The idea is to make the THz emission collected by PTFE-lenses from all the points along filament interfere constructively for a distinct THz frequency. So one need to insert additional π-retardation for the part of the THz radiation that has previously interfered destructively. Ideally, there should be also a non-zero on-axis value in THz power distribution in this case. Thus, instead of the metallic iris diaphragm we place a 1 mm-thick (π-retardation for ~0.7 THz) PTFE-sheet with ~3 mm diameter aperture or a set of 2 such screens with a 25 mm distance between them (the distance corresponding to π walk-off in the air). Scheme is depicted on Fig.1 (b).

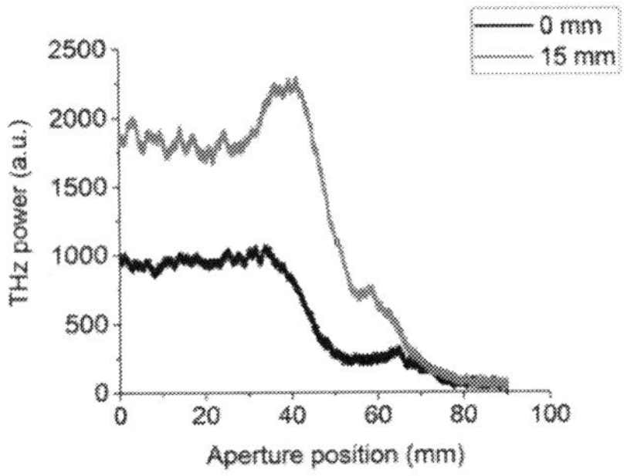

Fig. 2. Terahertz power dependence on iris diaphragm position along the filament created by 2.3 mJ laser-pulse for two positions of second harmonic crystal (SHG).

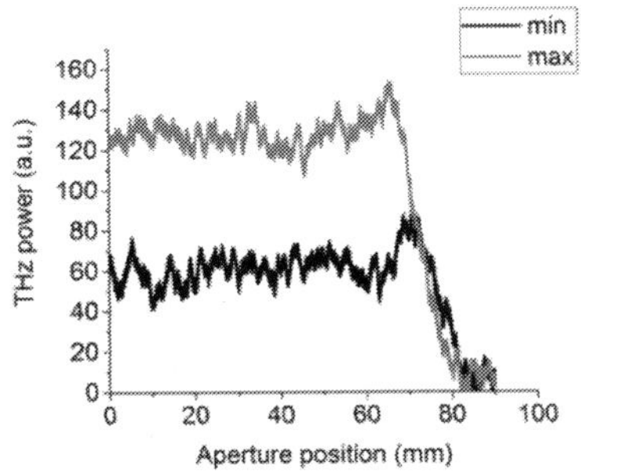

Fig. 3. Terahertz power dependence on iris diaphragm position along the filament created by 1.0 mJ laser pulse for two positions of second harmonic crystal (SHG).

For THz power measurements, a Golay cell (Tydex GC-1P) is used. For THz field measurements a standard electro-optical sampling scheme with ZnTe crystal (3x3x1 mm^3, <110> cut) is applied.

We measure THz power dependence on the iris diaphragm position along the two-color filament for different laser pulse energies from 1.0 to 2.3 mJ and for different initial phase difference between fundamental and second harmonics (SHG positions). We see strong influence on overall THz power due to constructive and destructive interference of THz emission from different parts of filament. For optimal initial phase and screening diaphragm position one can achieve significant increase in THz power. Examples of such dependencies for 2.3 mJ laser pulse with co-polarized harmonics are given on Fig. 2.

However, for short filament (low laser pulse energy) the effect of phase walk-off along plasma channel is sufficiently decreasing. See dependencies for 1.0 mJ laser pulse with co-polarized harmonics on Fig. 3.

For the experiment on spectral modulation, we collected waveforms of THz pulses and THz power value for different positions of π-retarder screen/set of screens. Thereafter waveforms are processed with fast Fourier transform to get spectra of THz pulses. Here one can see significant narrowing of spectrum for some position of the screens with peak amplitude on 0.7 THz frequency, that corresponds to the thickness of the PTFE-sheets used. There is also a bright peak on the doubled frequency — 1.4 THz. However, spectral change is accompanied with a dramatic drop in emitted THz power, so that even the amplitude on peak frequency is lower than that in spectrum with no screens placed. On Fig. 4 are given spectra of THz pulses for different positions of the set of two PTFE-screens: 0 mm position corresponds to undisturbed emission (screens placed before the whole filament), 40 mm position corresponds to significant influence of the π-retarders on emitted THz radiation.

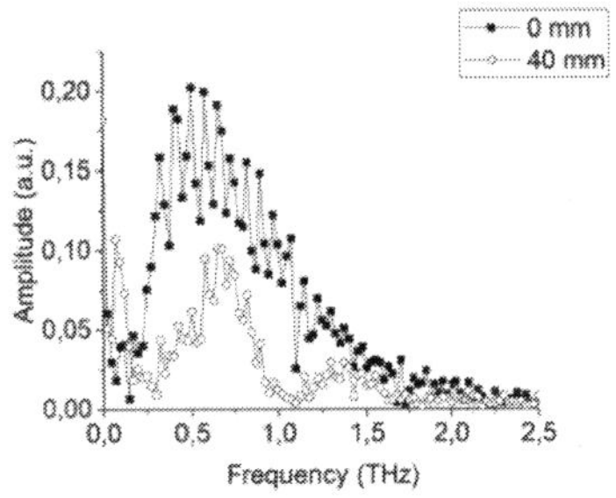

Fig. 4. Terahertz pulse spectra for two PTFE-screens positions.

References

1. *Zhang, X.-C., Shkurinov, A., Zhang, Y.* Extreme terahertz science // Nature Photonics. 2017. V. 11, No. 1. P. 16–18.

2. *Hamster, H., Sullivan, A., Gordon, S., White, W., Falcone, R. W.* Subpicosecond, Electromagnetic Pulses from Intence Laser-Plasma Interaction // Phys. Rev. Lett. 1993. V. 71, No. 17. P. 2725–2728.

3. *Cook, D. J., Hochstrasser R. M.* Intense terahertz pulses by four-wave rectification in air // Opt. Lett. 2000. V. 25, No. 18. P. 1210–1212

4. *You, Y. S., Oh, T. I., Kim, K. Y.* Off-Axis Phase-Matched Terahertz Emission from Two-Color Laser-Induced Plasma Filaments // Phys. Rev. Lett. 2012. V. 109, No. 18. P. 183902.

5. *Zhang, Z., Chen, Y., Chen, M., Zhang, Z., Yu, J., Sheng, Z., Zhang, J.* Controllable Terahertz Radiation from a Linear-Dipole Array Formed by a Two-Color Laser Filament in Air // Phys. Rev. Lett. 2016. V. 117, No. 24. P. 243901.

6. *Wang, T.-J., Chen, Y., Marceau, C., Théberge, F., Châteauneuf, M., Dubois, J., Chin, S. L.* High energy terahertz emission from two-color laser-induced filamentation in air with pump pulse duration control // Appl. Phys. Lett. 2009. V. 95, No. 18. P. 131108.

THz sources based on frequency conversion of multi-line molecular lasers in nonlinear crystals and on optically pumped molecular lasers

A. A. Ionin, I. O. Kinyaevskiy, Yu. M. Klimachev, A. A. Kotkov, A. M. Sagitova

P.N. Lebedev Physical Institute of the Russian Academy of Sciences, Moscow, Russia, umk@sci.lebedev.ru

At present, the main method for generating terahertz radiation is down-conversion of laser radiation in nonlinear crystals. However, at converting the radiation frequency from the visible or near infrared to the terahertz range, the efficiency of the process is limited by the Manley-Row ratio, i.e. the ratio of the photon energies of the converted and incident light on the crystal. Therefore, for applications that require high power of terahertz radiation, the use of a CO_2 laser operating at wavelength of 10 μm is more preferable than using solid-state near-IR lasers. At the present time in the scientific literature one can find several papers on the conversion of the frequency of the CO_2 laser radiation in nonlinear crystals to the terahertz range (see, for example, [1-5]). Note that the conversion efficiency of the super high-power CO_2 laser system (UCLA, Neptune Laboratory) under collinear interaction in the GaSe crystal turned out to be lower than for the non-collinear interaction in the GaAs crystal [2], that can be explained by a number of factors: the use of GaSe crystals with low optical quality and realized non-optimal type of interaction. In practice, CO_2 lasers with much lower energy are usually used, and the use of non-collinear phase matching becomes less effective in this case (see, for example, [3]). In turn, the implementation of collinear phase matching allows one the use of a longer length of nonlinear crystal and an increase conversion efficiency. Thus, in [4], at crystal length of 47 mm, the average power of pulsed-periodic terahertz radiation reached 0.26 mW with efficiency of 10^{-4}, which is three orders of magnitude greater than in [1]. The use of a doped GaSe crystal instead of a pure one, made it possible to increase the conversion efficiency by additional factor of 2 [5]. It should be noted that in papers [1-5] two CO_2 lasers or a two-channel CO_2 laser with independent resonators on the same active medium were used. Usually, the generation of the CO_2 laser radiation occurs only on one spectral line, although a fundamental possibility of generating a CO_2 laser on two or more spectral lines has already been demonstrated (see, for example, the review [6]). The use of pulsed laser emission acting synchronously on two or more spectral lines will allow us to realize a collinear three-frequency interaction in our RFBR project, simplify the optical scheme of the terahertz radiation generator, thanks to the automatic spatial-temporal matching of laser pulses, and reduce optical losses due to a decrease in the number of used elements, that increase the reliability of the entire system. Such a variant of the optical scheme was realized, for example, in [1] mentioned above. However, it should be emphasized that it will be necessary to solve a set of problems in order to ensure real synchronism of the generation of radiation pulses on laser spectral lines both in the Q-switched mode and in the of active mode locking. In addition, it will be necessary to investigate the influence of the parameters of the electric-discharge pump, the quality factor of the laser cavity, and the composition of the active medium of the laser on its spectral characteristics.

Other promising sources of terahertz radiation are optically pumped lasers operating on those molecular gases in which the radiating transition is either purely rotational (for diatomic molecules, molecules of the asymmetric and symmetrical top type except NH_3), or by inversion or inversion-rotational lasers (for NH_3) [7]. It was on inversion-rotational transitions of NH_3 molecule that laser radiation was generated by French colleagues on our project during optical pumping of ammonia by a quantum-cascade laser [8]. Typically, a CO_2 laser is used as the pumping source of terahertz gas lasers [7, 9-10] as the most convenient and well-studied source of radiation. In contrast to a CO_2 laser, a molecular CO laser is capable of operating in the mid-IR range almost simultaneously on hundreds of spectral lines in the wavelength range from 2.5 to 8.7 μm, which makes it possible to significantly increase the number of excited molecules and, accordingly, to enrich spectrum of radiation in the terahertz range.

Gas Lasers Laboratory of P.N. Lebedev Physical Institute of the Russian Academy of Sciences is one of the leading teams in Russia in the field of research of molecular lasers with electric discharge pumping. Unique wide-aperture pulsed electron beam sustained discharge laser devices with the possibility of cryogenic cooling of the laser medium working on vibrational-rotational transitions of CO, CO_2, and N_2O molecules and also on isotopically substituted modifications of these molecules have been created and investigated in the laboratory. An investigation was made of NH_3 laser with optical pumping by CO_2 laser radiation. Compact slab CO laser system with pumping by a transverse capacitive radio-frequency (RF) discharge with the possibility of cryogenic cooling of electrodes was developed. At such facilities, the mode of generation of a slab CO laser was realized for the first time, both on the fundamental vibrational transitions and on the first-overtone vibrational transitions of the CO molecule.

Within the framework of this project, it is planned to use laser systems that will be developed on the basis of existing laser devices pumped with an active medium by a pulsed electron beam sustained discharge [11], a pulsed-periodic RF discharge [12] and a DC discharge [13]. In addition, the members of the scientific team have experience in the implementation Q-switching modes and of the active mode-locking of molecular CO [14] and CO_2 lasers [15]. Our studies have been carried out on the frequency conversion of molecular gas lasers in various nonlinear crystals, some of which are expected to be used in carrying out work under this project.

The most close to the tasks of the project are the results of generation of the sum and difference frequencies of CO and CO_2 laser emission in nonlinear GaSe and $AgGaSe_2$ crystals [13, 16], as well as $PbIn_6Te_{10}$ [17] into wavelength interval from 2.5 to 16.6 μm. A large number (more than 1000) of the spectral lines of a CO laser makes it possible to realize single-stage frequency converters by generating the difference frequencies of both bands (fundamental and first-overtone) and covering the long-wave part of the mid-IR band up to 30 μm. The generation of difference frequencies between closely spaced spectral lines of a CO laser from one or another radiation band allows one to cover the entire terahertz spectral range [18]. In the paper [19], based on an analysis of the nonlinear optical properties of the new $PbIn_6Te_{10}$ crystal, the possibility of conversion of radiation on the lines of CO and CO_2 lasers into the spectral interval of wavelengths from 130 to 1000 μm (from 0.3 to 2.3 THz) is shown.

The research was supported by the RFBR (Project No. 18-52-16019)

References

1. *V.V. Bezotosnyi, E.A. Cheshev, M.V. Gorbunkov et al.*, Coherent THz repetitive pulse generation in a GaSe crystal by dual-wavelength Nd:YLF laser // Physics Procedia 72, 405 (2015).

2. *S.Ya. Tochitsky, J.E. Ralph, C. Sung, C. Joshi*, Generation of megawatt-power terahertz pulses by noncollinear difference-frequency mixing in GaAs // J. of Appl. Phys. 98, 026101 (2005).

3. *Y. Lu, X. Wang, L. Miao, et. al.*, Efficient and widely step-tunable terahertz generation with a dual-wavelenght CO_2 laser // Appl. Phys. B 103, 387 (2011).

4. *Y. Jiang, Yu.J. Ding*, Efficient terahertz generation from two collinearly propagating CO_2 laser pulses // Appl. Phys. Lett. 91, 091108 (2007).

5. *D.-D. Sie, D. Guo, L.-M. Zhang et al.*, Frequency conversion of a nanosecond CO_2 laser to the THz range in doped GaSe crystals // Fundamental problems of modern materials science, 9 (4) 486 (2012) (in Rissian).

6. *D.J. Biswas, A.K. Nath, U. Nundy, U.K. Chatterjee*, Multiline CO2 lasers and their uses // Prog. Quant. Electr. 11, 1 (1990).

7. *A.A. Vedenov, G.D. Myl'nikov, D.N. Sobolenko*, Generation of coherent far-infrared radiation using lasers // Physics-Uspekhi (Advances in Physical Sciences). 25, 833 (1982).

8. *A. Pagies, G. Ducournau, J.-F. Lampin*, Low-threshold terahertz molecular laser optically pumped by a quantum cascade laser // APL Photonics. 1, 031302 (2016).

9. *L.F.L. Costa, J.C.S. Moraes, F.C. Cruz, et al.*, Infrared and far-infrared spectroscopy of $^{13}CH_3OH$: TeraHertz laser lines and assignments // Applied Physics B. 86(4) 703 (2007).

10. *F.C. Cruz, T. Nogueira, L.F.L. Costa, et al.*, Continuous and Pulsed THz generation with molecular gas lasers and photoconductive antennas gated by femtosecond pulses // AIP Conference Proceedings 992, 383 (2008).

11. *A. Ionin, Yu Klimachev, A. Kotkov, et al.*, Multiline Laser Probing for Active Media CO:He, CO:N_2, and CO:O_2 in Wide-Aperture Pulsed Amplifier // Journal of Russian Laser Research. 27(1) 33 (2006).

12. *A.A. Ionin, Yu.V. Kochetkov, A.Yu. Kozlov, et al.*, Q-switched slab RF discharge CO laser // Laser Phys. Lett. 14, 055001 (2017).

13. *A. A. Ionin, I. O. Kinyaevskiy, Yu. M. Klimachev, and A. A. Kotkov*, Frequency Conversion of Radiation of IR Molecular Gas Lasers in Nonlinear Crystals (A Review) // Optics and Spectroscopy. 119, 356 (2015).

14. *O.V. Budilova, A.A. Ionin, I.O. Kinyaevskiy, et al.*, Mode-locked and Q-switched carbon monoxide laser system // Optics Communications 345, 163 (2015).

15. *A.A. Ionin, I.O. Kinyaevskiy, Yu.M. Klimachev et al.*, CO_2 laser with active mode-locking or Q-switching // VI International Scientific Youth School-Conference "Modern Problems of Physics and Technology", Theses of reports, Part 1, M: NRNU MEPhI, 322 (2017). (in Rissian).

16. *O.V. Budilova, A.A.Ionin, I.O.Kinyaevskiy, et al.*, Ultra-broadband hybrid infrared laser system // Optics Communications 363, 26 (2016).

17. *A.A. Ionin, I.O. Kinyaevskiy, Yu.M. Klimachev, et al.*, Frequency conversion of molecular gas lasers in $PbIn_6Te_{10}$ crystal within mid-IR range // Optics Letters. 41, 2390 (2016).

18. *Yu.M. Andreev, A.A. Ionin, I.O. Kinyaevskiy, et al.*, Frequency doubling and mixing of the radiation of carbon monoxide lasers in nonlinear $ZnGeP_2$ and GaSe crystals // Journal of Optical Technology. 78, 102 (2011).

19. *Yu.M. Andreev, V.V. Badikov, A.A. Ionin, et al.*, Optical properties of $PbIn_6Te_{10}$ in the long-wave IR // Laser Physics Letters. 13, 125405 (2016).

Temperature variation in the process of terahertz wave generation by intense laser pulses

G. Kh. Kitaeva[1], E.V. Moiseenko[2], and A.V. Shepelev[3]

[1]*Lomonosov Moscow State University, Moscow, Russia*
[2] *Nuclear Safety Institute of the Russian Academy of Sciences, Moscow, Russia,moi@ibrae.ac.ru*
[3] *I.M.Gubkin Russian State University of Oil and Gas, Moscow, Russia*

To reach satisfactory energy efficiency of the optical-to-terahertz frequency conversion, femtosecond and nanosecond pulsed periodic lasers are used. This considerably affects the lattice temperature and can lead to the dramatic decrease in conversion efficiency.

Nevertheless, the calculations done during the development of a conversion method are often based on the parameters of the non-linear medium taken without an account for its heating up by focused laser radiation. This can result in unpredicted effects.

We have developed a method for the account for non-linear medium heating based on first-principle analytical study of thermo-optical processes that take place in nonlinear-optical crystal structures under periodic femtosecond or nanosecond pumping. The developed approach will allow the experimentalists evaluate the temperature and thermo-optical anomalies just inside a pumped volume in every particular case.

It is generally considered that the refractive index of a crystal is uniquely defined by the crystal temperature and its stress-strain state. In case of short laser pulses, the situation is more complicated: 1) for times less than the highest relaxation time of photo-excited carriers and phonon subsystem, the concept of crystal temperature cannot be defined. In this time interval, even if the dielectric constant can be correctly defined at each moment, the computation of the refraction coefficient is a kinetics-type problem; 2) if the time interval exceeds the relaxation time, but is shorter than the sound propagation time, it is necessary to take into account the temperature waves; 3) if the time interval exceeds the sound propagation time, the standard parabolic equation for temperature is quite correct. We demonstrate that with a short pump pulse and a high pulse repetition frequency, the parabolic equation gives the correct results.

We have developed analytical and numerical methods for calculating the temperature for pulsed-periodic pumping. Numerical methods are not considered here. For the established periodic regime, an analytical solution for a specific experimental scheme can be obtained using known solutions from existing reference materials, replacing the time dependence.

In experiments for measuring *dn/dT* in some cases it is not this value that is directly determined, but the temperature coefficient of the optical path length change *ds/dT*, which describes the thermally induced distortion of a beam during its passing through a sample. Transversal inhomogeneity of crystal heating will lead to variation in change of optical path length for radiation propagating at different distances from the pumping beam optical axis. In turn, this will result in inclination, spherical deformation ("thermooptical lens" effect) or more complicated distortion of optical pumping wave front. Spatially inhomogeneous depolarization of radiation is also possible

Some concrete results ($LiNbO_3$ crystals) [1].
Case A. A cylinder-shaped crystal with radius 0.5 cm,

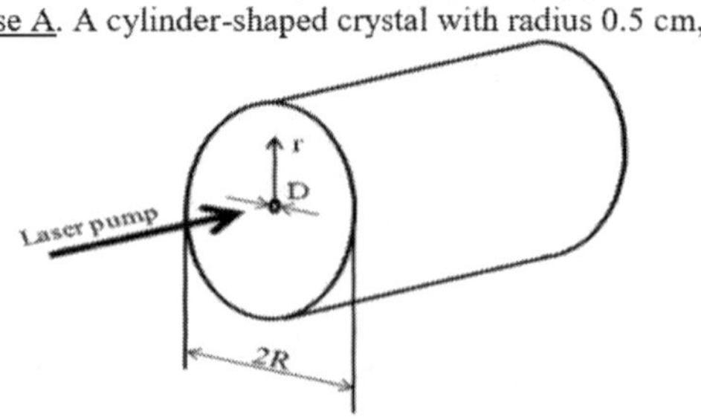

(a)

Fig.1a

much smaller than its doubled length (**Fig.1(a)**). Pumping radiation with Gaussian profile of diameter 100 μm propagates along the cylinder axis. On the crystal surface the heat exchange with air at 300 K

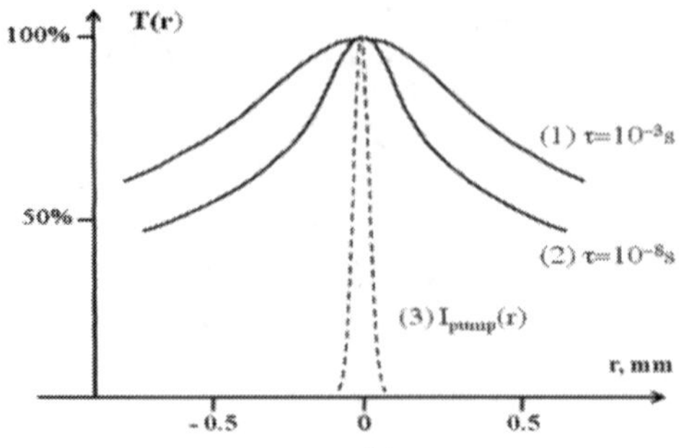

(b)

Fig.1(b)

temperature takes place according to Newton's law. The heat exchange coefficient is taken 8 W/m^2K. The graph of the corresponding dependence on the transversal coordinate *r* is shown in **Fig.1(b)**. With an increase in the repetition period of the pump pulses from 10^{-8} seconds to 10^{-3} seconds, the width of the profile of the temperature distribution increases approximately twice.

Case B. Cherenkov geometry, **Fig.2a**. A specially line-focused laser beam generally propagates as near as possible to one of the side facets of the 8 mm slab-shaped crystal equipped with a special prismatic massive Si-coupler. Here a pump laser beam (average power 8 W) propagates at the distance of 0.5 mm from the working side surface.

Case B1. The non-working surface of crystal is thermally insulated, the temperature of the Si-prism is 100 *K*, the repetition time frequency of the laser pulses is 3 kHz. In **Fig.2b (a) and (b),** the result of the numerical calculations of the temperature and dT/dr dependence on the distance r is shown; the scaled pump intensity profile is shown also there. The value of d^2T/dr^2 on the axis is 7 K/mm^2.

Case B2. The same as Case B1, but the temperature of the Si-prism is 300 *K*, the repetition time frequency of the laser pulses is 80 MHz, non-working surface experiences heat with air. **Fig.2c (a) and (b)** presents the spatial dependences of the temperature and dT/dr. The value of d^2T/dr^2 is 33 K/mm^2. d^2T/dr^2 on the axis of the laser beam is 7 K/mm^2.

Case B3. The same as in Case B2, but the non-working lateral surface contacts with a metallic cold loop. In **Fig.2d** the result of the numerical calculation of the temperature dependence on the distance r is shown. The value of d^2T/dr^2 is 18 K/mm^2.

The work was done under financial support of the Russian Science Foundation (Grant No. 17-12-01134).

References

1. *Kitaeva, G.Kh.., Moiseenko, E.V., Shepelev, A.V.* Temperature variation induced by the pulsed-periodic laser pumping under terahertz wave generation // Las. Phys. Lett. 2017, V.4, No.9, P.095401.

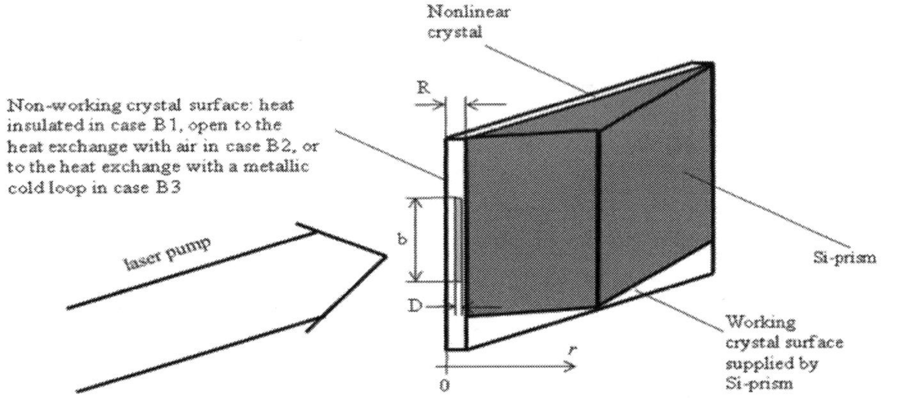

Fig.2a

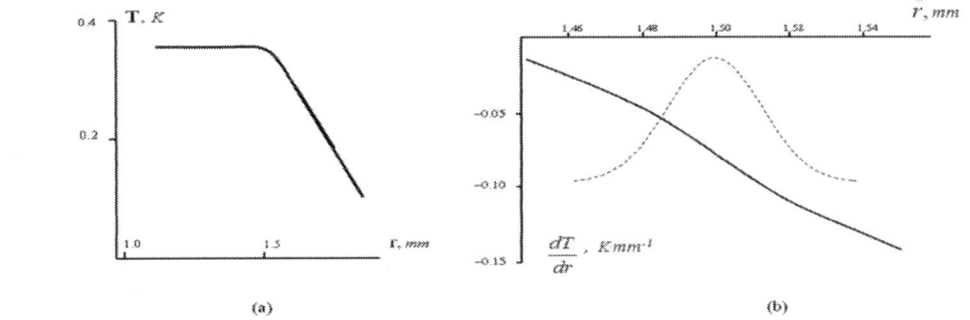

Fig.2b

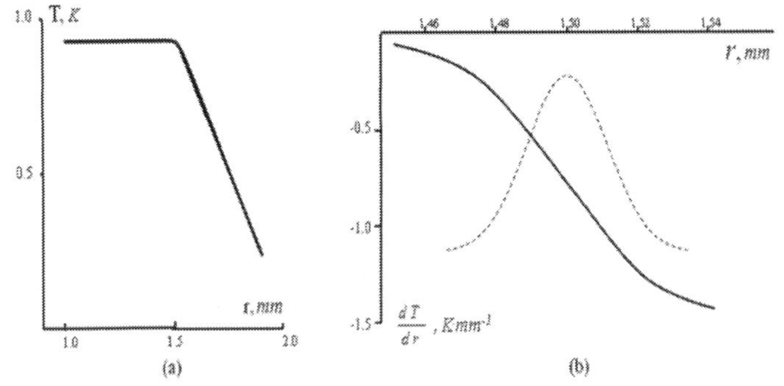

Fig.2c

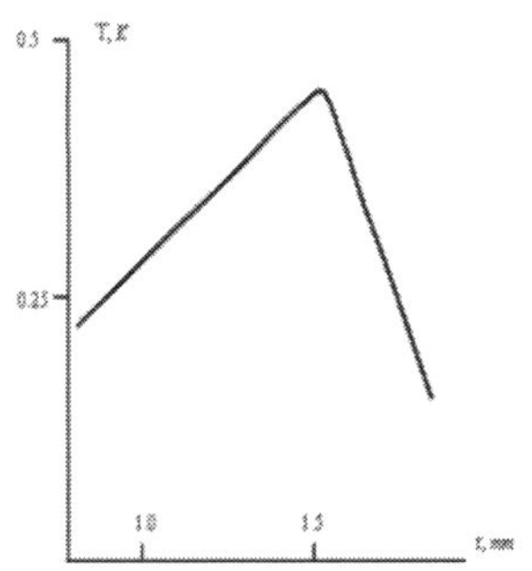

Fig.2d

EPJ Web of Conferences **195**, 03006 (2018) https://doi.org/10.1051/epjconf/201819503006
TERA-2018

Role of surface plasmons in laser-induced THz generation from metals

I. V. Oladyshkin, D. A. Fadeev, V. A. Mironov

Institute of Applied Physics of the Russian Academy of Sciences, Nizhny Novgorod, Russia
oladyshkin@gmail.com

Introduction

The paper is devoted to the analysis and preliminary theoretical interpretation of the last experimental results in laser-induced THz generation from metal structures. The main attention will be paid to metal gratings where resonant excitation of surface plasmons by the incident laser pulse is possible. Despite not a record efficiency level, study of optical-to-THz conversion on metal surfaces is relevant because of future applications in structural diagnostics and due to the fundamental interest in transformation mechanisms. Now there is no clear microscopic theory which is able to interpret the process of THz generation by metal gratings; in particular, the role of surface plasmons is not understood.

The first experimental measurements of THz signal from metal grating were performed more than 10 years ago [1]. In general, the strong influence of surface plasmon excitation on the THz generation was experimentally proved for different experimental conditions [1-3]. In the typical case a smooth resonant dependence of THz signal energy on the incidence angle is observed, where the resonant angle corresponds to the synchronism between the surface plasmon (on the optical frequency) and one of the diffraction waves [1, 2]. In the same time, one of the most complicated experimental features is non-quadratic dependence of the THz radiation energy on the optical fluence.

Mechanisms of plasmon-assisted THz generation

There are several possible microscopic mechanisms of plasmon-enhanced THz generation which were proposed before. The first one is based on the emission and acceleration of electrons above the surface, the second one – on the ponderomotive action of the plasmon field inside the metal, the third one takes into account thermal effects in the electronic gas.

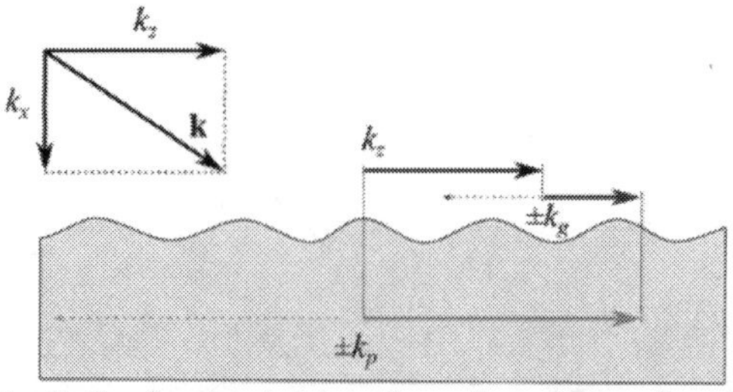

Fig. 1. Synchronism between the surface plasmon k_p and incident optical wave $\mathbf{k}$ in the case of periodical grating with wavenumber k_g: $k_z + k_g = k_p$

The first mechanism was previously experimentally refuted in [3], where the authors irradiated structured gold foil and observed THz emission from both incidence and rare side of the sample. Since electron emission and their acceleration into the substrate are impossible, the theory considering emitted electrons cannot be applicable in this case. The second mechanism can be also neglected since thermal nonlinear effects estimated to be stronger than ponderomotive-like ones in the case of metals [4]. So, here we will focus on the third (thermal) mechanism.

Thermal mechanism of THz generation for the case of flat metal surfaces was first proposed in [4], where analytical and numerical analysis was performed. Here we consider the same generation mechanism taking into account the possibility of surface plasmon excitation as an additional mechanism of the laser radiation absorption. Numerical model is based on Maxwell equations and hydrodynamical equations which describe electron gas motion and include heating by the laser field and heat diffusion.

Resonance THz generation

Well-known condition of synchronism between the surface plasmon wave and the optical wave above the metal grating is shown in the Fig. 1. In the first series of numerical calculations we study behavior of the THz response of the metal surface varying the grating period while the incidence angle was fixed (30°).

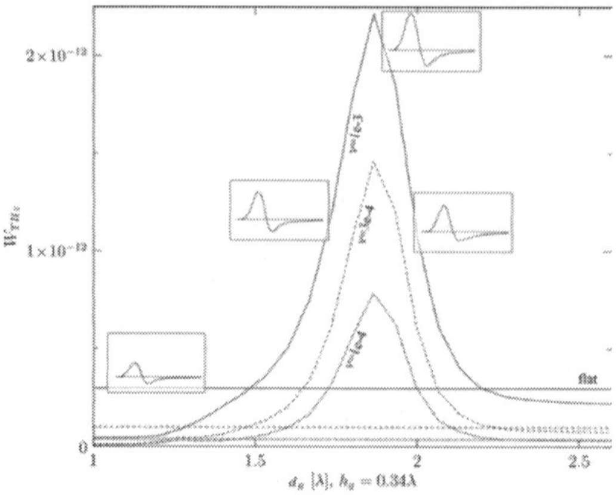

Fig. 2. Calculated dependence of the THz pulse energy on the grating period (angle of incidence = 30°, three different values of collision frequency are chosen). Insets: THz waveforms for different values of grating period.

Results presented in the Fig. 2 clearly indicate resonant character of the dependence (three curves correspond to three values of the scattering rate in metal, the grating period is measured in wavelengths of the incident optical pulse). Characteristic width of resonant curves equals to 10-15% of the resonant period which is in good agreement with experimental data from [2]. With the rise of collision frequency characteristic width also slightly increases.

Optimal grating depth

In the second series of numerical calculation we studied dependence of the conversion efficiency on the grating depth in the resonant point (see Fig. 3).

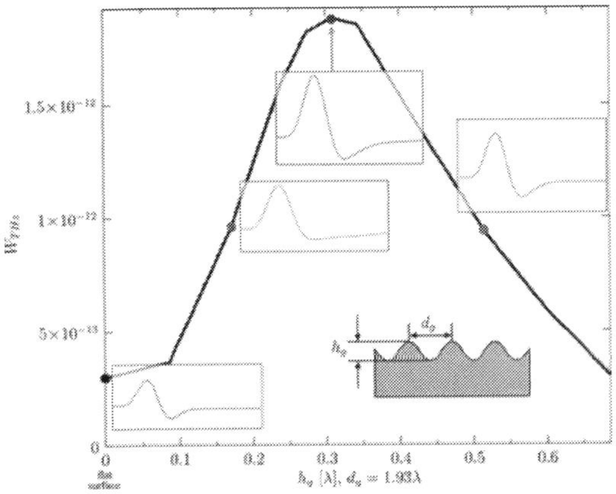

Fig. 3. Calculated dependence of the THz pulse energy on the grating depth (measured in the incident wavelengths). Angle of incidence equals to 30°. Insets: THz waveforms for different values of grating depth.

Obtained results are in good qualitative agreement with the experimental data from [3] where it was found that there is some optimal depth for THz generation (of about 150 nm for the optical wavelength 785 nm and incidence angle ≈45°). The further growth of the grating depth decreases THz signal. Quantitative comparison with the experiment requires modeling of the concrete grating profile (rectangular in most of cases). Calculated optimal value is about 0.3 wavelengths. The physical reason of this fact is a finite radiative lifetime of the surface plasmon which becomes shorter than the absorption time for too deep surface modulations.

To summarize, we perform numerical modeling of the THz signal generation process during the laser pulse reflection from the metal with periodically modified boundary. The key mechanism of low-frequency radiation generation is the medium polarization under the action of electronic temperature gradients. Here surface plasmon resonance can be treated an additional channel for the laser pulse absorption, so the resonance in absorption leads to rise of THz signal energy. Obtained numerical results (see Fig. 2 and Fig. 3) give a preliminary interpretation of the recent experimental data [2, 3]

Acknowledgements

This work was supported by the Program of basic research of the Presidium of the Russian Academy of Sciences "Nonlinear dynamics: fundamental problems and applications". I.O. is grateful for support to the Foundation for the advancement of theoretical physics and mathematics "BASIS"

References

1. *Welsh, G. H., Hunt, N. T., Wynne, K.* Terahertz-pulse emission through laser excitation of surface plasmons in a metal grating // Phys. Rev. Lett. 2007. V. 98. No. 2. P. 026803.

2. *Ramanandan, G. K. P., Ramakrishnan, G., Kumar, N., Adam, A. J. L., Planken, P. C. M.*, Emission of terahertz pulses from nanostructured metal surfaces // J. of Phys. D: Appl. Phys. 2014. V. 47, No. 37. P. 374003.

3. *Garwe, F., Schmidt, A., Zieger, G., May, T., Wynne, K., Hübner, U., Zeisberger, M., Paa, W., Stafast, H., Meyer, H.-G.*, Bi-directional terahertz emission from gold-coated nanogratings by excitation via femtosecond laser pulses // Appl. Phys. B. 2011. V.102, No. 3. P. 551—554.

4. *Oladyshkin, I. V., Fadeev, D. A., Mironov, V. A.*, Thermal mechanism of laser induced THz generation from a metal surface // J. of Optics 2015. V. 17, No. 7. P. 075502

Interplay effects of carrier-envelope phase and plasmon resonances in terahertz generation by ionizing ultrashort optical pulses

I. V. Osovitskaya[1,2], V. A. Kostin[1,2], N. V. Vvedenskii[1,2]

[1]Institute of Applied Physics, Russian Academy of Sciences, Nizhny Novgorod, Russia, oivnn121@mail.ru
[2]University of Nizhny Novgorod, Nizhny Novgorod, Russia

The direct conversion of femtosecond ionizing laser pulses into low-frequency radiation attracts considerable attention due to associated possibilities for creating sources of powerful broadband terahertz (THz) pulses. Previous studies of this phenomenon were mainly related to two limiting cases: (i) few-cycle (extremely short) ionizing pulses with several optical field cycles and (ii) multicycle (quasimonochromatic) pulses with a large number of field cycles. In the first case, the amplitude and energy of THz oscillations strongly depend on the carrier-envelope phase (CEP) in the ionizing pulse [1–3], whereas for multicycle pulses the CEP dependence is unnoticeable [4–6]. The spectral and mode structures of generated radiation also differ in these two cases. For the few-cycle pulses, the plasmonic oscillations corresponding to the dipole geometric resonance are excited and emitted from the laser-produced plasma filament to the surrounding space. For multicycle pulses, the radiation is produced by a relaxing current directed along the plasma filament, and the spectrum width and central frequency are determined by the collision frequency of electrons with heavy particles. Thus, the shape of low-frequency spectrum strongly depends on the pump pulse duration, and there is a range of durations where transition between different spectral shapes occurs. In this range of durations, the spectrum shape also significantly depends on the CEP, as was apparently observed in experiment [7]. Here, we study this intermediate case based on the universal model that considers the motion of the plasma electrons under forces both linear and quadratic with respect to the optical field strength.

The calculation of the THz radiation employs the two-stage approach [3, 5] which assumes that the characteristic frequency f_{THz} of the generated THz radiation is small compared with the inverse pump duration $1/\tau$: $f_{\mathrm{THz}}\tau << 1$. During the short (femtosecond) first stage, a gas medium is ionized and the free-electron currents are excited in the produced plasma by the optical pump pulse. In the longer (picosecond) second stage, there is a self-consistent evolution of these free-electron currents and the electromagnetic fields generated by them. Such separation into stages is convenient since one can neglect low-frequency plasma fields and assume the electric and magnetic fields given in the first stage, and one can consider fields and currents weak enough and neglect the nonlinear terms in the second stage.

Equations solved in the first stage are obtained directly from the kinetic equation

$$\frac{\partial f}{\partial t}+\mathbf{v}\cdot\frac{\partial f}{\partial \mathbf{r}}-\frac{e}{m}\left(\mathbf{E}+\frac{\mathbf{v}}{c}\times\mathbf{B}\right)\cdot\frac{\partial f}{\partial \mathbf{v}}=I$$

for the single-electron distribution function $f(\mathbf{r}, \mathbf{v}, t)$ (over positions $\mathbf{r}$ and velocities $\mathbf{v}$) with the ionization source term $I = (N_m - N)w_{\mathbf{v}}(\mathbf{E}, \mathbf{v})$. Here e is the elementary charge, m is the electron mass, c is the speed of light, $\mathbf{E}(\mathbf{r}, t)$ and $\mathbf{B}(\mathbf{r}, t)$ are the electric and magnetic fields, $w_{\mathbf{v}}(\mathbf{E}, \mathbf{v})$ is the differential (with respect to the initial photoelectron velocities) ionization probability per unit time in the static field $\mathbf{E}$, N_m is the initial density of neutral particles, and $N(\mathbf{r}, t)$ is the plasma density. Assuming the plasma being cold and the average velocity of newly-born electrons being zero, $\int w_{\mathbf{v}}(\mathbf{E}, \mathbf{v})\mathbf{v}\, d^3\mathbf{v} = 0$, one obtains equations for the plasma density, the electron density perturbation $n(\mathbf{r}, t) = \int f(\mathbf{r}, \mathbf{v}, t)\, d^3\mathbf{v} - N(\mathbf{r}, t)$, and the electron current density $\mathbf{j}(\mathbf{r}, t) = -e\int f(\mathbf{r}, \mathbf{v}, t)\mathbf{v}\, d^3\mathbf{v}$ as follows:

$$\frac{\partial N}{\partial t}=(N_m-N)\,w(\mathbf{E}),\quad \frac{\partial n}{\partial t}=\frac{\nabla\cdot\mathbf{j}}{e},\quad \frac{\partial \mathbf{j}}{\partial t}=\frac{Ne^2}{m}\mathbf{E}+\mathbf{S},\ (1)$$

$$\mathbf{S}=\frac{1}{e}\left[(\mathbf{j}\cdot\nabla)\frac{\mathbf{j}}{N+n}+\frac{\mathbf{j}}{N+n}(\nabla\cdot\mathbf{j})\right]+\frac{ne^2}{m}\mathbf{E}-\frac{e}{mc}\mathbf{j}\times\mathbf{B},\ (2)$$

where $w(\mathbf{E}) = \int w_{\mathbf{v}}(\mathbf{E}, \mathbf{v})\, d^3\mathbf{v}$ is the ionization probability per unit time. For numerical calculations below, we use the atmospheric density value $N_m = 3 \cdot 10^{19}$ cm^{-3} and the empirical formula for the hydrogen atom [8]:

$$w(\mathbf{E}) = 4\omega_a(E_a/|\mathbf{E}|)\exp(-2E_a/3|\mathbf{E}| - 12|\mathbf{E}|/E_a),$$

where $\omega_a = 4.13 \cdot 10^{16}$ s^{-1} and $E_a = 5.14 \cdot 10^9$ V/cm are the atomic units of frequency and electric field. To find fields and currents, we develop the perturbation approach for Eqs. (1) and (2) assuming that $n << N$ and all characteristic spatial scales (optical wavelength, size of plasma inhomogeneity) are large compared to the amplitude of electron oscillations.

In the first stage, we solve Eqs. (1) and (2) for the given fields $\mathbf{E}$ and $\mathbf{B}$ of the optical Bessel beam defined by the magnetic Hertz vector $\mathbf{\Pi}^m = \hat{\mathbf{y}}\int_{-\infty}^{\infty} A_\omega J_0(k_\perp r)e^{-i\omega\xi}\, d\omega/2\pi$, where θ is the focusing angle of the Bessel beam, $k_\perp = \sin\theta\,\omega/c$ is the transverse wavenumber, $\xi = t - z/V$ is the shifted time, $V = c/\cos\theta$ is the phase velocity, r is the distance to the beam axis (the z axis of the Cartesian and cylindrical coordinate systems), J_0 is the zero-order Bessel function of the first kind, A_ω is the Fourier spectrum of the y component of the Hertz vector at the beam axis (since $E_x = c^{-2}\cos\theta\, d^2\Pi_y^m/d\xi^2$, the spectrum of E_x at the axis is $-\cos\theta\,(\omega/c)^2 A_\omega$). In numerical calculations, the temporal profile of the pulse at the axis is chosen as

$$\Pi_y^m = -\left[(8\pi c^3 I_0)^{1/2}/(\omega_0\cos\theta)^2\right]\sin^4(\pi\xi/2\tau)\cos(\omega_0\xi+\psi)$$

for $|\xi + \tau| < \tau$, $\Pi_y^m = 0$ for $|\xi + \tau| \geq \tau$, where ω_0 is the central frequency of the ionizing field (which corresponds to the wavelength $2\pi c/\omega_0 = 800$ nm in calculations), I_0 is the maximum field intensity at the axis, ψ is the CEP. Thus, the fields depend on the shifted time ξ and transverse coordinates (radial coordinate r and azimuth angle φ in the cylindrical coordinate system. Another simplification is related to the cylindrical symmetry of the beam: The calculations show that the plasma density depends weakly on φ for small enough angles θ, and this dependence can be neglected. The current density in the first order of perturbation theory has a dipole structure, i.e., its components in cylindrical coordinates have one azimuthal variation and are proportional to cos φ or sin φ. The current density in the second order is a superposition of the symmetric term (the components do not depend on φ) and the term with two variations (components are proportional to cos 2φ or sin 2φ).

We integrate the equations for the first and second orders inside the cylindrical area $r < a = \mu_0 c/\omega_0 \sin\theta$, and determine the plasma density transverse profile remaining after the pump pulse has passed, $N_R(r) = N(r, \xi = 0)$, and the residual current density (RCD) distribution $\mathbf{j}_R(r, \varphi) = \mathbf{j}(r, \varphi, \xi = 0)$, which is a superposition of components with 0, 1, and 2 azimuthal variations; here $\mu_0 \approx 2.405$ is the first zero of J_0. We assume the field is so weak outside this cylindrical domain that the plasma and currents produced there do not significant affect the generated radiation. Then, the calculated distributions are used as parameters and initial conditions for the second stage, where the Maxwell equations and the linear equation for the electron current density in a cold collision plasma are solved,

$$\nabla\times\mathbf{E}=-\frac{1}{c}\frac{\partial\mathbf{B}}{\partial\xi},\ \nabla\times\mathbf{B}=\frac{1}{c}\frac{\partial\mathbf{E}}{\partial\xi}+\frac{4\pi}{c}\mathbf{j},\ \frac{\partial\mathbf{j}}{\partial\xi}+\nu\mathbf{j}=\frac{\omega_p^2}{4\pi}\mathbf{E},$$

where $\omega_p(r) = [4\pi N_R(r)e^2/m]^{1/2}$ is the plasma frequency and ν is the collision frequency of electrons with heavy particles ($5 \cdot 10^{12}$ s^{-1} in calculations). The initial conditions for fields are $\mathbf{E}|_{\xi=0} = 0$, $\mathbf{B}|_{\xi=0} = 0$; the RCD found in the first stage serves as the initial condition for the electron current density, $\mathbf{j}|_{\xi=0} = \mathbf{j}_R$. At the boundary $r = a$, the radiation conditions are set under assumption that there is no plasma in the external area, $N(r > a) \equiv 0$. We solve this initial problem separately for each of three azimuthal components of the RCD by applying the Laplace transform with respect to the variable ξ. After that, we calculate the spectral density $w_{\text{rad}}(f)$ and the radiated energy $W_{\text{rad}} = \int_0^{f_{\max}} w_{\text{rad}}(f)df$ per unit length along the z axis in THz range, where $f_{\max}$ is the upper bound of the THz range under consideration. The spectral density can be expressed in terms of the Laplace images of the longitudinal field components at the boundary of the computational area $w_{\text{rad}} = w_{\text{rad}}^{[0]} + w_{\text{rad}}^{[1]} + w_{\text{rad}}^{[2]}$, where

$$w_{\text{rad}}^{[M]}(f)=\frac{c^2}{2\pi^2 f\sin^2\theta}\frac{\left|\widetilde{E}_z^{[M]}(a,0)\right|^2+\left|\widetilde{B}_z^{[M]}(a,\pi/2M)\right|^2}{\left|H_M^{(1)}(2\pi\sin\theta\, fa/c)\right|^2}$$

is the spectral density of the radiated energy of the component with M azimuthal variations, $\widetilde{E}_z^{[M]}(r,\varphi)$ and $\widetilde{B}_z^{[M]}(r,\varphi)$ are the Laplace images of the longitudinal components of the electric and magnetic fields, respectively, for the Laplace variable $p \to 2\pi i f + 0$, and $H_M^{(1)}$ is the Hankel function of the first kind and order M.

We solve numerically the above equations for a wide range of focusing angles $1° \le \theta \le 30°$, durations 4 fs $\le \tau_p = [2\arcsin(2^{-1/8})/\pi]\tau \le 50$ fs, and intensities 10^{14} W/cm$^2 \le I_0 \le 10^{15}$ W/cm^2 of ionizing pulses for different CEPs $0 \le \psi \le \pi$. The dipole RCD strongly depends on ψ for any pumps considered, whereas the distribution of the plasma density and the RCD terms with 0 and 2 azimuthal variations practically do not depend on ψ even for extremely short durations τ_p. Although the dipole RCD proves to be the weakest, for few-cycle pulses its contribution to THz radiation can be determinant. However, the dipole component decrease rapidly with increasing duration, and the contributions from different azimuthal components to THz radiation are already comparable at not too long durations. Under these conditions, the THz spectrum shape can vary greatly as ψ changes. With a further increase in the duration, the dipole component becomes negligibly small, and the dependence on the CEP disappears.

This work was supported by the Russian Foundation for Basic Research (analytical study, Grants Nos. 16-32-60166, 18-02-01150 and 18-32-00951) and Russian Science Foundation (numerical calculations, Grant No. 18-11-00210).

References

1. *Löffler, T., Jacob, F., Roskos, H. G.* Generation of terahertz pulses by photo-ionization of air // Appl. Phys. Lett. 2000. V. 77, No. 3. P. 453–455.

2. *D'Amico, C., Houard, A., Franco, M.* Conical forward THz emission from femtosecond-laser-beam filamentation in air // Phys. Rev. Lett. 2007. V. 98, No. 23. P. 235002.

3. *Kostin, V. A., Vvedenskii, N. V.* Ionization-induced conversion of ultrashort Bessel beam to terahertz pulse // Opt. Lett. 2010. V. 35. No. 2. P. 247–249.

4. *Kreß, M., Löffler, T., Thomson, M. D.* Determination of the carrier-envelope phase of few-cycle laser pulses with terahertz-emission spectroscopy // Nat. Phys. 2006. V. 2, No. 5. P. 327–331.

5. *Gildenburg, V. B., Vvedenskii, N. V.* Optical-to-THz wave conversion via excitation of plasma oscillations in the tunneling-ionization process // Phys. Rev. Lett. 2007. V. 98, No. 24. P. 245002.

6. *Silaev, A.A., Vvedenskii, N. V.* Residual-current excitation in plasmas produced by few-cycle laser pulses // Phys. Rev. Lett. 2009. V. 102, No. 11. P. 115005.

7. *Bai, Ya, Song, L., Xu, R., Li, C., Liu, P., Zeng, Z., Zhang, Z., Lu, H.* Waveform-controlled terahertz radiation from the air filament produced by few-cycle laser pulses // Phys. Rev. Lett. 2012. V. 108, No. 25. P. 255004.

8. *Tong, X. M., Lin, C. D.* Empirical formula for static field ionization rates of atoms and molecules by lasers in the barrier-suppression regime // J. Phys. B. 2005. V. 38. No. 15. P. 2593–2600.

https://doi.org/10.1051/epjconf/201819503008

New Approach to Generation and Amplification of the THz Radiation in Plasma Created by Intense Two-Color Laser Fields

A. M. Popov[1,2,3], A. V. Bogatskaya[1,3]

[1]D. V. Skobeltsyn Institute of Nuclear Physics, Moscow State University, Moscow, Russia, alexander.m.popov@gmail.com
[2]Department of Physics, Moscow State University, 119991, Moscow, Russia
[3]P. N. Lebedev Physical Institute, RAS, Moscow Russia

Construction of new types of THz radiation sources is nowadays of sufficient fundamental and practical interest. This interest is motivated by the prospects of using THz radiation for THz imaging in a lot applications in science and technology [1]. However, till now THz frequency band is still insufficiently studied in comparison with infrared and microwave ones. Among others, THz generators based on the gas ionization by high-intensity femtosecond lasers are widely studied now. Such a way of THz generation allows to obtain radiation with a continuous spectrum throughout the terahertz bandwidth that is of importance for different spectroscopy applications. Typically, optical-to-THz conversion results from the four wave mixing process (FWM) in high-intensity femtosecond two color laser field ($\omega+2\omega$). Unfortunately, the efficiency of such conversion remains rather small - about 0.001- 0.01%. In this respect, problems of amplification of THz pulses as well as the development of new more efficient methods of optical to terahertz energy conversion are of significant importance. In this paper we propose an alternative way to enhance significantly the process of optical-to-THz conversion by using the resonance in four-wave mixing process. Such resonant enhancement is possible in the aluminum vapor irradiated by Ti-Sa laser pulse and its second harmonic via the aluminum atom excited state $4s(^2S_{1/2})$ with the energy $E_{4s}\approx 3.14$ eV close to the frequency of the Ti-Sa second harmonic $\hbar(2\omega)\approx 3.1$ eV. Schematically this process is shown at fig.1 We demonstrate that the THz output can be increased up to two orders of magnitude in comparison with the non-resonant process.

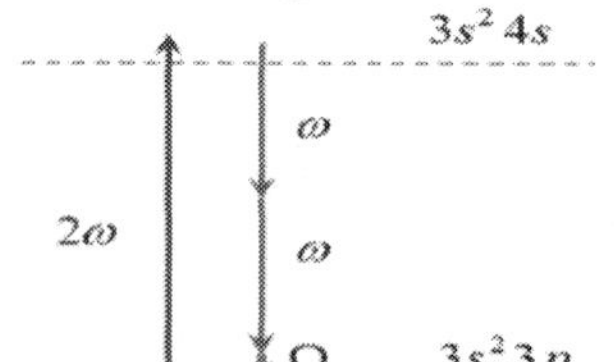

Fig 1. Nearly resonance four-wave-mixing process in aluminum atom

It is known that irradiation of gas by intense two color ($\omega+2\omega$) laser pulse results in low-frequency polarization of the medium $P_\Omega \sim \chi^{(3)}E_\omega^2 E_{2\omega}$ and THz emission [2] which arises from the four-wave mixing (FWM) process ($\Omega=\omega'+\omega''-2\omega'''$) and in the limit of perturbation theory is governed by the cubic susceptibility $\chi^{(3)}(\omega'+\omega''-2\omega'')$ (ω', ω'' are the frequencies from the spectral width of the pulse with frequency ω while ω''' is the frequency from the spectral width of second harmonic 2ω.). For femtosecond laser pulses this low-frequency spontaneous emission appears in the THz and far IR frequency band. In this paper we plan to study low frequency spontaneous background from gas or plasma irradiated by two-color laser field which could be further amplified according to the mechanism proposed in [3].

Below we analyze the resonant FWM process and the formation of the spontaneous THz background for xenon and aluminum atoms using the approach developed in [4] that is based on the accurate consideration of quantum system interaction with vacuum quantized field modes in the first order of perturbation theory while the intense laser field is considered classically beyond the perturbation theory.

The electric field strength in the two-color pulse was chosen in a form

$$E(t)=f(t)\times(E_\omega\cos(\omega t)+E_{2\omega}\cos(2\omega t)),$$

where ω the frequency of the Ti-Sa laser and $f(t)$ is the envelope being the same for both pulses. It was chosen in the sine-squared form with the half-duration $\tau=880$ atomic units (~20 fs), that corresponds to 8 optical cycles of the Ti-Sa laser or 16 optical cycles for its second harmonic.

The study of atomic emission was performed in the range of intensities for both fundamental and second harmonic $\sim 10^{-5}-2\times10^{-3}$ at. un. and the intensity of second harmonic is assumed to be several time less than the intensity of the fundamental one. On the one hand, the increment of the intensity gives rise to more efficient FWM process, on the other hand the ionization and the depletion of the ground state will reduce the probability of the resonance FWM.

Typical spontaneous emission spectra obtained in our simulations for aluminum atom are presented at figure 2. Curves 1, 2 and 3 represent the spectra obtained for Ti-Sa laser irradiation, its second harmonic as well as combined two-color action respectively. First of all, the peaks of Rayleigh scattering at $\hbar\omega_{k\lambda}=1.55$ eV for fundamental frequency and $\hbar\omega_{k\lambda}=3.1$ eV for the second harmonic are observed. The last peak has the triple structure (triplet Mollow) due to the near-resonant transition to 4s state. There are also some additional peaks appearing for the Ti-Sa laser resulting from different processes. Most important to highlight that for two-color laser pulse one more low-frequency peak with a maximum at $\hbar\Omega_{k\lambda}\approx 0.2$ eV is well pronounced. This peak manifests the FWM process when the second harmonic quantum absorption and two quanta of Ti-Sa laser stimulated emission is accomplished by spontaneous emission of THz - far IR quantum (see figure 1).

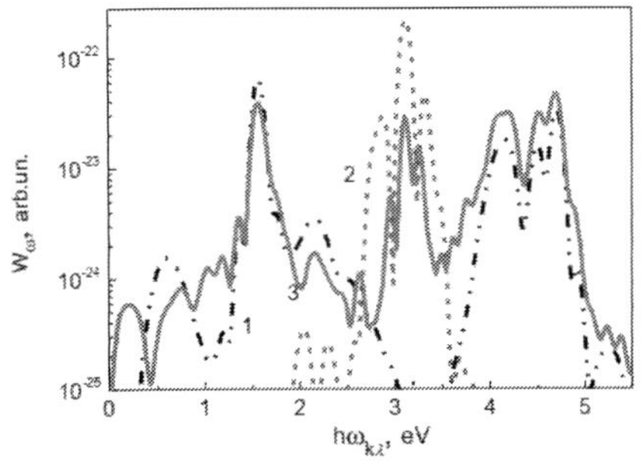

Fig. 2. Spectrum of spontaneous emission of aluminum atom irradiated by Ti-Sa laser pulse with $I_\omega = 1.4\times10^{-3}$ (in atomic units) (1), its second harmonic with $I_{2\omega} = 10^{-4}$ (2) and by two-colour pulse (3) with $I_\omega = 1.4\times10^{-3}$ and $I_{2\omega} = 10^{-4}$.

The detailed analysis of the low-frequency emission in the near resonant FWM process leads to the conclusion that optimal values for THz emission exist near the intensities $I_\omega = 1.4\times10^{-3}$ and $I_{2\omega} = 4\times10^{-4}$.

The resonant character of FWM can dramatically increase THz emission in comparison with the non-resonant case. To demonstrate it, the simulation of THz background from the xenon atom was also analyzed for nearly the same laser radiation parameters [5]. Comparative analysis of these data leads to the conclusion that resonant character of FWM provides the enhancement of the emission in THz bandwidth up to two orders of value.

Further amplification of a seed low-frequency pulse can be realized in the nonequilibrium plasma channel formed independently by the UV laser pulse. Earlier it was demonstrated in [6] that such a plasma produced in the process of multiphoton ionization of heavy rare gases by powerful UV laser pulse is characterized by population inversion that can be used for amplification of THz-band radiation. This amplification is possible if the energy of photoelectron peak is located in the interval of increasing with energy (faster than the linear dependence) electron transport cross section.

Possible realization of such a concept is performed at figure 3. Terahertz spontaneous signal emerges in the region of intense two-color laser action and then propagates through the amplifying zone which is formed by KrF laser pulse. In the frames of our concept the presence of two different lasers is mandatory because we require high-intensities for the two-color pulse to reach high level of nonlinearity for FWM process but at the same time we do not need high level of ionization in the amplifying plasma channel as domination of electron-electron collisions will cause the fast relaxation of the nonequelibrium electron energy distribution function (EEDF) and destroy the population inversion..

To conclude, in this paper we demonstrate that the resonance between quanta of the laser radiation and energy transition from the ground to one of the excited states of the atomic system can result in the dramatic enhancement of the four-wave mixing process in two-colour laser field ($\omega + 2\omega$) providing more efficient optical to THz conversion scheme. Possibility of further amplification of the THz background is discussed. Also, the new concept allowing to enhance the THz output from the two-color scheme in a nonequilibrium plasma channel formed by UV laser was proposed.

This work was supported by the Russian Science Foundation (project no. 18-72- 00125).

References

1. *Mittleman, D. M.* Twenty years of terahertz imaging // Opt. Express 2018.V. 26, P. 9417

2. *Cook, D. J. and Hochstrasser, R. M.* Intense terahertz pulses by four-wave rectification in air // Opt. Lett. 2000. V. 25, P. 1210

3. *Bogatskaya, A. V., Volkova, E. A., Popov, A. M.* Generation and amplification of sub-THz radiation in a plasma of rare gases formed by two-color femtosecond laser pulse // Las. Phys. Lett. 2018. V. 15, P. 065301

4. *Bogatskaya, A. V., Volkova, E. A., Popov, A. M.* Spontaneous transitions in atomic system in the presence of high intensity laser field // Europhysics Letters 2016. V. 116, P. 14003

5. *Bogatskaya, A. V., Popov, A. M.* New approach to the problem of THz generation by intense two-color laser fields // Laser Phys. 2018. V. 28 submitted

6. *Bogatskaya, A. V., Popov, A. M.* On the possibility of the amplification of subterahertz electromagnetic radiation in a plasma channel created by a high-intensity ultrashort laser pulse // JETP Lett. 2013. V. 37. P. 388-392

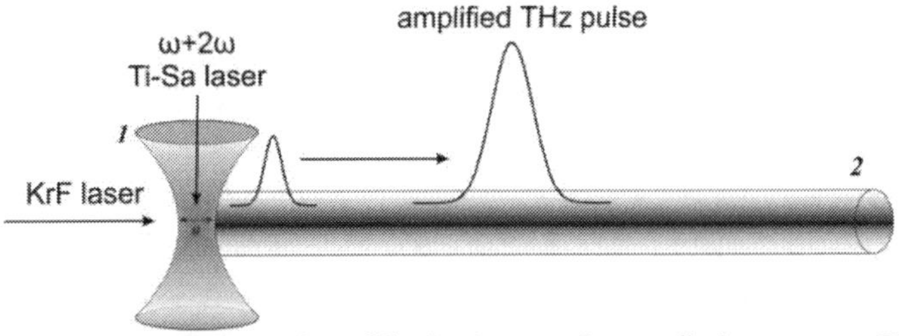

Fig. 3. Principal scheme of THz pulse generation and amplification in case of perpendicular geometry. THz signal is produced in the region of focal waist of two-color Ti-Sa laser action (channel 1) and then enhanced in the 2nd channel formed in xenon by KrF laser pulse.

Terahertz radiation in two-color laser fields: from single atom to extended gas response

S. Yu. Stremoukhov[1,2], A.V. Andreev[1]

[1]Faculty of Physics, Lomonosov Moscow State University, Leninskie Gory, 1, build.2, 119991, Moscow, Russia, sustrem@gmail.com
[2]National Research Centre "Kurchatov Institute", pl. Akademika Kurchatova, 1, 123182, Moscow, Russia

One of the promising methods of the THz radiation generation is the use of the two-color laser field focused into the gas media [1, 2]. The polarization of the low frequency THz field generated in gas is linear [2], although, in some conditions, it is elliptical due to the modulation of the laser phase and its polarization in the gas plasma [3]. At the same time, for some practical applications, it is critical to have the elliptically polarized THz radiation [4]. For its generation certain methods [2, 5] mostly based on the mixing of the elliptically polarized laser fundamental radiation (ω) with linear (or elliptically polarized) second harmonic radiation (2ω) exist, the generated elliptical THz radiation lies in low frequency region (<5THz).

Here, we demonstrate the method of generation of the highly elliptical high-frequency THz radiation (>25 THz) by the extended gas media interacting with the two-color laser field having linearly polarized components but with a nonzero angle between them [6]. The consistent quantum-mechanical description is given. The two spatial levels are introduced. Initially, the time-dependent Schrodinger equation for single atom "placed" in a finite number of positions in the extended gas (i.e. being under the action of the two-color laser field having various intensities and relative phases between the pulses [7]) is solved. Then, we calculate the response of extended system by changing the spatial parameters of the gas (mainly, the gas length, l) and the laser beam (mainly, beam width, σ)).

For the calculation of the properties of the THz radiation emitted by a single atom we have used the non-perturbative theory [8, 9] and the model of the Ar atom described in [7, 10]. In simulations we assume that the components of the two-color laser field have the fixed intensity ($I_\omega = 6.77\times10^{12}\ W\cdot cm^{-2}$ and $I_{2\omega} = 5.85\times10^{11}\ W\cdot cm^{-2}$), fixed temporal width of the laser field components ($\tau_{1,2} = 30\ fs$), zero delay time between the pulses ($\Delta t_{1,2}=0$) and relative phase equal to $\varphi_0=\pi/2$. At these parameters of laser field the ionization-free regime of interaction occurs [10]. In this way we obtain a series of photoemission spectra for different angles between the polarizations of two-color laser field components (θ) and different delay times between the pulses.

Single atom numerical calculations demonstrate that at frequences $>25\ THz$ there are some regions of high ellipticity (up to circular), the maximum value of the ellipticity and the widths of these regions decrease along with an increase in the THz frequency [6].

Basing on the single-atom response we have calculated the response of the extended gas media. Previously we have used the formula (4) from [7] for describing the linearly polarized THz laser field spatial distribution in an aria around the gas. In that calculations we have summarized two projections onto perpendicular axes of THz fields emitted from individual atoms and have extracted total intensity of the THz radiation at given points of the area. In current calculations of the elliptically polarized THz field response each atom emits elliptically polarized radiation which could be expanded on two projections on the perpendicular axes $E_{xj} = |E_{xj}| e^{i\alpha_{xj}}$ and $E_{yj} = |E_{yj}| e^{i\alpha_{yj}}$ (see Fig. 1).

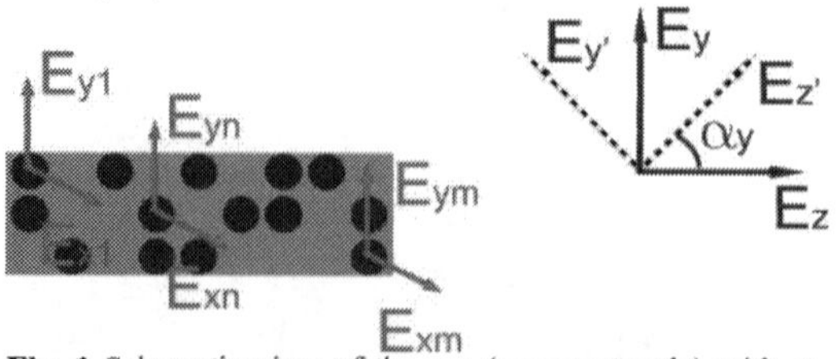

Fig. 1 Schematic view of the gas (gray rectangle) with atoms (black circles) emitting the THz field (red arrows indicate two projections). By the black solid arrows and black dashed arrows indicated the two projections of the field formed by the sum of projections of the single color field on the one axes. α_y denotes the rotation angle.

By using formulae (4) from [7] two projections of the summarized field could be found at given point of the area for each projections ($x-$ and $y-$) of the THz field separately (see Fig.1 with shown projections only for $y-$ component of the THz field, the other two projections for $x-$ component of the THz field lies in the perpendicular plane). As a result, two projections of the single atom THz field response are transformed at four ones when the problem of extended gas media response is under consideration. But the turning of two perpendicular axes on the angle can decrease the number of projections. Indeed, if we turn $yz-$ axes on the angle α, the projections of the THz field on the new $y'z'-$ axes become

$$E_{y'} = E_y \cos\alpha + E_z \sin\alpha,$$
$$E_{z'} = -E_y \sin\alpha + E_z \cos\alpha.$$

If the rotation angle will be chosen by the following way

$$\tan\alpha_y = \frac{E_z}{E_y},$$

then $E_{z'} = 0$ and $E_{y'}$ will have the form of

$$E_{y'} = E_y \cos\left(\arctan\left(\frac{E_z}{E_y}\right)\right) + E_z \sin\left(\arctan\left(\frac{E_z}{E_y}\right)\right).$$

As a result, only one complex projection have non-zero value. Providing the following procedure with other two projections (with are corresponding to the x−component of the atoms THz components) we will have only one projection of the THz field which has the form of

$$E_{x'} = E_x \cos\left(\arctan\left(\frac{E_z}{E_x}\right)\right) + E_z \sin\left(\arctan\left(\frac{E_z}{E_x}\right)\right).$$

So, there are only two complex projections of the THz field at each point of space. Using them and providing some standard procedures ones can extract properties of the THz field (intensity, ellipticity, polarization angle).

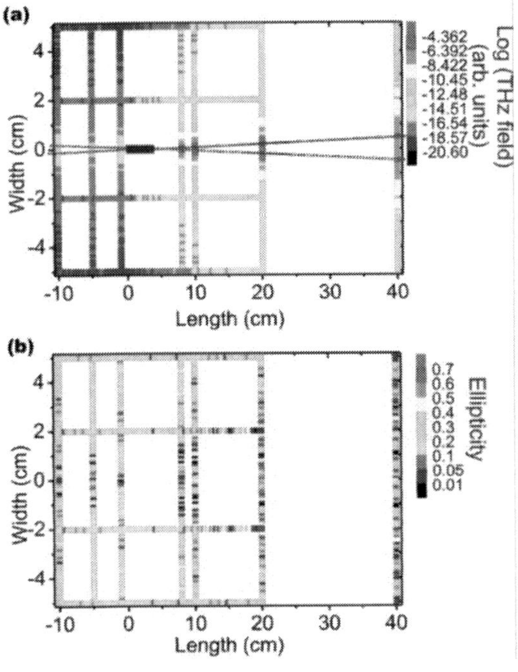

Fig. 2 Spatial distribution of the THz field strengths (a) and ellipticities (b) calculated for ν=29 THz for l=6 cm and σ=0.1 cm. The black rectangle is the emitted medium.

The typical THz field distributions calculated for $l=6$ *cm*, $\sigma=0.1$ *cm*, $\theta=\pi/4$ and $\nu=29$ *THz* are presented in Fig. 2. The THz field distributions demonstrate the cone structure mainly. These distributions also contain the conical structure in the backward direction [11], which is possible due to the complete orders in "emitted atoms" locations in the gas (with a fixed period). The backward emission is less intensive compared to the forward one. Fig. 2b demonstrated the ellipticity distribution of the generated radiation.

We have investigated three possibilities to control the value of ellipticity (and the THz field strength) of the generated radiation. The first one is to change the relative phase between the pulses of the two-color laser field at the left side of the gas (the same method was previously investigated in [2]).

The second way to an effective control of the value of the ellipticity of the generated THz radiation is to change the spatial geometry of the interaction (the length of the gas l and the width of the laser beam σ).

The third method of the effective control of the ellipticity value of the generated THz radiation is by changing the spatial profile of the laser beam, i.e. by using an iris and a mask which is completely opposite to the iris: the iris transmits, in this case, the central part of the laser pulses, and the mask transmits the periphery one [6].

The work was partially supported by the Russian Foundation for Basic Research under Projects Nos. 18-02-00528, 18-02-00743, 18-02-40014.

References

1. *Cook, D. J., Hochstrasser, R. M.* Intense terahertz pulses by four-wave rectification in air // Opt. Lett. 2000. V. 25, p. 1210.

2. *Dai, J., Karpowicz, N., Zhang ,X. C.* Coherent Polarization Control of Terahertz Waves Generated from Two-Color Laser-Induced Gas Plasma // Phys. Rev. Lett. 2009. V. 103, p. 023001.

3. *Xu, J., Galan, J., Ramian, G., Savvidis, P., Scopatz, A., Birge, R. R., Allen, S. J., Plaxco, K.* Terahertz Circular Dichroism Spectroscopy of Biomolecules // Proc. SPIE. 2004. 5268, 19–26.

4. *Hughes, S., Citrin, D.S.* Interaction of terahertz transients and broadband optical pulses in quantum wells // J. Opt. Soc. Am. B. 2000. V. 17, p. 128.

5. *Sato, M., Higuchi, T., Kanda, N., Konishi, K., Yoshioka, K., Suzuki, T., Misawa, K., Kuwata-Gonokami, M.* Terahertz polarization pulse shaping with arbitrary field control // Nat. Photonics. 2013. V. 7, p. 724.

6. *Stremoukhov, S., Andreev, A.* Quantum-mechanical fingerprints in generation of elliptical terahertz radiation by extended media interacting with two-color laser field // J. Opt. Soc. Am. B. 2017. V. 34(2), p. 232-237.

7. *Stremoukhov, S. Yu., Andreev, A. V.* Spatial variations of the intensity of THz radiation emitted by extended media in two-color laser fields // Laser Phys. Lett. 2015. V. 12, p. 015402.

8. *Andreev, A.V., Stremoukhov, S. Yu., Shoutova, O. A.* Light-induced anisotropy of atomic response: prospects for emission spectrum control // Eur. Phys. J. D. 2012. V. 66, 16.

9. *Stremoukhov, S., Andreev, A., Vodungbo, B., Salières,P., Mahieu, B., Lambert, G.* Origin of ellipticity of high-order harmonics generated by a two-color laser field in the cross-polarized configuration // Phys. Rev. A. 2016. V. 94, p. 013855.

10. *Andreev, A. V., Stremoukhov, S. Yu.* Terahertz-radiation generation in the ionization-free regime of light-atom interaction // Phys. Rev. A. 2013. V. 87, p. 053416.

11. *Jahangiri, F., Hashida, M., Nagashima, T., Tokita, S., Hangyo, M., Sakabe, S.* Intense terahertz emission from atomic cluster plasma produced by intense femtosecond laser pulses // Appl. Phys. Lett. 2011. V. 99, p. 261503.

Backward terahertz emission from two-color laser induced plasma spark

Aleksandr Ushakov[1,2], Pavel Chizhov[1], Vladimir Bukin[1], Nikolay Panov[2], Daniil Shipilo[2], Olga Kosareva[2], Andrei Savel'ev[2], Sergey Garnov[1]

[1]A.M. Prokhorov General Physics Institute RAS, Moscow, Russian, ushakov.aleksandr@physics.msu.ru
[2]M.V. Lomonosov Moscow State University, Moscow, Russia

Introduction

Femtosecond gas plasma created by focusing of two-color laser pulses is one of the promising laser-driven THz sources [1]. Investigation of spatial distribution of THz radiation from plasma is of great interest [2-4]. Several theoretical works have predicted an essential backward emission of terahertz radiation from two-color femtosecond plasma under tight focusing conditions [5]. We proved experimentally the presence of such radiation previously in [6]. However, to get a clear insight into mechanisms leading to backward emission one has to investigate characteristic features of this radiation. Here we perform its spectral and energetic investigation.

Experimental setup

A Ti:Sa femtosecond laser (800 nm, 40 fs, 2.8 mJ, 1 kHz rep. rate, 12 mm dia.) is used. Laser radiation is partially converted into second harmonic in BBO crystal (10x10x0.2 mm^3, I-type). To provide efficient THz generation polarizations of fundamental and second harmonic pulses are made collinear by a dual-wavelength phase plate and adjusted for optimal temporal shift by a delay compensator plate. Two-color femtosecond laser pulse is focused in ambient air by a parabolic mirror with 0.8 inch focal length to create plasma.

To perform forward THz radiation measurements we use two PTFE-lenses (5 cm diameter, 6 and 10 cm focal lengths). The first lens collimates radiation and the second focuses it on a THz detector. We use Golay cell for power measurements and electro-optical sampling in ZnTe crystal (10x10x0.5 mm^3, <110> cut) for spectral analysis.

The parabolic mirror that is used for laser beam focusing collimates backward THz radiation. Thereafter it is partially reflected on 90° by a metal plate with an aperture cut for passing of the laser beam. Then THz radiation is focused by a PTFE-lens on the THz detector. For power measurements it is 5 cm diameter 6 cm focal length lens. For electro-optical sampling it is 5 cm diameter 4 cm focal length lens with a 1 mm aperture for passing of a probe laser pulse.

Results

We have obtained waveforms (see Fig.1(a)) and corresponding spectra of THz radiation spreading from two-color femtosecond laser plasma both in forward and backward directions. The backward pulse spectrum is somewhat narrower than the forward one.

A numerical simulation based on a simple interferometric model is performed to support experimental data [6, 7]. During the calculations we assume the length of plasma channel to be 250 μm; the length of plasma channel recorded in our experiment is (256 ± 6) μm (see inset in Fig. 1b). The backward P_{bwd} and forward P_{fwd} spectral powers of THz radiation are calculated with taking into account the collecting angles of the parabolic mirror and PTFE lenses; the transmission of the metal plate with ∅15-mm hole is accounted. The spectral powers P_{bwd} and P_{fwd} are found for the THz frequencies ν_{THz} from 0.05 to 3 THz with the step 0.05 THz. The ratio $E(\nu_{THz})=P_{bwd}(\nu_{THz})/P_{fwd}(\nu_{THz})$ is the estimation the transmission function from the forward THz spectrum $S_{fwd}(\nu_{THz})$ to the backward one $S_{bwd}(\nu_{THz})$:

$$E(\nu_{THz})=P_{bwd}(\nu_{THz})/P_{fwd}(\nu_{THz}) \quad (1)$$

The ratio $E(\nu_{THz})$ has the order of unity only for $\nu_{THz} \leq 0.5$ THz (see Fig. 1b), so one can expect the shift of backward THz spectrum to the lower frequencies within the agreement with the experiment. We reconstruct the backward THz spectrum from the experimentally measured forward one for calculated function $E(\nu_{THz})$. The adequate semi-quantitative agreement between measured and reconstructed spectra supports the conclusion about the low-frequency bias in the backward THz spectrum. The qualitative explanation of this effect is the following: for the spectral THz components with the wavelength much longer than the plasma channel length of about 250 μm (i.e. $\nu_{THz} \ll 1.2$ THz in good agreement with the previous estimation $\nu_{THz} \leq 0.5$ THz) the plasma spark is closer to point-source as compared with shorter THz wavelength components. This point-source emits the low-frequency THz radiation both in forward and backward directions. We note, that the THz spectrum of two-color optical breakdown spreads up to ~50 THz, and the EO system can measure its only small part up to 3 THz. However, the transmission function, which determines the backward THz spectrum mostly, has the width of ~0.5 THz (see Fig. 1(b)) within the EO system detection range. Therefore, we measure the whole spectrum of the backward THz emission from two-color microplasma.

They also demonstrate narrowing of the spectrum of radiation emitted in backward direction. This feature arises from destructive interference from local emitters along plasma spark. The less the wavelength is, the more periods of radiation correspond to the plasma length. Thus, overall interference is more destructive. However, the experimentally observed narrowing is less dramatic as compared to the simulations.

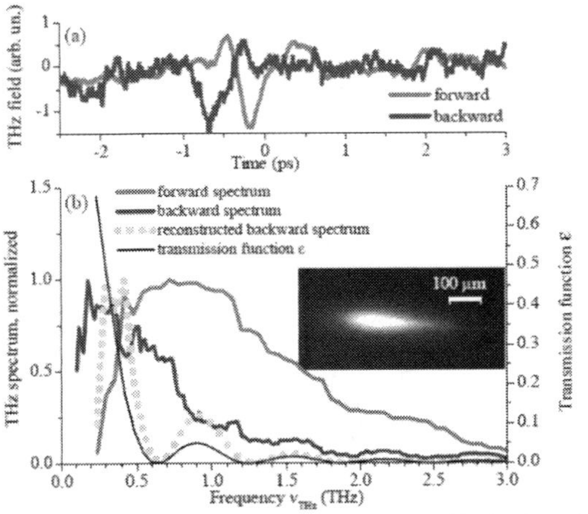

Fig. 1. (a) Waveforms of THz field measured in backward and forward directions. (b) Spectra of experimentally measured forward (red line) and backward (blue line) THz emission with plasma channel, transmission function received from formula 1 (black line), numerically calculated backward THz emission (green dotted line).

We also get backward/forward emission energy dependencies on the energy of the two-color laser pump pulse (see Fig.2).

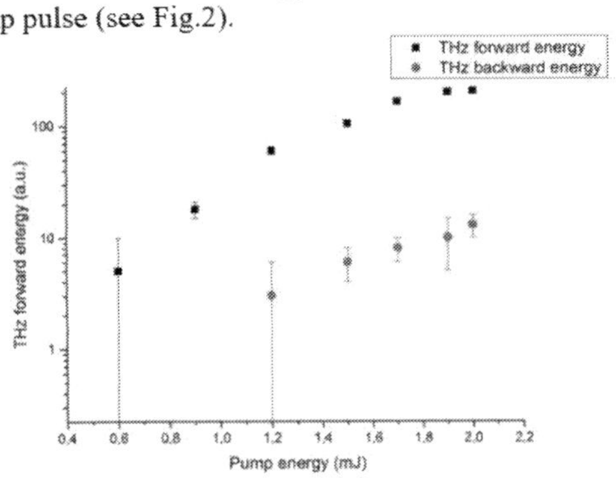

Fig. 2. Energy of forward and backward THz radiation dependence on laser pump pulse energy.

Conclusions

We have performed investigation of spectral parameters of forward and backward terahertz radiation from optical breakdown plasma created by tightly focused two-color femtosecond laser pulses. It appears that backward THz emission spectrum is narrower as compared to that of THz radiation co-directed with the optical pump. This result is supported by numerical simulation with simple interferometric model.

Dependencies of forward/backward THz energy on laser pump energy are measured. This can be useful for indirect control of forward spreading radiation by monitoring of backward one.

References

1. *Cook, D. J., Hochstrasser R. M.,* Intense terahertz pulses by four-wave rectification in air // Opt. Lett. 2000. V. 25, No. 16. P. 1210-1212.

2. *Zhong H., Karpowicz N., Zhang X.-C.* Terahertz emission profile from laser-induced air plasma // Appl. Phys. Lett. 2006. V. 88, No 26. P. 261103.

3. *Andreeva V. A., Kosareva O. G., Panov N. A., Shipilo D. E., Solyankin P. M., Esaulkov M. N., Gonzalez De Alaiza Martinez P., Shkurinov A. P., Makarov V. A., Berge L., Chin S. L.,* Ultrabroad terahertz spectrum generation from an air-based filament plasma // Phys. Rev. Lett. 2016. V. 116, No 6. P. 63902.

4. *Shkurinov A. P., Sinko A. S., Solyankin P. M., Borodin A. V., Esaulkov M. N., Annenkov V. V., Kotelnikov I. A., Timofeev I. V., Zhang X.-C.* Impact of the dipole contribution on the terahertz emission of air-based plasma induced by tightly focused femtosecond laser pulses // Phys. Rev. E. 2017. V. 95, No 4. P. 43209.

5. *Panov N., Andreeva V., Kosareva O., Shkurinov A., Makarov V. A., Bergé L., Chin S. L.* Directionality of terahertz radiation emitted from an array of femtosecond filaments in gases // Laser Phys. Lett. 2014. V. 11, No12. P. 125401.

6. *Ushakov A. A., Matoba M., Nemoto N., Kanda N., Konishi K., Chizhov P. A., Panov N. A., Shipilo D. E., Bukin V. V., Kuwata-Gonokami M., Yumoto J., Kosareva O. G., Garnov S. V., Savel'ev A. B.* Backward terahertz radiation from a two-color femtosecond laser filament // JETP Letters. 2017. V. 106, No 11. P. 706–708.

7. *Panov N. A., Kosareva O. G., Andreeva V. A., Savel'ev A. B., Uryupina D. S., Volkov R. V., Makarov V. A., Shkurinov A. P.* Angular distribution of the terahertz radiation intensity from the plasma channel of a femtosecond filament // JETP Letters. 2011. V. 93, No 11. P. 715-718.

*** Formatting Issue - Best Available Paper/Graphic ***

EPJ Web of Conferences **195**, 03011 (2018) https://doi.org/10.1051/epjconf/201819503011
TERA-2018

Resonant Effects in Terahertz Generation with Laser-Induced Gas Plasmas

I. Thiele [1,2], B. Zhou [3], A. Nguyen [4], E. Smetanina [1,5], R. Nuter [1], P. González de Alaiza Martínez [1], K. J. Kaltenecker [3], J. Déchard [4], L. Bergé [4], P. U. Jepsen [3], and S. Skupin [1,6]

[1]Univ. Bordeaux - CNRS - CEA, Centre Lasers Intenses et Applications, Talence, France
[2]Department of Physics, Chalmers University of Technology, Göteborg, Sweden
[3]DTU Fotonik, Technical University of Denmark, Kongens Lyngby, Denmark
[4]CEA/DAM Ile-de-France, Bruyères-le-Châtel, 91297 Arpajon, France
[5]Department of Physics, University of Gothenburg, Göteborg, Sweden
[6]Institut Lumière Matière, Université Lyon 1 - CNRS, Villeurbanne, France, stefan.skupin@univ-lyon1.fr

Research on intense terahertz (THz) electromagnetic sources has received an increasing attention owing to numerous applications, for example, in time-domain spectroscopy, biomedical imaging or security screening [1]. Among the various techniques employed to generate THz radiation, focusing intense two-color femtosecond pulses in air or noble gases provides interesting features like absence of material damage, large generated bandwidth (up to ~100 THz) and high amplitudes of the emitted THz pulses (> 100 MV/m) [2]. First reported by Cook et al. [3], THz emission from intense two-color pulses was initially attributed to optical rectification via third-order nonlinearity. However, it was shown later that the plasma built-up by tunneling photoionization is necessary to explain the high amplitudes of the THz field [4], and a quasi-dc plasma current generated by the temporally asymmetric two-color field is responsible for THz emission [5].

Numerous experimental results show that the laser-induced free electron density has a strong impact on the THz emission [4,6,7]. While it is frequently observed that a larger free electron density leads to broader THz spectra, the origin of the effect remains controversial. In [6,7], homogeneous plasma oscillations were proposed as an explanation, even though those oscillations are in principle non-radiative [8,9]. Moreover, nonlinear propagation effects have been held responsible for THz spectral broadening as well [10]

On the other hand, the gas plasma produced by the fs laser pulse is a finite conducting structure with a lifetime largely exceeding the fs time scale. Thus, one can expect that the gas plasma features plasmonic resonances which may have a strong impact on the THz emission propertie [11]. However, no direct evidence of plasmonic effects in laser-induced gas-plasmas was observed so far: To make an evidence of plasmonic effects, those need to be distinguished from nonlinear propagation effects. Also from the theoretical point of view capturing plasmonic effects is not trivial: plasmonic effects require at least a full two-dimensional Maxwell-consistent description, and reduced models like the unidirectional pulse propagation equation [12], which are frequently used to describe plasma-based THz generation [5,7,10], are by construction not capable of capturing such resonant effects.

In this work, we consider the two-color-laser-induced plasma as a plasmonic structure, and investigate under which conditions such plasmonic perspective is important. In the context of plasmonic nanoantennas (or metamaterials), e.g, for second-harmonic generation, tailoring plasmonic resonances by tuning the shape of the plasmonic particle is a standard approach. Therefore, we follow a similar strategy and modify the usually prolate spheroidal plasma shape. Going to tri-axial ellipsiods which can be achieved by using elliptically shaped laser beams turns out to be already sufficient [13]. Depending on whether the laser polarization is oriented along the long beam axis (quasi transverse electric, qTE) or along the short beam axis (quasi transverse magnetic, qTM), plasmonic resonances are triggered or not (see Fig. 1).

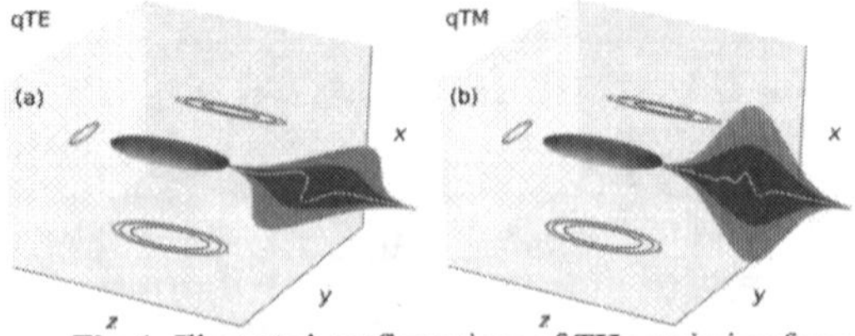

Fig. 1. Illustrated configurations of THz emission from an ellipsoidal plasma induced by a two-color Gaussian laser pulse~(funamental in red, second-harmonic in purple) with strongly elliptical beam shape propagating along z. The laser electric field is y-polarized (along the long axis of the beam, qTE) in (a) and x-polarized (along the short axis, qTM) in (b). The plasma is sketched as blue surface. Simulated forward emitted THz pulses are presented as white lines demonstrating a significantly shorter pulse duration for qTM polarization, which can be attributed to triggering a plasmonic resonance.

While nonlinear propagation effects are in both cases equally present, any difference between the THz emission spectra in this two cases is linked to plasmonic effects. We demonstrate experimental results which reveal a significant difference: THz pulses are shorter and have a broader emission spectrum when the plasma is excited by the laser field in the direction with the short focal beam width and plasma width (see Fig. 2). A simple analytical model allows us to link the broadening to a leaky mode. It turns out that the resonance has a strong impact on the spectrum whenever electrons are excited along a direction where the plasma size is smaller than the plasma wavelength. Finally, direct three-dimensional (3D) Maxwell consistent simulations in tightly focused geometry confirm that these plasmonic resonances indeed broaden the emitted THz spectrum significantly.

EPJ Web of Conferences **195**, 03011 (2018) https://doi.org/10.1051/epjconf/201819503011
TERA-2018

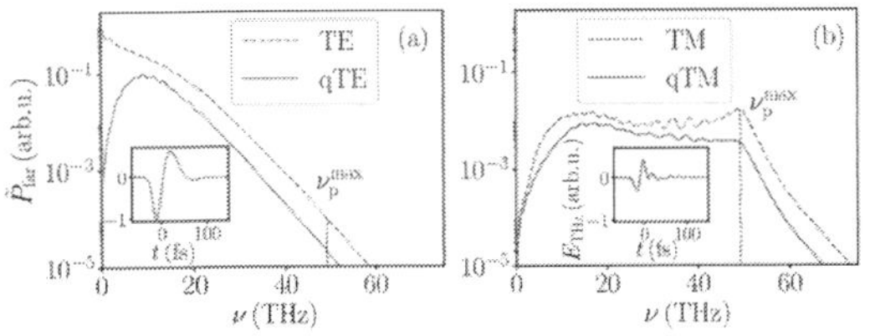

Fig. 2. Experimental THz spectra for qTE (a) and qTM (b) polarization (see text for details). Corresponding on-axis THz waveforms are shown as insets. The dashed lines specify the estimated maximum plasma frequency.

References

[1] M. Tonouchi. Cutting-edge terahertz technology. Nature Photon., 1:97, 2007.

[2] K. Kim et al. High-Power Broad-band Terahertz Generation via Two-Color Photo-ionization in Gases. IEEE J. Quant. Electron., 48:797, 2012.

[3] D. Cook et al. Intense terahertz pulses by four-wave rectification in air. Opt. Lett., 25:1210, 2000.

[4] K. Kim et al. Coherent control of terahertz supercontinuum generation in ultrafast laser-gas interactions. Nature Photon., 2:605, 2008.

[5] L. Bergé et al. 3D numerical simulations of THz generation by two-color laser filaments. Phys. Rev. Lett., 110:073901, 2013.

[6] H. Hamster et al. "Short-pulse terahertz radiation from high-intensity-laser-produced plasmas," Phys. Rev. E, 49:671, 1994.

[7] V. A. Andreeva et al. Ultrabroad terahertz spectrum generation from an air-based filament plasma," Phys. Rev. Lett., 116:063902, 2016.

[8] V. T. Tikhonchuk, Comment on "generation of electromag-netic pulses from plasma channels induced by femtosecond light strings", Phys. Rev. Lett., 89:209301, 2002.

[9] I. Thiele et al. Theory of terahertz emission from femtosecond-laser-induced microplasmas, Phys. Rev. E, 94:063202, 2016.

[10] E. Cabrera-Granado et al. "Spectral self-action of thz emission from ionizing two-color laser pulses in gases, New J. Phys., 17:023060, 2015.

[11] V. A. Kostin at al. Dc to ac field conversion due to leaky-wave excitation in a plasma slab behind an ionization front, New J. Phys., 17:033029, 2015.

[12] M. Kolesik at al., Nonlinear optical pulse propagation simulation: From Maxwell's to unidirectional equations, Phys. Rev. E, 70:036604, 2004.

[13] I. Thiele et al. Terahertz emission from laser-driven gas-plasmas: a plasmonic point of view. arXiv:1803.06889 [physics.optics], 2018.

EPJ Web of Conferences **195**, 03012 (2018) https://doi.org/10.1051/epjconf/201819503012
TERA-2018

Study of the cluster formation dynamics and its affect on generation of THz and X-Ray radiation in the expanding gas jet.

A. V. Balakin,[1,2] M. S. Dzhidzhoev,[1] V. M. Gorgienko,[1] I. A. Zhvaniya,[1] I. E. Ivanov,[1] N. A. Kuzechkin,[2,1] P. M. Solyankin,[2,1] A. P. Shkurinov[1,2,3]

[1]Faculty of Physics & International Laser Center, Lomonosov Moscow State University, Moscow 119991, Russia
[2]Institute on Laser and Information Technologies of RAS, Branch of the FSRC "Crystallography and Photonics", RAS, Shatura, Moscow Region 140700, Russia
[3]The National University of Science and Technology MISiS, Moscow 119049, Russia

Presently the interaction of intense femtosecond laser pulses with cluster target is a subject of active researches. Laser-cluster interaction proceeds with high efficiency and is accompanied with a plenty of nonlinear effects [1-3]. That makes it very attractive and promising as the basis for elaboration of a source for generation of intense coherent electromagnetic pulses in a wide spectral range from X-ray up to THz.

For detailed understanding and interpretation of the processes occuring in the gas-cluster nanoplasma, as well as for further optimization of their effectivity, a detailed information about properties of the gas-cluster jet is required. This information include distributions of the clusters size, cluster concentration, condensation degree and average atomic density along the spatial coordinates. In general, the process of the clusters formation in the gas jet is quite complex and possesses the probabilistic nature. The efficient method to get detailed information regarding the properties and parameters of the gas-cluster jet is the straight numerical simulation of the clusterization process.

In this work, we have carried out the numerical simulation and computing of dynamic picture of clusterization of argon atoms passed through the supersonic conical nozzle and future propagation of the jet into vacuum at a distance of up to 60 mm below the nozzle throat. The result of numerical simulation has demonstrated that ratio between argon monomers, small and large clusters fractions dramatically changes both as along the jet propagation direction, and as across the jet when the distance from the nozzle throat increase.

Some results of the numerical simulation of cluster formation process are shown on Fig. 1,2. The following parameters of the conical nozzle which was applied in our experiments were used for simulation: throat diameter 0.7 mm, output diameter 4.7 mm, and the nozzle length 24.7 mm. Backing pressure of argon was 2 Mpa. Fig.1 depicts spatial distribution of the average argon atoms density inside and outside of the nozzle. The direction along the axis of symmetry of the nozzle is denoted as an axis of abscissa (X-axis), that coincides with the jet propagation direction as well. The nozzle throat position is noted as X=0 mm, and X=24.7 mm corresponds to position of the nozzle output edge. Radial direction, which is perpendicular to the axis of the symmetry, is denoted as Y-axis. There are five stream lines (SL) indicated by dashed lines on the Fig. 1. Distributions of mean cluster size and mean cluster concentration through the stream lines as a function of distance from the nozzle are presented on Fig.2.

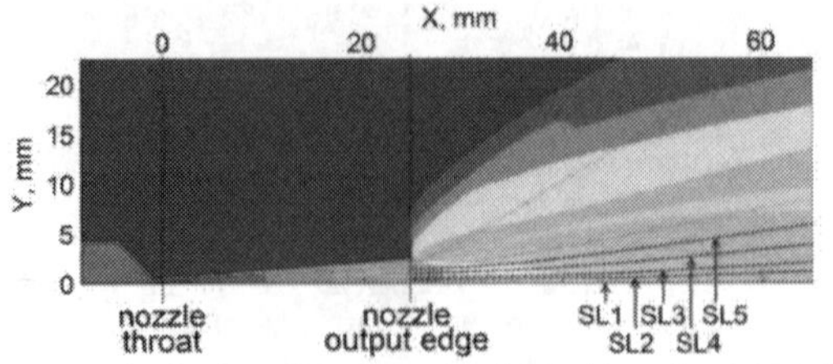

Fig. 1. Distribution of Ar atom density inside and outside of the conical nozzle (Y = 0 corresponds to the axis of symmetry of the nozzle, SL1-SL5 represent five stream lines). The magnitude of the density is represented by a color scale: ~5 kg/m^3 (red), ~0.3 kg/m^3 (yellow), ~10^{-2} kg/m^3 (green), ~3x10^{-4} kg/m^3 (blue), and ~10^{-5} kg/m^3 (dark blue).

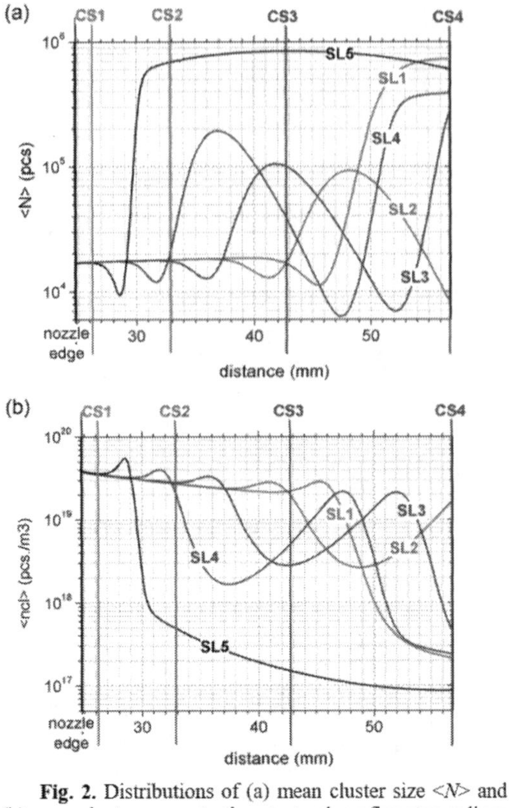

Fig. 2. Distributions of (a) mean cluster size <*N*> and (b) mean cluster concentration <n_{cl}>, along five stream lines (SL1-SL5) as a function of the distance from the nozzle throat.

The graphs shown on Fig. 1,2 clearly demonstrate that the distance from the nozzle throat is an important parameter which defines the properties of the gas-cluster jet. We suggest that this fact can be effectively used for efficient control over X-ray and THz emission yields produced from argon gas-cluster jet under irradiation with high-intense laser pulses.

In our experiments on laser-cluster interaction we used the setup described in details in our previous paper [4], which was modified to provide a possibility of focusing the laser beam into the gas-cluster jet at various distances below the output edge of conical supersonic nozzle. The focus point was located at the axis of symmetry of the nozzle and could be positioned discretely in the range between 1.5 mm to 32.3 mm below the nozzle edge that corresponds to the locations of CS1-CS4. Our laser system provided pulses with energy up to 30 mJ at 10 Hz repetition rate, central wavelength was 810 nm, the pulse duration could be tuned in the range of 50–600 fs by chirping the laser pulse in a vacuum grating compressor.

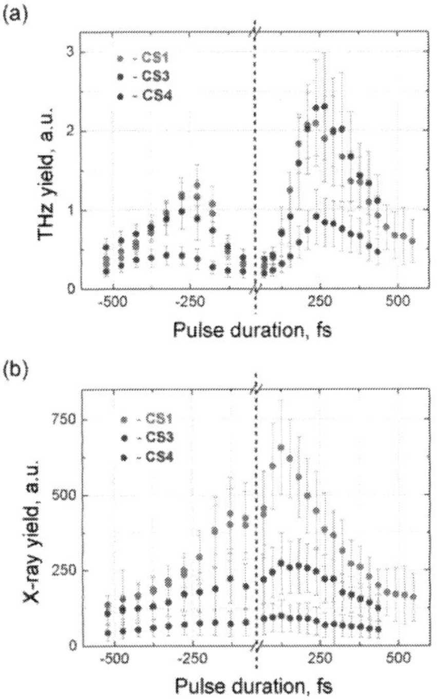

Fig. 3. X-Ray (a) and THz (b) yields from an Ar gas-clusters jet as a function of the temporal duration of the excitation laser pulses recorded at the cross-sections of CS1 (red dots), CS3 (blue dots), and CS4 (purple dots). The vertical dotted line in the center represents the pulse duration for a Fourier limited pulse.

Let us consider the experimental results. THz and X-Ray yields from Ar gas-clusters jet as a function of temporal duration of exciting laser pulse are presented on Fig. 3. Measurements have been performed for both positively (right side) and negatively (left side) chirped optical pulses. Pulse duration of around 55fs corresponds to Fourier-limited pulse and is indicated by related vertical line on Fig. 3. The symbols of different colors depict experimental data measured for three cases of position of laser beam focus relatively to the nozzle, and correspond to CS1, CS3 and CS4 cross-sections.

It can be seen from Fig. 3 that THz yield strongly decreases in the region of minimum pulse durations whereas X-Ray emission intensity demonstrates maximum values under these conditions. Maximal value X-ray signal monotonously decreases as the distance between excitation area and the nozzle throat increases from CS1 to CS4. In contrast with this behavoir, THz signal demonstrate roughly equal values in the CS1 and CS3, but it decreases in the CS4 position. As it can be seen from Fig. 1,2 there are mainly small-size clusters fractions are contained in cross-section CS1, and predominantly large-size clusters are in cross-section CS4.

In conclusion, we carried out numerical simulations of the cluster formation process in a supersonic jet of Ar. We have shown that the distance from the nozzle output edge is an important parameter describing the properties of the gas-cluster target, which should be taken into account when developing the effective sources of THz and X-ray radiation based on cluster nanoplasma.

This work was supported by the Ministry of Science and Higher Education within the State assignment FSRC «Crystallography and Photonics» RAS, by the RFBR under Grants 18-52-16016 and 17-02-01217, by the Ministry of Education and Science of the Russian Federation in the framework of the Increase Competitiveness Program of NUST "MISiS" (no. K2-2017-003).

References

1. *Krainov V., and Smirnov M.* Cluster beams in the super-intense femtosecond laser pulse // Phys. Rep. 2002. V. 370, P. 237-331.

2. *Nagashima T., Hirayama H., Shibuya K., Hangyo M., Hashida M., Tokita S., and Sakabe S.* Terahertz pulse radiation from argon clusters irradiated with intense femtosecond laser pulses // Opt. Express 2009 V. 17, P. 8907

3. *Alexeev I., Antonsen T.M., Kim K.Y., and Milchberg H.M.* Self-focusing of intense laser pulses in a clustered gas // Phys. Rev. Lett. 2003 V. 90, P. 103402

4. *Balakin A.V., Dzhidzhoev M.S., Gordienko V.M., Esaulkov M.N., Zhvaniya I.A., Ivanov K.A., Kotelnikov I.A., Kuzechkin N.A., Ozheredov I.A., Panchenko V.Y., Savel'ev A.B., Smirnov M.B., Solyankin P.M., Shkurinov A.P.* IEEE Trans. Terahertz Sci. Technol. 2017 V. 7, P. 70-79

Terahertz generation from single and multiple filaments in air

O. Kosareva[1,2,3], N. Panov[1,2], D. Shipilo[1,2], V. Andreeva[1,4], T.-J. Wang[5], Y. Chen[6], W. Liu[3], A. Savel'ev[1], and A. Shkurinov[1]

[1]Lomonosov Moscow State University, 1/2 Leninskie gori, Moscow, Russia, kosareva@physics.msu.ru
[2]Lebedev Physical Institute of Russian Academy of Sciences, Moscow, Russia
[3]Institute of Modern Optics, Nankai University, Tianjin, China
[4]School of Mathematics, University of Minnesota, Minneapolis, Minnesota, USA
[5]Shanghai Institute of Optics and Fine Mechanics, Chinese Academy of Sciences, Shanghai, China
[6]Shanghai Jiao Tong University, Shanghai, China

Terahertz (THz) photonics is entering an era of strong electromagnetic fields in the THz frequency range [1]. Optical breakdown plasma takes a specific place among THz sources. With the appearance of compact commercial terawatt laser systems and the actualization of remote sensing tasks, two-color optical breakdown in air has turned into a powerful mechanism of broadband THz generation at the observation point [2]. There are several problems associated with THz remote sensing applications. First, the directional diagram of emitted THz radiation usually has a conical structure [3]. Nevertheless, as the filament length increases, the divergence angle decreases [4]. Second, the limitation of air-breakdown THz applications to remote-sensing tasks is due to low signal-to-noise ratio in the amplitude analysis of THz radiation reflected from the object under study. Therefore, one can use polarization analysis of useful signals whose effective application requires thorough control of the state of polarization of THz radiation. Third, further optimization of the THz yield from two-color filaments is required to approach the signal level obtained from optical rectification in crystals [1]. Control over the relative temporal (and spatial) properties of the optical pump fields, e.g. via the pulse chirp, offers an important means to control and optimize the optical-to-THz conversion process. With increasing pulse peak power and transition to multiple filamentation more overall THz energy is generated however the proper regularization of the high-peak-power beam is needed to provide for the maximum THz yield in the narrow angle.

In this paper we present a series of studies for the control and optimization of THz directionality, polarization and energy yield.

We observe linear-to-elliptical transformation of THz radiation through independent control of initially linear polarization directions of ω and 2ω beams. Strongly elliptical THz radiation appears abruptly at the threshold angle of ~85° between the ω and 2ω light fields. The physical reason for THz beam ellipticity was identified by us as the nonlinear response of the birefringent medium induced by an 800 nm pump. We found THz ellipticity and energy yield using a vectorial nonparaxial propagation equation (Fig. 1).

We have demonstrated and justified the mechanism which dictates the optimum chirp to maximize the THz yield from a two-color filament. We have experimentally measured the THz energy yield using 800-nm pulses in a single filament regime, being chirped from 35 fs up to 250 fs both positively and negatively. We found that the positively chirped pulse proves to be a more advantageous choice since the ω–2ω pulse overlap is larger by ~15 fs as compared with the transform-limited pulse (Fig. 2).

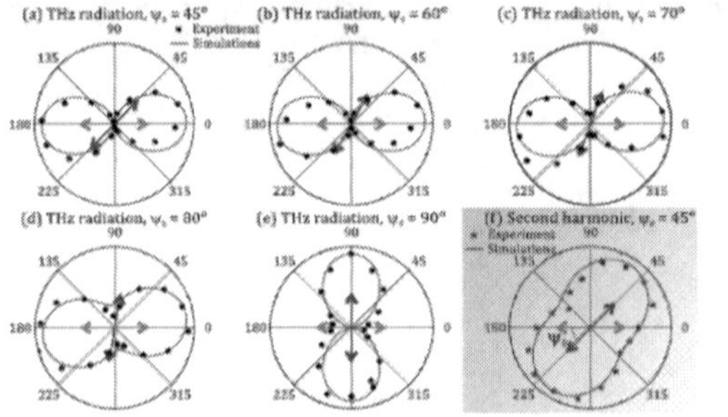

Fig. 1 (a)–(e) THz energy after transmission through THz analyzer, in dependence on its rotation angle, for different angles ψ_0, for different angles ψ_0 between polarization directions of the 800 nm (ω) and 400 nm (2ω) beams. (f) Second-harmonic energy after its transmission through polarization analyzer. Red (blue) arrow indicates the incident polarization of the ω (2ω) beam.

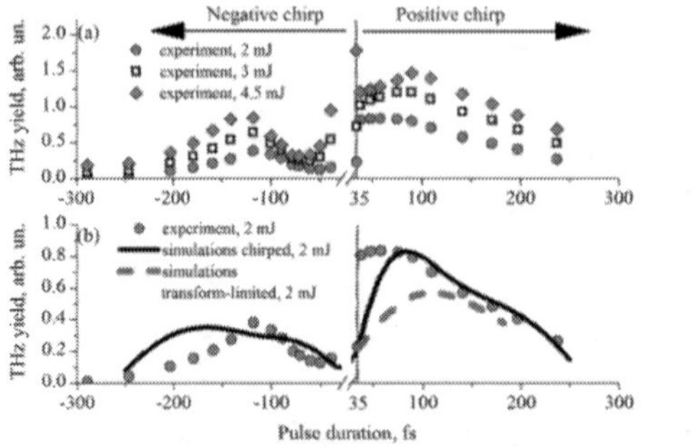

Fig. 2 (a) The experimentally measured dependence of the THz energy on the pump (800 nm) chirped pulse duration. The pump energy is fixed for each chirped pulse duration variation in the range from –300 fs (FWHM, negatively chirped) to +250 fs (FWHM, positively chirped). (b) The experimental vs. simulated dependence of the THz energy on the 800-nm 2-mJ chirped pulse duration. Green dashed curve shows simulated THz yield for the transform-limited pulse.

We have shown that a *2D* array of multiple plasma filaments in atmospheric density gases leads to the efficient increase of terahertz radiation emitted in the forward direction. As the number of filaments in an $N \times N$ array is augmented, the THz energy increase per unit polar angle in the central cone exceeds the standard gain factor N^2 [4,5]. The two-dimensional multiple filament array can be produced from an actual non-ideal elliptical beam using a phase plate to introduce a phase barrier and block the interaction between the neighboring filaments. Such regularized structure of plasma channels is used to control the directional diagram of THz radiation from a sub-terawatt-peak-power 800 nm + 400 nm femtosecond laser pulse.

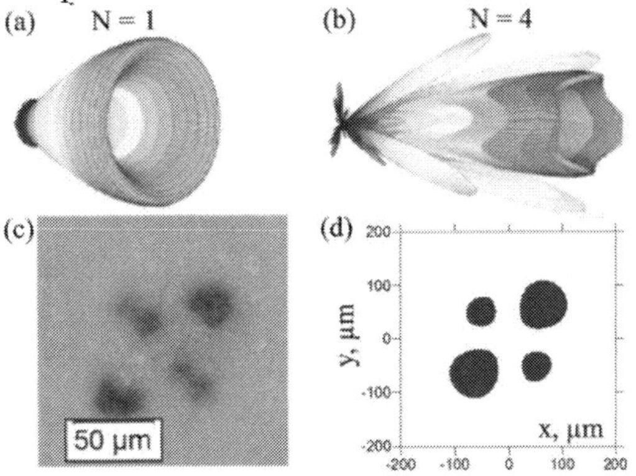

Fig. 3. (a, b) Full 3D angular THz intensity distribution from an array of N multiple filaments with the length $L = 1$ cm. (c) Single-laser-shot burnt photo paper images after the geometrical focus of 1.1-mJ 35-fs 800 nm + 400 nm elliptical beam transported through a phase mask in air. The reproduction of the plasma channels in 3*D*+time carrier resolved simulations is given in (d).

This work is financially supported by Russian Science Foundation (RSF, Grant No. 16-42-01060), the National key research and development program (2018YFB0504400), "111 Project (B16027)" the National Basic Research Program of China (2014CB339801); the National Natural Science Foundation of China (11474202, 11774228, 11721091); the Strategic Priority Research Program of the Chinese Academy of Sciences (XDB16000000); Key Project from Bureau of International Cooperation Chinese Academy of Sciences (181231KYSB20160045).

References

1. *X.-C. Zhang, A. Shkurinov, and Y. Zhang.* Extreme terahertz science // Nat. Photonics 2017. V. 11, p. 16.
2. *T.-J. Wang, S. Yuan, Y. Chen, J.-F. Daigle, C. Marceau, F. Théberge, M. Châteauneuf, J. Dubois, and S. L. Chin.* Toward remote high energy terahertz generation // Appl. Phys. Lett. 2010. V. 97, p. 111108.
3. *V. Andreeva, O. Kosareva, N. Panov, D. Shipilo, P. Solyankin, M. Esaulkov, P. G. de Alaiza Martinez, A. Shkurinov, V. Makarov, L. Bergé, and S. L. Chin.* Ultrabroad terahertz spectrum generation from an air-based filament plasma // Phys. Rev. Lett. 116, 063902 (2016).
4. *N. Panov, V. Andreeva, O. Kosareva, A. Shkurinov, V. A. Makarov, L. Bergé and S. L. Chin.* Directionality of terahertz radiation emitted from an array of femtosecond filaments in gases // Laser Phys. Lett. 2014. V. 11, p. 125401.
5. *S. I. Mitryukovskiy, Y. Liu, B. Prade, A. Houard, and A. Mysyrowicz.* Coherent synthesis of terahertz radiation from femtosecond laser filaments in air // Appl. Phys. Lett. 2013. V. 102, p. 221107.

Laser-plasma generation of tunable ultrashort pulses in terahertz and mid-infrared ranges

A. A. Silaev[1,2], V. A. Kostin[1,2], I. D. Laryushin[1,2], N. V. Vvedenskii[1,2]

[1]Institute of Applied Physics, Russian Academy of Sciences, Nizhny Novgorod, Russia, silaev@appl.sci-nnov.ru
[2]University of Nizhny Novgorod, Nizhny Novgorod, Russia

The generation of tunable ultrashort pulses in the terahertz (THz) and mid-infrared (mid-IR) ranges is an important problem of both fundamental interest and for many applications. In particular, such sources are required for the observation of ultrafast electronic and vibrational dynamics in a wide class of materials [1]. Such sources are also of interest because of the possibilities of spectroscopy and diagnostics of various media [2]. Apart from this, the use of intense ultrashort pulses in the mid-IR range with controlled phase for the generation of high harmonics makes it possible to reach the short-wave region of the soft X-ray range and to obtain ultrashort attosecond pulses [3].

In this paper we report on analytical and numerical investigation of generation of tunable ultrashort pulses in THz and mid-IR ranges based on the ionization-induced multiwave mixing of fields of intense two-color femtosecond pulses. If the ratio of the frequencies $\omega_{1,2}$ of quasi-monochromatic components of the two-color ionizing field is close to an irreducible rational fraction $\omega_2 : \omega_1 = a : b$, where $a + b$ is a not too large odd integer, the ionization-induced multiwave mixing in the produced plasma can result in the efficient excitation of the current of free electrons at a low (compared to the frequencies of the ionizing field) combination frequency responsible for generation of tunable ultrashort pulses in THz and mid-IR ranges [4, 5]. A strong dependence of the ionization rate on the field strength is responsible for a very short duration of generated pulses, which is determined by the ionization duration (characteristic time of increase in the plasma density). As a result, the generated pulses can be few-cycle or even subcycle, and the phase difference between the carrier and envelope in these pulses is determined by the phase shift between single-color components in the ionizing field.

As the most important and practically interesting example of ionization-induced generation of ultrashort pulses in the mid-IR range, we consider here the two-color ionizing pulses containing quasi-monochromatic components whose frequency ratio $\omega_2 : \omega_1$ is close to 2. Such two-color pulses usually consist of the pump field and an additional field, which is obtained using either frequency doubling crystals and frequency-selective elements creating detuning frequency [6] or an optical parametric amplifier generating radiation near half the frequency [7]. We assume that the time-dependent linearly polarized two-color ionizing field is given by

$$\mathbf{E}(t) = [F_1(t)\cos(\omega_1 t) + F_2(t)\cos(\omega_2 t + \varphi)]\mathbf{x}_0, \quad (1)$$

where $F_{1,2}(t)$ are the slow envelopes of single-color quasimonochromatic components, φ is the phase shift, and $\mathbf{x}_0$ is the unit vector. We consider Gaussian envelopes $F_{1,2} = (8\pi I_{1,2}/c)^{1/2}\exp(-t^2/2\tau^2)$ with the maximum intensities $I_{1,2}$, where $\tau = \tau_p/(4\ln 2)^{1/2}$, τ_p is the intensity full width at half-maximum, and c is the speed of light. Let ω_1 be the lowest of two frequencies, $\omega_2 = 2\omega_1 + \Delta\omega$, and $|\Delta\omega| << \omega_1$. We also assume that the duration of the generated pulse is much smaller than the characteristic time of current relaxation in plasma and the period of plasma natural oscillations. As a result, the field (1) can be considered as given and all dissipative terms in equations for low-frequency (at frequencies much lower than ω_1) current density $\mathbf{j}_{LF}$ can be neglected.

The performed *ab initio* simulations of the excitation of the electron current are based on the solution of the three-dimensional time-dependent Schrödinger equation (3D TDSE) [4, 5]. This equation allows describing all stages of the dynamics of the electron wavefunction $\psi(\mathbf{r},t)$ of the hydrogen atom in a strong external field $\mathbf{E}(t)$ The initial condition corresponds to the ground ($1s$) state. The time derivative of the electron current density is expressed in terms of as

$$\frac{\partial \mathbf{j}}{\partial t} = \frac{e^2 N_m}{m}\left[\mathbf{E} + \left\langle \psi \,|\, \frac{e\mathbf{r}}{r^3} \,|\, \psi \right\rangle\right], \quad (3)$$

where N_m is the density of neutral atoms before the ionization, e and m are electron charge and mass, respectively. The temporal profile of $\partial \mathbf{j}_{LF}/\partial t$ is obtained from $\partial \mathbf{j}/\partial t$ by means of the ideal low-pass filter (the cutoff frequency of the filter corresponds to the lowest frequency above $|\Delta\omega|$ at which the spectral density $\partial \mathbf{j}/\partial t$ reaches a local minimum).

Spectra and time profiles of $\partial \mathbf{j}_{LF}/\partial t$ calculated from the numerical solution of Eqs. (2) and (3) are shown in Figs. 1 and 2, respectively, for various values of the detuning frequency $\Delta f = \Delta\omega/2\pi$. As is seen in Fig. 1, the spectrum contains a pronounced bell-shaped peak near Δf. The width of the peak is much larger than the widths of the spectra of quasi-monochromatic components of the ionizing field and depends only slightly on Δf, whereas the height of the peak depends on both the magnitude and sign of Δf. Pulses generated at negative detuning are more intense than those generated at positive detuning.

This asymmetry with respect to the sign of detuning can be very large, and the spectral peak at positive detuning can be an order of magnitude lower than the peak at negative detuning. The peaks have

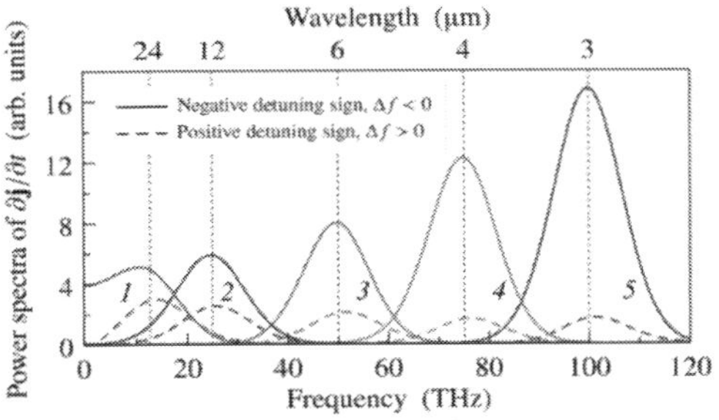

Fig. 1. Power spectra of $\partial \mathbf{j}/\partial t$ found from the numerical solution of the 3D TDSE (2) for the two-color ionizing field (1) at various (dashed lines) positive and (solid lines) negative detuning frequencies $\Delta f = \Delta\omega/2\pi =$ (1) ±12.5, (2) ±25, (3) ±50, (4) ±75, and (5) ±100 THz. The frequency ω_1 corresponds to the wavelength 800 nm, components intensities $I_1 = 2{\cdot}10^{14}$ W/cm², $I_2 = I_1/10$, pulse duration $\tau_p = 100$ fs, and phase shift $\varphi = 0$.

smooth flanks, and the corresponding time dependences also have fairly smooth envelopes, as is seen in Fig. 2 plotted for two values of the phase shift. Changing φ by $\pi/2$ results in a phase shift of $\pi/2$ in the generated pulse. The durations of pulses at different detuning values are approximately equal and are much smaller than the duration of the initial ionizing field, which corresponds to the generation of few-cycle or subcycle pulses (depending on the detuning frequency).

For the interpretation of the results of quantum-mechanical simulation, we use an analytical approach based on the equation for the plasma density $N(t)$ and the classical equation for the current density $\mathbf{j}(t)$ of free electrons in a cold collisionless plasma with a varying number of particles

$$\frac{\partial N}{\partial t} = (N_m - N)w(E), \quad \frac{\partial \mathbf{j}}{\partial t} = \frac{e^2 N}{m}\mathbf{E} \qquad (4)$$

with the initial condition $N = 0$ (at $t \to -\infty$) [4, 5, 7]. Here, $w(E)$ is the probability of tunnel ionization per unit time, and it is assumed that the ponderomotive energy of the electron in this field is much higher than the ionization energy. The use of perturbation theory with respect to the corresponding weak field (F_1 or F_2) allows to obtain closed-form analytical expression for $\partial \mathbf{j}_{LF}/\partial t$ and explain the origin of the asymmetry and other characteristic features of the excitation of low-frequency current. Particularly, it is shown that the duration of the excited current is determined by the ionization duration τ_i (the characteristic time of plasma formation), which is much shorter than the duration of the ionizing field because of the steep dependence of the ionization rate on the field strength. At quite large detuning frequencies, $|\Delta f|\tau_i \geq 1$, a pulse is generated at the frequency Δf, and its phase is determined by φ. At smaller detunings $|\Delta f|\tau_i < 1$, a subcycle pulse is generated, and φ affects both the phase and amplitude of the generated pulse. The central frequency of the generated pulse is about $1/\tau_i$, and this value specifies the lower limit of the possible frequency tuning range. In this case, a significant residual current density, i.e., a static component (zeroth harmonic) of $\partial \mathbf{j}_{LF}/\partial t$

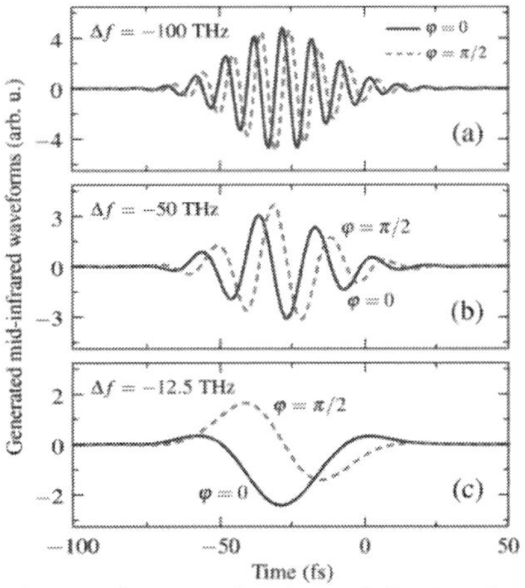

Waveforms of generated pulses of the low-frequency current $\mathbf{x}_0{\cdot}(-\partial \mathbf{j}_{LF}/\partial t)$ found from the numerical solution of 3D TDSE at the same parameters as in Fig. 1, but at two phase shifts $\varphi = 0$ and $\pi/2$ and three negative detuning frequencies $\Delta f =$ (a) −100, (b) −50, and (c) −12.5 THz.

depending on φ can be excited. The presence of this component can be seen in the results of numerical calculations presented in Fig. 1 (solid line 1) and in Fig. 2c. This residual current density is responsible for the generation of THz radiation [4, 6, 7], and Eqs. (10)–(12) determine the high-frequency part of the spectrum of this THz radiation, which extends to frequencies about $1/\tau_i$.

This work was supported by the Russian Foundation for Basic Research (Grant Nos. 16-32-60200, 16-32-60166, 18-02-01150, 18-32-00951), by the Grants Council under the President of the Russian Federation (Grant No. MK-2135.2017.2), and the Russian Science Foundation (numerical calculations, Grant No. 18-11-00210).

References

1. *Krausz, F, Stockman M. I.* Attosecond metrology: from electron capture to future signal processing // Nature Photonics. 2014. V.8. P.205.
2. *Cossel, K.C. et al.* Gas-phase broadband spectroscopy using active sources: progress, status, and applications // J. Opt. Soc. Am. B. 2017. V.34. P. 104-129.
3. *Zenghu, C., Corkum, P.B., Leone, S.R.* Attosecond optics and technology: progress to date and future prospects // J. Opt. Soc. Am. B. 2016. V. 33. P. 1081-1097.
4. *Kostin, V. A., Laryushin, I. D., Silaev, A. A., Vvedenskii, N. V.* Ionization-induced multiwave mixing: terahertz generation with two-color laser pulses of various frequency ratios // Phys. Rev. Lett. 2016. V. 117, No. 3. P. 035003.
5. *Silaev, A. A., Kostin, V. A., Laryushin, I. D., Vvedenskii, N. V.* Ionization mechanism of the generation of tunable ultrashort pulses in the mid-infrared range // JETP Letters. 2018. V. 107, No. 3. P.151-156.
6. *Theberge, F. et al.* Generation of tunable and broadband far-infrared laser pulses during two-color filamentation // Phys. Rev. A 2010. V. 81, No.3. P. 033821.
7. *Vvedenskii, N. V., et al.* Two-color laser-plasma generation of terahertz radiation using a frequency-tunable half harmonic of a femtosecond pulse // Phys. Rev. Lett. 2014. 112, No.5. P. 055004.

Quantum-mechanical simulations of low-frequency current excitation during ionization of many-electron atoms by intense laser pulses

A. A. Silaev[1,2], A. A. Romanov[1,2], N. V. Vvedenskii[1,2]

[1]Institute of Applied Physics, Russian Academy of Sciences, Nizhny Novgorod, Russia, silaev@appl.sci-nnov.ru
[2]University of Nizhny Novgorod, Nizhny Novgorod, Russia

Great progress in computer technology opens up the possibility for *ab initio* numerical study of various ionization-stimulated phenomena arising from the interaction of atoms and molecules with intense laser pulses. These include the excitation of a low-frequency current (with frequencies much lower than the optical one) in a plasma created by a laser pulse. The interest to this phenomenon is associated with the possibility of its use for the generation of coherent and broadband radiation in the terahertz and mid-infrared ranges and possibility of tuning the spectrum of generated radiation [1-11]. The low-frequency current density is excited effitiently under the asymmetry of ionizing pulses [1-10] or ionized molecules [11]. Among the various methods of generating low-frequency current, the most well-known one is based on the use of two-color ionizing laser pulses [1-8]. The low-frequency current is generated at some combination frequency of two-component laser pulse for a whole set of ratios between component frequencies [5-8].

The efficient excitation of high-power terahertz and mid-infrared radiation is associated with the need to use of high-intensity laser pulses, which produce fast ionization of a gas medium and accelerate the freed electrons up to high energies [2, 5, 7, 9, 10]. The action of such intense pulses on atoms or molecules (most of which are multielectron quantum systems) may be accompanied by a dynamic response of various electron orbitals, which was not taken into account in previous similar studies. This response can significantly influence the magnitude and waveform of the low-frequency current.

This paper is devoted to direct numerical simulations of ionization-induced excitation of low-frequency electron current density in multielectron atoms of noble gases. Our approach is based on the time-dependent density functional theory (TDDFT), which has recently been increasingly used in atomic, molecular, and strong-field physics [12]. The basis of this approach is the system of time-dependent Kohn-Sham (TDKS) equations, which takes into account the interaction of electrons with the atomic nucleus and the external electric field of the laser pulse, as well as the electron-electron interaction. In the atomic system of units (in which $|e| = \hbar = m_e = 1$, where $\hbar$ is the reduced Planck constant, $e = -|e|$ is the charge and m_e is the mass of the electron) these equations are:

$$i\frac{\partial}{\partial t}\psi_n(\mathbf{r},t) = \hat{H}\psi_n(\mathbf{r},t), \quad n = 1,\ldots,N,$$
$$\hat{H} = -\frac{1}{2}\nabla^2 - \frac{Z}{r} + \mathbf{r}\mathbf{E}(t) + V_{\mathrm{ee}}[\rho(\mathbf{r},t)]. \qquad (1)$$

Here, ψ_n is the wave function of the n-th TDKS orbital, Z is the nuclear charge, N is the number of electrons, $\rho(\mathbf{r},t)$ is the electron density, and $V_{ee}[\rho(\mathbf{r},t)]$ is the potential of electron-electron interaction. The initial Kohn-Sham orbitals correspond to electronic configuration of the unperturbed atom. For considered noble-gas atoms, the atomic orbitals are initially occupied by the pairs of electrons with opposite spin. Due to the weak influence of the magnetic field, the zero spin polarization is conserved during interaction with the laser pulse. The electron density is related to TDKS orbitals by equality

$$\rho(\mathbf{r},t) = \sum_{n=1}^{N} |\psi_n(\mathbf{r},t)|^2. \qquad (2)$$

The electron-electron interaction potential consists of the Hartree potential $V_{\mathrm{H}}[\rho(\mathbf{r},t)] = \int d^3r'\rho(\mathbf{r}')/|\mathbf{r}-\mathbf{r}'|$, describing the electron repulsion in the framework of the mean field and the exchange-correlation potential for which the spin unpolarized form of LB94 approximation [12] is used.

Using TDKS orbitals and Ehrenfest's theorem we find the time-dependent dipole acceleration of the atomic system:

$$\mathbf{a}(t) = -N\mathbf{E}(t) - \int d^3r \frac{Z\mathbf{r}}{r^3}\rho(\mathbf{r},t). \qquad (3)$$

The obtained dipole acceleration is proportional to the time derivative of macroscopic electron current density,

$$\frac{\partial \mathbf{j}(t)}{\partial t} = -N_m \mathbf{a}(t) \qquad (4)$$

where N_m is the gas density before the start of the ionization process.

Based on the developed computer code for solving the 3D TDKS, we calculate the current density $\mathbf{j}(t)$ excited by two-color laser pulses containing the main and second harmonics in a wide range of intensities and durations of the laser-pulse components. The found electron current density $\mathbf{j}(t)$ has a very wide frequency spectrum. Besides the low-frequency part, it contains a large component at the frequency of the ionizing laser pulse and its low-order harmonics. Also it contains a wide plateau ending with a pronounced cutoff in the frequency region, which corresponds to the extreme UV or soft-x-ray radiation [13]. We extract the low-frequency component from the full current density and study differences in its excitation for different noble-gas atoms. The results obtained on the basis of TDDFT are compared with the calculations of the single-active electron (SAE) approximation. In the framework of SAE approximation all orbitals except highest-occupied orbital are "frozen", and the field of the parent ion is described by a static potential well. It is shown that taking into account the dynamics of all the electrons can lead to a change in the optimal relative phase between the components of the two-color laser pulse and in the corresponding maximum low-frequency current. Based on the results obtained, the range of applicability of SAE approximation is determined.

This work was supported by the Russian Foundation for Basic Research (Grant Nos. 16-32-60200, 18-02-01150, 18-32-00951) and the Russian Science Foundation (numerical calculations, Grant No. 18-72-00192).

References

1. *Clough B., Dai J., Zhang, X.-C.* Laser air photonics: beyond the terahertz gap // Mater. Today. 2012. V. 15, No.1-2. P. 50-58.

2. *Vvedenskii, N. V., et al.* Two-color laser-plasma generation of terahertz radiation using a frequency-tunable half harmonic of a femtosecond pulse // Phys. Rev. Lett. 2014. 112, No.5. P. 055004.

3. *Balčiūnas, T. D. et al.* CEP-stable tunable THz-emission originating from laser-waveform-controlled sub-cycle plasma-electron bursts // Opt. Express. 2015. V. 23, No. 12. P. 15278-15289.

4. *Theberge, F. et al.* Generation of tunable and broadband far-infrared laser pulses during two-color filamentation // Phys. Rev. A 2010. V. 81, No.3. P. 033821.

5. *Kostin, V. A., Laryushin, I. D., Silaev, A. A., Vvedenskii, N. V.* Ionization-induced multiwave mixing: terahertz generation with two-color laser pulses of various frequency ratios // Phys. Rev. Lett. 2016. V. 117, No. 3. P. 035003.

6. *Zhang, L.-L. et al.* Observation of terahertz radiation via the two-color laser scheme with uncommon frequency ratios // Phys. Rev. Lett. 2017. V.119, No. 23. P. 235001.

7. *Silaev, A. A., Kostin, V. A., Laryushin, I. D., Vvedenskii, N. V.* Ionization mechanism of the generation of tunable ultrashort pulses in the mid-infrared range // JETP Letters. 2018. V. 107, No. 3. P.151-156.

8. *Kostin V.A., Vvedenskii N.V.* Generation of Few- and Subcycle Radiation in Midinfrared-to-Deep-Ultraviolet Range During Plasma Production by Multicolor Femtosecond Pulses // Phys. Rev. Lett. V. 120, No. 6. P. 065002.

9. *Silaev, A.A., Vvedenskii, N. V.* Residual-current excitation in plasmas produced by few-cycle laser pulses // Phys. Rev. Lett. 2009. V. 102, No. 11. P. 115005.

10. *Silaev, A. A., Vvedenskii N. V.* Analytical description of generation of the residual current density in the plasma produced by a few-cycle laser pulse // Phys. Plasmas. 2015. V. 22, No. 5. P. 053103.

11. *Alexandrov, L. N., Emelin, M. Yu., Ryabikin M. Yu.* Unidirectional current excitation in tunneling ionization of asymmetric molecules // Phys. Rev. A. 2013. V. 87, No. 1. P. 013414.

12. *Ullrich, C. A.* Time-dependent density-functional theory: concepts and applications // Oxford University Press. 2012.

13. *Krausz, F., Ivanov, M.* Attosecond physics // Rev. Mod. Phys. 2009. V. 81, No. 1. P. 163.

Quantum-cascade lasers of mid-IR spectral range: epitaxy, diagnostics and device characteristics

A. V. Babichev[1], A. G. Gladyshev[2], E. S. Kolodeznyi[1], A. S. Kurochkin[1], G. S. Sokolovskii[3], V. E. Bougrov[1], L. Ya. Karachinsky[1,2,3], I. I. Novikov[1,2,3], V. V. Dudelev[3], V. N. Nevedomsky[3], S. O. Slipchenko[3], A. V. Lutetskiy[3], A. N. Sofronov[4], D.A. Firsov[4], L.E. Vorobjev[4], N. A. Pikhtin[3], A. Bousseksou[5], A. Yu. Egorov[1]

[1]ITMO University, St. Petersburg, Russia, a.babichev@corp.ifmo.ru
[2]Connector Optics LLC, St. Petersburg, Russia
[3]Ioffe Institute, St. Petersburg, Russia
[4]Peter the Great St. Petersburg Polytechnic University, St. Petersburg, Russia
[5]Center of Nanoscience and Nanotechnology (C2N), Universit´e, Paris Sud and Paris-Saclay, Orsay cedex, France

Since the first practical realization of quantum-cascade lasers (QCL) the considerable progress has been achieved in the theory, epitaxy, and laser formation. Growth of QCL lasers by metal-organic chemical vapour deposition instead of molecular beam epitaxy (MBE) has been demonstrated [1,2]. The using of MBE increases the efficiency of QCL mainly due to the possibility to grow high quality identical multiperiod superlattices. Smooth control of epitaxial growth rate during the growth of QCL heterostructures [3,4] and/or using of cell with increased stability of the fluxes of elements of Group III [5,6] allow increasing the identity of QCL cascades. In this paper we report on epitaxy, diagnostics and device characteristics of QCL heterostructures grown using the industrial MBE system Riber 49 utilizing the ABI 1000 cells with a cylindrical crucible.

The first design (type 1) with the active region based on bound-to-continuum design [7,8] was suggested for lasing at ~ 6 μm. Before forming the stages of the laser on an InP substrate, an 100 nm $In_{0.53}Ga_{0.47}As$ buffer layer doped by silicon (Si) and lattice matched to InP was grown. The heterostructure contains 60 stages formed by the strain-compensated $In_{0.44}Al_{0.56}As/In_{0.60}Ga_{0.40}As$ heteropair that provided the necessary elastic balance in the QCL stages and a band offset ΔE_c of more than 630 meV at the heterointerface. The $In_{0.52}Al_{0.48}As$ top cladding layer and $In_{0.53}Ga_{0.47}As$ contact layer doped by Si have thickness of 1500 and 50 nm, respectively.

The active region design with the escape of charge carriers due to two-phonon resonance was used for 8 μm spectral range QCL heterostructure and consisted of 50 stages based on the $In_{0.53}Ga_{0.47}As/Al_{0.48}In_{0.52}As$ heteropair lattice matched to InP substrate. It was developed three waveguide types in 8 μm spectral range QCL heterostructures. First of them (type 2A) consisted of bottom cladding based on InP (2 μm thick) and $In_{0.53}Ga_{0.47}As$ (0.5 μm thick) layers, which were grown on a substrate in succession, the active region including 50 stages and $In_{0.53}Ga_{0.47}As$ layers of top cladding with 1×10^{17} cm^{-3} and 1×10^{19} cm^{-3} doping levels and with thickness of 0.1 μm and 0.02 μm, respectively [9,10]. Second of them (type 2B) was based on the bottom $In_{0.53}Ga_{0.47}As$ waveguide layer (0.5 μm thick), active region layers, top cladding layers based on InP (0.75 μm thick), waveguide based $In_{0.53}Ga_{0.47}As$ and contact $In_{0.53}Ga_{0.47}As$ layers with 1×10^{17} cm^{-3} and 1×10^{19} cm^{-3} doping level with thicknesses of 0.1 μm and 0.02 μm, respectively. Third of them (type 2C) consisted of bottom $In_{0.53}Ga_{0.47}As$ waveguide layer (0.5 μm thick), active region layers, top cladding layers based on InP (3.9 μm thick) and waveguide based on $In_{0.53}Ga_{0.47}As$ and contact $In_{0.53}Ga_{0.47}As$ layers with 1×10^{17} cm^{-3} and 1×10^{19} cm^{-3} doping level with thicknesses of 0.1 μm and 0.02 μm, respectively.

The three-phonon resonance design [5] was used for 9 μm spectral range QCL heterostructure (type 3) and consisted of 40 periods based on $In_{0.53}Ga_{0.47}As$ quantum wells with $Al_{0.48}In_{0.52}As$ barrier layers. Heterostructure consisted of $In_{0.53}Ga_{0.47}As$: Si (with 1×10^{18} cm^{-3} doping level) current spreading layer with a thickness of 50 nm, bottom cladding $In_{0.53}Ga_{0.47}As$: Si layer with 5×10^{16} cm^{-3} doping level and 4 μm thick active region layers, top cladding layers (4 μm thick) based on step-doped $Al_{0.48}In_{0.52}As$:Si ($1 \times 10^{17}/1 \times 10^{18}$ cm^{-3}) and 0.2 μm thick contact layer of $In_{0.53}Ga_{0.47}As$: Si (5×10^{18} cm^{-3}).

The heterostructure of the dual-frequency QCL (type 4) consisted of $In_{0.53}Ga_{0.47}As$: Si (with 1×10^{18} cm^{-3} doping level) current spreading layer with a thickness of 50 nm, bottom cladding $In_{0.53}Ga_{0.47}As$: Si layer with 5×10^{16} cm^{-3} doping level (2.1 μm thick), sequence of 15 stages of type 2 active region, sequence of 36 stages of type 3 active region, sequence of 15 stages of type 2 active region, top cladding layers based on $In_{0.53}Ga_{0.47}As$ (1.95 μm thick) and $In_{0.52}Al_{0.48}As$ (2 μm thick) and 0.02 μm thick contact layer of $In_{0.53}Ga_{0.47}As$: Si with 1×10^{19} cm^{-3} doping level.

X-ray diffraction measurements showed narrow satellite peaks, which indicate high structural perfection of the multi-stage QCL heterostructures. The spectra of X-ray diffraction were measured with high resolution in the vicinity of the InP symmetrical (004) reflection using a PANanalytical X'PertPro diffractometer with parallel geometry of the X-ray radiation beam. Additional analysis of high-resolution transmission electron microscopy (TEM) proved the absent of extended defects and that all layers have planar heteroboundaries. The TEM studies were conducted at a JEM2100F transmission electron microscope with an accelerating voltage of 200 keV.

The heterostructures of QCL were processed in four-cleaved and ridge geometries and mounted on copper heatsink using the indium solder to study the spontaneous and stimulated emission.

The QCL of type 1 based on strained heteropair $In_{0.44}Al_{0.56}As/In_{0.60}Ga_{0.40}As$ with $Al_{0.48}In_{0.52}As$ top cladding has shown the room-temperature lasing at threshold current density of 4.8 kA/cm^2 with the peak emission wavelength about 5.8 µm [3]. The estimated output optical power from one facet was about 150 mW in pulse mode.

The QCL of type 2A (waveguide with air cladding) demonstrated the lasing at room temperature with the threshold current density about 5.1 kA/cm^2 [9]. The lasing spectra demonstrate maximum close to 8 µm. The same QCL with full top metallization demonstrate the lasing up to 250 K with the threshold current density about 5 kA/cm^2 [11].

The QCL of type 2B (waveguide with thin top cladding) demonstrated the lasing at room temperature and the threshold current density about 5.6 kA/cm^2. The lasing wavelength was close to 7.8 µm. The estimated output peak optical power measured at 300 K was about 100 mW.

The QCL of type 2C (waveguide with thick top cladding) demonstrated the lasing at room temperature with wavelength 8.0 µm and the threshold current density about 4.9 kA/cm^2. The estimated output peak optical power measured at 300 K was about 200 mW.

For type 3 QCL the lasing up to 140 K was demonstrated. Threshold current density determined at 87 K was about 3 kA/cm^2. The estimated output optical power is about 20-35 mW, measured from one facet in pulse mode [5].

The heterostructure of dual-frequency QCL based on type 3 design was created and its structural and optical quality was studied. The design of waveguide with uniform pumping of both active regions [12] was implemented in this heterostructure. The realization of waveguide based on 2 active region one emitting at 7.6 µm wavelength and separated by other active region emitting at wavelengths of 9.6 µm allows to reach the lasing close to 7.6 µm and well pronounce of spontaneous emission near 9.6 µm.

To summarize, we presented the results of the study of QCL with different active region inside 5-10 µm spectral range. Room temperature lasing with output optical power about a few hundreds of milliwatt was demonstrated. The further optimization should be related with usage of thick InP top cladding and decreasing of injector doping level aimed to reach low threshold current density that is necessary for the high-power continuous-wave lasing.

This was supported by the Ministry of Education and Science of the Russian Federation within the Federal target program "Research and Development in Priority Areas of the Science and Technology Complex of Russia for 2014–2020", code no. 2016-14-579-0009, unique identifier RFMEFI57816X0204.

References

1. *Wang, C. A., Schwarz, B., Siriani, D. F., Missaggia, L. J., Connors, M. K., Mansuripur, T. S., Calawa, D. R., McNulty, D., Nickerson, M., Donnelly, J. P., Creedon, K.* MOVPE growth of LWIR AlInAs/GaInAs/InP quantum cascade lasers: Impact of growth and material quality on laser performance // IEEE J. Sel. Top. Quantum Electron. 2017. V. 23, No. 6. P. 1-13.

2. *Zasavitskii, I. I., Kovbasa, N. Y., Raspopov, N. A., Lobintsov, A. V., Kurnyavko, Y. V., Gorlachuk, P. V., Krysa A.B., Revin, D. G.* // A GaInAs/AlInAs quantum cascade laser with an emission wavelength of 5.6 µm // Quantum Electron. 2018. V. 48, No. 5. P. 472–475.

3. *Babichev, A. V., Bousseksou, A., Pikhtin, N. A., Tarasov, I. S., Nikitina, E. V., Sofronov, A. N., Firsov D. A., Vorobjev L. E., Novikov I. I., Karachinsky L. Y., Egorov, A. Y.* Room-temperature operation of quantum cascade lasers at a wavelength of 5.8 µm // Semiconductors 2016. V. 50, No. 10. P. 1299–1303.

4. *Mamutin, V. V., Vasilyev, A. P., Lyutetskiy, A. V., Ilyinskaya, N. D., Zadiranov, Y. M., Sofronov, A. N., Firsov, D. A., Vorobjev, L. E., Maleev, N. A., Ustinov, V. M.* On the Fabrication and Study of Lattice-Matched Heterostructures for Quantum Cascade Lasers // Semiconductors 2018. V. 52, No. 7. P. 950–953.

5. *Babichev, A. V., Gusev, G. A., Sofronov, A. N., Firsov, D. A., Vorobjev, L. E., Usikova, A. A., Zadiranov, Yu. M., Il'inskaya, N. D., Nevedomsky, V. N., Dudelev, V. V., Sokolovskii, G. S., Gladyshev, A. G., Karachinsky, L. Ya., Novikov, I. I., Egorov A. Yu.* Lasing of quantum-cascade laser at 9.6 µm wavelength // Tech. Phys. 2018. V. 88, No. 10, in press.

6. *Babichev, A. V., Gladyshev, A. G., Filimonov, A. V., Nevedomskii, V. N., Kurochkin, A. S., Kolodeznyi, E. S., Sokolovskii, G. S., Bugrov, V. E., Karachinsky, L. Ya., Novikov, I. I., Bousseksou, A., Egorov, A. Yu.* Heterostructures for quantum-cascade lasers of the wavelength range of 7–8 µm // Tech. Phys. Lett. 2017. V. 43, No. 7. P. 666–669.

7. *Faist, J., Beck, M., Aellen, T., Gini, E.* Quantum-cascade lasers based on a bound-to-continuum transition // Appl. Phys. Lett. 2001. V. 78, No. 2. P. 147–149.

8. *Egorov, A. Y., Babichev, A. V., Karachinsky, L. Y., Novikov, I. I., Nikitina, E. V., Tchernycheva, M., Tchernycheva, M., Sofronov, A. N., Firsov, D. A., Vorobjev, L. E., Pikhtin, N. A., Tarasov, I. S.* Lasing of multiperiod quantum-cascade lasers in the spectral range of (5.6–5.8)-µm under current pumping // Semiconductors 2015. V. 49, No. 11. P. 1527–1530.

9. *Babichev, A. V., Gladyshev, A. G., Kurochkin, A. S., Kolodeznyi, E. S., Sokolovskii, G. S., Bougrov, V. E., Karachinsky L. Ya., Novikov I. I., Bousseksou A., Egorov, A. Y.* Room Temperature Lasing of Multi-Stage Quantum-Cascade Lasers at 8 µm Wavelength // Semiconductors 2018. V. 52, No. 8. P. 1082–1085.

10. *Babichev, A. V., Gladyshev, A. G., Filimonov, A. V., Nevedomskii, V. N., Kurochkin, A. S., Kolodeznyi, E. S., Sokolovskii, G. S., Bugrov, V. E., Karachinsky, L. Ya., Novikov, I. I., Bousseksou, A., Egorov, A. Y. Heterostructures for quantum-cascade lasers of the wavelength range of 7–8 µm* // Tech. Phys. Lett. 2017. V. 43, No. 7. P. 666–669.

11. *Kurochkin, A. S., Babichev, A. V., Denisov, D. V., Karachinsky, L. Y., Novikov, I. I., Sofronov, A. N., Firsov, D. A., Vorobjev, L. E., Bousseksou, A., Egorov, A. Yu.* Quantum-cascade lasers in the 7-8 µm spectral range with full top metallization // J. Phys.: Conf. Ser. 2018. V. 993, No. 1. P. 012031.

12. *Babichev, A. V., Kurochkin, A. S., Kolodeznyi, E. C., Filimonov, A. V., Usikova, A. A., Nevedomsky, V. N., Gladyshev, A. G., Karachinsky, L. Ya., Novikov, I. I., Egorov, A. Yu.* Heterostructures of Single-Wavelength and Dual-Wavelength Quantum-Cascade Lasers // Semiconductors. 2018. V. 52, No. 6. P. 745–749.

https://doi.org/10.1051/epjconf/201819504002

Terahertz quantum cascade lasers with silver- and gold-based waveguides

R. A. Khabibullin[1], N. V. Shchavruk[1], D.S. Ponomarev[1], D.V. Ushakov[2], A.A. Afonenko[2], O.Yu. Volkov[3], V.V. Pavlovskiy[3], A.A. Dubinov[4]

[1]V.G. Mokerov Institute of Ultra High Frequency Semiconductor Electronics of RAS, Moscow, Russia, khabibullin@isvch.ru
[2]Belarusian State University, Minsk, Belarus
[3]Institute of radio-engineering and electronics of RAS, Moscow, Russia
[4]Institute for Physics of Microstructures RAS, Nizhny Novgorod, Russia

For the practical use of THz quantum cascade lasers (QCLs) in a large field of applications, including real-time THz imaging and THz spectroscopy, it is necessary to develop low loss waveguides [1].The most efficient in the THz region of the spectrum is a double metal waveguide (DMW) design in which the active region is placed between two metal layers. The mode confinement factor in such waveguides is $\Gamma \sim 1$, which is much higher than in plasmon waveguides ($\Gamma \sim 0.3$), which effectively work for mid-infrared QCLs. However, QCLs with DMW are complex in fabrication [2] and require preliminary theoretical and experimental studies of the behavior of the dielectric constant and the loss coefficient of both metals and semiconductors in the THz spectral region.

For the development of more efficient laser schemes for THz QCLs, as well as a low loss waveguide design, the information is needed on the loss in THz QCLs over a wide range of temperatures and frequencies. It is shown that the loss in THz QCLs plays an important role in the internal tuning of the radiation frequency [3]. In this work, we propose to use a double metal waveguide based on silver (Ag), for reducing the losses of the waveguide. In comparison with Au, Ag has a higher electrical and thermal conductivities, which should lead to low losses in the Ag-based waveguide. For this purpose, we investigate the spectra of mode losses in THz QCL with DMW based on Au and Ag.

Metal film of Ti/Au (30/1000 nm) and Ti/Ag (30/1000 nm) are deposited on SI-GaAs, and their resistivities are measured using Van der Pauw method in the temperature range from 4.2 to 300 K. The dielectric constants and loss coefficients (α_{met}) are calculated on the basis of measurements of the resistivity of metals using the Drude model. As shown in Fig. 1, the loss coefficient on the Au- and Ag-based DMW increases with increasing frequency and temperature. Using Ag for DMW allows to reduce α_{met} by more than 2 cm^{-1} comparing with DMW based on Au.

Calculations of the coefficient of total losses, including losses on the DMW, resonator mirrors, optical phonons and free charge carriers, are performed for a 10-μm waveguide. It is demonstrated that, taking into account the absorption of THz radiation by free carriers and optical phonons (see the inset of Fig. 2), the spectrum of total mode losses has a wide minimum in the region of 3-6 THz, which shifts to the high-frequency region of the spectrum with increasing temperature (see Fig. 2). The minimum losses in an Au-based waveguide with an increase of temperature from 100 to 300 K increase from 8 to 27 cm^{-1}. The use of Ag-based DMW allows to reduce losses by 2–4 cm^{-1} in comparison with Au-based DMW.

We have designed the active region of the THz QCL based on three tunnel-coupled quantum wells $GaAs/Al_{0.15}Ga_{0.85}As$ with a resonance-phonon depopulation scheme [4]. Numerical calculations of energy levels, matrix elements of dipole transitions, the degree of subbands populations and gain spectra are carried out depending on the applied electric field and temperature. It is shown that the maximum gain is realized under phonon resonance conditions at a frequency of 3.37 THz at an electric field strength of F = 12.3 kV/cm [5].

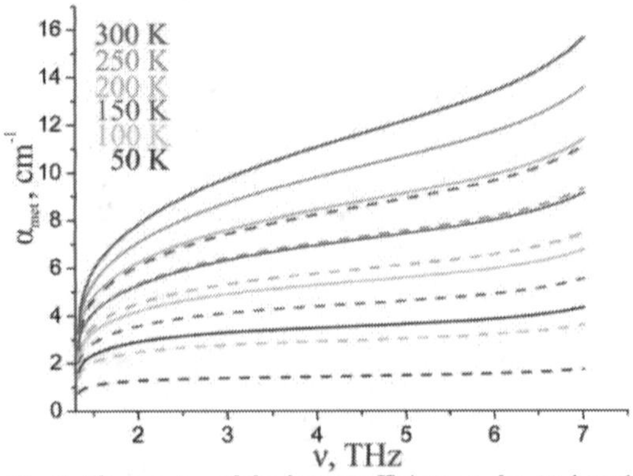

Fig. 1. The spectra of the loss coefficient on the Au-based DMW (solid lines) and Ag-based DMW (dashed lines) at temperatures from 50 to 300 K.

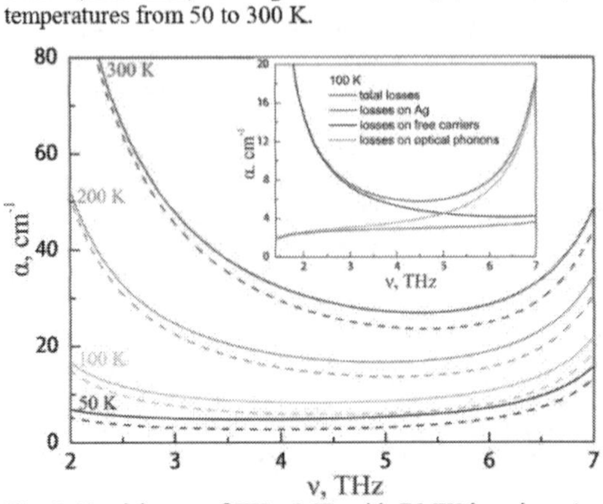

Fig. 2. Total losses of THz QCL with DMW based on Au and Ag for different temperatures. The inset shows the components for the total loss coefficient for THz QCL based on Ag DMW at 100 K.

Based on the proposed design, the laser heterostructures are grown by molecular beam epitaxy with 10 μm active region thickness [6]. We fabricate ridge structures with Au-based DMW [7]. The device is die bonded to a copper submount with wire bonding to ridge structures. Measurements are made with an abutted to laser facet sapphire hyperhemispherical lens.

The lasing threshold of the fabricated THz QCL is achieved at a current of 0.85–0.86 A for 20 K. The influence of temperature on the threshold current and output power of the fabricated THz QCL emitting near ~ 3.3 THz with a maximum operating temperature of T_{max} ~ 84 K is investigated. By fitting the dependence of threshold current J_{th} on temperature to the empirical expression $J_{th} \sim \exp(T/T_0)$, we obtain the characteristic temperature T_0 = 20 K for fabricated THz QCL. It is seen that with an increase in temperature from 40 to 58 K, an insignificant decrease in the THz QCL radiation power by ~ 35%, which allows the pumping of nitrogen vapors to be used to cool the laser. The characteristic energy of E_a = 23 meV, which is necessary for the temperature activation of LO phonon emission in the case of electron recombination from the upper laser level to the lower one, is determined from the Arrhenius temperature dependence of the output power.

In measured at 58 K spectrums, there are three equidistant spectral lines corresponding to the longitudinal Fabry-Perot modes with $\Delta f = c/2nL$ = 40 GHz for L=1 mm. The mode content varies with the bias current. There are three radiation modes (3.24, 3.28, 3.32 THz) at a pumping current of 0.940 A. Generation at 3.32 THz ceases when the current increases to 1.005 A, and only two high-frequency modes remain at 1.058 A. The maximum amplitude of the total radiation power is 28 μW in the 3.25–3.32 THz range at a temperature of 42 K [8].

We investigate the post-growth processing for THz QCL with Ag-based DMW. This processing includes the Ag–Ag thermocompression bonding of QCL heterostructures with a doped n+-GaAs substrate, mechanical lapping and selective wet etching of the substrate, and dry etching of QCL ridge structures through a Ti/Ag/Au metallization mask 100 μm wide. Reactive-ion-etching recipe with an inductively coupled plasma source in a BCl_3/Ar gas mixture is selected to obtain vertical walls of the QCL ridge structure with minimum Ti/Ag/Au mask sputtering (Fig. 6).

Conclusion

The coefficient of total losses of THz QCL with DMW based on Au and Ag have been analyzed. We have designed and fabricated THz QCLs with Au-DMW based on $GaAs/Al_{0.15}Ga_{0.85}As$ three-quantum well active module with resonant-phonon depopulation scheme. We investigate the dependence of threshold current and lasing output power on temperature. The dependence of the radiation spectrum on the amplitude of the excitation current pulse was investigated. The postgrowth processing for THz QCL with Ag-based DMW are studied.

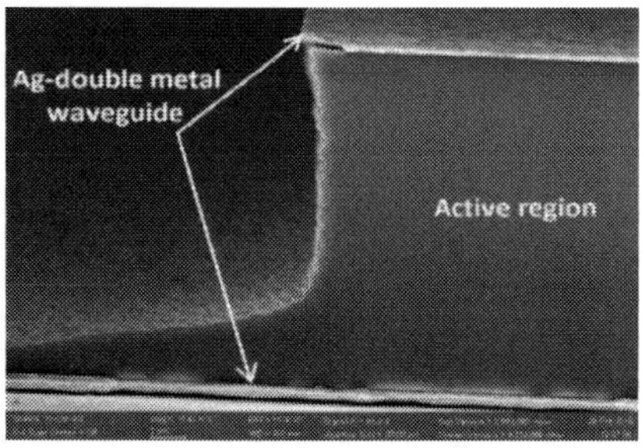

Fig. 3. SEM image of THz QCL with Ag-Ag double metal waveguide.

Acknowledgements

This work was supported by the Russian Science Foundation (Project #18-19-00493).

References

1. *Han, Y. J., Li, L.H., Zhu, J., Valavanis, A., Freeman, J.R., Chen, L., Rosamond, M., Dean, P., Davies, A.G., Linfield, E.H.* Silver-based surface plasmon waveguide for terahertz quantum cascade lasers // Opt. Express. 2018. V. 26, No.4. P. 3814-3827.

2. *Khabibullin, R. A., Shchavruk, N. V., Pavlov, A. Y., Ponomarev, D. S., Tomosh, K. N., R. R. Galiev, Maltsev, P. P., Zhukov, A. E., Cirlin, G. E., Zubov, F. I., Alferov, Z. I.* Fabrication of a terahertz quantum-cascade laser with a double metal waveguide based on multilayer GaAs/AlGaAs heterostructures // Semiconductors. 2016. V. 50, No. 10. P. 1377–1382.

3. *Schrottke, L., Lu, X., Roben, B., Biermann, K., Wienold, M., Richter, H., Hubers, H.W., Grahn. H.T.* Intrinsic frequency tuning of terahertz quantum-cascade lasers // J. Appl. Phys. 2018. V. 123, No. 21. P. 213102.

4. *Khabibullin, R. A., Shchavruk, N. V., Ponomarev, D. S., Ushakov, D. V., Afonenko, A. A., Vasil'evskii, I. S., Zaycev, A. A., Danilov, A. I., Volkov, O. Yu., Pavlovskiy, V. V., Maremyanin, K. V., Gavrilenko, V. I.* Temperature dependences of the threshold current and output power of a quantum-cascade laser emitting at 3.3 THz // Semiconductors. 2018. V. 52, No. 11. P. 1381–1386.

5. *Khabibullin, R. A., Shchavruk, N. V., Klochkov, A. N., Glinskiy, I. A., Zenchenko, N. V., Ponomarev, D. S., Maltsev, P. P., Zaycev, A. A., Zubov, F. I., Zhukov, A. E., Cirlin, G. E., Alferov, Zh. I.* Energy spectrum and thermal properties of a terahertz quantum-cascade laser based on the resonant-phonon depopulation scheme // Semiconductors. 2017. V. 51, No. 4. P. 514–519.

6. *Reznik, R. R., Kryzhanovskaya, N. V., Zubov, F. I., Zhukov, A. E., Khabibullin, R. A., Morozov, S. V., Cirlin G. E.* MBE growth, structural and optical properties of multilayer heterostructures for quantum-cascade laser // Journal of Physics: Conf. Series. 2017, V. 917. P. 052012

7. *Ikonnikov, A. V., Marem'yanin, K. V., Morozov, S. V., Gavrilenko, V. I., Pavlov, A. Yu., Shchavruk, N. V., Khabibullin, R. A., Reznik, R. R., Cirlin, G. E., Zubov, F. I., Zhukov, A. E., Alferov, Zh. I.* Terahertz radiation generation in multilayer quantum-cascade heterostructures // Technical Physics Letters. 2017, V. 43, No. 4. P. 358-361.

8. *Volkov, O. Yu., Dyuzhikov, I. N., Logunov, M. V., Nikitov, S. A., Pavlovskii, V. V., Shchavruk, N. V., Pavlov, A. Yu., Khabibullin, R. A.* Analysis of terahertz radiation spectra in multilayer GaAs/AlGaAs heterostructures // Journal of Communications Technology and Electronics. 2018. Vol. 63, No. 9. P. 1042–1046.

EPJ Web of Conferences **195**, 04003 (2018) https://doi.org/10.1051/epjconf/201819504003
TERA-2018

Overview of Techniques for THz QCL phase-locking

A. Khudchenko[1,2], D.G.Pavelev[3], V.L Vaks[4], A.M. Baryshev[1,5]

[1]Kapteyn Astronomical Institute/NOVA, University of Groningen, Groningen, Netherlands, a.khudchenko@sron.nl
[2]Kotel'nikov Institute of Radio Engineering and Electronics RAS, Moscow, Russia
[3]Lobachevsky State University, Nizhny Novgorod, Russia
[4]Institute for Physics of Microstructures RAS, Nizhny Novgorod, Russia
[5]Astro Space Center, Lebedev Physical Institute of Russian Academy of Science, Moscow, Russia

Since the first demonstration of a THz Quantum Cascade Laser (QCL) in 2002 [1] it was rapidly improved towards practical applications, for example THz imaging [2] and molecular spectroscopy [3] with high resolution.

At the moment, QCL is one of the most attractive continuous wave (CW) sources in the frequency range of 3-6 THz for heterodyne spectroscopy. In the terahertz astronomy QCLs are used as LO sources in heterodyne receivers for projects SOFIA [4] and GUSTO [5] and are considered for space observatories such as Millimetron [6], LOCUS [7] and OST [8]. Frequency stability of LO is absolutely crucial for a heterodyne instrument because it influences the quality of scientific data delivered by the receiver. The most reliable option to stabilize the LO frequency is phase-locking. Phase-locking of a THz QCL has been demonstrated by various groups using various methods. Here we make an overview of these techniques.

The first frequency stabilization was demonstrated by Betz *et al.* in 2005 [9] by frequency-locking (and partial phase-locking) of 3 THz 1 mW QCL to a far-infrared (FIR) gas laser using GaAs Schottky mixer. A while later, in 2006 Baryshev *et al.* [10] showed phase-locking of two-mode QCL, using an inter-mode beat signal generated on hot-electron bolometer mixer (HEB). Both these experiments were important demonstration of QCL stabilization, though they were difficulties for practical applications.

Afterwards, in 2009 Rabanus *et al.* [11] and Khosropanach *et al.* [12] showed independently stable phase-locking for single-tone QCL, in [11] it was used 1.5 THz 0.3 mW laser and in [12] – 2.7 THz 0.38mW one. The key moment is that both groups used HEB mixer to generate the beat signal for phase locking loop (PLL) system. The reference RF signal in [11] was generated by a high power multiplying chain able to pump an HEB mixer, while in [12] was used a frequency comb generated by superlattice diode mixer [13] pumped by 182 GHz microwave source. A little later, in 2010 Consolino *et al.* [14] realized similar approach for phase-locking of 2.5 THz 1 mW QCL, using HEB mixer, though the reference RF reference was a COMB signal generated by the Cherenkov effect in a lithium niobate waveguide effected by Ti:sapphire femtosecond laser. These three phase-locking approaches [11][12][14] can be combined in one group show schematically on fig.1, utilizing HEB mixer, which is known to be the most sensitive mixer from 1.5 to 6 THz. Nether the less, presence of HEB an in the locking scheme means arranging a separate 4K cryostat. It is not an issue for lab experiments, but it makes serious complications for on-bard application of a locking system.

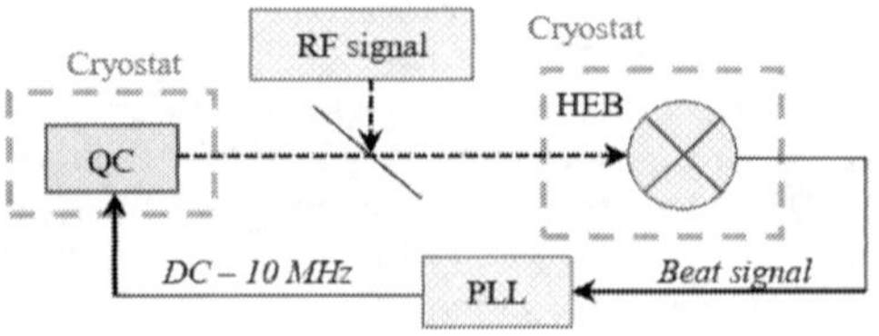

Fig. 1. Simplified block-diagram of QCL phase-locking described in papers [11], [12] and [14]. In all the cases the HEB mixer is used to provide a beat signal between the QLC and a reference signal. The RF signals in all the approaches are generated in different way.

In 2010 Barbiere *et al.* [15] demonstrated another approach of QCL phase-locking. In this experiment, a mode-locked femtosecond laser and a QCL radiation (2.7 THz, power of 25 mW) were applied to electro-optic detection system. This photo-mixing system generated a mixing product between a QCL frequency and the n_{th} harmonic of repetition rate of femtosecond laser of 90 MHz. The photocurrent beat signal frequency was 30 MHz. The electro-optic detection system is based on ZnTe crystal modulating amplitude of femtosecond pulses with the QCL frequency and on high speed (bandwidth of 300 MHz) silicon photodiodes. Similar technique was used a year later in 2011 by Ravaro *et al.* [16] to phase-lock 10mW QCL radiating at 2.5 THz (see fig. 2).

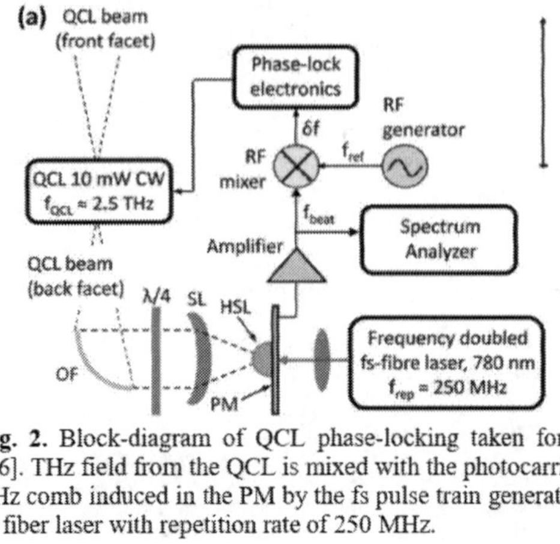

Fig. 2. Block-diagram of QCL phase-locking taken form [16]. THz field from the QCL is mixed with the photocarrier THz comb induced in the PM by the fs pulse train generated in fiber laser with repetition rate of 250 MHz.

The main new element relative to [15] is a GaAs photomixer. A big advantage of scheme in [15] and [16] is that only a room-temperature-operated elements are used. A fiber laser technology and the semiconductor photomixer are much more light and compact than superconducting bolometer mixers.

Another way of QCL phase-locking has been described in 2013 by Hayton *et al.* [17]. The authors phase locked a 3.4 THz 1 mW distributed feedback (DFB) laser to the 18th harmonic of a 190.7 GHz reference source using a room temperature GaAs/AlAs superlattice diode. Next year Khudchenko *et al.* [18] employed the same technique for 4.7 THz 0.25 mW QCL. The key element of this scheme (see fig. 3) is the harmonic mixer receiving QCL signal and mixing it with n_{th} harmonic of a reference microwave signal of about 10 mW power at frequency around 200 GHz. The product is used by PLL system to control phase and frequency of QCL. All the elements of the locking loop are working at room temperature. A very similar approach, but using Schottky harmonic mixer instead of super-lattice one, was been published in 2015 by Danylov *et al.* [19] and Bulcha *et al.* [20]. In [20] it was phase locked a 2.5 THz laser, and in [19] – 2.3 THz and 2.9 THz lasers of power correspondingly. All the experiments in [17], [18], [19] and [20] can be reflected by very simple scheme shown in fig. 3.

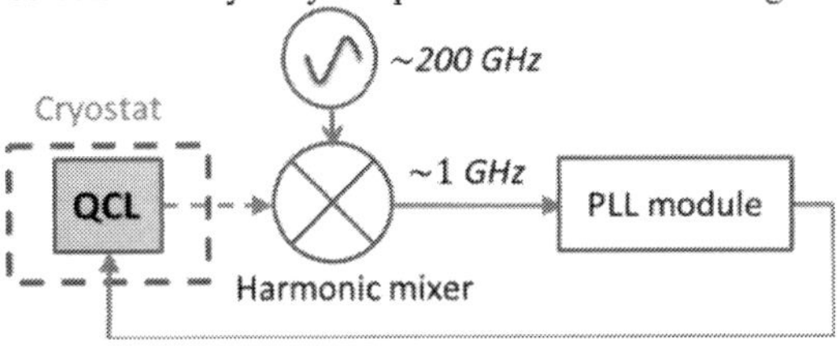

Fig. 3. Simplified block-diagram of QCL phase-locking shown in [17], [18], [19] and [20].

The latest conceptually unique method of QCL phase locking was presented in 2017 by Freeman *et al.* [21]. The 2 THz 1.7 mW QCL was locked not by an active feed-back system but by injection locking method. Employing infrared frequency combs and InGaAs photomixers a reference signal of 100 nW was generated at frequency very close to the QCL frequency. This signal was injected into the QCL and synchronized its radiation.

To summarize, the most promising are four techniques: [15]/[16]; [17]/[18] ; [19]/[20] and [21].

However, only in [18] it was shown phase locking at 4.7 THz, all the other QCLs are below 3 THz. Though, one of the most interesting applications for QCL is a role of LO in on-board heterodyne receiver at 4.7 THz to observe atomic oxygen line [4-8].

References

[1] *R. Köhler, A. Tredicucci, et. al.* // Terahertz semiconductor heterostructure laser // Nature 417, 156–159, 2002.
[2] *A. W. M. Lee, Q. Qin, et al.* // "Real-time terahertz imaging over a standoff distance (>25 meters) // Appl. Phys. Lett. ,89, 141125, 2006.
[3] *H.-W. Hübers, S. G. Pavlov, et. al.* // High-resolution gas phase spectroscopy with a distributed feedback terahertz quantum cascade laser // Appl. Phys. Lett. 89, 061115 2006.
[4] *H. Richter, M. Weinold, et al* // 4.7-THz Local Oscillator for the GREAT Heterodyne Spectrometer on SOFIA // IEEE Trans. on TST 5, 539-535, 2015.
[5] *https://www.sron.nl/missions-astrophysics/gusto*
[6] *http://millimetron.ru/index.php/en/*
[7] *http://www.locussatellite.com/*
[8] *J. Fortney et al. https://arxiv.org/abs/1803.07730*
[9] *A. L. Betz, R.T. Boreiko, et al.* // Frequency and phase-lock control of a 3 THz quantum cascade laser // Opt. Lett. 30, 1837-1839, 2005
[10] *A. Baryshev, J. N. Hovenier, A. J. L. Adam, I. Kaalynas, J. R. Gao, T. O. Klaassen, B. S. Williams, S. Kumar, Q. Hu, and J. L. Reno* // Phase locking and spectral linewidth of a two-mode terahertz quantum cascade laser // Appl. Phys. Lett. 89, 031115 , 2006.
[11] *D. Rabanus, U. U. Graf, M. Philipp, O. Ricken, J. Stutzki, B. Vowinkel, M. C. Wiedner, C. Walther, M. Fischer, and J. Faist* // Phase locking of a 1.5 Terahertz quantum cascade laser and use as a local oscillator in a heterodyne HEB receiver // Opt. Express 17, 1159-1168, 2009.
[12] *P. Khosropanah, A. Baryshev, W. Zhang, W. Jellema, J. N. Hovenier, J. R. Gao, T. M. Klapwijk, D. G. Paveliev, B. S. Williams, S. Kumar, Q. Hu, J. L. Reno, B. Klein, and J. L. Hesler* // Phase locking of a 2.7 THz quantum cascade laser to a microwave reference // Opt. Lett. 34, 2958–2960 , 2009.
[13] *D. G. Paveliev, Yu. I. Koschurinov, V. M. Ustinov, A. E. Zhukov, F. Lewen, C. Endres, A. M. Baryshev, P. Khosropanah, W. Zhang, K. F. Renk, B. I. Stahl, A. Semenov, and H.-W. Huebers* // conf. proc., ISSTT, 319, 2008.
[14] *L. Consolino, A. Taschin, P. Bartolini, S. Bartalini, P. Cancio, A. Tredicucci, H. E. Beere, D. A. Ritchie, R. Torre, M. S. Vitiello, and P. De Natale* // Phase-locking to a free-space terahertz comb for metrological-grade terahertz lasers // Nat. Commun. 3, 1040.
[15] *S. Barbieri, P. Gellie, G. Santarelli, L. Ding, W. Maineult, C. Sirtori, R. Colombelli, H. Beere, and D. Ritchie* // Phase-locking of a 2.7, THz quantum cascade laser to a mode-locked erbium-doped fibre laser // Nat. Photonics 4, 636–640, 2010.
[16] *M. Ravaro, C. Manquest, C. Sirtori, S. Barbieri, G. Santarelli, K. Blary, J. F. Lampin, S. P. Khanna, and E. H. Linfield* // Phase-locking of a 2.5 THz quantum cascade laser to a frequency comb using a GaAs photomixer //", Opt. Lett., vol. 36, pp. 3969-3971, 2011.
[17] *D. J. Hayton, A. Khudchenko, D. G. Pavelyev, J. N. Hovenier, A. Baryshev, J. R. Gao, T. Y. Kao, Q. Hu, J. L. Reno, and V. Vaks* // Phase-locking of a 3.4-THz Quantum Cascade Laser using a harmonic super-lattice mixer // Appl. Phys. Lett. 103, 051115.
[18] *A. Khudchenko, D. J. Hayton, D. G. Pavelyev, A. Baryshev, J. R. Gao, T. Y. Kao, Q. Hu, J. L. Reno, V. Vaks* // Phase locking a 4.7 THz Quantum Cascade Laser using a Super-lattice Diode as Harmonic Mixer // conf. proc. IRMMW-THz, 2014.
[19] *A. Danilov, N. Erickson, A. Light, and J. Waldman, // Phase locking of 2.324 and 2.959 terahertz quantum cascade lasers using a Schottky diode harmonic mixer* // Opt. Lett., vol. 40, pp. 5090–5092, 2015.
[20] *B.T. Bulcha, J.L. Hesler, A. Valavanis, V. Drakinskiy, J. Stake, R. Dong, J.X. Zhu, P. Dean, L.H. Li, A.G. Davies, E.H. Linfield, N.S. Barke. // Phase locking of a 2.5 THz Quantum Cascade Laser to a microwave reference using THz Schottky mixer* // conf. proc., IRMMW-THz, 2015.
[21] *J. R. Freeman, L. Ponnampalam, H. Shams, R. A. Mohandas, C. C. Renaud, P. Dean, L. Li, A. G. Davies, A. J. Seeds, and E. H. Linfield* // Injection locking of a terahertz quantum cascade laser to a telecommunications wavelength frequency comb // Optica, Vol. 4, No.9, pp. 1059-1064, 201

Characterization of the THz quantum cascade laser using fast superconducting hot electron bolometer

Yu. V. Lobanov[1,2], Yu. B. Vakhtomin[1,2], I. V. Pentin[1,2], R. A. Khabibullin[3], N. V. Shchavruk[3], K. V. Smirnov[1,2,4]

[1]LLC "Superconducting Nanotechnology" (SCONTEL), Moscow, Russia, lobanov@scontel.ru
[2] Moscow State Pedagogical University, Moscow, Russia
[3] V.G. Mokerov Institute of Ultra High Frequency Semiconductor Electronics of RAS, Moscow, Russia
[4]National Research University Higher School of Economics, Moscow Institute of Electronics and Mathematics, Moscow, Russia

Invention and demonstration of solid-state terahertz-range quantum cascade lasers (THz QCLs), followed by inevitable development of the fabrication technology, brought in a new era for the terahertz-range related fundamental and practical applications, such as spectroscopy and astrophysics, imaging and communication. With increase of the THz QCL research papers (with details found elsewhere, e.g., [1-4]), optimization of the lasing characteristics, improvement of its performance and development of the laser characterization techniques attracts a special attention among the researchers. There is a wide variety of different techniques for the detection of THz QCL. However, most of them are based on incoherent thermal detection with the slow time-constant of thermal process ~ 1 ms. In this paper, we present a new characterization method of THz QCL operated in the pulsed mode. In the proposed method the THz QCL is biased by applying a smooth gradual (i.e., with a "slow-rising" front edge) voltage pulse, and the laser output power was measured using NbN superconducting hot electron bolometer (HEB) within a single voltage pulse front edge, which was possible due to sufficiently short response time (of the order of 1 ns) of the HEB. The HEB which we used was similar to that described elsewhere [5]. The primary advantage of this method over the most common THz detection techniques derives from its ability to measure THz QCL output power-current relationship within a single voltage pulse. In addition, the use of a smooth gradual voltage pulse makes it possible to avoid strong fluctuations of the current flowing through the THz QCL at the beginning and the end of the driving pulse in the case of using a driving pulse with sharp edges [6]. In our setup, THz QCL was mounted onto a cold plate of a closed-cycle refrigerator, whose base temperature was varied in a range of 5 – 60 K. Principally, a very short measurement time set by duration of the driving pulse and its repetition rate (~ 5 μs and 200 Hz, respectively), prevents the QCL from overheating by the driving current. Amplitude of the driving pulse was chosen to be slightly higher than the laser threshold voltage. Fig. 1 depicts a driving pulse along with the measured lasing power which is given in arbitrary units. THz radiation was focused by a couple of off-axis parabolic mirrors onto the detector block. HEB was installed onto a cold plate of a wet helium cryostat. DC voltage and a resistive heater were used to bias the HEB to its optimal bias point where a high responsivity is obtained. THz QCL power incident on the detector was kept within detector's dynamic range. The HEB signal was amplified using a low noise cold amplifier and an R&S oscilloscope was used for the signal analysis.

THz QCL based on four tunnel-coupled quantum wells $GaAs/Al_{0.15}Ga_{0.85}As$ with a resonance-phonon depopulation scheme emitting near 2.3 THz was investigated by the proposed method. The laser was processed into a double metal waveguide based on Au with dimensions 1 mm × 100 μm ×10 μm using the conventional technology described in [7]. The laser is die bonded to a copper submount with wire bonding to ridge structures. No coatings or lens were deposited on the laser facet. A THz QCL emits radiation when the value of voltage exceeds the laser threshold level (i.e. when the laser gain coefficient is exactly balanced by the sum of waveguide and mirror losses). Applying a smooth gradual voltage pulse it becomes possible to smoothly approach the threshold condition which was examined at different operating temperatures of the QCL. As deduced from data given in Fig. 1, the laser threshold and peak power are achieved at voltages of 10.7 V and 10.8 V at 5 K, respectively. A further increase in the laser voltage is accompanied by a sharp drop in the optical power until lasing ceases at 11.3 V. After a sharp increase in output power beyond the laser threshold,

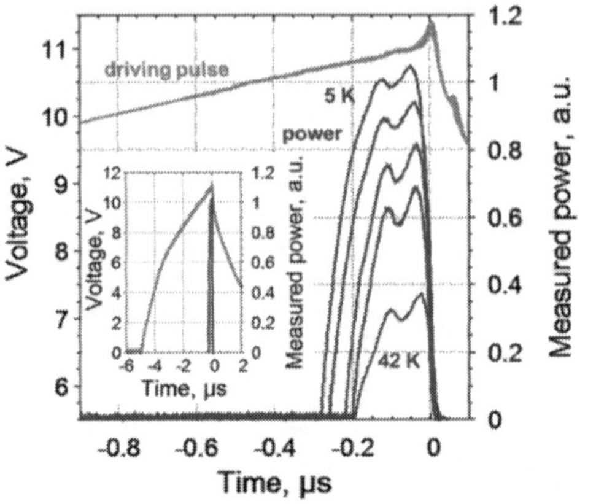

Fig. 1. QCL driving voltage pulse and measured lasing power registered with HEB. With increase of the QCL operating temperature from 5 to 60 K lasing power has decreased. In the figure, measured power at 5, 14, 36, 41, and 42 K is shown. The lasing pulse width has also decreased from ~ 250 ns at 5 K down to ~ 100 ns at 56 K. Inset shows rise and decay of the voltage pulse and relative position of the emitted THz pulse during the driving pulse

a nonmonotonic behavior of the output power-current dependence is observed. We suppose that this is due to the mode hopping effect with an increase of applied electric field. Commonly, the gain maximum shifts to higher frequencies with an increase in applied electric field, which is related to transformation of the laser levels positions [8]. The influence of temperature on the laser parameters is one of the most important characteristic of THz QCL. The temperature dependence of the threshold current J_{th} was measured by the proposed method. Using a phenomenological relationship for the temperature dependence:

$$J_{th} = J_1 + e^{T/T_0}, \quad (1)$$

we determined the characteristic temperature T_0 to be ~ 26 K. Since in our experiment we have measured rather voltage, the current was obtained using a mathematically derived fitting parameter *AR* such that $J_{th} = V_{th}/AR$, with $AR = 0.0217$. $J_{th}(T)$ and its fit according to (1) are depicted in Fig. 2. Following a paper by Albo & Hu [9] we have also examined *P(T)* dependence of the QCL power and experimental results are shown in Fig. 3. The presence of second plateaus in *P(T)* in the temperature range from 43 to 52 K is not traditional for this characteristic.

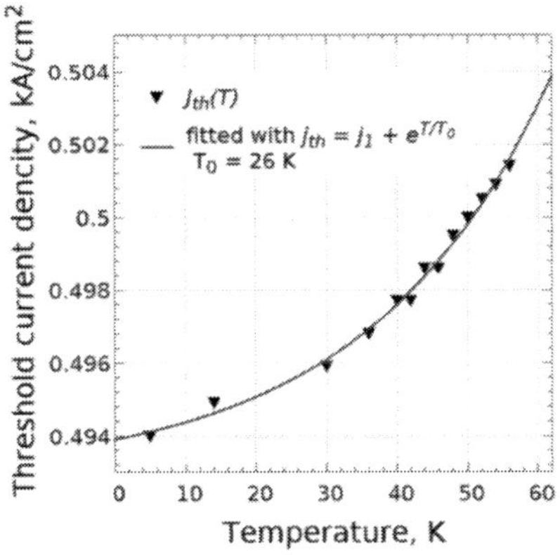

Fig. 2. Threshold current dependence on the QCL operating temperature. Experimental points are fitted according to (1)

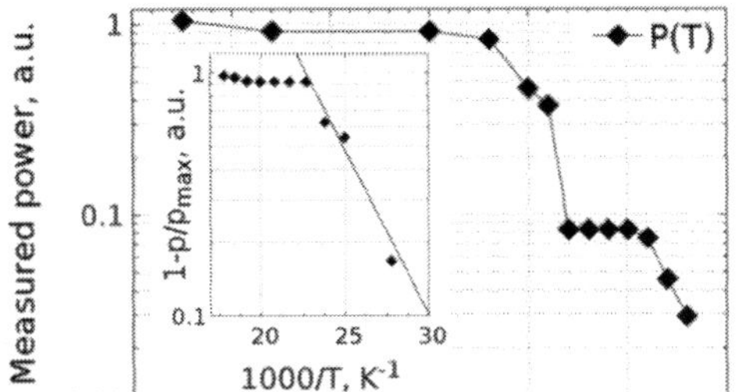

Fig. 3. Dependence of the laser output power (normalized) on the QCL operating temperature. Inset shows the Arrhenius plot with a fitting line in the temperature range 30-42 K, from which an activation energy E_a of ~ 26 meV has been derived using (2)

Typically *P(T)* gradually decreases with increasing temperature and at high temperatures the dominant mechanism of scattering is thermally activated LO-phonon scattering [9]. The formation of the second plateau in *P(T)* is most likely caused by the leakage of electrons from lower laser level to the continuum. Thus, a partial compensation of "negative" thermally activated LO-phonon scattering and "positive" leakage of electrons from lower laser level is obtained. Nevertheless, Arrhenius plot of the *P(T)* dependence (inset in Fig. 3) is well fitted to:

$$\frac{P_{out}(T)}{P_{out}^{max}} = 1 - \alpha e^{-E_a/kT} \quad (2)$$

in the 30-42 K temperature range, providing an activation energy E_a of ~ 26 meV which is reasonably close to a calculated value of ~ 28 meV.

In conclusion, we have proposed and demonstrated a new approach for THz-QCL characterization allowing $P(V_{bias})$, *P(T)*, and $I_{th}(T)$ analysis. Obtained results are in a reasonable agreement with previous results; however, further study of these dependencies may reveal additional information regarding the QCL structure. Exploiting fast response of our detector and tunability of the implemented pulsed gradual biasing a thorough investigation of the QCL emittance depending on the driving parameters could be carried out.

This research has been partially supported by the RSF grant No. 17-72-30036 (development and implementation of the experimental technique, data analysis) and RFBR grant No. 17-02-00070 A (design and processing of the QCL chip, data analysis).

References

1. *Williams, B. S.* Terahertz quantum-cascade lasers // Nat. Photonics 2007. V. 1. P. 517-525.
2. *Scalari, G., et al.*, THz and sub-THz quantum cascade lasers // Laser & Photon. Rev. 2009. V. 3, No. 1–2. P. 45–66.
3. *Sirtori, C., et al.*, Wave engineering with THz quantum cascade lasers Nat. Photonics 2013. V. 7. P. 691–701.
4. *Liang, G., et al.*, Recent Developments of Terahertz Quantum Cascade Lasers // IEEE J. Sel. Top. Quant. 2017. V. 23, Iss. 4. P 1-18.
5. *Seliverstov, S., et al.*, Fast and sensitive terahertz direct detector based on superconducting antenna-coupled hot electron bolometer // IEEE Trans. on Appl. Supercond. 2015. V. 25, No. 3. P. 2300304-1 - 2300304 5.
6. *Scheuring, A., et al.*, Transient Analysis of THz-QCL Pulses Using NbN and YBCO Superconducting Detectors // IEEE Trans. THz Sci. Techn. 2013. V.3, No. 2. P. 172-179.
7. *Khabibullin, R. A., et al.*, Fabrication of a terahertz quantum-cascade laser with a double metal waveguide based on multilayer GaAs/AlGaAs heterostructures // Semiconductors. 2016. V. 50, No. 10. P. 1377–1382.
8. *Khabibullin, R. A., et al.*, Temperature dependences of the threshold current and output power of a quantum-cascade laser emitting at 3.3 THz // Semiconductors. 2018. V. 52, No. 11. P. 1381–1386.
9. *Albo, A., Hu, Q.* Investigating temperature degradation in THz quantum cascade lasers by examination of temperature dependence of output power // Appl. Phys. Lett. 2015. V. 106. P. 131108-1 – 131108-5.

EPJ Web of Conferences **195**, 04005 (2018) https://doi.org/10.1051/epjconf/201819504005
TERA-2018

Polymer waveguides for THz QCL radiation delivery and filtering

Nazarov Maxim[1], A. Shilov[2], Z. Margushev[3], K. Bzheumikhov[3], I. Ozheredov[4], A. Angeluts[4], A. Sotsky[2], A. Shkurinov[4]

[1]Kurchatov Institute National Research Center, Moscow, Russia, nazarovmax@mail.ru
[2]State Institution of Higher Education "A.A. Kuleshov State University", Mogilev, Republic of Belarus
[3]Institute of Computer Science and Problems of Regional Management of RAS, Nalchik, Russia
[4]Department of Physics and International Laser Center, Lomonosov Moscow State University, Russia

The development of suitable waveguides with low attenuation will allow terahertz technologies to rise to a mass applications level. In the known reviews of perspective fibers for the THz range [1,2] the necessary requirements of flexibility, single-mode regime and low attenuation are still not met simultaneously. For each case a compromise between the requested properties must be found. We demonstrate a possibility to meet all the requirements in a capillary waveguide structure in the case of the THz quantum-cascade laser (QCL) radiation delivery. The tube lattice fiber (TLF) formed by one ring of the eight identical capillaries shown in Fig.1(c) is the best compromise for THz range. The advantages of this structure over other known types of waveguides are: a simple manufacturing process, the single-mode regime, small size and flexibility [3,4]. Discrete transparent spectral bands with sharp slopes inherent to this structure in some cases became an advantage of the TLF. We present manufacturing technology and theory allowing to optimize the TLF for the delivery of the narrow band QCL radiation at f=3 THz frequency.

The authors have extensive experience in the manufacturing of photonic crystal fibers from glass. Despite the difficulties in applying this technology to polymeric materials, we use original approaches to fabricate specified geometric parameters of waveguide structures in the multi-stage "stack and draw" process. As a result, a polypropylene flexible waveguides were manufactured in different configurations (Fig.1). Transmission properties of those TLF types are compared as well in the report as the most transparent polymer materials.

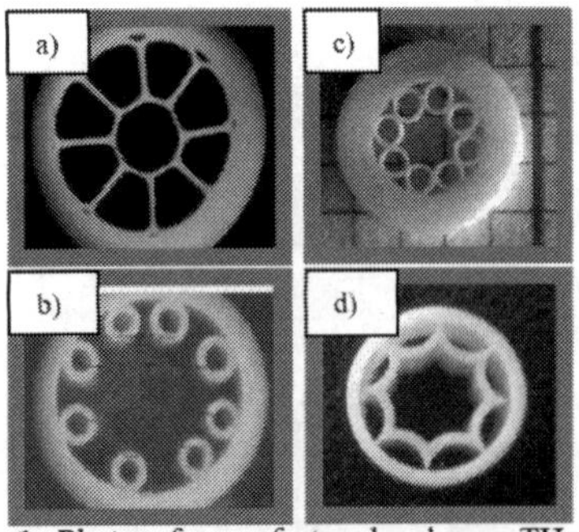

Fig. 1. Photo of manufactured polymer THz waveguides types

To create the most transparent and flexible waveguide for a fixed frequency, we studied in detail the influence of the core radius R_{core} and capillary wall thickness d (Fig.2(a)) on the shape and position of the transparency frequency windows. To characterize the fabricated TLF, it is necessary to take into account its imperfection. The wall thickness of each of the 8 capillaries in the "cladding" d_i has a random deviation of δd from the average value d. With the improvement of manufacturing technology, the size dispersion $\delta d/d \approx 0.14$ should be minimized, it is now the limiting factor to decrease d/R value and finally to decrease total TLF losses. The data in Fig.2(c) refer to polypropylene (PP) 8 capillary structure. Identical capillaries are tightly inserted into the tube cover. The ratio of the inner radius of the cover R to the outer radius of the capillaries r is given as follows: $r = R/(1+1/\sin(\pi/8))$, $R_{core} = R - 2R/(1+1/\sin(\pi/8)) = 0.4465\text{R}$ Here R, r and R_{core} are explained on Fig. 2a). Note that number of capillaries can differ from 8.

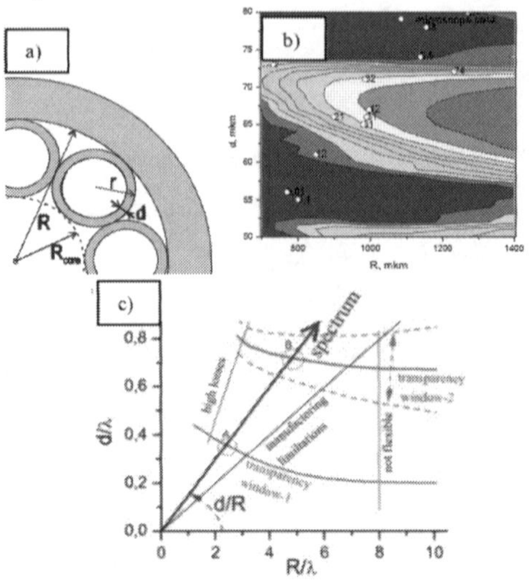

Fig. 2. a) TLF sizes, b) calculated (color area) and measured (circles) 30 cm TLF transmittance (in %) *vs* d and R, c) Optimization strategy and physical limitations for 8 capillary dielectric THz waveguide. Optimal wall thickness d *vs* air core radius R for first two transparency windows, both axes are normalized to the wavelength λ. Markers A and B indicate a particular parameters set.

The used TLF structure is optimized by two most important values - the wall thickness d (which determines the position of the transmission windows) and the core radius R, (affecting attenuation value, flexibility and modal regime). All structure proportions are given by the ratio d/R of the initial tubes, that indicates how thick the wall relative to other TLF sizes is. In fact the maximum achievable transparency of TLF is determined by this ratio, it should be as low as possible till wall quality and TLF flexibility are satisfactory- Fig 2c).

Considerations for optimal *d* should account for a quasiperiodic dependence of losses *versus d*. The actual spectral position of each transparency window depends on the size of the waveguide core *R*. We choose the second transparency window as the most promising of the technically available ones. To account for the deviation from analytical estimation [4], special calculations were carried out using the methods developed in [**Error! Bookmark not defined.**]. The relationship between *R* and *d* for the center of the first and second transparency windows for a structure of 8 polypropylene (PP, n=1.5) capillaries are shown in Fig. 2c). Since polymers refraction are almost constant in THz range, for TLF analysis variation of *d* is identical to variation of $1/\lambda$, in d/λ scale we obtain universal curve (Fig. 2c), some difference appear between $R/\lambda=const$ and $d/R=const$ cases for *d* or λ variation [**Error! Bookmark not defined.**].

The measurements

To couple QCL radiation into the TLF we focused a free space radiation with a lens into a 1 mm spot on the input end – Fig. 3a). At the output end of the 30 cm long sample we measured the radiation by the Golay cell, THz camera or interferometer. For bending loss measurements we moved the output end of the waveguide together with the Golay cell.

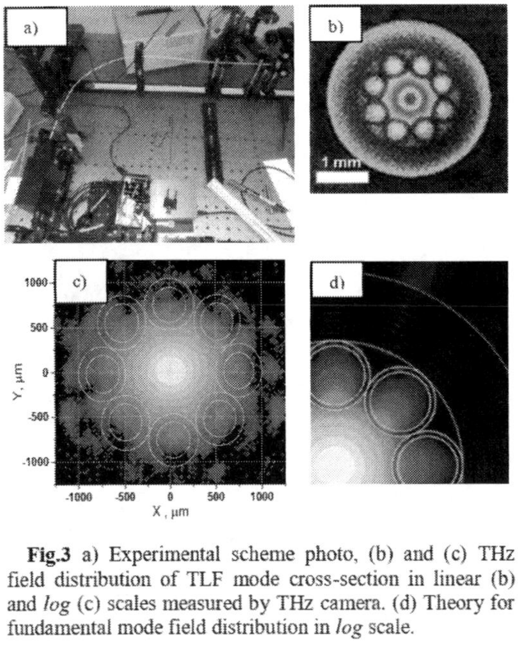

Fig.3 a) Experimental scheme photo, (b) and (c) THz field distribution of TLF mode cross-section in linear (b) and *log* (c) scales measured by THz camera. (d) Theory for fundamental mode field distribution in *log* scale.

More rigorously the shape and bandwidth of the transparency window can be measured in spectral domain by THz-TDS method for thicker walls and lower frequency – see Fig.4, for the same $\delta d/d$ and d/R average values. Using scalability of d/λ, one can compare frequency-domain spectra with wall-thickness domain spectra. The transparency window bandwidth and shape for fixed values of $\delta d/d$ and d/R are in good agreement with the theory. Note, that *d* and λ values indicated by markers *B"* in Fig.4 correspond to the second transparency window center indicated by marker *B* in Fig.2(c), in all cases $d/\lambda\approx0.75$.

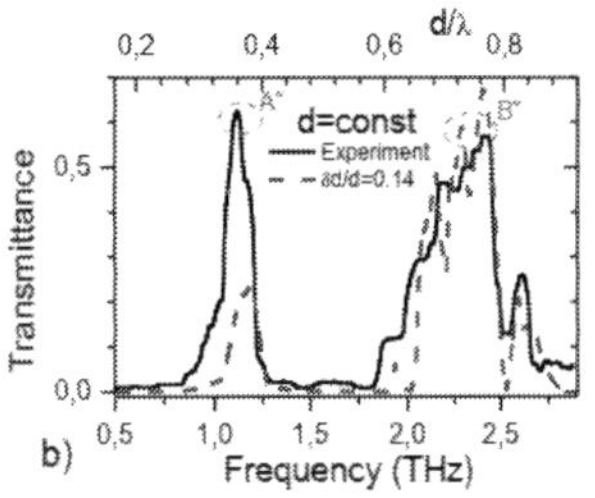

Fig. 4 Fixed d=95 µm and R=600 µm, TDS transmittance spectra - solid magenta line; dashed blue line – theory for $\delta d/d$=0.14.

In conclusion, we have demonstrated a purely dielectric, flexible, single mode, narrowband waveguide for the 3 THz frequency. In comparison with the previously known studies of TLF [5] we increased the used frequency from 0.3-1.5 THz to 3 THz, decreased the transversal TLF size from 1 cm to 3 mm. For filtering out undesirable frequencies of THz QCL radiation and selecting desired QCL frequency a set of waveguides can be used due to a steep slope of their transparency window. At the same time, the key application of the TLF seems to be a way to deliver single mode THz radiation to hardly accessible areas. A good agreement of the experimental data and theoretical calculations allows accounting for the existing technical limitations.

The work was supported by RFBR grants 18-52-00040, 17-00-00270, 16-29-11789 and BRFFI grant F18R-143.

References

1. S. Atakaramians, S. Afshar, V. T. M. Monro, D. Abbott, Adv. Opt. Photon. **5**, 169 (2013).
2. A. Bahr, B. P. Pal, G. P. Agrawal, R. K. Varshney, and M. A. Rahman, IEEE J. Sel. Topics Quantum Electron.**22**, 850015 (2016).
3. V. Setti, L. Vincetti, and A. Argyros, Opt. Express **21**, 3388 (2013).
4 M.M. Nazarov, A.V. Shilov, K. A. Bzheumikhov, Z. Ch. Margushev, V. I. Sokolov, A. B. Sotsky, and A. P. Shkurinov, IEEE Trans. THz Sci. Technol. **8**, 183 (2018).
5W. Lu, S. Lou, A. Argyros, IEEE J. of Sel.Top. In Quantum Electronics, **22**, 4401607 (2016)

Temporal stability and absolute composition issues in molecular beam epitaxy of AlGaAs/GaAs THz QCL

I. S. Vasil'evskii[1], A. N. Vinichenko[1], M. M. Grekhov[1], V. V. Saraykin[2], A. N. Klochkov[2], N. I. Kargin[1], R. A. Khabibullin[2], S. S. Pushkarev[2]

[1]National Research Nuclear University «MEPhI», Moscow, Russia, ivasilevskii@mail.ru
[2]V.G. Mokerov Institute of Ultra High Frequency Semiconductor Electronics of RAS, Moscow, Russia

The operation of the terahertz quantum cascade lasers (THz QCL) is strongly dependent on the repeatable fabrication of the well and barrier layers with the certain thicknesses throughout the whole active region stack epitaxial growth. This emphasizes the importance of the strict control of the growth rates and the stability of Si and group III effusion cell fluxes during growth [1]. It was reported [2] that two THz QCLs based on nominally identical multilayer heterostructures $Al_{0.15}Ga_{0.85}As$/GaAs emit at the different frequencies of 2.59 and 2.75 THz because of unintentional small deviations in the GaAs and AlAs growth rates (4 and 1.6 % respectively). Authors [3] determined the thickness tolerance for working lasing heterostructures to be minimally above 2% while the structures with thickness deviation 4.3 and 6.5 % are not lasing. In [4] the Ga cell temperature has to be increased to maintain a GaAs constant growth rate while the Al cell temperature remains nearly constant because variation in AlAs growth rate can be neglected. By using such growth rate compensation technique the two nominally identical structures approximately 10 mkm overall thick were found to show thickness difference of ~ 1 %. Thus calibration procedures and accurate analysis techniques become crucial to provide enhanced metrology possibilities.

In this work we discuss the following problems which have to be solved within the QCL growth metrology: precise determination of Al content in $Al_xGa_{1-x}As$ ternary, composition profiling, robust and rapid express-analysis for AlGaAs/GaAs quantum well (QW) layer thickness and composition, cell regime corrections for proper III fluxes stabilization.

The origins of III group flow long-term drift are material rearrangement and depletion in the effusion cells and thermometry drift. The latter can arise if environmental temperature is changed and thermocouple loop hasn't reference junction.

For the absolute $Al_xGa_{1-x}As$ composition measurement the different techniques have been used: X-ray diffractometry (XRD), photoluminescence spectroscopy (PL), secondary ion mass spectroscopy (SIMS), and Rutherford ion backscattering (RBS). SIMS measurements performed in cluster ion mode with $CsMe^+$ provide good accuracy for relative composition variation. Absolute content determination by different techniques, however, gives spreading within $\Delta x \sim \pm 0.015$. Discrepancy for XRD techniques x determination arises from inaccuracy of C_{11} and especially C_{12} elastic coefficient for AlAs in the different databases. RBS are known as the most direct technique but it ensures absolute value within $\Delta x \sim \pm 0.01$ because Al haven't resolved separately but matches the area of GaAs substrate signal, so areal fit used rather than peak fit in this case. PL technique is quite sensitive [5, 6] but $E_g(x)$ dependencies have notable differences within $\Delta x \sim \pm 0.01$. PL is also can produce complicated spectrum due to doping and possible defects in $Al_xGa_{1-x}As$ (Fig. 1).

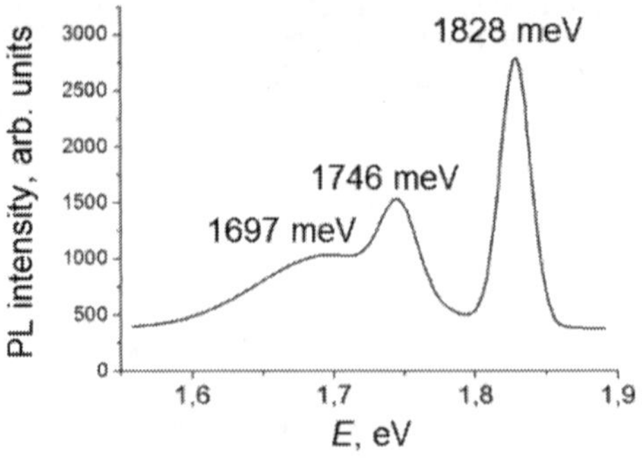

Fig. 1. PL spectra of $Al_xGa_{1-x}As$ at $T = 77$ K

Series of experiments performed on Compact 21 T epitaxy on 2 μm thick slightly Si doped $Al_xGa_{1-x}As$ (structure # 80) reveal a valuable drift of Al composition up to 1.2 %·μm^{-1} for the range of $x \sim 0.15$–0.24. After long-term cell regime stabilization and termal isolation of rear thermocouple terminals the variation has been decreased to ~ 0.1 %·μm^{-1} (Fig. 2), that could be sufficient for whole QCL structure growth.

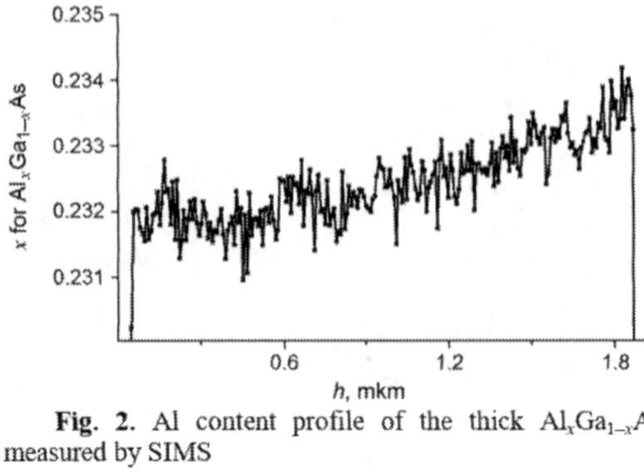

Fig. 2. Al content profile of the thick $Al_xGa_{1-x}As$ measured by SIMS

Another issue arises for superlattice structures $Al_xGa_{1-x}As$/GaAs, where both thickness and composition precision are needed. We have confirmed high efficiency of the calibration structures containing both $Al_xGa_{1-x}As$ layer and $Al_xGa_{1-x}As$ (96 Å)/GaAs (85 Å) short-period superlattice (SPSL) for the combination of metrological techniques. After proper calibration of composition and layer thicknesses, easy and fast analysis becomes possible by express technique as room temperature PL (Fig. 3). Sample # 486 has additional bulk $Al_xGa_{1-x}As$ layer that produces peak at 1.624 eV. Band structure calculation performed by Nextnano with conventional semiconductors parameters [7] shows good compliance with experiment. The technique sensitivity is about 0.6 meV energy shift per 1 Å GaAs thickness variation (Fig. 4). The absolute thickness determination accuracy for SPSL sample may be sufficiently higher by using XRD due to multiple satellite peaks and reaches 0.02 Å.

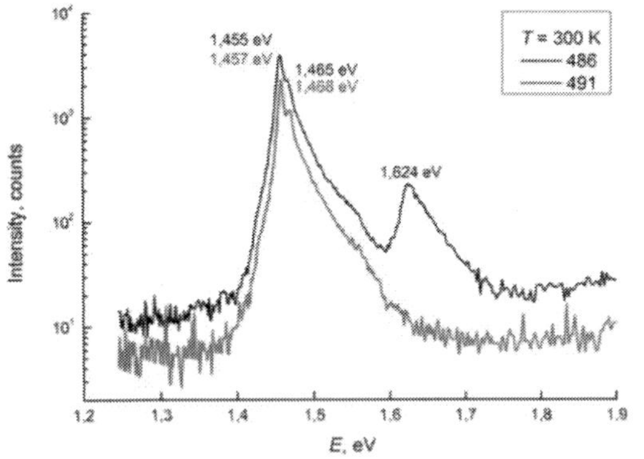

Fig. 3. Room temperature PL spectra of SPSL $Al_xGa_{1-x}As/GaAs$ samples

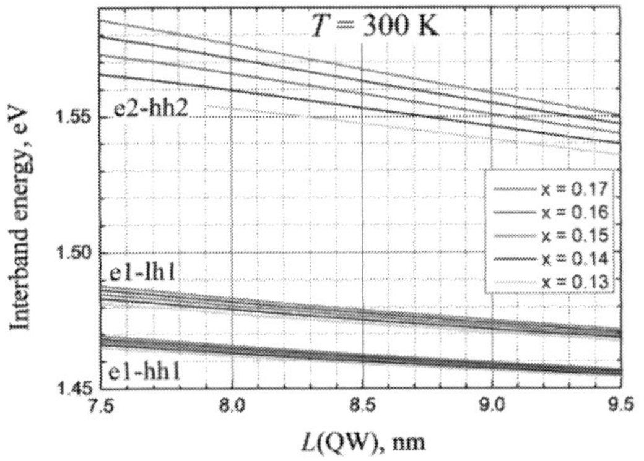

Fig. 4. Nomogramm of interband optical transitions in SPSL $Al_xGa_{1-x}As$ (96 Å)/GaAs (L_{QW})

Low temperature photoluminescence spectra of the two similar SPSL structures (samples # 486 and # 491) shown on Fig. 5 reveal excellent matching of e1-lh1 and e1-hh1 transitions in the different structures at 1.54–1.55 eV. PL measurements demonstrate high intensity of transitions and narrow well-resolved exciton peaks. The peaks beyond the quantum well energy correspond to GaAs bulk exciton (1.515 eV) and phonon replica of GaAs fundamental transition (1.473 eV). The hidden peak at 1.63–1.64 seeming as shoulder may be caused by e2-hh2 transition. Photoluminescence analysis based on e2-hh2 transition energy determination is more accurate in comparison with e1-transitions mainly in regard to x estimation, as one may see from Fig. 4. The explanation is that e2 energy level position is more sensitive to the height of quantum well barrier which depends on $Al_xGa_{1-x}As$ content.

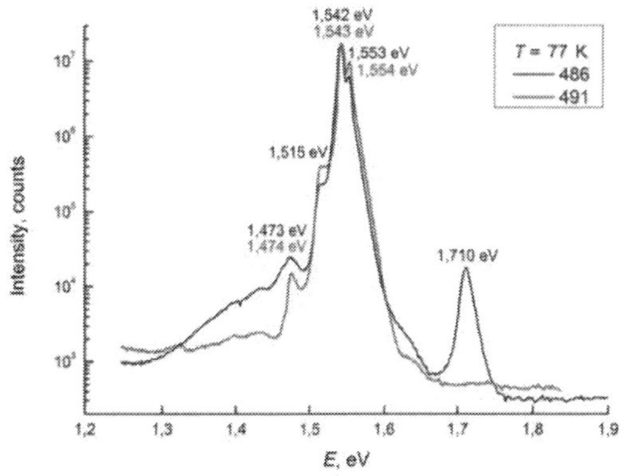

Fig. 5. Low temperature PL spectra of SPSL $Al_xGa_{1-x}As/GaAs$ samples at $T = 77$ K

As a result, different analysis techniques had been applied for $Al_xGa_{1-x}As$ composition analysis and spatial profile variation. XRD and PL techniques are collated and accuracy of the both are determined on short period $GaAs/Al_xGa_{1-x}As$ superlattice samples with layer arrangement, providing robust and rapid further control for QCL growth runs.

The study was partially supported by the Russian Federation President's Council on Grants (project MK-2342.2017.2) and by RFBR according to the research project № 17-02-00070 A.

References

1. *Beere, H. E., Fowler, J.,C., Alton, J., Linfield, E.,H., Ritchie, D. A., Köhler, R., Tredicucci, A., Scalari, G., Ajili, L., Faist, J., Barbieri, S.* MBE growth of terahertz quantum cascade lasers // Journal of Crystal Growth. 2005. V. 278. P. 756–764.

2. *Khanna, S. P., Chakraborty, S., Lachab, M., Hinchcliffe, N. M., Linfield, E. H., Davies, A. G.* The growth and measurement of terahertz quantum cascade lasers // Physica E. 2008. V. 40. P. 1859–1861.

3. *Roch, T., Andrews, A. M., Fasching, G., Benz, A., Schrenk, W., Unterrainer, K., Strasser, G.* High-quality MBE growth of $Al_xGa_{1-x}As$-based THz quantum cascade lasers // Central European Journal of Physics. 2007. V. 5, No. 2. P. 244–251.

4. *Li, L. H., Zhu, J. X., Chen, L., Davies, A. G., Lin field, E. H.* The MBE growth and optimization of high performance terahertz frequency quantum cascade lasers // Optics Express. 2015. V. 23, No. 3. P. 2720–2729.

5. *Galiev, G.B., Klimov, E.A., Klochkov, A.N., Lavruhin, D.V., Pushkarev, S.S., Maltsev, P.P., Vasil'evskii, I.S.* Specific features of the photoluminescence of HEMT nanoheterostructures containing a composite InAlAs/InGaAs/InAs/InGaAs/InAlAs quantum well // Semiconductors. 2015. V. 49, No. 2. P. 234–241.

6. *Lavrukhin, D. V., Khabibullin, R. A., Ponomarev, D. S., Maltsev, P. P.* Photoluminescence of heterostructures containing an $In_xGa_{1-x}As$ quantum well with a high in content at different excitation powers // Semiconductors. 2015. V. 49, No. 9. P. 1218–1221.

7. *Vurgaftman, I., Meyer, J. A., Ram-Mohan, L. A.* Band parameters for III–V compound semiconductors and their alloys // Journal of Applied Physics. 2001. V. 89, No. 11. P. 5815–5875.

EPJ Web of Conferences **195**, 04007 (2018) https://doi.org/10.1051/epjconf/201819504007
TERA-2018

3 THz quantum-cascade laser with metallic waveguide based on resonant-phonon depopulation scheme

F. I. Zubov[1], A. V. Ikonnikov[2], K. V. Maremyanin[2], S. V. Morozov[2], V. I. Gavrilenko[2], A. Yu. Pavlov[3], N. V. Shchavruk[3], R. A. Khabibulin[3], R. R. Reznik[1], G. E. Cirlin[1], A. E. Zhukov[1], A.A. Dubinov[2], Zh. I. Alferov[1]

[1]St Petersburg National Research Academic University of the Russian Academy of Sciences, St Petersburg, Russia, fedyazu@mail.ru
[2]Institute for Physics of Microstructures RAS, Nizhny Novgorod, Russia
[3]V.G. Mokerov Institute of Ultra High Frequency Semiconductor Electronics of RAS, Moscow, Russia

1. Introduction

Terahertz radiation, which frequency belongs to 0.1-10 THz range, is of great importance in various fields of science and engineering. One of the most important applications of THz radiation is the objects visualization and species detection. As this emission penetrates well through clothes and biological tissues and reflects from metals, it can be used in inspection systems for detection of hidden enclosures. At the same time, in contrast to X-ray, THz radiation is safe for living body. Promising sources of this radiation are quantum cascade lasers (QCLs), provided by their compact size, low current and voltage requirements, high brightness as well as relatively low cost. The first THz QCL was fabricated in 2002 [1]. As of today their output optical power reaches more than 200 mW in CW [2] and 1 W in pulse pumping regime [3], wavelength tuning range makes about 3 GHz for CW operation [4] and maximum lasing temperature reaches 200 K (for direct THz-emitted sources) [5]. However, QCL technology is in its infant stage in Russia.

In the present work we have fabricated and tested THz QCL having metal-metal waveguide. Heterostructure included 228 $Al_{0.15}Ga_{0.85}As/GaAs$ cascades with resonant phonon based triple quantum well design. Here we present results of studies of voltage-current and light-current characteristics as well as cyclotron resonance, which allowed us to measure the lasing frequency to be about 3 THz.

2. Experiment details

Heterostructure was grown in a molecular beam epitaxy reactor (Riber 21) on a semi-insulating GaAs (100) substrate. The active region was a 228-period superlattice $Al_{0.15}Ga_{0.85}As/GaAs$. Each cascade involved three tunnel-coupled quantum wells (QW) and had resonant-phonon design. The thicknesses of the layers in the cascade in Angstroms are <u>43</u>/89/<u>24.6</u>/81.5/<u>41</u>/160 (one period has 439.1 Å nominal thickness), where underlined are the dimensions of $Al_{0.15}Ga_{0.85}As$ layers. The complete width of the heterostructure was about 10 μm. Schematic band diagram at an applied bias corresponding to the lasing threshold is shown in Fig. 1a. Active transition is a transition between levels 2 and 3 with a nominal radiation frequency of 3 THz and calculated oscillator strength 0.3 at a bias field of 12 kV/cm^2. Transition 4-1 corresponds to passing of electrons between cascades (injection/extraction) and is carried out with a resonant optical phonon emission.

Growth rate was about 0.8 monolayers per second. The reactor was equipped with high speed shutters with 150 ms actuation, thus spreading of the interfaces did not exceed 0.4 Å. Layer-by-layer structure description is presented in Fig. 1b. Contact layers and the middle QW were doped with silicon concentrations of 5 x 10^{18} cm^{-3} and 5 x 10^{16} cm^{-3}, respectively. The growth was conducted in stabilized As-flow conditions; deposition temperature was controlled by IR pyrometer. The surface condition was monitored *in situ* by high-energy electron diffraction, and as grown wafer was examined by X-ray diffraction and photoluminescence technique, which spectra were compared with simulated one.

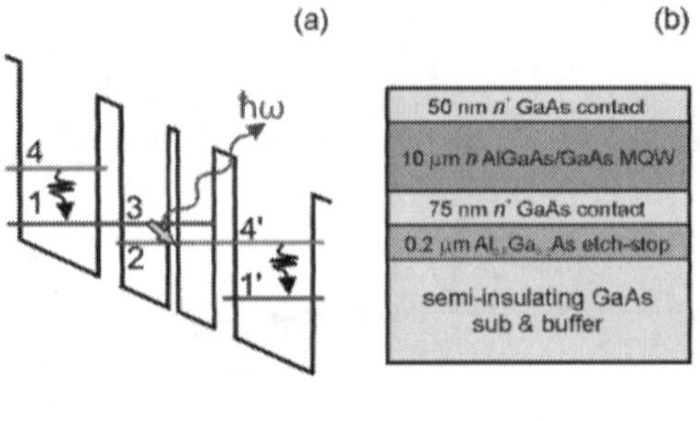

Fig. 1. (a) Schematic band diagram of QCL, (b) layer-by-layer structure description

Next metal-metal waveguide was fabricated that comprised the active region [6]. Postgrowth processing included thermal-compression bonding of the initial wafer epi-side to receptor n$^+$-GaAs substrate using Au-Au interlayer. QCL ridges of 100 microns width were fabricated by means of etching of the initial wafer down to the active region as well as electron lithography and dry etching of the active region through the upper metallization mask.

After thinning of the receptor to about 150 μm the wafer was cleaved to create 1.5 mm long resonators. Next chips were soldered the receptor side down to Cu heatsinks – the bottom contacts to the devices. In turn Au-wires were soldered to the ridges. The devices characteristics (voltage-current, light-current and cyclotron resonance spectra) were measured in pulse pumping regime at a liquid He temperature. To measure integral radiation intensity Ge:Ga photodetector was used. To determine QCL lasing frequency we used variable rejecting filter based on HgTe/CdHgTe QW, which attenuation spectrum depends on magnetic field.

3. Results and discussion

Results of studies of the synthesized structure by high resolution X-ray diffractometry are shown in Fig. 2. The figure also presents calculated rocking curve. Measured superlattice period is 436 Å, differs only by 0.7% from its nominal value. Low full width at half maximum (FWHM) of satellite peaks (15-19″) indicates a high accuracy of the cascades thicknesses within the structure as well as low roughness of the interfaces. Atomic-force microscopy studies indicated the high planarity of the surface even after 10 μm growth. The widths of terraces were 0.7-1.0 μm and RMS roughness was estimated about 0.2 nm.

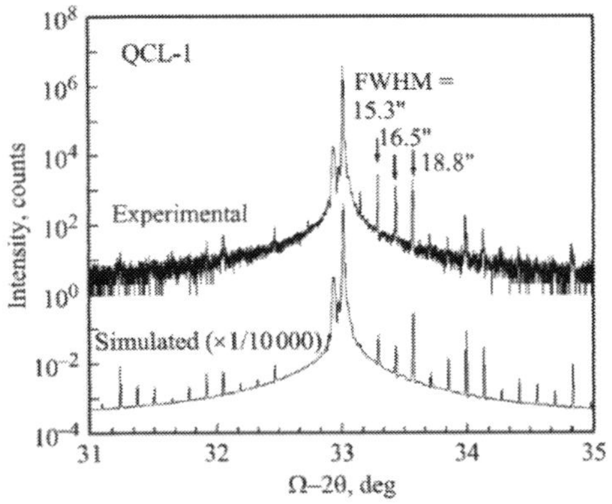

Fig. 2. X-ray rocking curve of the QCL structure near the GaAs (004) reflection and the corresponding simulated curve

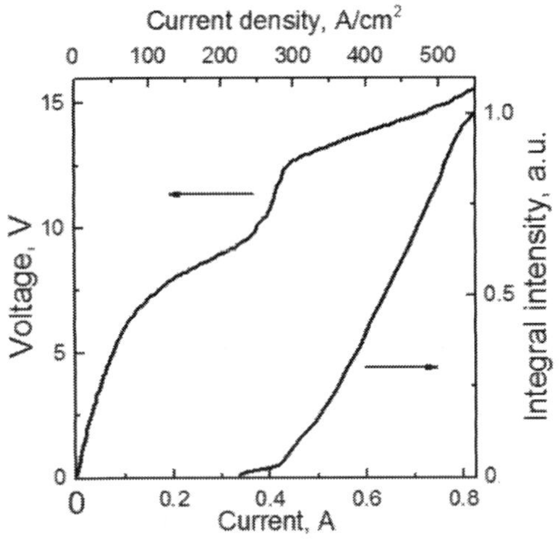

Fig. 3. Voltage-current (solid line) and light-current (dotted line) of the studied THz QCL

The dependence of the radiation intensity on the pumping current of the studied devices had characteristic threshold growth starting from 0.4 A (see Fig. 3), what is typical to the onset of laser generation. The voltage, corresponding to the kink in the light-current curve, about 11 V, is consistent with the calculated value of electric field, at which the levels are aligned, i.e. when the lasing should start (see Fig. 1a). Measurements of the radiation intensity transmitted through the variable filter based on HgTe/CdHgTe QW allowed to determine the lasing frequency of the QCL. The measured value about 3 THz coincides with the nominal frequency.

4. Conclusion

Thus, using fully domestic technological capacities we fabricated THz QCL with the double metal waveguide and the active region based on triple tunnel-coupled QWs with resonant-phonon depopulation scheme of the lower active level. High quality of the synthesized heterostructures was confirmed by high resolution X-ray diffractometry and atomic-force microscopy. Pulsed lasing was demonstrated at a liquid He temperature at 300 A/cm^2 threshold current density and 12 V voltage. The radiation frequency was estimated to be 3 THz, which is consistent with designed value.

Acknowledgements

This work was supported by the Russian Science Foundation (Project #18-19-00493).

References

1. *Köhler, R., Tredicucci, A., Beltram, F., Beere, H. E., Linfield, E. H., Giles Davies, A., Ritchie, D. A., Iotti, R. C., Rossi F.* Terahertz semiconductor-heterostructure laser // Nature. 2002. V. 417. P. 156–159.
2. *Wang, X., Shen, C., Jiang, T., Zhan, Z., Deng, Q., Li, W., Wu, W., Yang, N., Chu, W., Duan, S.* High-power terahertz quantum cascade lasers with ~0.23 W in continuous wave mode // AIP Advances. 2016. V. 6, No. 7. P. 075210.
3. *Li, H., Laffaille, P., Gacemi, D., Apfel, M., Sirtori, C., Leonardon, J., Santarelli, G., Rösch, M., Scalari, G., Beck, M., Faist, J., Hänsel, W., Holzwarth, R., Barbieri, S.* Dynamics of ultra-broadband terahertz quantum cascade lasers for comb operation // Opt. Express. 2015. V. 23, No. 26. P. 33270-33294.
4. *Hempel, M., Röben, B., Schrottke, L., Hübers, H.-W., Grahn, H. T.* Fast continuous tuning of terahertz quantum-cascade lasers by rear-facet illumination // Appl. Phys. Lett. 2016. V. 108, No. 19. P. 191106.
5. *Fathololoumi, S., Dupont, E., Chan, C. W. I., Wasilewski, Z. R., Laframboise, S. R., Ban, D., Matyas, A., Jirauschek, C., Hu, Q., Liu, H. C.* Terahertz quantum cascade lasers operating up to ~ 200 K with optimized oscillator strength and improved injection tunneling // Opt. Express. 2012. V. 20, No. 4. P. 3866-3876.
6. *Ikonnikov, A. V., Marem'yanin, K. V., Morozov, S. V., Gavrilenko, V. I., Pavlov, A. Yu., Shchavruk, N. V., Khabibullin, R. A., Reznik, R. R., Cirlin, G. E., Zubov, F. I., Zhukov, A. E., Alferov, Zh. I.* Terahertz radiation generation in multilayer quantum-cascade heterostructures // Tech. Phys. Lett. 2017 V. 43, No. 4, P. 362-365.

EPJ Web of Conferences **195**, 05001 (2018) https://doi.org/10.1051/epjconf/201819505001
TERA-2018

Imaging of powerful terahertz beams

I. V. Andreev[1,2], V. M. Muravev[1,2], A. R. Khisameeva[1,2], G. E. Tsydynzhapov[1,2], I. V. Kukushkin[1,2]

[1]MWAVE LLC, Moscow, Russia,
[2]Institute of Solid State Physics RAS, Chernogolovka, Russia,
ivan.andreev@terasense.com

THz frequency range is a field of constantly growing interest within modern applied physics. Non-destructive imaging, time-domain and frequency domain spectroscopy in THz range become important up-to-date techniques of material characterization [1]. Recently considerable interest was attracted by various THz personnel screening and security systems. Powerful THz sources, such as gyrotrons, could be used as a focused heating source for selective melting with controllable temperature distribution in glass and semiconductor industry [2], or for plasma heating into ITER tokamak. Lots of existing applications require quasi-optical setup, which demands proper beam alignment and focusing. Sometimes it becomes a complicated task since sub-THz radiation is invisible to the eye. We developed THz square cameras, which allows for real-time beam profiling of various vacuum-tube and semiconductor-based THz sources and could greatly simplify adjustment of quasi-optical setups. Cameras are suitable for both vacuum-tube (BWO, gyrotrons) and semiconductor-based (IMPATT and Gunn diodes) THz sources operating in range of 50–700 GHz.

Our group have developed a new semiconductor-based technology for detection of THz radiation in a broad range of 50–700 GHz. Detectors are fabricated from GaAs/AlGaAs quantum well heterostructures with high two-dimensional (2D) electron conductivity. Detection principle is rest on coupling of incident sub-THz radiation to so-called relativistic plasmons in 2D electron system [3] and subsequent rectification of plasmon ac potential on nonlinear defects, which provides a dc signal at the output of the detector. Single detector itself has a noise equivalent power (NEP) of 0.5 nW/Hz$^{1/2}$ and ultrafast response time of 120 ps. GaAs detectors are fully compatible with common semiconductor processing lines, which makes them comparatively inexpensive, and allows to manufacture detectors on industrial scale and with reproducible properties.

Further to our research, we are developing a graphene-based detector in THz frequency range. Plasma oscillations in graphene have typical frequencies in THz range, which allows graphene to be used as a detector in this range of electromagnetic radiation. This material has the highest value or 2D electron mobility that provides high 2D conductivity and allows to observe weakly damped relativistic plasma oscillations. In contrast to common two-dimensional plasmons, the damping of relativistic plasma mode [3] is not determined by the inverse time of scattering of charge carriers. Thanks to weak damping, these plasma oscillations can be excited even at room temperature when ordinary plasmons are strongly suppressed. Compared to existing analogs, the plasmon detector based on graphene should have the highest responsivity and fastest response time.

Fig. 1. Visual appearance of a Tera-4096 square imaging camera (64x64 pixels, 96x96 mm^2 active area size).

Ownership of the detector technology allowed us to develop a line of THz imaging cameras (Fig. 1). Camera consists of a focal plane array of detectors (pixels), combined with amplification and readout circuitry. Cameras have a modular design, being assembled from blocks, which provides a very good scalability in pixel number and imaging area size. Most common camera sizes are 16x16, 32x32 and 64x64 pixels. High frame rate (up to 90 fps) makes possible to perform a real-time THz imaging. Typical camera parameters are summarized in the table below.

Frequency range	50 – 700 GHz
NEP	1 nW/Hz$^{1/2}$
Frame rate	0.2 – 90 fps
Typical pixel number	16x16, 32x32, 64x64
Pixel size	1.5x1.5 or 3x3 mm^2
Features	Mini-USB interface, TTL sync out

As well as square cameras with frame rate not more than 100 fps, TeraSense also manufactures fast linear cameras with frame rate up to 5000 lines per second. Linear cameras are targeted at conveyor-belt based scanning applications.

TeraSense cameras proved itself as a flexible solution in a number of applications, such as non-destructive testing and quality control in industrial manufacturing lines [4], especially in area of humidity control in dry materials, mail inspection and security screening. Another promising application of THz imaging cameras is real-time beam profiling in various quasi-optical setups [5].

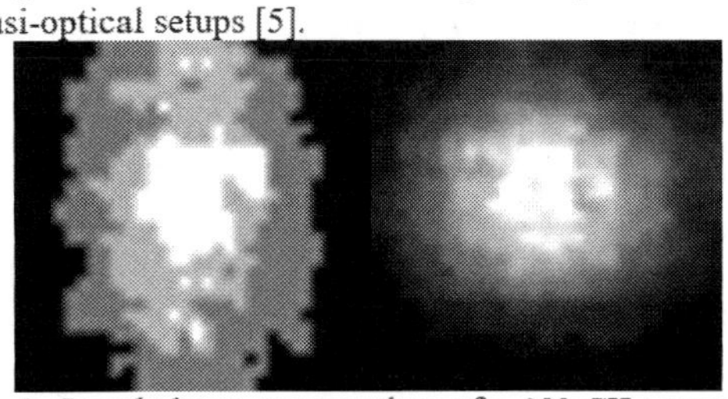

Fig. 2. Sample beam cross-sections of a 100 GHz source, taken with Tera-1024 32x32 pixel camera for two different output geometries. Left: WR8 rectangular waveguide. Right: conical horn.

Square cameras could capture a cross-section power distribution of a beam less than in one second. This simplifies beam alignment procedure, makes possible to control angular pattern of the source, and solve other tasks arising in everyday life of experimental physicists and engineers working with quasi-optical THz setups. Cameras are broadband sensitive in 50–700 GHz frequency range with several sensitivity peaks, which positions could be adjusted and optimized at the manufacturing stage. Minimal detectable power depends on the frame rate and is about $1 \mu W/cm^2$ at 1 fps, which allows cameras to operate even with low-power sources. Spatial resolution (discretization) of the beam cross-section image is limited by the pixel size and could be 1.5 or 3 mm. Fig. 2 demonstrates sample beam cross-section images of the same source for various output geometries. Cameras could be used for alignment not only continuous-wave, but pulsed beams as well. Operating with the pulsed beam, camera resembles mean beam power distribution. Camera pixels are tolerant to powerful beams, allowing to work not only with semiconductor-based sources, but with high-power vacuum-tube sources as well.

This work was partially financially supported by U.M.N.I.C. grant (contract No. 10667ГУ/2016).

References

1. *Zhang, H., Sfarra, S., Saluja, K., Peeters, J., Fleuret, J., Duan, Y., Fernandes H., Avdelidis N., Ibarra-Castanedo C., Maldague, X.* Non-destructive investigation of paintings on canvas by continuous wave terahertz imaging and flash thermography // Journal of Nondestructive Evaluation. 2017. V. 36, No. 2. P. 34.

2. *Sklyarevich V., Shevelev M.* Method for the rapid thermal treatment of glass and glass-like materials using microwave radiation. // U.S. Patent No. 6,408,649. 25 Jun. 2002.

3. *Muravev, V. M., Gusikhin, P. A., Andreev, I. V., Kukushkin, I. V.* Novel relativistic plasma excitations in a gated two-dimensional electron system // Physical review letters. 2015. V. 114. No. 10. P. 106805.

4. *Zhang, H., Sfarra, S., Sarasini, F., Santulli, C., Fernandes, H., Avdelidis, N. P., Ibarra-Castanedo C., & Maldague, X. P.* Thermographic non-destructive evaluation for natural fiber-reinforced composite laminates // Applied Sciences. 2018. V. 8. No. 2. P. 240.

5. *Yıldırım, İ. O., Özkan, V. A., İdikut, F., Takan, T., Şahin, A. B., & Altan, H.* Characterization of a terahertz wave scanned imaging system for threat detection at stand-off distances // Optical and Quantum Electronics. 2016. V. 48. No. 7. P. 367.

Optical-Mechanical Analogy Approach for the Purposes of Detection of IR-MW Radiation

A.V. Bogatskaya[1,2,3], N.V. Klenov[1,3,4,5], M.V. Tereshonok[3,5] and A.M. Popov[1,2,4]

[1]D. V. Skobeltsyn Institute of Nuclear Physics, Moscow State University, Moscow, Russia, annabogatskaya@gmail.com
[2]P. N. Lebedev Physical Institute, RAS, Moscow Russia
[3]Moscow Technical University of Communication and Informatics, 111024, Moscow, Russia
[4]Department of Physics, Moscow State University, 119991, Moscow, Russia
[5] Moscow Technological University (MIREA), 119454, Moscow, Russia

Here we propose a generalized physical approach providing possible technical realizations of a number of urgent problems related to the interaction of electromagnetic radiation with plasma and conducting media. Proposed approach is based on the use of well-known optical-mechanical analogy which exploit mathematical identity of the stationary Schroedinger equation in quantum mechanics for the particle motion in potential field and Helmholtz equation in wave theory.

Let us consider spatially inhomogeneous non-magnetic medium characterized by the susceptibility $\chi(\vec{r})$, or permittivity $\varepsilon(\vec{r}) = 1 + 4\pi\chi(\vec{r})$. In the case of monochromatic field $\vec{E} = \vec{E}_0(\vec{r})\exp(-i\omega t)$ ($\vec{H} = \vec{H}_0(\vec{r})\exp(-i\omega t)$, ω is the radiation frequency) Maxwell equations for electric and magnetic field strength can be written as:

$$rot\,\vec{E} = \frac{1}{c} i\omega\vec{H}, \quad div(\varepsilon\vec{E}) = 0,$$
$$rot\,\vec{H} = -\frac{\varepsilon}{c} i\omega\vec{E}, \quad div\vec{H} = 0. \tag{1}$$

From (1) one can obtain the following equation for electric field strength $\vec{E}$:

$$\Delta\vec{E} + \nabla\left(\frac{1}{\varepsilon}(\vec{E}\nabla)\varepsilon\right) + \frac{\varepsilon\omega^2}{c^2}\vec{E} = 0. \tag{2}$$

For the case when permittivity depends only on one spatial coordinate $\varepsilon = \varepsilon(z)$ and wave field propagates alone this direction the equation (2) transforms to the well-known Helmholtz equation for the spatial distribution of electric field strength E:

$$\frac{d^2E}{dz^2} + k_0^2(1 + 4\pi\chi(z))E = 0. \tag{3}$$

with $k_0^2 = \omega^2/c^2$. Here electric field propagates in the direction perpendicular to z-axis.

Equation (3) is mathematically equivalent to the stationary Schrödinger equation in quantum mechanics for the particle wave function $\psi(z)$ in the potential field $V(z)$:

$$\frac{d^2\psi}{dz^2} + \kappa_0^2\left(1 - \frac{V(z)}{\zeta}\right)\psi = 0, \tag{4}$$

where $\kappa_0^2 = 2m\zeta/\hbar^2$ is the wave vector of the particle with energy ζ. Direct comparison of eq. (3) and (4) leads to the conclusion that potential function $V(z)$ in quantum mechanics is similar to the susceptibility in electromagnetic theory $(2m/\hbar^2)V(z) \to (1-\varepsilon)\cdot(\omega/c)^2 = -4\pi\chi(z)\cdot(\omega/c)^2$. Thus the eigenvalue problem for the Hamiltonian in quantum theory turns out to be mathematically identical to the problem of calculating the stationary distribution of the electric field strength in a wave. The medium with $\varepsilon > 0$ can be associated with an attractive potential $V(z) < 0$ (potential well) while the medium with $\varepsilon < 0$ acts as potential barrier $V(z) > 0$. In particular, the transport of the electron flux in heterostructures is mathematically identical to the problem propagation of electromagnetic waves through inhomogeneous media.

If the potential curve $V(z)$ has the piecewise-continuous structure (Fig. 1), both the ψ-function and its derivative $d\psi/dx$ should be continuous functions in the potential breaking points. Similar boundary conditions appear to exist in electromagnetic theory: the tangential components of $\vec{E}, \vec{H}$ should also be continuous functions at the interface regions. Using Maxwell equations one can rewrite the boundary conditions as the continuity of tangential components of $\vec{E}$ and its derivative. For the normal incidence when only tangential components of $\vec{E}, \vec{H}$ have the non-zero values these boundary conditions are equivalent to boundary conditions for the wave function in quantum mechanics. The above conclusion known as an optical-mechanical analogy in quantum theory gives rise to a lot of practical applications and transfer the quantum theory problem solutions to optics and vice versa.

We perform an analysis of the possibility of penetration of electromagnetic waves through opaque media using the analogy of tunneling of a quantum-particle flux through a potential barrier with a height greater than its kinetic energy. As an example, we consider plasma sheath surrounding the hypersonic vehicle as a potential barrier and analyze the overcoming of radiocommunication blackout problem [1].

Really, for the collisionless plasma the permittivity reads $\varepsilon_p = 1 - \omega_p^2/\omega^2$, where $\omega_p^2 = 4\pi e^2 n_e/m$ is the plasma frequency squared and n_e is the electron density. From this point of view the plasma sheath appearing around the hypersonic vehicle during the flight looks like a potential barrier for the target transmission frequencies less than plasma frequency.

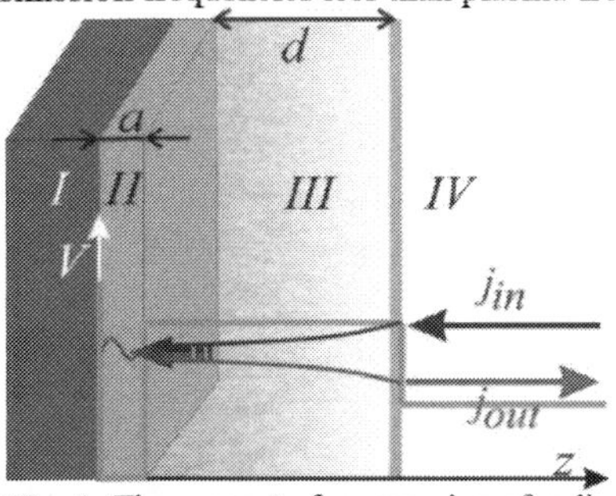

Fig. 1. The concept of overcoming of radio communication blackout: profile of the "potential barrier" $V(z) \leftrightarrow (1-\varepsilon(z))$ containing a vehicle surface (I), dielectric layer (thickness a) with embedded antenna (II) and plasma sheath (III). (IV) corresponds to the region of infinite motion of the electromagnetic wave (atmospheric air).

The main idea is to embed a «resonator» (it can be a dielectric layer with rather large value of permittivity ε_d) between the surface on the vehicle and plasma sheath which is supposed to provide an effective tunnelling of the signal to the receiving antenna. Calculations (see Fig.2) show sharp increase of tunnelling probability of electromagnetic signal if the frequency of incident radiation coincides with the eigen-frequency of resonator (dielectric layer) which is determined by formula $\omega_n \approx \frac{\pi c}{a\sqrt{\varepsilon_d}} n$, $n = 1,2,3\ldots$. Here we introduce the filling factor $F(f)$ which represents the degree of resonator filling by the incoming radiation flux:

$$F(f) = \max\left\{|E_d|^2 / |E_{a+}|^2\right\}, \qquad (5)$$

where $|\vec{E}_d|^2$ and $|\vec{E}_{a+}|^2$ are the squared absolute values of electric field strength in dielectric layer and air correspondingly.

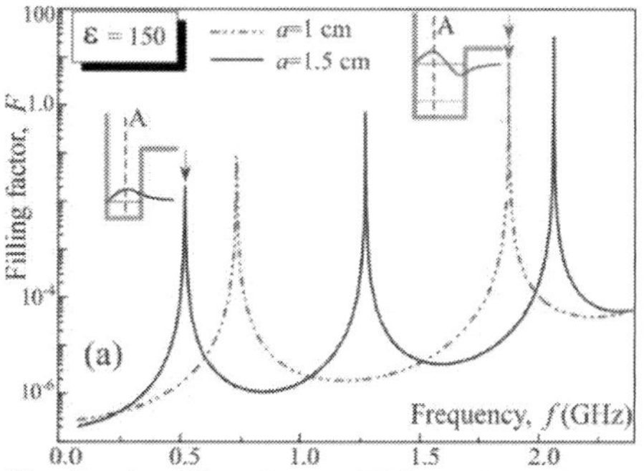

Fig. 2. The filling factor $F(f)$ in dependence on the transmitted signal frequency (normal incidence on vehicle surface). Calculations are made for $d = 10$ cm and $\varepsilon_d = 150$. The inset to figure schematically represents parts of the corresponding wave functions of stationary states within the framework of the optical-mechanical analogy.

The role of collisions in plasma as well as cases of normal and oblique incidence of radiofrequency waves on the vehicle surface are studied in [1].

One more promising application of the suggested approach implies a method for increasing the efficiency of bolometric photodetectors. The results of calculations show that by choosing thickness and dielectric constant of the substrate-resonator under a thin layer of metal (superconductor) one could bring the fraction of absorbed radiation in the detector to a value close to unity at a certain frequency of the infrared range [2].

This work was supported by the Russian Science Foundation (project no. 18-72- 00125).

References

1. *Bogatskaya A. V., Klenov N. V., Tereshonok M. V., Adjemov S. S., Popov A. M.* // J. Phys. D. 2018. V. 51, P. 185602.

2. *Bogatskaya A. V., Klenov N. V., Tereshonok M. V., Popov A. M.* // *Technical Physics Letters* 2018. V.44, № 8, p. 667–670

Efficient electron cooling in Cold Electron Bolometers

A.V. Gordeeva[1,2], A.L. Pankratov[1,2], V.O. Zbrozhek[1],

A.V. Blagodatkin[1,2], L.S. Revin[1,2], D.A. Pimanov[1], L.S. Kuzmin[1,3]

[1]Nizhny Novgorod State Technical University n.a. R.E. Alekseev, Nizhny Novgorod, Russia, anna.gord@list.ru
[2]Institute for Physics of Microstructures of RAS, Nizhny Novgorod, Russia
[3]Chalmers University of Technology, Gothenburg, Sweden

We have reached experimentally a theoretical minimum of electron cooling in Cold Electron Bolometer at base temperature 300 mK. The efficiency of the cooling depends on how the removal of hot quasiparticles from the vicinity of the tunnel barrier is organized. By double stock and traps for hot quasiparticles we have managed to make the removal of the hot quasiparticles as efficient as 99%. We also explain such high efficiency by the absence of Andreev reflections due to hybrid superconductor/ferromagnet structure for absorber.

The superconductor-insulator-normal metal (SIN) tunnel junctions are known for their ability to remove heat from the electron system of the normal metal electrode. Two SIN junctions, connected in SINIS structure, remove the heat twice more efficiently than a single junction [1]. This property is intensively used in Cold-Electron Bolometers [2,3], where normal metal serves as an absorber of incoming signal.

The maximum cooling by 200 mK was demonstrated in several groups, for example, from 300 mK to 100 mK in [4]. However, the theory predicts that the electron system can be cooled even lower down to 70 mK.

The efficiency of the cooling depends on how the removal of hot quasi-particles from the vicinity of the tunnel barrier is organized. In SINIS structures if hot quasi-particles stay near the barrier in S-electrodes, some power may return back to the normal metal through the phonon system or due to backtunneling.

We present three types of cold electron bolometers with different cooling efficiency. We start from the first design, when about 30% of the heat, removed through the SIN junctions, return back to the absorber. Then we describe the second and the third designs with improvements, that made the return power to decrease from 30% to 6% and finally to just 0.5%. Thus, we remove the hot quasiparticles from the vicinity of the tunnel barrier in the most effective possible way.

The tunneling current in SIN junction is given by the well known expression:

$$I = 1/(eR_T)\int \nu(\varepsilon)\,[f((\varepsilon - eV)/\tau_E) - f(\varepsilon/\tau_S)], \quad (1)$$

where V - voltage drop across one SIN junction, R_T - normal state resistance of SIN junction, $\nu(\varepsilon)$ - density of states in superconductor, $f(\varepsilon)$ - Fermi-Dirac distribution, τ_E and τ_S are electron temperatures in normal metal and in superconductor, respectively. We use the expression (1) to restore electron temperature in the normal absorber together with the heat balance equation, as described, for example, in [5]. The current I, voltage V and R_T are determined experimentally, τ_S does not influence the result and can be set to the phonon temperature. Thus, only τ_E remains unknown quantity in this expression.

The electron temperatures, found from (1), are shown in Fig. 1 and Fig. 2 on right axis. Normal resistance of the measured samples varies from 0.7 to 1.6 kΩ. The samples are made of aluminum with critical temperature 1.24 K.

In Fig.1 we compare a sample with a single connection of superconducting electrode to the thermal bath with a sample, which has two connections to thermal bath for each superconducting electrode. One can see a dramatic increase in electron cooling for the second sample. The better cooling efficiency is seen on IV-curves as increasing superconducting gap. We have measured the critical temperatures of superconducting electrodes for all our samples and found that it was 1.47 K for the first sample and 1.24 K for the second one, i.e. it would have opposite effect. Therefore, the change in IV-curves is due to different electron temperatures in the absorber.

We note that the first sample was expected to show good cooling properties as well due to implementation of normal metal traps below the superconducting electrodes. But as appears, the double stock works much more effectively than the single stock with traps.

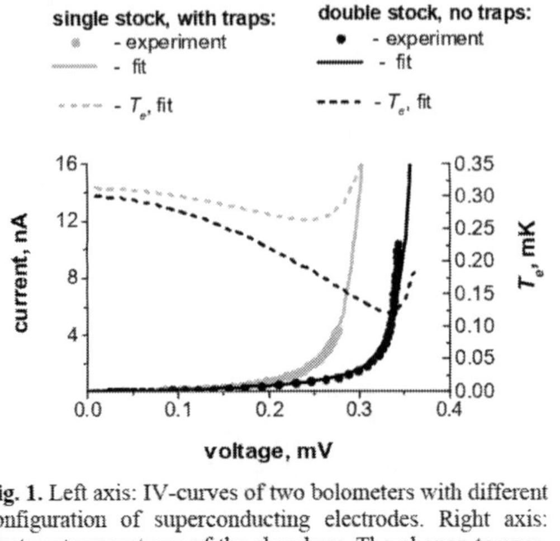

Fig. 1. Left axis: IV-curves of two bolometers with different configuration of superconducting electrodes. Right axis: electron temperatures of the absorbers. The phonon temperature is 300 mK for both bolometers.

In Fig. 2 we compare the second sample from Fig. 1 with a sample, which has both double stock and traps. Again we can see a significant improvement in cooling properties for the latter sample (black colour).

Now let us point out another possible reason of reaching the minimal electron temperatures, namely suppression of Andreev reflection. It was shown in [6] that two-particle current due to Andreev reflection generates Joule heating in normal electrode, restricting the cooling effect at low temperatures. In our samples we use sublayer of Fe in order to suppress superconductivity in aluminum, used as a normal metal absorber. We suppose that this thin layer of ferromagnet reduces the Andreev currents in our samples.

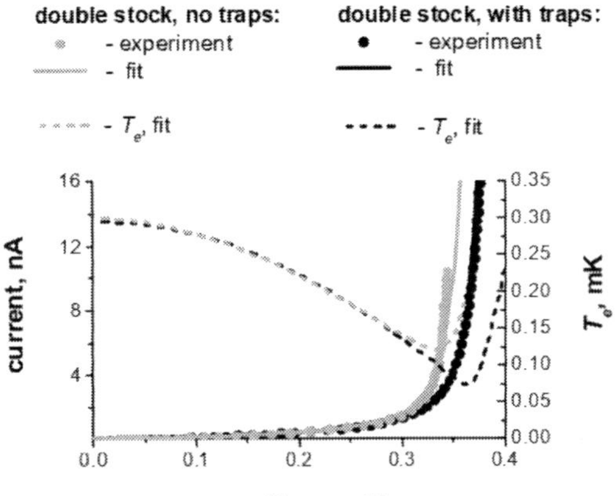

Fig. 2. Left axis: IV-curves of two bolometers without and with normal metal traps below the superconducting electrodes. Right axis: electron temperatures of the absorbers. The phonon temperature is 300 mK for both bolometers.

Thus, by double stock, by traps for hot quasiparticles and by suppression of Andreev current we manage to make the self cooling as efficient as possible. The suppression of Andreev current is organized using hybrid superconductor/ferromagnet structure for the absorber.

Authors would like to thank M. Tarasov and A. Gunbina for help in the sample fabrication. The samples were fabricated in the Chalmers nanotechnology Center. The facilities of the Center of Cryogenic Nanoelectronics of NNSTU and Common Research Center "Physics and technology of micro- and nanostructures" of IPM RAS were used.

The work is supported by Russian Science Foundation (Project 16-19-10468).

References

1. *Nahum M., Eiles T. M. and Martinis J. M.* Electronic micro-refrigerator based on a NIS tunnel junction // Appl. Phys. Lett. 1994. V. 65. P. 3123.

2. *Kuzmin L.* Ultimate Cold-Electron Bolometer with Strong Electrothermal Feedback // Proceedings of SPIE. 2004. V. 5498. P. 349.

3. *Kuzmin L.* An Array of Cold-Electron Bolometers with SIN Tunnel Junctions and JFET readout for Cosmology Instruments // Journal of Physics: Conference Series. 2008. V. 97, P. 012310.

4. *Leivo, M.M., Pekola, J.P. and Averin, D.V.* Microrefrigeration by NIS tunnel junctions // Czech J Phys. 1996. V. 46(Suppl 5): P. 2763.

5. *Gordeeva,A.V., et al.* Observation of photon noise by cold-electron bolometers // Appl. Phys. Lett. 2017. V. 110. P. 162603

6. *Vasenko A.S., Bezuglyi E.V., Courtois H., Hekking F.W.J.* Electron cooling by diffusive normal metal–superconductor tunnel junctions // Phys. Rev. B. 2010. V. 81. P. 094513.

Terahertz Electro-Optic Sampling in Crystals with High Natural Birefringence

I.E. Ilyakov[1,2,3], G.Kh. Kitaeva[4], B.V. Shishkin[1,2], and R.A. Akhmedzhanov[1]

[1]Institute of Applied Physics RAS, Nizhny Novgorod 603950, Russia, i.ilyakov@hzdr.de
[2]N.I. Lobachevsky State University of Nizhny Novgorod, Nizhny Novgorod 603950, Russia
[3]Helmholtz-Zentrum Dresden-Rossendorf, Bautzner Landstr. 400, 01328 Dresden, Germany
[4]M.V. Lomonosov Moscow State University, Faculty of Physics, Moscow 119991, Russia

The femtosecond lasers allows one to generate via various nonlinear processes pulsed terahertz wave radiation (0.1-10 THz) with unique characteristics: subpicosecond duration containing only one-two periods of oscillation and extremely wide spectrum. These terahertz pulses are in demand for different applications which is supported also by sensitive optical methods of coherent detection. In recent years, great progress was made in the laser-to-terahertz pulse conversion efficiency using highly nonlinear organic materials and lithium niobate (LN) crystal structures [1, 2]. In spite of high conversion efficiencies obtained during generation process there is a lack of data about efficient use of this crystals and crystal structures as detectors. The main reason is that their application in the most popular standard electro-optic (EO) detection set up based on Pockels effect [3] is hindered by decoherence of the laser polarization components because of high natural birefringence of these crystals [4]. In this work we perform the detection methods suited to be used with EO crystals with various natural birefringence and test experimentally their applicability with different EO crystals and crystal structures.

When the laser pulse interacts with THz in EO crystal not only its polarization ellipticity but also its energy could be changed [5, 6]. These changes depend on the amplitude and phase of the THz overlapped with the laser pulse and can be used for terahertz time-domain measurements [5, 6]. The method based on energy changes can separately measure the energies of the orthogonal polarization components of the laser pulse and, in contrast to techniques based on ellipticity changes, does not require maintaining coherence between the polarization components and thus, can be used with birefringent crystals.

The sensitivity of both the "energy-based" method and the standard "ellipticity-based" one can be improved if the laser pulses with the edge-cut spectrum are used. As we shown theoretically [7, 8] the relative changes of spectral energy density and ellipticity at the frequencies corresponding to the edge-cut of the laser pulse spectrum could be much higher than when a laser pulse with Gaussian spectrum is used. To realize this in experiment an edge filter can be positioned before the EO crystal to produce the high gradient of the spectral energy density of the laser pulse and the laser pulse after interaction with THz in the EO crystal can be decomposed in the frequency spectrum by a diffraction grating or can be passed through another edge filter transmitting only the optical waves with frequencies corresponding to the slope of the first filter [7, 8].

These techniques were tested in our experiments with a number of crystals [7-10]: with ZnTe and GaP - commonly used for EO sampling, and with birefringent ones – periodically-poled LN (PPLN) and DSTMS crystals. All of them demonstrated an applicability of being used with the suggested methods.

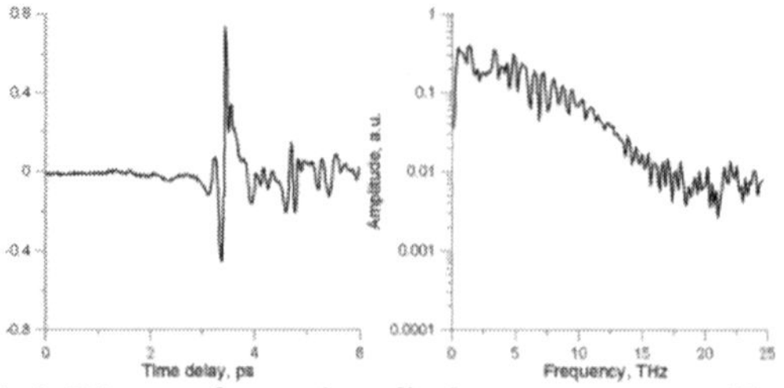

Fig.1. THz wave form and amplitude spectrum measured by DSTMS crystal.

In its easiest configuration the technique based on the total laser pulse energy changes demands even fewer optical elements than the standard one. It was tested as for narrow band detection with PPLN crystal, which allows one to achieve optical-THz synchronism at various terahertz frequencies and laser wavelengths. Application of this method with DSTMS crystal led to significant broadening of the detection bandwidth. The THz waves with frequencies up to 15 THz could be measured by the laser pulses with duration of 50 fs (Figure 1). However, for the same crystal – ZnTe or GaP this technique demonstrates the lower sensitivity than the standard one.

To improve the sensitivity of measurements the techniques with edge-pass filters were used [7, 8]. At high frequencies they demonstrated more than an order higher sensitivity comparing with the standard detection technique when the same ZnTe and GaP detection crystals were used. Further improvement of the detection sensitivity and bandwidth was achieved with the PPLN (Figure 2) and DSTMS crystals. The use of DSTMS crystal also demonstrated higher bandwidth comparing with the use of GaP crystal.

The sandwich structures with LN core demonstrated previously the high optical-terahertz conversion efficiency [11]. The use of these structures for noncollinear EO sampling was demonstrated previously basing on the spatial modulation of the laser power density [12]. However, due to the less sensitivity of this technique the signal was lower

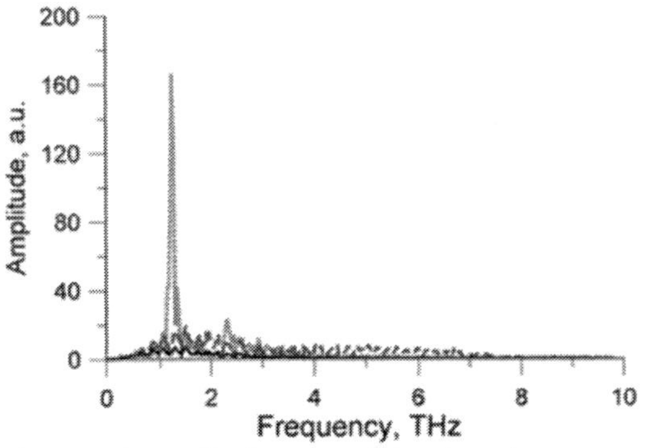

Fig.2. Fourier amplitude spectra of time dependences measured by the standard measurement scheme with the GaP crystal (black-solid line); by the scheme with edge filters and the GaP crystal (blue-dashed line) and with the PPLN crystal (red-dotted line, highest curve).

comparing with the standard one. Our future plan is to combine such structures with the detection methods described above. Due to the noncollinear detection scheme with a Si prism coupling it is possible to achieve a broadband optical-THz synchronism in LN and reduce the wavelength absorption. The detection sensitivity significantly depends from the probe pulse width. However, for the suggesting method with the edge filters we can take only the part of the optical beam after the crystal and reduce by this way the effective transverse area of the optical-THz interaction. As is seen from the Figure 3, the detection bandwidth and detection sensitivity could be much higher than the ones of the standard detection technique with the ZnTe crystals of various thicknesses. The calculation results of the detection sensitivity at different probe beam diameters will be also presented in the report.

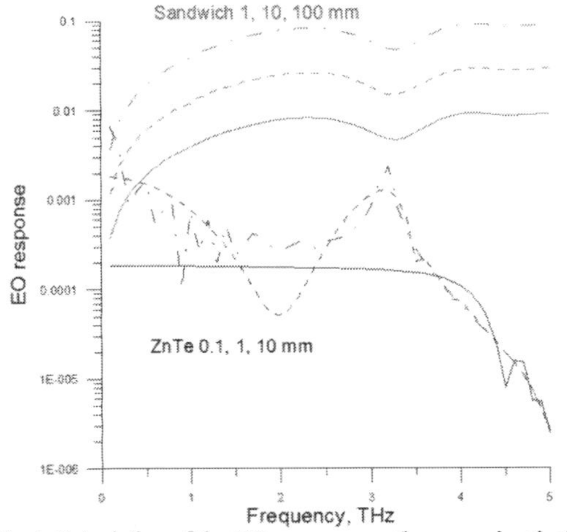

Fig.4. Calculation of the EO response on the same signal of the standard detection technique with the ZnTe crystal and energy detection technique with LN sandwich structure of various lengths (probe beam width 15 μm).

Acknowledgment. This work was supported by the Russian Science Foundation Grant No. 18-19-00486.

References

1. *Bodrov, S. B., Ilyakov, I. E., Shishkin, B. V., and Stepanov, A. N.* Efficient terahertz generation by optical rectification in Si-LiNbO3-air-metal sandwich structure with variable air gap // Appl. Phys. Lett. 2012. V. 100, N. 20. P. 201114.

2. *Vicario, C., Jazbinsek, M., Ovchinnikov, A. V., Chefonov, O. V., Ashitkov, S. I., Agranat, M. B., and Hauri, C. P.* High efficiency THz generation in DSTMS, DAST and OH1 pumped by Cr:forsterite laser // Opt. Exp. 2015. V. 23. P. 4573-4580.

3. *Wu, Q. and Zhang, X.-C.* Free-space electro-optic sampling of terahertz beams // Appl. Phys. Lett. 1995. V. 67. P. 3523-3525.

4. *Han, P. Y., Tani, M., Pan, F., and Zhang, X.-C.* Use of the organic crystal DAST for terahertz beam applications // Opt. Lett. 2000. V. 25. P. 675-677.

5. *Kitaeva, G. Kh., Kovalev, S. P., Naumova, I. I., Akhmedzhanov, R. A., Ilyakov, I. E., Shishkin, B. V., and Suvorov E. V.* Quasi-phase-matched probe-energy electro-optic sampling as a method of narrowband terahertz detection // Appl. Phys. Lett. 2010. V. 96. P. 071106.

6. *Ilyakov, I. E., Kitaeva, G. Kh., Shishkin, B. V., and Akhmedzhanov, R. A.* Laser pulse amplitude changes induced by terahertz waves under linear electro-optic effect // Appl. Phys. Lett. 2014. V. 104. P. 151107.

7. *Ilyakov, I. E., Kitaeva, G. Kh., Shishkin, B. V., and Akhmedzhanov, R. A.* Terahertz wave electro-optic measurements with optical spectral filtering // Appl. Phys. Lett. 2015. V. 106. P. 121101.

8. *Ilyakov, I. E., Kitaeva, G. Kh., Shishkin, B. V., and Akhmedzhanov, R. A.* Terahertz time-domain electro-optic measurements by femtosecond laser pulses with an edge-cut spectrum // Opt. Lett. 2016. V. 41. P. 2998–3001.

9. *Ilyakov, I.E., Kitaeva, G.Kh., Shishkin, B.V., and Akhmedzhanov, R.A.* Electro-optic sampling of terahertz waves by laser pulses with an edge-cut spectrum in birefringent crystal // Opt. Lett. 2017. V. 42, No. 9. P. 1704–1707.

10. *Ilyakov, I.E., Kitaeva, G.Kh., Shishkin, B.V., and Akhmedzhanov, R.A.* The use of DSTMS crystal for broadband terahertz electro-optic sampling based on laser pulse amplitude changes // Laser Phys. Lett. 2018. V. 15, No. 12. P. 125401.

11. *Bodrov, S. B., Ilyakov, I. E., Shishkin, B. V., and Stepanov, A. N.* Efficient terahertz generation by optical rectification in Si-LiNbO3-air-metal sandwich structure with variable air gap // Appl. Phys. Lett. 2012. V. 100, No. 20. P. 201114.

12. *Tani, M., Kinoshita, T., Nagase, T., Horita, K., Que, C. T., Estacio, E., Yamamoto, K., and Bakunov, M. I.* Non-ellipsometric detection of terahertz radiation using heterodyne EO sampling in the Cherenkov velocity matching scheme // Opt. Exp. 2013. V. 21. P. 9277-9288.

Optical-Terahertz Biphotons

G.Kh. Kitaeva

Lomonosov Moscow State University, Moscow, Russia, gkitaeva@physics.msu.ru

The quantum correlated pairs of photons, usually named as biphotons, are the well known quantum optical objects, comprehensively studied and exploited in modern quantum information schemes [1]. Biphotons are easily generated as pure quantum states in the process of parametric down-conversion (PDC) of optical laser radiation in a quadratic nonlinear medium, and serve further as a basis for preparation of different types of entangled photons and squeezed vacuum states. During PDC each pump photon of frequency ω_p decays into a pair of photons: a signal photon of frequency $\omega_s \geq \omega_p/2$ and an idler photon of lower frequency $\omega_i \leq \omega_p/2$. Formally, only one restriction is imposed on specific values of ω_i and ω_s. It follows from the energy conversation law $\hbar\omega_i + \hbar\omega_s = \hbar\omega_p$, so that the biphotons can be generated in a wide spectral range. Up to now PDC processes are studied mostly in the frequency-degenerate cases, when $\omega_s = \omega_i = \omega_p/2$, or slightly non-degenerate ones, when the difference between signal and idler frequencies is less than the spectral band of optical transmission of the nonlinear medium. We propose to consider properties of biphotons with extremely different frequencies, when the signal photon frequency is just close to ω_p, while the idler photon frequency hits the terahertz range. This regime can be referred as strongly frequency non-degenerate (SFND) PDC.

The SFND regime is interesting with new possibilities that appear in the generation of correlated photons from different spectral ranges, generation of localized long-wave excitations in matter that are bound to optical delocalized states, construction of quantum single-photon sources of terahertz frequencies, quantum calibration of the terahertz wave sources and detectors, and other applications. However, direct transfer of the results and methods, obtained up to now for the PDC in the optical range, to the mixed optical-terahertz case is a quite challenging task. The main problems are caused by the presence of classical thermal field fluctuations at THz frequencies, inherent absorption of the non-linear medium at THz frequencies, and high angular diversity of the generated THz idler photons. Results of the theoretical treatment of correlation functions and frequency-angular distributions of signal and idler photons emitted via spontaneous (low-gain) PDC are presented here accounting the peculiar properties of the SFND regime.

Calculations were made using the nonlinear Kirchhoff law, which expresses the second-order moments of the output PDC-generated fields in terms of the second-order moments of the input fields via elements of the scattering matrix of a nonlinear medium. According to its definition, the scattering matrix $\hat{U}$ consists of coefficients that connect linearly first-order moments of the both signal and idler input and output fields. In particular,

$$\langle a_s^+(L)\rangle = \sum_{s'} U_{ss'}\langle a_{s'}^+(0)\rangle + \sum_{i'} U_{si'}\langle a_{i'}(0)\rangle. \quad (1)$$

Here and below $\langle a_m(z)\rangle$ is considered as an average of the field operator which describes propagation of one plane mode of frequency ω_m ($m = i, s$) along the layer-cut medium at some angle ϑ_m with respect to the layer's normal. The nonlinear Kirchhoff-type law was formulated by D.N. Klyshko within the logic of fluctuation-dissipation approach for parametric light scattering by polaritons [2]. Applying this approach, one does not need to solve exact Heisenberg equations for field operators with the introduction of special noise operators responsible for the absorbing reservoir. It is enough to consider equations for spatially varying *average* values of the field operators. These equations have the same structure as the classical wave equations for slowly varying field amplitudes in an absorptive crystal. They were considered accounting the multi-mode character of the parametric interaction caused by a Gaussian transverse-limited pump beam. Explicit expressions for the multimode scattering matrix elements were obtained for the case of spontaneous PDC.

By substituting the scattering matrix elements into the nonlinear Kirchhoff law the second-order moments of the output fields were obtained. The results show that occupations of output modes $\langle N_m\rangle \equiv \langle a_m^+(L)a_m(L)\rangle$ consist of several different contributions. The first-type contribution (which is true spontaneous) appears due to inherent radiation of the pumped nonlinear medium,

$$\langle N_s^{Spon}\rangle = \kappa_s\left(\sum_i |U_{si}|^2\right) + (\kappa_s - 1)\left(\sum_i |U_{si}|^2\right)\langle N_T\rangle \quad (2)$$

in each signal mode, and

$$\langle N_i^{Spon}\rangle = \left(\sum_s |U_{si}|^2\right) + (1-\kappa_i)\left(\sum_s |U_{si}|^2\right)\langle N_T\rangle + \\ + \left(1 - e^{-2\mu_i L}\right)\langle N_T\rangle \quad (3)$$

in each idler mode. The first terms in Eqs. (2)-(3) contain the pure quantum parts of the nonlinear emission. They are proportional to the sums over all modes of opposite type which make their contributions with the corresponding nonlinear efficiencies $|U_{si}|^2$, whereas

$$U_{si} = iL\gamma_{psi}\, e^{(i\Delta k_{\|} - \mu_i)L/2}\,\frac{\sinh(i\Delta k_{\|} - \mu_i)L/2}{(i\Delta k_{\|} - \mu_i)L/2}. \quad (4)$$

Here, $\mu_m \equiv \alpha_m / 2\cos\vartheta_m$ depends on the intensity absorption coefficient α_m, $\Delta k_{\|}$ is a longitudinal projection of the wavevector mismatch, and $\gamma_{psi} \equiv \frac{2\pi\omega_s\omega_i}{c^2\sqrt{k_{sz}k_{iz}}} E_p \chi^{(2)}$ is a specific gain per unit length which depicts efficiency of the parametric interaction between each three plane modes taken at pump, signal and idler frequencies. $\chi^{(2)}$ is an effective value the medium second-order susceptibility. The pump plane mode components are taken with the amplitudes $E_p(\mathbf{k}_p^{\perp}) \sim w_p{}^2 \exp(-w_p{}^2 \mathbf{k}_p^{\perp 2}/4)$ for transverse wave vectors $\mathbf{k}_p^{\perp}$; w_p is a Gaussian beam waist. The angular diversity of the idler modes that are parametrically connected with each optical signal mode was analyzed on the basis of Eq. (4). It was shown, that the solid angle occupied by spontaneously emitted idler photons rapidly increases when the idler wave frequency goes down and the idler wave vector becomes comparable with the allowed wave vector mismatch. However, by proper selection of the transverse ($\sim 2w_p$) and longitudinal ($\sim L$) dimensions of the interaction volume, the idler radiation divergence can be kept at a moderate level.

The second terms in expressions for spontaneous emission (2)-(3) are responsible for parametric frequency conversion (in Eq.(2)) or amplification (in Eq.(3)) of classical thermal fluctuations at an idler frequency. They are proportional to the mean number of thermal photons $\langle N_T \rangle = [\exp(\hbar\omega_i / k_B T_{cr}) - 1]^{-1}$ at a temperature T_{cr} of the nonlinear medium. The loss-factors κ_s and κ_i account different influence of THz absorption on the efficiencies of parametric processes with thermal and quantum field fluctuations. With a high degree of accuracy they can be taken outside the summation procedures as coefficients that depend only on the THz absorption $\mu_i L$,

$$\kappa_s = 2\left(\mu_i L - 1 + e^{-\mu_i L}\right) / \left(1 - e^{-\mu_i L}\right)^2, \quad \kappa_i = \kappa_s(-\mu_i L). \quad (5)$$

In case of negligibly small absorption the nonlinear contributions of thermal fluctuations to the inherent emission of the medium disappear. The last term in Eq.(3) for the number of emitted idler photons describes the non-parametric thermal radiation by the linear Kirchhoff law.

Apart from the spontaneous emission there can be contributions from the signal or idler radiation incident on the medium. For example, if there is any external idler radiation which is characterized by an effective temperature T_{ext} or with the spectral brightness distributed over all the idler modes with the corresponding mode occupations $\langle N_i \rangle$, the additional photons

$$\left\langle N_s^{Ind} \right\rangle = \sum_i |U_{si}|^2 \langle N_i \rangle \quad (6)$$

are detected at the output in the signal channel. The method of quantum calibration of spectral brightness of an external terahertz radiation is based on comparison between the spontaneous and induced signals [3]. In case of the pure SPDC registration scheme there are no special sources of the external input radiation. However, the equilibrium radiation of the medium environment should be taken into account.

In most cases the temperatures of the medium and of the environment are the same, and a noticeable contribution of external thermal radiation persists only at idler frequencies. As a result, the detected output signal and idler photon numbers are

$$\begin{aligned} \langle N_s \rangle &= \left\langle N_s^{Spon} \right\rangle + \left\langle N_s^{Therm.Ind} \right\rangle = \kappa_s \left(\langle N_T \rangle + 1\right) \sum_i |U_{si}|^2, \\ \langle N_i \rangle &= \left\langle N_i^{Spon} \right\rangle + \left\langle N_i^{Therm.Ind} \right\rangle = \left(\langle N_T \rangle + 1\right) \sum_s |U_{si}|^2 + \langle N_T \rangle. \end{aligned} \quad (7)$$

Following the nonlinear Kirchhoff law, the second-order moment, which describes correlation of idler and signal fields, is calculated in this case as

$$\left\langle a_s^+(L) a_i^+(L) \right\rangle = \left(\langle N_T \rangle + 1\right) \sum_{s'} U_{is'} U_{ss'} \approx \left(\langle N_T \rangle + 1\right) U_{is}. \quad (8)$$

In experimental quantum optics the *intensity* correlation function

$$\langle I_s I_i \rangle \sim \left\langle : \sum_s N_s \sum_i N_i : \right\rangle = \sum_{s,i,s',i'} \left\langle a_s^+(L) a_i^+(L) a_{s'}(L) a_{i'}(L) \right\rangle$$

is frequently measured as an indicator of non-classicality of the generated field states. Consider two ideal detectors in signal and idler channels connected by a correlation scheme which measures the normalized intensity correlation function. The readings of such scheme should depend on the nonlinear medium parameters as

$$\frac{\left\langle : \sum_s N_s \sum_i N_i : \right\rangle}{\sum_i \langle N_i \rangle \sum_s \langle N_s \rangle} = 1 + \frac{\left| \sum_{s,i} U_{is} \right|^2}{\kappa_s \sum_{s,i} |U_{is}|^2} \left(\sum_{s,i} |U_{is}|^2 + \sum_i \langle N_T \rangle / \left(\langle N_T \rangle + 1\right) \right)^{-1} \quad (9)$$

In the case of properly cooled and transparent medium, low gain and low number of involved modes this ratio is sufficiently more than its classical value 1. However, when absorption, number of modes and the medium temperature are increased, the quantum excess above 1 can become vanishingly small. Final Eq. (9) shows the influence of each of these parameters and can be used as a guide in constructing PDC schemes for generating optical-terahertz biphotons.

The work was done under financial support of the Russian Science Foundation (Grant No. 17-12-01134).

References

1. *Bouwmeester D., Ekert A., Zeilinger A.* The Physics of Quantum Information: Quantum Cryptography, Quantum Teleportation, and Quantum Computation // New York: Springer. 2000.
2. *Klyshko, D.N.* Photons and Nonlinear Optics // New York: Gordon and Breach. 1988.
3. *Kitaeva, G. Kh., Penin, A. N., Tuchak, A. N., Shepelev, A.V., Yakunin, P. V.* Measurement of the spectral brightness of terahertz radiation in the process of nonlinear_optical detection // JETP Lett. 2010. V. 92, No. 5, P. 291–294.

https://doi.org/10.1051/epjconf/201819505006

Cold-electron bolometer as a photon-noise-limited detector with on-chip electron self-cooling

L.S. Kuzmin[1,2], A.L. Pankratov[1,3], A.V. Gordeeva[1,3], V.O. Zbrozhek[1],

A.V. Blagodatkin[1,3], L.S. Revin[1,3]

[1]Nizhny Novgorod State Technical University n.a. R.E. Alekseev, Nizhny Novgorod, Russia
[2]Chalmers University of Technology, Gothenburg, Sweden, leonid.kuzmin@chalmers.se
[3]Institute for Physics of Microstructures of RAS, Nizhny Novgorod, Russia

After discovering the electron cooling by Superconductor-Insulator-Normal metal (SIN) tunnel junctions, significant efforts were applied to obtain an effective on-chip cooler for mm/IR detectors. We have developed a Cold-Electron Bolometer (CEB) with effective direct electron self-cooling of the absorber [1], [2]. A photon-noise-limited pixel, consisting of an array of CEBs with self-cooling, is realized for OLIMPO Balloon Telescope for incoming power P_{in} up to 60 pW at phonon temperature T_{ph} = 310 mK. Operation of the bolometer at electron temperature less than phonon temperature significantly increases its sensitivity so that the noise equivalent power goes beyond the photon-noise-limited mode, which means that the internal bolometer noise is smaller than the noise of incoming signal [3]. We demonstrate the electron cooling from 310 to 120 mK without signal and from 410 to 225 mK for Pin = 60 pW at 350 GHz. The proposed technology is a potential replacement for the high-cost dilution refrigerators for space applications.

The cold electron bolometer is a SINIS structure, integrated in a planar antenna, designed to absorb terahertz radiation at certain frequency. Its SEM image is shown in Fig.1. The normal metal is made of aluminum with suppressed superconductivity, the tunnel barrier is formed by oxidation of aluminum and the superconducting electrodes are made of clean aluminum. All layers are deposited on a Si substrate by method of shadow evaporation. The volume of the normal absorber is just 0.02 μm^3.

A single CEB can absorb up to 0.3-0.5 pW of power before it saturates. In order to fulfill the requirements of high power load for OLIMPO mission, we connected 192 single CEBs in one array. Each bolometer is integrated in a dipole antenna. The array is tuned to have its maximal absorption at 350 GHz. The experiments have shown that the efficiency of absorption for fabricated samples is more than 50%.

The cold electron bolometer is well described by the heat balance equation for normal absorber with resistance R_N:

$$P_{abs} + I^2R_N + V^2/R_{leak} = 2*P_{COOL} + \sum V(T_e^2 - T_{ph}^2), (1)$$

where I - tunnel current, V - voltage drop across one SIN junction, P_{abs} - absorbed power, R_{leak} - leakage resistance of SIN junctions, P_{COOL} - cooling power of SIN junctions, $\sum$ - electron-phonon constant, V - volume of the absorber, T_e and T_{ph} are electron and phonon temperatures of the absorber, respectively.

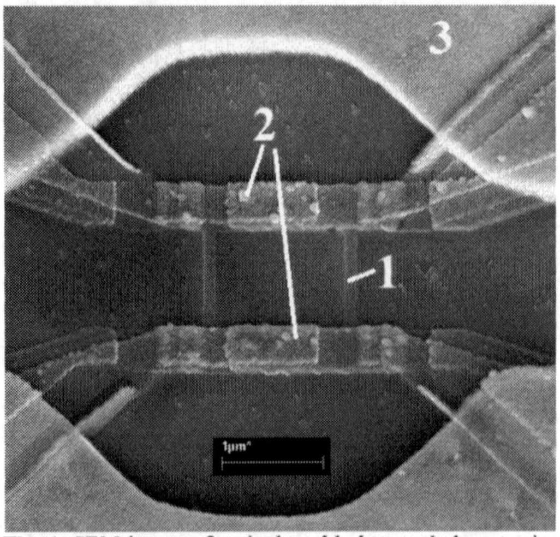

Fig. 1. SEM image of a single cold electron bolometer integrated in a gold dipole antenna. 1 - normal absorber, 2 - SIN junction, 3 - antenna.

The fabricated samples have been cooled in a dilution cryostat down to 310 mK and irradiated by a black body source through a set of quasi-optical filters for 350 GHz. The voltage response and voltage noise at several black body temperatures from 2K to 46 K have been measured, using room temperature ultralow noise amplifiers AD745.

We have compared the experimental IV-curves with Eq. (1) and found that our samples can be fitted very well with the following parameters: critical temperature of superconductor 1.24 K, phonon temperature 310 mK, $\sum$=1.3 nW/K^5/μm^3, normal resistance of SIN junction 1.6 kΩ, R_{leak}=25 MΩ, R_N=50 Ω.

The main characteristic of any detector is the noise equivalent power (NEP). The modern bolometers for astronomical applications are required to have intrinsic NEP less than the photon NEP of incoming signal, i.e. the detector has to be limited by the photon noise. We have shown in a set of our experiments, that we indeed can see the photon noise in our detectors in a broad range of incoming powers from 20 to 60 pW.

The NEP of the bolometer array is shown in Fig.2 for the case of zero incoming power. One can see good agreement between measured and theoretical values. The optimal operational point of the bolometer can be chosen at the minimal NEP around 12 mV. At this voltage the largest contribution to the bolometer NEP comes from the noise of SIN contacts, which is the sum of shot noise of current through a tunnel junction, noise of heat flow and correlation between them. Next largest NEP component is amplifier NEP. Whereas NEP due to electron-phonon interaction is significantly smaller than other NEP components and can be even disregarded.

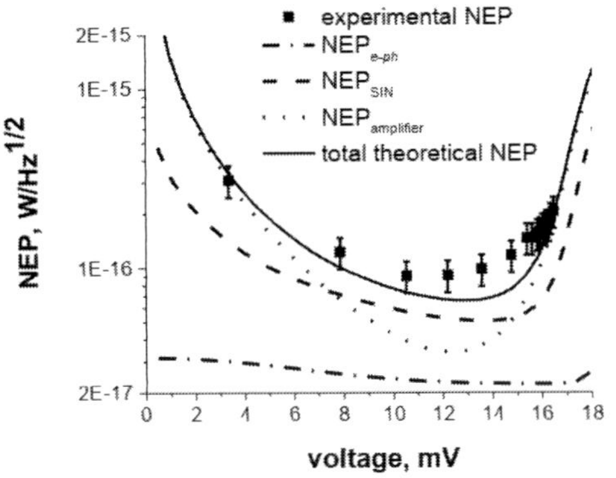

Fig. 2. Noise equivalent power versus voltage on the bolometer array with zero power load. The phonon temperature 310 mK.

In Fig. 3 we show NEP of the same bolometer array but for high power load 60 pW, which corresponds to 32 pW of absorbed power. Now we have one more NEP component - photon noise of absorbed power (straight gray line in Fig. 3). One can see that at voltages 11-12 mV the photon NEP goes a bit higher that all other NEP components together, which means that the detector is photon limited.

Analyzing our data, we found that for black body temperatures above 20 K, the phonon temperature of the sample also increases and can be higher than the cryostat temperature. For example for Fig. 3 the phonon temperature was 380 mK. But even with this overheating the samples show photon limited mode of operation.

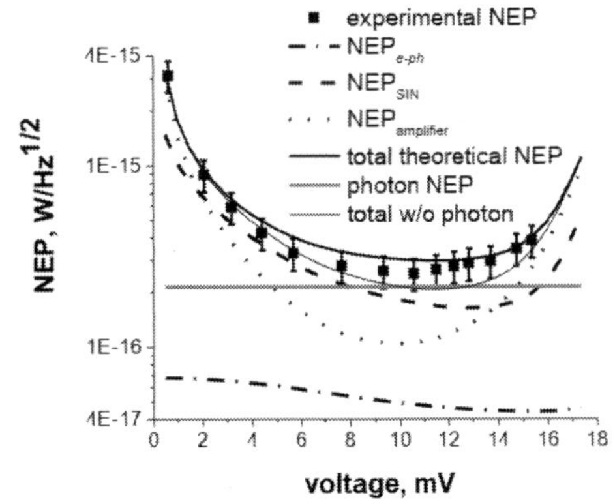

Fig. 3. Noise equivalent power versus voltage on the bolometer array. Radiated power 60 pW. Absorbed power 32 pW.

Authors would like to thank M. Tarasov and A. Gunbina for help in the sample fabrication, E. Skorokhodov for making SEM images of the samples. The samples were fabricated in the Chalmers anotechnology Center. The facilities of the Center of Cryogenic Nanoelectronics of NNSTU and Common Research Center "Physics and technology of micro- and nanostructures" of IPM RAS were used.

The work is supported by Russian Science Foundation (Project 16-19-10468).

References

1. *Kuzmin, L.* An Array of Cold-Electron Bolometers with SIN Tunnel Junctions and JFET readout for Cosmology Instruments // Journal of Physics: Conference Series, 8th European Conference on Applied Superconductivity. 2008. V. 97.

2. *Kuzmin, L; Golubev, D.* On the concept of an optimal hot-electron bolometer with NIS tunnel junctions // Physica C-Superconductivity and Its Applications. 2002. V. 372. P. 378-382.

3. *Gordeeva,A.V., et al.* Observation of photon noise by cold-electron bolometers // Appl. Phys. Lett. 2017. V. 110. P. 162603.

https://doi.org/10.1051/epjconf/201819505007

Terahertz Time-Domain Spectrometer with Precision Delay Line Encoder

A. A. Mamrashev[1], F. A. Minakov[1,2], L. V. Maximov[1,2], N. A. Nikolaev[1], P. L. Chapovsky[1,2]

[1]Institute of Automation and Electrometry SB RAS, Novosibirsk, Russia, mamrashev@iae.nsk.su
[2]Novosibirsk State University, Novosibirsk, Russia

Introduction

Accuracy of optical delay line is crucial for terahertz time-domain spectroscopy (THz-TDS). Typically such delay line consists of a motorized translation stage with a mounted retroreflector. Random and systematic errors in positioning of the translation stage result in distortion of sampled terahertz pulses and hence terahertz spectra.

It is intuitive that random misplacements in the sampling position (registration jitter) translate into an error in the amplitude of the measured terahertz signal proportional to its derivative. This results in a higher noise floor in the terahertz spectra [1,2]. It was recently found that periodic sampling errors lead to creation of spurious mirror spectra around the error's frequency [3]. In a typical terahertz measurement a series of sampled terahertz pulses or spectra are averaged to achieve better signal to noise ratio. However, even small delays in the sampling start times can actually increase the noise level in the averaged signal [4]. So it was suggested to align terahertz pulses in the time domain before averaging using algorithms from [5].

In the custom-made THz-TDS developed at IA&E SB RAS [6] we used motorized translation stage 8MT173-50-20 (Standa Ltd, Lithuania) with the full movement range of 50 mm (time range of 333 ps) and the step of 1.25 μm (time step of 8.34 fs). In order to independently measure translation stage position in this study we installed an optical encoder Resolute RL32BAT001B50 with RTLA absolute scale (Renishaw, UK) (see Fig. 1). The scale length was 100 mm and its accuracy was 0.5 μm. The encoder resolution was 1 nm. Thus we detected delay line inaccuracies and corrected them.

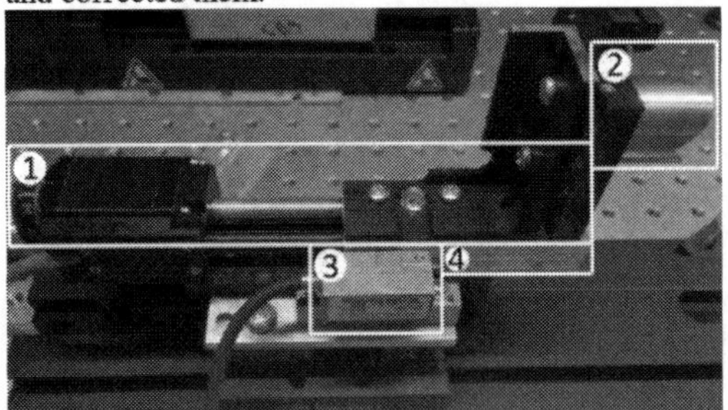

Fig. 1. Upgraded optical delay line consisting of the motorized translation stage (1), the retroreflector (2), the optical encoder (3) and the high accuracy scale (4)

Results

We compared position of the translation stage (L_{st}) with the position measured by the optical encoder (L_{en}). We studied their difference for four translation stages by installing the encoder on each (Fig. 2). It was found that three systematic errors are associated with the measurements: growing offset of the position, periodic error, and drift of the starting position.

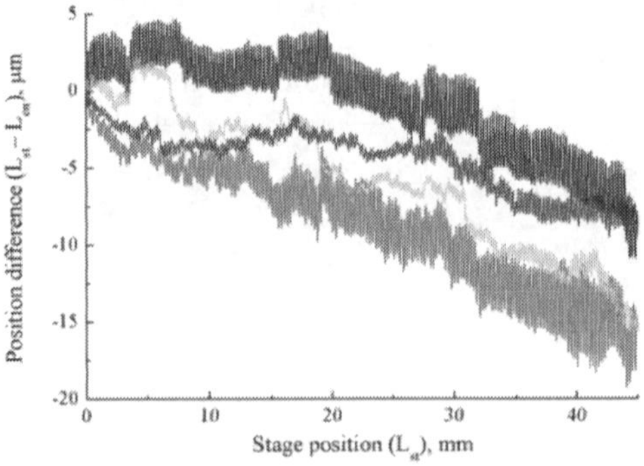

Fig. 2. Difference between the positions of the translation stage (L_{st}) and the optical encoder measurements (L_{en}). The data are presented for four different stages

First, it can be seen from the Fig. 2 that all four translation stages exhibit growing position offset that can roughly be approximated by a linear function. Abrupt changes of the offset may be explained by mechanical defects in the translation stage screw. Total offset varies from –8 to –15 μm for different stages. Such offset manifests itself in a compressed terahertz pulse and consequently stretched terahertz spectrum, i. e. having linearly increasing frequency shift. In our experiments it leads to a total shift of up to 400 MHz at 1 THz. Correction of such systematic error is crucial in the measurements of narrow-band absorption peaks such as spectra of nuclear spin isomers of water vapor [7].

Second, we notice periodic sinusoidal sampling error that becomes apparent after zooming in on the finer structure of the difference between translation stage position and encoder measurements (see Fig. 3).

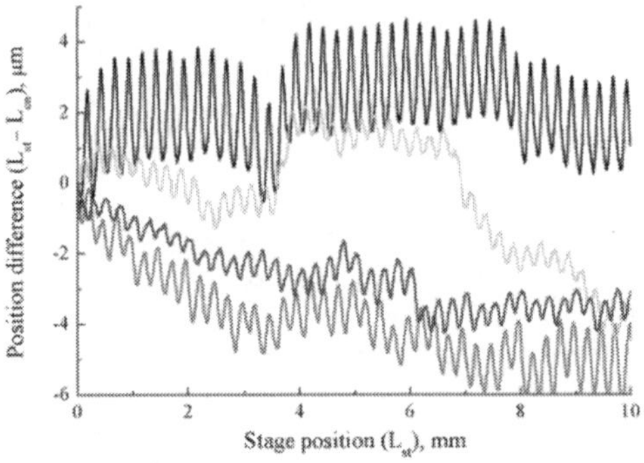

Fig. 3. Difference between the positions of the translation stage (L_{st}) and the optical encoder measurements (L_{en}). The data are presented for four different stages

The period of the oscillations is 250 μm that corresponds well with a single revolution of the stage screw. The amplitude of the oscillation varies from 0.3 to 1.6 μm for different stages. According to [3] such periodic sampling errors result in the presence of additive spurious mirror copies of the main pulse spectrum around error's frequency and its harmonics. In our case oscillation period of 250 μm corresponds to the frequency of 600 GHz. The amplitude of the spurious spectra is proportional to the amplitude of the oscillations. In our measurements it results in the proportionality coefficient as low as 10^{-3} which is comparable with the signal-to-noise ratio making this error difficult to detect.

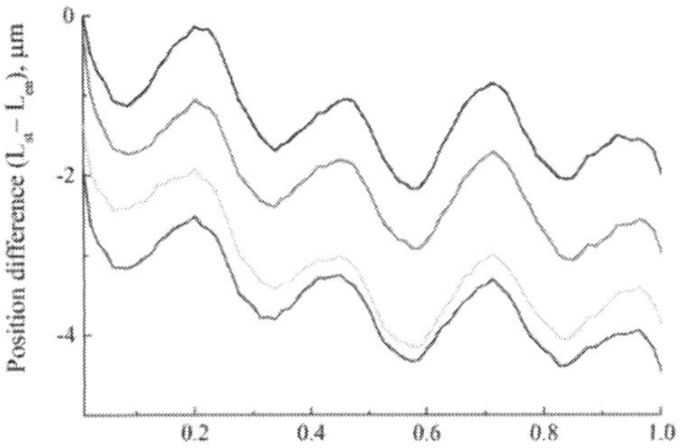

Fig. 3. Difference between the positions of the translation stage (L_{st}) and the optical encoder measurements (L_{en}). The data are presented for four consecutive measurements of the same translation stage

Finally, we find that the starting position of the same translation stage in the consecutive measurements drifts by less than 1 μm and the drift magnitude gets smaller each time. It results in time delays of the measured terahertz pulses or equivalently in phase shifts in their spectra. This error requires correction before averaging of the series of consecutive terahertz measurements. Otherwise, it may lead to an increased error in the averaged signal.

In conclusion, we found several systematic errors in the optical delay line of THz-TDS by independently measuring its absolute position with precision optical encoder. We found that correction of the position offset of the delay line is crucial in gas analysis. Correction of the starting position drift should be applied before averaging the series of terahertz pulses measured consecutively. Periodic sampling error resulted in a small additive spectral error which can only be found in certain spectra such as ones provided by high contrast band pass filters.

Acknowledgement

The study was financially supported by the Russian Science Foundation (RSF), project 17-12-01418.

References

1. *Jahn D. et al.* On the influence of delay line uncertainty in THz time-domain spectroscopy // Journal of Infrared, Millimeter, and Terahertz Waves. 2016. Vol. 37, No. 6. P. 605–613.

2. *Cohen N. et al.* Experimental signature of registration noise in pulsed terahertz systems // Fluctuation and Noise Letters. 2006. Vol. 6, No. 1. P. L77–L84.

3. *Rehn A. et al.* Periodic sampling errors in terahertz time-domain measurements // Optics Express. 2017. Vol. 25, No. 6. P. 6712.

4. *Humphreys D.A., Naftaly M., Molloy J.F.* Effect of time-delay errors on THz spectroscopy dynamic range // 39th International Conference on Infrared, Millimeter, and Terahertz Waves (IRMMW-THz 2014). IEEE, 2014.

5. *Coakley K.J., Hale P.* Alignment of noisy signals // IEEE Transactions on Instrumentation and Measurement. 2001. Vol. 50, No. 1. P. 141–149.

6. *Antsygin V.D. et al.* Effect of a Magnetic Field on Wideband Terahertz Generation on the Surface of Semiconductors // IEEE Trans. Terahertz Sci. Technol. 2015. Vol. 5, No. 4. P. 673–679.

7. *Mamrashev A.A. et al.* Detection of Nuclear Spin Isomers of Water Molecules by Terahertz Time-Domain Spectroscopy // IEEE Transactions on Terahertz Science and Technology. 2018. Vol. 8, No. 1. P. 13–18.

EPJ Web of Conferences **195**, 05008 (2018)
TERA-2018 https://doi.org/10.1051/epjconf/201819505008

Temperature dependence of signal spectra generated via spontaneous parametric down-conversion in strongly frequency non-degenerate regime

T.I. Novikova, K.A. Kuznetsov, G.Kh. Kitaeva
MSU Physics Faculty, Moscow, Russia, tanya.novik00@mail.ru

The process of spontaneous parametric down-conversion (SPDC) represents the process of decay of photons of monochromatic radiation (pumping) with a frequency ω_p on pairs of photons due to interaction with quantum fluctuations of the electromagnetic field in a medium with a non-zero quadratic susceptibility $\chi^{(2)}$. As a result, radiation occurred inside the medium, consists of pairs of correlated photons. We propose to consider properties of biphotons with extremely different frequencies, when the signal photon frequency is very close to ω_p, while the idler photon frequency hits the terahertz range. This regime can be referred as strongly frequency non-degenerate parametric down-conversion (SFND PDC). The frequency-angular distributions of the power of signal photons generated in the Stokes and anti-Stokes ranges are measured. We study the dependence of the power of signal photons generated in Stokes and anti-Stokes ranges from the temperature of the non-linear $LiNbO_3$ crystal.

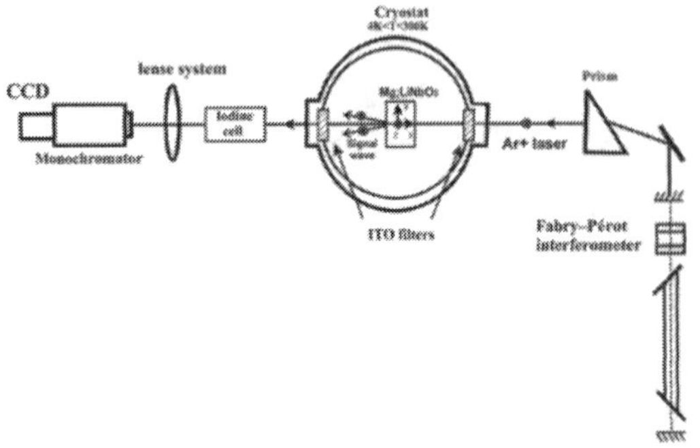

Fig. 1. Experimental setup

In our experiment we use argon laser with generation at a wavelength of 514.5 nm as a pumping source for SPDC. The design of this laser provides the possibility of narrowing the generation line to 100 MHz using the Fabry-Perot interferometer that allow to use narrowband iodine vapor filter for later exclusion of pump radiation. Using the prism we create two beams. One of them allows measuring the pump power without disturbing the main beam, which is a pumping source for 5 mol% MgO-doped $LiNbO_3$ crystal. This crystal is an excellent material for optical frequency converters and it is widely used in quasi-phase-matched conversion applications, because of its large optical nonlinearity. The length of the crystal was L=1.5 cm.

In order to study the influence of thermal fluctuations on the PDC signal the nonlinear crystal has been placed into the cryostat (SCONTEL, Russia) and held at 5 - 300 K. Also, the cryostat in this case has been equipped with a second window for optical pump and signal radiation removal. Stray light from the optical pump was blocked with an iodine cell heated at 75°C. The cryostat windows diameter was D ~ 20 mm. Input and output window of the cryostat were closed by filters, based on indium tin-doped oxide (ITO). They do not allow external thermal terahertz radiation to get into the cryostat. Filters save us from external terahertz thermal photons. We use three lens system in order to focus all signal radiation on the entrance slit of monochromator and get the frequency-angular distributions of emitted photons from CCD camera.

According to the nonlinear Kirchhoff law the power of Stokes and Anti-Stokes radiated photons depends on the crystal temperature in the following way:

$$P_S \sim 1+\langle N_T\rangle \quad (1)$$

$$P_{AS} \sim \langle N_T\rangle \quad (2)$$

where $\langle N_T\rangle = \dfrac{1}{e^{\frac{\hbar\omega}{kT}}-1}$ - the mean number of thermal photons per mode. So, emitted signal photons in Stokes range arise due to up-conversion of thermal and quantum fluctuations. But anti-Stokes signal photons arise due to up-conversion of thermal fluctuations only. One can see the power of Anti-Stokes photons tends to zero while the temperature tends to zero. During the experiment we cooled the crystal up to 4 K and observed a total disappearance of the Anti-Stokes branch.

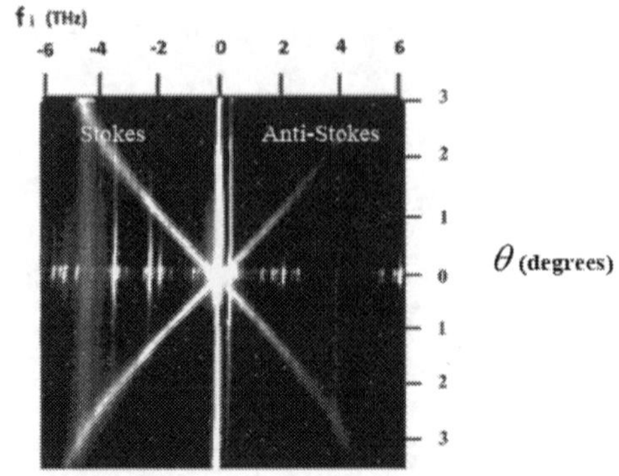

Fig.2.
Frequency-angular spectrum of PDC at T=297K

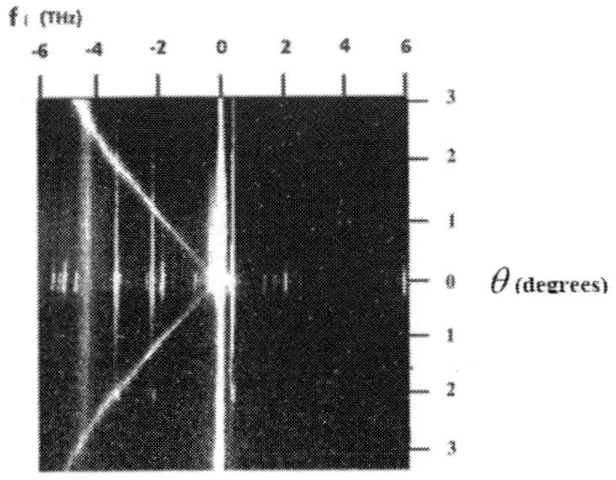

Fig.3. Frequency-angular spectrum of PDC at T=4K

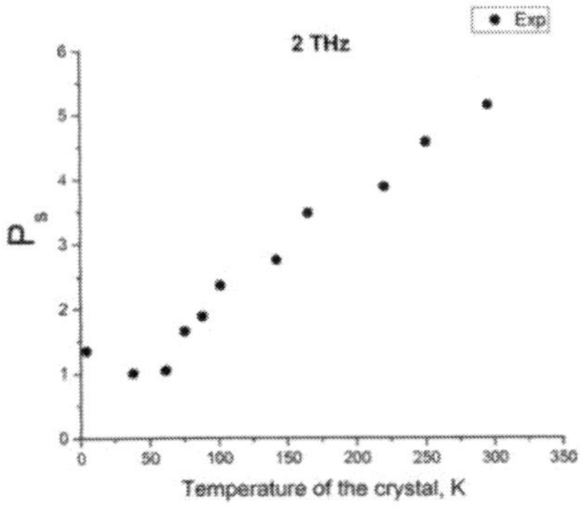

Fig.4. Experimental data for 2 THz

Using the nonlinear Kirchhoff law, formulated by D.N. Klyshko [1], with account of crystal absorption in the THz idler-frequency range and presence of classical thermal field fluctuations at THz frequencies, it is logical to offer such a theoretical model to describe the obtained results:

$$P_s^{SPDC} = C_{\omega_i}\left(1+\langle N_T\rangle\right)\int S_{\omega}^{SPDC}(\vartheta_i,\varphi_i)d\vartheta_i d\varphi_i d\vartheta_s d\varphi_s \quad (3)$$

where

$$S_{\omega}^{SPDC}(\vartheta_i,\varphi_i) = \left(Lw_p\chi^{(2)}\right)^2 \omega_i^3 tg\vartheta_i e^{-\frac{w_p^2(\mathbf{k}_{s\perp}+\mathbf{k}_{i\perp})^2}{2}} * \left[\frac{1+(i\Delta k_z-\mu_i)L-e^{(i\Delta k_z-\mu_i)L}}{(\Delta k_z+i\mu_i)^2L^2}+c.c.\right] \quad (4)$$

Here, $\mu_m \equiv \alpha_m / 2\cos\vartheta_m$ depends on the intensity absorption coefficient α_m, Δk_z is a longitudinal projection of the wave-vector mismatch, w_p is a Gaussian beam waist, L is the length of our crystal, ϑ_i is the angle of idler wave inside the crystal.

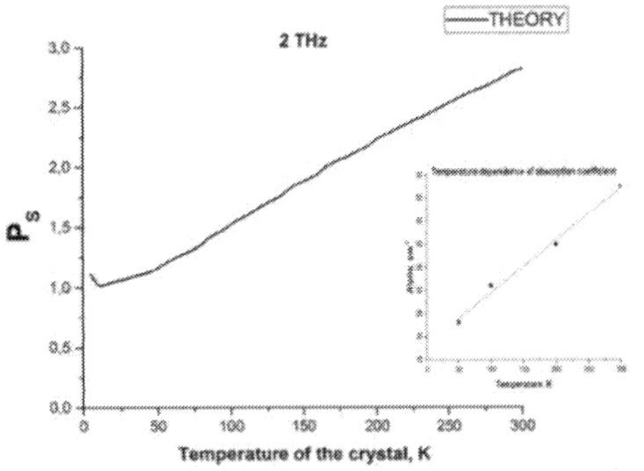

Fig.4. Calculation for 2 THz

The absorption coefficient α_m was taken from the [2]. For crystal absorption in the THz idler-frequency range there were given several values for different temperatures. These data were linearly extrapolated in our calculations. From this extrapolation we took numerical values for our temperatures near 5K. So, considering theoretical values of P_s^{SPDC} we should take into account the approximation error.

This work was done under financial support of the Russian Science Foundation (Grant No. 17-12-01134).

References

1. Klyshko D.N. Photons and Nonlinear Optics // New York: Gordon and Breach. 1988.
2. *Xiaojun Wu, Chun Zhou, Wenqian Ronny Huang, Frederike Ahr, and Franz X. Kärtner.* Temperature dependent refractive index and absorption coefficient of congruent lithium niobate crystals in the terahertz range // OPTICS EXPRESS. 2015. V. 23, No.23. P. 29729–29737.

Wideband metamaterial-based array of SINIS bolometers

A.S. Sobolev[1,2], B. Beiranvand[1], A.M. Chekushkin[1,2], A.V. Kudryashov[1], M.A. Tarasov[1,3], R.A.Yusupov[1,2], A. Gunbina[4], V.F. Vdovin[4], V. Edelman[5]

[1] Moscow Institute for Physics and Technology, Moscow, Russia, sobolev@hitech.cplire.ru
[2] Kotel'nikov Institute of Radio Engineering and Electronics, Moscow, Russia
[3] Chalmers University of Technology, Gothenburg, Sweden
[4] Institute of Applied Physics, N.Novgorod, Russia
[5] Kapitza Institute for Physical Problems RAS

Introduction

Recently we have developed and experimentally studied a wideband 2D-array of periodically arranged electrically small rings, each containing Superconductor-Insulator-Normal metal-Insulator-Superconductor (SINIS) bolometers. Earlier such bolometers in a regular annular antenna array [1] demonstrated voltage responsivity up to 10^9 V/W with a bandwidth of 5-10% in a 350 GHz frequency band. Such detector array can operate in background noise limited condition for incident powers up to 80 pW [2,3]. The typical diameter of the rings was $\lambda/2$, where λ is an effective wavelength on the silicon-vacuum interface related to free-space wavelength λ_0 by

$$\lambda = \frac{\lambda_0}{\sqrt{\frac{\epsilon+1}{2}}} \quad (1)$$

Wavevector **k** of a normally incident plane wave has zero tangential component. For this reason a periodic array with the elementary unit cell size of the order of λ is seen by a normally incident plane wave as a homogeneous thin film with effective impedance Z_{FSS}. The condition of normal incidence is satisfied in quasioptical systems when the detectors array is placed into the focus of a lens where the wase of the Gaussian beam is. Large spatial period of the array reduces the operational frequency bandwidth. In the present work we report on the studies of the wideband array of cold-electron bolometers designed for the range 300-450 GHz and consisting of periodically arranged rings, each containing four bolometers. This periodic array with the unit cell size less than $\lambda/10$ acts as a distributed metamaterial absorber, which is seen as homogeneous metamaterial film by the wave incident at any angle. This fact makes the array applicable for integration into waveguides where all three orthogonal k-vector components of the propagating wave are nonzero. Small unit cell size ensures higher density of bolometers and therefore increases the bandwidth and the dynamic range of a single pixel.

Numerical model and optimization

The chip with a wideband array is designed for mounting on a silicone lens similar to integrated lens antennas. Figure 1 shows a layout of the unit cell with dimension 37um x 38 um in XY plane. A structure made of golden thin film is fabricated on a silicone substrate and the four cold-electron bolometers are integrated into it. The Floquet ports are placed at the boundaries corresponding to the minimum and the maximum of z-coordinate.

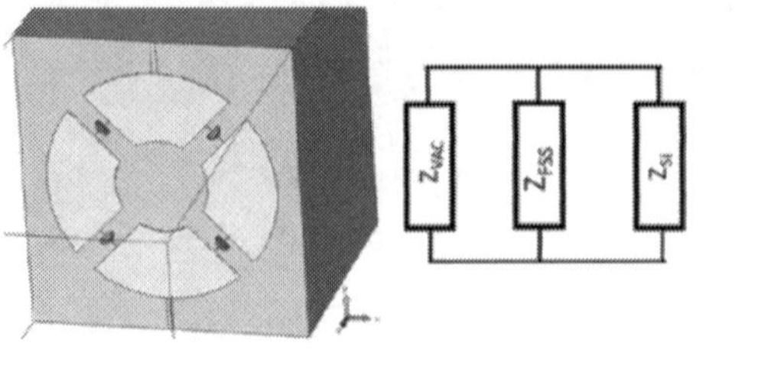

Fig.1. Unit cell design and the lumped elements equivalent schematic

Most of the RF-power coming from the silicone lens is absorbed by the SINIS-bolometers, but some fraction of it is transmitted further into the vacuum half-space because of no backshot. The equivalent microwave schematics of this setup can be represented as a parallel connection of the three lumped impedances: surface impedance Z_{FSS} of the array is shunted by impedance Z_{VAC} = 377 Ω and Z_{SI}=377/√11,7 Ω of vacuum and silicon half-spaces. From this simple schematic one can derive the following formula for the power P_{abs} transmitted from Z_{SI} to Z_{SI}:

$$P_{abs} = \frac{Z_{VAC}}{Z_{FSS}+Z_{VAC}} \cdot \frac{4Z_{SI}Z_{VAC}Z_{FSS}(Z_{FSS}+Z_{VAC})}{(Z_{SI}Z_{FSS}+Z_{SI}Z_{VAC}+Z_{VAC}Z_{FSS})^2} \quad (2)$$

Dependence of P_{abs} as a function of Z_{FSS} is plotted in figure 2. It achieves maximum P_{abs} = -1 dB at $Z_{FSS}\approx$ 80 Ω. This value of Z_{FSS} is slightly less than that of parallel connection of Z_{SI} and Z_{VAC}.

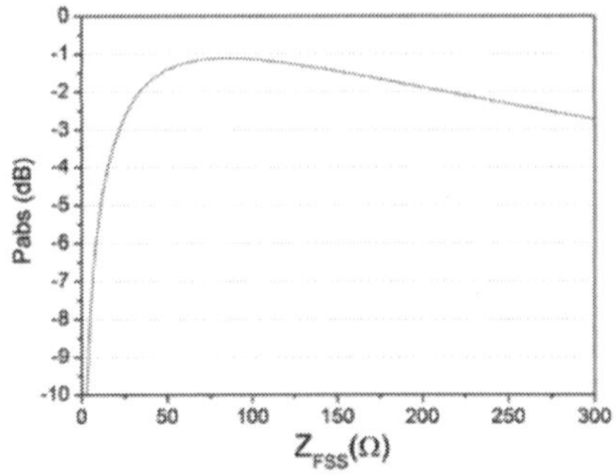

Fig.2. Absorbed power P_{abs} as a function of Z_{FSS}. Maximum P_{abs} corresponds to $Z_{FSS}\approx$ 80 Ω

During numerical simulations we considered the two lowest port modes with orthogonal polarization and constant distribution of the electric and the magnetic fields. The the E-field of the first Y-polarized mode is directed along the narrow DC-biasing lines that connect adjacent unit cells together and create series inductance, which tunes the capacitance of the SINIS-bolometers out. For the second X-polarized mode Z_{FSS} is determined by the planar capacitance between the rings of the adjacent unit cells.

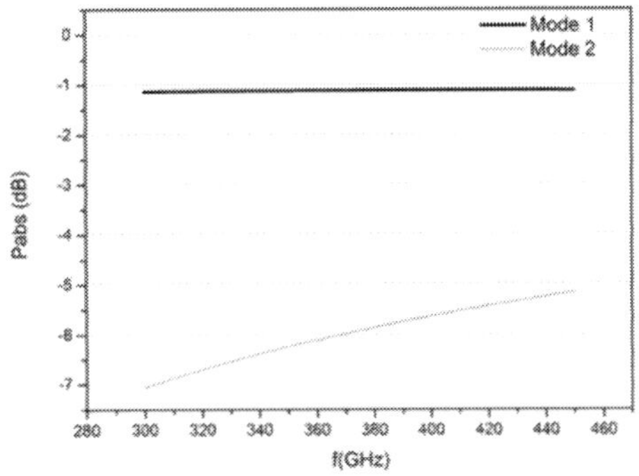

Fig.3. Simulated absorbtion P_{abs} of the SINIS-array fo the two modes with orthogonal polarization

Figure 3 shows the simulated absorbtion of both modes for the array with the optimized unit cell design. and the optimal capacitance C=20 fF and the resistance R=100 Ω of each SINIS-bolometer. According to numerical simulations based on the FDTD-method P_{abs} for the first mode is expected to be close to the -1 dB limit in the whole frequency range 300-450 GHz as like it was predicted by formula (2). The capacitive reactance of the bolometer array seen by the second mode results in much lower P_{abs} ranging from -7 dB to -5 dB.

Experimental test

During experimental tests the spectral response of the sample with the array of SINIS-bolometers was measured in He3 sorption cryostat with optical window and a set of cold neutral density filters that were acting as cold attenuators to reduce the power of room temperature background blackbody radiation. A tuneable external narrowband submm source of radiation was illuminating the 10x10 unit cells array that covered the Airy spot in the focus of the lens. An additional absorber was placed behind the array at some distance from the vacuum side to get rid of the standing waves and make more uniform and wideband frequency response characteristics. The experimentally measured spectral response in the range 240-370 GHz is shown in figure 4. The response is wide and quite nouniform. The nonuniformity may come from the standing waves existing between the cold neutral sensity filters and optical windows. Also the spectral response can be influenced by spurios reflections from some parts of the mesurement setup close by the sample holder.

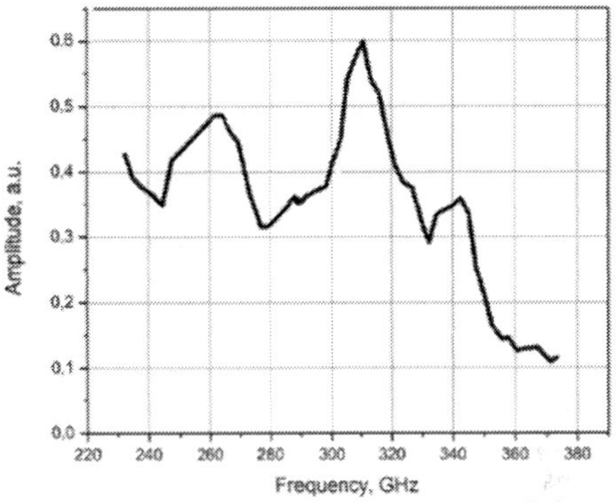

Fig.4. Experimentally measured spectral response of the SINIS-bolometers array

From the measurements with a blackbody source The voltage responsivity of the array with 100 rings measured from a blackbody source was as much as 1.3*10^9 V/W

The work was supported by the RFBR grants #18-07-00785 and #18-07-00743.

References

1. *S.Mahashabde, A.Sobolev, M.Tarasov, G.Tsydanzhapov, L.Kuzmin,* Planar frequency selective bolometric array at 350 GHz // IEEE Trans. Terahertz Sci. and Technol., vol. 5, No 1, pp.37-43, 2015.

2. *Mahashabde, S. Sobolev, A.S., Bengtsson, A.; Andren, D.; Tarasov, M.A.; Salatino, M.; de Bernardis, P.; Masi, S.; Kuzmin, L.S.* A Frequency Selective Surface Based Focal Plane Receiver for the OLIMPO Balloon-Borne Telescope // IEEE Transactions on Terahertz Science and Technology, Volume 5, Issue 1, pp.145-152, 2015

3. *S. Mahashabde, M. A. Tarasov, M. Salatino, A. Sobolev, S. Masi, L. S. Kuzmin,and P. de Bernardis* A distributed-absorber cold-electron bolometer single pixel at 95 GHz // Appl. Phys. Let., 107, 092602, 2015

EPJ Web of Conferences **195**, 05010 (2018) https://doi.org/10.1051/epjconf/201819505010
TERA-2018

Arrays of annular cryogenic antennas with SINIS bolometers and cryogenic receivers for SubTHz observatories

M. Tarasov[1], A. Gunbina[2,5], M.Mansfeld[2,5], G. Yakopov[4], A.Chekushkin[1], R. Yusupov[1], S.Lemzyakov[6], V. Edelman[6], and V. Vdovin[2,3,5],

[1]V.Kotelnikov Institute of Radio Engineering an d Electronics RAS
[2]Institute of Applied Physics RAS
[3]P.N. Lebedev Physical Institute of the Russian Academy of Sciences
[4]Special Astrophysical Observatory RAS
[5]Nizhny Novgorod State Technical University n.a. R.E. Alekseev
[6]P.Kapitza Institute for Physical Problems RAS

Series and parallel arrays of annular antennas optimized for 350 GHz frequency band with SINIS (Superconductor - Insulator - Normal metal - Insulator-Superconductor) bolometers and cryogenically cooled receivers for series of modern SubTHz observatories have been designed, fabricated and studied. The detectors are matched to incoming beam by an immersion sapphire lens. In the low background power case a voltage responsivity approaches $1.5*10^9$ V/W corresponding to NEP=10^{-17} W/Hz$^{1/2}$. The current responsivity for parallel array is $4*10^4$ A/W and shot noise limited intrinsic noise equivalent power is NEP=$4*10^{-18}$ W/Hz$^{1/2}$. At the background radiation temperature level of 2.7 K the responsivity to variations of radiation temperature is over 75 μV/K. The intrinsic time constant of the bolometer array is 0.1-0.2 μs.

Introduction

Observation of the CMB radiation anisotropy and some other objects of cold Universe are the priority tasks for ambitious projects in radioastronomy where some new observatories are under development: international space observatory Millimetron, international observatory Suffa (Uzbekistan) and QTT observatory, China. Technical requirements for receivers of listed observatories are rather contradictive, high sensitivity should be combined with a wide dynamic range and a low time constant. Room-temperature semiconducting electronics practically expired potential for improvement and superconducting devices offer promising potential. Only superconductors (TES, KID and other) and cryogenically cooled electronics gives possibility to provide new observatories by proper instruments. A realistic alternative to transition edge sensors (TES) can be SINIS bolometers that provide high responsivity, wide dynamic range and high speed. In present contribution we study performance of such bolometers integrated in series and parallel arrays.

Design and fabrication

Bolometers were integrated in annular ring antennas designed for central frequency of 345 GHz. Antennas were arranged in series and parallel arrays to increase saturation power and dynamic range. The layout of SINIS bolometers was modified to increase the volume of superconducting electrodes up to 100 times (area from 2 μm^2 in our previous designs [1, 2] to 250 μm^2) to reduce overheating by strong dilution of hot quasiparticles in such big volume. Another layout modification is in eliminating of additional layer of so-called thin gold that suppressed superconductivity and connecting Al electrodes directly to thick TiAuPd antennas that also improve cooling of superconducting electrodes.

Measurement setup

Our previous arrangement of experiment was based on a back-to-back horn that reduces a role of antenna dimensions on overall spectral characteristics and beampattern. In present research we have developed results of previous results [1, 2] and used immersion sapphire lens instead of horn. Such substrate lens provides direct illumination of array, avoid substrate modes, and increase the gain of planar antenna. Top view of dilution cryostat [3] with 3 such lenses is presented in Fig. 1. Each sample was connected to room-temperature electronics through cold resistors that provide suppression of interferences and noise. The sapphire hemisphere lens diameter is 8 mm. Total extension for hyperhemisphere should be 1.5 mm, in this case antenna on the chip 0.28 mm thick is placed in the second focus of elliptical lens. Such elliptical lens convert a spherical wave into a planar wave.

Cryogenic fast radiation source comprising NiCr film on sapphire substrate was equipped with a thermometer and placed on 0.5 K temperature stage. Bandpass and lowpass filters between such source and sample determine the bandwidth of irradiating power.

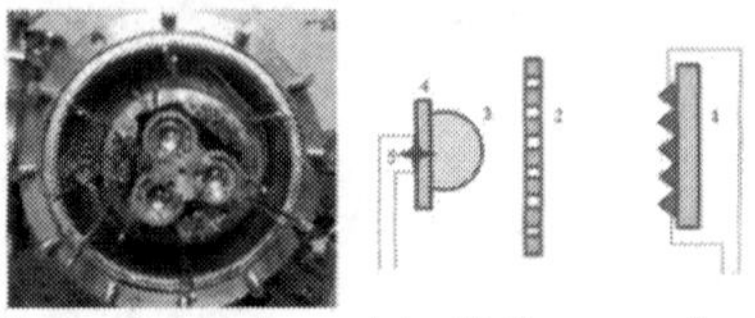

Fig. 1. Photo of opened cryostat with three samples on three sapphire lenses (left), and schematics of sensitivity measurements (right): 1 fast radiation source, 2 bandpass and lowpass filters, 3 hemisphere lens, 4 extension substrate, 5– bolometer array

Experimental results

The asymptotic resistance for single ring is 1700 Ω, for series array of 25 rings it is 39 кΩ, and for 25 parallel rings it is 65 Ω. Responsivity of bolometers was measured by varying the radiation source temperature from initial 0.5 K by applying DC

 https://doi.org/10.1051/epjconf/201819505010

current through NiCr film on sapphire substrate suspended on thin wires. We compared voltage response of series array for variation of bath temperature and for variation of radiation temperature (see Fig. 2). Dependencies are identic that is a clear proof of thermal mechanism of response contrary to our results with previous generation of samples with lower volume of superconducting electrodes and different geometry of antennas. Position of maximum is slightly shifted from about 1 nA at low radiation power up to about 1.6 nA for higher power. Estimated power level calculated for receiving of single mode.

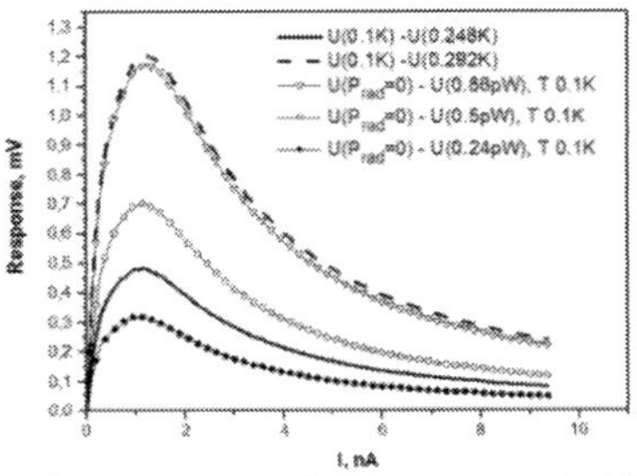

Fig. 2. Voltage response for temperature (0.248 K and 0.292 K) and radiation power

Illustration for saturation and dynamic range is presented in Fig. 3 for series array at bias currents 1 nA and 1.6 nA and for single ring response voltage multiplied by factor of 25 for comparison. Both dependencies can be fitted by a simple relation presented as a line. For radiation power of 2 pW and bias current 1.6 nA the responsivity dV/dP is reduced only by 30% compared to low radiation power case. Small reducing of response for single ring can be due to its position at the edge of chip with lower illumination power. We also compared response of series array for the same signal on the bath temperature (Fig. 4). Three times increase of temperature from 0.1 to 0.3 K leads to decrease of maximum response only by a factor of 1.5. A figure of merit for astronomic applications is radiation temperature responsivity (Fig. 5). At the background power level of 2.7 K the responsivity to variations of radiation temperature is over 75 μV/K. For amplifier noise limited case of 7 nV/Hz$^{1/2}$, the detection of relative radiation temperature variation is at the level of $4*10^{-5}$.

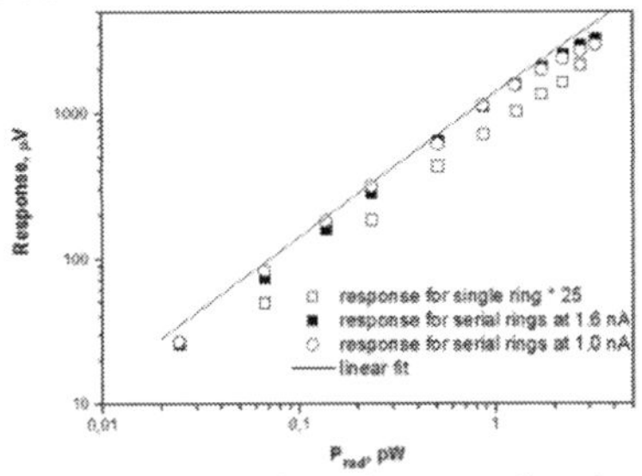

Fig. 3. Response at 0.1 K for series array of 24 rings, single ring x25, and linear fitting

The time constant of such detector (Fig. 6) have been measured by response for the heating pulses of 85 V amplitude and duration of 1 μs. Fitting by exponent of signal increase after the end of heating pulse shows that the overall time constant of receiver is $\tau_{sum} \approx 0.55$ μs. The main impact to this time constant is due to the readout amplifier that is $\tau_{amp} \approx 0.4$ μs measured separately with test signal from optocoupler pair. From these data we can estimate the intrinsic time constant of the bolometer as 0.1-0.2 μs.

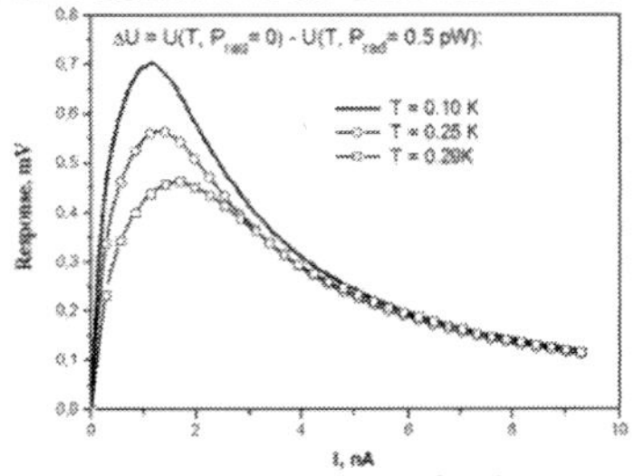

Fig. 4. Response of 25 series array for the same radiation power of 0.5 pW measured at bath temperatures 0.1 K, 0.25 K, and 0.29 K

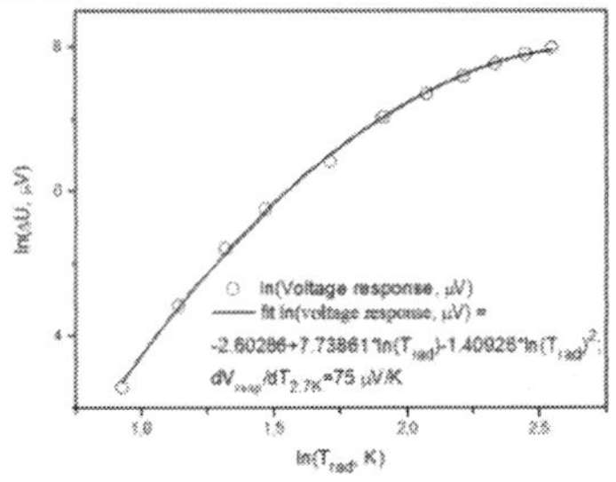

Fig. 5. Logarithmic plot of voltage response for series array at 0.1 K. Radiation temperature responsivity is 75 μV/K at background temperature of 2.7 K

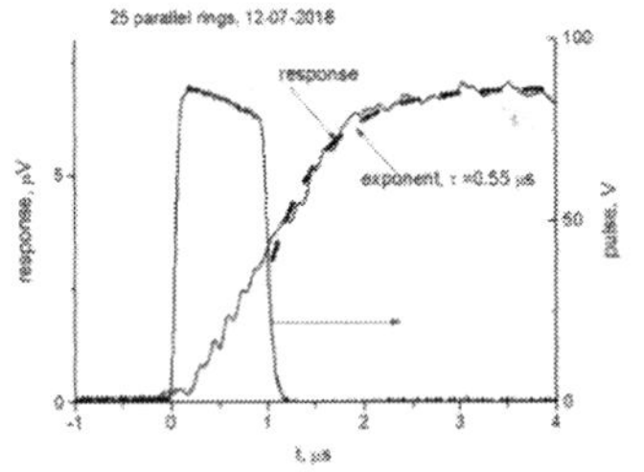

Fig. 6. Probing pulse applied to radiation source (right axis), and output voltage response (left axis) together with fitting by exponent (dashed)

References

1. *S. Mahashabde, A. Sobolev, A. Bengtsson, et al.* A frequency selective surface based focal plane receiver for the OLIMPO balloon-borne telescope //IEEE Trans. on Terahertz Science and Technology. 2015.V.5,N1. P.145.
2. *S. Mahashabde, A. Sobolev, M. Tarasov, et al.* Planar frequency selective bolometric array at 350 GHz //IEEE Trans. on TST. 2015.V.5, No 1. P. 37-43.
3. *V.S. Edelman.* A dilution microcryostat-insert //Instrum. and Exp. Tech. 2009. V. 52, No 2. P. 301-307.
4. *G.M. Rebeiz.* Millimeter-wave and Terahertz integrated circuit antennas // Proc. of the IEEE. 1992. V. 80, No 11, P. 1748-1770

Acknowledgements

The work was supported by the IAP RAS state program (project No. 0035-2014-0021) and by the Basic Research Program of the Presidium of RAS No. 32 Project "PRIZE"

 https://doi.org/10.1051/epjconf/201819505011

Technology for NbN HEB based multipixel matrix of THz range

Ivan Tretyakov[1], N. Kaurova[1], S. Raybchun[1] and G.N. Goltsman[1]

[1]Department of Physics, Moscow State University of Education, 29 Malaya Pirogovskaya St, Moscow, 119435, Russia

***Abstract*—The influence of homogeneity disorder degree of the thin superconducting NbN film across of Si wafer on characteristics of the Hot Electron Bolometers (HEB) has been investigated. Our experiments have been carried out near the superconducting transition and far below it. The high homogeneity disorder degree of the NbN film has been achieved by preparing the Si substrate surface. The fabricated HEBs all have almost identical R (T) characteristics with a dispersion of Tc and the normal resistance R_{300} of not more than 0.15K and 2 Ω, respectively. The quality of the devises allows us to demonstrate clearly the influence of non-equilibrium processes in the S'SS' system on the device performance. Our fabrication technology also allows creating multiplex heterodyne and direct detector matrices based the HEB devices.**

[1] INTRODUCTION

The disordered thin superconducting NbN film has found an application as a sensitive element in bolometric detectors of terahertz and infrared ranges, such as HEB devices [1], single-photon SSPD detectors [2]. Modern instruments of observational astronomy SOFIA [3] and GUSTO [4] require the use of matrix detectors for the terahertz range. Each pixel of the matrix – an HEB - is now considered as a separate detector, requiring individual adjustment of the bias voltage and local oscillator (LO) power. With the matrix size of several dozen pixels, this approach is impractical from the point of view of Allan's time. A solution to the problem can become unified and not requiring individual adjustment HEB pixels. Modern HEBs as heterodyne detectors have practically reached their sensitivity limit [5,6]. However, the task of fabricating a relatively large number of similar HEBs remains unresolved. In this paper, we present the capabilities of our technology of NbN HEB fabricating on Si substrates that allows us to obtain detectors with very close R (T) characteristics for geometrically identical detectors within one batch. The key aspects of our technology are the surface preparing process of the high-resistance Si substrate, as well as the process of cleaning the NbN contact areas before the Au deposition.

[2] RESULTS

For our experiments the NbN film is deposited by the AJA ORION 8 unit for 69 sec. to achieve a thickness of 5 nm. Fig. 1 shows a schematic of the HEB. The detector consists of a superconducting NbN film embedded into Ti/Au terminals of a planar antenna deposited onto a Si substrate. The detector inner part consists of an NbN film between the antenna inner terminals, the so called NbN bridge with a given width W and length L, and a multilayer structure of the NbN/Ti/Au antenna –outer terminals. The multilayer NbN/Ti/Au has a critical temperature Tc2 lower than the Tc1 of the NbN bridge.

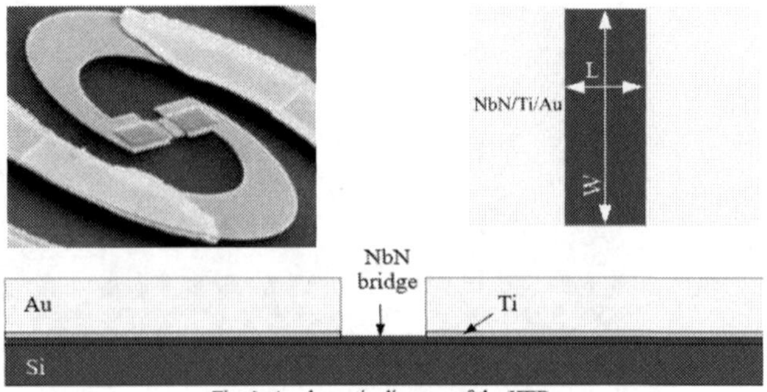

Fig. 1. A schematic diagram of the HEB.

The cleaning process of the Si substrate surface allows us to fabricate devices with a Tc1 deviation not more than 0.1K, for example, the *in-situ* technology presented in [7] gives a Tc1 dispersion of about 0,7-0,8 K. Fig. 2 presents a family of R (T) characteristics of NbN HEBs. The similar shape of the R (T) curves demonstrates HEBs fabricated within *in-situ* NbN/Au technology for HEB contacts [6]. The similarity of the Tc1 for the fabricated NbN HEBs across Si wafer gives the possibility to operate them at one common physical temperature as a single direct detector, or as a heterodyne detector with the same local oscillator power.

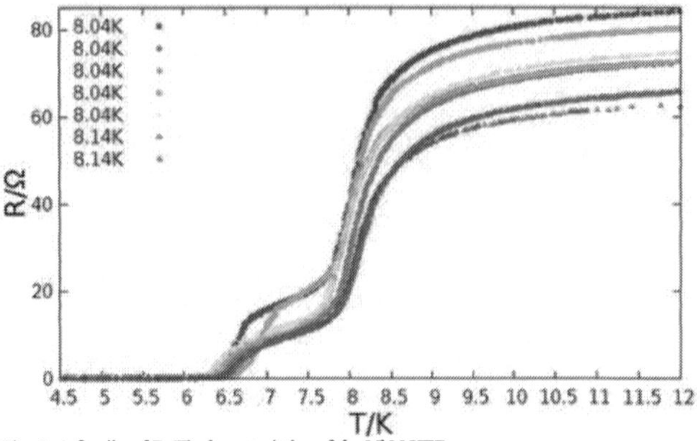

Fig. 2. A family of R (T) characteristics of the NbN HEBs.

The NbN film under Ti/Au antenna ports was cleaned in Ar and O2 plasmas. The idea of cleaning was first proposed and applied for HEBs in [5]. This process made it possible to minimize the contact resistance between NbN and Ti/Au, and also to reduce the dispersion of the normal resistances to not more than 2 Ω for detectors with the same L/W ratio. The quality of the electrical contact between NbN and Ti/Au explains the appearance of second transition on R (T) due to proximity effect. The HEBs resistance at Tc1<T<Tc2 is built into the NbN superconducting bridge and is caused by the conversion of the normal electron current to the current of Cooper pairs over a length of oder ξ [7]. Moreover, the charge conversion process also could influence the coordinate

dependence of the energy gap near Tc1 [8]. It leads to a smearing of the superconducting transition and an increase of ΔTc1.

Fig. 3 shows a family of the IV curves of detector 1498_2 # 9 taken at the bath temperature ranging from 5 K to T>Tc1. Fig. 3 we interprete as the evolution of the HEB IV curves upon transition from the S'SS' state [8], when the NbN bridge and the multilayer NbN/Ti/Au are in the superconducting state to the NSN [8] state where the NbN bridge is superconducting but multilayer NbN/Ti/Au is in the normal state.

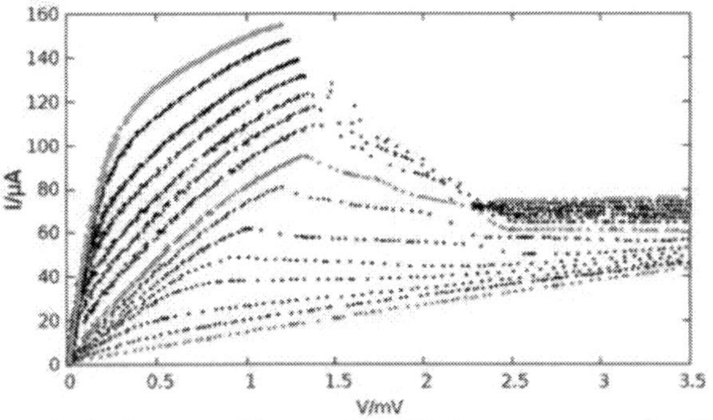

Fig. 3. A family of IV curves of detector 1498_2 # 9 in the temperature range from 5 K to T> Tc1.

An HEB as a direct detector of THz radiation operates at the edge of the superconducting transition, being in the NSN state. In the mixer mode, an HEB operates at T<<Tc1, being in the S'SS' state. The deviation in the measured minimal NEP of our HEB detectors is less than 15% at level of $5*10^{-13}$ W/√Hz. The dispersion of the HEBs optimal bias current and voltage is 3-5 μA and 0.1 mV respectively. Such a small dispersion also indicates the identity of the HEBs. The measured values of the NEP are worse than the calculated value $1\text{-}2*10^{-14}$ W/√Hz, which can be attributed to the influence of the processes at the metal/superconductor interface, which reduce the dR/dT of the detector and give an additional contribution to the detector noise.

Fig. 4 presents the results of our study of HEB detectors as mixers in the S'SS' state. These mixers were fabricated using *in-situ* technology [6]. However, the surface of the Si substrate was not prepared in a special way before NbN deposition. The L/W bridge ratio was kept constant at 0.1-0.12 for all devices. Fig. 4 presents the dependence of the HEB noise temperature Tn measured at 2.5THz on the width of the NbN bridge. As can be seen from Fig. 4, there is a certain optimum volume of the NbN bridge with a minimum dispersion and a minimum Tn of 650K. The dispersion of Tn for devices with the same L and W fabricated from the same NbN film clearly demonstrates the influence of homogeneity disorder degree of the thin superconducting NbN film across of Si wafer on the characteristics of final devices. The optimal W indicates some optimal ratio of the NbN bridge heat capacity C, the resistance coordinate dependences ρ(L) and the distribution of the order parameter Δ(L) presented in Fig. 20a of [9]. For large W the optimal value the volume of the NbN bridge increases also as its C, and accordingly the voltage responsivity Sv falls. The decrease of W of the NbN bridge leads to a proportional decrease of the ρ(L) and the "hump" Δ(L) widths in Fig. 20a of [9].

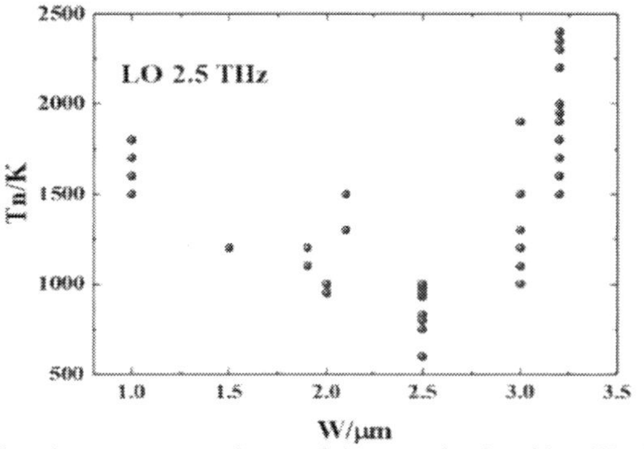

Fig. 4. The noise temperature Tn of an HEB bolometer as function of the width of the its superconducting NbN bridge.

Since, the absolute value of the "hump" Δ(L) width decreases, taking into account the diffusion length Le for a thin film of NbN in the resistive state [6, 10], some electrons absorbing the signal power, can escape into metal contacts not through the Andreev reflection but through tunneling, carrying energy with them - without a contribution to the deviation ρ(L) and the the voltage responsivity of the entire system will fall.

[3] CONCLUSION

We demonstrated a fabrication technology of NbN HEB on Si substrates which allows us to obtain detectors with almost identical R (T) characteristics and explain the optimal dimensions of the bridge for a low Tn. This technology allows the simple creating of the multiplex heterodyne and direct detector matrix based on HEB devices.

ACKNOWLEDGEMENTS

The research has been carried out with the support of the Russian Science Foundation (project No. 17-72-30036

References

[1] E. M. Gershenzon, G. N. Goltsman, I. G. Gogidze, Y. P. Gusev, A. I. Elantiev, B. S. Karasik, and A. D. Semenov, "Millimeter and submillimeter range mixer based on electronic heating of superconducting films in the resistive state," Superconductivity, vol. 3, no. 10, pp. 1582–1597, 1990

[2] A. Semenov, G. N. Gol'tsman, R. Sobolewski, "Hot-Electron Effect in Superconductors and Its Applications for Radiation Sensors", LLE Review, V 87, pp 134-152, 2002.

[3] www.sofia.usra.edu

[4] www.sron.nl/missions-astrophysics/gusto

[5] J. J. A. Baselmans, M. Hajenius, J. R. Gao, T. M. Klapwijk, P. A. J. de Korte, B. Voronov, and G. Gol'tsman, "Doubling of sensitivity and bandwidth in phonon cooled hot electron bolometer mixers," Appl. Phys. Lett., vol. 84, pp. 1958–1960, 2004.

[6] I. Tretyakov, S. Ryabchun, M. Finkel, A. Maslennikova, N. Kaurova, A. Lobastova, B. Voronov, and G. Goltsman, "Low noise and wide bandwidth of NbN hot-electron bolometer mixers," Appl. Phys. Lett., vol. 98, p. 033507, 2011.

[7] M. Shcherbatenko, I. Tretyakov, Y. Lobanov, S. N. Maslennikov, N. Kaurova, M. Finkel, B. Voronov, G. Goltsman, and T. M. Klapwijk, "Nonequilibrium interpretation of DC properties of NbN superconducting hot electron bolometers," Appl. Phys. Lett., vol. 109, p. 132602, 2016.

[8] N. Vercruyssen, T.G.A. Verhagen, M.G. Flokstra, J.P. Pekola, and T.M. Klapwijk, Phys.Rev. B85, 224503, 2012.

[9] T. M. Klapwijk and A. V. Semenov, "Engineering Physics of Superconducting Hot-Electron Bolometer Mixers," IEEE Transactions on Terahertz Science and Technology, vol/ 7, issue 6, pp. 627 - 648, 2017.

[10] A.D. Semenov, G.N. Goltsman, and A.A. Korneev, Physica C 351, 349, 2001.

Response of carbon nanotube film transistor to the THz radiation

V. V. Belosevich[1], I. A. Gayduchenko[2], N. A. Titova[2], E. S. Zhukova[1], G. N. Goltsman[2], G. E. Fedorov[1]

[1]Moscow Institute of Physics and Technology, Moscow, Russia, vsevolod.belosevich@phystech.edu
[2]Moscow State Pedagogical University, Moscow, Russia

Nowadays radiation in teraherz frequency range has many applications in different areas of science and industry. Despite years of research, the THz region of electromagnetic spectrum still has lack of sources and detectors. Almost all created high sensitive detectors require cooling to liquid nitrogen or even helium temperatures.

In 1996 M.Dyakonov and M.Shur have shown theoretically [1] that a short channel High Electron Mobility Transistor demonstrate a resonance response to electromagnetic radiation at the plasma oscillation frequencies of the two-dimensional electrons in the device. The described device implements Field-Effect-Transistor (FET) configuration and consists of transport channel framed by two contacts (source and drain) and a top-gate contact. As a consequence of the high plasma wave velocity and small FET dimensions, the plasma wave frequencies are in the terahertz range. The plasma waves are accompanied by a variation of a dipole moment created by charges in the FET channel and the mirror image charges in the gate.

In that view devices based on carbon nanostructures look promising because of operation with zero bias and their high electron drift mobility. In carbon nanonutubes (CNT) it reaches 100'000cm2/Vs [2].

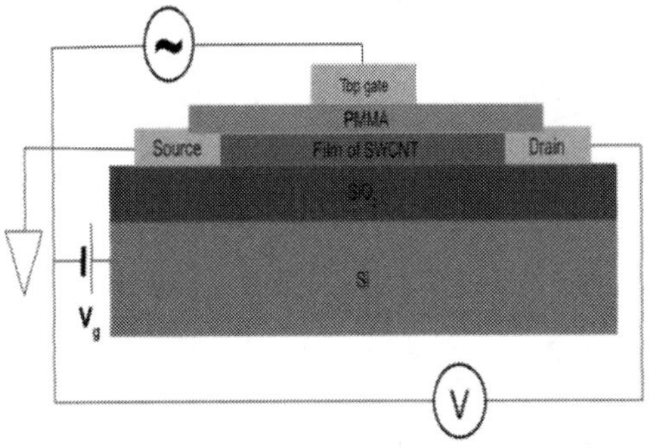

Fig. 1. Scheme of internal structure of the sample

In this work we report on first studies of efficiency of such detectors (Fig.1). Using CVD method we grew our CNT films on oxidized silicon substrate that serves as a back gate of the transistor. After that using photo- and electron- lithography we made Ti/Au contacts and put over a layer of overexposed PMMA resist as dielectric [3]. On the top of the PMMA we plated one more contact as top-gate. Source contact and top gate are part of a broadband log-spiral antenna that couple the radiation (see figure 1). Applying back gate voltage we control concentration and type of charge carriers in transport channel. We observed a strong photoresponse in a broad spectral range from 130 to 900GHz.

In order to charecterize the device we first sweep potential of bottom gate and measure conductance as a function of the back gate voltage applying source-drain voltage of 10meV (see Figure 2). All measurements were carried out at three temperatures (room temperature, temperature of liquid nitrogen, temperature of liquid helium). We further used these dependencies to calculate Seeback coefficient and compare its dependence on Vg to the gate dependence of the device responsivity (see figure 4).

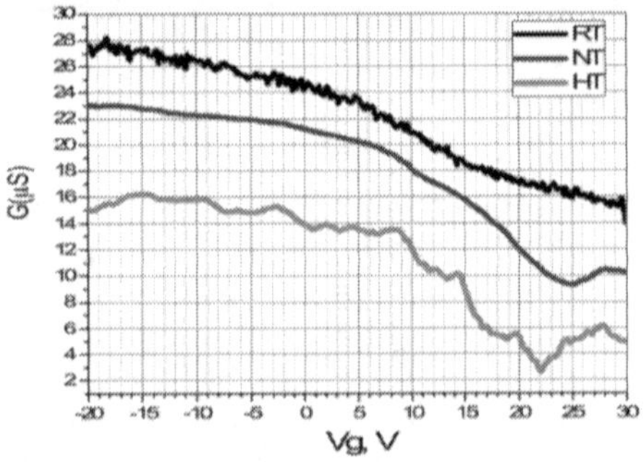

Fig. 2. Transistor characteristic of the tested sample measured at three temperatures (300K, 77K, 8K)

As seen from the figure 2 the conductance decreases with cooling down, This indicates that there are more semiconducting tubes in the film than metallic, which corresponds with the probabilistic estimate obtained from the conditions on the chirality vector components. [4]

At the lowest temperature on figure 2 (8 K) it is possible to observe numerous irregularities. This happens due to different reasons but poor dielectric quality that introduces lots of local potential fluctuations makes this pattern non-reproducible. Charge traps formed by organic molecules of PMMA make noise during the low-temperature experiment.

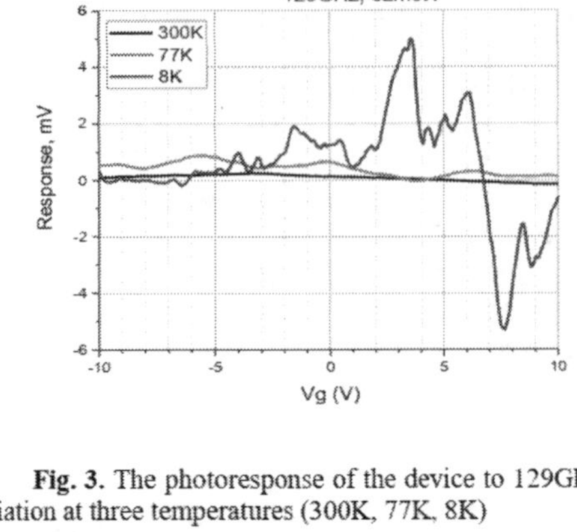

Fig. 3. The photoresponse of the device to 129GHz radiation at three temperatures (300K, 77K, 8K)

 https://doi.org/10.1051/epjconf/201819505012

The experimental curves of photoreponse to different temperatures are shown on figure 3. It can be seen that the response grows up with cooling down the device. On the figure 4 responsivity curves are shown. As can be seen from the figure 4 all curves have a common trend regardless to the frequency of radiation. The main mechanism involved in rectifying the THz radiation was identified.

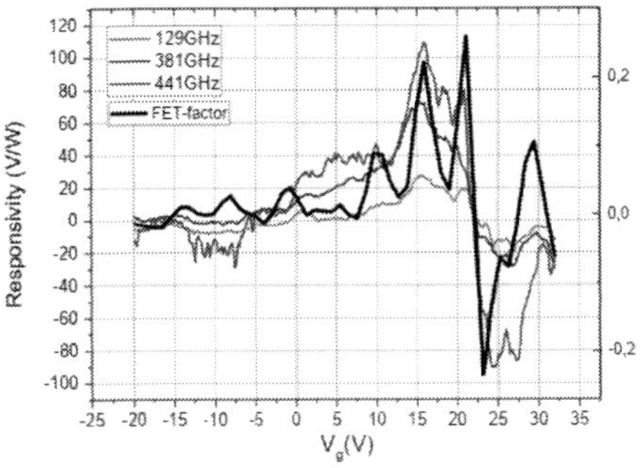

Fig. 4. The responsivity of the device to different frequencies radiation at liquid helium temperature

Thus, we showed that FET device with carbon nanotube film as a transport channel detect THz radiation in all temperature ranges. Previously CVD graphene device in the Dyakonov-Shur configuration was shown to have 1V/W responsivity [5] and graphene encapsulated between two slabs of hexagonal boron nitride 20V/W [6]. In comparison with this CNT device shows responsivity up to 100V/W to radiation of THz frequency range. This fact makes CNT devices promising for industry and interesting for scientific research.

References

1. *Dyakonov, M. I., Shur, M. S.* Detection, mixing, and frequency multiplication of terahertz radiation by two-dimensional electronic fluid // IEEE Transactions on Electron Devices 1996. V. 43. P. 380–387.

2. *Dürkop T.,Getty S. A. ,Cobas Enrique, and Fuhrer M. S.* Extraordinary Mobility in Semiconducting Carbon Nanotubes // *Nano Letters* 2004 *4* (1), 35-39

3. *Sanne A., Movva H. C. P., Kang S., McClellan C., Corbet C. M., Banerjee S. K.* Poly(methyl methacrylate) as a self-assembled gate dielectric for graphene field-effect transistors // Appl. Phys. Lett 2014. 104 (8), 83106.

4. *Charlier J. C. , Blase X., Roche S.* Electronic and transport properties of nanotubes // B: Reviews of Modern Physics 2007. V.79 Iss 2.

5. *Fedorov G.E. , Gaiduchenko I.A., Golikov A.D., Obraztsova E.D., Rybin M.G., Voronov B.M., Coquillat D., Diakonova N., Knap W., Goltsman G.N.* Response of Graphene Based Gated Nanodevices Exposed to THz Radiation // EPJ Web of Conferences 103, 10003 (2015).

6. *Bandurin D.A., Gaiduchenko I.A., Cao Y., Moskotin M., Principi A., Grigorieva I.V., Goltsman G.N., Fedorov G.E., Svintsov D.* Dual origin of room temperature sub-terahertz photoresponse in graphene field effect transistors // Applied Physics Letters 112, 141101 (2018).

Spectral measurements of THz radiation emitted from intrinsic Josephson junction stacks

V.P. Koshelets[1], N.V. Kinev[1], A.B. Ermakov[1], F. Rudau[2], R. Wieland[2], D. Koelle[2], R. Kleiner[2], and H.B. Wang[3]

[1] Kotel'nikov Institute of Radio Engineering and Electronics RAS, Moscow 125009, Russia; valery@hitech.cplire.ru
[2] Physikalisches Institut and Center for Collective Quantum Phenomena in LISA+, Universität Tübingen, D-72076 Tübingen, Germany;
[3] Research Institute of Superconductor Electronics, Nanjing University, Nanjing 210093, China.

A superconducting integrated receiver (SIR) comprises all of the elements needed for heterodyne detection on a single chip. Recently, the SIR was successfully implemented for the first spectral measurements of terahertz (THz) radiation emitted from intrinsic Josephson junction stacks (BSCCO mesa) at frequencies up to 750 GHz; a linewidth below 10 MHz has been recorded in the high bias regime. In this report the results of the spectral measurements of THz radiation emitted from intrinsic Josephson junction stacks are summarized; recent results of spectrometric gas detection using THz radiation from a BSCCO mesa are presented.

In recent years, coherent THz emission has been obtained from stacks of intrinsic Josephson junctions (IJJs), created naturally in the BSCCO unit cell with the CuO layers forming the superconducting electrodes and the BiO and SrO layers forming the barrier layer [1, 2]; a 1- μm-thick crystal consists of about 670 IJJs. Terahertz emission from BSCCO mesa has been obtained both at a low bias (where the temperature distribution in the stack is almost homogeneous) and a high bias regime (where an over-heated part and a cold part of the sample coexist) [3, 4].

Coherent emission above 1 THz by intrinsic Josephson BSCCO junction stacks with improved cooling has been demonstrated [5, 6]. Due to the variable size of the hot spot and the temperature rise caused by the self-heating, the emission frequency can be tuned over a wide range of up to 700 GHz [5]. So far, emitted by one device power up to 30 μW was obtained [7, 8]. These are very encouraging results, although for most practical application spectral properties of the novel oscillators are vitally important.

The spectral characteristics of the oscillator were studied using the high-sensitivity super-heterodyne SIR, which was developed at Kotel'nikov IREE [9 - 12]. Such a receiver is intended to perform spectral studies of the electromagnetic radiation in the frequency range 450–700 GHz and successfully used for measuring the profiles of the spectral lines of the gas-molecule radiation and absorption and for the spectral study of any external terahertz oscillator radiating in the operation frequency range of the receiver. The best noise temperature of the SIR is 120 K and its spectral resolution is better than 0.1 MHz, which exceeds the resolution of modern terahertz-range Fourier spectrometers by several orders of magnitude.

Two configurations for the oscillator and receiver location were used: in the first case the oscillator was located in a cryostat of the SIR in the vicinity of the mixing unit [11]; in the second case the oscillator and the receiver were located in independent cryostats with Mylar quasioptical windows. The SIR operates at temperature of about 4.5 K, whereas the optimal BSCCO-oscillator temperature is 20–50 K. The spectral lines of the oscillator radiation are recorded by the SIR and displayed on the spectrum-analyzer screen in the intermediate-frequency range 4–8 GHz. The spectrum analyzer allows one to average the signal, read it by a computer, and perform other necessary digital operations for the spectrum analysis and processing.

Application of the SIR has allowed to measure radiation emitted from intrinsic Josephson junction stacks in both regimes with spectral resolution better than 1 MHz [4].While at low bias we found that linewidth is not smaller than 500 MHz, at high bias, emission linewidth turned out to be in the range 10–100 MHz (see Fig. 1a). We attribute this to the hot spot acting as a synchronizing element; a linewidth as narrow as 7 MHz has been recorded at high bias [12], see Fig. 1b.

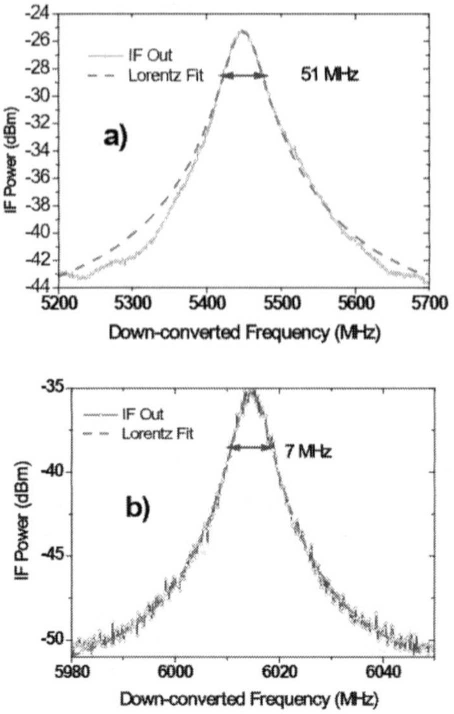

Fig. 1. Down-converted spectrum of the BSCCO at 527 (a) 476 GHz (b) in the high bias regime; dashed line is a Lorenzian fit with full linewidth 51 and 7 MHz respectively.

Typical dependencies of the linewidth on the BSCCO frequency both in the low-bias and high-bias regimes that were measured by the SIR are presented in the Fig. 2. Important to note that the tuning of the BSCCO oscillator frequency is continuous over the range; that was confirmed by fine tuning of the SIR LO frequency. Actually for the presented data the lowest measured frequency of about 550 GHz was limited by the BSCCO mesa, while losses in the Nb interconnection lines of the SIR restrict the measurements at frequencies higher than 730 GHz.

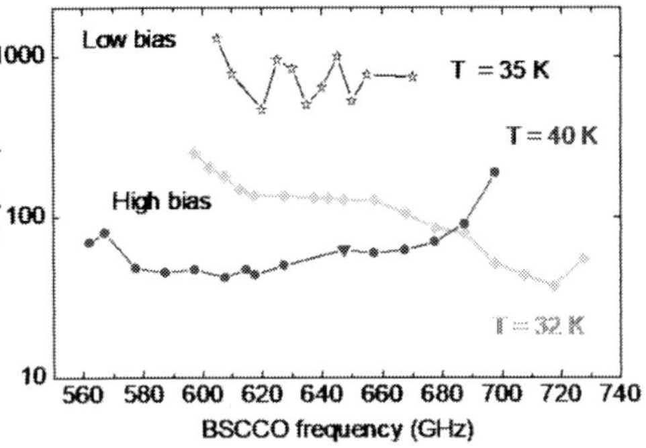

Fig. 2. Typical dependence of the BSCCO linewidth on the frequency (measured for the different sample temperatures — solid (open) symbols are for high (low) bias regime.

A combination of the BSCCO mesa and the SIR was used to accurately measure the terahertz absorption spectra of ammonia and water vapor [13]. In this experiment, the bias current through the BSCCO emitter is kept at a constant value, tuned to the respective gas-line frequency, and intermediate-frequency (IF) spectra are taken using the SIR.

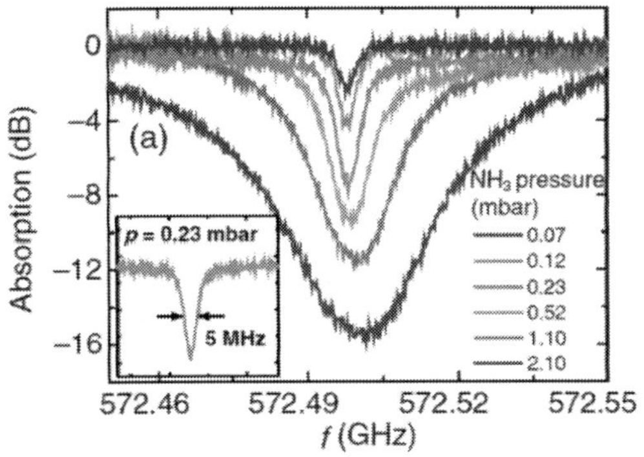

Fig. 3. Terahertz absorption spectra of ammonia (mixed with water; 10% solution) at different pressures by a BSCCO emitter and a SIR terahertz detector. The emitter and the SIR are operated at T_b = 4.2 K. The empty cell curve is subtracted from the data. Spectrum of ammonia vapor at p = 0.23 mbar, with an absorption linewidth of 5 MHz is presented as an inset.

Measurements are done at T_b = 4.2 K, where the linewidth of radiation of the emitter is 60 MHz at 572.5 GHz (NH_3). With this setup, high resolution THz absorption spectra of ammonia achieved in a narrow frequency band lower than the linewidth of the BSCCO emitter; see Fig. 3 [13]. The emitter, operated in a free-running mode, is stable enough to permit this kind of measurement. The absorption lines of ammonia are highly broadened at large gas pressures, with the linewidth decreasing with decreasing pressure. For ammonia, at p = 0.23 mbar, which is the order of the concentration of trace gases, a clear absorption dip with an absorption linewidth of about 5 MHz is observed, as shown in the inset of Fig. 3. At p = 0.07 mbar, the measured linewidth is 4 MHz (although affected by Doppler broadening). The observed rotation frequencies of ammonia (572.498 GHz) coincide with literature values. Further improvement of the lowest resolvable linewidth is possible by using proper feedback techniques for the oscillator [10].

The work was supported by the Russian Foundation for Basic Research, grant No. 17-52-12051, and by the Deutsche Forschungsgemeinschaft, Project KL-930/13-2.

References

1. *Kleiner, R., Muller, P., Kunkell, G., Steinmeyer, F.* Intrinsic Josephson effects in Bi Sr CaCu O single crystals // Phys. Rev.Lett. 1992. V. 68, P. 2394–2397.

2. *Ozyuzer, L., et al.* Emission of coherent THz radiation from superconductors,"Science, 2007. V. 318, No. 5854. P. 1291–1293.

3. *Wang, H. B., et al.* Coherent terahertz emission of intrinsic Josephson junction stacks in the hot spot regime // Phys. Rev. Lett. 2010. V. 105, No. 5. Art. ID 057002.

4. *Li, M. Y., et al.* Linewidth dependence of coherent terahertz emission from BiSrCaCuO intrinsic Josephson junction stacks in the hotspot regime // 2012. Phys. Rev. B. V. 86. Art. ID 060505.

5. *Ji, M., et al.* BiSrCaCu O intrinsic Josephson junction stacks with improved cooling: Coherent emission above 1 THz // Appl. Phys. Lett. 2014. V. 105 Art. ID 122602.

6. *Kashiwagi, T., et al.* A high-Tc intrinsic Josephson junction emitter tunable from 0.5 to 2.4 terahertz // Appl. Phys. Lett. 2015. V. 107 P. 082601.

7. *An, D. Y., et al.* Terahertz emission and detection both based on high-Tc superconductors: Towards an integrated receiver // Appl. Phys. Lett. 2013. V. 102. Art. ID 092601.

8. *Sekimoto, S., et al.* Continuous 30 W terahertz source by a high-Tc superconductor mesa structure // Appl. Phys. Lett. 2013. V. 103. Art. ID 282601.

9. *Koshelets, V. P., and Shitov, S. V.* Integrated superconducting receivers // Supercond. Sci. Technol. 2000. V. 13. No. 5. P. R53–R69.

10. *Koshelets, V. P., et al.* Superconducting integrated THz receivers: Development and applications // Proc. SPIE. 2010. P. 78540J-1–78540J-13.

11. *Kinev, N. V., Filippenko, L. V., Li, M. Y., Yuan, J., Wang, H. B., and Koshelets, V. P.* Spectral properties of a terahertz oscillator based on the BiSrCCuO mesastructure // Radiophysics and Quantum Electronics, 2014, V. 56, No. 8–9, P. 582-590.

12. *Koshelets, V. P., et al.* Superconducting integrated terahertz spectrometers // IEEE Trans. Terahertz Sci. Technol. 2015. V. 5. P. 687.

13. *Sun, H., et al.* Terahertz Spectroscopy of Dilute Gases Using BiSrCaCuO Intrinsic Josephson Junction Stacks // Phys. Rev. Applied. 2017. V. 8, P. 054005.

SubTHz Arrays of Planar Antennas with SINIS bolometers for BTA

G. Yakopov[4], M. Tarasov[1], A. Gunbina[2,5], M.Mansfeld[2,5], R. Yusupov[1], V. Edelman[6], and V. Vdovin[2,3,5]

[1]V.Kotelnikov Institute of Radio Engineering an d Electronics RAS
[2]Institute of Applied Physics RAS
[3]P.N. Lebedev Physical Institute of the Russian Academy of Sciences
[4]Special Astrophysical Observatory RAS
[5]Nizhny Novgorod State Technical University n.a. R.E. Alekseev
[6]P.Kapitza Institute for Physical Problems RAS

We have investigated the optical and spectral response of a 350 GHz receiver made of annular antenna array with SINIS bolometer elements. Receiver is intended for using at Big Alt-Azimuthal Telescope (BTA). Samples were measured in a dilution cryostat at temperatures in the range 80-400 mK. Voltage responsivity approaches 3*10^9 V/W. Spectral bandwidth in 230-380 GHz range for single array was measured using Backward Wave Oscillator source.

BTA SAO RAS

The BTA-6 is a 6-metre aperture optical telescope at the Special Astrophysical Observatory located in the Zelenchuksky District on the north side of the Caucasus Mountains at an altitude of 2070 m above sea level. Photo and schematic picture of BTA is presented in Fig.1. The mounting of a cryostat with receiving array is supposed in the Nasmyth focus. The equivalent focus length is 184 meters and diameter of diffraction image is 60 mm. For matching receiving array with incoming radiation we can use horn with large aperture (up to 60 mm) or lens with the same diameter or decreasing of focal length of 24 times by using a large lens.

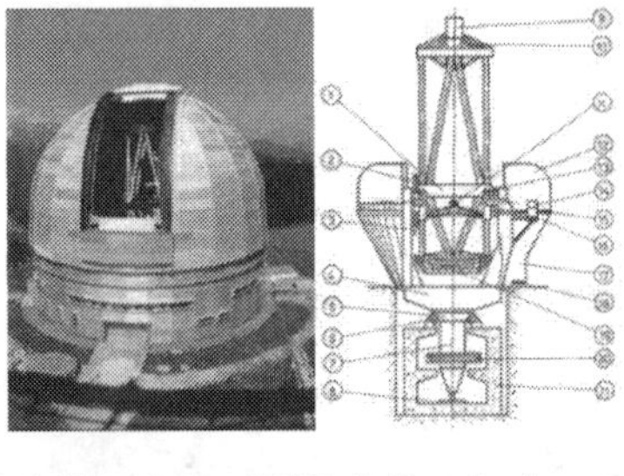

Fig.1. The photo of BTA (left) and schematic picture (right): 1- middle unit; 2 - worm gear; 3 - elevator; 4 - rotation support platform; 5 - spherical support of vertical axis; 6 - spherical pads; 7 - vertical axis; 8 - lower bearing; 9 - observer's cage; 10 - prime focus unit; 11 - flat mirror; 12 - platform cover; 13 - oil pads; 14 - main spectrograph; 15 - observing platform; 16 - support for spectrograph; 17 - pier; 18 - 2m camera mirror of main spectrograph; 19 - primary mirror cell; 20 - spur and worm gears; 21 - reinforced concrete.

Matching of incoming signal with receiver array

Receiving array can be matched with incoming signal using arrays of horns, quasi-optical lenses or distributed absorbers. The main advantages and disadvantages of using feedhorns and filled arrays are represented in [1] and in table.

Feedhorns	Filled array
Advantages	**Advantages**
a) Provides maximum efficiency for detection of a point source with known position; b) the bolometer angular response is restricted to the telescope, giving good stray-light rejection; c) the susceptibility to electromagnetic interference can be controlled — the horn plus integrating cavity act as a Faraday enclosure; d) the number of detectors needed to fill a given array field of view is minimized	a) Provide a higher efficiency for mapping observations; b) they allow full sampling of the instantaneous field of view of the array by use of pixels of 0.5Fλ or smaller, making jiggling unnecessary; c) they yield, for a 0.5Fλ array, a slightly narrower beam profile on the sky for a given telescope size owing to the stronger illumination of the outer parts of the telescope
Disadvantages	**Disadvantages**
a) In order to achieve full spatial sampling of the sky, even for a region smaller than the array field of view, jiggling or scanning are needed, which complicates the observing modes; b) the efficiency for mapping is considerably less than the ideal value	a) the background power per pixel is lower than for the larger feedhorn coupled detectors, typically by a factor of 4 –5, yielding a photon noise NEP that is lower by a factor of 2 or more, and thus more difficult to achieve; b) the detectors are much more vulnerable to stray light because of the very broad pixel angular response — by a factor of $\pi F^2/4$, assuming a pixel beam solid angle of π steradians; c) the vulnerability to electromagnetic interference is also greater owing to the naked array architecture; d) the need for more detectors to fill a given field size

In case of using immersion lens close to elliptical lens [2] allows to remove substrate modes and to increase gain coefficient of planar antenna. The spherical wave transforms in plane wave by mounting the source in the second focus.

We suggest two types of arrays for using at BTA: half-wave antenna arrays (with back-to-back horn) and metamaterial arrays (with lens). A single element is an annular antenna with two (or more) SINIS-bolometers (Fig.2). Schematic image of matching such arrays with incoming radiation are presented in Fig.3.

A metamaterial is made of a periodic array of subwavelength metallic resonators that are collectively coupled to the free space excitation. In the case of small antennas the matrix can be made more wideband and much smaller that allows placing it in the waist of a single-mode horn or in the focus of immersion lens.

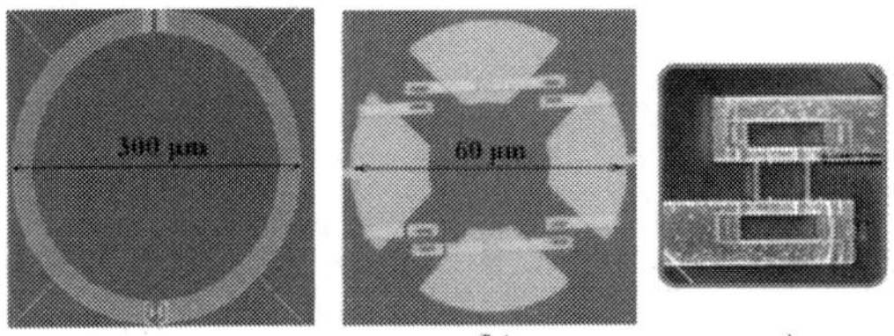

a) b) c)

Fig.2. The single elements of investigated structures for 345 GHz receiver arrays: a) Half-wave antenna, b) Metamaterial, c) SINIS-bolometer

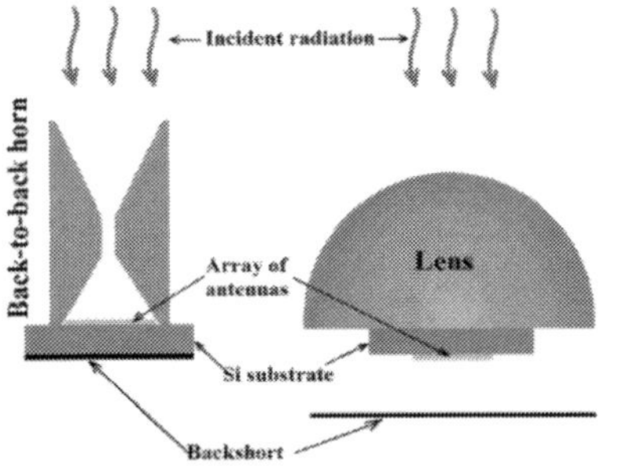

Fig.3. Schematic image of measurements such arrays: half-wave antenna array with back-to-back horn (left) and metamaterial array with lens (right)

Experimental setup and results

Quasistatic optical response to incoming radiation was measured in dilution cryostat [3] using cold black body (BB) source made of NiCr film on sapphire substrate. Source was equipped with thermometer, connected to current source and mounted on 2.7 K stage. For spectral response measurements we use a 230-380 GHz Backward Wave Oscillator (BWO) illuminating the antennas array via optical window and 3 neutral density filters with transmission below -10 dB placed at radiation shields 100 K, 3 K, 0.3 K temperature stages. Two channels were measured simultaneously by lock-in amplifier, one for signal from bolometer, and another from pyroelectric detector that is monitoring level of the incoming power. Schematic pictures of experimental setup are presented in Fig.4. Results of spectral and optical response measurements are presented in Fig.5,6.

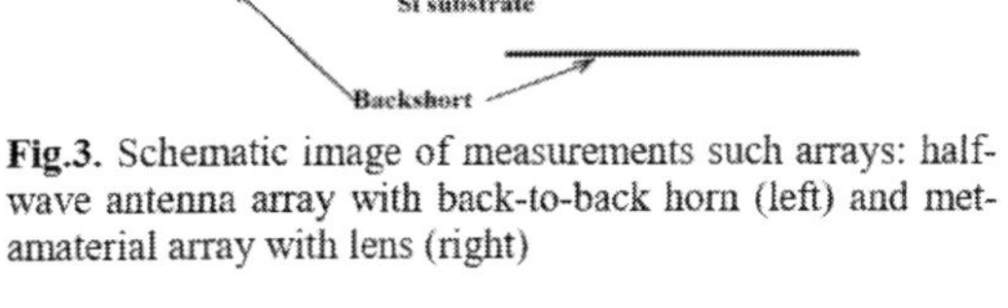

Fig.4. Experimental setup with cold BB (left) and with BWO source (right)

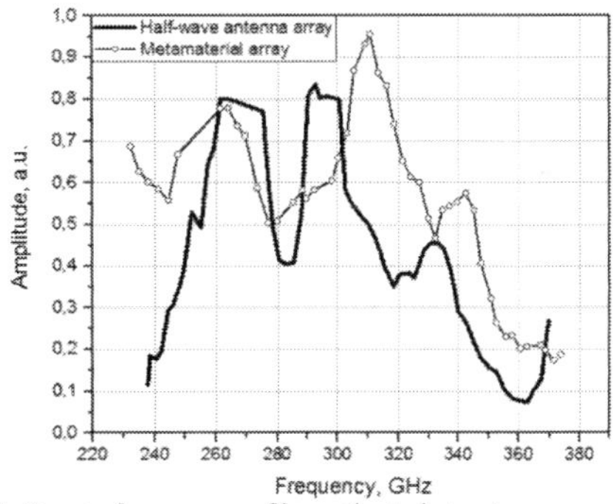

Fig.5. Spectral response of investigated structures

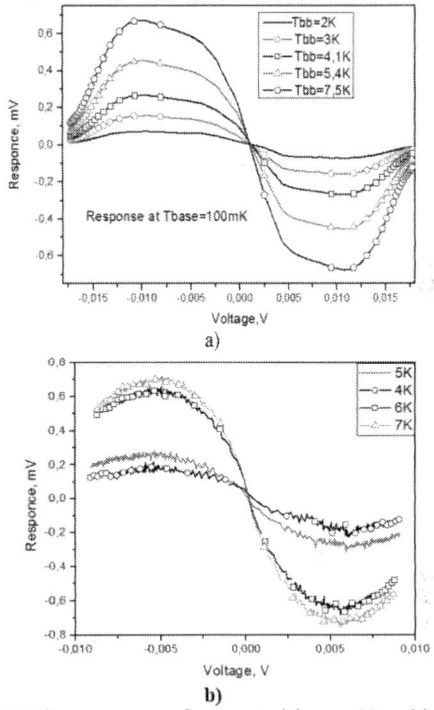

Fig.6. Voltage response of metamaterial array (a) and half-wave antenna array (b) to blackbody radiation

References

1. *M.Griffin, J.Bock, W.Gear*, Relative performance of filled and feedhorn-coupled focal-plane architecture, Applied Optics, v/ 41, No31, 6543-6554 (2002)

2. *D.F.Filipovic*, Double-slot antennas on extended hemispherical and elliptical silicon dielectric lenses, IEEE Trans. Microwave Theor. and Techn., vol. 41, No10, pp.1738-1749 (1993).

3. *V. S. Edelman*, A dilution microcryostat-insert, Instrum. Exp. Tech. **52**, N2, 301-307 (2009), DOI: 10.1134/S002044120902033X.

Acknowledgements

The work was supported by the IAP RAS state program (project No. 0035-2014-0021) and by the Basic Research Program of the Presidium of the Russian Academy of Sciences No. 32 "Nanostructures: Physics, Chemistry, Biology, Technology Basics". Project "PRIZE"

Modelisation of a gas phase polarization induced by a 200 GHz chirped pulse

R. Bocquet[1], D. Fontanari[1], C. Bray[1], G. Mouret[1], A. Cuisset[1], G. Dhont[1], K. Hickson[2] and F. Hindle[1]

[1]LPCA, Université du Littoral Côte d'Opale, 59140 Dunkerque, France, robin.bocquet@univ-littoral.fr

[2]Institut des Sciences Moléculaires, Université de Bordeaux, Talence, F-33400, France

Introduction

A millimeter wave chirped pulse spectrometer (mmWCP) has been developed to study chemical reactivity at low temperature using the CRESU technique **[1]**. The project goal is to detect and quantify gas species created during reactivity at thermodynamic conditions close to interstellar media, requiring the combination of the 2 techniques cited above. The first part of the project, undertaken at the LPCA, is the realization and the development of the mmWCP operating between 190 - 210 GHz **[2]**. One of the interests of this instrument is its capability to simultaneously and rapidly measure numerous molecular transitions of different species in gas phase by probing individual rotational transitions. However, during the trials of the experiment, we were confronted with the difficulty of determining the rotational line intensities according to their position within the chirped pulse. For this reason, we have undertaken a theoretical study to get obtain an analytical expression for the detected signal as a function of the line position in the pulse.

Experimental set up

A mmWCP instrument operates in 2 phases. The first is the polarization of the gas under study by the CP source. The second phase is the recording the Free Induction Decay (FID) signal after extinction of the CP source.

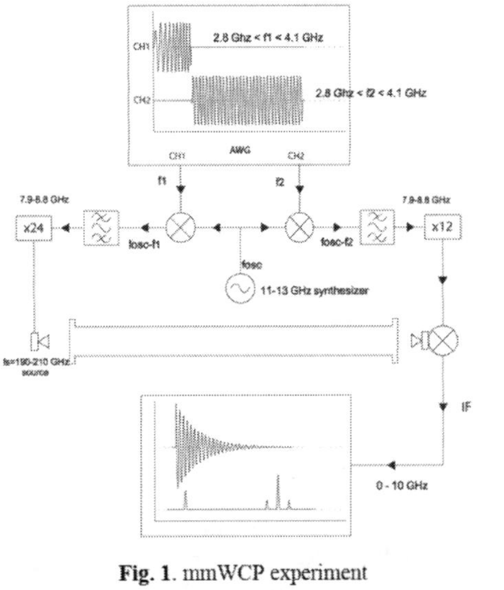

Fig. 1. mmWCP experiment

The mmWCP is generated by mixing the AWG CH1 with a fixed oscillator before being frequency-multiplied to the 190 to 210 GHz range. It is launched into free space using a diagonal horn antenna and propagated through the sample chamber. A second diagonal horn antenna is used to couple the molecular FID signal to the sub-harmonic mixer. The other ports of the mixer are used to provide the local oscillator (LO) signal and extract the intermediate frequency (IF). The IF signal is recorded using a fast oscilloscope while being triggered by a timing signal generated by the AWG. A large number of measurement cycles are averaged before applying a FFT to obtain the spectrum. The experiment can detect line intensities of the order of 10^{-26} cm^{-1}/molec.cm^{-2} in few seconds and has the possibility to scan over 10 GHz in one shot **[2]**.

Theory

Polarization phase - The measured molecular signal is proportional to the gas polarization at the end of the chirped pulse T_c. At this stage we used the optical Bloch equations **[3, 4]** to describe the evolution of an isolated molecular system for which 2 energy levels |a> and |b> are probed. We will suppose that the system is submitted to an electric field **E** with a linear time dependent pulsation:

The optical Bloch equations are then given by :

$$\begin{aligned} \dot{z}(t) &= -(\tau_2 + i\omega_0) z(t) - i\varepsilon(t) W(t) \\ \dot{W}(t) &= -\tau_1 (W(t) - W_{eq}) + \frac{1}{2i}(\varepsilon^*(t) z(t) - \varepsilon(t) z^*(t)) \end{aligned} \quad (1)$$

$$\varepsilon(t) = \Omega_0 e^{-i[(\omega_0 - c\Delta\omega)t + \frac{\Delta\omega}{2T_c}t^2]} \quad (2)$$

$$\Omega_0 = \frac{\mu_{ab}.E_0}{\hbar} \text{ and } \omega(t) = \omega_0 - c\Delta\omega + \frac{\Delta\omega}{T_c}t \quad (3)$$

where z(t) is the polarization, W(t) the difference population of the 2 levels, W_{eq} the difference at the thermodynamic equilibrium, τ_1 and τ_2 are the inverse of relaxation times of levels and dipoles respectively, Ω_0 is the Rabi frequency, T_c is the duration of the chirp, c is a unitless parameter describing the molecular resonance position in the chirped pulse ($0<c<1$ and $\omega(cT_c) = \omega_0$) and $\Delta\omega$ is the pulsation extension of the chirp.

In order to find a solution, at least in a first approximation, we solve in the equations (1) the one containing $\dot{z}(t)$ while considering W(t) given, and the other one while considering z(t) given. Two uncoupled first-order differential equations are therefore obtained providing a general formulation of the solution. Then, considering the initial conditions (z(0)=0 and W(0)=W_{eq}), we have

$$\begin{aligned} z(t) &= -i\int_0^t e^{-(i\omega_0 + \tau_2)(t-x)} \varepsilon(x) W(x)\, dx \\ W(t) &= W_{eq} + \frac{1}{2i}\int_0^t e^{-\tau_1(t-x)} (\varepsilon^*(x) z(x) - \varepsilon(x) z^*(x))\, dx \end{aligned} \quad (4)$$

We can write the 2 functions as series:

$$z(t) = z^{(0)}(t) + z^{(1)}(t) + z^{(2)}(t) + \ldots$$
$$W(t) = W^{(0)}(t) + W^{(1)}(t) + W^{(2)}(t) + \ldots$$

and find the solutions by successive approximations:

$$\begin{aligned} z^{(n+1)}(t) &= -i\int_0^t e^{-(i\omega_0+\tau_2)(t-x)}\varepsilon(x)W^{(n)}(x)dx \\ W^{(n+1)}(t) &= \frac{1}{2i}\int_0^t e^{-\tau_1(t-x)}(\varepsilon^*(x)z^{(n)}(x)-\varepsilon(x)z^{(n)*}(x))dx \end{aligned} \quad (5)$$

Taking the initial conditions ($z(0)=0$ and $W(0)=W_{eq}$), only the odd terms of the z-series and the even terms of the W-series remain. We will work with $z^{(1)}(t)$ and $W^{(0)}(t)=W_{eq}$ and do a changing variable to simplify the equations (rotating frame):

$$\xi^{(1)}(t)=z^{(1)}(t)e^{i[(\omega_0-c\Delta\omega)t+\frac{\Delta\omega}{2T_c}t^2]} \quad (6)$$

Then we can show that for $t=T_c$:

$$\xi^{(1)}(T_c)\approx\rho_{app}e^{i\theta_{app}}+\tilde{\xi} \quad (7)$$

with

$$\begin{aligned} \rho_{app} &= \frac{\sqrt{2\pi}\,\Omega_0 W_{eq}}{\sqrt{\Delta\omega}}\sqrt{T_c}\exp[\frac{-T_c}{T_2}(1-c)] \\ \theta_{app} &= (1-c)^2\frac{\Delta\omega}{2}T_c-\frac{3}{4}\pi \\ \tilde{\xi} &= \frac{W_{eq}\Omega_0}{\Delta\omega}(\frac{1}{1-c}+\frac{1}{c}\exp[i\frac{\Delta\omega}{2}T_c(1-2c)])\exp[\frac{-T_c}{T_2}(1-c)] \end{aligned} \quad (8)$$

ρ_{app} (which is positive) represents the amplitude of $\zeta^{(1}(T_c)$, θ_{app} its phase and $\tilde{\xi}$ induces a small oscillation. In particular ρ_{app}, tends to 0 when T_c tends to 0, increases until it reaches a maximum when $T_c=\frac{T_2}{2(1-c)}$ with value $\frac{\sqrt{2\pi}}{e}\frac{\sqrt{T_c}}{\Delta\omega}W_{eq}\Omega_0$ and then it decreases toward 0 as $T_c\to+\infty$. The approximation which retains ρ_{app} is already very good to describe the system behavior.

Free Induction Decay (FID) – In this phase, the Rabi frequency vanishes and the optical Bloch equations simplify **[3]**. Using an heterodyne detection scheme, we can demonstrate that the output signal is :

$$S(t')=B\sqrt{\frac{T_c}{\Delta\omega}}e^{\frac{-T_c}{T_2}(1-c)}[e^{\frac{-t'}{T_2}-\frac{t'^2\Delta\omega_D^2}{4}}\cos(\omega_{IF}t'+\Phi)] \quad (9)$$

where $t'=t-T_c$ and B is a constant depending of the experiment (amplifiers, mixers ...). Applying a FFT, we can access the signal amplitude at the intermediate frequency:

$$A_{IF}(\nu_{IF})=\frac{B}{\sqrt{\Delta\omega}}\sqrt{T_c}\exp[\frac{-T_c}{T_2}(1-c)] \quad (10)$$

Experimental results

We compared the theoretical model we obtained with experimental data recorded with carbonyl sulfide gas (OCS). We chose the 17->16 rotational line centered at 206.745 GHz. The output signal amplitude at the IF have been measured for different gas pressure and different positions of the molecular resonance in the CP described by the c parameter. The figure 2 compares experimental measurements and analytical modelisations for 3 different c values. A very good agreement is obtained between experimental data (blue points) and model (red curve) for a 1 GHz extension chirp and for a 100 µbar gas pressure (collisional width larger than Doppler width).

Conclusion

We used a mathematical method by successive approximations to solve the optical Bloch equations and to get an analytical formula giving the output signal of a chirped pulse experiment. This will be useful for absolute intensity lines measurements and to optimize the pulse duration to obtain the best signal to noise ratio.

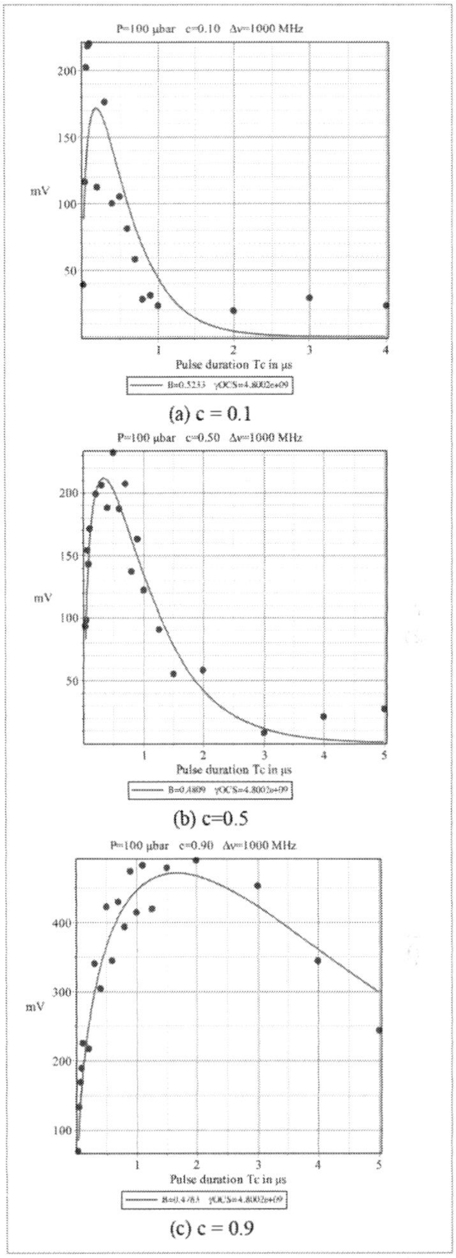

Fig. 2. Output signal amplitude $A_{IF}(\nu_{IF})$, function of pulse duration T_c, recorded for OCS at 206.745 GHz with line resonance position c as parameter. Comparison of experimental data (blue points) and model (red curve) from equation (10)

References

1. *K. M. Hickson, P. Caubet and J.-C. Loison, J. Phys. Chem. Lett, 4, 2843, 2013.*
2. *F. Hindle, C. Bray, K. Hickson, D. Fontanari, M. Mouelhi, A. Cuisset, G. Mouret, and R. Bocquet, Journal of Infrared, Millimeter, and Terahertz Waves, vol. 39, no. 1, pp. 105–119, 2018.*
3. *P. Meystre and M. Sargent III, Elements of quantum optics 3rd edition. Springer, 1999.*
4. *J. C. McGurk, T. G. Schmalz, and W. H. Flygare, J. Chem. Phys, vol. 60, no. 11, pp. 4181–4188, 1974.*

Terahertz induced optical second harmonic generation from dielectric interfaces: mechanism and application

S.B. Bodrov[1,2] Yu.A. Sergeev[2], A.I. Korytin[2], M.Y. Emelin[2], M.Y. Ryabikin[2] and A.N. Stepanov[2]

[1]University of Nizhny Novgorod, Nizhny Novgorod 603950 Russia, bosbor@ufp.appl.sci-nnov.ru
[2]Institute of Applied Physics of the Russian Academy of Sciences, Nizhny Novgorod 603950 Russia

Optical second harmonic (SH) generation from interfaces of different materials is widely used for surface studies. The surface breaks the symmetry allowing SH generation even in isotropic media. The quality of the surface as well as presence of an additional layers on the surface significantly influence on SH generation. Also, it was shown that application of constant external electric field changes the magnitude of the generated SH signal [1]. In Ref. [2] it was proposed to use a weak THz electric field to modulate SH generation from surface of nontransparent optical material (silicon) as a method to measure THz field waveform. Here we propose to use THz induced SH generated by femtosecond laser radiation to detect optically hidden layers (glue) placed not only on the surface but also into a bulk of transparent materials (quartz). We showed that existence of terahertz radiation influences on SH generation only from the interfaces (glue-quartz, air-quartz) and not from the bulk of the quartz. To explain this effect a model based on interaction of electromagnetic radiation with an electron in coulomb potential biased by embedded surface electric field was proposed and analyzed.

A Ti:Sapphire laser system (1 mJ, 1 kHz, 70 fs) was used in our experiment. The optical beam was divided into two parts (see Fig. 1). The first pulse generated THz radiation in a LiNbO3 crystal by the tilted-front technique [3]. The THz pulse (the waveform is shown in Fig. 2(a)) was focused on the sample collinear with a probe (weak) femtosecond optical pulse. The sample was fabricated by cementing of two 0.9 mm fused quartz plates by glue with thickness of about several μm. The generated optical SH was detected by PMT in reflection and transmittance setups. In the first setup the sample was placed at 45° to incident probe pulse and PMT detected reflected optical SH. In the second setup the probe optical pulse propagated normally to the incident surface of the sample and PMT detected transmitted optical SH.

Figure 2(b) shows the SH signal as a function of delay time between optical and THz pulses in the transmittance setup (the similar signal but two orders less magnitude was measured in the reflection setup). Due to different velocities of THz and optical pulses, their interaction occurs at different points inside the sample. The measured SH signal at $t = 0$ corresponds to the signal from the entrance surface of the sample, at $t = 1.4$ ps – from the interface glue-quartz, and at $t = 2.7$ ps – from the output boundary of the sample. The dependence obtained in Fig. 2 (b) demonstrates the possibility of determining the presence of internal boundaries in the material that are not visible in the optical range. With the known values of the refractive indexes for optical and THz radiation, one can find a depth of internal inhomogeneity.

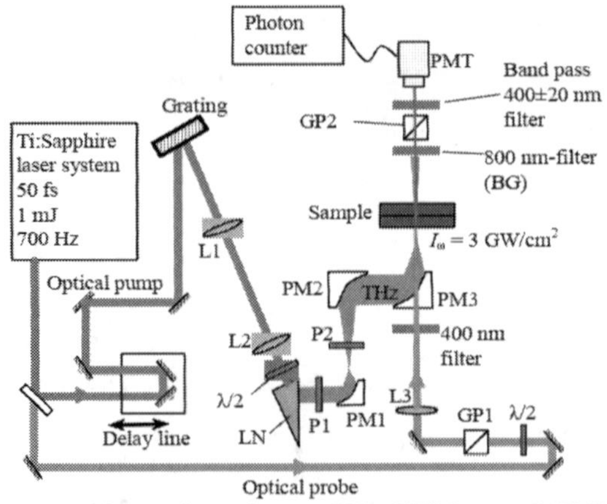

Fig. 1. (a) Experimental setup. LN – $LiNbO_3$ crystal; PM1, PM2 and PM3 – parabolic mirrors; GP1,GP2 – Glan prisms; PMT – photo multiplier tube; P1, P2 – THz polarizers.

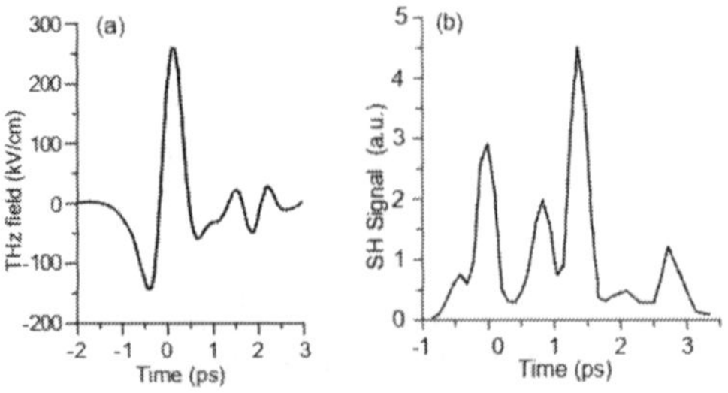

Fig. 2. (a) THz waveform. (b) THz induced optical SH generation from cemented quarts plates.

The fundamental feature of the dependence in Fig. 2(b) is absence (or negligible small value) of SH signal from the bulk. Indeed, the SH in the bulk of isotropic material can be generated by the fundamental harmonic via cubic nonlinearity in the presents of external constant (or low-frequency) field. Thus, the optical and THz pulses should intersect in space. Due to material dispersion the optical and THz pulses interact in the bulk on a walk-off distance $L_w = c\tau / (n_{THz} - n_{opt})$, where τ – THz pulse duration, n_{THz} and n_{opt} – terahertz and optical refraction coefficients, respectively. Taking $\tau = 500$ fs we have for fused quartz $L_w = 300$ μm. Changing the time delay results to moving the position of interaction region inside the sample. If the thickness of the sample (1.8

mm) is more than the walk-off length (as in our experiment) the dependence of SH signal on time delay should have step-like profile with rising and falling times $\tau \approx 0.5$ ps (with possible modulation due to short ~ 10 μm coherent length between first and second optical harmonics). Such signal was not measured in the experiment. At the same time, the signal tracing the form of THz intensity was well observed that indicates the presents of a source of the second harmonic only near the interfaces.

To qualitatively explain the above-discussed experimental result we considered a nonlinear response of hydrogen-like atom in the field of optical and THz radiation. The presence of the interface was modeled by applying additional constant field with corresponding orientation. The simulation was based on 2D and 3D numerical solution of Schrödinger equation. It was shown that near the surface the SH response has a resonance defined by the value of the constant field. Switching-off of the constant field (that corresponds to moving atom into the bulk) leads to a decrease in the second harmonic by several orders of magnitude in agreement with experimental data. This single-particle theory gives a good qualitative explanation of experimental results. For more accuracy (and better quantitative agreement with experimental data) one should take into account the number of atoms in the bulk and near the surface as well as the dynamics of their radiation.

This work was supported in part by the RFBR, research project No. 18-02-00968.

References

[1] O. A. Aktsipetrov, A. A. Fedyanin, A. V. Melnikov, E. D. Mishina, A. N. Rubtsov, M. H. Anderson, P. T. Wilson, M. ter Beek, X. F. Hu, J. I. Dadap, and M. C. Downer. Dc-electric-field-induced and low-frequency electromodulation second-harmonic generation spectroscopy of Si(001)–SiO2 interfaces // Phys. Rev. B. 1999. V. 60. P. 8924.

[2] A. Nahata and T.F. Heinz. Detection of freely propagating terahertz radiation by use of optical second-harmonic generation. // Opt. Lett. 1998. V. 23. P. 67.

[3] S.B. Bodrov, A.A. Murzanev, Y.A. Sergeev, Y.A. Malkov, and A.N. Stepanov. Terahertz generation by tilted-front laser pulses in weakly and strongly nonlinear regimes. // Appl. Phys. Lett. 2013. V. 103. P. 251103.

Coherent control of electron-nuclear states of rare-earth ions in crystals using radio-frequency and microwave radiation

M. R. Gafurov, G.V. Mamin, E.I. Baibekov, I. N. Kurkin, F.F. Murzakhanov, S.B. Orlinskii

Kazan Federal University, Kazan, Russian Federation, mgafurov@gmail.com

Electron and/or nuclear spin states are well-known for their use as potential qubits for quantum computing [1]. Single-qubit operation on an electron spin-1/2 qubit can be performed by using microwave (mw) pulse with frequency matching the energy of the transition between the two spin states. Double resonance techniques allow one to manipulate more than two quantum states simultaneously during the pulse sequence. Thus, certain two-qubit operations can be implemented [2].

We have performed electron-nuclear double resonance (ENDOR) and double electron-electron resonance (DEER) manipulations of electronic and nuclear spin states of trivalent gadolinium impurity ions in $CaWO_4$ host crystal. The sample of $CaWO_4:Gd^{3+}$ (0.01 at. %) single crystal was grown in Kazan Federal University by the Czochralski method. Experimental studies were done by using helium flow cryostats at X-band Bruker Elexsys E580 spectrometer (T = 5-25 K, microwave frequency $\nu \approx 9.7$ GHz). Pulsed electron paramagnetic resonance spectrum of Gd^{3+} ion in $CaWO_4$ host crystal (magnetic field **B** $\parallel$ *c* axis) is shown in Fig. 1.

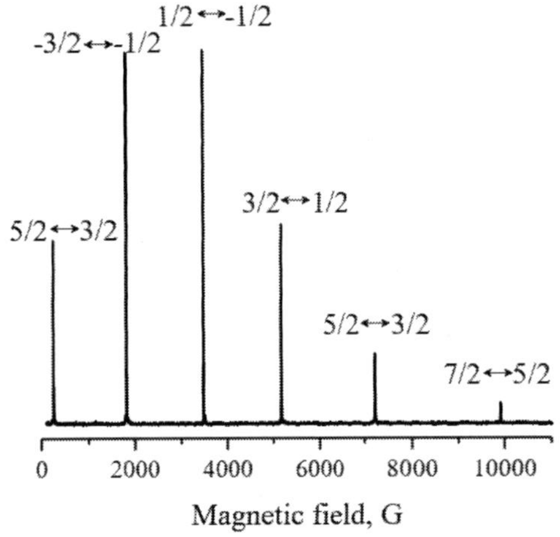

Fig. 1. Pulsed EPR of Gd^{3+} ion in $CaWO_4$ crystal, **B** $\parallel$ *c* axis

It consists of six resonance lines corresponding to various $S_{z1} \rightarrow S_{z2}$ transitions between the states of the ground spin-7/2 multiplet. Containing 8 lowest-energy spin states ($S_z = -7/2 \div 7/2$), Gd^{3+} ion can be viewed as model three-qubit system. Single-qubit manipulations in this system have been demonstrated in our previous work [3]. There, mw pulse excited one of the six transitions (chosen by appropriately tuned magnetic field), and the resulting Rabi oscillations were recorded in the time domain of the mw pulse.

Since the impurity gadolinium ions contain odd ^{155}Gd and ^{157}Gd isotopes with nuclear spin $I = 3/2$ (natural abundance 14.8% and 15.6%, respectively), one can use pulsed ENDOR in order to manipulate both the electron and nuclear spin states of these isotopes simultaneously. The total basis of 8·4 = 32 electro-nuclear states would expand the possible quantum algorithm to 5-qubit operations. In order to demonstrate the control of both the electron and nuclear spin states simultaneously, we used Davies ENDOR pulse sequence [4]: an initial mw π pulse is followed by the radiofrequency (rf) π pulse, then the remaining electron state is probed by the mw two-pulse spin-echo sequence. If the rf pulse frequency matches the nuclear transition, the final echo intensity changes. An example of ENDOR spectrum corresponding to 5/2→3/2 electronic transition near $B = 244$ G swept in the range 0-50 MHz of the rf pulse is shown in Fig. 2. Here, four peaks corresponding to certain electronuclear transitions (two pairs corresponding to ^{155}Gd and ^{157}Gd nuclear spin flips) are present. The positions of the peaks agree well with our theoretical estimations.

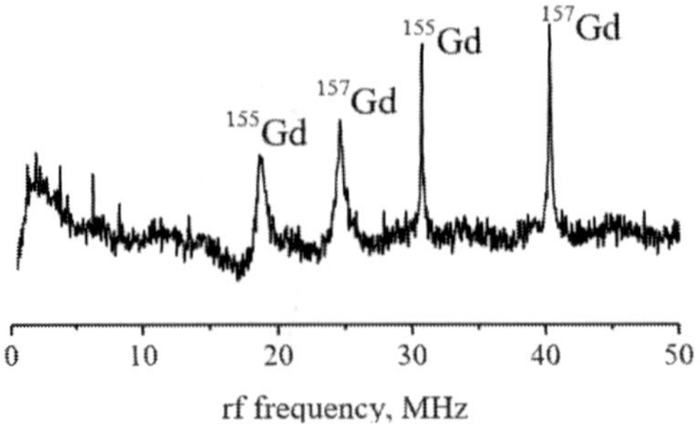

Fig. 2. Pulsed Davies ENDOR of Gd^{3+} in $CaWO_4$ crystal, *B* = 244 G is directed along the crystal *c* axis, *T* = 20 K

Since the positions of the lines in Fig. 1 are orientation-dependent [5], one can choose an appropriate direction of **B** with respect to the crystallographic axes, so that two electronic transitions lie within the 800 MHz bandwidth of the two-frequency DEER equipment. EPR spectrum in certain orientation (13 degrees between **B** and *c* axis) is shown in Fig. 3. There, two high-field transitions, 5/2→3/2 and 7/2→5/2, involving three electronic levels, are drawn together, the corresponding resonance fields differ by ≈ 100 G. Then, the following two-frequency pulse sequence has been applied: $\pi/2(\nu_1) - \pi/2(\nu_1) - \pi(\nu_2) - \pi/2(\nu_1)$, where the indices in brackets denote the mw pulse frequency. There, ν_1 = 9.74 GHz corresponded to the resonant transition 5/2→3/2 at 7210 G, while ν_2 was swept in the range 9.3-10 GHz. When ν_2 did not match any transition, a usual three-pulse stimulated echo for 5/2→3/2 transition was recorded. However, when ν_2 matched either 5/2→3/2 or 7/2→5/2 transition energy, the third (flipping) π pulse inversed the populations of the two involved spin states, so that the echo intensity was decreased.

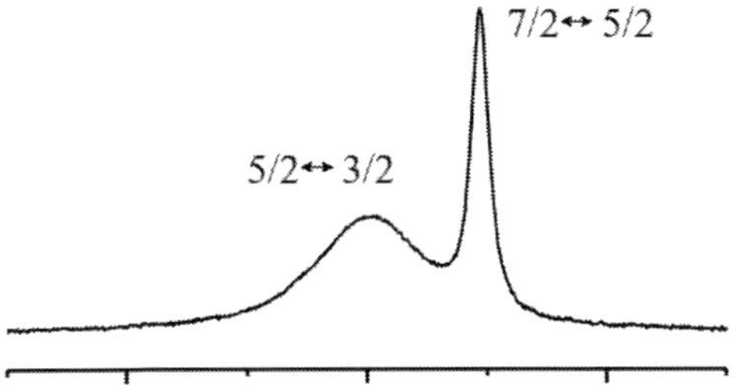

Fig. 3. A part of pulsed EPR spectrum of Gd:CaWO$_4$ when **B** is directed at an angle of 13 degrees with respect to crystal *c* axis

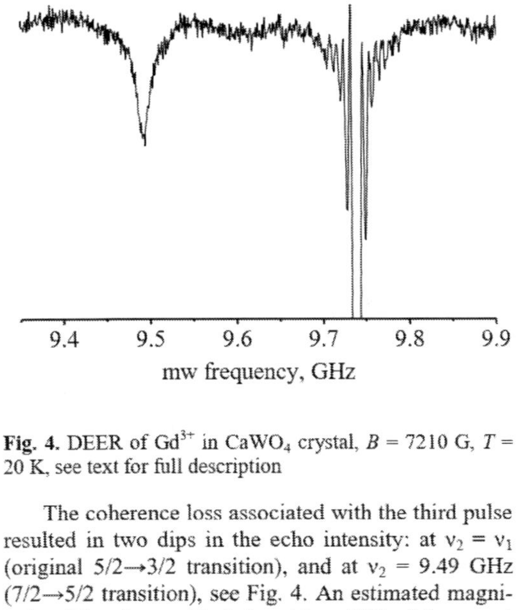

Fig. 4. DEER of Gd^{3+} in CaWO$_4$ crystal, B = 7210 G, T = 20 K, see text for full description

The coherence loss associated with the third pulse resulted in two dips in the echo intensity: at $\nu_2 = \nu_1$ (original 5/2→3/2 transition), and at ν_2 = 9.49 GHz (7/2→5/2 transition), see Fig. 4. An estimated magnitude of the observed hole burnt is ~ 10% of the one at $\nu_2 = \nu_1$, this figure is attributed to the inhomogeneous broadening of the resonance lines. In order to measure the corresponding coherence time of the three-state ensemble (a qutrit), we have recorded the transient nutation of the 7/2→5/2 coherence. There, the duration of the third ν_2 pulse resonant with 7/2→5/2 transition was varied in the range 0-4 μs, while the duration of the $\pi(\nu_2)$ pulse was ≈ 180 ns. The resultant echo intensity vs. the pulse duration is shown in Fig. 5.

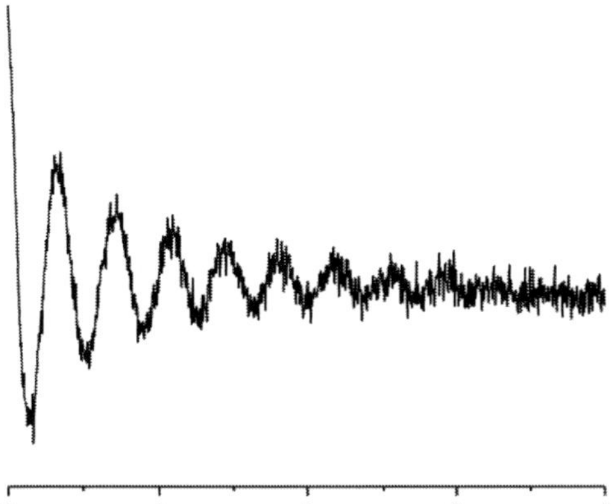

Fig. 5. Oscillations of the 5/2→3/2 transition coherence implemented by pulsed DEER, see text for full description

Finally, we have demonstrated electron-electron and electron-nuclear spin manipulations of Gd^{3+} ion in CaWO$_4$ crystal. The results suggest that the studied system is perspective for multiqubit implementation in quantum computing.

This work was financially supported by the Russian Science Foundation (Project no. 17-72-20053).

References

1. *DiVincenzo, D. P.* Quantum computation // *Science* 1995. V. 270. P. 255–261.

2. *Jelezko, F., Gaebel, T., Popa, I., Domhan, M., Gruber, A., Wrachtrup, J.* Observation of coherent oscillation of a single nuclear spin and realization of a two-qubit conditional quantum gate // *Physical Review Letters* 2004. V. 93. P. 130501.

3. *Baibekov, E. I., Gafurov, M. R., Zverev, D. G., Kurkin, I. N., Rodionov, A. A., Malkin, B. Z., Barbara, B.* Coherent spin dynamics in a gadolinium-doped CaWO$_4$ crystal // *Physical Review B* 2017. V. 95. P. 064427.

4. *Davies, E. R.* A new pulse ENDOR technique // *Physics Letters A* 1974. V. 47. P. 1–2.

5. *Vazhenin, V.A., Guseva, V.B., Artemov, M.Y.* Effects of averaging of spin packets of interacting resonances in Gd^{3+} EPR spectra of scheelites // *Physics of the Solid State* 2003. V. 45. P. 2271–2275.

Terahertz time-domain spectroscopy of astrophysical ice analogs: A pilot study

A. A. Gavdush[1,2], B. M. Giuliano[3], B. Müller[3], G. A. Komandin[1], M. E. Palumbo[4], G. A. Baratta[4], C. Sciré[4], S. O. Yurchenko[2], K. I. Zaytsev[1,2], A. V. Ivlev[3], and P. Caselli[3]

[1]Prokhorov General Physics Institute of the Russian Academy of Sciences, Moscow, Russia, arsenii.a.gavdush@gmail.com
[2]Bauman Moscow State Technical University, Moscow, Russia
[3]Max-Planck-Institute für extraterrestrische Physik, Garching bei München, München, Germany
[4]Osservatorio Astrofisico di Catania, Catania, Italy

Ices are one of the most common kinds of dense matter in the Universe. There are more than 180 various molecular astrophysical ices, including H2O, CO, CO2, CH3OH, etc. [1]. Surface chemical processes on icy mantles of dust particles play an important role in formation of new molecules, including biologically relevant organic species [2, 3]. Therefore, thorough studies of the ice properties may shed light to such astrophysical problem, as understanding the origin of the volatiles and organics – the building blocks of life – in space and then on the early Earth.

Analysis of astronomical observations in different ranges of electromagnetic spectrum is the main tool for experimental studies of ices in the Universe. These observations employ mostly a mid-infrared (IR) range of wavelengths [1, 4], but also cover far-IR and terahertz (THz) ranges, provided by such space telescopes, as Hershel [5] and SOFIA [6]. A cornerstone of proper interpretation of these observations is their comparison to IR and THz spectra of astrophysical ice analogs obtained in laboratory conditions. Recent works were mainly dedicated to the analysis in mid-IR frequency region [1], with only a few works concerned on THz spectroscopy [7, 8]. However, in spite of its importance, the presently available data related to spectroscopy in IR and THz range are still incomplete or even have not been performed for a lot of ice compounds, to our knowledge.

In the present work, we apply previously developed method [9-13] for reconstruction of both real and imaginary parts of dielectric permittivity, using the THz time-domain spectroscopy (TDS). This technique allows us to measure optical properties of astrophysical ice analogs in a broad range of wavelengths, that is required for detailed study of radiative transfer in ice mantles of astrophysical dust particles. We developed an experimental setup allowing simultaneous growth and THz TDS of thin ice films of different molecular compositions.

The experimental setup for the spectroscopic measurements is composed closed-cycle He cryostat and THz time-domain pulsed spectrometer (TPS). The cryostat is mounted on a motor controlled translational stage to tune its position with respect to the THz beam. The cryocooler vacuum chamber is equipped with four ports for optical access and for the gas inlet. Optical windows and a substrate for the ice growing are made of high-resistivity float-zone silicon (HRFZ-Si) with different thickness, the smallest one for the substrate, in order to suppress multiple resonances in the THz spectra. The pumping station that operates with gas mixture for ice sample deposition is composed of a turbomolecular pump combined with a backing rotary pump. The THz TPS used for the present work has been basically purchased from the Batop GmbH, and it is coupled with two photoconductive antennas made of low-temperature grown gallium arsenide and a femtosecond laser with a pulse repetition rate of 100 MHz and an average input power of 65 mW. Further technical details on an experimental setup one can find in the upcoming work at Astronomy & Astrophysics journal [11].

Growth of the ices on the substrate is organized using standard technique using vapor deposition on the cold substrate in the vacuum chamber. To attain ice thickness of a mm order, fast deposition regime is employed. The first experiments were performed with ice of CO, deposited at 14 K in a several deposition regimes, with 4, 5 and 6 min deposition steps, and up to the total 30 min deposition time.

The reconstruction procedure of the optical properties of ice layers is organized in the same manner as in previous works [9-13]. The signal propagated through the empty vacuum chamber at 14 K was used as the reference one for analysis of the sample signal propagated through the chamber with grown ice layers on the substrate. Then, the time-domain signals (reference ballistic and sample satellite pulses) were analyzed to have initial estimations of ice layers' thicknesses and a real part of the sample refractive index, to enhance the accuracy of data post processing. Reliability of the employed approach is confirmed by independent measurements of the sample thickness [15], and observed linear increase of the total initial thickness of ices in different deposition regimes. In result [14], the optical properties of CO ice were measured in the range of 0.3–2.0 THz, with a clearly expressed Lorentz-like peak centered near 1.5 THz, which stands in a good agreement with previous studies [16, 17].

This work was partly supported by the Russian Foundation for Basic Research (RFBR), Project #18-32-00816.

References

1. *Boogert, A. C. A., Gerakines, P. A., and Whittet, D. C. B.* Observations of the icy universe // Annual Review of Astronomy and Astrophysics 2015. Vol. 53, P. 541–581.

2. *Öberg K. I.* Photochemistry and Astrochemistry: Photochemical Pathways to Interstellar Complex Organic

Molecules // Chemical Reviews 2016. V. 116, No 17 P. 9631–9663.

3. *McGuire, B. A., Carroll, P. B., Loomis, R. A.., Finneran, I. A., Jewell, P. R., Remijan, A. J., Blake, G. A.* Discovery of the interstellar chiral molecule propylene oxide (CH3CHCH2O) // Science 2016. V. 352, No. 6292. P. 1449–1452.

4. *van Dishoeck, E. F.* Astrochemistry of dust, ice and gas: introduction and overview // Faraday Discuss. 2014. V. 168, P. 9-47.

5. *Doyle, D., Pilbratt, G. and Tauber, J.* The Herschel and Planck Space Telescopes // Proceedings of the IEEE 2009. V. 97, No. 8. P. 1403-1411.

6 *Gehrz, R. D., Becklin, E. E., de Pater, I., Lester, D. F., Roellig, T. L. and Woodward, C. E.* A new window on the cosmos: The Stratospheric Observatory for Infrared Astronomy (SOFIA) // Advances in Space Research 2009. V. 44, No. 4. P. 413-432.

7. *Allodi, M. A., Ioppolo, S., Kelley, M. J., McGuire, B. A., and Blake, G. A.* The structure and dynamics of carbon dioxide and water containing ices investigated via THz and mid-IR spectroscopy // Physical Chemistry Chemical Physics 2014. V. 16, No. 8. P. 3442-3455.

8. *McGuire, B. A., Ioppolo, S., Allodi, M. A., and Blake, G. A.* THz time-domain spectroscopy of mixed CO2-CH3OH interstellar ice analogs // Physical Chemistry Chemical Physics 2016. V. 18, No. 30. P. 20199-20207.

9 *Zaytsev, K. I., Gavdush, A. A., Lebedev, S. P., Karasik, V. E., and Yurchenko S. O.* A Method of Studying Spectral Optical Characteristics of a Homogeneous Medium by Means of Terahertz Time-Domain Spectroscopy // Optics and Spectroscopy 2015. V. 118, No. 4. P. 552–562.

10. *Zaytsev, K. I., Gavdush, A. A., Chernomyrdin, N. V., and Yurchenko, S. O.* Highly Accurate In Vivo Terahertz Spectroscopy of Healthy Skin: Variation of Refractive Index and Absorption Coefficient along the Human Body // IEEE Transactions on Terahertz Science and Technology 2015. V. 5, No. 5. P. 817-827.

11. *Yakovlev, E. V., Zaytsev, K. I., Dolganova, I. N., Yurchenko, S. O.* Non-Destructive Evaluation of Polymer Composite Materials at the Manufacturing Stage Using Terahertz Pulsed Spectroscopy // IEEE Transactions on Terahertz Science and Technology 2015. V. 5, No. 5. P. 810 – 816.

12. *Zaytsev, K. I., Kudrin, K. G., Karasik, V. E., Reshetov, I. V., Yurchenko S. O.* In vivo terahertz spectroscopy of pigmentary skin nevi: Pilot study of non-invasive early diagnosis of dysplasia // Applied Physics Letters 2015. V. 106, No. 5 P. 053702.

13. *Zaytsev, K. I., Gavdush, A. A., Karasik, V. E., Alekhnovich, V. I., Nosov, P. A., Lazarev, V. A., Reshetov, I. V., Yurchenko, S. O.* Accuracy of sample material parameters reconstruction using terahertz pulsed spectroscopy // Journal of Applied Physics 2014. V. 115, P. 193105.

14. *Giuliano, B. M., Gavdush, A. A., Müller, B., Zaytsev, K. I., Palumbo, M. E., Baratta, G. A., Sciré, C., Komandin, G. A., Yurchenko, S. O., Ivlev, A. V., and Caselli, P.* Broadband spectroscopy of astrophysical ice analogs. I. Direct measurement of complex refractive index using terahertz pulsed spectroscopy // Astronomy & Astrophysics manuscript.

15. *Baratta, G. A., Palumbo, M. E.* Infrared optical constants of CO and CO2 thin icy films // J. Opt. Soc. Am. A 1998. V. 15, No. 12. P. 3076-3085.

16. *Anderson, A., Leroi, G. E.* Far-Infrared Spectra of Crystalline Nitrogen and Carbon Monoxide // The Journal of Chem. Phys. 1966. V. 45, P. 4359.

17. *Ron, A., Schnepp, O.* Lattice Vibrations of the Solids N_2, CO_2, and CO // The Journal of Chem. Phys. 1967. V. 46, No. 10 P. 3991-3998.

EPJ Web of Conferences **195**, 06005 (2018)
TERA-2018

https://doi.org/10.1051/epjconf/201819506005

Accurate broadband THz molecular spectroscopy

Krupnov A.F., Belov S.P., Tretyakov M.Yu., Golubiatnikov G.Yu., Parshin V.V., Koshelev M.A., Serov E.A., Vilkov I.N., Makarov D.S., Bubnov G.M., Leonov I.I., Chernova A.I., Andriyanov A.F., Shkaev A.P.

Institute of Applied Physics RAS, Nizhny Novgorod, Russia, koma@ipfran.ru

Teraherz (THz) molecular spectroscopy is a powerful instrument for solving both fundamental and applied problems. Molecular physics, remote sensing, chemical and isotopic analysis, medicine (breath analysis) are among the applied problems which can be solved using molecular spectroscopy methods.

Sensitivity, resolution and operating pressure range are the most crucial parameters of any spectrometer. High resolution of gas spectrometers allows observing the lines that uniquely as fingerprints characterize molecules, opening up vast possibilities, on the one hand, for the outlook at the molecular level, and, on the other hand, for qualitative and quantitative analysis of gas mixtures. High sensitivity of the spectrometer allows the significant expand of the range of the studied objects. The higher the sensitivity, the greater the number of lines which can be observed in the experiment. Finally, the wider the pressure range, the extensively the knowledge about intermolecular interaction which is required for adequate interpretation and modeling of the molecular spectra. However, use of only type one spectrometer cannot guarantee the data reliability. Use of several instruments of the different principal of operation allows obtaining more information on molecular spectra with higher accuracy and reliability.

In this report we present a complex of broadband THz/subTHz spectrometers (see [1] and references therein) with complementary abilities developed in IAP RAS. The complex consists of three spectrometers with different methods of spectra recording covering the widest pressure range from a part of mTorr to a few atmospheres. All of them utilize backward wave oscillator (BWO) as a radiation source which frequency is stabilized by a phase lock loop (PLL) against a harmonic of a microwave frequency synthesizer and cover a frequency range from 30 GHz to about 1 THz.

The first spectrometer is a conventional video spectrometer [2] operating in a pressure range from a part of mTorr to about 1 Torr. The principle of its operation consists in detection of a change of radiation power passed through the gas cell filled with the gas investigated. A spectrometer cell is a 2 m long and 113 mm diameter stainless-steel tube with polished internal cell surface. Optical vacuum windows are made of high-density polyethylene and are cone-shaped to reduce the standing-wave amplitude between them. The spectrometer allows accurate measurement of line frequency, intensity, pressure broadening and shifting parameters. For precise measurement of line positions with subkHz uncertainty, method of nonlinear coherent spectroscopy (Lamb-dip method) was implemented in the spectrometer. A typical line recording obtained using video spectrometer is presented in Fig. 1.

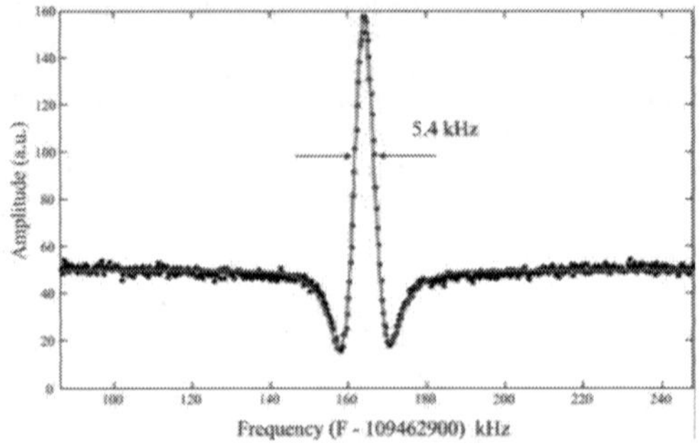

Fig. 1. Lamb-dip observed in OCS rotational transition J=9-8. Frequency detuning is shown from 109463.063 MHz. Collisional line width (HWHM) obtained from the fit to the Lorentz shape is 6.56(3) kHz. Doppler width is 81 kHz.

The second instrument of the spectroscopic complex is a spectrometer with radio-acoustic detector of absorption signal [3]. The output signal of the RAD spectrometer is proportional to the radiation power absorbed by the gas. It brings us to two main advantages of the RAD method: i) almost zero baseline (comparing to a video spectrometer); ii) the spectrometer sensitivity is increased as radiation power increases. The operating pressure range of the spectrometer is in interval from 10 mTorr to 10 Torr. The spectrometer is the best suited for the line shape analysis and measuring pressure broadening and shifting parameters of the lines and their speed-dependence in a wide temperature range. One more advantage of the RAD spectrometer is high enough sensitivity at rather compact size of the gas cell (10 cm long and 2 cm in diameter). It allows easy shielding of the gas cell from the ambient magnetic fields that is important for studying the paramagnetic lines, e.g., for the lines of oxygen spectra. Typical recordings of oxygen lines obtained using RAD spectrometer are shown in Fig. 2.

At the present level of sensitivity of THz spectrometers most of spectroscopic studies are aimed at refinement of the spectroscopic information. Expanding the range of the studied objects requires significant improvement of the sensitivity. In THz range radio-acoustic detection is seemed to be the only method allowing reaching high sensitivity at reasonable recording time by increasing radiation power. It was demonstrated in our earlier study [4] that the use of the high power coherent THz radiation of the gyrotron allows increase of the RAD spectrometer sensitivity by at least three orders of magnitude. This opens up possibility of investigating, for example, the forbidden spectra of molecules.

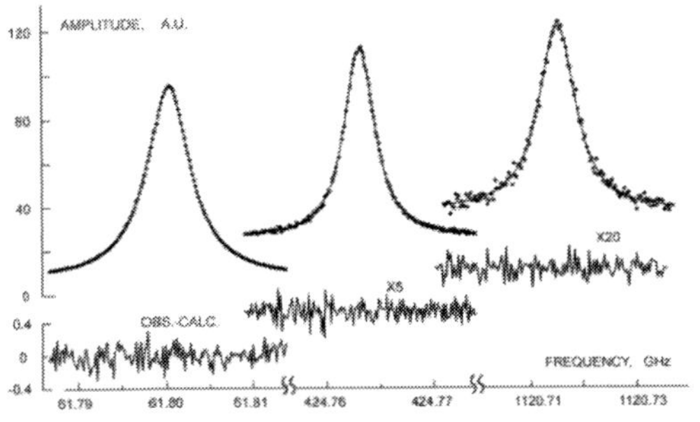

Fig. 2. Lines of the oxygen molecule obtained using RAD spectrometer (from left to right): a) fine-structure line 11+; b) rotational line (N,J)=(3,2)-(1,2); c) rotational line (N,J)=(7,6)-(5,6). The result of the line fitting to Voigt profile is shown by solid line. The residue of the fit is shown in the lower part of the figure.

In third, the resonator spectrometer [5,6] operating in a pressure range from 10 Torr up to a few atmospheres gas absorption is measured sequentially at each eigen-frequency as a change of the Q-factor of the Fabry-Perot resonator filled with the studied and non-absorbed gases, respectively. The PLL system provides fast phase-continuous frequency scan, resulted in significant progress in the spectrometer sensitivity (which now reaches $4\cdot10^{-9}$ cm^{-1} in terms of the absorption coefficient). A part of the atmospheric absorption spectra observed using resonator spectrometer is presented in Fig. 3. It is worth noting, that the line intensity measurements performed using the resonator spectrometer are supposed to be the most accurate comparing to other methods.

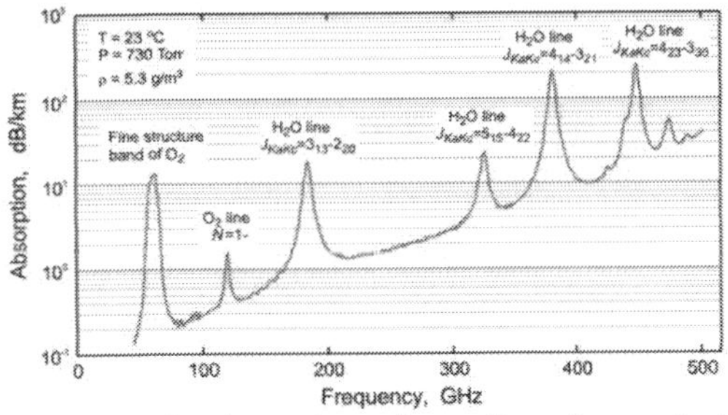

Fig. 3. Broadband experimental recording of atmospheric air spectrum obtained using resonator spectrometer.

Use of the complex of THz spectrometers allows determination of possible systematic errors. This is confirmed by a series of complementary experiments in which parameters of spectral lines measured by one spectrometer were revised by a spectrometer of another type. Coincidence of the values of the parameters measured by different spectrometers means high accuracy and reliability of the data obtained. As an example, results of the pressure broadening study for 118-GHz oxygen line [7] (RAD and resonator) and 22-GHz water line [8] (RAD and video) are presented in Fig. 4.

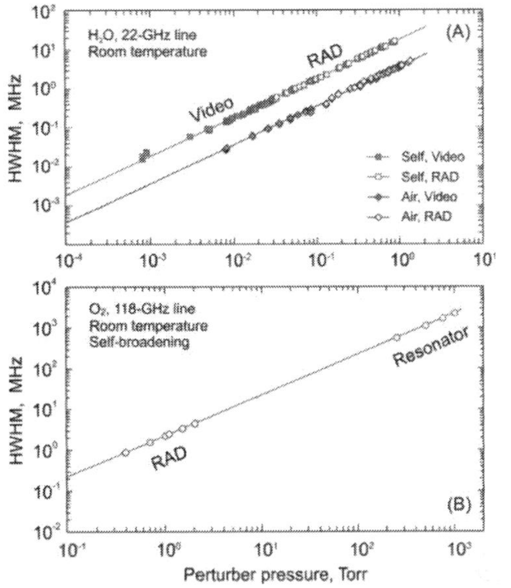

Fig. 4. Pressure dependence of the width of the (A) 22-GHz water line measured using video and RAD spectrometers and (B) 118-GHz oxygen line measured using RAD and resonator spectrometers.

Improvement of the RAD spectrometer sensitivity is supported by RSF (project 17-19-01602). Studies of molecular spectra using complex of spectrometers are supported in parts by RFBR projects 18-05-00698 (atmospheric pressure recordings) and 18-02-00705 (line shape analysis).

References

1. *Krupnov A.F., et al.* Accurate broadband rotational BWO-based spectroscopy // J. Molec. Spectrosc. 2012. V. 280 P.110-118.

2. *Golubiatnikov G.Yu., et al.* Precision Sub-Doppler Millimeter and Submillimeter Lamb-Dip Spectrometer // Radiophysics and Quantum Electronics. 2014. V 56. N. 8–9. P. 599–609.

3. *Tretyakov M.Yu., et al.* Precise Measurements of Collision Parameters of Spectral Lines with a Spectrometer with Radioacoustic Detection of Absorption in the Millimeter and Submillimeter Ranges // Instrum. Exp. Tech. 2008. V. 51. P. 78–88.

4. *Koshelev M.A., et al.* Molecular Gas Spectroscopy Using Radioacoustic Detection and High-Power Coherent Subterahertz Radiation Sources // J. Molec. Spectrosc. 2017. V. 331 P. 9–16.

5. *Parshin V.V., et al.* Modern resonator spectroscopy at submillimeter wavelengths. // IEEE Sensors Journal. 2013 V.13. N. 1. P. 18-23.

6. *Koshelev M.A., et al.* New frontiers of modern resonator spectroscopy // submitted in IEEE Trans. THz Sci. Technol. 2018.

7. *Koshelev M.A., et al.* Accurate modeling of the diagnostic 118-GHz oxygen line for remote sensing of the atmosphere // J. Quant. Spectrosc. Radiative Transfer 2017. V. 196 P. 78-86.

8. *Koshelev M.A., et al.* Line shape parameters of the 22-GHz water line for accurate modeling in atmospheric applications // J. Quant. Spectrosc. Radiative Transfer 2018. V. 205 P. 51-58.

Pulsed magnets with high field intensity for laser-plasma experiments and TDS spectroscopy

T. Krapivnitskaia, A. Luchinin, V. Malyshev, M. Morozkin, M. Starodubtsev, M. Proyavin, A. Fokin, and M. Glyavin

Institute of Applied Physics, 46 Ulyanov Street, 603950 Nizhny Novgorod, Russia

The studies of different physical processes in presence of strong magnetic fields are one of the ever-lasting problems in experimental physics. Development of techniques and investigation of complex dynamics leads to the implementation of magnetic systems with fields more than 20 T. The laser-plasma experiments [1] and time-domain spectroscopy [2] are one of the outstanding applications for strong magnetic fields. Conventional magnetic systems based on large superconducting magnets are not convenient for ultrafast spectroscopy due to remote access to the magnet coil and unavoidable dispersive elements. The solution is to use the small-scale pulsed systems, which provide easy alignment and strong field at the same time.

Magnets for time-domain spectroscopy

Terahertz time-domain spectroscopy (THz-TDS) is ideally suited for the study of coherent low-energy dynamics in condensed matter, determining the complex conductivity of material. The terahertz frequency range is rich in collective and elementary excitations in solids, many of which have magnetic origins such as cyclotron resonance, electron spin resonance, and magnon excitations. An external magnetic field provides a powerful tool for tuning these magnetic THz excitations through Landau and Zeeman quantization of electrons' orbital and spin energies. Probing these excitations through THz-TDS in high magnetic fields can provide valuable insight into the dynamics of interacting electrons. Therefore, there is a need for developing a source of strong magnetic field, suitable for TDS.

The IAP RAS has great experience in development and production of pulsed magnetic systems, which were mainly used for high-frequency gyrotrons [3]. The record magnetic field of 50 T in a bore of 9 mm (Fig. 1) was produced for the gyrotron with 1.3 THz frequency [4] along with a number of 30 T magnets with 2 cm bore. The developed technology is based on copper magnets encased in fiberglass body with liquid nitrogen as a coolant. The cooling to cryogenic temperatures reduces the resistance of copper, lowers the heating of the coil during the pulse and provides better pulse-to-pulse stability. Resulting coil operated with 1 Hz repetition rate and successfully produced more than 3000 pulses with field > 30 T and more than 100 pulses with field > 45 T without any degradation. The magnets of this type surely can find their application in experiments with time-domain spectroscopy in strong magnetic fields.

Fig. 1. The 50 T Pulsed magnet

Magnets for laser-plasma experiments

The studies on the formation of different plasma structures in strong magnetic field are of great interest for astrophysics and inertial fusion. Different astrophysical objects could be modeled in such down-scaled laboratory experiments. Such astrophysical objects characterizes by the perturbation of the magnetic field structure and development of different magneto-hydrodynamic instabilities, which could not be studied analytically and should be studied through laboratory modeling [5].

The specially designed magnetic coil provides the needed conditions for investigation of plasma dynamics in magnetic field. The principal scheme of the setup and coil photo are presented in Fig. 2. To satisfy the requirements for the plasma formation and diagnostics the coil must have three mutually perpendicular apertures of free space inside the coil, that provide the possibility of positioning the target, irradiating the target with powerful nanosecond laser beam and observing the dynamics of plasma expansion perpendicular to the main axis of the direction of propagation of the plasma stream.

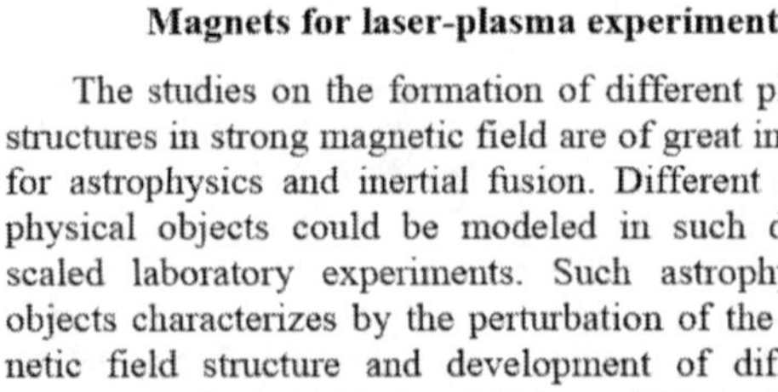

Fig. 2a. General view of the experimental setup.

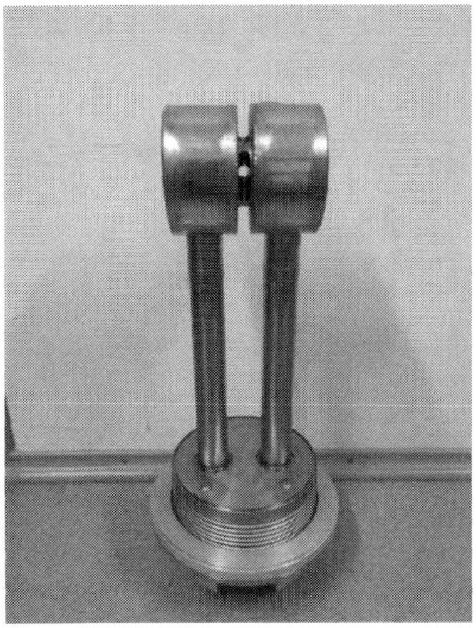

Fig. 2b. The pulsed magnet photo.

In order to satisfy the requirements, the Helmholtz magnetic coil was developed and manufactured by the IAP RAS. The coil current is controlled by a capacitor discharge unit with maximum current up to 8 kA, ensuring the maximum magnetic field of 15 T. The discharge of the coil occurred over times of the order of several milliseconds. The field was considered stationary at time scales of the order of several microseconds, much larger than the times of the investigated plasma dynamics of the order of 100 ns. The magnetic field was practically uniform in the axial and radial directions at scales of the order of 20 mm, which exceeded the characteristic scales of the investigating plasma (~ 10 mm). The repetition of the magnetic field amplitude and profile was ensured by cryogenic cooling of the system with liquid nitrogen. Typically, the coil can operate with repetition rate of 1 shot per 10 minutes. The construction of the magnet allowed switching of the direction of both coils, providing the co- and counter-directional setup of the magnetic field and opening up the possibility of investigation of plasma dynamics in different conditions. The measured profiles of magnetic field for both configurations are presented on Fig. 3. The magnet was already used in successful experiment, carried out on PEARL facility [6].

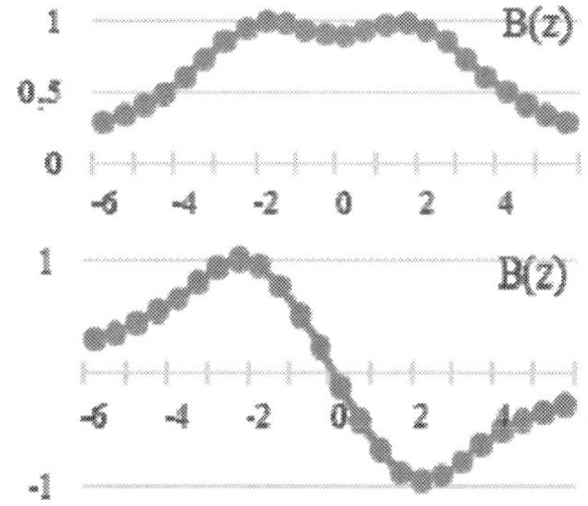

Fig. 3. Magnetic field profiles for co and counter-directional setup of coils.

Conclusion

The developed pulse magnets with fields up to 50 T and complex magnetic field structure open up the possibilities for investigation of properties of materials and laboratory modeling of astrophysical phenomena.

The work supported in part by the Russian Scientific Foundation under the project 14-12-00887.

References

1. *Albertazzi, B. et al.* Laboratory formation of a scaled protostellar jet by coaligned poloidal magnetic field // Science, 2014, V. 346, Vol. 346, No 6207, P. 325-328.

2. *G. Timothy Noe et al.* Single-shot terahertz time-domain spectroscopy in pulsed high magnetic fields // Optics Express, 2016, V. 24, No 26, P. 30328-30337.

3. M. Glyavin et al. A pulse magnetic field generator for terahertz gyrodevices // Instruments and Experimental Techniques, 2011, V. 54, No. 1, P. 77-80

4. Glyavin M.Y., Luchinin A.G., Golubiatnikov G.Y. Generation of 1.5-kW, 1-THz Coherent Radiation from a Gyrotron with a Pulsed Magnetic Field // Phys. Rev. Lett. 2008. V. 100, No 1. P. 15101.

5. M. Starodubtsev et al. Laboratory investigation of magnetized laser plasmas expansion into the vacuum, ICONO/LAT 2016, Minsk, Belarus

6. Kurbatov, E.P. et al. Comparison of Dimensionless Parameters in Astrophysical MHD Systems and in Laboratory Experiments // Astronomy reports, 2018, V. 62, No. 8, P. 483–491

Single-color pump-probe setup at the NovoFEL facility for measurements of carrier relaxation dynamics in semiconductors

V. D. Kukotenko[1,3], Y. Y. Choporova[1,2], Boris A. Knyazev[1,2], V. V. Gerasimov[1], R. K. Zhukavin[4], K. A. Kovalevsky[4]

[1]Budker Institute of Nuclear Physics, Novosibirsk, Russia, kukotenkovaleria@gmail.com
[2]Novosibirsk State University, Novosibirsk, Russia
[3]Novosibirsk State Technical University, Novosibirsk, Russia
[4]Institute for Physics of Microstructures, Nizhny Novgorod, Russia

The understanding of carrier dynamics in semiconductors is of prime importance for applications but also interesting from a fundamental point of view. As a basic example the speed of semiconductor-based photodetectors is determined by the capture time of free charge carriers. When a semiconductor like Si or Ge is doped with shallow impurities (As, Ga, Sb) the excess electrons or holes behave almost as in a hydrogen atom. However, Bohr radii and energy levels are scaled according to the effective masses and dielectric constants of their host lattice. The perturbation from there ground state and their relaxation back to equilibrium can be directly investigated by the "pump-probe" method.

To selectively excite excitation lines a monochromatic radiation source is needed. For impurities in Ge and Si these lines are in the terahertz frequency range. Temporal information is obtained using short pulses and the time resolution is determined by the pulse duration. The Novosibirsk free-electron laser (NovoFEL) facility is ideally suited for this purpose. NovoFEL generates monochromatic radiation with a tunable wavelength in the range of 6-240 μm. The radiation is emitted in a train of 40 - 100 picosecond pulses with a repetition rate of 5.6 MHz. The spectral width is typically 1%.

The idea of the pump-probe method follows a basic principle: a pump beam with a wavelength equal to a certain energy transition excites a sample. During the passage of the pump pulses, the sample becomes partially transparent as carriers are lifted into an excited state. The relaxation back to the ground state occurs on a characteristic timescale, the lifetime. A second beam, the probe beam, measures the recovery of the transmittance in time as its optical path is changed with respect to the pump pulse.

We use a wire grid polarizer to divide the initial beam into a pump beam and a probing beam. The optical path difference between both pulses can be varied by an optical delay line up to 5 ns. Both beams are then focused onto the sample by off-axis parabolic mirrors (Fig. 1). A sample, e.g. germanium doped with arsenic (Ge:As), is placed in a liquid-He flow cryostat. The temperature of the sample can be varied from 4-200 K. As a detector, we use two Golay cells for the probe and the pump beam. The probe signal is a function of the delay between the two beams and called the pump-probe signal.

The second golay cell in the pump beam acts as a reference signal. To increase the signal-to-noise ratio, the continuous train of the NovoFEL radiation allows us to use a synchronous detection circuit (Lock-in), typically with a frequency of 15 Hz. This frequency is given by the maximum sensitivity of the Golay cells.

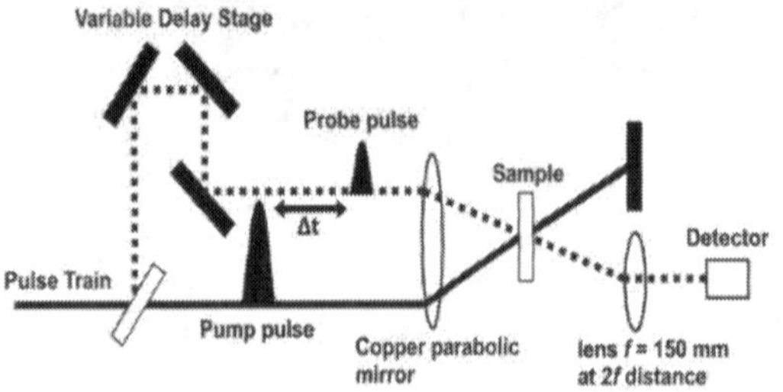

Fig. 1. Optical scheme of the "pump-probe" system at the NovoFEL.

A typical pump probe signal for a Ge:As sample is shown in figure 2. Here, the FEL radiation was tuned to the 1s → $2p_0$ transition. The decay of the signal can be fitted by a single exponential function. The exponential time constant τ corresponds to the relaxation time of the energy transition and was found to be τ = 1035 ± 12 ps. The temperature was kept constant at T = 5 K, the average pump radiation power was 90 mW, the probe radiation power was 2.5 mW, and the wavelength was λ = 131.2 μm.

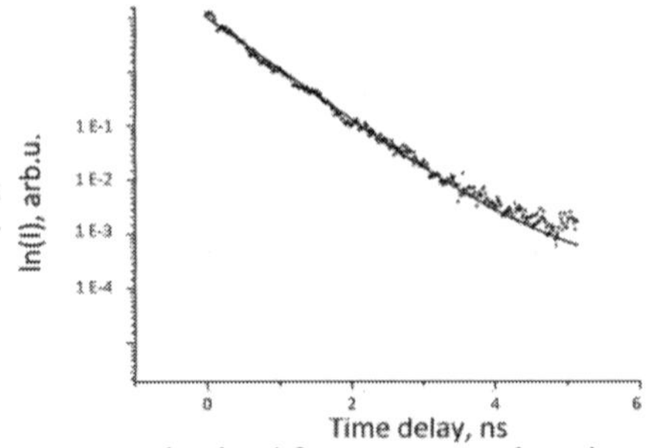

Fig. 2. Pump probe signal for a Ge:As sample at the wavelength was λ = 131.2 μm.

References

1. *N. Deßmann, S. G. Pavlov.* Time-resolved electronic capture in n-type germanium doped with antimony// Physical review. 2014. V. 89, No. 3. P. 1–8.
2. *R. Kh. Zhukavin, K.A. Kovalevsky.* The times of low-temperature intracenter relaxation of small donors in Ge// Letters to JETP. 2017. V. 789, No. 9. P. 555 – 560.

Generation of terahertz pulses from the island films of topological insulator $Bi_{2-x}Sb_xTe_{3-y}Se_y$

K.A. Kuznetsov[1], G.Kh. Kitaeva[1], P.I. Kuznetsov[2], and G.G. Yakushcheva[2]

[1]M.V. Lomonosov Moscow State University, 119991 Moscow Russia, kirill-spdc@yandex.ru
[2]Kotelnikov IRE RAS (Fryazino branch), 141190 Fryazino, Russia

A great interest to topological matter is stimulated by the study of the THz response to laser radiation. Angle-resolved photoemission spectroscopy (ARPES) indicates the presence of Dirac electrons in many topological insulators (TI) [1] Besides of fundamental aspects, TI's are very perspective objects for practical purposes both THz detectors as THz emission sources.

In this work, we perform a study of the THz emission properties of $Bi_{2-x}Sb_xTe_{3-y}Se_y$ thick films with thickness of hundreds nm and one island film with total thickness about 40 nm grown by MOCVD method on sapphire substrate. Rhombohedral $Bi_{2-x}Sb_xTe_{3-y}Se_y$ films were grown on (0001) Al2O3 substrates with thin (10 nm) ZnTe buffer layer with orientation (111) at atmospheric pressure of hydrogen in horizontal quartz reactor. An example of AFM scan of thin island film surface is shown in Fig. 1.

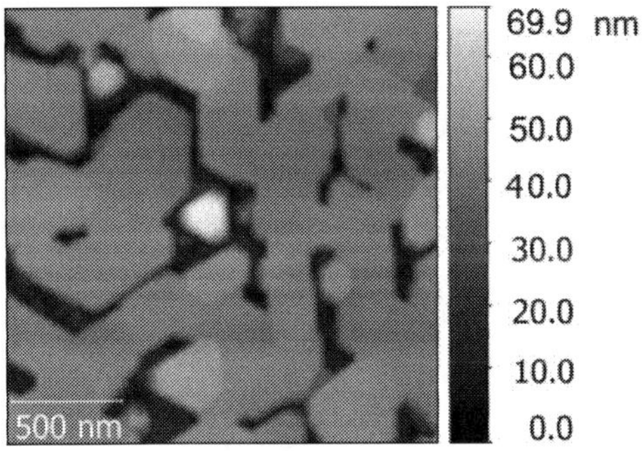

Fig. 1. AFM scan of the TI island film's surface

We use experimental setup for the terahertz emission time-domain spectroscopy (TDS) in the backward geometry. The pump is realized by the Erbium fiber laser in the Q-switched mode locking near a threshold regime at 1.56 μmm wavelength. Er^+-laser laser operated in the picosecond mode, generating optical pulses of 2.5 ps duration with the repetition rate of 70 MHz. The pump bam was divided into two unequal parts with a mean power of 100 and 20 mW, respectively. The low-power beam was focused on a photoconductive dipole antenna. The high-power beam, after propagation through the delay line and an optical chopper, was focused by the through a hole in the parabolic mirror onto the TI sample. The angle of incidence was 15°. The sample was in contact with copper electrodes (the distance between them was 0.5 mm) so that external voltage could be applied. The THz radiation generated in the sample in the backward direction was collected by four parabolic mirrors and focused on the commercial THz antenna. The antenna was oriented to register vertical components of the THz field collinear with the polarization of the pump. Thus, the electric field of THz waves generated in the direction opposite to pumping was measured at different times after the arrival of the pump pulse.

We have shown that the intensity of the THz signal from the island film is 25 times greater than that in "thick" TI samples. In order to demonstrate the amplification effect for THz radiation using an external electric field, we applied a voltage between the electrodes in contact with a surface of the island film. Fig.2 shows the waveforms of THz radiation from an island film with a bias voltage. When the polarity of the bias voltage changed, the THz signal also changed its phase by 180 degrees. Thus, for the first time, the mode of the radiating antenna in a topological insulator was demonstrated.

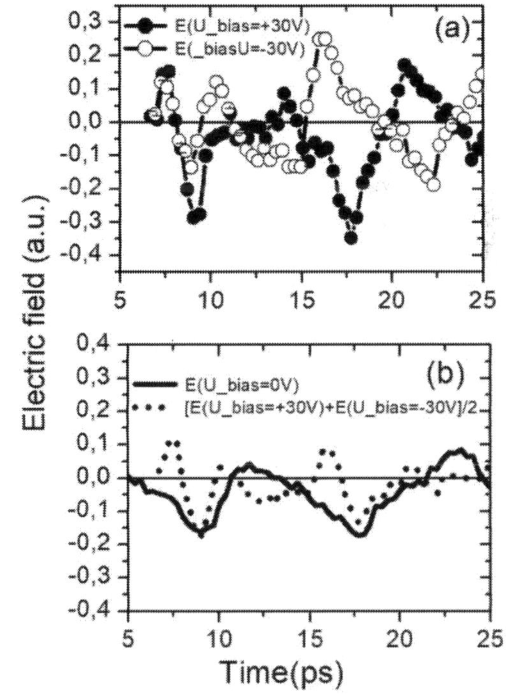

Fig. 2. Waveforms of the THz pulses

The observed waveforms contain some additional signal, which does not change its polarity with the voltage. To reveal this part we have summarized the signals, observed with +30 V and -30V voltages. Half of this sum is shown in Fig.2b (see the dotted line). The opposite-sign parts of the signals from Fig.2a almost canceled each other, so that one could expect to observe the residual constant part of the same shape as the zero-voltage signal. However, this is not true. For comparison, we repeat the zero-voltage signal together with this residual signal. One can see that the antenna pulses contain additional parts corresponding to higher modulation frequencies. This fact is illustrated in Fig.3 where the spectra of the both waveforms are shown.

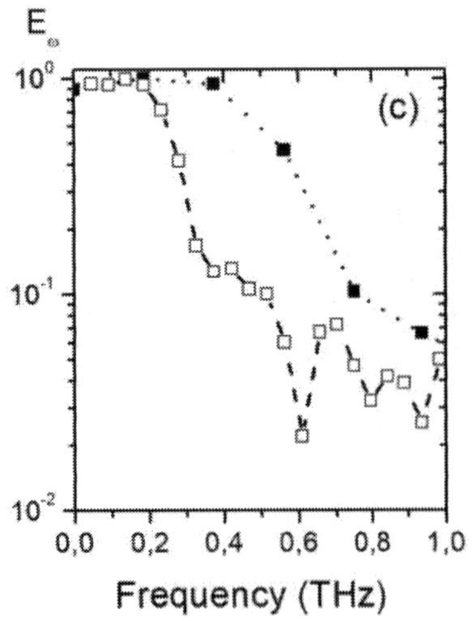

Fig. 3 Spectrum of zero-voltage signal (hollow squares) and spectrum of half-sum of signals at external bias (filled squares).

The spectrum of the residual part of the antenna signal is about two-times broader than the spectra of all the other signals, i.e. the zero-voltage signal, and the polarity-sensitive parts of +30 V and -30V voltages signals. The peak frequency corresponds to times that are of order or even less than the pump pulse duration. It means that some carrier relaxation effects proceed at considerably shorter relaxation times, than it is usually assumed to describe the THz emission of free TI samples. We believe that this effect arises from the relaxation of fast non-equilibrium charge carriers accelerated by an external field. Probably electrons are captured by the traps that are p-type impurities with energy above the bulk valence band maximum in the energy of 3-6 meV or other defects.

The authors are grateful to Dr. T.V. Murzina, for discussions and help. The study of THz emission was performed under the support of the RFBR grant № 16-02-258a. The growth and characterization of the $Bi_{2-x}Sb_xTe_{3-e}Se_y$ samples was supported by grant 17-19-01057 of the Russian Science Foundation.

References

1. *M. Z. Hasan and C. L. Kan.* Colloquium: Topological insulators // Rev. Mod. Phys. 2010, V.82,P. 3045.

Aligned planar-wire zero-index metamaterial for terahertz frequency range

E.A. Litvinov[1], P. S. Demchenko[1], E.B. Sheklanova[2], M.K. Khodzitsky[1]

[1]Terahertz Biomedicine Laboratory, Saint-Petersburg, Russia, thzbiomedicine@corp.ifmo.ru
[2] International Laboratory "Nonlinear optical molecular crystals and microlasers", Saint-Petersburg, Russia

Introduction

Demonstrations of specific class of materials with extremely small (near zero) values of permittivity have attracted attention of scientific community[1]. These unique materials have the refractive index equal to zero what causes almost constant spatial distribution of the electric field of the propagating wave inside these media[2].

Considering its extreme properties, ENZ-material can also be used as delay line that brings no phase delay in propagating wave[3], as ideal coupler between waveguides of the same diameter and can find effective application in cloaking[4]. It should be mentioned that ENZ-, MNZ- ("mu near zero") and LH- ('left-handed") materials can be designed using transformation optics approximation[5–7].

ENZ or MNZ coating may be used to radically enhance transmission through subwavelength aperture in planar ideally opaque screen which allow to overcome the diffraction limit. In general, isolated subwavelength aperture can't provide enough transfer of energy, but with ENZ coating the transfer increases by several orders of magnitude[8].

This work is dedicated to zero-index (ZI) terahertz metamaterial designed with one-layer aligned planar aluminum wires on dielectric PET substrate and acting like epsilon-near-zero (ENZ) medium.

Metal stripes on dielectric substrate can be easily manufactured. The idea is that metal stripes may have permittivity described by Drude model in terahertz frequency range, what gives the opportunity to design ENZ-material.

Numerical simulation of the structure

To prove that such structure can act like ENZ-material, number of numerical simulations was done with the periodic boundary conditions. Simulations were done also in two steps: the first is for the unit cell and the second is for the one-layered structure.

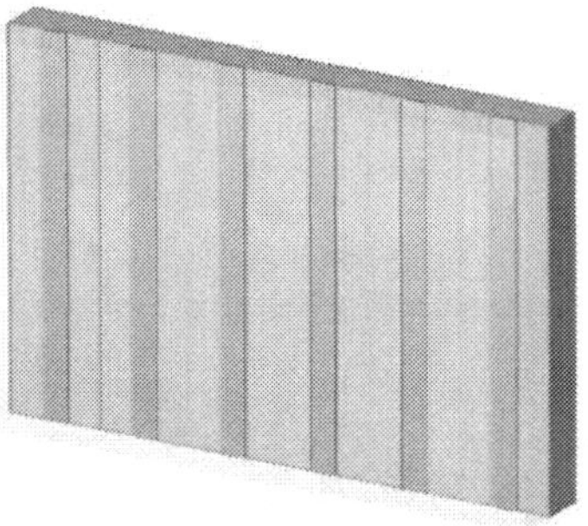

Fig. 1. The investigated structure: dielectric PET (Polyethylene terephthalate) substrate and parallel aluminum stripes.

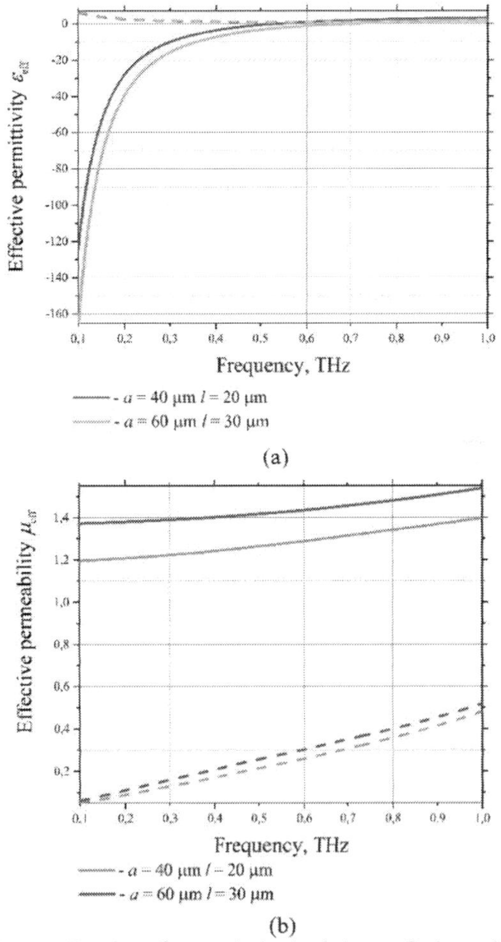

Fig. 1. Results of numerical simulations of the striped metamaterial (a is the lattice constant, l is the width of the stripe). 'Electric' plasma frequency for l = 20 μm is 0,641 THz and for l = 30 μm is 0,562 THz.

Experimental verification

The material was made by laser ablation method using LaserGraver LG2500 SP47. Two samples were used for the experiment: ones with l = 20 μm a = 40 μm and l = 30 μm a = 60 μm. Examined samples were measured using time-domain terahertz spectroscopy in 0,1-1 THz frequency range in transmission mode.

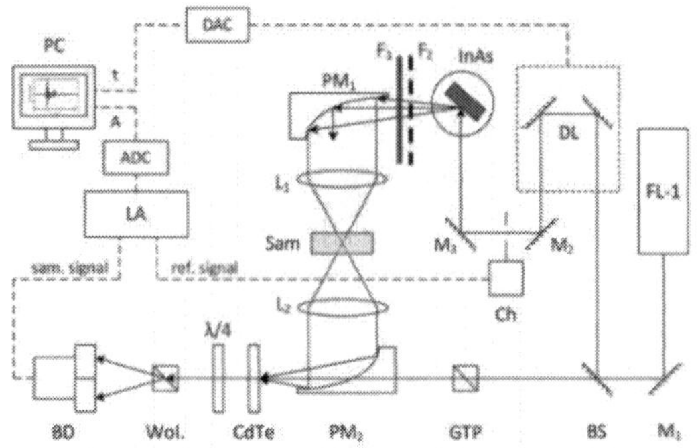

Figure. 1. Experimental THz TDS setup[9] in transmission mode (FL-1 – femtosecond laser based on potassium-yttrium tungstate crystal activated with ytterbium (Yb: KYW), generating femtosecond pulses; F1,2 – teflon filters for IR wavelength range cutting off, BS – beamsplitter, DL – optical delay line, M1,2,3 – mirrors, Sam – investigated sample, Wol. – Wollaston prism, CdTe – electro-optical cadmium-telluric crystal, BD – balanced detector, LA – lock-in amplifier, PC – personal computer, GTP – Glan-Taylor prism, PM1,2 – parabolic mirrors, Ch – chopper, DAC – digital to analog converter, ADC – analog to digital converter)

The experiment showed that samples have the dispersion of refractive index according to Drude model (Figure 8). In a little area near plasma frequencies samples behave as ENZ-material. It should be mentioned that below plasma frequencies medium shows left-handed properties in materials previously designed[10,11].

(a)

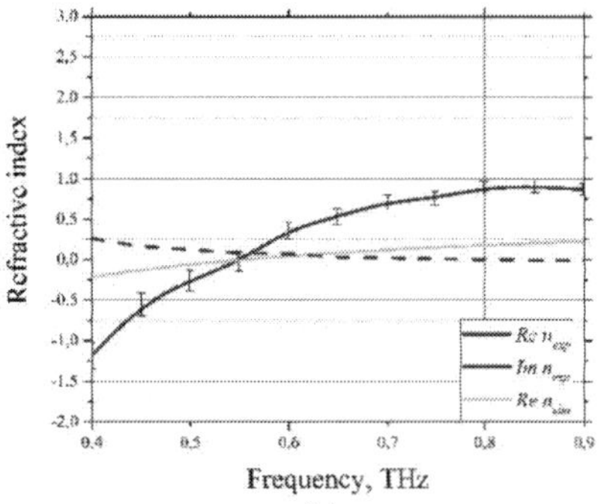

(b)

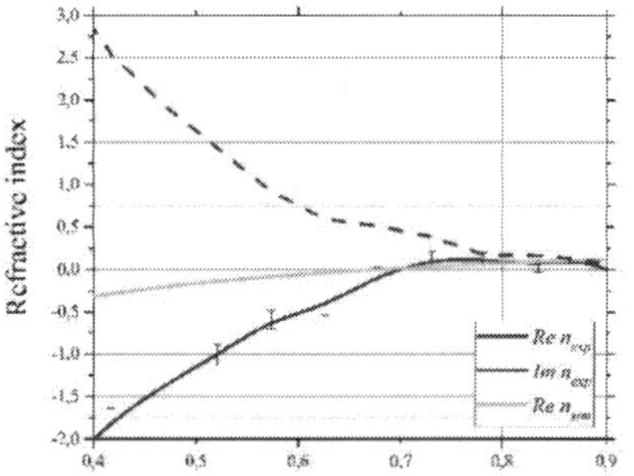

Figure 2. Experimental (blue) and calculated (cyan) effective permittivity and permeability of two striped metamaterial samples. (a) is the data on s ample with $l = 20$ μm $a = 40$ μm and (b) is the data on sample with $l = 30$ μm $a = 60$ μm. Plasma frequencies are approximately 0,556 THz for (a) and 0,648 THz for (b).

Conclusion

This paper showed the possibility of designing ZI-materials for terahertz frequency range using one-layered copper wire or aluminum stripe lattices. Copper-wired medium can be manufactured using electrochemical etching or electron-beam lithography. Aluminum-striped material was made from metallized PET substrate using laser ablation. ENZ properties of metal-striped materials were showed using numerical simulation and verified experimentally. Two striped-material samples were about to have effective plasma frequency in terahertz frequency range and showed left-handed properties below the plasma frequency.

References

[1] Engheta, N. and Liberal, I., "The rise of near-zero technologies," Science **358**(6370), 1540–1541 (2017).

[2] Ziolkowski, R. W., "Propagation in and scattering from a matched metamaterial having a zero index of refraction," Physical Review E **70**(4) (2004).

[3] Enoch, S., Tayeb, G., Sabouroux, P., Guérin, N. and Vincent, P., "A Metamaterial for Directive Emission," Physical Review Letters **89**(21) (2002).

[4] Alu, A., Engheta, N., Erentok, A. and Ziolkowski, R., "Single-Negative, Double-Negative, and Low-index Metamaterials and their Electromagnetic Applications," IEEE Antennas and Propagation Magazine **49**(1), 23–36 (2007).

[5] Islam, S. S., Faruque, M. R. I. and Islam, M. T., "An Object-Independent ENZ Metamaterial-Based Wideband Electromagnetic Cloak," Scientific Reports **6**(1) (2016).

[6] Javani, M. H. and Stockman, M. I., "Real and Imaginary Properties of Epsilon-Near-Zero Materials," Physical Review Letters **117**(10) (2016).

[7] Kyoung, J., Park, D. J., Byun, S. J., Lee, J., Choi, S. B. and Park, S., "Epsilon-Near-Zero meta-lens for high resolution wide-field imaging," Optics Express **22**(26), 31875 (2014).

[8] Shi, K. and Lu, Z., "Field-effect optical modulation based on epsilon-near-zero conductive oxide," Optics Communications **370**, 22–28 (2016).

[9] Bespalov, V. G., Gorodetskiĭ, A. A., Denisyuk, I. Y., Kozlov, S. A., Krylov, V. N., Lukomskiĭ, G. V., Petrov, N. V. and Putilin, S. é., "Methods of generating superbroadband terahertz pulses with femtosecond lasers," Journal of Optical Technology **75**(10), 636 (2008).

[10] Khodzitsky, M. K., Tarapov, S. I., Belozorov, D. P., Pogorily, A. M., Tovstolytkin, A. I., Belous, A. G. and Solopan, S. A., "Negative permittivity and left-handed behavior of doped manganites in millimeter waveband," Applied Physics Letters **97**(13), 131912 (2010).

[11] Khodzitsky, M. K., Kharchenko, A. A., Strashevskyi, A. V. and Tarapov, S. I., "Left-Handed Properties of Metal-Ferrite Composites Placed into Waveguide in Millimetric Wave Range," Telecommunications and Radio Engineering **68**(7), 561–566 (2009).

Angle-Susceptible Sensing Metasurface in Terahertz Regime

N. A. Nikolaev[1, 2], S. A. Kuznetsov[3, 4], M. Beruete[5]

[1]Institute of Automation and Electrometry SB RAS, Novosibirsk, Russia, Nazar@iae.nsk.su
[2]Institute of Laser Physics SB RAS, Novosibirsk, Russia
[3]Rzhanov Institute of Semiconductor Physics SB RAS, Novosibirsk Branch "TDIAM", Novosibirsk, Russia
[4]Novosibirsk State University, Novosibirsk, Russia
[5]Public University of Navarra, Pamplona, Spain

Introduction

Nowadays thin-film coatings and structures are widely used in advanced industrial and scientific applications that makes the tasks of thin-film sensing highly demanded in practice. Last decade, stimulated by progress in terahertz (THz) instrumentation, a keen interest has been attracted to the THz spectral range to develop its potential for detecting and measuring properties of thin films. The THz radiation can be an alternative to visible and IR waves when examining optically opaque coatings. Meanwhile, due to a relatively large wavelength λ, the conventional spectroscopic methods (TDS-, FDS-, FTIR-, BWO-based) are ill-suited for direct characterization of films with the thickness d of about 2–4 orders of magnitude smaller than λ. This problem can be solved with metamaterials, in particular, with plasmonic metasurfaces (PMSs) [1–5]. The plasmonic resonance exhibits a high sensitivity of its spectral response to the dielectric environment due to a strong field localization what makes possible measuring of analyte layers satisfying $d \ll \lambda$ condition.

The traditional approach of THz thin-film sensing with PMSs is based on detecting a frequency shift of the resonance when the analyte is deposited onto the PMS. In this work, we present the idea to substitute THz spectral measurements for tracking the PMS response at a fixed wavelength upon changing the incidence angle θ of the exciting THz beam. This concept works well for the PMS with a narrowband resonance sensitive to θ.

The results of the numerical investigations and experimental study of such PMS designed as a single-layer array of hexagon-shaped annular slots (Fig. 1) with angle-susceptible resonant transmission near 0.85 THz are presented.

Metasurface design

Originally, the proposed PMS was intended for band-pass filtering applications in experiments on studying sub-THz emission from hot dense plasma during relativistic electron beam-plasma interaction [6, 7]. The listed PMS dimensions provide a resonant pass band centered at 0.5 THz with the relative bandwidth of 63%. Later, when inspecting the angular performance of the PMS of this kind, a higher-frequency "spurious" narrow-band resonance susceptible to the angle of incidence was revealed in the vicinity of 0.8–0.9 THz. A detailed study of this resonance for the purpose of its applicability to thin-film sensing became the essence of the current work.

To pattern the PMS, we employed a contact photolithography technique [8, 9] which was specifically adapted to flexible solid film substrates, such as PP, whose industrial production does not allow obtaining a liquid material suitable for posterior film deposition via spin coating.

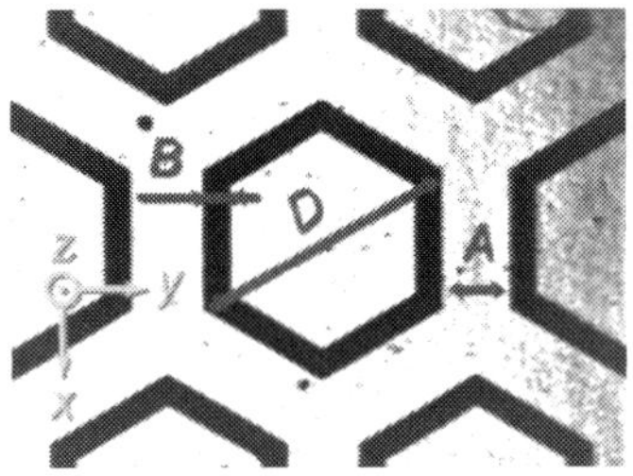

Fig. 1. Reflection-mode microphotograph of the fabricated plasmonic metasurface. The metallic area is light-colored. The slots are patterned in a 0.35 μm thick aluminum layer deposited on a 15 μm thick polypropylene film. D = 220 μm, B = 20 μm, A = 20 μm. The coordinate system xyz is introduced to show the structure orientation relative to the incident wave vector $k = (k \cdot sin\theta \cdot cos\phi,\ k \cdot sin\theta \cdot sin\phi,\ k \cdot cos\theta)$.

Experimental setup

The transmission spectra of the designed metasurface were measured on a custom-made THz time-domain spectrometer (TDS) described in Refs. [10, 11]. The PMS prototype under study was mounted on a manually controlled rotary stage capable of tilting the prototype both in TE and TM planes. The measurements were carried out by varying the incidence angle θ of the probing THz beam within 0°–55° with 5° incremental step. The THz-TDS signals were acquired with the time resolution of 125 fs in the range of 100 ps. This corresponds to the resolution of ~10 GHz in the frequency spectrum. The transmission was calculated as the ratio of the Fourier-transform spectra for the time-domain signals registered with and without the PMS prototype. The resulting experimental spectra were averaged over four independent measurements.

Results

Fig. 2 illustrates the typical spectral behavior of the angle-susceptible PMS resonance for different thicknesses d of the analyte overlayer plotted at a fixed angle of incidence θ. A photoresistive material was used as the analyte, whose dielectric permittivity $\varepsilon = 2.7\text{–}j \cdot 0.25$ was retrieved from direct transmission measurements of a 100 μm thick liquid cell.

To elucidate the idea of replacing THz spectral measurements by tracking the PMS response at a fixed λ upon changing the incidence angle θ, the PMS transmittance as a function of θ is plotted in Fig. 3. The latter corresponds to the frequency of 830 GHz chosen as an example.

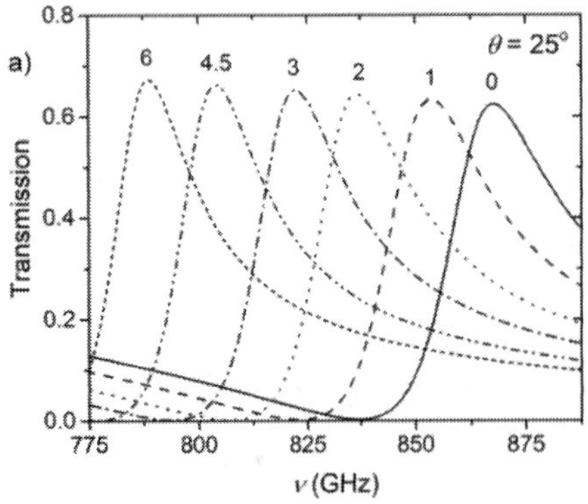

Fig. 2. Transmission spectra for the PMS covered with the analyte of different thicknesses d simulated at θ=25°.

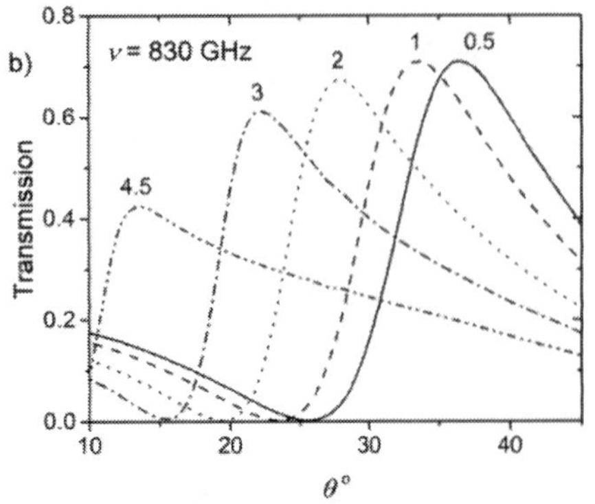

Fig. 3. PMS transmission at 830 GHz as the function of the incident angle θ modelled for different thicknesses d of the analyte overlayer. The numbers indicate the values of d in micrometers.

The presented data remarkably demonstrate the expected effect of shifting the angular peak due to thickening the analyte. Note, the resonance peak amplitude decreases and its bandwidth increases when the analyte thickness d changes from 0.5 to 4.5 μm. This means that the method sensitivity degrades for the thicknesses larger than 4.5 μm. Nevertheless, the peak remains almost stable in amplitude for the thicknesses $d < 1$ μm and the angles of both maximum and minimum transmission can be used to evaluate the analyte thickness. Thus, the proposed approach is estimated to be promising for THz sensing of sub-μm-thick analyte layers, thereby capable of detecting the level of $d/\lambda \sim 10^{-3}$ at least.

Acknowledgements

The work is partially supported by the Russian Foundation for Basic Research (Project #17-32-80039).

The authors gratefully acknowledge the Shared Equipment Center "Spectroscopy and Optics" of the Institute of Automation and Electrometry, SB RAS for the provided instrumental support of THz measurements.

References

1. *M. Tonouchi* Cutting-edge terahertz technology // *Nature photonics*. 2007. V. 1, No. 2. P. 97-105.
2. *M. Perenzoni and D. J. Paul, Eds. Physics and applications of Terahertz radiation,* Dordrecht, Netherlands: Springer, 2014.
3. *D. M. Mittleman* Perspective: Terahertz science and technology // *J. Appl. Phys.* 2017. V. 122, No. 23, P. 230901.
4. *F. F. Sizov* Infrared and terahertz in biomedicine // *Semiconductor Physics, Quantum Electronics & Optoelectronics*. 2017. V. 20, No. 3. P. 273-283.
5. *H. J. Song and T. Nagatsuma, Eds. Handbook of terahertz technologies: devices and applications,* Boca Raton, U.S.: CRC press, 2015.
6. *A. V. Arzhannikov, A. V. Burdakov, V. S. Burmasov, et al.* Observation of spectral composition and polarization of sub-terahertz emission from dense plasma during relativistic electron beam-plasma interaction // *Phys. Plasmas*. 2014. V. 21, No. 8. P. 082106.
7. *A. V. Burdakov, A. V. Arzhannikov, V. S. Burmasov et al.* Microwave generation during 100 keV electron beam relaxation in GOL-3 // *Fus. Sci. Technol.* 2013. V. 63, No. 1T. P. 286-288.
8. *M. Navarro-Cia, S. A. Kuznetsov, M. Aznabet et al.* Route for bulk millimeter wave and terahertz metamaterial design // *IEEE J. Quantum Electron.* 2011. V. 47, No. 3. P. 375-385.
9. *S. A. Kuznetsov, A. V. Arzhannikov, V. V. Kubarev et al.* Development and characterization of quasi-optical mesh filters and metastructures for subterahertz and terahertz applications // *Key Eng. Mat.* 2010. V. 437, P. 276-280.
10. *Mamrashev A. A., Maximov L. V., Nikolaev N. A., Chapovsky P. L.* Detection of nuclear spin isomers of water molecules by terahertz time-domain spectroscopy // *IEEE transactions on terahertz science and technology*. 2018. V. 8, No. 1. P. 13–18.
11. *Antsygin V.D., Mamrashev A.A., Nikolaev N.A., Potaturkin O.I., Bekker T.B., Solntsev V.P.* Optical properties of borate crystals in terahertz region // *Optics Communications*. 2013. V. 309. P. 333–337.

EPJ Web of Conferences **195**, 06011 (2018) https://doi.org/10.1051/epjconf/201819506011
TERA-2018

Far IR continuum absorption of $H_2{}^{16}O$ and $H_2{}^{18}O$

T.A. Odintsova[1], M.Yu Tretyakov[1], A.O. Zibarova[1,2], O. Pirali[3], P. Roy[3] and A. Campargue[4]

[1]Institute of Applied Physics, Russian Academy of Sciences, Nizhniy Novgorod, Russia, odintsova@ipfran.ru
[2]Lobachevsky State University of Nizhny Novgorod, Nizhny Novgorod, Russia
[3]SOLEIL Synchrotron, L'Orme des Merisiers, Saint-Aubin 91192, Gif-Sur-Yvette, France
[4]Univ. Grenoble Alpes, CNRS, LIPhy, 38000 Grenoble, France

Water vapor is the main greenhouses gas and it is responsible for about 70% of solar radiation absorption by the Earth atmosphere. Water vapor absorption consists of resonance absorption (sum of water lines) and continuum absorption (smoothly varying with frequency). Even though continuum is several orders of magnitude weaker than absorption at the center of absorption lines, its integrated contribution to total atmospheric absorption is significant and exceeds contribution of other greenhouse gases, such as CO_2 and CH_4. Thus, appropriate calculation of radiation balance, modeling of global climate changes and atmosphere remote sensing require accounting of the water vapor continuum.

Nowadays the semi-empirical MTCKD model [1] is the standard model implemented in radiative transfer codes to describe the water absorption continuum. This model is a far wing model constrained on experimental data obtained in the IR frequency range mainly at room or elevated temperatures. Extrapolation of this model to other frequency ranges (including terahertz region) and atmospheric temperatures are hazardous and might deviate from new experimental data [2].

Appropriate description of water vapor continuum absorption requires development of physically based model. In spite of long history of theoretical and experimental investigation of the continuum, it remains poorly characterized and understood. Its nature is not fully settled yet. Nevertheless, it was established, that the continuum originates from non-ideality of water vapor and its magnitude is quadratic function of pressure (self continuum). There are several possible sources of continuum: intermediate wings of resonance lines, bound and quasi-bound water dimer absorption and collision induced absorption.

Our investigation of the continuum absorption in the frequency range of water molecule rotational band is motivated by: (i) filling the spectral gap on the continuum data, and (ii) determination of water dimer and intermediate wings contributions to the continuum.

The weakness of the continuum makes to use a spectrometer with a high sensitivity and a high stability. In the current study the continuum was measured using high resolution Fourier transform spectrometer IFS-125HR equipped with multipass gas absorption cell with total pathlength of 151 m, that allowed to distinguish weak continuum absorption even at few mbar of water vapor pressure. The measurements were performed at the AILES beamline of SOLEIL synchrotron. The high power and high stability of the synchrotron radiation allows increasing sensitivity of the spectrometer up to 15 times in comparison with standard laboratory sources. To cover the frequency range of the rotational band, the coherent (15-35 cm^{-1}) and standard (50-500 cm^{-1}) radiation modes of the synchrotron were used. In order to refine the magnitude and clarify the physical origin of the continuum, spectra of the two major water isotopologues, $H_2{}^{16}O$ and $H_2{}^{18}O$, were considered. Recordings at several water vapor pressures were used to check the expected quadratic pressure dependence of the continuum.

The continuum was obtained by difference between experimental absorption α_{EXP} and modeled resonance absorption α_{WML}:

$$\alpha_C = \alpha_{EXP} - \alpha_{WML}.$$

Resonance absorption was modeled as sum of all significant lines of 7 most abundant water isotopologues:

$$\alpha_{WML}(\nu) = \sum_{j=1}^{7}\sum_{i} S_i^j \cdot \Phi(\nu),$$

where $\Phi(\nu)$ is Van Vleck - Weisskopf profile with commonly accepted wings cut off at 25 cm^{-1} and line shape parameters (including line strength S_i, center frequency and collisional broadening coefficient) from HITRAN 2016 database [3].

The continuum cross-section was determined as

$$C(\nu,T) = \frac{\alpha_C(\nu)\cdot k_B T}{P^2}$$

where T is temperature, k_B - Boltzmann constant and P - water vapor pressure.

Water vapor absorption spectra in the frequency range of 15-35 cm^{-1} were recorded at two pressures (11 and 16 mbar of natural water and 13 and 16 mbar of water sample enriched with O^{18}). The coincidence of the continuum cross-sections corresponding to two pressures is a validation of the pressure squared dependence. Fig.1. compares our results to literature measurements [4-10] in the range up to 60 cm^{-1} together with a calculated spectral function of $(H_2{}^{16}O)_2$ dimer [11]. The experimental data show a good agreement with the frequency dependence of the function. The $H_2{}^{18}O$ continuum shows a similar spectrum in this range (Fig.2.). These facts confirm the dominating contribution of water dimer absorption to the observed continuum. It should be mentioned, that the MTCKD model does not account satisfactorily of the observed frequency dependence in this region (see Fig. 1).

In the 50-500 cm^{-1} range, $H_2{}^{16}O$ and $H_2{}^{18}O$ continuum spectra were recorded for a number of pressures between 2 and 6 mbar. The very good quadratic pressure dependence of the measured continuum was obtained. The continuum absorption corresponding to $H_2{}^{16}O$ and $H_2{}^{18}O$ molecules are similar in shape and magnitude. Non-smooth frequency dependence of the $H_2{}^{16}O$ and $H_2{}^{18}O$ continua is observed between retrievals performed in nearby micro-windows (Fig.3). Inadequate modern line shape modeling in the range of intermediate wing (detuning of 5-10 cm^{-1} from line center) is believed to be responsible of the obtained fluctuations. We believe that investigation of isotopic dependence of the continuum could help to determine contribution of intermediate wings of resonance lines.

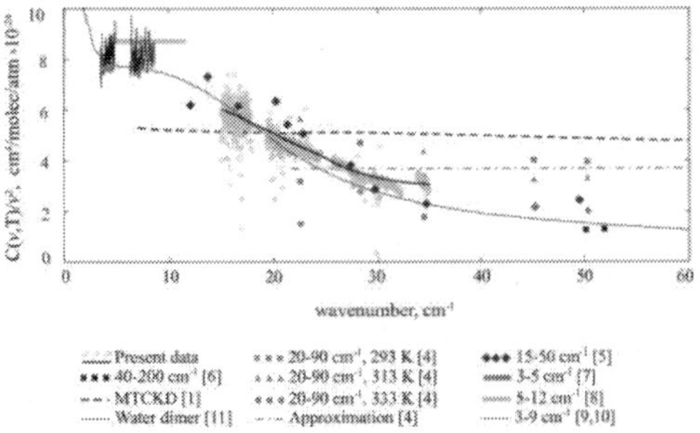

Fig. 1. Calculated water dimer spectral function [11], water vapor ($H_2{}^{16}O$) self-continuum cross-section normalized by frequency squared: obtained data, results of previous measurements.

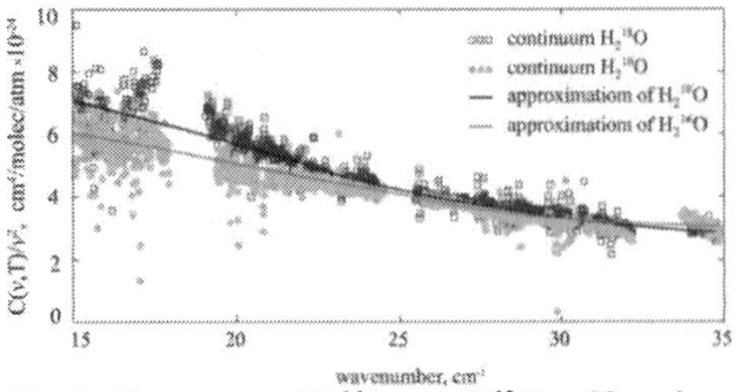

Fig. 2. Water vapor ($H_2{}^{16}O$ and $H_2{}^{18}O$) self-continuum cross-section normalized by frequency squared.

The new data on $H_2{}^{16}O$ continuum are in a good agreement with results of our previous measurements in the range of 40-200 cm^{-1} [6] and with results obtained by Burch in the range of 350-800 cm^{-1} [5]. New data filled the 200-350 cm^{-1} "gap" where the continuum was not measured before (Fig.3). The data confirmed that the widely used in atmospheric applications MT-CKD model [1] overestimates the continuum amplitude by about 30% in the whole range of the rotational band of H_2O (Fig.3).

This work was partially supported by grant № 18-02-00705 of Russian Foundation for Basic Research.

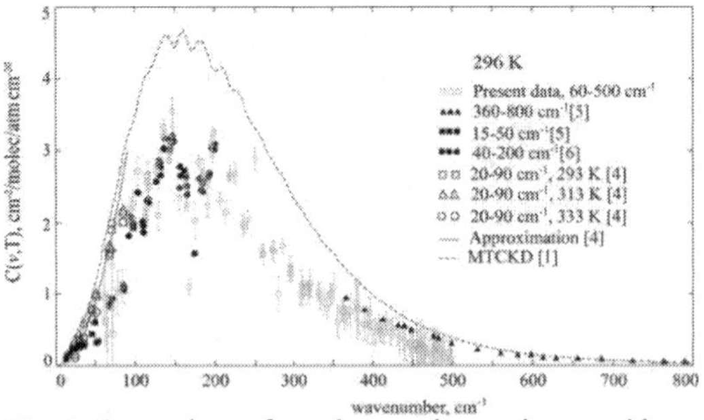

Fig. 3. Comparison of our data on the continuum with results of previous experimental studies [4-6] and MTCKD model [1].

References

1. *Mlawer E. J., et al.* Development and recent evaluation of the MT_CKD model of con-tinuum absorption // Philos. Trans. R. Soc. A. 2012. No. 370. P. 2520 – 56.

2. *Campargue A., et al.* Accurate laboratory determination of the near infrared water vapor self-continuum. A test of the MT_CKD model. J. Geophys. Res. Atmos. 2016. No. 13. P. 180 - 203.

3. *Gordon I. E., et al.* The HITRAN2016 Molecular Spectroscopic Database. J. Quant. Spectrosc. Rad. Trans. 2017. No. 203. P. 3–69.

4. *Podobedov V. B.,et al.* New measurements of the water vapor continuum in the region from 0.3 to 2.7 THz. J. Quant. Spectrosc. Rad. Trans. 2008. No. 109. P. 458 – 67.

5. *Burch D.E.* In: Continuum absorption by H2O. 1982. Report No. AFGL-TR-81-0300

6. *Odintsova T. A. et al.* Water Vapor Continuum in the Range of Rotational Spectrum of H2O Molecule: New Experimental Data and Their Comparative Analysis. J. Quant. Spectrosc. Rad. Trans. 2017. No.187. P. 116–123

7. *Koshelev M. A., et al.* Millimeter Wave Continuum Absorption in Moist Nitrogen at Temperatures 261–328 K. J. Quant. Spectrosc. Rad. Trans. 2011. V. 112. No. 270. P. 4 – 12.

8. *Kuhn T., et al.* Water vapor continuum: absorption measurements at 350 GHz and model calculations. J. Quant. Spectrosc. Rad. Trans. 2002. No. 74. P. 545 – 62.

9. *Tretyakov M. Yu., et al.* Water Dimer Rotationally Resolved Millimeter-Wave Spectrum Observation at Room Temperature. Phys. Rev. Lett. 2013. No. 110. P. 093001

10. *Serov E. A., et al.* Rotationally Resolved Water Dimer Spectra in Atmospheric Air and Pure Water Vapour in the 188–258 GHz Range. Phys. Chem. Chem. Phys. 2014. V. 16. No. 47. P. 26221 – 33.

11. *Scribano Y., Leforestier C.* Contribution of water dimer absorption to the millimeter and far infrared atmospheric water continuum. J. Chem. Phys. 2007. No. 126. P. 234301.

 https://doi.org/10.1051/epjconf/201819506012

Oxide nonlinear crystals: optical properties and phase-matching for terahertz wave generation

O. I. Potaturkin[1], V. D. Antsygin[1], A. A. Mamrashev[1], N. A. Nikolaev[1], Yu. M. Andreev[2,3], G. V. Lanskii[2,3], V. A. Svetlichnyi[3], K. A. Kokh[4]

[1]Institute of Automation and Electrometry SB RAS, Novosibirsk, Russia, potaturkin@iae.nsk.su
[2]Institute of Monitoring of Climatic and Ecological Systems SB RAS, Tomsk, Russia
[3]Siberian Physical Technical Institute of Tomsk State University, Tomsk, Russia
[4]Institute of Geology and Mineralogy SB RAS, Novosibirsk, 630090, Russia

Introduction

Oxide nonlinear crystals such as beta-barium borate (β-BaB_2O_4 or β-BBO), potassium titanyl phosphate ($KTiOPO_4$ or KTP), potassium titanyl arsenate ($KTiOAsO_4$ or KTA) are widely used for efficient harmonic, difference frequency and optical parametric generation/oscillation in the visible and near-infrared ranges [1]. The crystals have wide transparency range stretching from 0.19 μm for β-BBO, 0.35 μm for KTP and KTA to near-IR or even mid-IR range. They also have moderate nonlinear coefficients and high optical damage threshold.

Recently these crystals attracted attention as prospective materials for terahertz applications. Several studies were devoted to measuring their optical properties (refractive index and absorption coefficient) in the terahertz domain [2–5]. Different schemes for IR-to-THz and THz-to-THz conversion were investigated [6–9]. Besides nonlinear crystals another attractive materials such as metals and semi-metals were considered for efficient conversion of near-IR femtosecond pulses to terahertz radiation on their surface [10].

In this study we measure terahertz optical properties and discuss possibilities of applying oxide nonlinear crystals, metals and semi-metals for terahertz wave generation.

Selected results

Absorption coefficient and refractive index dispersions in the THz range for all axes of nonlinear oxide crystals were measured at the room (RT), liquid nitrogen (LN), and liquid helium (LH) temperatures within the range of 0.2-2.5 THz using custom-made terahertz time-domain spectrometer (THz-TDS) described elsewhere [2].

Absorption coefficient and refractive index of β-BBO crystals for ordinary and extraordinary waves were measured at RT and LN temperatures of 293 K and 81 K, respectively. Sellmeier equations for both *o*- and *e*-waves were formulated. Dispersion equations for at RT valid in the range of 187–10000 μm:

$$n_o^2 = 2.040 + \frac{0.816\lambda^2}{\lambda^2 - 12815},\; n_e^2 = 2.478 + \frac{0.160\lambda^2}{\lambda^2 - 15598} \quad (1)$$

Dispersion equations at LN temperature valid in the range of 182–3530 μm for *o*- and 144–2727 μm for e-wave:

$$n_o^2 = 4.612 + \frac{3.117\lambda^2}{\lambda^2 - 16175},\; n_e^2 = 5.970 + \frac{0.800\lambda^2}{\lambda^2 - 16659} \quad (2)$$

Using resulting dispersion curves we were able to find phase-matching (PM) for down-conversion or difference frequency generation of near-IR radiation into the THz range was found possible by *o–o→e*, *e–e→o* and original *e–e→e* types (Fig. 1) of three-wave interactions.

It was found that despite the significant decrease in the absorption coefficient at LN temperature, refractive index components and birefringence did note change a lot keeping PM conditions favorable for efficient down-conversion into the THz range.

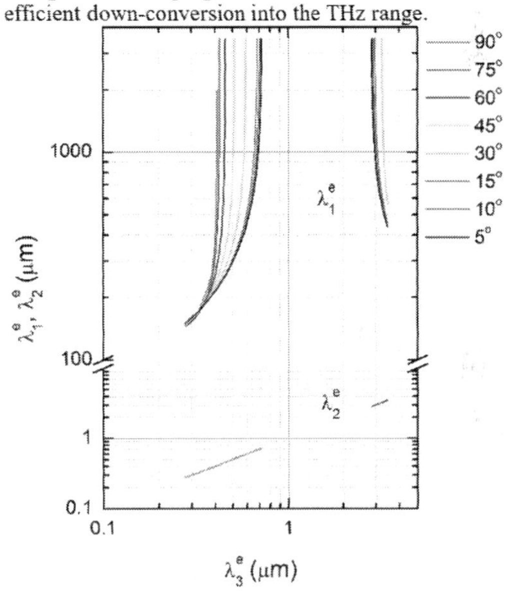

Fig. 1. Phase-matching curves for type I (*e–e→e*) three-wave interactions in β-BBO at RT. θ angles are shown in the figure inset

In a similar manner, terahertz optical properties of KTP crystals were measured at three temperatures. The refractive index for all three optical axes of KTP crystal was approximated in the form of Sellmeier equations. Here is a set of equations for LH temperature of 5.3 K where λ is in μm:

$$n_x^2 = 8.67715 + \frac{1.34263\lambda^2}{\lambda^2 - 11458}$$
$$n_y^2 = 8.86222 + \frac{1.22979\lambda^2}{\lambda^2 - 12045} \quad (3)$$
$$n_z^2 = 13.18625 + \frac{0.86157\lambda^2}{\lambda^2 - 21447}$$

Using dispersion equations for the visible, infrared and terahertz ranges we obtained comprehensive results for phase-matched difference frequency generation in KTP crystals at different temperatures and different wavelengths presented in the Fig. 2.

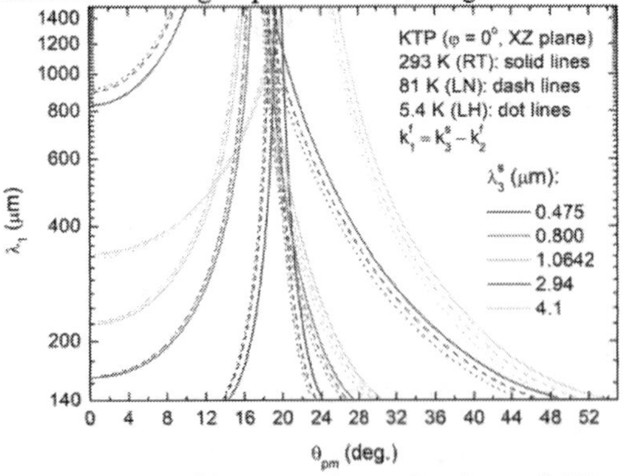

Fig. 2. Phase-matching curves for $s - f \to f$ type of difference frequency generation in the ***XZ*** principle plane of KTP crystal at RT (bold lines), LN (dash lines) and LH (dot lines) temperatures for different λ from 0.475 to 4.1 μm

KTA refractive index measured at RT was approximated in the form of the following Sellmeier equations in the range from 132 μm (n_x), 155 μm (n_y), 313 μm (n_z) to 3200 μm:

$$n_x^2 = 9.70796 + \frac{1.24017\lambda^2}{\lambda^2 - 12935}$$
$$n_y^2 = 9.62652 + \frac{1.44084\lambda^2}{\lambda^2 - 17674} \quad (4)$$
$$n_z^2 = 16.6706 + \frac{0.12511\lambda^2}{\lambda^2 - 85811}$$

Analogous phase-matching curves for DFG of Nd:YAG laser in KTA crystals is shown in the Fig. 3.

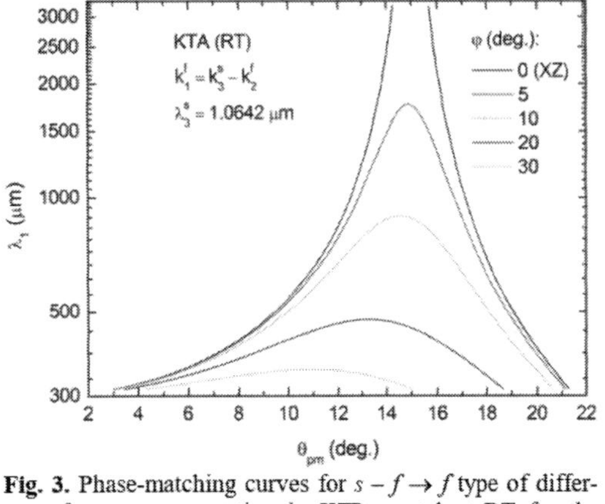

Fig. 3. Phase-matching curves for $s - f \to f$ type of difference frequency generation in KTP crystal at RT for the wavelength of Nd:YAG lasers

A common feature of all three abovementioned crystals was that cooling them down did not significantly affect phase-matching conditions. At the same time their absorption coefficient lowered drastically making cooled crystals more effective for terahertz wave generation.

The crystals were tested as narrowband THz radiation sources for a homemade spectrometer described elsewhere [11] having a long (>1 km) measuring path which consists of a transceiver adopted from a lidar complex [12].

We also studied terahertz generation on the surface of metals (Ti, Ni, Au, Cu) and their alloys ($Ti_{50.5}Ni_{49.5}$) pumped by femtosecond Ti:Sapphire laser. The power of generated radiation was measured using Golay cell. Fig. 4 shows that the output signal is linearly depends on the pump pulse energy. We also found that the efficiency of optical-to-terahertz conversion is proportional to the material conductivity.

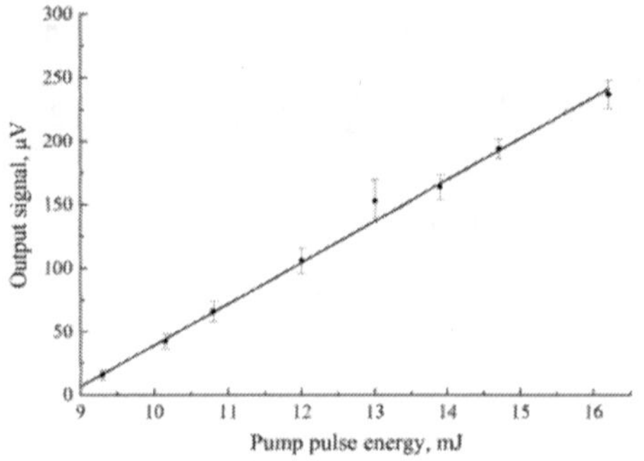

Fig. 4. Dependence of Golay cell signal proportional to the terahertz power generated on the surface of $Ti_{50.5}Ni_{49.5}$ alloy on the pump pulse energy

References

1. *Nikogosyan D.* Nonlinear optical crystals: a complete survey. New York: Springer, 2005.
2. *Antsygin V.D. et al.* Optical properties of borate crystals in terahertz region // Optics Communications. 2013. Vol. 309. P. 333–337.
3. *Antsygin V.D. et al.* Terahertz optical properties of potassium titanyl phosphate crystals // Optics Express. 2014. Vol. 22, Is. 21. P. 25436.
4. *Andreev Y.M. et al.* Observation of a different birefringence order at optical and THz frequencies in LBO crystal // Optical Materials. 2017. Vol. 66. P. 94–97.
5. *Nikolaev N.A. et al.* Terahertz optical properties of LBO crystal upon cooling to liquid nitrogen temperature // Quantum Electronics. 2018. Vol. 48, Is. 1. P. 19–21.
6. *Huang J.-G. et al.* Phase matching in RT KTP crystal for down-conversion into the THz range // Laser Phys. Lett. 2018. Vol. 15, Is. 7. P. 075401.
7. *Mamrashev A.A. et al.* Optical Properties of KTP Crystals and Their Potential for Terahertz Generation // Crystals. 2018. Vol. 8, Is. 8. P. 310.
8. *Nikolaev N.A. et al.* Optical properties of β-BBO and potential for THz applications // Journal of Physics: Conference Series. 2018. Vol. 951. P. 012003.
9. *Nikolaev N.A. et al.* Temperature dependence of terahertz optical properties of LBO and perspectives of applications in down-converters // Journal of Physics: Conference Series. 2018. Vol. 951. P. 012005.
10. *Mironov V.A., Oladyshkin I.V., Fadeev D.A.* Optical-to-THz radiation conversion on a semi-metal surface // Quantum Electronics. 2016. Vol. 46, Is. 8. P. 753–758
11. *Huang J.-G. et al.* Remote imaging by nanosecond terahertz spectrometer with standoff detector // Russian Physics Journal. 2018. Vol. 60, Is. 9. P 1638-1643.
12. *Andreev Yu. M., Geiko P.P., Sherstov I.V.* Development and testing of the lidar gas analyzing complex // Proc. SPIE. 1999. Vol. 3983, P. 386-394.

EPJ Web of Conferences **195**, 06013 (2018)
TERA-2018

https://doi.org/10.1051/epjconf/201819506013

Terahertz induced optical second harmonic generation from silicon surface

S.B. Bodrov[1,2], A.I. Korytin[1], Yu.A. Sergeev[1], A.N. Stepanov[1]

[1]Institute of Applied Physics of the Russian Academy of Sciences, Nizhny Novgorod, Russia, yas@ufp.appl.sci-nnov.ru
[2]University of Nizhny Novgorod, Nizhny Novgorod, Russia

Second harmonic generation (SHG) of optical radiation from surface is one effective technique for surface diagnostics [1]. It is well known, that in bulk media with inversion center the second harmonic generation is prohibited in the dipole approximation. However, the surface breaks the symmetry and allows SHG. In this regard, the second harmonic (SH) signal is extremely sensitive to the state of the surface. The SHG technique allows to measure inhomogeneities of the surface, dynamics of fast processes, such as fast melting under the action of femtosecond laser pulses, contains information about the surface states of the material. An application of the external field on the surface can significantly change the second harmonic signal [2,3].

In this work, we investigated influence of intense THz field on SHG from the silicon surface (111).

A Ti:Sp femtosecond laser system at 795 nm with energy of 0.7 mJ, duration of 70 fs and repetition rate of 700 Hz was used in the experiment. Optical radiation was divided into two beams (see Fig.1). The pump beam was used to generate THz radiation by the tilted intensity front technique in a LiNbO3 crystal [4,5]. The terahertz radiation was focused on the sample using a system of off-axis parabolas. The maximum THz field in focus reached a value of 300 kV/cm. For the attenuation of THz fields were used two THz polarizer. A probe laser pulse focused on the sample in the spot of about 100 μm (smaller than the size of the THz spot). The second harmonic radiation generated from the surface was detected by the PMT connected to the pulse counting system. The polarization properties of the SH was controlled by polarizer GP2. The special filters was used to block laser radiation.

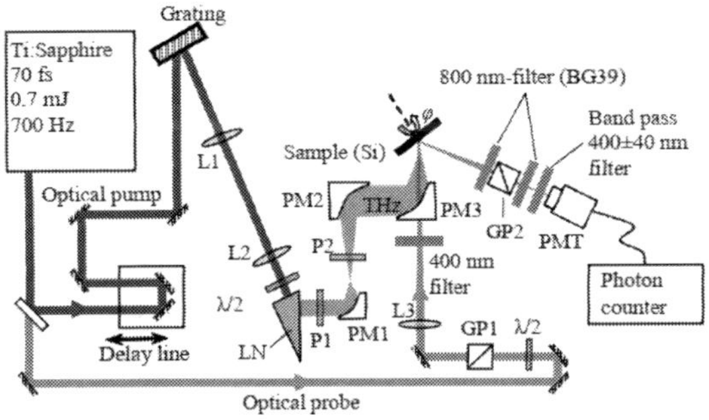

Fig. 1. Experimental setup (L1, L2, L3- lenses; PM1, PM2, PM3 – parabolic mirrors; LN – LiNbO3; GP1, GP2 – Glan-Taylor prism; P1,P2 – THz polarizer)

In the course of the experiment, the generation of second harmonic at different combinations of polarization of optical and THz radiation was investigated. Under application of THz field, the SH signal was significantly increased when second harmonic and THz electric fields were parallel. In the case when THz, first and second harmonic electric fields have *s* polarization, the SH signal increased by 30 times. The dependences of the SH signal on the energy of optical and THz pulses, on the delay time between them, and the dependence on the azimuthal angle (rotation angle of the sample) were investigated. In Fig. 2 an example of angular dependence is presented, where all fields (THz field, field of the first and second harmonics) have *p* polarization.

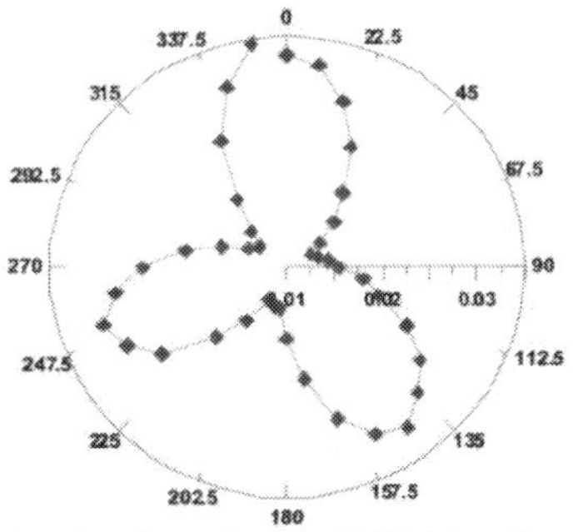

Fig. 2. Angular dependence of SH intensity: optics P-polarization, THz field P-polarization, SH P-polarization

To describe the experimental results, the phenomenological theory of second harmonic generation from (111) silicon surface was developed with account of presents of THz field. By comparison with experimental data, it was found that THz induced source of SH is in phase with the source of SH without THz field.

References

1. *Bloembergen, N., Chang, R. K., Jha, S. S., Lee, C. H.* Second-harmonic generation of light in reflection from media with inversion symmetry // Phys. Rev. 1966. V. 16, No. 22. P. 986-989.

2. *Nahata, A., Heinz, T.F.* Detection of freely propagating terahertz radiation by use of optical second-harmonic generation // Opt. Letters, 1998. V. 23, No. 1. P. 67-69.

3. *Lee, C. H., Chang, R. K., Bloembergen N.* Nonlinear electroreflection in silicon and silver// Phys. Rev. Lett. 1967. V. 18, No 5. P. 167-170.

4. *Fulop, J. A., Palfalvi, L., Almasi, G., Heblin, J.* Design of high-energy terahertz sources based on optical rectification // Opt. Express, 2010. V. 18, No. 12. P. 12311-12327

5. *Bodrov, S. B., Murzanev, A. A., Sergeev, Y.A., Malkov, Y.A., Stepanov A.N.* Terahertz generation by tilted-front laser pulses in weakly and strongly nonlinear regimes // Applied Physics Letters. 2013. V.103, No. 25. P. 251103

Dielectrics for output windows of medium power gyrotrons

Evgeny Serov[1], V. V. Parshin[1], S. V. Egorov[1], A. N. Konovalov[2], A. I. Makarov[1], K. V. Vlasova[1]

[1]Institute of Applied Physics of the Russian Academy of Sciences, Nizhny Novgorod, Russia, serov@ipfran.ru
[2]Quartz technologies Ltd, Shilovo, Ryazan' region, Russia

For achieving optimal characteristics of gyrotron it is necessary to choose a suitable material for output window. Balance of relatively low dielectric losses, slowly dependent on temperature, and relatively high thermal conductivity are required for good heat removal. Certainly, modern low loss CVD diamond windows meet these requirements, but their price is too expensive for wide application.

In this paper we present an analysis of properties of several dielectric materials: boron nitride (BN), which is currently widely used for gyrotron output windows production, crystal Quartz, and Spinel.

Basing on experimentally measured dielectric losses of these materials and on their thermal properties we made a calculation of maximal achievable output power that can be transmitted through disks made of such materials.

Boron nitride

Most of currently produced medium power (~10–100 kW) gyrotrons have output windows made of hexagonal pyrolytic BN. Its dielectric properties in sub-THz range were studied in our previous works [1,2]. Usual samples of BN have $\tan\delta \approx 10^{-3}$ at 140 GHz which is higher in comparison with other investigated materials. But the loss tangent temperature dependence is rather weak: it increases by only about 40% with temperature increasing from 25 to 700°C.

Another important characteristic is thermal conductivity of BN. Usual value used as estimate in practice is 45 W/m·K for the direction parallel the layers. However, according to the available publications [3,4], thermal conductivity of BN can reach at least 220 W/m·K for samples with high crystal perfection.

Spinel

Spinel ($MgAl_2O_4$) is ceramic material with relatively low dielectric loss which is resistant to high temperatures. In our work we studied spinel samples obtained by microwave heating of ultrafine powder components [5].

Measurements of dielectric parameters (n and $\tan\delta$) were made using high-Q resonator technique (see [1] and Refs. therein) in the frequency range 50–300 GHz at temperatures 20–200°C. Unlike BN, spinel demonstrates rather strong temperature dependence of dielectric loss. It is illustrated in Fig. 1. At temperature of about 400°C dielectric losses in spinel become higher than that in BN.

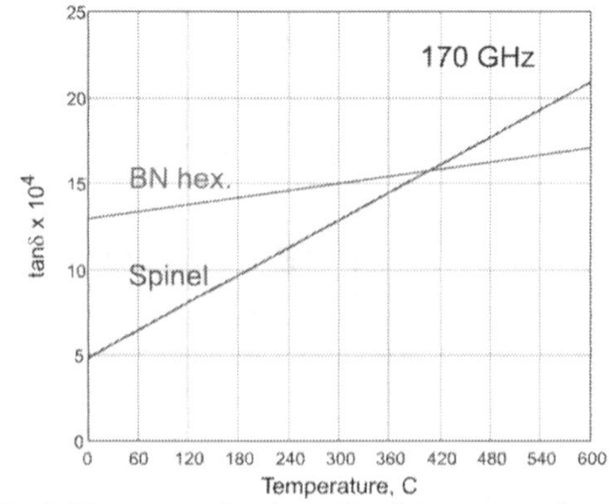

Fig. 1. Linear approximations of $\tan\delta$ temperature dependencies at f = 170 GHz based on experimental data for spinel and hexagonal BN.

Thermal properties of spinel were taken from [6]. In comparison with BN it has several times lower thermal conductivity (18 W/m·K at room temperature).

Quartz

Crystal quartz produced by Quartz Technologies Ltd was grown by the method of hydrothermal synthesis on the own raw material base of the deposit "Zhelannoe". The total concentration of impurities including alkali, gas-liquid and mineral elements does not exceed 2-3 ppm (chemical composition studies were carried out at the G.G. Devyatykh Institute of High-Purity Chemistry, ANZAPLAN analytical laboratories and QSIL laboratories). It has extremely low absorption coefficient in the optical range ($\leq 10^{-6}$cm^{-1} at $\lambda = 1\,\mu$m).

Investigation of quartz sample dielectric properties was made in frequency range 65–200 GHz at room temperature. Due to specific geometric sizes of the sample only n_o component and corresponding value of $\tan\delta$ were measured. Fig. 2 demonstrates experimentally measured frequency dependence of dielectric losses in quartz along the main axis. The losses in this sample are practically the lowest ones in comparison with earlier measured samples [1].

Thermal conductivity in quartz was taken from paper [7]: 6 W/m·K for [010] direction and 13 W/m·K for [001] direction at room temperature.

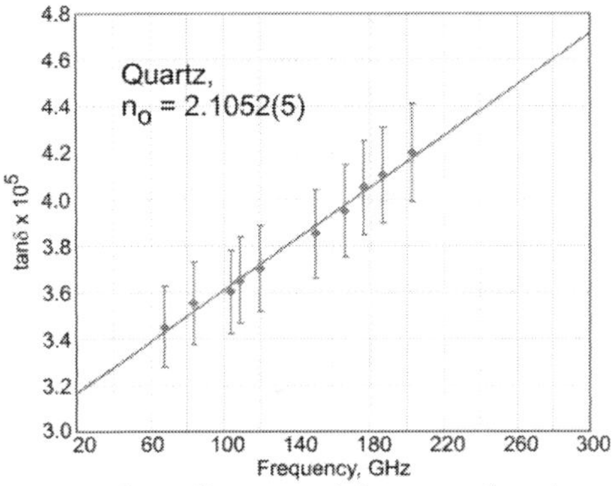

Fig. 2 Experimentally measured frequency dependence of dielectric losses in quartz for the direction parallel to the main crystal axis. Line is the linear regression.

Thermal calculation and discussion

Using data on dielectric losses and thermal properties of the samples we made a calculation of temperature distribution for a disks used as gyrotron output windows in the continuous-wave operation (CW). The calculation was made in a similar way as in [8]. We used Newton's law of cooling as the boundary condition on the perimeter. The coolant temperature was 30°C, and the coefficient of heat removal from the window perimeter was 4 W/cm^2·K. Heat flows from the window-vacuum and window-air boundaries were neglected.

The maximal output power for gyrotron windows made of BN and spinel was found as maximal beam power for which the temperature of the disk did not exceed 600°C. The thermal conductivity of BN was taken as 45 W/m·K. For quartz window the calculation is more estimative because we do not have information about the temperature dependence of tanδ. Besides, it is unlikely that quartz disk will withstand a large temperature gradient arising when the temperature of the disk center rises to 600°C. Thus, in our estimation of maximal power we limited maximal temperature of quartz to 300°C. For more strict calculation additional factors should be taken into account.

The window thickness was resonant, i.e., the optical thickness was equal to an integer number of half-wavelengths. The number of the half-wavelengths was chosen for the reason of the window strength. For disks with diameters of about 100 mm, the thicknesses ranged from 3 to 4 mm, and for disks with diameters of 30 mm (for f ~ 30 GHz), the thicknesses was 2-2.5 mm. Fig. 3 shows the maximal power transmitted through the window.

On the basis of Fig. 3 we can conclude that $MgAl_2O_4$ spinel is not optimal material for gyrotron output windows. Crystal quartz is a more suitable one, comparable to boron nitride in terms of maximum power level and even surpassing it at high frequencies. However, some problems associated with the use of crystal quartz may occur due to difficulty of soldering and possible mechanical stresses. Additional data on temperature dependence of tanδ are required.

Comparison of Fig. 3 (above) and Fig. 5 from [8] leads us to the conclusion that 6H-SiC disk can transmit several times more powerful beam than BN or quartz disks. But if we use BN with high crystal perfection and high thermal conductivity then such material can also provide several hundred kilowatts level for the CW mode in the sub-THz range.

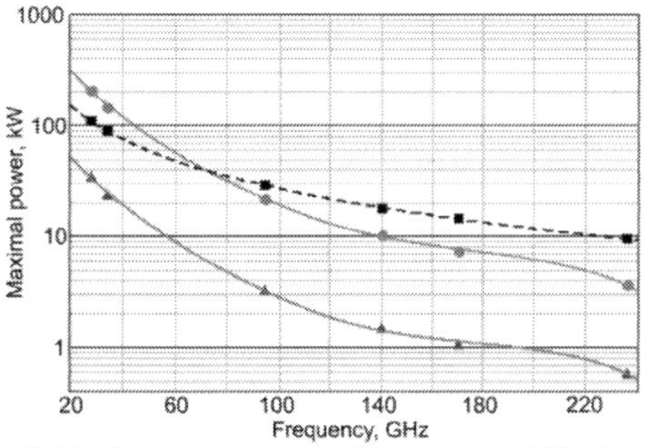

Fig. 3 Maximal output power for hexagonal BN (circles), spinel (triangles) and quartz (squares). Curves are smooth approximations. See text for details.

Acknowledgements

The work was supported by the Russian Foundation for Basic Research and the Government of the Nizhny Novgorod region, project No. 18-42-520015.

References

1. *Parshin, V. V., Tretyakov, M. Yu., Koshelev, M. A., Serov, E.A.* Instrumental complex and the results of precise measurements of millimeter- and submillimeter-wave propagation in condensed media and the atmosphere // Radiophys. & Quant. Electron. 2009, V. 52, No. 8. P. 525–535.

2. *Parshin, V. V.* Dielectric materials for gyrotron output windows // International Journal of Infrared and Millimeter Waves. 1994, V. 15, No. 2. P. 339–348.

3. *Sichel, E. K., Miller, R. E., Abrahams, M. S., Buiocchi, C. J.* Heat capacity and thermal conductivity of hexagonal pyrolytic boron nitride // Phys. Rev. B. 1976, V. 13, No. 10. P. 4607–4611.

4. *Duclaux, L., Nysten, B., Issi, J-P.* Structure and low-temperature thermal conductivity of pyrolytic boron nitride // Phys. Rev. B. 1992, V. 46, No. 6. P. 3362–3367.

5. *Egorov, S. V., Bykov, Yu. V., Eremeev, A. G., Sorokin, A. A., Serov, E.A., Parshin, V. V., Balabanov, S. S., Belyaev, A. V., Novikova, A. V., Permin, D. A.* Millimeter-Wavelength Radiation Used to Sinter Radiotransparent $MgAl_2O_4$ Ceramics // Radiophys. & Quant. Electron. 2017, V. 59, No. 8–9. P. 690–697.

6. *Harris, D. C., Linda F. Johnson, L. F., Seaver, R., Lewis, T. Turri, G., Bass, M., Zelmon, D. E., Haynes, D. N.* Optical and thermal properties of spinel with revised (increased) absorption at 4 to 5 μm wavelengths and comparison with sapphire // Optical Engineering. 2013, V. 52, No.8. 087113.

7. *Kanamori, H., Fujii, N., Mizutani, H.* Thermal Diffusivity Measurement of Rock-Forming Minerals from 300° to 1100° K // Journal of Geophysical Research. 1968, V. 73, No. 2. P. 595–605.

8. *Parshin, V., Serov, E., Denisov, G., Garin, B., Denisyuk, R., V'yuginov, V., Klevtsov, V., Travin, N.* Silicon carbide for high-power applications at MM and THz ranges // Diamond & Related Materials. 2017, V. 80, P. 1–4.

Effect of MIT in epitaxial VO_2 films on THz transmittance

D. I. Sharovarov[1,2], F. Ya. Akbar[1], D. P. Lelyuk[3], A. M. Makarevich[4], O. V. Boytsova[1,5], A. R. Kaul[2,4]

[1]Department of Material Science, Lomonosov Moscow State University, Moscow, Russia, dmitrii.sharovarov@gmail.com
[2]Oxyfilm LLC, Moscow, Russia
[3]Institute for Theoretical and Applied Electromagnetics RAS, Moscow, Russia
[4]Department of Chemistry, Lomonosov Moscow State University, Moscow, Russia
[5]Kurnakov Institute of General and Inorganic Chemistry RAS, Moscow, Russia

The development of THz technologies (communication, imaging, spectroscopy etc.) requires the design of materials for high-speed modulation of radiation in THz range. For such applications, the thin films based on oxides showed a metal-insulator transition (MIT) are interesting due to large different of optical transmittance before and after critical point. Among the oxide materials with MIT vanadium dioxide, which has a transition temperature the closest to the room temperature (68°C for bulk samples), is of particular interest, with the MIT characterized by a record speed (<1 ps) and a large amplitude (a change in the conductivity is 10^5 times for single crystals). It is known that the MIT transition in VO_2 is accompanied by a first-order phase transition, while the crystalline structure from the monoclinic (crystal type MoO_2) with the properties of the semiconductor transforms into a tetragonal (rutile type) with metallic conductivity. The electronic transition in the vanadium dioxide can be caused by the temperature, the electric field and laser radiation. A set of these unique properties makes vanadium dioxide a promising key component of optoelectronic devices (switches, modulators, lenses, etc.) for the THz range controlled by thermal action, electric voltage or laser pulse.

Recently the efforts of numerous researches were directed to the vanadium dioxide in the form of thin epitaxial films of thickness (100-500 nm) on optically transparent single-crystal substrates (Al_2O_3, TiO_2). The epitaxial VO_2 films exhibit the sharp transition with an amplitude changes in conductivity up to four orders of magnitude and showed above 90% change of THz transmission.

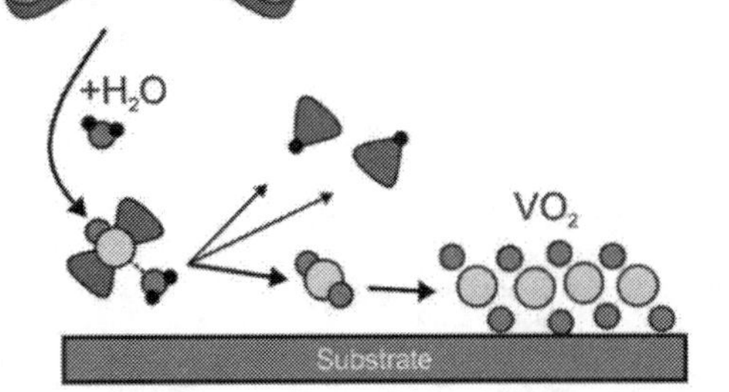

Fig. 1. The scheme of reaction between $VO(hfa)_2$ molecule and water resulted to formation of VO_2.

The properties of vanadium dioxide are very sensitive to the concentration of defects, morphology, contact between crystallites, impurities of other elements. The synthesis of epitaxial films of vanadium dioxide, extremely limited by a narrow region of homogeneity and a high sensitivity to the partial pressure of oxygen, is a complex problem, which resulted in a low reproducibility of the composition and properties of films obtained in different scientific groups.

The method of chemical vapor deposition is a promising method for the production (on an industrial scale) of vanadium dioxide films demonstrating the properties of an optical switch in the THz range, however, provided a simple and effective process for obtaining high-quality samples that are not contaminated with impurity phases. In the case of materials sensitive to the oxygen partial pressure, the use of water vapor for destruction of the vanadium containing molecules makes possible to obtain stoichiometric VO_2 due to stabilization of vanadium oxidation state +4 and to carry out the deposition process at low temperatures (> 350°C) without carbon contamination of the film. In our work we performed the deposition of epitaxial $(001)VO_2$ films on $r\text{-}Al_2O_3$ substrates (diameter 1-3 inches) by gas-phase reaction between vapors of vanadyl diketonates $VO(dik)_2$ ($dik^- = acac^-$, thd^-, hfa^-) and water in argon atmosphere at temperature range 350-600°C.

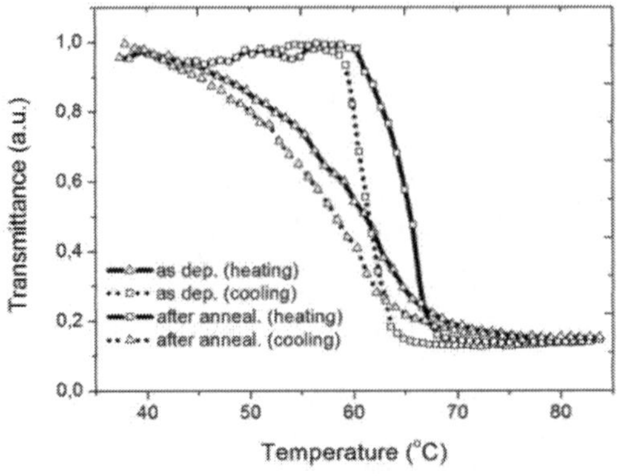

Fig. 2. The THz transmittance of $VO_2/r\text{-}Al_2O_3$ (as deposited and after annealing) films at different temperature.

Among the many factors that influences on the nature of the MIT in vanadium dioxide, the film microstructure plays one the most important role. Fine-grained films that have a large number of partially oxidized grain boundaries show a diffuse hysteresis loop and a smaller amplitude of the resistance change. The increase of the grain size can be achieved either by increasing the substrate temperature, or by using an additional heat treatment. In the first case, the formation of a partially oxidized grain surface is not prevented, whereas additional annealing successfully solves this problem due to the peritectic decomposition of the oxidized phases occurs to VO_2 and trace amounts of the melting phases promoting recrystallization process. The samples obtained by MOCVD process were annealed under conditions of controlled oxygen partial pressure at temperature range 550-650°C resulted in increasing of grain size, changing of morphology, electrical and optical properties. It was investigated the films composition, structure, morphology and physical properties.

Measurements of the THz transmission as a function of temperature were carried out on a pulsed THz spectrometer. It was demonstrated the effect of films synthesis on the contrast of transmission in the dielectric and conducting state. The shape of the hysteresis loop differs from the resistance curve and is somewhat shifted downward in temperature associated with percolation of conductive grains into the insulation matrix. For large-scale substrates the maps of THz amplitude change were analyzed.

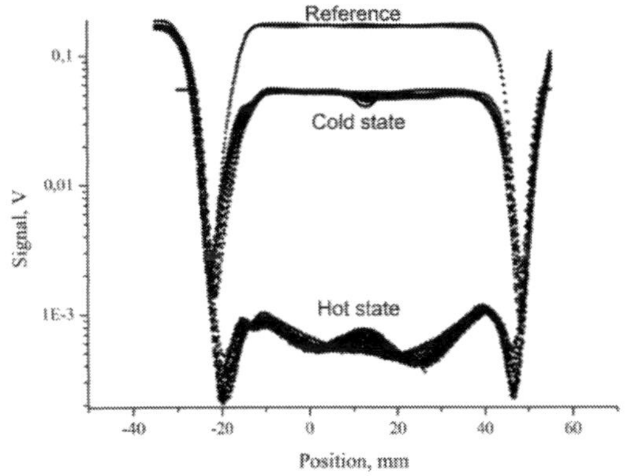

Fig. 3. The THz transmittance of 3-inch VO_2 films on r-Al_2O_3 at different position from the center (10 mm).

Moreover most of practical applications require non-trivial shape and design of functional optical elements, then it is important to have possibility to realize the chemical lithography operations that accurately was demonstrated in this work.

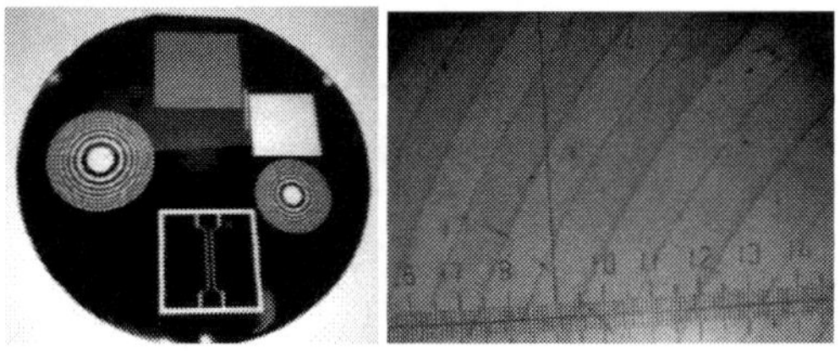

Fig. 4. The different patterns on 3-inch VO_2 film/Al_2O_3 obtained by photolitographic etching process.

The authors thank M.N. Esaulkov, P.M. Solyankin and A.P. Shkurinov (Institute on Laser and Information Technologies of RAS) for measurements of films THz transmittance. The work was supported by RSF 18-73-10212.

Investigation of the properties of a 3-level broadband antireflective structure on silicon by THz time-domain spectroscopy

I. A. Tzibizov[1], G. I. Kropotov[1], V.S. Pavelyev[2,3], K. N. Tukmakov[2,3], A. S. Reshetnikov[2]

[1]Tydex LLC, 16, Domostroitelnaya st., 194292, St. Petersburg, Russia
[2]Samara University, 34, Moskovskoe shosse, Samara, 443086, Russian Federation
[3]Image Processing Systems Institute of the Russian Academy of Sciences, 151, Molodogvardejskaya st., Samara, 443001, Russian Federation

The realization of a broadband antireflective coating in the THz range has been an urgent problem for many years. At present, only narrow-band antireflection coatings made on the basis of a single-layer coating are commercially available [1].

Here we want to present the results of the study of properties of a 3-level broadband antireflective (BBAR) structure on low-resistance silicon by THz time-domain spectroscopy (THz-TDS). BBAR structure was developed as a part of the manufacturing of a broadband THz radiation absorber, which can be used as a sensitive element of THz radiation detectors [2].

Since low-resistance silicon has a high absorption in the THz range, it was decided to use it as a substrate for a THz. To increase the amount of absorption, it is necessary to reduce losses associated with Fresnel reflection. For this, it was decided to create an antireflection coating on one side of the silicon substrate [3, 4].

To provide broadband antireflection effect, on the one side of the substrate a 3-level structure has been realized by use of reactive ion etching (Bosch process) [3, 5].

Sample device was fabricated from p-type ρ=0.54 Ohm·cm silicon substrate with a diameter of 50 mm and thickness of 500 μm (Fig. 1).

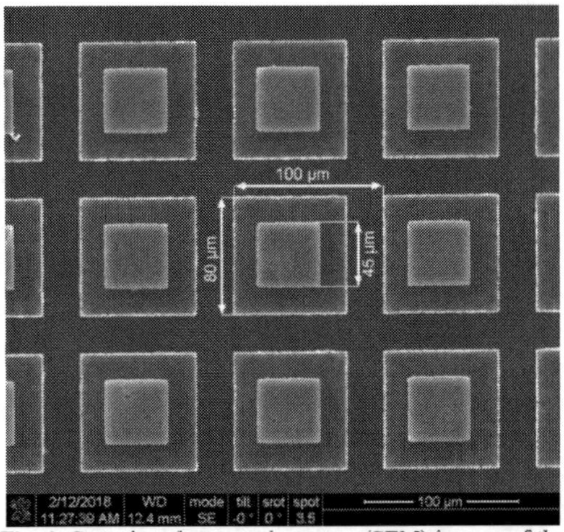

Fig. 1. Scanning electron microscope (SEM) images of the produced sample

To characterize the antireflection properties of the produced sample, the transmission and reflection measurements were carried out on the Tera K8 terahertz time-domain spectroscopy (THz-TDS) system (produced by Menlo Systems Gmbh).

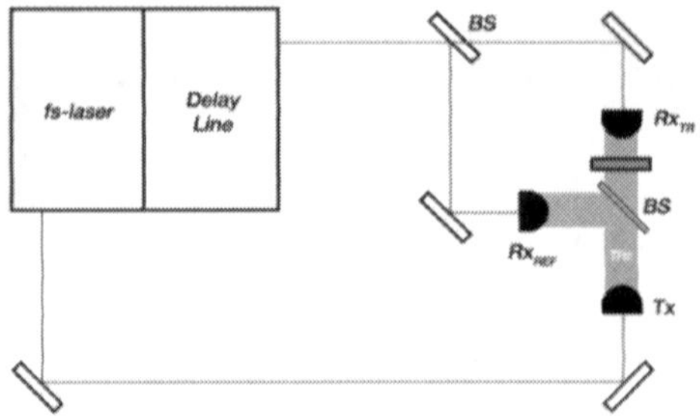

Fig. 2. Experimental setup

Tera K8 spectrometer operates in 0.2-2.5 THz range. Spectral resolution – 5 GHz. The experimental setup is shown on Fig. 2. Both, bare silicon substrate and fabricated sample with 3-level antireflection structure were measured. Reflectance was measured under the normal angle of incidence. The value of Fresnel reflection decreased from 28-30% to less than 3% in 0.5 – 2 THz range (Fig. 3).

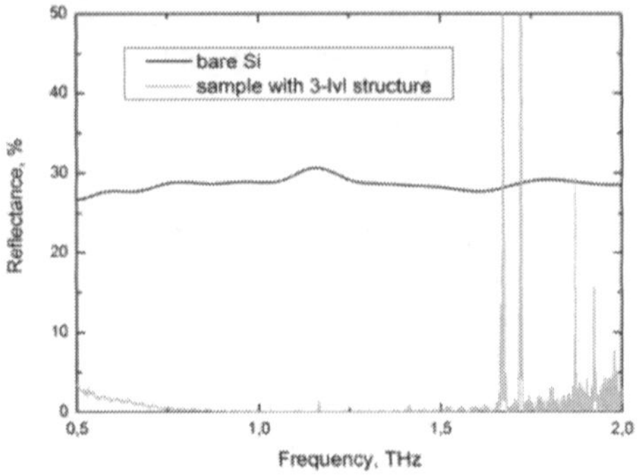

Fig. 3. Reflectance spectra of the fabricated sample

It can be concluded that fabricated structure can be used for reducing of reflection losses at the air-silicon interface of transmissive silicon optical elements in the broad range of the THz range frequencies. Also, it could be optimized for the certain frequency range.

References

1. *A. J. Gatesman, J. Waldman, M. Ji, C. Musante, and S. Yngvesson*, An Anti-Reflection Coating for Silicon Optics at Terahertz Frequencies // IEEE Microwave and Guided Wave Letters, 10, 7, pp 264 – 266 (2010).

2. *J. Lehman, A. Sanders, L. Hanssen, B. Wilthan, J. Zeng, C. Jensen*, Very black infrared detector from vertically aligned carbon nanotubes and electric-field poling of lithium tantalite // Nano Letters, 10, 3, P. 3261 (2010).

3. *A.N. Agafonov, B.O. Volodkin, A.K. Kaveev, B.A. Knyazev, G.I. Kropotov, V.S. Pavel'ev, V.A. Soifer, K.N. Tukmakov, E.V.Tsygankova, Yu.Yu. Choporova*, Silicon diffractive optical elements for high-power monochromatic terahertz radiation, Optoelectronics, Instrumentation and Data Processing, 49, pp. 189-195 (2013)

4. *C. Brückner, T. Käsebier, B. Pradarutti, S. Riehemann et al.*, Broadband antireflective structures applied to high resistive float zone silicon in the THz spectral range, Optics Express, 17, 5, P. 3064 (2009).

5. *A.N. Agafonov, B.O. Volodkin, D.G. Kachalov, B.A. Knyazev, G.I. Kropotov, K.N. Tukmakov, V.S. Pavelyev, D.I. Tsypishka, Yu.Yu. Choporova, A.K. Kaveev*, Focusing of Novosibirsk Free Electron Laser (NovoFEL) radiation into paraxial segment, Journal of Modern Optics, 63, 11,pp. 1051-1054 (2016)

https://doi.org/10.1051/epjconf/201819506017

Recent results on THz gyrotron-based molecular spectroscopy

Golubiatnikov G.Yu., Koshelev M.A., Tsvetkov A.I., Fokin A.P., Glyavin M.Yu., and Tretyakov M.Yu.

Institute of Applied Physics RAS, Nizhny Novgorod, Russia, koma@ipfran.ru

Method of opto-acoustic (also called radio- or photo-acoustic) molecular spectroscopy is a powerful tool for both fundamental studies and applications. In this method the result of the interaction of the radiation with matter is detected by the change of the parameters of the matter and not of the radiation. It is likely the only method allowing reaching high sensitivity at reasonable recording time by increasing radiation power. This was particularly demonstrated in the teraherz (THz) region in our recent work [1] using spectrometer with radio-acoustic detection of absorption (RAD spectrometer) [2] and radiation of the free-running 263-GHz gyrotron operating in continuous wave regime with up to 1 kW output power [3].

Further improvement of the spectrometer sensitivity and its use for spectroscopy requires precise control of the gyrotron radiation frequency. A phase-locked loop (PLL) against a signal of the reference oscillator was used in order to control the gyrotron modulating anode voltage [4]. The PLL system, which block diagram is presented in Fig. 3, is similar to that used for backward wave oscillator (BWO) [5]. A specially designed fast voltage control unit can vary voltage within 0-1 kV with a speed of somewhat faster 1 kV/μs. Preliminary testing of the control system showed a modulation bandwidth of 150 kHz, defined by the time constant of the gyrotron anode circuit.

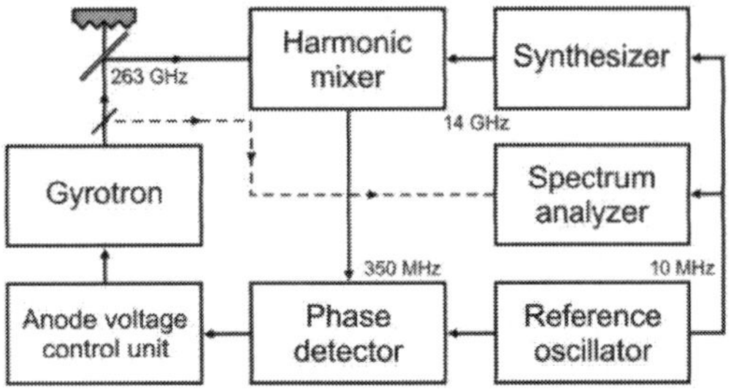

Fig. 1. Block diagram of the phase-locked loop units for a sub-THz gyrotron.

The principle of its operation is following. Part of the output radiation of the 263-GHz gyrotron is transmitted to the harmonic mixer, where it is mixed with the a harmonic of the signal from a microwave synthesizer with a frequency of about 14 GHz. The resulting signal at an intermediate frequency of 350 MHz is then directed to the phase detector for a frequency-phase comparison with the harmonic of a quartz clock serving as the reference oscillator. The error signal from the phase detector is then used as a control signal for the anode voltage control unit. In addition, the radiation spectrum was analized by Keysight spectrum analyzer N9010A equipped with an external mixer. Phase-locking the gyrotron frequency reduced the radiation spectrum width from 0.5 MHz for a free-running gyrotron down to about 1 Hz for the stabilized gyrotron (Fig. 2) which corresponds to relative frequency stability $\Delta f/f \sim 3\cdot10^{-12}$ for a measurement time of a few seconds.

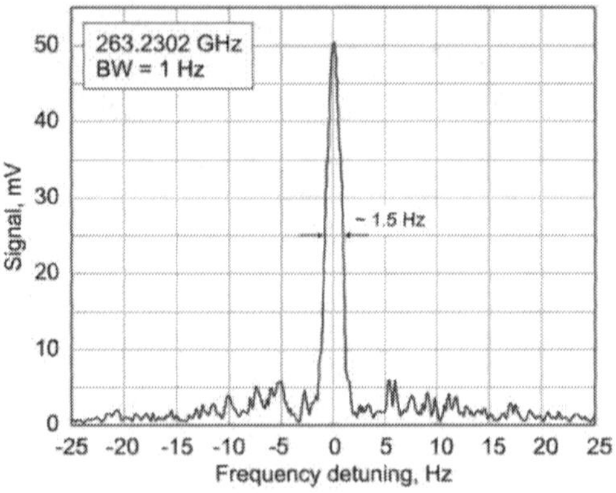

Fig. 2. Frequency spectrum of the gyrotron observed at 263.2302 GHz with phase-locked loop. Bandwidth of the spectrum analyzer was 1 Hz.

Our recent spectroscopic study using RAD method and a free running gyrotron facility [1] revealed the necessity for modulation of the gyrotron frequency aimed at reduction of the spectrometer baseline. It became possible by implementation of the PLL system of the gyrotron. Typical recordings of SO_2 line obtained using RAD spectrometer and gyrotron radiation at approximately 1 W power are shown in Fig. 3. Modulation of either radiation power (Fig. 3A) or its frequency along with synchronous signal detection at the first (Fig. 3B) and second (Fig. 3C) harmonic of the modulation frequency were applied for spectra recording. The line asymmetry due to baseline is clearly observed from the experimental spectra in case of power modulation in Fig. 3A. The situation is even worse if a weaker line is studied. Use of the frequency modulation decreases influence of the baseline on the line shape which is seen in Figs. 3B and 3C.

Experimental spectrum of a mixture of SO_2 (0.027 Torr) and Ar (0.26 Torr) was registered in the frequency range 263-264 GHz using RAD method and gyrotron radiation of about 0.5 W power (Fig. 4A). Calculated spectrum shown in Fig. 4B was modeled using Lorentz profile and line parameters from HITRAN and JPL catalogs.

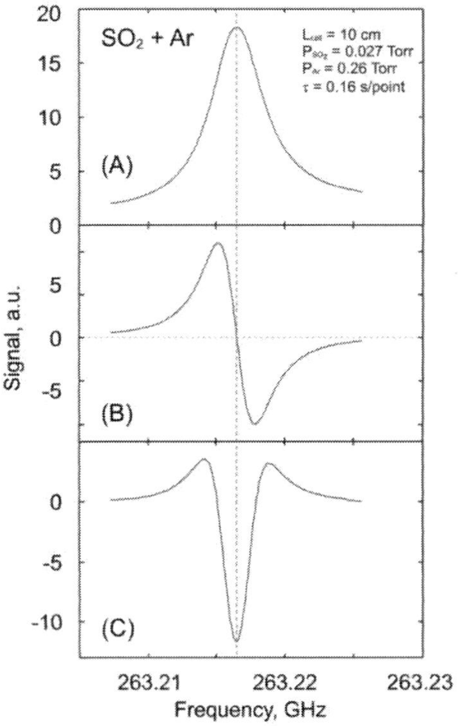

Fig. 3. Experimental recordings of the $44_{6,38} - 45_{5,41}$ line of $^{32}S^{16}O_2$ in a mixture with argon obtained using RAD spectrometer and gyrotron radiation (~ 1 W). (A) - modulation of the radiation power; (B) and (C) - modulation of the radiation frequency and signal detection at first (B) and second (C) harmonic of the modulation frequency (180 Hz).

It follows from the comparison of these two spectra that only some of the observed lines of the well-studied spectrum of SO_2 can be assigned to the tabulated ones. Futher analysis of the experimental spectra revealed that unassigned lines are narrower than the assigned ones. This may indicate that these lines arise from an interaction of absorbing molecules with radiation at second, third, etc. harmonics of the fundamental frequency. Calculated specta in the 526-528 GHz and 789-791 GHz ranges are shown in Fig. 4B and 4C, respectively (note frequency scales normalized by the corresponding harmonic number). Comparison of the calculated spectra with the experimental one confirmes presence of radiation at second and third harmonics of the gyrotron fundamental frequency. Using known parameters of SO_2 lines we estimated relative fraction of the radiation power coming to the gas cell at second harmonic as ~ 1% of the full power and ~ 0.1% for the third harmonic. Routine sensitivity of the spectrometer was determined from spectra analysis as $\sim 8 \cdot 10^{-10}$ cm^{-1} (at 2-s time constant), which is limited by the power saturation effect of the molecular transition.

Results were confirmed by the high resolution spectra obtained using a BWO-based video spectrometer.

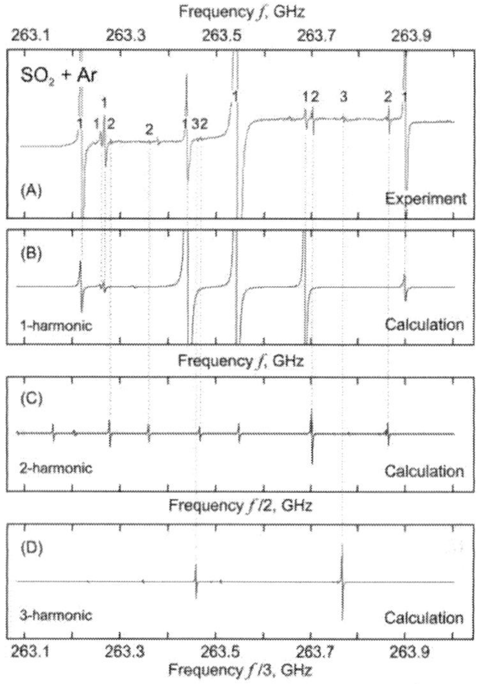

Fig. 4. (A) Experimental spectrum of a mixture of SO_2 and Ar obtained using RAD and gyrotron radiation of ~ 0.5 W power. Lines belonging to the 1st, 2nd and 3rd harmonics of radiation frequency are denoted by the corresponding numbers. (B,C and D) Calculated spectra of SO_2. Frequency scales are normalized by the corresponding harmonic number.

The study was supported by Russian Science Foundation (project 17-19-01602).

References

1. *Koshelev M.A., et al.* Molecular Gas Spectroscopy Using Radioacoustic Detection and High-Power Coherent Subterahertz Radiation Sources // J. Molec. Spectrosc. 2017. V. 331, P. 9-16.

2. *Tretyakov M.Yu., et al.* Precise Measurements of Collision Parameters of Spectral Lines with a Spectrometer with Radioacoustic Detection of Absorption in the Millimeter and Submillimeter Ranges // Instrum. Exp. Tech. 2008. V. 51. P. 78–88.

3. *Glyavin M.Yu., et al.* Experimental tests of a 263 GHz gyrotron for spectroscopic applications and diagnostics of various media // Rev. Sci. Instr. 2015. V. 86 N. 5 P. 054705.

4. *Fokin A. et al.* High-power sub-terahertz source with a record frequency stability at up to 1 Hz // Scientific Reports. 2018. V. 8. P. 4317.

5. *Krupnov A.F.* Phase Lock-In of MM/SUBMM Backward Wave Oscillators: Development, Evolution, and Applications // Int. J. IR MM Waves. 2001. V. 22. P. 1-18.

H_2O Molecules Hosted By A Crystalline Matrix – New State Of Water?

Elena S. Zhukova[1], M. A. Belyanchikov[1], M. Savinov[2], P. Bednyakov[2], V. G. Thomas[3], L. S. Kadyrov[1], E. A. Simchuk[1], Z. V. Bedran[1], V. I. Torgashev[4], A. Dudka[5], M. Dressel[1,6], B. P. Gorshunov[1].

[1] Moscow Institute of Physics and Technology, Dolgoprudny, Moscow Region, 141700 Russia, zhukova.es@mipt.ru
[2] Institute of Physics AS CR, Na Slovance 2, 18221 Praha 8, Czech Republic
[3] Institute of Geology and Mineralogy, RAS, 630090 Novosibirsk, Russia
[4] Faculty of Physics, Southern Federal University, 344090 Rostov-on-Don, Russia
[5] Shubnikov Institute of Crystallography, Federal Scientific Research Centre "Crystallography and Photonics", Russian Academy of Sciences, 119333 Moscow, Russia
[6] 1. Physikalisches Institut, Universität Stuttgart, 70569 Stuttgart, Germany

Along with the liquid and solid phases of water, there are many natural and artificial systems and objects - biological, geological, meteorological, physical, chemical, etc., in which water molecules are found in nano-confined state. As in the case of any other material, in such conditions water must acquire qualitatively new characteristics and properties that it did not possess in its macroscopic liquid state. Currently, there is a surge of active research on such properties, which is stimulated by the importance of obtaining new knowledge about the environment, the need to solve actual problems of medicine and public health, the creation of qualitatively new devices, mechanisms and technologies, materials with new characteristics that are promising for use in bio-sensors, diagnostics and in many other applications. It is known that the properties of liquid water remain less understood in comparison with other liquids, which is largely due to the ability of H_2O molecules to form hydrogen bonds with each other. Under the conditions of a nano-confinement these properties will become even more exotic, diverse and complex, especially in complex systems such as biological. Therefore, the most effective would be to study first as simple system as possible. We have studied an array of *single* H_2O molecules distributed over a periodical matrix of nano-sized cages inside the crystal lattice of dielectrics, see Fig.1. The advantages of such objects are the absence of short-range intermolecular hydrogen bonds (the distance between the pores is 5-10 Angstroms), the complete certainty of the spatial arrangement of the cages, and hence the H_2O molecules, as well as the known type and magnitude of the interaction of molecules with the environment (weak van der Waals bonds) and between themselves (long-range electric dipole-dipole interaction). Importantly, these factors are essential from the point of view of simplifying and hanging the accuracy of the model analysis (molecular dynamics, density functional methods, ...) of the properties of the subsystem of water molecules.

Using broad-band spectroscopy (frequencies from ~1 Hz up to the visible, temperatures from 300 K down to 0.3 K) we have studied specific single-particle and collective excitations of water molecular ensemble hosted by dielectric crystalline matrices of hexagonal beryl $Be_3Al_2Si_6O_{18}$ and orthorhombic cordierite $(Mg,Fe)_2Al_4Si_5O_{18}$. Experiments on dehydrated crystals allowed to identify spectral features connected exclusively with the response of water molecular subsystem. We observe well-known H_2O intramolecular modes together with librational and rattling-like translational excitations of separate water molecules confined within the nano-cages. Using density-functional and molecular dynamics approaches allowed us to identify the origin of the single-particle modes. In crystals of beryl containing both, regular and heavy water molecules we discovered an incipient ferroelectric state within water molecular subsystem that manifests itself in the form of a Curie-Weiss temperature dependence of quasi-static permittivity

$$\varepsilon'(T) = \varepsilon_\infty + C(T - T_C)^{-1},$$

where ε_{inf} is the high-frequency dielectric constant, C and T_C are the Curie constant and Curie temperature, respectively. Below ≈10 K the behavior ε'(T) saturates that can be described by the Barrett expression [1]

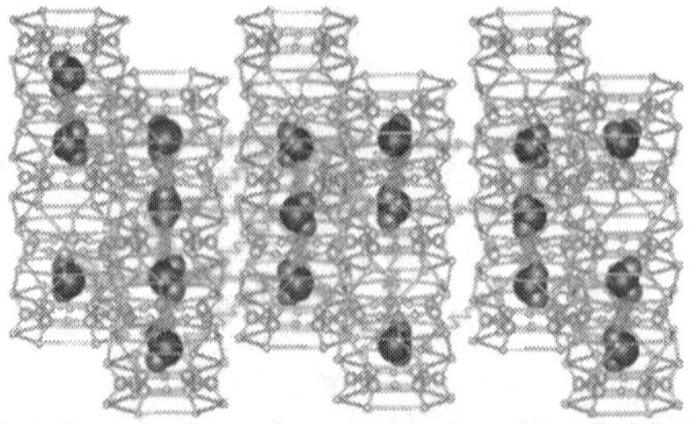

Fig. 1. Schematic view of a crystal lattice of beryl with nano-sized cages formed by ions of crystal lattice. The cages are arranged in channels that are elongated along crystallographic c-axis. Single water molecules are captured within the cages during crystal growth. The distance between H_2O molecules along the channels is about 5 Angstroms and between H_2O molecules in neighboring channels about 10 Angstroms. Orange lines denote intermolecular electric dipole-dipole interaction.

$$\varepsilon' = \varepsilon_\infty + \frac{C}{\frac{T_1}{2}\coth\frac{T_1}{2T} - T_C}$$

where C is constant and T_1 gives the energy scale of quantum fluctuations that are suppressing the macroscopic ferroelectric phase transition. We suggest that

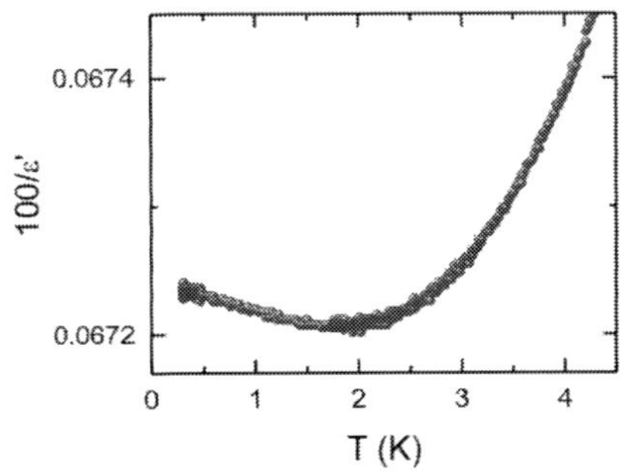

quantum tunneling is responsible for the suppression of ferroelectric phase transition, as confirmed by neutron experiments [2].

Fig. 2. Temperature-dependent inverse dielectric permittivity of an array of nano-confined water molecules in beryl crystalline matrix measured at 1 kHz.

As seen from Fig.2, the inverse dielectric permittivity displays weak minimum at the lowest temperatures. This effect can be considered as either coming from onset of short-range spatial ferroelectric (antiferroelectric) correlations between the dipole moments or signs of quantum critical behavior of water molecular electric dipoles. Further experiments are needed to clarify this point.

The specific potential profile experienced by the water molecules localized within the nano-cages of cordierite leads to emergence of an overdamped temperature dependent relaxation at radiofrequencies and a rich set of excitations in the terahertz range. The origin of both types of excitations needs detailed investigation.

In conclusion, spectroscopic studies of an array of dipole-dipole interacting water molecules localized within nano-sized cages of crystal lattice of dielectrics demonstrate a rich set of excitations that are not observed in liquid water. The corresponding H_2O molecular network can thus be considered as a model system whose studies will allow to proceed to investigations of more complicated water-containing objects.

The work was supported by Russian Foundation for Basic Research program Project No 18-32-00286.

References

1. *Barrett, J. H.* Dielectric constant in perovskite type crystals // Phys. Rev. 1952. V. 86, P. 118–120.
3. *Kolesnikov A. I.* et al. Quantum tunneling of water in beryl: a new state of the water molecule // Phys. Rev. Lett. 2016. V. 116, P. 167802.

Millimeter-Wave Spectroscopy of Weakly Bound Molecular Complexes and Small Clusters

L. A. Surin[1,2]

[1]Institute of Spectroscopy, Troitsk, Moscow, Russia, surin@isan.troitsk.ru
[2]I. Physikalisches Institut, University of Cologne, Cologne, Germany

Currently, there are a number of international projects on a full-scale study of the infrared and submillimeter-wave (SMMW) radiation in space, such as the Space Observatory "Herschel" (mission is completed, analysis of the data is in progress), the Stratospheric Observatory "SOFIA" based on a Boeing-747SP wide-body aircraft, astronomical interferometer of MMW and SMMW telescopes in the Atacama Desert of northern Chile "ALMA", and many others. The demand of high-level spectroscopic data, including the parameters of interaction between abundant in the interstellar space molecules and atoms, is extremely high, because it is necessary for the interpretation of the information coming from the radio telescopes. Many laboratory and theoretical research directed on the understanding of the interaction between ammonia (NH_3), water (H_2O), carbon monoxide (CO), nitrogen (N_2), etc. and their main partners by the collisions in the interstellar medium - molecular hydrogen H_2 and atomic helium He [1, 2]. These data are then used to test intermolecular interaction potentials, and they provide the necessary background for the development for the astrophysical models.

Another application, in which the intermolecular interactions with He and H_2 play an important role, is the doping of helium or *para*-hydrogen clusters with a probe molecule to perform high resolution spectroscopy. Such experiments are directed on the detection of superfluid response of molecular rotation in the He and *para*-H_2 clusters [3].

Typically, microwave (MW) spectra provide information about ground state *end-over-end* pure rotational energy levels of the complexes, MMW spectra give access to their internal rotor states, and SMMW or THz wavelengths probe intermolecular vibrations. Gas phase high resolution spectroscopic data give unambiguous information on the dynamics, composition, and structure of the complexes, because their bound states are very sensitive to the interaction potential.

The overview of the spectroscopic methods as well as recent results on molecular complexes and small clusters are presented.

OROTRON Spectrometer with Supersonic Jet

In OROTRON, the MMW radiation is produced by interaction of an electron beam with the electromagnetic field of an open Fabry-Perot resonator. The principal scheme of the spectrometer is shown in Fig. 1. In order to allow the measurements in the molecular jet expansion, the entire OROTRON is placed into a vacuum chamber. The high vacuum part of the OROTRON includes the cathode, the collector and the fixed plane mirror with a periodic structure ruled on its surface. This part is separated from the rest of the resonator by a thin mica window, which is transparent to the MMW radiation. The second part of the OROTRON consists of a movable spherical mirror. These two parts are connected to each other by a hollow brass cylinder with two openings for the propagation of a molecular jet through the resonator.

Absorption inside the resonator is sensitively detected by measuring the appropriate change of the electron current in the collector circuit. The principle of the method is that a change in the quality factor of the resonator due to the absorption in the gas will cause a change in the velocity distribution of the electrons and thus in their passage to the collector [4].

The double resonance option strongly helps in the identification of complex spectra. If one of the two transitions in a three level system is saturated, the populations of the levels and, correspondingly, the radiation absorption intensity in the second transition change. Therefore, the absorption of radiation from an additional source in the first transition can be detected in a change in the signal of the spectrometer tuned to the second transition. As shown in Fig. 1, the OROTRON spectrometer assumes the role of signal detection, while the microwave pump radiation is delivered by a microwave synthesizer followed by a high power broadband amplifier (2-110 GHz). The molecular beam travels perpendicular to both MMW and MW radiation beams and interacts simultaneously with them.

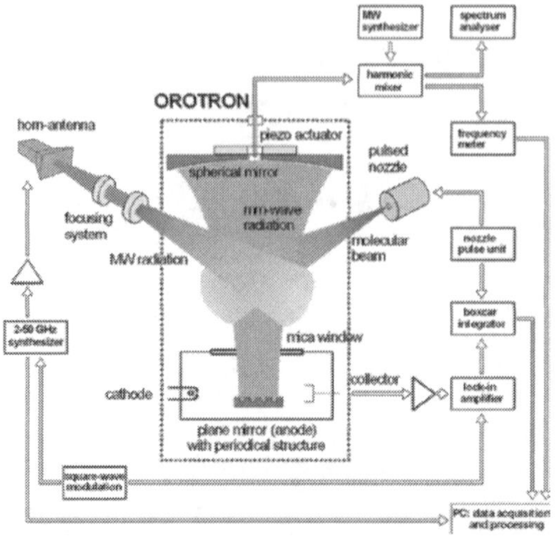

Fig. 1. Schematic diagram of the OROTRON jet spectrometer (100-150 GHz) with MW–MMW double resonance arrangement.

This technique was applied to study many weakly bound van der Waals systems [5]. Fig. 2 shows one example of a 20 MHz survey for $R(0)$ transitions of the He_8–$^{12}C^{18}O$ cluster [3].

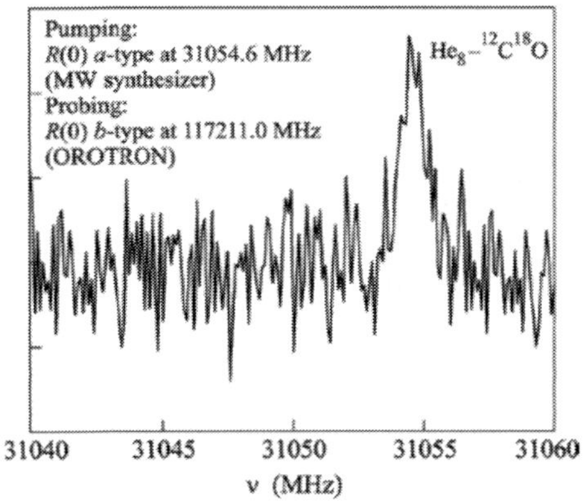

Fig. 2. Record of the $R(0)$ transition in the He_8–$^{12}C^{18}O$ cluster by the double resonance method with OROTRON.

Supersonic Jet Spectrometer for Terahertz Applctions (SuJeSTA)

For measurements of the weakly bound van der Waals NH_3–H_2 complexes the SuJeSTA spectrometer (150-800 GHz) [6] was applied. Details of the current experimental setup used in the Astrophysics group of the University of Kassel for carbon clusters experiments are shown in Fig. 3. A microwave synthesizer is used as a radiation source, followed by a frequency multiplier. The SMMW radiation intersects a pulsed supersonic jet ~5 cm downstream in a perpendicular configuration.

A multi-pass optics is used to further enhance the signal-to-noise ratio. The absorption signal is detected by a low-noise liquid-He cooled hot electron bolometer. The frequency of the synthesizer is modulated at 40 kHz, and phase sensitive detection of the signal is achieved by a lock-in amplifier in 2*f*-mode operation. In addition to the frequency modulation of the radiation source, an on–off modulation of the jet is used for background subtraction through a pair of boxcar integrators. The combination of frequency and source modulation substantially improves the S/N ratio and at the same time suppresses standing wave effects.

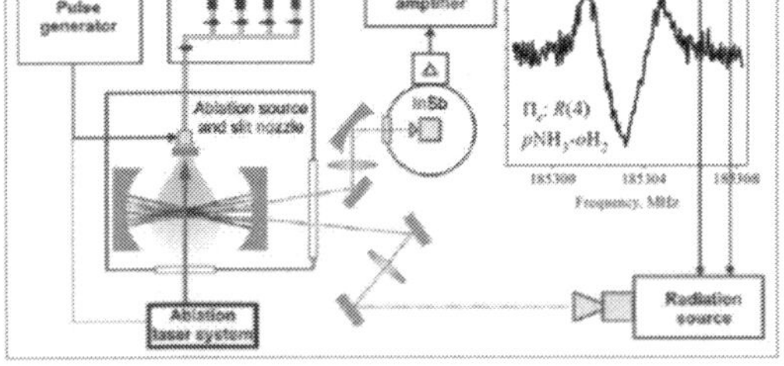

Fig. 3. SuJeSTA spectrometer (150-800 GHz) with multi-pass cell and recorded line of the pNH_3–oH_2 complex.

Double Resonance Spectroscopy with Cryogenic Ion Trap Apparatus

A novel action spectroscopic method for rotational spectroscopy of weakly bound complexes, which has been recently developed in the Spectroscopy group of the University of Cologne. This method uses a double resonance consisting of a rotational transition (MMW) followed by a predissociating transition (IR), and uses the final destruction of the complex as the spectroscopic action signal. This approach was applied to obtain high-resolution rotational data of the He–HCO^+ complex. HCO^+ is particularly interesting because it is probably the most abundant molecular ion in dense cores of interstellar molecular clouds and its complex with He is regarded as reaction intermediate of the proton exchange reaction.

The experiments have been carried out in the 22-pole ion trap machine COLTRAP [7]. The spectroscopic parameters of the ground state have been determined to microwave precision, thus providing the most reliable reference data for testing available He–HCO^+ interaction potentials.

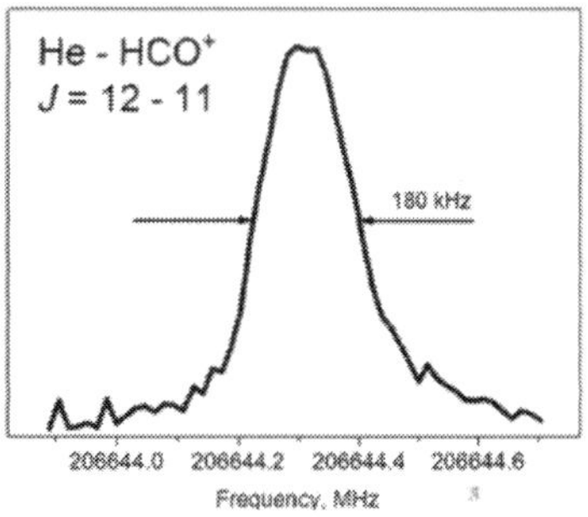

Fig. 4. Example for the double resonance technique applied to the He–HCO^+ complex.

The author acknowledges the Russian Science Foundation (Grant 17-12-01395).

References

1. *Green S., Thaddeus P.* Rotational excitation of CO by collisions with He, H, and H_2, under conditions in interstellar clouds // Astrophys. J. 1976. V. 205, P. 766-785.
2. *Chefdeville S., Stoecklin T., Naulin C., et. al.* Experimental and theoretical analysis of low-energy CO + H_2 inelastic collisions // Astrophys. J. Lett. 2015. V. 799, L9.
3. *Surin L.* Nonclassical rotations of single molecule in small helium and hydrogen clusters: manifestation of "microscopic superfluidity" // JETP Lett. 2013. V. 97, P. 57-65.
4. *Surin L.A., Dumesh B.S., Lewen F., Roth D.A., et. al.* Millimeter-wave intracavity-jet OROTRON-spectrometer for investigation of van der Waals complexes // Rev. Sci. Instrum. 2001. V. 72, P. 2535-2542.
5. *Dumesh B.S., Surin L.A.* Unusual rotations in helium and hydrogen nanoclusters and "nanoscopic" superfluidity // Physics-Uspekhi. 2006. V. 49, P. 1113-1129.
6. *Surin L.A., Tarabukin I.V., Schlemmer S., Breier A.A., Giesen T. F., et. al.* Rotational spectroscopy of the NH_3–H_2 complex // Astrophys. J. 2017. V. 838, Art. 27.
7. *Salomon T., Töpfer M., Schreier P., Schlemmer S., Asvany O., Kohguchi H., Surin L.* Double Resonance Rotational Spectroscopy of He–HCO^+ // Phys. Chem. Chem. Phys. 2018.

Nonlinear quantum interferometry in terahertz spectroscopy

E.I. Malkova[1], S.P. Kovalev[2], K.A. Kuznetsov[1], G.Kh. Kitaeva[1]

[1]M.V. Lomonosov Moscow State University, Moscow, Russia, 119991
[2]Helmholz Zentrum Dresden Rossendorf, Dresden, Germany, 01328

At present time, biphoton fields are widely studied both in quantum optics and spectroscopy. There are many experimental and theoretical works devoted to nonlinear interferometers [1-5]. However, there is a lack of studies of generation for optical-terahertz biphotons in nonlinear interferometers. The goal of this work to analyze the possibility of observation for interference of the optical-terahertz biphoton fields in Young and Mach-Zehnder schemes, and calculating the angular-frequency intensity patterns at the output of the interferometers.

Earlier a special approach was applied for measuring of the dielectric function imaginary part at THz frequencies, based on the analysis of visibility of three-wave interference under spontaneous parametric down-conversion (SPDC) in the Young scheme [6]. The experiments were performed using SPDC-based three-wave interferometric technique in geometry of near-forward Raman scattering by phonon polaritons. This technique provides measuring absorption coefficients at frequencies of infrared and terahertz phonon polaritons using crystal samples of any arbitrary length with respect to propagation depth of absorbed waves.

In this paper, we calculated the frequency-angular spectrum of the biphoton field at the output of the nonlinear Mach-Zehnder interferometer in the regime of strongly frequency-non-degenerate parametric light scattering. The lithium niobate crystal doped by magnesium ($LiNbO_3$: Mg) was considered as a model of a nonlinear interferometer (Fig.1), as two plates of thickness L with a layer L ' of the other medium being placed between them. The scattering geometry was chosen as follows: x(y,y) xz.

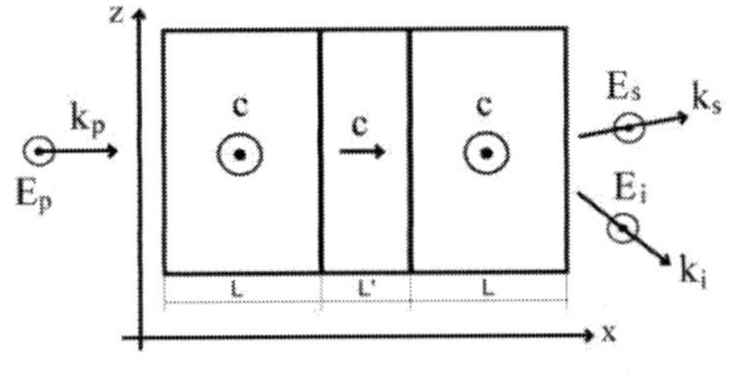

Fig. 1. The geometry of the nonlinear Mach-Zehnder interferometer, c - the optical axes of lithium niobate crystals.

It is well known that frequency-angular SPDC spectrum from one $LiNbO_3$ crystal can be written as:

$$I_1(\theta_s,\omega_i) \propto \left(\frac{\sin(\delta_x/2)}{\delta_x/2}\right)^2 \cdot \left(\frac{\sin(\delta_z/2)}{\delta_z/2}\right)^2 \quad (1)$$

where δ_x, δ_z – longitudinal and transversal wave mismatches, respectively. According to [4], the modulation function of the middle layer was calculated as:

$$m(\theta_s,\omega_i) = \cos^2\left(\frac{\delta_x+\delta'_x}{2}\right) \quad (2)$$

The resulting interference pattern is described by the following expression:

$$I(\theta_s,\omega_i) \propto I_1(\theta_s,\omega_i) m(\theta_s,\omega_i) \quad (3)$$

Two lithium niobate plates of 0.023 cm thickness with Mg content 5.1 mol% MgO were oriented as in Fig.1. The axes of the crystals are perpendicular to the wave pump vector, so eee- interaction was realized. In order to avoid the total internal reflection of idler wave at the boundary of a lithium niobate crystal, a material with a sufficiently high refractive index (~6) should be used as an intermediate layer. Between the plates of the interferometer is placed another plate of $LiNbO_3$: Mg with a thickness of 0,007 cm with a Mg content of 4.1 mol% MgO. The optical axis of this layer was parallel to the wave pump vector.

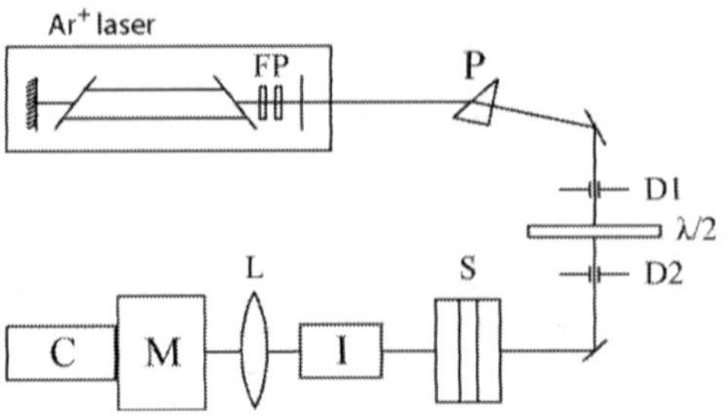

Fig. 2. Schematics of the experimental setup.

Measurements were carried out using Ar+-laser generating a single-longitudinal mode at a wavelength of 514.5 nm. To enable the subsequent attenuation of the pump by an iodine chamber, an additional Fabri-Perot interferometer was placed in the laser resonator, separating one longitudinal mode and significantly reducing the spectral width of the generation so that it does not exceed the spectral band of the absorption line of the iodine cell. The radiation power of one longitudinal mode was about 100 mW. To rotate the polarization of the pump we used a λ/2 plate. Heated up to temperature of 75°C an iodine vapor cell has many absorption lines near the generation frequency of the argon laser, so an iodine cell was used to filter the pump radiation. The signal radiation is eight orders of magnitude weaker than the pump radiation and is shifted from the pump frequency by 0.1-7 THz, so the use of an iodine cell makes it possible to detect the frequency-angular spectrum of the signal radiation

 https://doi.org/10.1051/epjconf/201819506020

without the parasite background of the laser pump. Using a system of three lenses, the signal radiation was focused on the entrance slit of the MDR-41 monochromator. CCD-camera Proscan HS-101H with a Hamamatsu matrix recorded resulting interference patterns; the size of one pixel of the matrix was 24 microns.

Calculated and measured frequency-angular spectra are shown in Fig.3, one can see a good agreement between them. We suppose, that interferometry of optical-terahertz biphoton fields in the scheme of the Mach-Zehnder can be used as a method of spectroscopy for various substances in the terahertz range.

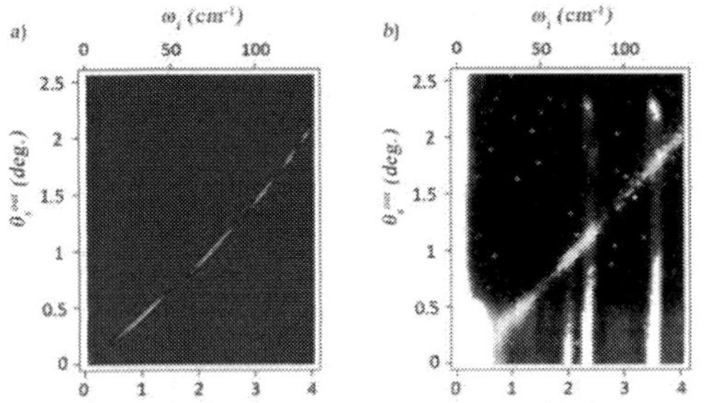

Fig. 3. a) calculated frequency-angular spectrum, b) experimental frequency-angular spectrum

This work was done under financial support of the Russian Science Foundation (Grant No. 17-12-01134).

References

1. *Chekhova M.V. and Ou Z.Y.* Nonlinear interferometers in quantum optics // *Advances in Optics and Photonics*. 2016. V. 8. No. 1. P. 104-155.

2. *A.V. Burlakov, Yu.B. Mamaeva, A.N. Penin, M.V. Chekhova* Three-wave interference with participation of polaritons// *J. Exp. Theor. Phys*. 2001. v.93. p.55

3. *Korystov D.Y., Kulik S.P., Penin A.N.* Rozhdestvenski hooks in two-photon parametric light scattering //, *JETP Letters*. 2001. V. 73, No. 5, P. 214-218.

4. *Korystov D.Y., Kulik S.P., Penin A.N.* Interferometry of spontaneous parametric light scattering // *Quantum Electronics*. 2000. V.30, P.922-926.

5. *Kalashnikov D.A., Paterova A.V., Kulik S.P., Krivitsky L.A.* Infrared Spectroscopy with Visible Light // *Nature Photonics*. 2016. V.10, P.98-102.

6. *Kuznetsov K.A., Kitaeva G.Kh, Kovalev S.P., Germansky S.A., Buryakov A.M., Tuchak A.N., Penin A.N.* Complex extraordinary dielectric function of Mg-doped lithium niobate crystals at terahertz frequencies // *Applied Physics B: Lasers and Optics*. 2016. V. 122. P. 223-232.

Terahertz Range Surface-Wave Bragg Resonators with Optimized Ratio between Ohmic and Radiative Losses

Andrey M. Malkin, N.S. Ginzburg, S.E. Fil'chenkov, A.S. Sergeev, and V.Yu. Zaslavsky

Institute of Applied Physics RAS, Nizhny Novgorod, Russia, malkin@appl.sci-nnov.ru

Finite grating area at the surface of a metal plate can form a highly selective surface-wave Bragg resonator which can be used for sensing in the THz range, among other applications [1]. For 1D and 2D structures, we describe surface modes confined at a metallic plate. For a given material conductivity and grating dimensions, an optimum corrugation depth can be determined that provides the maximum value of the fundamental mode quality factor due to the interplay between radiative and Ohmic losses.

Surface electromagnetic waves confined at the interface of metal and dielectric have been widely studied [2] since they possess a number of unique features making them useful for various applications, including antenna applications Raman scattering, etc. Since the evanescent waves are decelerated, Cherenkov devices based on interaction of electron beams with surface modes of the corrugated structures is of special interest in the short wavelength radiation bands, including the terahertz band [3]. These devices allow to solve the problem of longitudinal and transverse mode selection in oversized electrodynamical systems.

In this paper, we consider a periodically corrugated section of a metallic plate as a surface-wave resonator (see Fig.1) in a dielectric media with a permittivity of ε. We obtain the equations of wave coupling with Ohmic losses taken into account and derive an integral equation describing the eigenmodes. Based on this equation, we find the optimum values of the corrugation depth providing the maximum Q factor of the fundamental mode dependent on the ε. We also discuss the 2D variant of a surface-wave resonator.

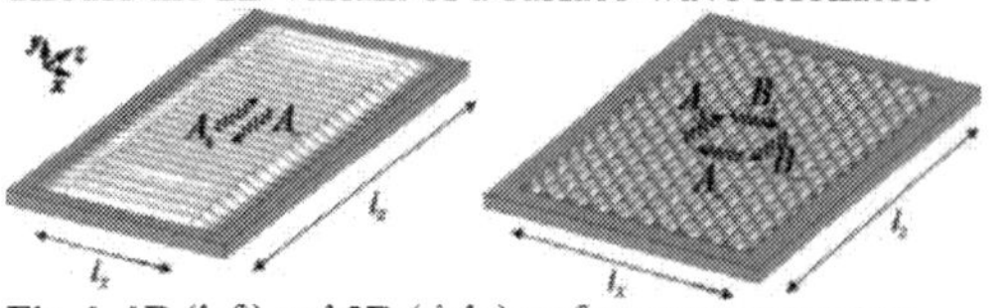

Fig. 1. 1D (left) and 2D (right) surface-wave resonators.

In the case when the depth of sinusoidal corrugation $b(z)=(b_0/2)\cos\bar{h}z$ is small in the scale of wavelength and of the structure period d the electromagnetic field of the surface mode existing near the 1D corrugated metal plate can be presented as a sum of two quasi-optical wavebeams with slowly varying amplitudes $A_\pm$ which can be characterized by the magnetic field

$$H_x=\mathrm{Re}\left[A_+(z,y,t)e^{i(\omega_0 t-h_0 z)}+A_-(z,y,t)e^{i(\omega_0 t+h_0 z)}\right],$$

where $h_0=\bar{h}/2$, $\omega_0=c\varepsilon^{-1/2}h_0$ is the Bragg frequency. Propagation of electromagnetic field in the vicinity of the Bragg resonance can be described by a system of coupled parabolic wave-beam equations which take into account the diffraction in y direction [2],

$$\pm\frac{\partial A_\pm}{\partial z}+\sqrt{\varepsilon}\frac{\partial A_\pm}{c\partial t}+\sigma\delta(y)A_\pm+\frac{i}{\bar{h}}\frac{\partial^2 A_\pm}{\partial y^2}=i\alpha\delta(y)A_\mp \quad (1)$$

At the edges of the corrugation we set non-reflection boundary conditions: $A_+(z=0)=0$, $A_-(z=l_z)=0$. Here $\alpha=\bar{h}b_0/8$ is the coupling parameter, $\delta(y)$ is the delta function. Parameter $\sigma=\sqrt{i/2}\varepsilon k d_{skin}$ describes the Ohmic losses in the metal corresponding to the Leontovich boundary condition, d_{skin} is the skin depth of the metal.

Equations (1) applied to a system infinite in z direction lead to the existence of a surface wave $A_\pm\sim e^{i\Gamma z-g_\pm y+i\Omega t}$ with a dispersion law which in the lossless case can be put down as

$$\varepsilon\left(\frac{\Omega}{c}\right)^2-\Gamma^2=\bar{h}^2\alpha^4 \quad (2)$$

Dispersion characteristic corresponding to (2) is shown in the lower part of Fig. 2.

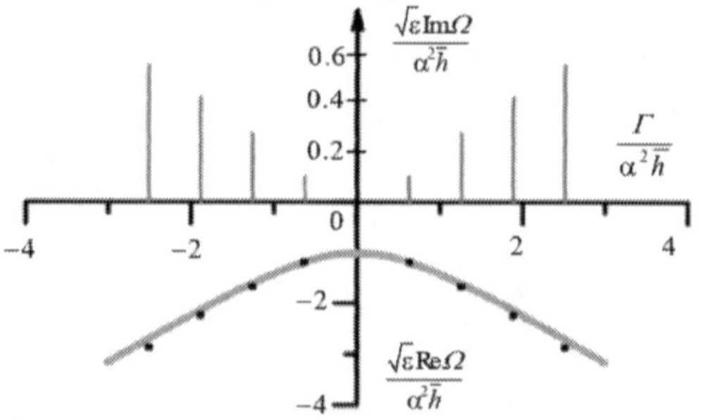

Fig. 2. Dispersion curve of the surface wave. Dots mark the surface-wave resonator modes at $\alpha^2\bar{h}l_z=5$, lines in the upper part indicate the losses of these modes.

In the case when the length l_z of the structure is finite, a set of longitudinal modes can be found. Seeking modes of (1) at an eigenfrequency of Ω, and using the Fourier transform, one can obtain an integral equation equivalent to system [5] which possesses an infinite discrete eigenmode spectrum. Frequencies and spatial profiles of eigenmodes can be found numerically. Real parts of the eigenfrequencies of the first several modes are marked with dots close to the dispersion in lower part of Fig.2 for $\alpha^2\bar{h}l_z=5$ while the imaginary parts of these modes are shown as vertical lines in the upper part of the same figure. It can be seen that the $\mathrm{Im}\,\Omega$ increases almost linearly. Quality factor $Q=\omega_0\left(2\,\mathrm{Im}\,\Omega\right)^{-1}$ of the fundamental mode is about 3 times higher than that the next (second mode). It demonstrates the high selective properties of surface wave resonator. Making use of the fact that according to the exact solution of (3), the longitudinal field structure of the mode tends to sine shape with increasing parameter $\alpha^2\bar{h}l_z$, an asymptotic formulas for the fundamental mode eigenfrequency and Q factor can be obtained in the following form:

$$\mathrm{Re}\,\omega_1 \approx \frac{c\bar{h}}{\sqrt{\varepsilon}}\left(\frac{1}{2}-\alpha^2\left(1+\frac{2\pi^2}{\alpha^4\bar{h}^2 l_z^2}\right)\right) \qquad (3)$$

$$Q=\frac{1}{Q_{diff}^{-1}+Q_{Ohm}^{-1}},\ Q_{diff}=\frac{\alpha^4\bar{h}^3 l_z^3}{5\pi^2};\ Q_{Ohm}=\frac{2}{\alpha\, Re\,\sigma}$$

Note the sensitivity of the eigenmode position to the value of ε. According to (3), the precision of dielectric sensing by such structure can be roughly estimated as Q^{-1}. It is important that the two Q terms in (3) demonstrate different behavior at varying parameter α that is proportional to the corrugation depth b_0. Total Q factor has a maximum in its α dependence for each l_z. This maximum can be found from expressions (4) as

$$Q_{max}=\left(\frac{2^4}{5^6\pi^2}\frac{l_z^3}{\bar{h}\varepsilon^4 d_{skn}^4}\right)^{1/5}\approx 0.16\frac{l_z^{3/5}}{\bar{h}^{1/5}\varepsilon^{4/5}d_{skn}^{4/5}}$$

reached at

$$\alpha_{opt}=\left(\frac{10\pi^2}{\bar{h}^4 l_z^3\varepsilon d_{skn}}\right)^{1/5}\approx\frac{2.5}{\bar{h}^{4/5}l_z^{3/5}\varepsilon^{1/5}d_{skn}^{1/5}}. \qquad (4)$$

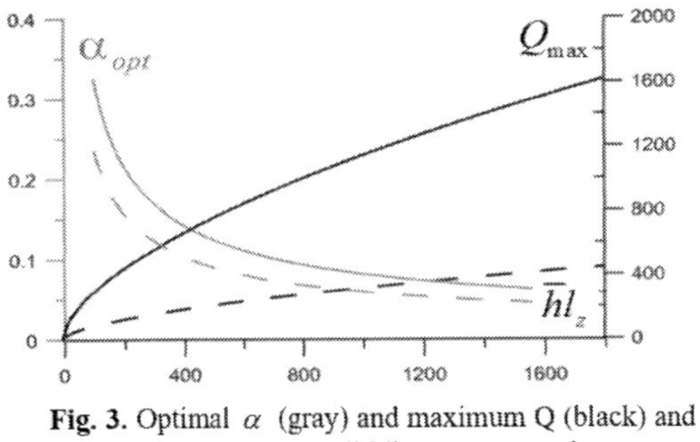

Fig. 3. Optimal α (gray) and maximum Q (black) and vs. length of the resonator. Solid lines correspond to copper structure in vacuum, dashed lines correspond $\varepsilon = 5$.

The optimum Q and α values vs. length of the resonator are depicted in Fig.3 for a structure with a Bragg frequency of about 1 THz (d=150 μm) made of copper ($d_{skn} \approx 0.065\ \mu m$) for a structure in vacuum (solid lines) and in a dielectric with $\varepsilon = 5$. According to (4), in this parameter area the small corrugation approximation is validated for structures of several dozen periods or longer. For instance, for a structure with a length of 100 periods ($\bar{h}l_z \approx 630$, see dotted line in Fig.5), a corrugation depth of 1/10 wavelength provides an optimal Q factor of about 1000 for copper at normal temperature. In a dielectric with $\varepsilon = 5$, Q value is much lower while the optimum corrugation depth remains almost the same.

Theory of 1D surface-wave resonators given above is applicable only at moderate (in the sense of Fresnel scales) transverse (x) sizes of the system, as it describes, in fact, a single transverse mode approximation. At larger x-dimensions of the system, various transverse modes with close Q factors would compete with each other. In [5], 2D surface-wave resonators were introduced to provide a means for synchronization of radiation from wide electron beams into the evanescent wave. For a 2D resonator (see Fig.1) with a corrugation of

$$l(z,x)=l_{2D}/4\left[\cos(\bar{h}_{2D}(z-x))+\cos(\bar{h}_{2D}(z+x))\right],$$

we use the time-domain simulation to find the fundamental mode. In this case, four partial quasi-optical wave-beams take part in formation of a surface mode, two of which ($A_\pm$) are propagating in $\pm z$ directions and the other two ($B_\pm$) are propagating in $\pm x$ directions. For 2D Bragg structures, Bragg frequency and wavenumber are defined as $\omega_0 = ch_0$, $h_0 = \bar{h}_{2D}$.

Wavebeam coupling is described by the following system of equations [5]:

$$\pm\frac{\partial A_\pm}{\partial z}+\frac{\partial A_\pm}{c\partial t}+\sigma\delta(y)A_\pm+\frac{i}{2k}\frac{\partial^2 A_\pm}{\partial y^2}=i\alpha_{2D}\delta(y)(B_+ + B_-),$$

$$\pm\frac{\partial B_\pm}{\partial x}+\frac{\partial B_\pm}{c\partial t}+\sigma\delta(y)B_\pm+\frac{i}{2k}\frac{\partial^2 B_\pm}{\partial y^2}=i\alpha_{2D}\delta(y)(A_+ + A_-).$$

Here $\alpha_{2D}=\bar{h}_{2D}l_{2D}/8$. To complete the 3D nonstationary model, we also set an "ideal metal" condition $\partial A_\pm/\partial y=0,\ \partial B_\pm/\partial y=0$ at some point $y=l_y$ with l_y much larger than the scale of decay of the fields. These equations were simulated with random initial conditions set at *t=0* for the parameters of $\alpha_{2D}^2 h_0 l_z = 4.4,\ \alpha_{2D}^2 h_0 l_x = 3.14$. In simulation, a single mode with highest quality factor survived at $\alpha_{2D}^2\omega_0 t > 100$. This fundamental mode is confined at the corrugation and possesses a bell-shaped structure in (z-x) plane. Q factor of this mode is about $15\alpha^{-2}$. Thus we have shown that a 2D surface-wave resonator with shallow corrugation is a highly selective structure with respect to three coordinates.

This work was supported by RSF grant No. 18-79-10252.

References

1. B. Ng, J. Wu, S.M. Hanham, et al., Spoof Plasmon Surfaces: A Novel Platform for THz Sensing // Advanced Optical Materials 1(8), 2013.

2. W.L. Barnes, A. Dereux, and T.W. Ebbesen, Surface plasmon subwavelength optics // Nature, v. 424, pp. 824–830, 2003.

3. N.S. Ginzburg, A.M. Malkin, A.S. Sergeev, V.Yu. Zaslavsky, Quasi-optical theory of relativistic submillimeter surface-wave oscillators // *Appl. Phys. Lett.* 2011, v. **99**, p. 121505.

4. N.S. Ginzburg, A.M. Malkin, A.S. Sergeev, V.Yu. Zaslavsky, Powerful surface-wave oscillators with two-dimensional periodic structures // *Appl. Phys. Lett.* 2012, v. **100**, p. 143510.

5. N.S. Ginzburg, A.M. Malkin, A.S. Sergeev, S.E. Fil'chenkov, V.Yu. Zaslavsky, "Highly selective surface-wave resonators for terahertz frequency range formed by metallic Bragg gratings", // *Phys. Lett. A*, v. 382, p. 925, 2018.

Study of influence of densification on control of conductivity and spectral characteristics of thin films of carbon nanotubes in terahertz frequency range

P. Demchenko[1], D. Gomon[1], I. Anoshkin[1,2], S. Smirnov[2], D. Lioubtchenko[1,2], M. Khodzitsky[1]

[1]Department of Photonics and Optical Information Technologies, ITMO University, Saint Petersburg, Russian, pert.s.demchenko@gmail.com
[2]Department of Micro and Nanosystems, KTH Royal Institute of Technology, Stockholm, Sweden.

The terahertz (THz) frequency range has a large number of potential applications. Terahertz radiation interacts with rotational and vibrational transitions of organic molecules, thus, this one can be used for spectroscopy of small organic molecules, for example glucose [1], water [2], macromolecules, cells [3], biological _tissues and medicines is possible. Also, the terahertz frequency band can be used in quality control of electrical circuits, drugs and product, astronomy, security systems and communications [4].

Each application requires components for generation, detection, polarization, amplification, and modulation of terahertz radiation. Currently, devices are being developed, the parameters and properties of which can be controlled. Several materials are capable of changing the properties and characteristics in the terahertz frequency range one of them is single-walled carbon nanotubes (SWCNTs). So, in our last article [5] we showed that the complex dielectric permittivity of a thin film of carbon nanotubes under the influence of optical pumping varies in the terahertz frequency range. Such thin layers of nanostructures readily absorb various liquids. Therefore, such samples can be processed in various ways. One of them is densification. This is the process of compacting the material. In this paper, we investigated the optical properties of thin SWCNT layers before and after densification, and how the optical properties change under the influence of infrared pumping at wavelength of 980 nm.

SWCNT film was made by chemical vapor deposition (CVD) [6] on nitrocellulose filter. This film was examined with scanning electron microscope (SEM) (Fig. 1(a)) and transmission electron microscope (TEM) to determine dispersion of length of nanotubes, their diameter and film's thickness.

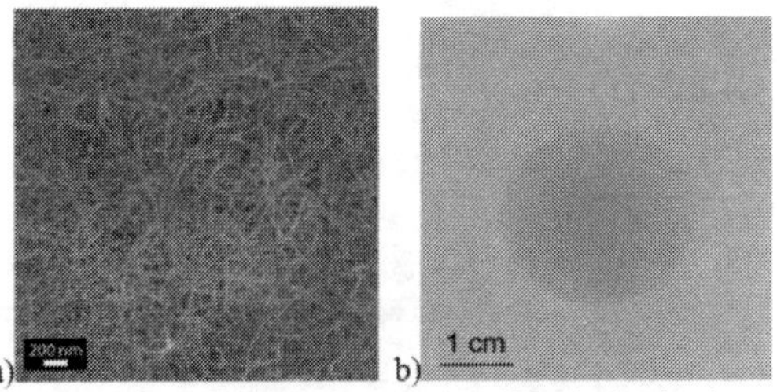

Fig. 1. a) Sample image of a SWCNT film obtained using a scanning electron microscope, b) Photo of sample

Characteristics of SWCNT film:

Nanotubes's length	~15 um
Diameter of nanotubes	1.3 – 2 nm
Film's thickness	55 nm

After that, SWCNT film was moved on teflon (PTFE) substrate with thickness of 350 um (Fig. 1(b)). The sample was studied using terahertz time-domain spectrometer (THz-TDS) [7] with additional 980 nm infrared laser for optical pumping (up to 0.7 W/cm^2). Scheme of THz-TDS is shown in Fig. 2.

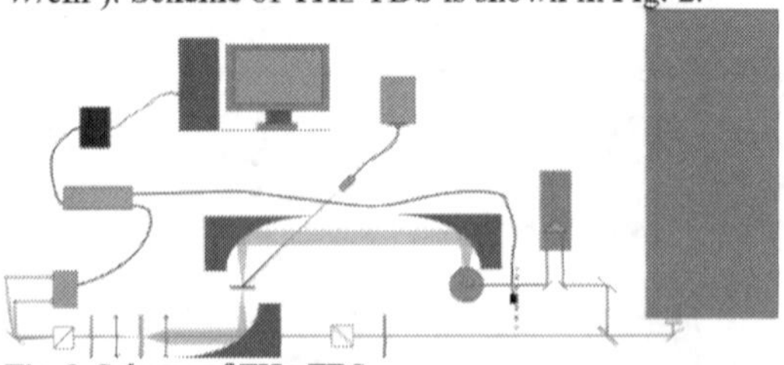

Fig. 2. Scheme of THz-TDS

THz waveforms were obtained before densification and after, with and without optical pumping (Fig. 3).

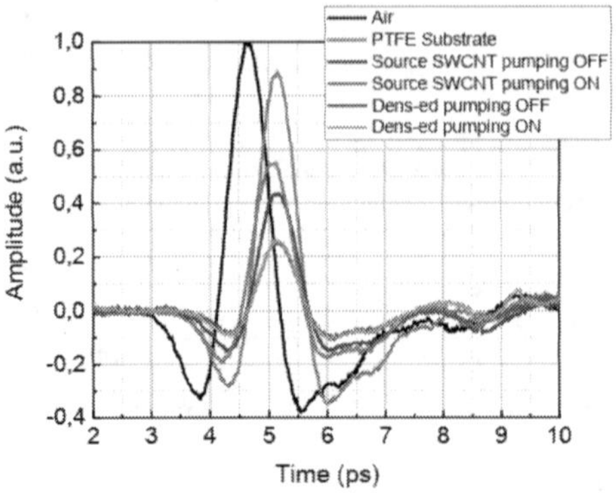

Fig. 3. Typical THz waveforms

To calculate a complex conductivity of the sample, the thin-film method [8] was used

$$\hat{\sigma}(f) = \frac{1}{Z_0}\left(\hat{n}_{Sub}(f)+1\right)\left(\frac{\hat{E}_0(f)}{\hat{E}(f)}-1\right), \qquad (1)$$

were Z_0 – impedance of free space, f - frequency, n_{Sub} – refractive index of substrate, $\hat{E}_0(f)$ and $\hat{E}(f)$ – the Fourier transforms of the detected THz wave transmitted through the SWCNT layer on the PTFE substrate, and the PTFE alone, respectively.

Spectra of complex conductivity calculated by the formula (1) are shown in Figure 4.

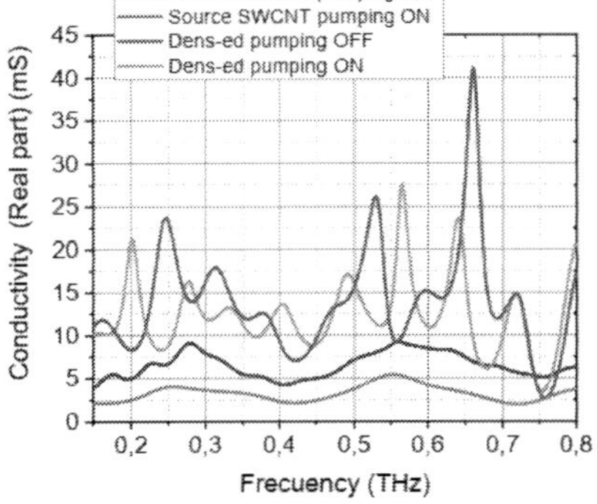

a)

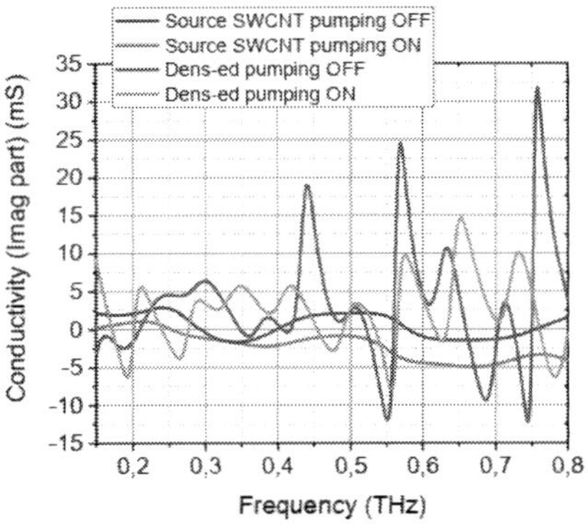

b)

Fig. 4. Real part (a) and Imaginary part (b) of complex conductivity of thin-film SWCNT

The sample before densification has the effect of changing the complex conductivity (as in article 4). One explanation for this effect is that the nanotubes in the original sample are relatively free of each other, like "wool." Upon irradiation of a sample, the SWCNT oscillates more strongly; therefore, the contact area between adjacent nanotubes decreases and the conductivity of the sample decreases. During the process of densification the distance between the nanotubes decreases, the overall density of the film increases, hence the contact area between the tubes increases and the conductivity increases. This is confirmed in our work. After densification, the conductivity increased and does not change upon optical pumping. In addition, peaks appeared on the spectra after sample processing, presumably due to the uneven compaction of a thin-film and a decrease in the signal-to-noise ratio for densified sample.

According to the result of the work, it was obtained that after densification, the sample began to pass less (increased losses / increased the real part of the conductivity), the real part of the refractive index slightly increased. At the same time, the ability to control explicitly is almost imperceptible. This can be used in terahertz antennas and integrated circuits for the terahertz frequency range, where part of the nanotubes are densified to attenuate control by optical pumping.

References

1. *Gusev S., Demchenko P., Cherkasova O., Fedorov V., Khodzitsky M.* Influence of Glucose Concentration on Blood Optical Properties in THz Frequency Range // Chinese Optics 2018. V 11. No. 2. P. 182–189.

2. *Borovkova M., Khodzitsky M., Demchenko P., Cherkasova O., Popov A., Meglinski I.* Terahertz time-domain spectroscopy for non-invasive assessment of water content in biological samples // Biomedical Optics Express 2018. V. 9. No.5. P. 2266.

3. *Borovkova M., Serebriakova M., Fedorov V., Sedykh E., Vaks V., Lichutin A., Salnikova A., Khodzitsky M,* Investigation of terahertz radiation influence on rat glial cells // Biomed. Opt. Express, BOE 2017. V. 8. No. 1. P. 273–280.

4. *Federici J., Moeller L.* Review of terahertz and subterahertz wireless communications // Journal of Applied Physics 2010. V. 107. No. 11. P. 111101.

5. *Smirnov S., Anoshkin I., Demchenko P., Gomon D., Lioubtchenko D., Khodzitsky M., Oberhammer J.* Optically controlled dielectric properties of single-walled carbon nanotubes for terahertz wave applications// Nanoscale 2018. V. 10. No.26. P. 12291–12296.

5. *Anoshkin I., Nasibulin A., Tian Y., Liu B., Jiang H., Kauppinen E.* Hybrid carbon source for single-walled carbon nanotube synthesis by aerosol CVD method // Carbon 2014. V. 78. P. 130–136.

6. *Bespalov, V., Gorodetskiĭ, A., Denisyuk, I., Kozlov, S., Krylov, V., Lukomskiĭ, G., Putilin, S.* Methods of generating superbroadband terahertz pulses with femtosecond lasers //Journal of Optical Technology. 2008. V. 75. No. 10. P. 636-642.

7. Tinkham M. Energy Gap Interpretation of Experiments on Infrared Transmission through Superconducting Films // Physical Review 1956. V. 104. No. 3. P. 845–846

EPJ Web of Conferences **195**, 06023 (2018) https://doi.org/10.1051/epjconf/201819506023
TERA-2018

Terahertz plasmonics: achievements and prospects

A. K. Nikitin[1,2], B. A. Knyazev[2,3], V. V. Gerasimov[2,3]

[1] Scientific and Technological Center of Unique Instrumentation, Russian Academy of Sciences, Moscow, Russia, alnikitin@mail.ru
[2] Novosibirsk State University, Novosibirsk, Russia
[3] Budker Institute of Nuclear Physics, Siberian Branch of Russian Academy of Sciences, Novosibirsk, Russia

I. Brief theory and definitions

Plasmons are collective oscillations of the free electron gas density. Just as light consists of photons, the plasma oscillation consists of plasmons. Plasmon coupled with photon forms another quasiparticle called a plasmon-polariton. Plasmon-polaritons propagating along a "metal-dielectric" interface are called surface plasmon-polaritons (SPPs). They represent themselves a complex of an evanescent *p*-polarized electromagnetic wave coupled with a wave of density of free charges on surface of a (semi) conductor.

The field of optics studying surface plasmons is often described as Plasmonics, by analogy with electronics [1]. "SPPs is the best known method for confining and guiding EM radiation into the smallest possible volume" [2].

By solving Maxwell's equations with the appropriate boundary conditions, one can find out that: 1) field of surface plasmons is *p*-polarized, in order to build polarization of the metal surface; 2) they can exist provided the metal dielectric constant is negative; 3) their dispersion relation (between the tangential component of the SPP wave number k_x and the angular frequency ω of their field) for a "metal-dielectric" interface has the form [2]:

$$k_x = \frac{\omega}{c} \cdot \sqrt{\frac{\varepsilon_M \cdot \varepsilon_D}{\varepsilon_M + \varepsilon_D}}, \qquad (1)$$

here c is the speed of light in free space; ε_M and ε_D are the dielectric constants of the metal and the dielectric.

The complex value of ε_M makes k_x a complex quality as well, with a real component k_x' that defines the dispersion of SPPs and an imaginary component k_x'' that is proportional to the Joule losses of the SPPs.

The dispersion curves for SPPs on a metal adjoining vacuum or a dielectric are presented in Fig.1. The dotted lines indicate so called "light lines" depicting the dispersion of plane waves in a relevant medium. As one can see, the real component of the SPP wave number is close to the light lines at low frequencies; SPPs have light-like characteristics in the terahertz (THz) range that is THz SPPs are weakly bounded to the interface and propagate a long distance (meters). The value of k_x' deviates from the respective light line as the frequency increases, which is accompanied by stronger confinement (in the dielectric) of the SPPs to the interface. Note, that as ω approaches the value $\omega_{SPP} = \omega_p / \sqrt{1+\varepsilon_D}$ (here ω_p is the plasma frequency of the metal) the value of $\varepsilon_M \to 0$ and the deviation of k_x' from k_0 (the wave number of a plane wave in vacuum) infinitely increases, while the propagation length of the SPPs is drastically reduced [2]. Thus one may state that in the low-frequency limit (THz range) SPPs acquire the photonic character, while in the high-frequency limit ($\omega \to \omega_{SPP}$) they resemble a mechanical wave (plasmonic character).

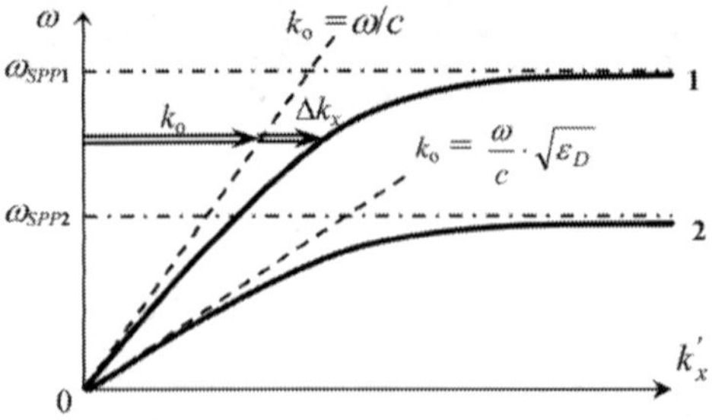

Fig. 1. The dispersion curves of SPPs on interfaces "metal – vacuum" (curve 1) and "metal – dielectric with ε_D" (curve 2).

From the expression for ω_{SPP} it follows that there are at least two ways for controlling confinement (as well as the propagation length) of SPPs to the surface: 1) to change the value of ω_p by, for example, replacing the metal with a semiconductor; 2) to cover the surface with a dielectric layer (curve 2 in Fig.1).

II. Achievements in THz plasmonics

Metals have very large permittivity at THz frequencies, typically 5 to 6 orders of magnitude larger than at optical frequencies. This reduces the penetration depth of the SPP field into the metal and results in a weak confinement of the field to the surface, as well as in an increase of the SPP propagation length L due to the reduction of ohmic loses in the metal.

According to the Drude model for the dielectric permittivity of metals values of $L \sim \lambda^2$ (here λ is the radiation wavelength). But already first experiments with SPPs excited by THz gas lasers revealed great discrepancy between the calculated and measured L values. In the experiments, THz SPPs ran distances an order or even two orders smaller than those predicted by the theory: real THz SPPs ran from a few to tens of centimeters [3]. As it was found out later, the main reason for this was not only the difference between the dielectric permeability of the metal surface and its bulk value, but also the significant radiative losses arising from scattering of SPPs on the surface inhomogeneities [4]. To reduce the losses it was suggested to cover the surface with a thin (~λ/100) dielectric layer shifting in this way the dispersion curve from the light line and thus making the "SPPs – bulk radiation" process less probable.

Later on it was proposed to employ the newly developed time-domain spectroscopy (TDS) technique for experimental studies of THz SPPs [5, 6]. Using the technique Prof. D. Grischkowsky with colleagues estimated average values of L and SPP field penetration depth δ into air; they also discovered the capability of THz SPPs to jump through cm air gaps between the adjacent metal plates. The TDS is beneficial in many respects when investigating bulk samples, but this method has obvious pitfalls of dispersion origin when SPPs are under study. Using monochromatic radiation produced by lasers, one can avoid problems with dispersion of broadband SPPs, uncertainty in phase shifts at the coupling and decoupling elements, as well as the necessity of the inverse Fourier transform in data processing.

Intensive studies of THz SPPs are conducted with the Novosibirsk free-electron laser (Novo FEL) at the Siberian Synchrotron and Terahertz Radiation Centre [7]. Here they managed to master the end-fire coupling method for effective conversion of FEL radiation into THz SPPs employing cylindrical segments as coupling elements; this made it possible to measure carefully field distribution along the SPP track, to study diffraction of THz SPPs on the rectangular sample edge, to detect bulk waves emitted from the SPP track, to investigate explicitly the process of SPP passing from one to another substrate via air gaps, to establish capability of THz SPPs to be reflected by plane mirrors and to be divided by plane beam splitters [4, 8, 9]. Based on these achievements we have developed and tested the device for measuring the propagation length of THz SPPs in which the input and output elements are fixed; this enables one to perform measurements with a high signal-to-noise ratio.

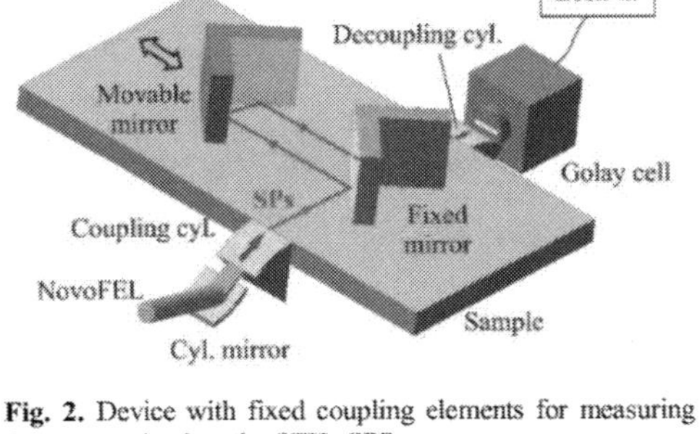

Fig. 2. Device with fixed coupling elements for measuring the propagation length of THz SPPs.

Recently one more way for controlling the value of ω_p, and thus the value of ω_{SPP}, was discovered; this method consists in creating periodic structures (holes, recesses, or objects) of subwavelength dimensions on the metal surface [10]. Such SPPs were called "spoof" or "designed" surface plasmons. The method allows confining THz SPPs not only on metals, but also on graphene, as well as on superconductors [1, 2]. Moreover, it turned out that conductive layers "decorated" with ordered holes have the ability to selectively transmit THz radiation with the transmission efficiency proportional to $(r/\lambda)^4$ (here r is the radius of each hole).

III. Prospects of THz plasmonics

We can distinguish the following promising areas of THz plasmonics development:

1) SPP communication channels based on extended metal-dielectric structures (wires) with rectangular or circular cross-section. The advantages of this method of information transmission are the following: first, the speed of THz SPPs is the maximum possible (it almost equals the speed of light); secondly, it leads to an extremely low group velocity dispersion, allowing essentially undistorted pulse propagation;

2) design of SPP interferometers that will make it possible to create surface plasmon Fourier spectrometers of the THz range;

3) intensive study of "spoof" surface plasmons (especially on graphene) and their interaction with bulk THz radiation;

4) generation of THz SPPs on semiconductor surfaces by radiation with $\omega\approx\omega_p$ condensing the light in the vicinity of the surface and thus paving the way to THz microscopy with super high resolution;

5) study of "localized" SPPs (LSPPs) in semiconductor micron-size particles which, in contrast to metal ones, are pervious for THz radiation. Exciting LSPPs one may concentrate the radiation energy in a very limited space.

The work was supported by the Russian Science Foundation (grants No. 18-12-00430 and 14-50-00080).

References

1. *Maier, S. A.* Plasmonics: Fundamentals and Applications. Springer, 2007. – 224 p.

2. *Gomez, R. J., Berrier A.* Fundamental aspects of surface plasmon polaritons at THz frequencies // In "Handbook of THz technology". Eds by Saeedkia D. Woodhead Publ. Series in Electronic and Optical Materials, 2013. – 688 p.

3. *Koteles, E. S., McNeill*, W. H. Far infrared surface plasmon propagation // Intern. J. Infrared Millim. Waves. 1981. V. 2. P. 361-371.

4. *Gerasimov, V.V., Knyazev, B.A., Lemzyakov, A.G., Nikitin, A.K., Zhizhin, G.N.* Growth of THz surface plasmon propagation length due to thin-layer dielectric coating // JOSA (B). 2016. V. 33. Is. 11. P. 2196-2203.

5. *Jeon, T.-I., Grischkowsky, D.* THz Zenneck surface wave (THz surface plasmon) propagation on a metal sheet // Appl. Phys. Lett. 2006. V. 88. 061113.

6. *Nazarov, M. M., Mukina, L. S., Shuvaev, A.V., Sapozhnikov, D. A., Shkurinov, A. P., Trofimov, V. A.* Excitation of surface EM waves studied by THz spectrochronography // Laser Phys. Lett. 2005. V. 2, No. 10. P. 471–475.

7. *Kulipanov, G. N., Bagryanskaya, E. G., Chesnokov, E. N. et al.* Novosibirsk free electron laser — facility description and recent experiments // IEEE Trans. on THz Science and Technology. 2015. V. 5. No. 5. P. 798-809.

8. *Gerasimov, V.V., Knyazev, B.A., Nikitin, A.K., Zhizhin, G.N.* Experimental investigations into capability of THz surface plasmons to bridge macroscopic air gaps // Optics Express. 2015. V. 23. No. 26. P. 33448-33459.

9. *Gerasimov, V.V., Knyazev, B.A., Nikitin, A.K.* Reflection of THz surface plasmon-polaritons by a fl at mirror // Quantum Electronics. 2017. V. 47. No. 1. P. 65–70.

10. *Pendry, J. B., Martin-Moreno, L., Garcia-Vidal, F. J.* Mimicking surface plasmons with structured surfaces // Science. 2004. V. 305. No. 5685. P. 847-848.

The first observation of the free induction signals of OH radicals in the terahertz region.

E.N.Chesnokov[1], L.N.Krasnoperov[2], V.V.Kubarev[3], P.V.Koshlyakov[1].

[1]Institute of Chemical Kinetics and Combustion, Novosibirsk, Russia, chesnok@kinetics.nsc.ru
[2]New-Jersey Institute of Technology., Newark,USA
[3] Institute of Nuclear Physics, Novosibirsk,Russia

Hydroxyl radical OH is the main oxidizing agent in combustion processes and in atmospheric chemistry. The rotational spectrum of OH is comparatively simple - there are less than 10 absorption lines in the therahertz region, corresponding to transitions between the lower rotational states of the radical. The FID signal was observed at the OH absorption line at 83.8 cm^{-1}, that corresponds to the transition from the lowest rotational state of OH. Hydroxyl radicals were generated in the following sequence of chemical reactions:

$$O_3 + (h\nu = 266\ nm) \rightarrow O(^1D) + O_2 \quad (1)$$

$$O(^1D) + H_2O \rightarrow 2\ OH \quad (2)$$

$$OH + OH \rightarrow H_2O + O \quad (3)$$

The concentration of OH radicals was ~ 10^{15} cm-3, lifetime ~ 300 .... 700 mks.

The pulse of UV laser (266 nm) was synchronized with the FEL pulses. Shape of the FEL pulse with FID signal after the optical cell was recorded by an ultrafast Schottky diode detector and 30GHz oscilloscope[1]. The The experiment was organized similarly to our work on ultrafast timedomain spectroscopy [2-18].

Fig.1 shows the first experimentally observed FID signal of a hydroxyl radical, accumulated at 500 mks after the UV pulse The moment t = 0 corresponds to the input FEL pulse. The signal from the laser pulse was accumulated separately, in the absence of radicals., and was subtracted .

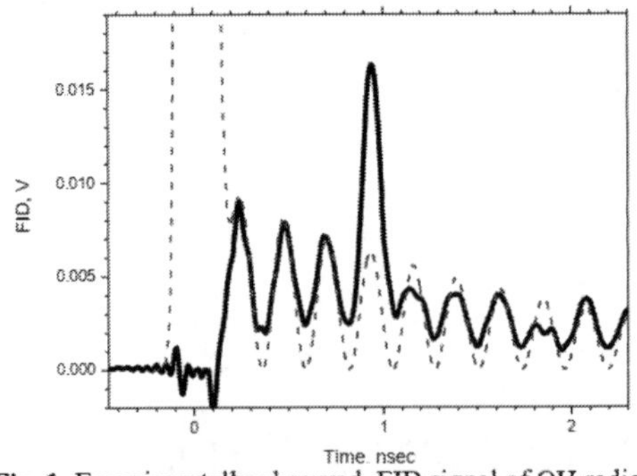

Fig. 1. Experimentally observed FID signal of OH radical. Exciting laser pulse was acc accumulated separately and subtracted. Dashed line – numeric modeling of the FID.

Absorption line of the OH radical is a Λ- doublet with a splitting of 4.3 GHz. The width of the FEL spectrum exceeds this splitting, therefore the laser simultaneously excites both components of the doublet. The FID signal contains characteristic beats with a period of 0.23 ns.

By changing delay between UV pulse and moment of measurements we can see the growing of OH concentration and at big delays decay of the OH concentration.

During the lifetime of the OH radical, many pulses of FEL pass through the cell, each of which initiates a FID signal. We could register a sequence of 10 pulses simultaneously, obtaining separate frames of the birth and death of the radical

The influence of the magnetic field on the FID signal was studied by numerical simulation. A longitudinal magnetic field leads to a rotation of the polarization plane of the signal. The angle of rotation depends on the time after the FEL pulse. The effect of a nonmonotonic rotation of the polarization was predicted - after a certain time after the pulse, the direction of rotation changes

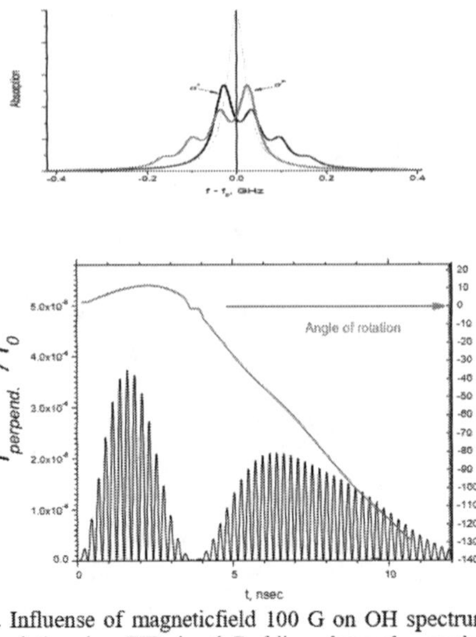

Fig. 2. Influense of magneticfield 100 G on OH spectrum (upper plot) and on FID signal. Red line shows the rotation angle of polarization plane of the FID vs.time.

This work was made with financial support of Russian Science Foundation (project 14-50-00080). The NovoFEL operation was supported by Ministry of Education and Science of The Russian Federation (project RFMEFI62117X0012).

References

1. *Chesnokov E. N., Kubarev V. V., Koshlyakov P. V., Kulipanov G. N.* Direct observation of the terahertz optical free induction decay of molecular rotation absorption lines in the sub-nanosecond time scale// Appl. Phys. Lett.2012, V. 101, 131109.

2. *Kubarev, V., Chesnokov, E., Koshlyakov, P.* Ultrafast high-resolution THz time-domain spectroscopy // Proc. IRMMW-THz-2012. 2012. Wollongong. Australia. Thu-C-2-4.

3. *Kubarev, V., Chesnokov, E., Koshlyakov, P.* Terahertz free induction decay of molecular transitions: direct observation and practical use // Proc. TERA-2012. 2012. Moscow. Russia. P. 68–69.

4. *Kubarev, V. V., Chesnokov, E. N., and Koshlyakov, P. V.* One-pulse high-resolution THz time-domain spectroscopy: development and applications // Proc. IRMMW-THz-2013. 2013. Mainz. Germany. Tu5-5.

5. *Kubarev, V. V.* Ultrafast high-resolution spectroscopy of separate NovoFEL pulses // Proc. IRMMW-THz-2013. 2013. Mainz. Germany. Mo14-4.

6. *Chesnokov, E. N., Kubarev, V. V., Koshlyakov, P. V., Kulipanov, G. N.* Very long terahertz free induction decay in gaseous hydrogen bromide // Laser Phys. Lett. 2013. V. 10. P. 055701-055703.

7. *Chesnokov, E. N., Kubarev, V. V., and Koshlyakov, P. V.* Rotation commensurate echo of asymmetric molecules - Molecular fingerprints in the time domain // Appl. Phys. Lett. 2014. V. 105. P. 261107-(1–4).

8. *Chesnokov, E. N., Kubarev, V. V., Koshlyakov, P. V., Getmanov, Ya. V., Shevchenko, O. A.* Non-Faraday rotation of the free induction decay in gaseous NO // Chemical Physics Letters. 2015. V. 636. P. 203–207.

9. *Chesnokov, E. N., Kubarev, V. V., Koshlyakov, P. V., Chichinin, A. I., Getmanov, Ya. V., Shevchenko, O. A.* The influence of magnetic field on the echo-like free induction decay in NO2 // Chemical Physics Letters. 2016. V. 662. P. 62–66.

10. *Chesnokov, E. N., Kubarev, V. V., Koshlyakov, P.V.* Observation of the Formation of 0-π Pulses in Rotation Spectra of HCN and HBr // Physics Procedia. 2016. V. 84. P. 135 – 141.

11. *Bragin A.V.* et al. Superconducting solenoid for superfast THz spectroscopy // Physics Procedia. 2016. V. 84. P. 82 – 85.

12. *Chesnokov, E. N., Kubarev, V. V., Koshlyakov, P. V., and Fedorov, V. V.* Heterodyne Method of Detection of Molecular Gas in the Terahertz Region Using the Beats Between Free Induction Decay Signals // IEEE Transactions on terahertz science and technology. 2017. V. 7, No. 2. P. 144–150.

INTERACTION OF HIGH-POWER TERAHERTZ RADIATION WITH METALLIC FILMS

O. V. Chefonov[1], A. V. Ovchinnikov[1], S. I. Ashitkov[1], S. A. Evlashin[2], P. S. Kondratenko[3], M. B. Agranat[1], V. E. Fortov[1]

[1]Joint Institute for High Temperatures of the Russian Academy of Sciences, Moscow, Russia, oleg.chefonov@gmail.com
[2]Center for Design, Manufacturing and Materials, Skolkovo Institute of Science and Technology, Moscow, Russia
[3]Nuclear Safety Institute of the Russian Academy of Sciences, Moscow, Russia

We report on the experimental observation of high-power terahertz (THz) radiation-induced damage in thin aluminum and nickel films with a thickness less than a terahertz skin-depth. The damage threshold fluence of thin metal films is evaluated for single pulse experiments.

The experiments were performed using a terawatt femtosecond Cr:forsterite laser system, that delivers 100-fs pulses with a bandwidth of 25 nm at a central wavelength of 1240 nm with energy above 80 mJ (energy stability less than 10%) at 10 Hz repetition rate [1]. THz pulses in the range of 1–3 THz were generated by optical rectification of femtosecond laser pulses in organic nonlinear crystal DSTMS (4-N,N-dimethylamino-4-N-methyl-stilbazolium 2,4,6-trimethylbenzene-sulfonate) of 8 mm in diameter [2, 3]. THz low pass filter was used for cutting wavelengths below 34 μm, including the pumping radiation at a wavelength of 1240 nm and its second harmonic at 620 nm generated in nonlinear crystal during optical rectification process. The attenuation coefficient of the THz low pass filter at wavelengths of 1240 nm and 620 nm was more than 10^8. A telescope 6:1 consisted of two off-axis parabolic mirrors with a reflected focus length of 25.4 and 152.4 mm was used to compensate THz beam divergence. The collimated THz beam traveling from the telescope was focused onto a sample using an 90° off-axis parabolic mirror with a reflected focus length of 50.8 mm and a diameter of 50.8 mm to a diffraction-limited spot size. Terahertz pulse energy was measured by means of a calibrated optoacoustic detector (Golay cell) and was up to ~110 μJ at the samples surface.

The samples were a 20-nm Al and Ni films deposited on a glass substrate of 180 μm in thickness using magnetron sputtering of a high-clean metal target (purity of 99.99%) at a rate of 1–1.91 nm/s in an argon atmosphere (purity of 99.999%). The scatter in films thickness were about 1–3 nm from the specified value.

Scanning electron microscope (SEM) images of thin metal film damage patterns at different fluences F of a single THz pulse are presented in Fig. 1 and Fig. 2.

The single-pulse damage threshold of the thin film was experimentally determined using a standard technique, where the squared radius (r_D [μm]) of observed damaged regions is plotted versus the logarithm of the numerical value of the energy of incident THz pulses W_{THz} [μJ] (Fig. 3). This dependence should be linear for a Gaussian beam. Thus, the threshold energy W_{th} could be derived from the line's intersection with an X-axis, while the line's slope determines the parameter of the Gaussian beam at the level of $1/e$.

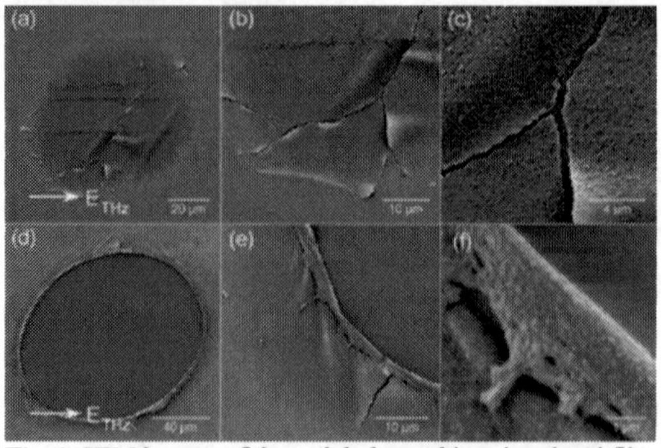

Fig. 1. SEM images of through holes and its edges in Al film produced by single THz pulses. (a), (b), (c): $F = 0.14$ J/cm^2; (d), (e), (f): $F = 0.3$ J/cm^2. The arrows indicate the direction of the THz electric field.

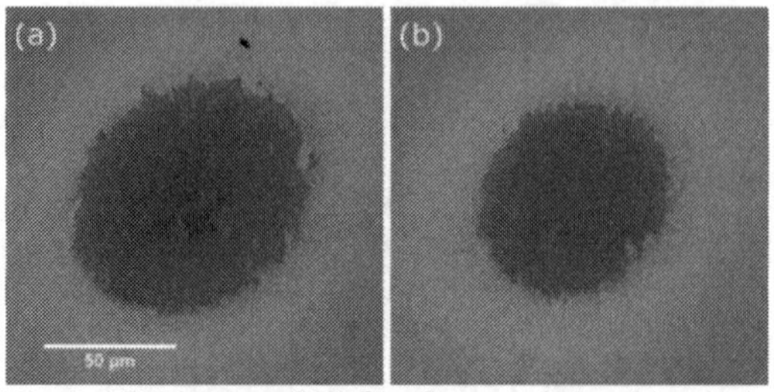

Fig. 2. SEM images of Ni film damages induced by single terahertz pulse at different fluences. (a): $F = 0.43$ J/cm^2, (b): $F = 0.39$ J/cm^2. The direction of the THz electric field is the same as in Fig. 1.

The experimental data in Fig. 3 are well approximated with a linear function (solid lines) indicating a Gaussian profile of the terahertz beam, high homogeneity of the used metal film and a distinct threshold behavior of its damage. The measured radius at the level of $1/e$ was of $r_0 \sim 90 \div 92$ μm, which gives a single-pulse damage threshold of the incident fluence of $F_{th} = W_{th}/(\pi r_0^2) \approx 0.15$ J/cm^2 for Al and 0.19 J/cm^2 for Ni at an electric-field strength about 15 MV/cm at the center of the focal spot. The value of electric-field strength was estimated using the energy of a THz pulse, its duration and a focused to a diffraction-limited spot size, assuming a Gaussian pulse shape [2].

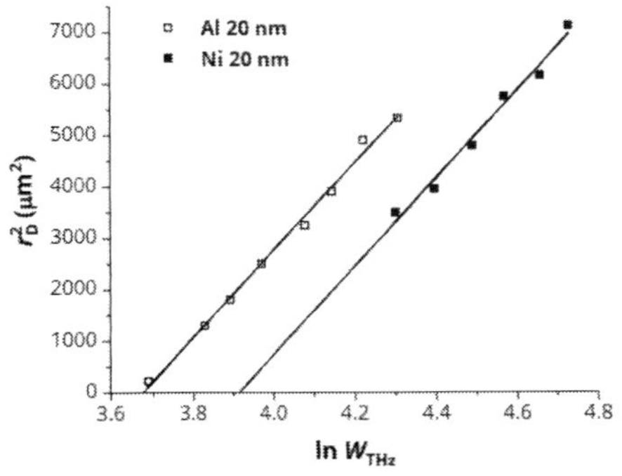

Fig. 4. Determination of the damage threshold in Al and Ni films irradiated with a single THz pulse; experimental data are shown with square symbols approximated with linear functions (solid lines).

SEM images of damage patterns in the Al and Ni films induced by multiple THz pulses at different fluences above the single-pulse damage threshold and number of pulses are shown in Fig. 4 and Fig. 5. The results were obtained at a pulse repetition rate of 10 Hz.

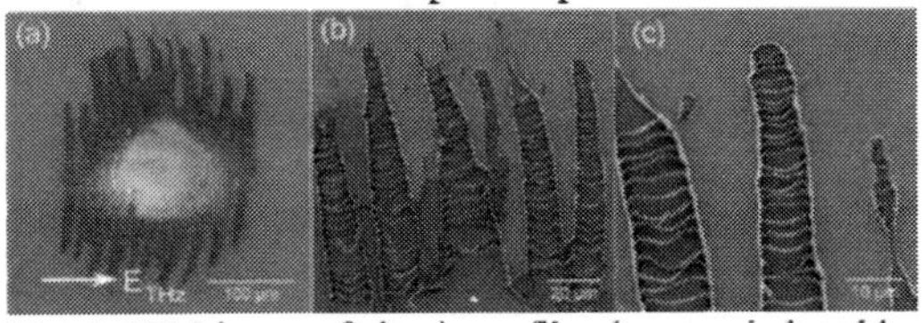

Fig. 4. SEM image of aluminum film damages induced by multiple THz pulses at fluence $F = 0.24$ J/cm^2, $N = 60$.

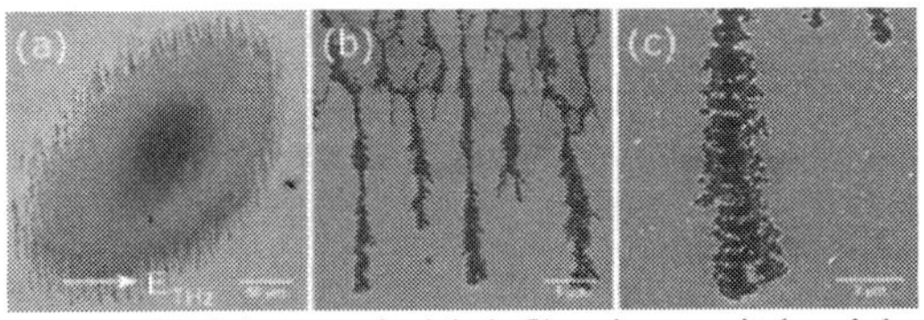

Fig. 5. SEM image of nickel film damage induced by multiple THz pulses at fluence $F = 0.43$ J/cm^2, $N = 40$.

We suppose the observed single THz pulse-driven damages in thin metal films to be formed as a result of absorption of the THz pulse energy by conduction electrons with the following generation of hot electrons that transfer energy to the lattice and heat it. Depending on the fluence the lattice heating after the THz pulse could result in film damage due to thermal expansion or melting and ablation processes. Due to differences in acoustical impedances at film boundaries the film center of mass gains momentum in the direction from the substrate. Besides, a tensile stress wave results in adhesive debonding. Delamination of the film occurs when exceeding the adhesion strength. With increasing fluence of the THz pulse the film expands into free space and breaks down.

Damage behavior of metal films irradiated with multiple THz pulses at fluences below the single pulse damage threshold (at the periphery, where according to Gaussian distribution the fluence is below the single-pulse damage threshold) is rather challenging to explain. We suppose that the THz pulse at the center of the focused spot (a region with the peak electric field) generates a microcrack as a result of induced stress. It might be supposed that for the subsequent THz pulses there is a local field enhancement in the gap of the induced microcrack. This results in a microcrack growth perpendicular to the electric-field vector and an increase in width in across-track direction. The microcrack growth occurs along with film ablation in microregions similar to the mechanism of the single pulse damage. As for the Al film, the peripheral damage pattern has the form of elongated surface discontinuities (a kind of channels) that are perpendicular to the electric field vector of the terahertz pulse. However, these "channels" in the Ni film differ in character (pattern) and sizes. For the Al film the observed damages have clearly defined plastic behaviour (the damage surface is smooth, there are smooth and rolled up edges, a rim at the edge of damage is 1–2 μm in width). For the Ni film the damage pattern is similar to fragile or quasi-fragile type (the damaged surface is rough and grainy, there are no smooth edges and rims). This could be associated with different physical and adhesion properties of the films (the Ni film has an increased adhesion as opposed to the Al film), and different processes of the film formation on a substrate during sputtering. Explain the difference in the size and periodicity of the structures formed on the surface of the films is not yet possible. A mechanism related to the generation of such channel-like damages and based on a new phenomenon of electrostriction in thin metal films was proposed in [2], while this phenomenon has not been and could not be observed in bulk metals due to a strong damping of the electric field.

This work is supported by Russian Science Foundation Grant No. 17-19-01261. The experiments were performed using the unique scientific facility "Terawatt Femtosecond Laser Complex" in the "Femtosecond Laser Complex" Center of the Joint Institute for High Temperatures of the Russian Academy of Sciences.

References

1. *Agranat, M. B., Ashitkov, S. I., Ivanov, A. A., Konyashchenko, A. V., Ovchinnikov, A. V., Fortov, V. E.* Terawatt femtosecond Cr:forsterite laser system // Quantum Electron. 2004. V. 34, No. 6. P. 506–508.

2. *Agranat, M. B., Chefonov, O. V., Ovchinnikov, A. V., Ashitkov, S. I., Fortov, V. E., Kondratenko, P. S.* Damage in a Thin Metal Film by High-Power Terahertz Radiation // Phys. Rev. Lett. 2018. V. 120, P. 085704.

3. *Chefonov, O. V., Ovchinnikov, A. V., Evlashin, S. A., Agranat, M. B.* Damage Threshold of Ni Thin Film by Terahertz Pulses // has been accepted for publication in Journal of Infrared, Millimeter, and Terahertz Waves.

EPJ Web of Conferences **195**, 07002 (2018) https://doi.org/10.1051/epjconf/201819507002
TERA-2018

Disappearance of Self-Focusing for Few-Cycle THz Pulses

S. A. Kozlov[1], A. A. Drozdov[1], M. A. Kniazev[1], D. A. Kislin[1], S. Choudhary[2,3], R. W. Boyd[2,3]

[1] ITMO University, Saint-Petersburg, Russia, kozlov@mail.ifmo.ru
[2] University of Ottawa, Ottawa, Canada
[3] The Institute of Optics, University of Rochester, Rochester, NY 14627-0186, USA

In the present work, we demonstrate that, for few-cycle wave packets with longitudinal dimension less the transverse size, the concept of critical power of self-focusing can lose its physical meaning because of the dominance of dispersion over diffraction. We present simple formulas for the estimation of the parameters of the field and medium under which self-focusing disappears and illustrate with numerical calculations the changes in the self-action phenomenon of light in this case.

Theoretical estimations

We have shown that when the nonlinearity of the optical medium competes primarily not with diffraction but with dispersion, the concept of critical power for self-focusing of radiation P_{cr} will begin to lose its original meaning. Such condition can be expressed in terms of laboratory parameters as:

$$\frac{l_0}{D_0} < \sqrt{c\omega_0 n(\omega_0)\beta_2}\,, \qquad (1)$$

where $l_0 = 2c\tau_0$ is longitudinal size of the wave packet and $D_0 = 2r_0$ is its transverse size, c is the speed of light in vacuum, τ_0 and r_0 are the duration and radius of the beam at the entrance of nonlinear medium, ω_0 is the central frequency of radiation, $n(\omega_0)$ is the linear refractive index of the medium at central frequency, and $\beta_2 = \left(\partial^2 k/\partial\omega^2\right)_{\omega_0}$, where k is the wave number.

For an extremely simple dispersion formula of $n(\omega) = N_0 + ac\omega_0^2$, where N_0 and a characterize the medium dispersion, the inequality (1) can be reduced to

$$\frac{l_0}{D_0} < \sqrt{6N_0\Delta n_{disp}}\,, \qquad (2)$$

where $\Delta n_{disp} = ac\omega_0^2$ is the change of the refractive index at the central wavelength due to dispersion.

Numerical illustrations

We consider boundary conditions (the electric field of radiation at the entrance to nonlinear medium) in the form of a Gaussian axisymmetric paraxial beam with a small number of oscillations ($N = \tau_0 / T_0$)

$$E(0,r,t) = E_0 \exp\left(-\frac{r^2}{r_0^2}\right)\exp\left(-\frac{t^2}{\tau_0^2}\right)\sin(\omega_0 t), \quad (3)$$

where $r = \sqrt{x^2 + y^2}$, and $T_0 = 2\pi / \omega_0$ is central period of electric field oscillations.

The pulse propagation was modeled in terms of the evolution of the instantaneous electric field [1, 2] and not its envelope. The use of the field equation allows on the correctly describe the generation of radiation at tripled and higher frequencies [1]. We show two-dimensional plots of the electric field distribution in Figures 1-3. The light-grey and dark-grey areas correspond to maximum positive and negative values of the electric field, while white lines correspond to zero field values, respectively.

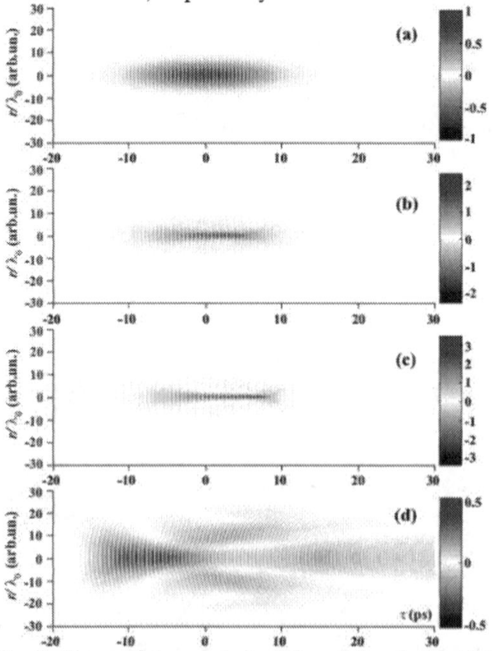

Fig. 1. The spatiotemporal dynamics of the electric THz field E of wave packet (3) in a MgO:LiNbO$_3$ crystal at distances (a) $z = 0$ cm, (b) $z = 8$ cm, (c) $z = 11$ cm, (d) $z = 25$ cm. Initial pulse parameters are $\lambda_0 = 300$ μm, $r_0 = 5\lambda_0$, $\tau_0 = 8$ ps, $I = 3 \times 10^8$ W/cm^2, $P_0 = 4P_{cr}$. We see that the beam shows self-focusing behavior at 8 and 11 cm and the beginning of breakup into filaments at 25 cm

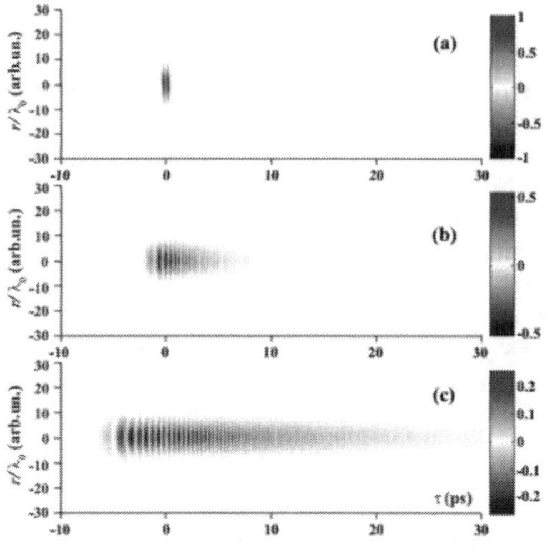

 https://doi.org/10.1051/epjconf/201819507002

Fig. 2. The spatiotemporal dynamics of the electric field for a single-cycle THz wave packet (3) in a MgO:LiNbO$_3$ crystal at distances (a) z = 0 cm, (b) z = 0.5 cm, (c) z = 2 cm. Initial pulse parameters are λ_0 = 300 μm, $r_0 = 5\lambda_0$, τ_0 = 0.3 ps, $I = 3 \times 10^8$ W/cm^2, $P_0 = 4P_{cr}$. In this case the beam does not display self-focusing

Figures 1 and 2 show typical results for the self-focusing dynamics for a many-cycle pulse (Fig 1) and a single-cycle pulse (Fig 2). The changes in self-focusing dynamics are illustrated for ultrashort pulses with a central wavelength of λ_0 = 300 μm for a number of field oscillations N = 16 ($2\tau_0$ = 16 ps) and N = 0.6 ($2\tau_0$ = 0.6 ps). Stoichiometric MgO:LiNbO$_3$ crystal with nonlinear refractive index $n_2 = 5.4 \times 10^{-12}$ cm^2/W [1] and dispersion parameters described in [1] was considered as the nonlinear dielectric medium. We also used the same ratio between initial power P_0 and critical power P_{cr} for self-focusing, $P_0 = 4P_{cr}$. As can be seen from Fig. 1, at initially stage of THz wave propagation its transverse compression occurs (Fig. 1 b). Then, a shock wave is formed in the generated filament (Fig. 1 c). At the distance of z = 11 cm (Fig. 1 c) a self-focused filament is formed with a transverse size 6.5 times smaller than the input value. The intensity of the radiation increased by a factor of 12 within the filament and reaches a value of 3.6×10^9 W/cm^2. The on-axis pulse duration is seen to decrease by a factor of 1.4 at the e^{-1} level.

With a decreasing of the number of oscillations down to just one (Fig. 2), the dispersion becomes so strong that self-focusing is no longer observed. For example, the axial pulse duration increased in 13 times (at e^{-1} level) already at a distance of z = 5 mm (Fig. 2 b). The curvature of the wave front due to diffraction is noticeable only at the leading edge of the beam. As one can see from Figures 1-2, reduction of N leads to decreasing of self-focusing efficiency because dispersion starts play a significant role in pulse evolution.

For a higher intensity of the input single-cycle THz beam (for example for $I = 1.48 \times 10^{10}$ W/cm^2 such that $P_0 = 200P_{cr}$), we observe an effect that is similar to the phenomenon of self-focusing of radiation (Fig. 1). In contrast to the case considered in Fig. 1, a significant increase of pulse duration is observed here at the initial stage of THz wave propagation because of the combined strong effect of dispersion and nonlinearity. Significant curvature of the wave front due to nonlinearity is also observed (Fig. 3 b,c). At the next step of THz wave propagation, transverse compression occurs due to the curvature. At the distance of z = 44 mm (Fig. 3 d), a filament is formed with transverse dimension 6 times smaller than the input value. However, the intensity of the radiation is decreased down to 5.3×10^9 W/cm^2. In Fig. 3 d we show additionally the result of propagation of the initial beam (Fig. 3 c) in linear dielectric medium 20 mm long with the same dispersion. It can be clearly seen that the focusing occurs in a linear medium as well due to the large curvature of the phase surfaces. Then usual dispersive-diffraction wave packet spreading occurs (Fig. 3 e).

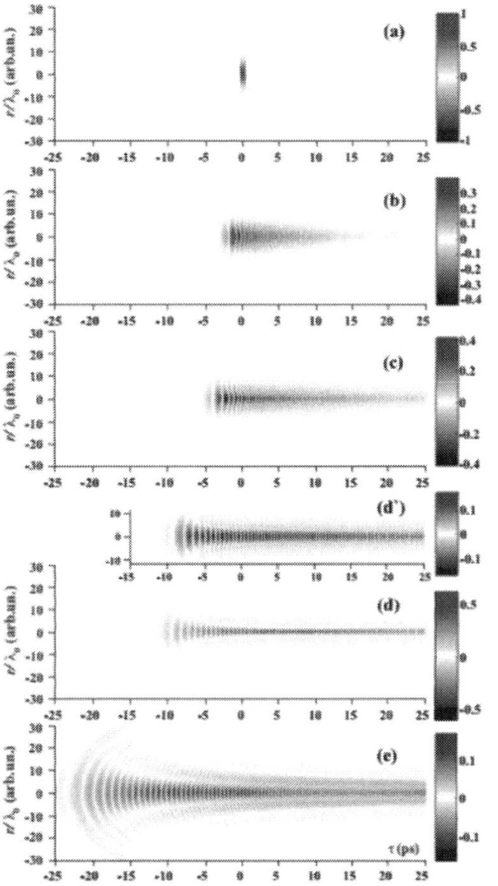

Fig. 3. The spatiotemporal dynamics of the THz electric field E of a single-cycle wave packet (3) in a MgO:LiNbO$_3$ crystal at distances (a) z = 0, (b) z = 9 mm, (c) z = 17 mm, (d) z = 44 mm, (e) z = 90 mm. Initial pulse parameters are λ_0 = 300 μm, $r_0 = 5\lambda_0$, τ_0 = 0.3 ps, $I = 1.48 \times 10^{10}$ W/cm^2, $P_0 = 200P_{cr}$. Case (d') corresponds to propagation of the initial beam though a 20-mm-long linear dielectric medium

Conclusion

A simple expression is derived and analyzed for the conditions under which the concept of a critical power for self-focusing loses its fundamental physical meaning due to the dominance of the process of dispersion over diffraction. It is shown that the power of single-cycle terahertz waves, when self-focusing occurs, can exceed the critical power by several hundred times. It is also shown that the nature of the compression of the single-cycle light beam is significantly different from the self-focusing of longer pulses.

References

1. *Drozdov, A. A., Kozlov, S. A., Sukhorukov, A. A., Kivshar, Y. S.* Self-phase modulation and frequency generation with few-cycle optical pulses in nonlinear dispersive media // Phys. Rev. A. 2012. V. 86, No. 5. P. 053822.

2. *Kozlov, S. A., Samartsev, V. V.* Fundamentals of femtosecond optics Woodhead. Cambridge. UK. 2013. P. 272.

EPJ Web of Conferences **195**, 07003 (2018) https://doi.org/10.1051/epjconf/201819507003
TERA-2018

Ultrafast multi-electron dynamics studied with THz-field streaking

M. Krikunova[1,2], E. Klimešová[1], O. Kulyk[1], T. Oelze[2], B. Schütte[3], T. Gebert[4], J. Andreasson[1,5]

[1]ELI Beamlines, Dolní Břežany by Prague, Czech Republic, maria.krikunova@eli-beams.eu
[2]Technische Universität Berlin, Berlin, Germany
[3]Max-Born Institute, Berlin, Germany,
[4]Max Planck Institute for the Structure and Dynamics of Matter, Hamburg, Germany,
[5]Chalmers University of Technology, Göteborg, Schweden

Temendous developments of novel ultrafast light sources based on High Harmonic Generation (HHG) or Free-Electron-Lasers (FELs) in the last decade have opened up new experimental regimes in the vacuum-ultraviolet (VUV), extreme-ultraviolet (XUV) down to X-ray spectral regions. Especially, the unique combination of coherent intense radiation with short pulse duration holds very promising perspectives for time-resolved *molecular movies* by obtaining information on molecular structure and function on atomic length scale with time resolution down to the attosecond domain. An understanding of the way how an intense X-ray light interacts with matter is of fundamental importance for studies of molecular structure and dynamics. Therefore this research topic is actively pursued by theoretical and experimental work within the photon science community.

The main objective of the Extreme Light Infrastructure – ELI-Beamlines project – in Dolní Břežany near Prague, Czech Republic is to establish an international user facility for fundamental and applied research with advanced laser sources. The research group for applications in molecular, biomedical, and material sciences works on the development of experimental capabilities using secondary sources that are driven by the unique ELI-Beamlines lasers. A central activity is the development of the multi-purpose user end-station (MAC) for experiments in atomic, molecular, and optical sciences and coherent diffractive imaging. The MAC user end-station is equipped with electron/ion time-of-flight spectrometers, Velocity Map Imaging detector, a detector for coherent diffractive imaging and state-of-the-art sample delivery systems to enable advanced photon science experiments on low density targets (atoms, molecules, clusters, nanometer size single intact free of substrate organic and inorganic particles and crystals, vacuum compatible tens of nanometers to a few micrometers thick liquid sheets formed with gas and by colliding jets technique, cylindrical GDVN jets and aerosols) in a broad range of the electromagnetic spectrum from THz to (soft) X-rays.

The THz-field driven streaking is a novel experimental technique to study the electronic response of many-electron systems to irradiation with intense ultra-short (soft) X-rays pulses [1]. The effect of light-field streaking can be understood as an additional momentum acquired by the free electrons in the presence of a dressing electric field [2]. By changing the time-delay between the ionizing XUV pulse with respect to the streaking field and measuring the electron kinetic energy spectra, a streaking spectrogram is obtained. In the streaking experiment half of the oscillation period of the streaking field has to be shorter than the temporal width of the electron distribution [2]. Therefore, to resolve the electron dynamics that is initiated by the femtosecond (soft) X-ray pulse, a THz-field is used [1].

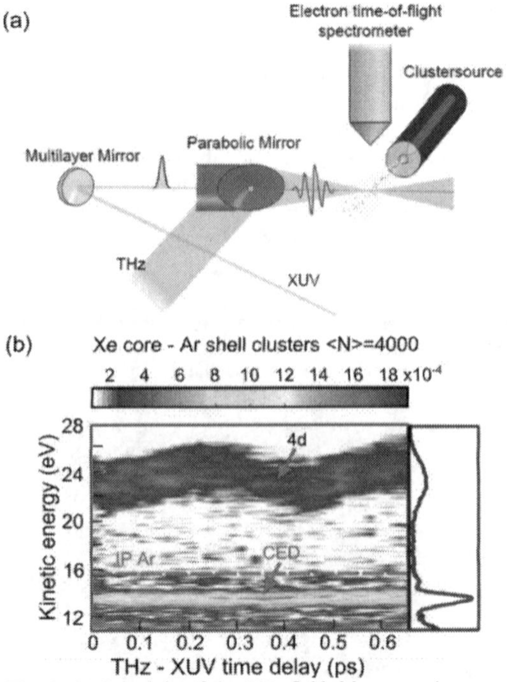

Fig. 1. (a) Principle of the THz-field driven streak-camera measurement. THz and XUV/(soft) X-ray pulses are collinearly overlapped and focused in the interaction region on atomic, molecular or nano-particle target. A series of photoelectron spectra, a so called streaking spectrogram, is measured with an electron-spectrometer as a function of relative time-delay between THz and XUV pulses. **(b)** THz-streaking spectrogram recorded in Xe-Ar clusters at Free-Electron-Laser (FEL) in Hamburg, FLASH. An observation of an un-modulated peak (marked as CED) clearly discloses an electron emission delayed by at least several picoseconds after ionization of Xe-Ar clusters with intense FEL pulses. Data are reproduced from [3].

Figure 1 (a) shows the typical experimental geometry of a THz-streaking experiment [1, 3, 4]. Figure 1 (b) shows an example of a streaking spectrogram recorded upon ionization of Xe-Ar clusters with intense FEL pulses (92 eV photon energy) [3]. In this experiment THz radiation is produced in the THz-undulator by the same electron bunch that is used to produce XUV pulse [1]. Note that electrons released from the 4d orbital of Xe show typical oscillatory behavior due to the shift in electron momentum induced by the instantaneous THz-field at the instance of the electron emission (Figure 1 (b)). Electrons marked as CED (correlated electronic decay), however, remain un-streaked, indicating that these electrons are emitted long after the THz pulse is terminated (i.e. on the time-scale of at least few picoseconds) [3].

For experiments at the MAC user end-station atomic, molecular, cluster or aerosol nano-particles will be available as targets. XUV pulses will be delivered by the HHG-source to the MAC end-station. Two types of THz-sources are planned. The first one is based on the optical rectification in a LiNbO3crystal. To achieve phase matching conditions a tilted-pulse-front geometry will be used [5, 6]. This source typically provides single cycle THz pulses in the frequency range between 0.1 and 1 THz. The expected maximal THz-field strength in the interaction region is ~10 MV/m [4]. The second source of THz radiation is based on difference frequency generation in an organic crystal [7]. This source provides narrow band multi-cycle THz-pulse tunable in the frequency range between 4 and 18 THz. Peak electric field of 370 MV/m has been achieved [7]. Using the same near-infrared laser to drive the HHG process and the THz-source will provide XUV and THz-pulses that are intrinsically synchronized with sub-cycle accuracy [4]. These developments will enable advanced experiments on the structure and dynamics of matter using a combination of THz and HHG beams.

Acknowledgements: This work was supported by the projects Advanced research using high intensity laser produced photons and particles (ADONIS) (CZ.02.1.01/0.0/0.0/16_019/0000789) and Structural dynamics of biomolecular systems (ELIBIO) (CZ.02.1.01/0.0/0.0/15_003/ 0000447) from the European Regional Development Fund and the Ministry of Education, Youth and Sports as part of targeted support from the National Programme of Sustainability II.

References

1. *Frühling, U., Wieland M., Gensch, M., Gebert, T., Schütte, B., Krikunova,M., Kalms, K., Budzyn, F., Grimm, O., Rossbach, J., Plönjes, E., Drescher, M.*. Single-shot terahertz-field-driven X-ray streak camera // Nature Photon. 2009. V. 3, P. 523–528.
2. *Kienberger, R., Goulielmakis, E., Uiberacker, M., Baltuska, A., Yakovlev, V., Bammerr, F., Scrinzi, A., Westerwalbesloh, T., Kleineberg, U., Heizmann, U., Drescher, M., Krauz, F.*. Atomic transient recorder // Nature. 2004. V. 427. P. 817–821.
3. *Oelze, T., Schütte, B., Müller, M., Müller, J.P., Wieland, M., Frühling, U., Drescher, M., Al-Shemmary, A., Golz, T., Stojanovic, N., Krikunova, M.* Correlated electronic decay in expanding clusters triggered by intense XUV pulses from a Free-Electron-Laser // Sci. Rep. 2017. V. 7, P. 40736.
4. *Schütte, B., Frühling, U., Wieland, M., Azima, A., Drescher, M.*. Electron wave packet sampling with laser-generated extreme ultraviolet and terahertz fields // Opt. Express 2011. V. 19, No. 20. P. 18833–18841.
5. *Hebling, J., Almasi, G., Kozma, I.Z.* Velocity matching by pulse front tilting for large-area THz-pulse generation // Geophys. Opt. Express 2002. V. 10, No. 21. P. 1161–1166.
6. *Yeh, K.L., Hoffmann, M.C., Hebling, J., Nelson, K.A.* Generation of 10 μJ ultrashort terahertz pulses by optical rectification // Appl. Phys. Lett. 2007. V. 90, No. 17. P. 171121.
7. *Liu, B., Bromberger, H., Cartella, A., Gebert, T., Först, M., Cavalleri, A.*. Generation of narrowband, high-intensity, carrier-envelope phase-stable pulses tunable between 4 and 18 THz // Opt. Lett. 2017. V. 42, No. 1. P. 129–131.

https://doi.org/10.1051/epjconf/201819507004

Experiments using extreme parameters of the NovoFEL radiation

V. V. Kubarev [1,2], Ya. V. Getmanov[1,2], O. A. Shevchenko[1], E. N. Chesnokov[3], P. V. Koshlyakov[3], L. N. Krasnoperov[4]

[1]Budker Institute of Nuclear Physics, Novosibirsk, Russia, V.V.Kubarev@inp.nsk.su
[2]Novosibirsk State University, Novosibirsk, Russia
[3]Voevodsky Institute of Chemical Kinetics and Combustion, Novosibirsk, Russia
[4]NJIT, New Jersey, USA

Three different types of experimental investigations using unique extreme NovoFEL parameters [1] are presented in the report. The first one is continuous terahertz laser discharge which was developed from an effective demonstration of the NovoFEL power [2] to detailed investigation of its physical process [3-5] and creation of a point-like atmospheric-pressure discharge with record temperature of 28 000 °K at plasma density of $2{\cdot}10^{17}$ cm^{-3} [6](Fig.1-2).

Second set of the experiments is ultrafast time-domain spectroscopy [7-18] used now both for a single-pulse diagnostics of the NovoFEL radiation [11] and in free-induction-decay molecular spectroscopy including a magnetic spectroscopy [15,17](Fig. 3) and spectral "cinema" of the generation/decay dynamics of shot-lived molecular OH-radicals (Fig. 4).

Third type of the experiments is high-resolution spectroscopy based on very good coherency between NovoFEL pulses and very narrow lines of the fine structure of its radiation [19,20] (Fig. 5). One-mode filtration by a set of three resonance Fabry-Perot interferometers allows creation a source of powerful ultramonochromatic tunable radiation for high-resolution molecular spectroscopy [20] (Fig.6).

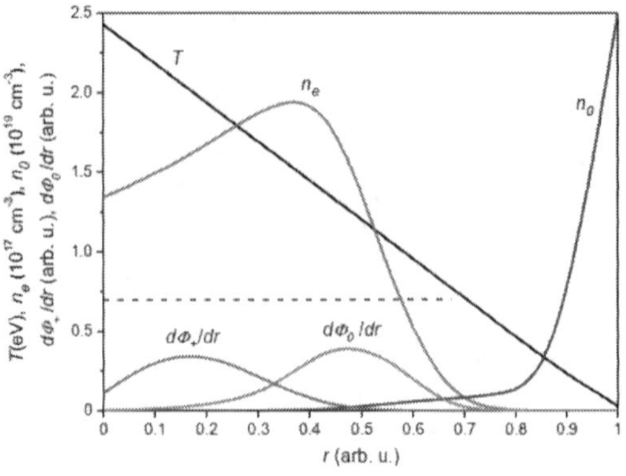

Fig. 2. Radial distributions of temperature *T*, plasma and gas densities n_e and n_0, and integral radiation of radial spherical layers for ionic $d\Phi_+/dr$ and atomic $d\Phi_0/dr$ lines of argon in the point-like high-temperature laser discharge. The dashed line indicates the critical plasma density (boundary of reflection of THz radiation from plasma)

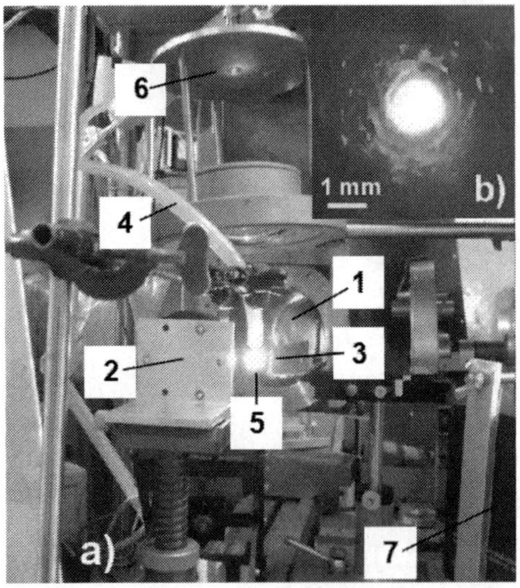

Fig. 1. Experimental setup for investigation of the THz laser discharge (a): 1 – output window of the NovoFEL optical beamline, 2 – main parabolic focusing mirror, 3 – stabilizing spherical mirror, 4 – gas tube, 5 – optical discharge, 6 – ultrafast optical photodiode, 7 – Mightex optical spectrometer. Insert (b) – photo of the point-like high-temperature laser discharge through a navy-blue glass filter

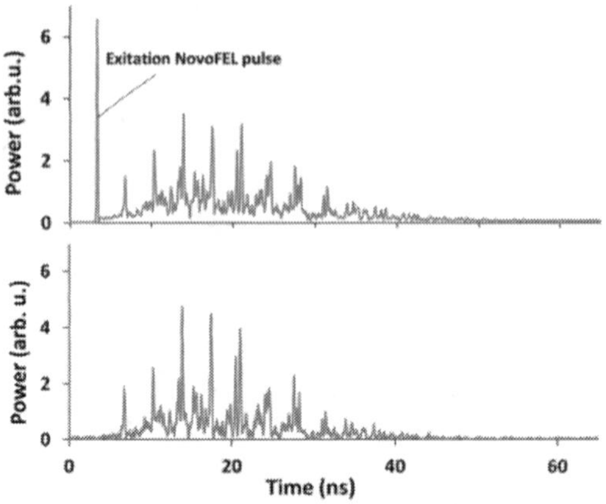

Fig. 3. Polarization separation of powerful excitation NovoFEL pulse and week FID signal of NO_2 molecules in magnetic field: upper – slightly misaligned cross input and output polarizers, lower – exactly aligned cross input and output polarizers (excitation pick is fully dumped)

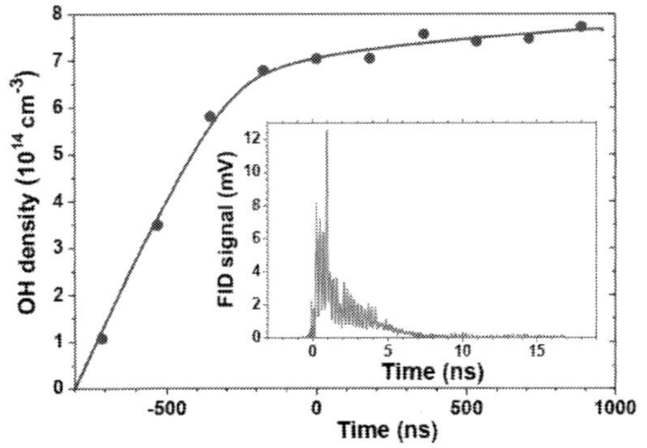

Fig. 4. Dynamics of OH-radical generation after shot pulse of UV radiation: points - treatment of the FID signals (insert) after series of ten NovoFEL excitation diagnostic pulses

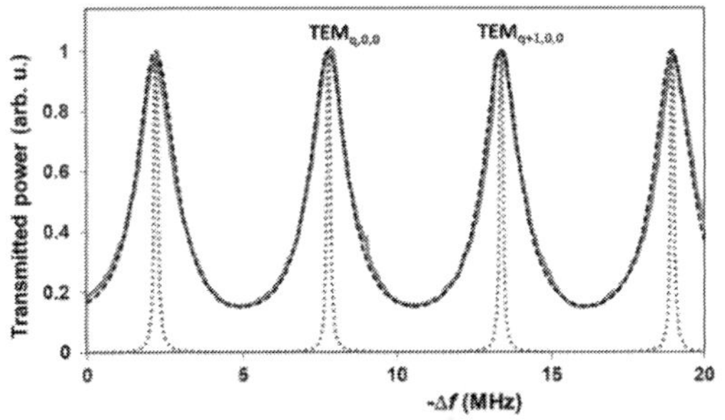

Fig. 5. Fine mode structure of the NovoFEL radiation measured by ultra-long waveguide vacuum mesh resonance FPI: solid red line – experiment, dash black line - instrumental function of the FPI, point blue line – reconstructed longitudinal mode structure of the NovoFEL

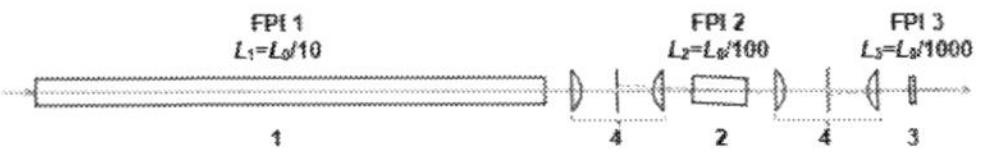

Fig. 6. Optical scheme of setup for one-mode filtration of the NovoFEL radiation: 1-3 – resonance FPI's (1 - ultra-long waveguide vacuum mesh resonance FPI), 4 – optical isolators

This work was made with financial support of Russian Science Foundation (project 14-50-00080). The NovoFEL operation was supported by Ministry of Education and Science of The Russian Federation (project RFMEFI62117X0012).

References

1. *Kulipanov, G. N., et al.* Novosibirsk free electron laser – facility description and recent experiments // IEEE Transaction on Terahertz Science and Technology. 2015. V. 5, No. 5. P. 798-809.

2. *Bolotin, V. P., et al.* Quasi-continuous sub-millimeter optical discharge on Novosibirsk free electron laser: experiments and elementary theory // Proc. IRMMW-THz-2005. 2005. Williamsburg. USA. P. 126–127.

3. *Kubarev, V. V.* Dynamics of the THz optical discharge // Proc. IRMMW-THz-2014. 2014. Tucson. USA. T2-A-16.2.

4. *Kubarev, V. V., Getmanov, Ya. V., Shevchenko, O. A.* Self-oscillations in terahertz laser discharge: Nature, parameters, and suppression // Proc. IRMMW-THz-2016. 2016. Copenhagen. Denmark, DOI:10.1109/IRMMW-THz.2016.7759003.

5. *Kubarev, V. V., Getmanov, Ya. V., Shevchenko, O. A., Koshlyakov, P. V.* Threshold conditions for terahertz laser discharge in atmospheric gases // Journal of Infrared, Millimeter, and Terahertz Waves. 2017. V 38, No. 6. P. 787-798.

6. *Kubarev, V. V., Getmanov, Ya. V., and Shevchenko, O. A.* Continuous point-like high-temperature laser discharge produced by terahertz free //AIP Advances. 2017. V. 7. P. 095123-(1-8).

7. *Kubarev, V., Chesnokov, E., Koshlyakov, P.* Ultrafast high-resolution THz time-domain spectroscopy // Proc. IRMMW-THz-2012. 2012. Wollongong. Australia. Thu-C-2-4.

8. *Kubarev, V., Chesnokov, E., Koshlyakov, P.* Terahertz free induction decay of molecular transitions: direct observation and practical use // Proc. TERA-2012. 2012. Moscow. Russia. P. 68–69.

9. *Chesnokov, E. N., Kubarev, V. V., Koshlyakov, P. V., and Kulipanov, G. N.* Direct observation of the terahertz optical free induction decay of molecular rotation absorption lines in the sub-nanosecond time scale // Appl. Phys. Lett. 2012. V. 101. P. 131109-(1–4).

10. *Kubarev, V. V., Chesnokov, E. N., and Koshlyakov, P. V.* One-pulse high-resolution THz time-domain spectroscopy: development and applications // Proc. IRMMW-THz-2013. 2013. Mainz. Germany. Tu5-5.

11. *Kubarev, V. V.* Ultrafast high-resolution spectroscopy of separate NovoFEL pulses // Proc. IRMMW-THz-2013. 2013. Mainz. Germany. Mo14-4.

12. *Chesnokov, E. N., Kubarev, V. V., Koshlyakov, P. V., Kulipanov, G. N.* Very long terahertz free induction decay in gaseous hydrogen bromide // Laser Phys. Lett. 2013. V. 10. P. 055701-055703.

13. *Chesnokov, E. N., Kubarev, V. V., and Koshlyakov, P. V.* Rotation commensurate echo of asymmetric molecules - Molecular fingerprints in the time domain // Appl. Phys. Lett. 2014. V. 105. P. 261107-(1–4).

14. *Chesnokov, E. N., Kubarev, V. V., Koshlyakov, P. V., Getmanov, Ya. V., Shevchenko, O. A.* Non-Faraday rotation of the free induction decay in gaseous NO // Chemical Physics Letters. 2015. V. 636. P. 203–207.

15. *Chesnokov, E. N., Kubarev, V. V., Koshlyakov, P. V., Chichinin, A. I., Getmanov, Ya. V., Shevchenko, O. A.* The influence of magnetic field on the echo-like free induction decay in NO_2 // Chemical Physics Letters. 2016. V. 662. P. 62–66.

16. *Chesnokov, E. N., Kubarev, V. V., Koshlyakov, P.V.* Observation of the Formation of 0-π Pulses in Rotation Spectra of HCN and HBr // Physics Procedia. 2016. V. 84. P. 135 – 141.

17. *Bragin A.V. et al.* Superconducting solenoid for superfast THz spectroscopy // Physics Procedia. 2016. V. 84. P. 82 – 85.

18. *Chesnokov, E. N., Kubarev, V. V., Koshlyakov, P. V., and Fedorov, V. V.* Heterodyne Method of Detection of Molecular Gas in the Terahertz Region Using the Beats Between Free Induction Decay Signals // IEEE Transactions on terahertz science and technology. 2017. V. 7, No. 2. P. 144–150.

19. *Kubarev, V. V.* Instabilities, coherency, and spectra of the NovoFEL radiation // EPJ Web of Conferences. 2017. V. 149. P. 05007(1-2).

20. *Kubarev, V. V., Getmanov, Ya. V.* NovoFEL as source of powerful ultramonochromatic tunable terahertz radiation // Proc. IRMMW-THz-2018. 2018. Nagoya. Japan. Fr-A2-1b-3.

Vector and mixed beams with orbital angular momentum

N. D. Osintseva[1], Yu. Yu. Choporova[1,2], O. E. Kameshkov[1,2], B. A. Knyazev[1,2], V. S. Pavelyev[3]

[1]Budker Institute of Nuclear Physics SB RAS, Novosibirsk, Russia, natalyaosintseva@gmail.com
[2]Novosibirsk State University, Novosibirsk, Russia
[3]Samara National Research University, Samara, Russia

Introduction

Vector beams with radial polarization are of great interest in surface plasmon science. Surface plasmon generation and propagation require polarization state which is perpendicular to the surface. Hence vector beams increase the generation efficiency of surface plasmon-polaritons on cylindrical metal-dielectric surfaces, such as wires [1].

In this work vector [2] and mixed beams with orbital angular momentum (OAM) [3] were generated by silicon binary diffractive optical elements (DOEs) in the THz range. The radiation was generated by the Novosibirsk free electron laser (NovoFEL), which has the possibility of smooth wavelength tuning from 6 to 220 μm [4]. Hence, we studied the efficiency of DOEs at different wavelengths. Using Fourier filtration diffraction efficiencies of 25-30% can be achieved (except of λ/2 and λ/4).

Vector beams

Three phase plates (Fig.1) transformed a linearly polarized Gaussian beam into Hermit-Gaussian (HG) modes (1,0) and (0,1) and into Bessel beam of first and second order. To reach a phase difference of π at a wavelength of 141 μm we used a profile height of $\Delta h = \lambda / 2(n-1) = 29.1\,\mu m$, where n=3.42 is the silicon refractive index [3]. The electric field of a HG beam (10 or 01) has opposite direction for each of its two maxima and hence, the polarization is opposite as well. Two orthogonally polarized HG beams were combined in an in-line Mach-Zehnder interferometer (Fig. 2). A phase singularity in the center leads to the generation of a vector Laguerre-Gaussian (LG) beam. To achieve an axially polarized beam a vertically (1,0) and a horizontally polarized (0,1) HG beam should be combined. A radially polarized beam is formed by the superposition of orthogonal polarization of both beams (Fig. 3).

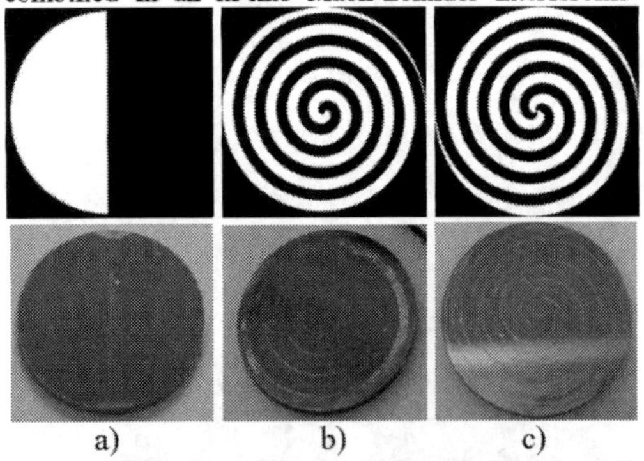

Fig. 1. Diffractive optical elements for transformation THz Gaussian beam into Hermit-Gaussian (left column) and vortex Bessel beams with topological charges l=1 and l=2 (middle and right columns respectively). The phase distribution (top line) and photo of silicon elements (bottom line). Black zones – phase is 2πN, white zones – πN, where N is integer.

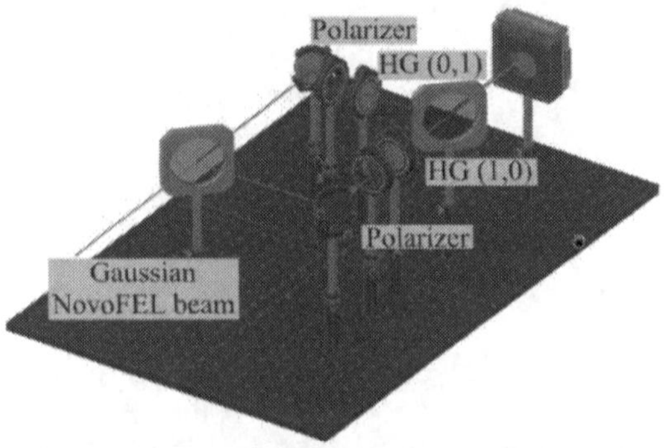

Fig. 2. The experimental scheme of generation a Laguerre-Gaussian vector beam. The Mach-Zehnder interferometer consists of two film beamsplitters and two mirrors. A polarizer and a DOE are situated in each arm f the interferometer.

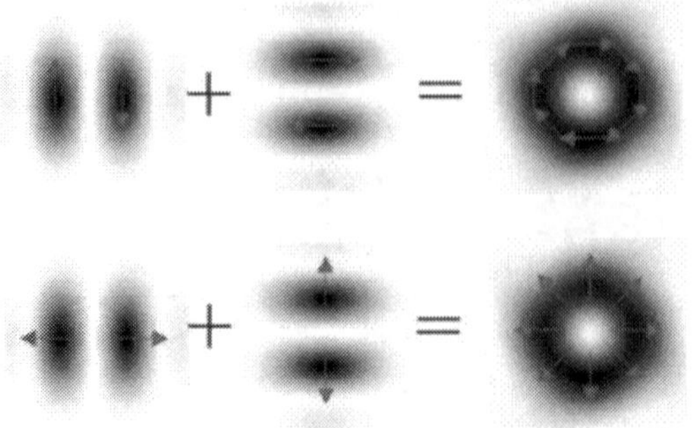

Fig. 3. Principal of generation of a vector beam. Maxima of an intensity distribution is shown with dark color. Red arrows show the direction of polarization. HG (1,0) and HG(0,1) are used to generate Laguerre-Gaussian beam with axial (top line) and radial (bottom line) polarization.

Mixed beams

Mixed beams were generated by combination of Bessel beams with different OAM or vortex beams [6]. A helical wavefront leads to an azimuthal phase dependence in the perpendicular plane. A helical wavefront is characterized by a topological charge, or helicity. It is equal to the number of times the phase changes from 0 to 2π in the perpendicular plane. Vortex Bessel beams with topological charges ± 1 and ± 2 were generated by phase binary axicons with spiral zones (Fig. 1) [7]. The sign of the topological charge can be changed by turning the axicon.

Interference patterns of mixed beams with different topological charges are shown in figure 4. The electric field amplitude can be written by the equation:

$$E(r,\varphi,z) = E_0(r,z)e^{i(k_z z + l\varphi)},$$

где $E_0(r,z)$ - complex amplitude, $r = \sqrt{x^2 + y^2}$, φ – azimuthal angle. Thus in a case of superposition of beams with topological charges of -1 and +1 the topological charge is equal to 0.

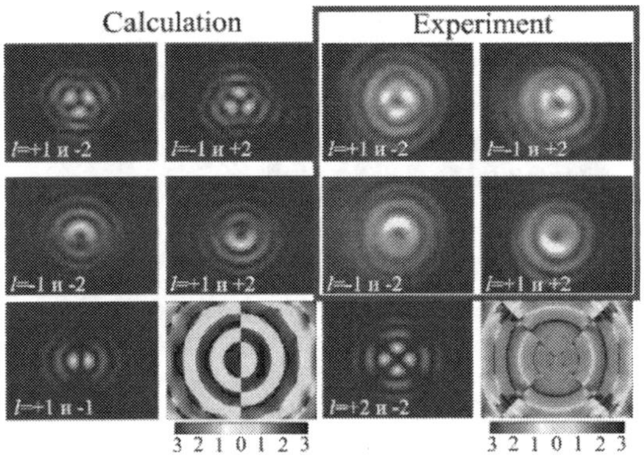

Fig. 4. Calculation and experimental results of mixed beam's generation. Bottom line: calculated intensity and phase distribution for mixed beams with equal values and different signs of topological charges.

Combinations such -1 and +2, or +1 and -2 are more complicated and requier detailed study. If total topological charge is different from the simple ± 1 vortex beam we can understand by using vector diffraction theory and plotting a Poynting vector (under development).

References

1. *Weibin, C., Zhan, Q.* Realization of an evanescent Bessel beam via surface plasmon interference excited by a radially polarized beam. // Opt. Lett. 2009. V. 34. P. 722-724.

2. *Oron, Ram, et al.* The formation of laser beams with pure azimuthal or radial polarization. // Applied Physics Letters. 2000. V. 77. No. 21. P. 3322-3324.

3. *Bouchal, Z., et al.* Orbital angular momentum of mixed vortex beams. // 15th Czech-Polish-Slovak Conference on Wave and Quantum Aspects of Contemporary Optics. 2007. Vol. 6609. International Society for Optics and Photonics.

4. *Volodkin, B. O., Choporova, Yu. Yu., Knyazev, B. A., et al.* Fabrication and characterization of diffractive phase plates for forming high-power terahertz vortex beams using free electron laser radiation. // Opt. Quant Electron. 2016. V. 48: No. 4. P. 223.

5. *Agafonov, A. N., et al.* Control of transverse mode spectrum of Novosibirsk free electron laser radiation. // Applied Optics. 2015. V. 54, No. 12. P. 3635-3639.

6. *Allen, L., Beijersbergen, M. W., Spreeuw, R. J. C., Woerdman, J. P.* Orbital angular momentum of light and the transformation of Laguerre–Gaussian laser modes. // Phys. Rev. 1992. V. 45. P. 8185–8189.

7. *Choporova, Yu. Yu., et al.* High-power terahertz non-diffractive bessel beams with angular orbital momentum: Generation and application. // IRMMW-THz 2015: 40th International Conference on Infrared, Millimeter, and Terahertz waves. Hong Kong, 23 - 28 August 2015. - Piscataway: IEEE, 2015. - Art.nr 7327684. - DOI 10.1109/IRMMW-THz.2015.7327684.

Terahertz optical elements for control of high-power laser irradiation

V.S. Pavelyev[1,2], A.N. Agafonov[1,2], B.O. Volodkin[1,2], K.N. Tukmakov[1,2], B.A. Knyazev[3,4], Yu.Yu. Choporova[3,4]

[1]Samara University, Samara, Russia, nano@ssau.ru
[2]Image Processing Systems Institute, Russian Academy of Sciences, Samara, Russia
[3]Budker Institute of Nuclear Physics of SB RAS, Novosibirsk, Russia
[4] Novosibirsk State University, Novosibirsk, Russia

Appearance of the sources of coherent and high power THz radiation [1] opened new horizons for investigations in this frequency range [2]. High attention is focused on silicon diffractive optical elements (DOE), which are used for the beam manipulation [3-7]. The lithographic etching of a silicon substrate has been used in [3-7] to fabricate binary relief of diffractive optical elements. Binary silicon element [4] coated with the antireflection coating remained intact upon exposure to an average radiation power density of 4 kW/cm^2; the peak power in a 100 ps pulse was almost 8 MW/cm^2. Experimental estimates of the diffraction efficiency of the elements coated with the antireflection coating [4] are in good agreement with theoretical estimates.

Such applications like imaging, material ablation, generation of continuous optical discharge, and even more exotic for the terahertz range application, namely the field ionization of individual atoms, require focusing of THz radiation [3-5], often with an enhanced focal depth [3].

Non-diffractive Bessel beams with angular orbital momentum (vortex beams) with different topological charges were formed by use of binary phase spiral axicons [6]. Binary phase axicon (BPA) with spiral zone structure and with aperture diameter of 100 mm (Fig. 1a,b) has been realized in [7] by technology similar to described in [3-6].

Fig. 1a. Calculated phase of binary diffractive optical element (black, 0; white, π)

Fig. 1b. Photo of realized element

The research has been performed at workstations of the free electron laser NOVOFEL (Budker Institute of Nuclear Physics of SB RAS, Novosibirsk). A high power radiation (in a routine regime, the average power is 50-150 W), a relatively narrow linewidth and the tunability of the radiation enable performing a wide variety of experiments. Vortex beam (Fig. 2) was formed from the NOVOFEL Gaussian beam (wavelength is of 129.5 μm) transformed by binary phase axicon (BPA) with spiral zone structures (Fig. 1) and with aperture diameter of 100 mm.

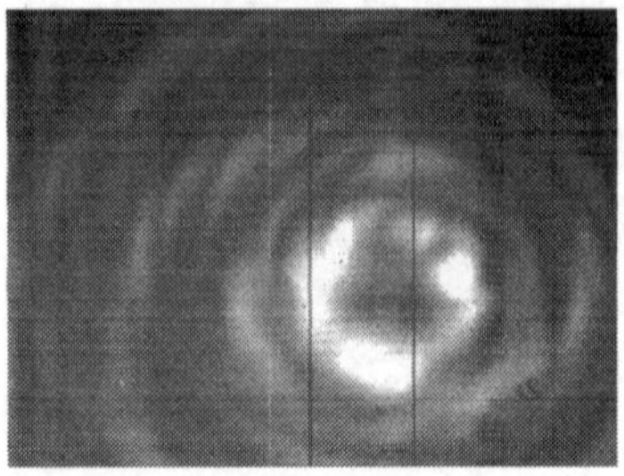

Fig. 2. Intensity distribution measured in the cross-section of beam formed by realized element (Fig. 1)

Vortex beams have great potential for use in free telecommunications and remote sensing [8].
However, lithographic etching has disadvantages in the case of multilevel elements, when an expensive and complicated procedure of alignment of photomask is required [8]. Binary (two-level) elements [3-7], in turn, have limited energy efficiency [8]. The laser ablation technology has been used in [9, 10] for fabrication of multilevel diffractive lens with high energy efficiency. However, diffractive optical elements are designed for working with monochromatic radiation of fixed wavelength [8] only. Fabrication of

terahertz reflective free-form elements for transformation of high-power beams of Free Electron Laser has been considered in [11]. Aluminium elements (spherical and cylindrical mirrors (Fig. 3), reflective axicons) were fabricated by technology of micromilling [11].

Fig. 3. Fabricated mirror

Realized optical elements were tested in the beam of NOVOFEL at the wavelength of 129.5 μm (Fig.4). The measured diffractive efficiency of the spherical mirror (>94%) is in good agreement with both numerical calculations and theoretical predictions.

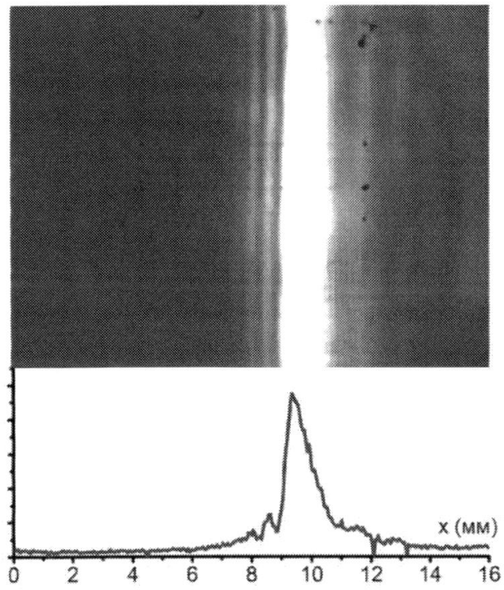

Fig. 4. Intensity distribution in the focal plane of cylindrical mirror

Acknowledgements

The elements were designed under support of the Ministry of Education and Science of the Russian Federation (project 16.7894.2017/6.7). The computer simulation of elements has been realized under support of RFBR grant 16-47-630560.

The terahertz beamline was designed under support of RSF grant 14-50-00080. The work was carried out in the collective research center supported by the Ministry of Education and Science of the Russian Federation, project RFMEFI62117X0012.

References

1. *Knyazev, B.A., Kulipanov, G.N., Vinokurov, N.A.* Novosibirsk terahertz free electron laser: instrumentation development and experimental achievements // Measurement Science and Technology. 2010. V. 21, No. 5. 054017, PP. 1-13
2. *Kulipanov, G.N., Bagryanskaya, E.G., Chesnokov, E.N., Choporova, Yu.Yu., Gerasimov, V.V., Getmanov, Ya.V., Kiselev, S.L., Knyazev, B.A., Kubarev, V.V., Peltek, S.E., Popik, V.M., Salikova, T.V., Scheglov, M.A., Seredniakov, S.S., Shevchenko, O.A., Skrinsky, A.N., Veber, S.L., Vinokurov, N.A.* Novosibirsk Free Electron Laser—Facility Description and Recent Experiments // IEEE Transactions on Terahertz Science and Technology. 2015. V. 5, No. 5, PP. 798-809
3. *Agafonov, A.N., Volodkin, B.O., Kachalov, D.G., Knyazev, B.A., Kropotov, G.I., Tukmakov, K.N., Pavelyev, V.S., Tsypishka, D.I., Choporova, Yu.Yu., Kaveev, A.K.* Focusing of Novosibirsk Free Electron Laser (NovoFEL) radiation into paraxial segment // Journal of Modern Optics. 2016. V. 63, No. 11, PP. 1051-1054.
4. *Agafonov, A.N., Volodkin, B.O., Kaveev, A.K., Knyazev, B.A., Kropotov, G.I., Pavel'ev, V.S., Soifer, V.A., Tukmakov, K.N., Tsygankova, E.V., Choporova, Yu.Yu.* Silicon diffractive optical elements for high-power monochromatic terahertz radiation // Optoelectronics, Instrumentation and Data Processing. 2013. V. 49, No. 2, PP. 189-195.
5. *Agafonov, A.N., Volodkin, B.O., Volotovsky, S.G., Kaveev, A.K., Knyazev, B.A., Kropotov, G.I., Tykmakov, K.N., Pavelyev, V.S., Tsygankova, E.V., Tsypishka, D.I., Choporova, Yu.Yu.* Silicon optics for focusing of terahertz laser radiation in a given two-dimensional domain// Computer Optics. 2013. V. 37, No. 4, PP. 464-470 (In Russ.).
6. *Knyazev, B.A., · Choporova, Yu.Yu., · Mitkov, M.S., · Pavelyev, V.S., · Volodkin, B.O.* Generation of Terahertz Surface Plasmon Polaritons Using Nondiffractive Bessel Beams with Orbital Angular Momentum // Physical Review Letters. 2015. V. 115, N 16, art 163901.
7. *Pavelyev, V.S., Volodkin, B.O., Tukmakov, K.N., Knyazev, B.A., Choporova, Yu.Yu.* Transmissive Diffractive Microoptics for High-Power THz Laser Radiation // AIP Conference Proceedings. 2018. V. 1989, No. 1, art. 020025.
8. *Computer Design of Diffractive Optics*, edited by V.A. Soifer. Cambridge International Science & Woodhead: 2012.
9. *Volodkin, B.O., Knyazev, B.A., Kononenko, V.V., Kononenko, T.V., Konov, V.I., Pavelyev, V.S., Soifer, V.A., Tukmakov, K.N., Choporova, Yu.Yu.* Fabrication of a multilevel THz Fresnel lens by femtosecond laser ablation // Quantum Electronics. 2015. V. 45, No. 10, PP. 933 –936.
10. *Minkevičius, L., Indrišiūnas, S., Šniaukas, R., Voisiat, B., Janonis, V., Tamošiūnas, V., Kašalynas, I., Račiukaitis, G., Valušis, G.* Terahertz multilevel phase Fresnel lenses fabricated by laser patterning of silicon // Optics Letters. 2017. V. 42, No. 10, PP. 1875-1878.
11. *Pavelyev, V., Agafonov, A., Ahmetova, E., Platonov, V., Knyazev, B.,* Reflective free-form optical elements for focusing of high-power THz radiation // Book of abstracts of International Conference "Synchrotron and Free electron laser Radiation: generation and application" (SFR-2018), PP. 41-42.

Nonlinear Transfer of Intense Few Cycle Terahertz Pulse Through Opaque Semiconductors

O. V. Chefonov[1], A. V. Ovchinnikov[1], M. B. Agranat[1], V. E. Fortov[1], E. S. Efimenko[2], A. N. Stepanov[2], T.Ozaki[3], X.Chai[3], X.Ropagnol[3], A.A.Ushakov[4] and A.B. Savel'ev[4]

[1]*Joint Institute for High Temperatures of the Russian Academy of Sciences, Moscow, Russia*
[2]*Institute of Applied Physics of the Russian Academy of Sciences , Nizhny Novgorod, Russia*
[3]*INRS-EMT, Advanced Laser Light Source, INRS, Varennes, Canada*
[4]*Lomonosov Moscow State University, Moscow, Russia*

An ultra-strong terahertz (THz) electric field has become available for experiments recently and this opened feasibility of nonlinear terahertz optics. Record-breaking results have been obtained using synchrotron radiation sources (pulse energy W_{THz} > 100μJ, electric field strength E_{THz} up to 20 MV/cm in the frequency range > 10 THz) or using optical rectification in nonlinear organic crystals (W_{THz} ~900 μJ, E_{THz} ~50 MV/cm in the frequency range ~ 1 THz).

Self-induced transparency of a medium under the action of terahertz radiation (or bleaching) was previously observed in semiconductors irradiated with a THz field of relatively low strength and increase in transmission was of a few percent only.

We studied the interaction of an intense THz pulse with the n-doped Si sample under the conditions inaccessible to quasi-stationary electric fields. In this work we observed that transmission by energy of the 700 fs THz pulse through n-doped Si sample amounts to as high as ~ 8% with field strength up to 5–7 MV/cm and then gradually decreases nearly twofold at higher electric fields of 10–20 MV/cm [1,2]. Analysis of the transmitted pulse waveform demonstrated its strong distortions and generation of higher frequency spectral components at 7–10 MV/cm, and finally a 300-fs single-cycle pulse was observed at a maximum field. Simulations showed that with E_{THz} ~5–7 MV/cm saturation of the electron-phonon collision rate occurs, and THz transmission saturates. Generation of a single-cycle pulse at even higher fields was associated with formation of a thin ionized layer at the front sample's surface during the first intense oscillation of the THz field cutting out subsequent field oscillations and decreasing total transmission.

We used the open-aperture Z-scan technique, based on gradual varying of the THz radiation intensity on the surface of the sample moved along the beam line by a stepping motor, to study nonlinear effects. THz radiation was generated by the optical rectification in the nonlinear organic crystal 4-N,N-dimethylamino-4'-N'methyl-stilbazolium 2,4,6-trimethylbenzenesulfonate (DSTMS), pumped by 100-fs femtosecond laser pulses at a central wavelength of 1240 nm delivered by a Cr:forsterite laser system [3,4]. A broadband THz filter (LPF8.8-47, Tydex) with a cutoff wavelength of 34 μm was placed after the crystal. Two off-axis parabolic mirrors were used as a 5:1 telescope to expand the THz beam. Finally, THz pulses with energy of 6.3 μJ were focused with an off-axis parabolic mirror with a 2" diameter and a focal length of 2" onto the sample to a spot with a radius of $w_0 \simeq$ 200 μm at $1/e^2$, measured using the knife edge technique. The duration τ_{FWHM} of the THz pulse measured with a first order autocorrelator (a THz Michelson-type interferometer) was 700 fs at full width at half-maximum (FWHM).

The Golay Cell GC-1D (Tydex) was used to measure THz pulse energy transmitted through the sample. The THz electric field in the time domain after the sample was measured using electro-optical (EO) sampling in a nonlinear bilayer 2.1-mm-thick GaP crystal with a 2-mm-thick GaP (100) and a 0.1-mm-thick GaP (110). Fields strengths at the crystal were low enough to ensure linearity of measurements. This scheme was placed instead of the Golay cell.

Maximal THz field strength reached at z=0 was estimated in two ways. First, we employed the well-known time-averaged Poynting vector approach using data on the energy, spot size and duration of the THz pulse. Second, we fulfilled EOS with a 100-μm thick GaP crystal placed at z=0 with femtosecond laser radiation at 1240 nm as a probe. The both measurements gave an estimate of the maximal THz field strength $E_{THz} \approx 21\pm1$ MV/cm at $W_{THz} \approx 105$ μJ.

The sample used was a commercially available n-doped silicon wafer of 245 μm thickness with carrier concentration and mobility of 8.7×10^{16} cm^{-3} and 800 cm^2 V^{-1} s^{-1}, respectively, obtained from Hall effect measurements.

Transmittance of the THz radiation was assessed as the ration of the THz pulse energy passed through the sample to the THz pulse energy without the sample (Fig.1). With increasing maximum electric-field strength above 1 MV/cm a complex nonlinear dynamic of transmission of the 245-μm thick n-doped Si with carrier concentration of 9×10^{16} cm^{-3} was observed. An increase of electron-phonon collision rate (due to an increase of "instantaneous" electron energy) results in the bleaching at electric fields below 5 MV/cm. The maximum transmission by energy was found out ~ 8% that is more than two orders of magnitude higher than with a low-intensity THz pulse. Saturation of the transmission is attributed to saturation of the collision rate at electron energies above 1.5 eV.

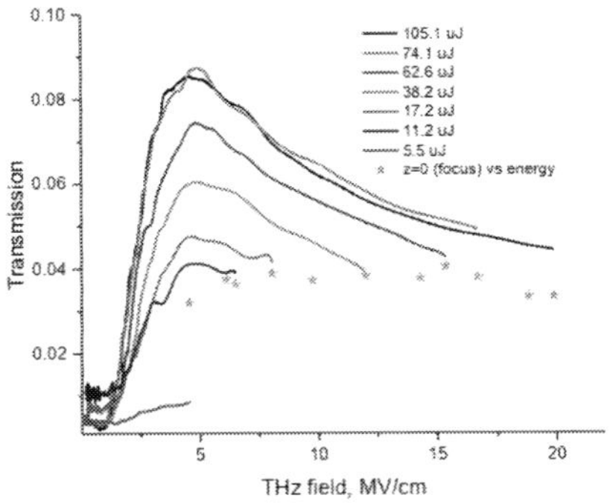

Fig.1. Experimental dependencies of the transmission η on the strength E_{THz} at different pulse energies.

The transmission gradually decreases at higher fields above 7 MV/cm that is related to the onset of ionization by electron impact. Ionization occurs in a thin surface layer during the first intense oscillation of the THz field. This results in efficient reflection and attenuation of subsequent field oscillations leading to formation of a single-cycle THz pulse with sharp rising edge and broad spectrum (Fig.2). This pulse propagates deeper into the sample under the bleaching conditions. The observed effect strongly depends on the spatial energy distribution over the THz beam cross section and, in our experiments, the maximum transmittance was observed out of the focal plane.

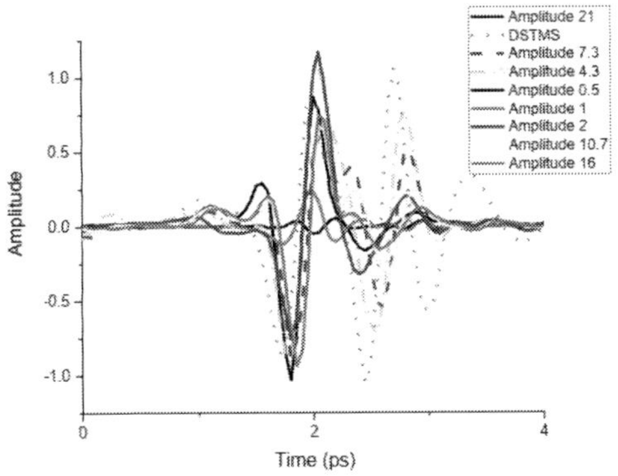

Fig.2. Waveforms of THz pulses without (DSTMS) and with the sample at different THz failed strengths.

To make deeper insight into the spectral-temporal changes of the THz pulse, preliminary numerical simulations of nonlinear propagation of an intense THz pulse through a silicon plate were performed. The Finite Difference Time Domain method was used to solve numerically the 1D Maxwell's equations for the electromagnetic field together with equations for plasma currents and carrier concentration. Dependencies of both the collision and impact ionization rates in Si on the E_{THz} were taken into account when calculating the current.

We also studied the nonlinear carrier dynamics in n-doped semiconductor InGaAs [5]. Intense THz pulses have pushed the nonlinear responses of doped InGaAs semiconductor into a new regime that involves interband carrier dynamics. Intraband heating dominates the overall THz responses and therefore doped semiconductor can still be a good candidate for the applications of saturable absorber.

The work was supported by Russian Science Foundation no. 17-19-01261. Experiments were performed using the unique scientific facility "Terawatt Femtosecond Laser Complex" in the Center "Femtosecond Laser Complex" of the Joint Institute for High Temperatures of the Russian Academy of Sciences.

References

1. *Chefonov O.V., Ovchinnikov A.V., Romashevskiy S.A., Chai X., Ozaki T., Savel'ev A.B., Agranat M.B., Fortov V. E.* Giant self-induced transparency of intense few-cycle terahertz pulses in n-doped silicon // Optics Lett. 2017. V.42, No 23. P.4889–4892.
2. *Chefonov O.V., Ovchinnikov A.V., Agranat M.B., Fortov V.E., Efimenko E.S., Stepanov A.N., Savel'ev A.B.* Nonlinear Transfer of Intense Few Cycle Terahertz Pulse Through Opaque n-doped Si // submitted to Phys.Rev.B.
3. *Vicario C., Ovchinnikov A. V., Ashitkov S. I., Agranat M. B., Fortov V. E., Hauri C. P.* Generation of 0.9-mJ THz pulses in DSTMS pumped by a Cr:Mg_2SiO_4 laser // Opt. Lett. 2014. V.39, No.23. P. 6632-6635.
4. *Vicario C., Jazbinsek M., Ovchinnikov A. V., Chefonov O. V., Ashitkov S. I., Agranat M. B., Hauri C. P.* High efficiency THz generation in DSTMS, DAST and OH1 pumped by Cr:forsterite laser // Opt. Express 2015. V.23, No.4. P.4573-4580.
5. *Chai X., Ropagnol X., Ovchinnikov A., Chefonov O., Ushakov A., Garcia-Rosas C. M., Isgandarov E., Agranat M., Ozaki T., Savel'ev A.* Observation of crossover from intraband to interband nonlinear terahertz optics // submitted to Optics Letters.

Relaxation of Coulomb States in Semiconductors Probed by FEL Radiation

R.Kh. Zhukavin[1], K.A. Kovalevsky[1], V.V. Tsyplenkov[1], S.G. Pavlov[2], H-W. Hübers[2,3], Yu.Yu. Choporova[4,5], B.A. Knyazev[4,5], J.M. Klopf[6], B. Redlich[7], N.V. Abrosimov[8], Yu.A. Astrov[9], V.N. Shastin[1]

[1]Institute for Physics of Microstructures, Nizhny Novgorod, Russia, zhur@ipmras.ru
[2]Institute of Optical Sensor Systems, German Aerospace Center (DLR), Berlin, Germany
[3]Department of Physics, Humboldt-Universität zu Berlin, Berlin, Germany
[4]Budker Institute of Nuclear Physics, Novosibirsk, Russia
[5]National Research Novosibirsk State University, Novosibirsk, Russia
[6]Helmholtz-Zentrum Dresden-Rossendorf, Dresden, Germany
[7]FELIX Laboratory, Radboud University, Institute of Molecules and Materials, Nijmegen, The Netherlands
[8]Leibniz-Institut für Kristallzüchtung (IKZ), Berlin, Germany
[9]Ioffe Institute, St. Petersburg, Russia

The typical semiconductors such as silicon and germanium with embedded donors and acceptors have been intensively studied since 1950th [1]. The main reason was the importance of doped semiconductors in the production of various electronic devices based on the properties of p-n junction or Schottky-junction. The new interest to Coulomb centers in semiconductors has arisen from several new directions including single-atom electronics, quantum logic and lasing based on impurity centers. For use of Coulomb centers in the frame of any of those fields one should know such properties of bound electron as energy spectrum and relaxation rates. The experimental methods being used for the investigations compose a very broad set of different techniques involving transport and optics and their combinations under various conditions. Last two decades the appearance of new user facilities based on THz free electron lasers (FELs) emitting high power short THz pulses provides the unique opportunity for studying of impurity centers in semiconductors under resonant excitation of chosen energy levels using single color pump-probe technique [2-4]. In particular, the lifetimes of different donors and acceptors in silicon have been measured to localize the time scale of the existing relaxation rates. More recently similar experiments have been performed to explore impurities in germanium. Here the authors present a review of pump-probe experiments devoted to lifetime measurements of excited donor states in silicon and germanium under the uniaxial stress.

As well known the relaxation of shallow donor electrons in semiconductors under low temperatures is governed by the emission of acoustic or optical phonons. The elemental semiconductors such as silicon and germanium have multi-valley structure of the conduction band that, together with appropriate energy gaps between states of the Coulomb centers, allow intervalley electron scattering processes. In silicon those are accompanied by the emission of f-type or g-type optical or acoustic phonons. The former corresponds to e.g. [100]-[010] transition while the latter to [100] – [-100] transition in the reciprocal space. The uniaxial stress applied along [100] crystal axis removes the degeneracy resulting in splitting of valleys and corresponding energy levels. Thus uniaxial stress can modify relaxation rates if changes the energy gaps. Fig. 1 demonstrates the energy diagram of phosphorus donor in silicon under the stress as an example. The excited states move parallel to their "parent" valleys while the singlet ground state $1s(A_1)$, which contains equal contributions from 6 valley under zero stress, moves nonlinearly losing the contribution of upper valleys.

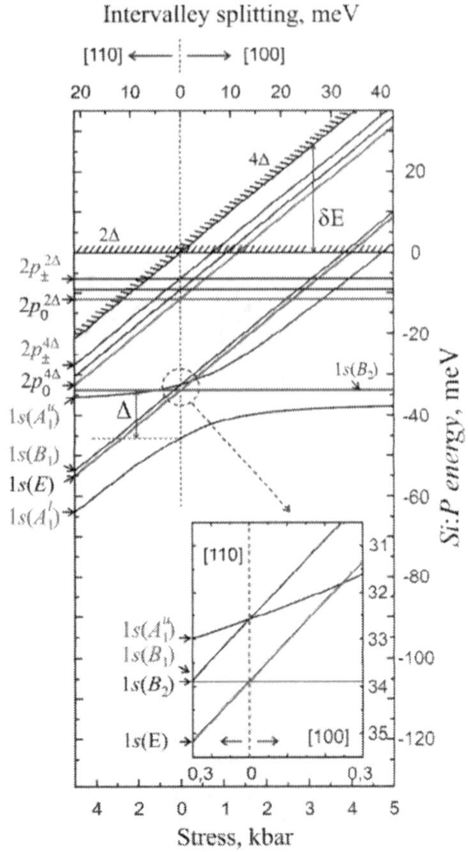

Fig. 1. Energy diagram for phosphorus donor in silicon as a function of uniaxial stress along [110] (left) or [100] (right) crystallographic orientation. Only lower excited states and ground states are shown as well as position of conduction band valleys 2Δ and 4Δ.

The silicon and germanium crystals doped by different donors were taken for sample preparation. The silicon samples were cut along 100 axis while germanium samples were cut along 111 axis. Those directions were chosen to provide the maximum valley splitting for given stress value. Typical donor concentrations were in the range of N_D=1-5×10^{15} cm^{-3}. The samples were shaped into parallelepipeds with dimensions of ~7×5×2 mm^3. Each sample was placed into the special module designed to apply uniaxial stress. The stress module containing one sample and individual stress value was mounted in an optical liquid helium flow cryostat and cooled to about 5 K. The achieved stress value in each module was estimated from the splitting of the donor levels as measured by absorption spectroscopy. Relaxation times of particular electronic states were derived from the time-resolved pump-probe technique. This involved the frequency-selective excitation of the investigated sample with an optical pulse, where the excitation photon energy corresponded to the energy of the transition between the probed electronic states. To measure the characteristic decay time τ, the pulse duration of optical excitation must be much shorter than τ and the time delay between pump and probe pulses must be appropriate to resolve τ. The overall data set was obtained using pump-probe experimental setups at several free electron laser facilities: FELIX (The Netherlands), FELBE (Germany) and NovoFEL (Russia). Details of the pump-probe setups are given elsewhere. Spectral width of the FEL emission on half width on half maximum (FWHM) was better than 1 % of a FEL wavelength. The typical micropulse duration was about 10 ps for FELIX and FELBE and 100 ps for NovoFEL.

Fig. 2 demonstrates the pump-probe signals obtained for Si:P stressed along [100] when pumped into the lower stress component of $2p_0$ level for two values of uniaxial stress – 80 and 650 bar. It is clear seen from the picture that higher stress value results in longer lifetime indicating suppression of the f-type phonon emission. Similar results have been obtained for other donor in silicon such as bismuth and arsenic. It has been obtained that typical lifetimes for donors in germanium are much longer (0.5-3 ns) than for acceptors in germanium (150-250 ps) as well as donors (100-200 ps) and acceptors (30-100 ps) in silicon. The stress applied along [111] axis for Ge:Sb results in increase of lifetime for $2p_{\pm}$ level from 1.8 ns to about 3.2 ns that can be explained by switching off the intervalley relaxation.

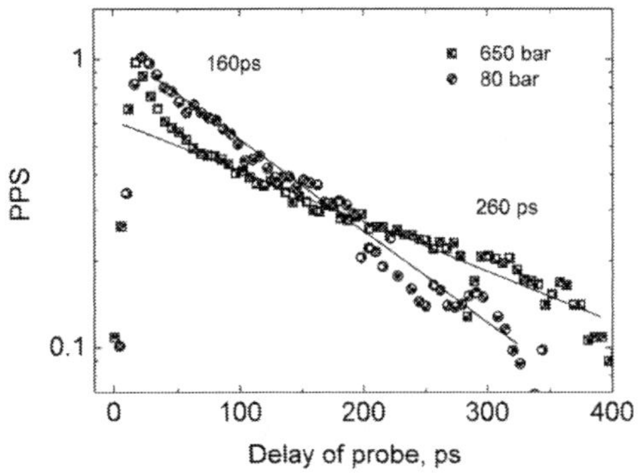

Fig. 2. Pump-probe signal for Si:P under two values of uniaxial stress (80 bar and 650 bar) along [100].

This work was partly supported by the joint German-Russian project of the Russian Foundation for Basic Research (RFBF Project No. 18-42-520064, 18-502-12077-DFG) and of the Deutsche Forschungsgemeinschaft (DFG No. 389056032).

References

1. *Ramdas, A. K., Rodriguez, S.* Spectroscopy of the solid-state analogues of the hydrogen atom: donors and acceptors in semiconductors, Rep. Prog. Phys. 1981, V. 4, P. 1291-1387.

2. *Hübers, H.-W., Pavlov, S. G., Lynch, S. A., Greenland, Th., Litvinenko, K. L., Murdin, B., Redlich, B., van der Meer, A. F. G., Riemann, H., Abrosimov, N. V., Becker, P., Pohl, H.-J., Zhukavin, R. Kh., Shastin, V. N.* Isotope effect on the lifetime of the $2p_0$ state in phosphorus-doped silicon // Phys. Rev. B. 2013. V. 88, P. 035201(1-5).

3. *Pavlov, S. G., Deßmann, N., Pohl, A., Shuman, V. B., Portsel, L. M., Lodygin, A. N., Astrov, Yu. A., Winnerl, S., Schneider, H., Stavrias N., van der Meer, A. F. G., Tsyplenkov V. V., Kovalevsky, K. A., Zhukavin, R. Kh., Shastin, V. N., Abrosimov, N. V., Hübers, H.-W.* Dynamics of nonequilibrium electrons on neutral center states of interstitial magnesium donors in silicon // Physical Review B. 2016. V. 94, P. 075208(1-8).

4. *Choporova, Yu. Yu., Gerasimov, V. V., Knyazev, B. A., Sergeev, S. M., Shevchenko, O. A., Zhukavin, R. Kh., Abrosimov, N. V., Kovalevsky, K. A., Ovchar, V. K., Hübers, H-W., Kulipanov, G. N., Shastin, V. N., Schneider, H., Vinokurov, N. A., Kulipanov, G. N.* First terahertz-range experiments on pump–probe setup at Novosibirsk free electron laser // 2016. Physics Procedia. V. 84, P. 152-156.

Impact of Scattering in Quasi-Ordered Structures on THz Imaging

I. N. Dolganova[1,2,3], N. V. Chernomyrdin[1,2,4], A. A. Kuznetsov[1], K. M. Malakhov[1], V. E. Karasik[1], K. I. Zaytsev[1,2,4]

[1]Bauman Moscow State Technical University, Moscow, Russia
[2]Sechenov First Moscow State Medical University, Moscow, Russia
[3]Institute of Solid State Physics of the Russian Academy of Sciences, Chernogolovka, Russia
[4]Prokhorov General Physics Institute of RAS, Moscow, Russia
in.dolganova@gmail.com

Terahertz science and technologies demonstrate a wide range of imaging applications in physics, biology, medicine, industry, etc. [1-7]. Scattering has one of the most essential impacts on imaging quality [1-4]. The mostly applied methods for estimation of scattering parameters and imaging losses are analytical diffuse and small-angle approximation of radiative transfer theory, Monte Carlo (MC), discrete ordinates, and T-matrix numerical methods, finite-difference time-domain method (FDTD), and experimental methods, e.g., using integrating sphere [8-11].

Nevertheless, one should pay attention to local ordering of scatterers in the considered media, since it could lead to significant changes of scattering signal [1-2].

We have studied the impact of quasi-ordered scattering materials on THz imaging, using a combined computational approach, based on computational electrodynamics, radiative scattering theory and Monte Carlo simulations. We assumed that the scattering effects could be accounted for by finding a single scattering phase function for clusters of particles, which form quasi-ordered scattering material, and using it in MC simulation of scattering, considering the whole medium as a random structure of such clusters. We applied FDTD technique for solution of Maxwell's equations in the near-field (NF) zone with the NF to far-field (FF) transformation based on the calculation of the diffraction integral. It helped us to determine scattering phase function and include it in MC numerical simulations in order to find scattered radiance $L(\theta)$ and the corresponding imaging modulation transfer functions (MTF).

The described approach was used to estimate an impact of quasi-ordered media [7,12-13].

We estimated imaging MTF in case of clothes scattering material. The numerical simulations and experimental measurements were performed for the following system parameters: imaging frequency of the THz source (backward-wave oscillator) was 0.25 THz, distance between the imaging system and the bar-pattern test-object was 4 m; for the scattering materials with groups of four cylindrical particles with period 0.3 mm dielectric permittivity of a single particle of diameter 0.2 mm was 2.9; for the scattering materials with groups of seven cylindrical particles with period 1.34 mm dielectric permittivity of a single particle of diameter 1.34 mm was 1.6; the thickness of scattering layers was 0.7 mm in the first case, and 4 mm in the second. The experimental scheme and the results are demonstrated in Fig. 1.

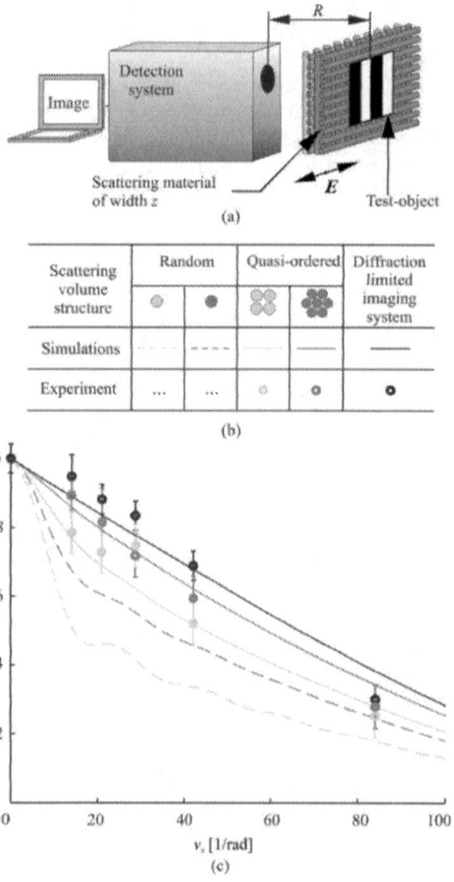

Fig. 1. (a) An experimental scheme for finding MTF using a set of test-objects; (c) and (d) experimental and numerical results, where black color corresponds to measurement without scattering layer.

We also calculated the MTF in case of random distribution of particles in the scattering materials and in case of quasi-ordering.

According to the numerical simulations [7], it is possible to determine particular combination of wavelength, particle system parameters, local period and dielectric parameters, for which imaging contrast of a remote object (and the corresponding MTF) would increase. On the contrary, almost opaque media could be achieved.

The demonstrated analysis could be effectively used for finding optimal wavelengths for imaging applications.

This work is supported by the Russian Science Foundation (RSF), Project # 14-19-01083.

References

1. *Fokina, I. N. et al.* Impact of structure geometry on scattering in partially-ordered media // Journal of Quantitative Spectroscopy & Radiative Transfer 2014. V. 149, P. 108–116.

2. *Dolganova, I. N. et al.* Combined terahertz imaging system for enhanced imaging quality // Optical and Quantum Electronics 2016. V. 48, No. 6, P. 325.

3. *Dolganova, I. N. et al.* A hybrid continuous-wave terahertz imaging system // Review of Scientic Instruments 2016. V. 86, P. 113704.

4. *Chernomyrdin, N. V. et al.* Reflection-mode continuous-wave 0.15lambda-resolution terahertz solid immersion microscopy of soft biological tissues // Applied Physics Letters 2018. V. 113, No. 11, P. 111102.

5. *Chernomyrdin, N. V. et al.* Solid immersion terahertz imaging with sub-
wavelength resolution // Applied Physics Letters 2017. V. 110, No. 22, P. 221109.

6. *Yakovlev, E. V. et al.* Non-Destructive Evaluation of Polymer Composite Materials at the Manufacturing Stage Using Terahertz Pulsed Spectroscopy // IEEE Transactions on Terahertz Science and Technology 2015. V. 5, No. 5, P. 810–816.

7. *Dolganova, I. N. et al.* The role of scattering in quasi-ordered structures for terahertz imaging: local order can increase an image quality // IEEE Transactions on Terahertz Science and Technology 2018. V. 8, No. 4, P. 403–409.

8. *Tuchin, V. V.* Tissue Optics: Light Scattering Methods and Instruments for Medical Diagnostics 2015. 3rd ed. Bellingham.

9. *Stamnes, K. et al.* Numerically stable algorithm for discrete-ordinate-method radiative transfer in multiple scattering and emitting layered media // Applied Optics 1988. V. 27, No. 12, P. 2502–2509.

10. *Waterman, P.* Matrix formulation of electromagnetic scattering // Proc. IEEE 1965. V. 53, No. 8, P. 805–812.

11. *Taylor, A. H.* The measurement of diffuse reflection factors and a new absolute reflectometer. // J. Opt. Soc. Amer. 1920. V. 115, No. 19, P. 9–23.

12. *Zaytsev, K. I. et al.* Accuracy of sample material parameters reconstruction using terahertz pulsed spectroscopy // Journal of Applied Physics 2014. V. 115, No. 19, P. 193105.

13. *Zaytsev, K. I. et al.* Invariant embedding technique for medium permittivity profile reconstruction using terahertz time-domain spectroscopy // Optical Engineering 2013. V. 52, No. 6, P. 068203.

Generation of vortex beamlet lattices via diffraction of Bessel vortex beams on 2D hole arrays: analytical and numerical calculations and comparison with experiments

O. E. Kameshkov[1,2], B. A. Knyazev[1,2], I.A. Kotelnikov[1,2]

[1]Budker Institute of Nuclear Physics of Siberian Branch Russian Academy of Sciences, Novosibirsk, Russia, o.kameshkov@g.nsu.ru
[2]Physics Department, Novosibirsk State University, Novosibirsk, Russia

The Novosibirsk Free Electron Laser (NovoFEL) [1] is a source of monochromatic and frequency-tunable terahertz radiation of high average power. At the entrance to a workstation, the beam of the laser is a Gaussian beam within good accuracy. However, it is necessary to transform the beam mode composition (focusing radiation with a defined intensity distribution). Diffraction optical elements (DOEs) were shown to excel refractive elements in the manipulation of high-power laser radiation in the experiments at the NovoFEL; DOEs can be considered as amplitude-phase masks (APMs) within the framework of physical optics [2-3]. Objects in some experiments, such as terahertz holography [4] and beam diffraction of complex mode composition on lattices [5], are also APMs.

For the experiments, a program in the Matlab environment with an easy-to-use interface has been written to simulate radiation transmission through optical systems consisting of a sequence of amplitude-phase elements. The calculations were performed within the framework of the scalar theory of diffraction. The software calculates the Rayleigh-Sommerfeld integral in the Fresnel approximation [6] using a combination of the impulse response method and the transfer function method, which ensures the solution correctness in the entire Fresnel diffraction region (see eq. (1)) [7].

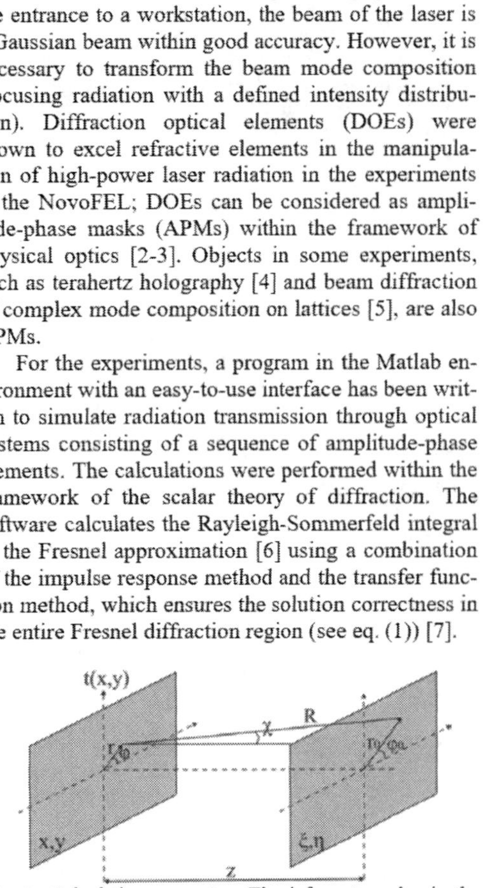

Fig. 1. Calculation geometry. The left rectangular is the amplitude-phase mask plane and the right rectangular is the observation plane.

$$E(z,\xi,\eta)=\frac{1}{i\lambda}\begin{cases}F^{-1}\left\{F\left[u(x,y)\right]F\left[h_f(z,x,y)\right]\right\} & z\geq\dfrac{L\cdot\Delta x}{\lambda}\\ F^{-1}\left\{F\left[u(x,y)\right]\cdot H_f\right\} & z\leq\dfrac{L\cdot\Delta x}{\lambda}\end{cases} \quad (1)$$

where

$$h_f(z,x-\xi,y-\eta)=\frac{\exp(ikz)}{z}\cdot\exp\left(ik\frac{(x-\xi)^2+(y-\eta)^2}{2z}\right)$$

is the impulse response function and its Fourier image is $H_f(\nu_x,\nu_y)=i\lambda\cdot\exp(ikz)\cdot\exp\left[i\pi\lambda z\left(\nu_x^2+\nu_y^2\right)\right]$, which is a transfer function, $u(x,y)=E(0,x,y)\cdot t(x,y)$, F is the Fourier transformation, L is the size of grid, and Δx is the grid spacing.

During a year, the program was widely applied to modeling of beam propagation in different optical systems used in experiments at the NovoFEL such as formation of beams of different mode compositions with DOEs (Hermite-Gaussian and Laguerre-Gaussian beams and Bessel beams with an orbital angular momentum), diffraction of beams on amplitude and phase gratings etc. In all cases, the results of experiments and modeling were in good agreement. Experiments on the diffraction of Bessel beam with an orbital angular momentum on periodic gratings (the quasi-Talbot effect) turned out to be the most interesting among these explorations. They were conducted for the first time.

We have investigated transmission of vortex Bessel beam formed by binary phase axicons through two-dimensional periodic gratings of round holes. It has been found that behind the grating, in the planes corresponding to the planes of self-images of the classical Talbot effect (see eq. (2)), a lattice of vortex ring beams appears, the topological charge of which corresponds to the charge of the incident beam:

$$L_T=\frac{2p^2}{\lambda}\left(N-1+\frac{n}{m}\right)=Z_T\cdot\left(N-1+\frac{n}{m}\right), \quad (2)$$

where N and $n<m$ are integers and p is the period of the grating.

After the experiments and numerical calculations, it became clear to us that the diffraction patterns observed in the Talbot planes had a beautiful geometry, which suggested existence of a beautiful analytical solution for their description. The problem was also solved analytically. The analytical expression obtained so far for the main Talbot plane is as follows:

$$E(\boldsymbol{x},z_N)=\frac{2\pi^2R^2}{\kappa}\sum_{m,n}\iint\frac{d^2k_\perp}{(2\pi)^2}\exp(i\boldsymbol{k}_\perp\cdot\boldsymbol{x}+i\ell\varphi_k)\cdot i^{-\ell}\times$$
$$\times\delta(k_\perp-\kappa)\cdot\delta\left(\boldsymbol{x}-\boldsymbol{p}_{mn}-\frac{k_\perp z_N}{k}\right), \quad (3)$$

where $\boldsymbol{x}=\{x,y\}$, $\boldsymbol{p}_{mn}=\{mp,np\}$, R is the hole radius, k is the wavenumber, $z_N=Z_T\cdot N$, ℓ is the orbital angular momentum, and κ is the radial wavenumber.

Hence, it is clear that the image has a form of circles with centers at the points $\boldsymbol{x}=\boldsymbol{p}_{mn}$ and radii $a_N=\frac{\kappa\cdot z_N}{k}=\kappa\frac{p^2}{\pi}N$. The results of the analytical studies of the problem in more detail will be published elsewhere.

Funding Information

The study of the vortex beams was carried out with the support of the RFBR grant (project 15-02-06444). The NovoFEL radiation transport beamline to the workstation was constructed with the support of the Russian Science Foundation (grant 14-50-00080). The work was carried out at the collective research center supported by the Ministry of Education and Science of the Russian Federation (project RFMEFI62117X0012).

Acknowledgments

The authors are grateful to G.N. Kulipanov, V. G. Serbo, and V. A. Soifer for stimulating discussions, V.S. Pavelyev and B. O. Volodkin for technical support, Yu. Yu. Choporova and N.D. Osintseva for participating in the experiments and useful discussions, and Ya. V.Getmanov, V. V. Kubarev, T. V.Salikova, M. A. Scheglov, O. A. Shevchenko, D. A. Skorokhod, and other members of the NovoFEL team for the invaluable support of the experiments.

References

1. *Kulipanov G.N., Bagryanskaya E.G., Chesnokov E.N., Choporova Yu.Yu., Gerasimov V.V.,Getmanov Ya.V., Kiselev S.L., Knyazev B.A., Kubarev V.V., Peltek S.E., Popik V.M., Salikova T.V.,Scheglov M.A., Seredniakov S.S., Shevchenko O.A., Skrinsky A.N., Veber S.L., Vinokurov N.A.* Novosibirsk free electron laser as a user facility // *IEEE Trans. on THz Sci. and Techn.* 2015. V. 5. №. 5. P. 798.

2. *Knyazev B.A., Cherkassky V.S., Choporova Yu.Yu., Gerasimov V.V., Vlasenko M.G., Dem'yanenko M.A., Esaev D.G.* Real-time imaging using a high-power monochromatic terahertz source: comparative description of imaging techniques with examples of application // *J Infrared Milli Terahz Waves*. 2011. V. 32, P. 1053.

3. *Agafonov A.N., Choporova Yu.Yu., Kaveev A.K., Knyazev B.A., Kropotov G.I., Pavelyev V.S.,Tukmakov K.N., Volodkin B.O.* Control of transverse mode spectrum of Novosibirsk free electron laser radiation // *Appl. Optics. 2015. V. 54, № 12. P. 3635.*

4. *Choporova Y.Y., Knyazev B.A., Mitkov M.S.* Classical holography in the terahertz range: Recording and reconstruction techniques // *IEEE Trans. on THz Sci. and Techn. 2015. V. 5. №. 5. C. 836.*

5. *Knyazev B.A., Kameshkov O.E., Vinokurov N.A., Cherkassky V.N., Choporova Yu.Yu., Pavelyev V.S.* Quasi-Talbot effect with vortex beams and formation of vortex beamlet arrays // Opt. Expr. 2018. V. 26. №. 11. P. 14174.

6. *Goodman, Joseph W.* Introduction to Fourier optics. *Roberts and Company Publishers,* 2005.

7. *Voelz, D.G.* Computational Fourier optics: a MATLAB tutorial. *Bellingham, WA: SPIE press*, 2011.

https://doi.org/10.1051/epjconf/201819508003

Diagnostic value of microwave imaging of dielectric tissues properties in patients with Dupuytren disease

Andrew K. Martusevich[1], A.G. Galka[1,2], S.Yu. Krasnova[1], S.V. Petrov[1], A.V. Novikov[1]

[1] Privolzhsky Research Medical University, Nizhny Novgorod, Russia, cryst-mart@yandex.ru
[2]Institute of applied Physics, Nizhny Novgorod, Russia

Dupuytren's disease - a pathology of connective tissue, leading to wrinkling of the Palmar aponeurosis and progressive deformation of the fingers, is a fairly common pathology. For example, in Germany about 1.9 million people suffer from Dupuytren's contracture, but in the U.S. 3% of the population has this pathology [1, 2]. In recent years, there has been a tendency to increase the frequency of this pathology. According to the Department of hand surgery in Moscow scientific research Institute of emergency care named after N.V. Sklifosovsky, the number of patients operated for Dupuytren's contracture, is about 20% of the total number of planned operations [3].

To date, the overwhelming majority of the authors consider surgical treatment of Dupuytren's disease to be the most radiacal and effective. Partial or total fasciectomy is considered as a gold standard of surgical treatment in most cases [4, 5]. However, despite the introduction of new technologies of surgical intervention, unsatisfactory treatment outcomes reach 30%, and as the contracture progresses, the results of operations deteriorate [6]. The number of relapses after surgical treatment remains high, which ranging from 26 to 80%, according to some foreign authors [2, 4].

In our opinion, one of the reasons for the poor results of surgical treatment of Dupuytren's contracture is the complexity of determining the boundaries of the lesion of Palmar aponeurosis, which determines the choice of rational surgical tactics. As a rule, these boundaries are determined visually by the surgeon directly during the operation. I. Zh. Osmonaliev et al. (2013) suggest using magnetic resonance imaging for this purpose without using the mode with suppression of the signal from adipose tissue [7]. According to the authors, the use of MRI allows more accurate visualization of the boundaries of the affected aponeurosis in patients with I-III degree of Dupuytren's contracture, and, respectively, to select a minimally invasive access and minimal traumatical method of excision of the affected tissues.

We believe that in modern economic conditions, the use of MRI to assess lesions in Dupuytren's contracture is justified due to the high cost of the survey. That is why the search for other objective methods of diagnosis of the area and boundaries of pathologically altered aponeurosis is very relevant and promising.

In this regard, in recent decades, studies have been conducted on the microwave diagnosis of the structure of biological tissues [8-11]. The method of resonance near-field microwave tomography, which allows to study the spatial distribution of the dielectric permeability and conductivity of living tissues with a resolution significantly lower than the wavelength λ, has broad prospects among non-invasive diagnostic methods. Unlike passive microwave sensing, near-field tomography requires a much smaller sensor (probe), and the resolution increases significantly [8-14].

The advantages of the method are confirmed by experimental studies. Thus, the assessment of skin electrodynamic properties in dermatoses indicated the diagnostic value of their study in microbial eczema and keratodermia [8, 9, 15]. Resonance near-field microwave diagnosis is potentially informative for the diagnosis of organ cancer (superficial or subepithelial localization), in determining the boundaries of the pathological focus [8, 15]. A.V. Arsenyev et al. (2011) investigated the level of functional activity of tissue growth zone in children, on the basis of which the presence of sexual characteristics of this process was established [16]. In addition, near-field microwave sensing makes it possible to quickly diagnose organ viability during transplantation [8]. Thus, with the help of near-field microwave sensing of the tissue structure, it is possible to obtain information about the biological object and the processes occurring in it [17-19]. At the same time, there is no data in the literature on the possibility of using microwave imaging in Dupuytren's contracture.

The aim of this study was to estimate the dielectric properties of fibrously altered tissues in patients with Dupuytren's contracture.

Material and methods

The study included 12 patients (male, mean age 53.9 years) with Dupuytren contracture of II-III degree according to R. Tubiana (1968), treated at the University hospital of the "PRMU" of Ministry of health of Russia. All patients were examined before surgical intervention.

The dielectric properties of the skin and subcutaneous structures were studied in different areas of the hand, including in the area of fibrous (altered and healthy tissues).

The dielectric characteristics of the biological tissues was evaluated by the method of resonance near-field microwave sensing. Near-field microwave sensing of tissues was performed using a special installation created at the Institute of Applied Physics of the Russian Academy of Sciences (Nizhny Novgorod), as well as specialized software that interfaces the installation with a PC and allows to calculate the effective part of the dielectric permittivity [8, 9]. This indicator was recorded and evaluated at depths of 2 to 5 mm using a series of probes.

The results were processed using the program Statistica 6.0.

The results and discussion

The study allowed to establish that in the field of healthy tissues the microwave profile of the skin corresponds to the physiological pattern formed by us on the basis of examination of healthy volunteers [10, 15]. It is revealed that the real part of the permittivity in the intact part of the Palmar aponeurosis increases monotonically with increasing sensing depth (Fig. 1).

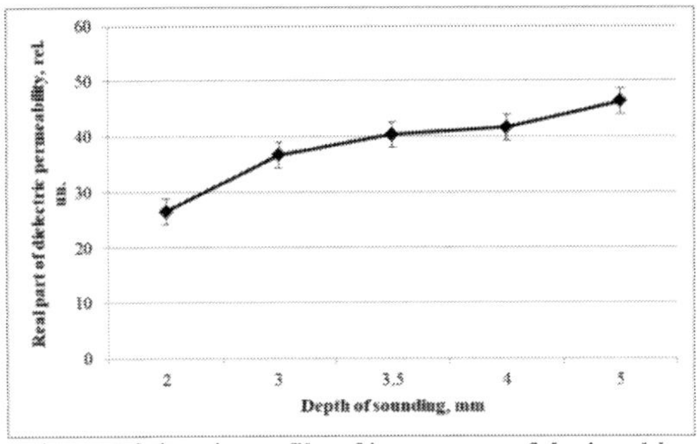

Fig. 1. Dielectric profile of intact area of the hand in patients with Dupuytren contracture

Then the microwave profiles of sub-skin tissues were compared at points 1 and 3 corresponding to intact and fibrous-modified areas (Fig. 2). It is established that fibrosis-changed tissues su-substantially differ in their dielectric parameters from the healthy, which leads to the significant transformation of the microwave profile in the area of measurement and control hardware of the Palmar aponeurosis are relatively to physiological pattern. The data obtained allow the conclusion that fibrosis in the tissue absorb microwave radiation, has extremely low values of the real part of dielectric permittivity. Such shifts are recorded at a depth of 2-3.5 mm, which approximately corresponds to the depth of pathologically altered tissues under Dupuytren's contracture [20, 21]. It should be emphasized that the sounding of deeper layers (4-5 m) does not detect significant deviations from the norm. This indicates the presence of intact morphological structures.

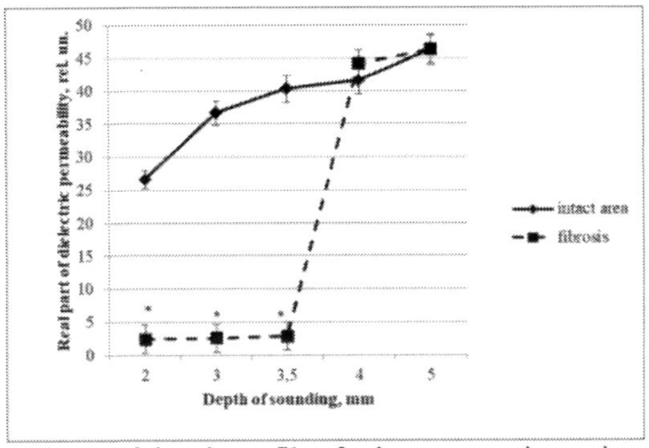

Fig. 2. Dielectric profile of subcutaneous tissues in intact and transformed area of the hand in patients with Dupuytren's contracture ("*" – statistical differences to intact area $p<0.05$)

Conclusion

The conducted studies allowed to form a microwave pattern of the actual part of the dielectric permeability in patients with Dupuytren's contracture in the area of healthy and fibrous-changed tissues, and a sharp decrease in this parameter was found in the zone of the pathological process at depths up to 3.5 mm.at the same time, in the area of healthy tissues, there were no features of dielectric properties compared with healthy volunteers. It is also shown that the fibrous-modified Palmar aponeurosis has a fairly uniform microwave structure, which allows us to count on the possibility of accurate visualization of its boundaries. This is crucial for the planning of surgery in patients with Dupuytren's contract.

References

1. *Brenner P., Krause-Bergmann A., Van V.H.* Dupuytren contracture in North Germany. Epidemiological study of 500 cases. Unfallchirurg 2001. V. 104, No 4. P. 303-311.
2. *Au-Yong I.T., Wildin C.J., Dias J.J., Page R.E.* A review of common practice in Dupuytren surgery. Tech. Hand Up Extrem. Surg. 2005. V. 9, No 4.178-187.
3. *Skoff H.D.* The surgical treatment of Dupuytrens contracture: a synthesis of techniques. Plast. Reconstr. Surg. 2004. V. 11, No P. 540-544.
4. *Kostrov A.V., Smirnov A.I., Yanin D.V. et al.* Resonanse near-field microwave diagnostics of non-homogenous mediums. Izvestiya RAS. Ser, Phys. 2005. V. 69, No 12. P. 1716-1720.
5. *Kostrov A.V., Smirnov A.I., Yanin D.V. et al.* The study of electrodynamics of biological tissues. Almanah of clinical medicine. 2008. No 17-2 P. 96-99.
6. *Martusevich A.K., Yanin L.D., Bogomolova E.B. et al.* Possibilitied and perspectives of the use of mocrowavw tomography in the estimation of skin state. Biomedical radioelectronics. 2017. No 12. P. 3-12.
7. *Reznik A.N., Yurasova N.V.* Near-field microwave tomography of biological objects // Zhurnal technicheskoi fiziki. 2004. V. 74, No 4. P. 108-116.
8. *Gaikovich K.P.* Subsurface near-field scanning tomography. Physical Review Letters 2007. V. 98, No 18: 183902.
9. *Raicu V., Kitagawa N., Irimajiri A.* A quantitative approach to the dielectric properties of the skin. Phys. Med. Biol. 2000. V. 45, No 2. P. L1-L4.
10. *Tirchin I.V. Method of optic biomedical vizualization: from subcelleular to ogranics to tissues.* Physics-Uspekhi. 2016. 59(5), c. 487-501.
11. *Tamura T., Tenhunen M., Lahtinen T. et al.* Modelling of the dielectric properties of normal and irradiated skin. Phys. Med. Biol. 1994. V. 39, No 6. P. 927–936.
12. *Berger A., Delbruck A., Brenner P., Hinzmann R.* Dupuytren's disease: pathobiochemistry and clinical management. Berlin, Heidelberg: Springer-Velard, 1994. 220 p.
13. *Warwick D., Tomas A., Bayat A.* Dupuytren's disease: overview of a common connective tissue disease with a focus on emerging treatment options. Int. Clin. Rheumatol. 2012.V. 7, No 3. P. 309-323.

Comparative study of dielectric properties of the skin of human and laboratory animals

Andrew K. Martusevich[1], A.G. Galka[1,2], S.Yu. Krasnova[1], D.V. Yanin[1,2], A.V. Kostrov[1]

[1] Privolzhsky Research Medical University, Nizhny Novgorod, Russia, cryst-mart@yandex.ru
[2]Institute of applied Physics, Nizhny Novgorod, Russia

Work on the study of electrical conductivity of tissues, including skin, first appeared about 40 years ago [12]. Since that time, the number of publications, despite their interesting results, is relatively small [7-11].

The skin, despite the superficial localization, being a difficult object for visualization, for a long time remained only the subject of histological examination [2, 5, 7, 9-12]. Existing methods (for example, optical coherence tomography, IR thermography, etc.) allow to study only the surface and the nearest surface structures of the skin [2, 5, 6, 10].

A more extensive methodological apparatus is available for monitoring skin vessels [5], while the deep structure of the latter is difficult for non-invasive study [1, 2, 5]. In this regard, the work on profiling of the skin by its dielectric properties attracts attention [1, 3, 4, 6, 8], but this information is isolated and jerky. This is, in particular, due to the lack of available diagnostic tools for assessing the dielectric characteristics of the skin and other tissues [3, 4].

In this regard, the aim of the study was to study the possibilities of near-field microwave sensing in assessing the structure of human and rat skin.

Material and methods

The study, which included a single microwave sounding, was performed in 20 practically healthy people and 20 healthy mature male Wistar rats.

Near-field microwave sensing of tissues was performed using a special installation created in the Institute of Applied Physics of the RAS (Nizhny Novgorod), as well as specialized software that interfaces the installation with a PC and allows to calculate the real part of the dielectric permeability [3]. The dielectric characteristics of the skin were evaluated at depths of 2 to 5 mm using a series of probes.

Measurement in all examined people was performed on the forearm at a single point, and in animals – at one point, localized in the middle part of the back, on a pre-epilated surface.

The obtained data were processed in the program package Statistica 6.1.

Results and discussion

It was found that the real part of the dielectic permeability of human skin increases monotonously with increasing depth of sounding (Fig. 1), showing a tendency to increase by 1.74 times when comparing the parameter values obtained at depths of 2 and 5 mm ($p<0.05$).

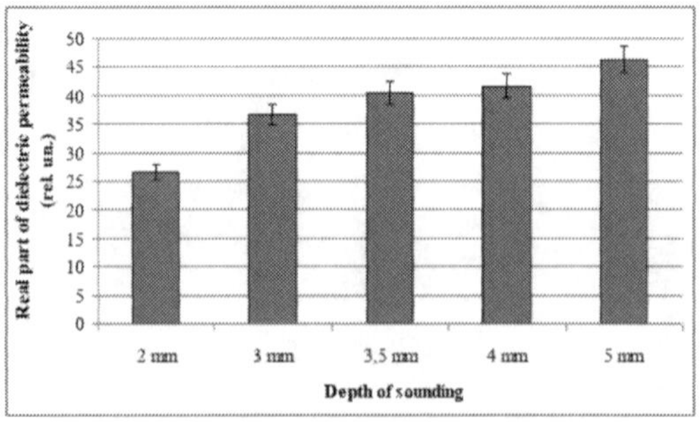

Fig. 1. The profile of dielectric permeability of the skin in healthy people (in rel. un.)

This is due to the fact that the value under consideration is cumulative, and each subsequent value includes the previous as well as the contribution made by tissues located from the previous to the current level of sensing. On the basis of the obtained data, a linear mathematical model of the change in the dielectrical permeability of the skin is constructed, which sufficiently describes its subsurface profile (determination coefficient - 0.94).

The linear regression equation, which allows to predict the value of dielectric permittivity at other sensing depths, is presented in the following form:

$$y = 6{,}4125 \bullet x + 15{,}844 \qquad (1)$$

Analysis of the dielectric properties of rat skin allowed to establish that the permeability of the latter is significantly, almost an order of magnitude lower than in humans, but the nature of the dependence of the considered parameter remains, demonstrating a monotonously increase in the value with a maximum at a depth of 5 mm (fig. 2).

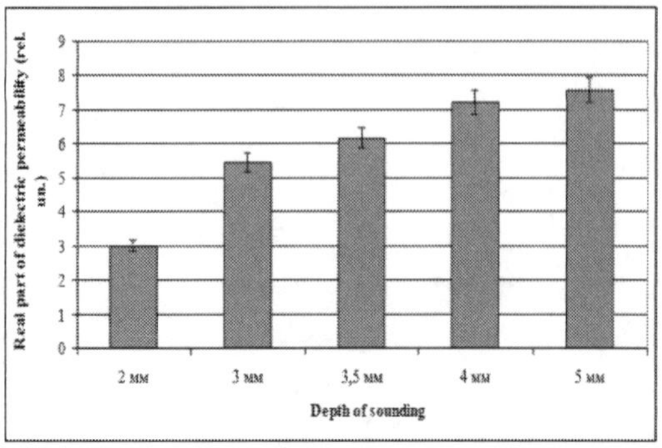

Fig. 2. The profile of dielectric permeability of the skin in healthy rats (in rel. un.)

At the same time, the level of the studied parameter at the minimum and maximum depths differs by 2.55 times ($p<0.01$), which is related to the assessment of deeper structures in rats during microwave profiling by a single date that performs sounding at a depth of 5 mm.

For the microwave profile of the skin of rats, we also formed the linear regression equation, as well as for humans, which fully reflects the experimentally obtained values of the dielectric permittivity (the determination coefficient is 0.91). This equation has the following form:

$$y = 1{,}55 \bullet x + 0{,}455 \quad (2)$$

Conclusion

The conducted studies allowed to establish a picture of the deep distribution of dielectrical permeability of the skin of healthy people and animals (Wistar rats), which can serve as a physiological microwave pattern for the study of subsurface tissues, including various layers of the skin and the nearest subcutaneous structures. It is shown that the real part of the dielectric permittivity at all the studied depths in humans is much higher than in rats, and monotonically increases with an increase in the sensing depth in the range from 2 to 5 mm in increments of 0.5 to 1 mm.

The stability of this pattern precedes the possibility of using the method of microwave profiling of the skin in the assessment of its structure in normal and local changes (benign and malignant neoplasms, burns, etc.), and the obtained equations can serve as a guide for the subsequent study of the dielectric characteristics of human cover tissues and mature rats in different experiments.

References

1. *Arsenyev A.V., Volchenko A.N., Likhacheva L.V., Pechersky V.I.* The use of the method of HF near-field sounding in diagnostics of bioobjects // Scientific and technical Gerald of information technologiesm mechanics and optics. 2011. No 2. P. 154-157.

2. *Gladkova N.D., Sergeev A.M.* PGuidelines on optic coherence tomography. Moscow: Fizmatlit, 2007. 295 p.

3. *Kostrov A.V., Smirnov A.I., Yanin D.V. et al.* Resonanse near-field microwave diagnostics of non-homogenous mediums. Izvestiya RAS. Ser, Phys. 2005. V. 69, No 12. P. 1716-1720.

4. *Reznik A.N., Yurasova N.V.* Near-field microwave tomography of biological objects // Zhurnal technicheskoi fiziki. 2004. V. 74, No 4. P. 108-116.

5. *Turchin I.V.* Methods of optiv biomedical visualization: from subcellular structures to tissues and ograns // Uspechi fizicheskih nauk. 2016. V. 186, No 5. P. 550–567.

6. *Gaikovich K.P.* Subsurface near-field scanning tomography // Physical Review Letters. 2007. Vol. 98, N 18. P.183902.

7. *Hayashi Y., Miura N., Shinyashiki N., Yagihara S.* Free water content and monitoring of healing processes of skin burns studied by microwave dielectric spectroscopy in vivo // Phys. Med. Biol. 2005. Vol. 50. N4. P. N8-N14.

8. *Raicu V., Kitagawa N., Irimajiri A.* A quantitative approach to the dielectric properties of the skin // Physics in Medicine and Biology. 2000. Vol. 45, N2. P. L1-L4.

9. *Schertlen R., Pivit F., Wiesbeck W.* Wound diagnostics with microwaves // Biomed. Tech. (Berlin). 2002. Vol. 47. Suppl 1, Pt. 2. P. 672-673.

10.*Semenov S.* Microwave tomography: Review of the progress towards clinical applications // Philos. Trans A Math Phys. Eng. Sci. 2009. Vol. 367, N 1900. P. 3021–3042.

11.*Sunaga T., Ikehira H., Furukawa S. et al.* Measurement of the electrical properties of human skin and the variation among subjects with certain skin conditions // Phys. Med. Biol. 2002. Vol. 47, N 1. P. N11-N15.

12.*Tamura T., Tenhunen M., Lahtinen T. et al.* Modelling of the dielectric properties of normal and irradiated skin // Phys. Med. Biol. 1994. Vol. 39, N 6. P. 927–936.

Microwave imaging of skin damage at experimental burns

Andrew K. Martusevich[1], A.G. Galka[1,2], S.Yu. Krasnova[1], A.G. Soloveva[1]

[1] Privolzhsky Research Medical University, Nizhny Novgorod, Russia, cryst-mart@yandex.ru
[2]Institute of applied Physics, Nizhny Novgorod, Russia

The ubiquitous prevalence of thermal injury predetermines not only the search and testing of innovative technologies for the treatment of severely burned, but also the improvement of the diagnostic apparatus of combustiology [1, 2, 4]. Currently, the clinical assessment of the local status has the greatest importance for the considered contingent of victims, including the one associated with the use of a number of empirical algorithms to determine the spatial characteristics of the injury ("rule of nines", etc.) [1, 2]. On the other hand, in combustiology there is a significant number of diagnostic difficulties associated with the specification of the boundaries of burn injury, tissue viability in the near-wound zone, wound uniformity, etc. [1, 2, 4]. Verification of the depth of skin and subcutaneous structures lesions should be singled out as a separate item [1-4]. In order to solve this complex of problems, in addition to the empirical approach prevailing in real clinical practice, the possibilities of thermal IR-imaging were studied [1, 3, 6]. It is shown that this technology is informative in a number of situations, but it allows us to judge only the state of the skin surface and the nearest underlying structures. Modern variants of ultrasound examination, which have high informative value and resolving ability in other pathology, do not allow to achieve the necessary contrast [3] in relation to the thermal injury.

An additional complicating factor in tissue imaging in combustiology is the presence of a physical barrier between the sensor and the surface of the skin (temporary and permanent wound coatings), which is not possible to eliminate for diagnostic manipulations (for example, when using biopo-coatings containing matrix with stem cells).) [1-3]. This is an obstacle for most methods of investigation of subsurface structure and blood tissues, in particular, for ultrasound examination. Therefore, it is necessary to search and test fundamentally different technologies of the assessment of the deep characteristics of the burn wound and the near-wound zone [2-4].

In this aspect, the method of near-field resonance microwave profiling, which has recently appeared in biomedicine and is based on the study of the dielectrical properties of tissues (dielectric permeability and conductivity) is very interesting [5]. Previous studies have shown that this technology is high informative in dermatology, allowing how to carry out primary diagnostics and differential diagnostics of various skin diseases and to monitor the efficacy of treatment, predicting the occurrence of the patient in the remission phase [5]. At the same time in combustiology the considered method was not applied earlier.

The **purpose of this study** is to evaluate the diagnostic capabilities of near-field microwave sensing in the estimation of the deep structure of the skin in the norm and in experimental burn wounds.

Material and methods

The study was performed on 30 male rats of the Wistar line, divided into 2 equal groups. The first group of animals (n=15) was a control group (no manipulations were performed, except for a single microwave sounding). Rats in the second group (n=15) was simulated thermal contact burns on 20% body square percent with our methodology (Peretyagin S.P. et al., 2009) [6]. The dielectrical properties of burn wound tissues were studied immediately after the application of burn and on 1 day after it.

Near-field microwave sensing of tissues was performed using a special device created at the Institute of Applied Physics of the Russian Academy of Sciences (Nizhny Novgorod), as well as specialized software, matching the installation with a PC and allowing the calculation of the real part of the dielectic permeability [5]. Dielectric characteristics of the skin were assessed at depths of 2 to 5 mm using a series of probes. All animals were measured at one point, localized in the middle part of the back, on a preepilated surface.

The results were processed using the program Statistica 6.0.

The results and discussion

The conducted research allowed to establish that it is possible to carry out the analysis of dielectric properties of skin of rats by means of the studied hardware-software complex.

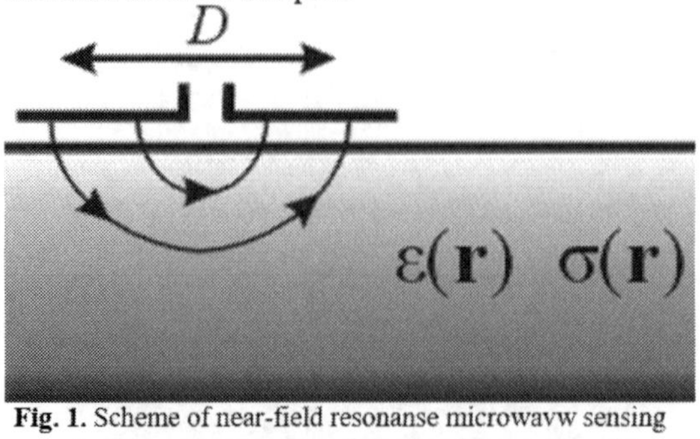

Fig. 1. Scheme of near-field resonanse microwavw sensing of subsurface structure (D – diameter of the sensor; ε(r) – dielecric permittivity; σ(r) – conductivity)

At the same time, it was possible to show that at the studied depths (2-5 mm) the level of dielectric conductivity of subsurface structures is at values less than 9 rel. un., monotonically increasing with increasing depth of sensing.

Given that each value is cumulative, i.e. includes the conductivity of the entire subsurface layer up to the specified depth (Fig. 1), the maximum level of the parameter is recorded at a distance of 5 mm from the surface of the skin. This is reflected in the character of the method of sounding as near-field (Fig. 1).

At the same time, the level of the studied parameter at the minimum and maximum depths differs by 2.55 times ($p<0.01$), which is associated with the assessment of deeper structures in rats during microwave profiling by one sensor performing sensing at a depth of 5 mm. This is due to the fact that each subsequent value of the dielectric conductivity includes the previous one in conjunction with the contribution made by tissues located from the previous to the current level of sensing [11, 19-21].

Based on the data obtained from intact rats, a linear mathematical model of changes in the dielectric permittivity of animal skin is constructed, which sufficiently describes its subsurface profile (determination coefficient – 0.91). The linear regression equation allows to predict the value of dielectric permittivity at other sensing depths. This model can be used to calculate the physiological level of dielectric conductivity of the subsurface structures of the skin of rats, used as a guide to identify its changes caused by various pathological processes.

The features of the deep structure of the dielectric properties of the skin and subcutaneous tissues in animals with modeled thermal injury (in the form of a thermal burn on the previously epilated surface of the back skin) were also studied. Evaluation of the dielectric permittivity of the subsurface tissues of the experimental burn wound was carried out by us immediately after the injury and on 1 day after its simulation. This allowed to form deep skin profiles by this parameter in the dynamics of the experiment and in comparison with intact biological tissue (Fig. 2).

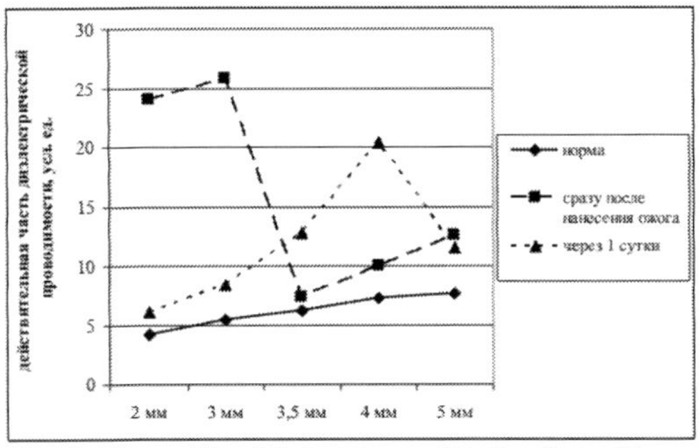

Fig. 2. The profile of dielectric permittivity of rats skin an subcutaneous tissues in normal conditions and after the burn

It was found that the dielectric properties of the burn wound differ significantly from the intact cover tissue. These shifts are characterized by an increase in the real part of the dielectric permittivity of the medium at both observation points. Thus, immediately after the application of thermal trauma, the greatest changes occur in the nearest subsurface layers of the skin (2-3 mm.), in which the figure under consideration repeatedly increases relative to intact rats (7.46 and 9.47 times at the depths of sensing 2 and 3 mm; $p<0.05$ for both cases), amounting to about 24.0 and 25.8 units, respectively. This may be due to the rapid intensive local heating of the tissues at a shallow depth immediately after exposure, whereas this effect has not yet spread to the deeper layers.

In 1 day after the burn the depth profile of the dielectric conductivity of the skin was significantly transformed (Fig. 3). During this period, there is a deepening of the lesion of subsurface structures, which is accompanied by a decrease in the level of the studied parameter at minimum depths (2-3 mm.) with its increase relative to the intact skin at a distance of 3.5-5 mm below the skin surface ($p<0.05$ for all cases). At the same time, the maximum dielectric conductivity was recorded at a depth of 4 mm, which in our experiments corresponded to the zone of greatest damage. This indicates a partial cooling of the surface layers of the skin with simultaneous overheating of more proximally lying and, consequently, a shift in the focus of damage to the deeper layers. The presented data indicate the possibility of monitoring the depth of thermal tissue damage, including creating an experimental basis for testing the processes of deepening burn in the post-traumatic period.

Conclusion

The studies have shown that the burn wound tissue shows a higher level of the real part of the dielectric permittivity in comparison with intact skin, and these changes have a temporal dynamics. So, immediately after the burn, the parameter change prevails in the surface layers of biological tissue, and after one day – in the deeper layers.

The stability of this pattern determines the possibility of using the method of microwave profiling of the skin in the assessment of its structure in normal and local changes (benign and malignant tumors, burns, etc.), and the equations can serve as a guide for the subsequent study of the dielectric characteristics of the cover tissues of mature Wistar rats in diverse experiments.

References

1. *Arai T.* Burns. Nihon Rinsho. 2016. V. 74, No 2. P. 231-235.

2. *Daigeler A, Kapalschinski N, Lehnhardt M.* Therapy of burns. Chirurg. 2015. V. 86, No 4. P. 389-401.

3. *Ida T., Iwazaki H., Kawaguchi Y. et al.* Burn depth assessments by photoacoustic imaging and laser Doppler imaging. Wound Repair Regen. 2016. V. 24, No 2. P. 349-355.

4. *Li H., Zhang J., Chen J. et al.* Integration of burn treatment and rehabilitation for a child with extremely severe burn. Zhonghua Shao Shang Za Zhi. 2015. V. 31, No 2. P. 130-134.

5. *Kostrov A.V., Smirnov A.I., Yanin D.V. et al.* Resonanse near-field microwave diagnostics of non-homogenous mediums. Izvestiya RAS. Ser, Phys. 2005. V. 69, No 12. P. 1716-1720.

6. *Peretyagin S.P., Martusevich A.K., Vazina I.R. et al.* Development of new method of combined thermal trauma modeling. Sovremennye techologgii v meditsine. 2011. No 2. P. 106-109.

Subsurface diagnostics of quasi-one-dimensional inhomogeneities using the method of near-field microwave sounding

D. V. Yanin, A. G. Galka, A. V. Kostrov, A.I. Smirnov

Federal Research Center Institute of Applied Physics of the Russian Academy of Sciences (IAP RAS), Nizhny Novgorod, Russian Federation, dyanin@appl.sci-nnov.ru

Potentialities of the near-field microwave sounding are studied for the case of model media with spatially localized inhomogeneities.

The method of resonance near-field sounding [1-3] is used successfully to study electrodynamic properties of various substances in the microwave range. Depending on the operating frequency, either oscillatory circuits with lumped parameters or distributed resonance systems can be used as measurement sensors. When a sample is introduced into the region occupied by the quasi-static field of a sensor, its resonance frequency and the Q-factor change. These variations can be used to judge about the electric and magnetic properties of the studied object. Using the sensors with different sounding depths, one can study the internal structure of the object without disrupting its integrity.

The near-field measuring system, which we use for diagnostics of inhomogeneous media, was a microwave resonator in the form of a section of a loop line closed on one end. A measuring capacity at the other end of this section is made up by two parallel wires with the radius r=0.5 mm, the length d_m=4 cm, and the varying distance d between the wires. Magnetic coupling loops were used to excite the resonator and receive its responsse. In this case, the maximum sounding depth was close to d. Eight sensors with sounding depths ranging from 5 mm to 25 mm were used to study the internal structure of the media. The eigen frequencies f_0 of the measuring systems were about 550 MHz, and their Q-factors, about 150.

The relationship between the resonant frequency f of the sensor and the parameters of the inhomogeneous medium is expressed as follows

$$\frac{f-f_0}{f_0}=\frac{d}{l}\left(1-\varepsilon_{\text{int}}\frac{\rho_1}{\rho_2}\right) \tag{1}$$

where ε_{int} - the integral permittivity of the inhomogeneous medium, ρ_1 and ρ_2 - wave resistances of two-wire lines corresponding to the resonator and the measuring capacitance, l - the length of the resonator. In the derivation of the expression (1) it was assumed that the electrical length of the measuring part is significantly less than the wavelength, $d_m \ll c/(f_0\varepsilon_{\text{int}})$ (c – the speed of light in vacuum). In the case of planar media, using the image method, you can obtain the following expression for ε_{int}.

$$\varepsilon_{\text{int}}=(\varepsilon_1+1)\left(2+\left(\frac{2\varepsilon_1}{\varepsilon_1+1}\right)\left(\frac{\varepsilon_1-\varepsilon_2}{\varepsilon_1+\varepsilon_2}\right)\frac{\ln\left(\frac{d^2+(2h)^2}{r^2+(2h)^2}\right)}{\ln\left(\frac{d}{r}\right)}\right)^{-1} \tag{2}$$

h - depth of inhomogeneity, ε_1 and ε_2 -dielectric permeability of the medium and inhomogeneity, respectively.

The medium studied in the experiments, in which the method of near-field microwave sounding was tested, was organic glass having the dielectric permittivity ε_1=3.5 with the rectangular inclusions of glass having the dielectric permittivity ε_2=7.8, which were situated at the depths h=4 and 5.5 mm. The length and thickness of the inhomogeneity were equal to L_n=10.5 cm and h_n=7 cm, respectively.

To spot the inhomogeneities, a probe with a sounding depth of 10 mm moved parallel to the flat surface of the organic glass. The value of the shift of the resonance frequency Δf was registered depending on the position x of the sensor (see Fig.1).

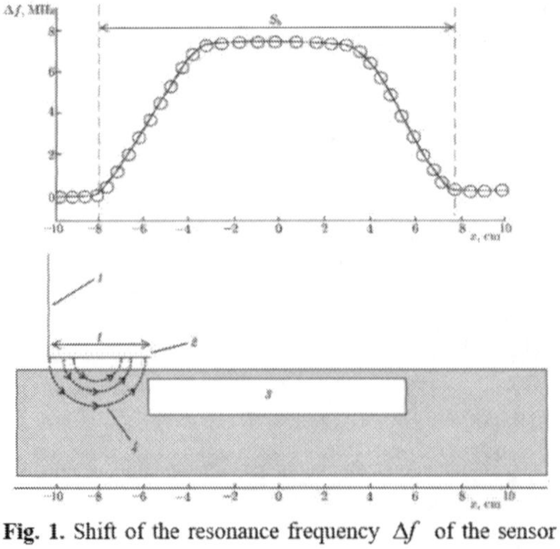

Fig. 1. Shift of the resonance frequency Δf of the sensor with a sounding depth of 10 mm as a function of the probe coordinate x for the case of scanning along the plane surface of the medium with a rectangular inhomogeneity.

The length of inhomogeneities is equal to the difference between the length s of the "swell" in the plot of the resonance frequency and the length d of the wires of the measuring capacity.

Further, the medium was studied using sensors with different sounding depths. The resonance frequency of the measuring systems was registered as the systems contacted the surface of the medium near the center of the inhomogeneity. Then, the dielectric permittivity ε_{ex} of the medium was determined for each probe using Eq. (1).

$$F(\varepsilon_1,\varepsilon_2,h)=\sqrt{\frac{1}{n}\sum_{i=1}^{n}\left(\varepsilon^{ex}(d_i)-\varepsilon_{\text{int}}(\varepsilon_1,\varepsilon_2,h,d_i)\right)^2} \tag{3}$$

To measure the medium parameters, we minimized function (3) being the r.m.s. deviation between $\varepsilon_{int}(\varepsilon_1, \varepsilon_2, h, d)$ for the plane-layered medium and the experimentally found values of the dielectric permittivity ε_{ex}. Table shows the actual and experimental values of the electrodynamic and geometric parameters of the studied media.

This paper deals with testing of the method of the near-field microwave sounding. The algorithm of solving of the inverse problem is presented, and the geometric and electrodynamic parameters of rectangular inhomogeneities are determined with good accuracy.

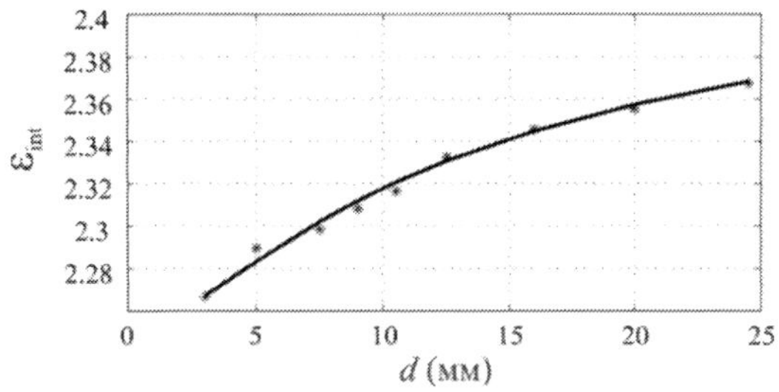

Fig. 2. Integral dielectric permittivity ε_{int} as a function of the distance d between the wires of the sensor measuring head in the case of studying an inhomogeneity in the form of a rectangular parallelepiped. The asterisks mark the experimental data, and the solid curve represents the theoretical dependence that corresponds to Eq. (2) for the found values of ε_1, ε_2 and h.

Model media 1	ε_1	ε_2	h	L_n
Experiment	3.5	7.3	5.0	11.4
Real value	3.5	7.8	5.5	10.5
Model media 2	ε_1	ε_2	h	L_n
Experiment	3.5	7.6	3.6	11.4
Real value	3.5	7.8	4.0	10.5

This work was supported by the Russian Foundation for Basic Research (project Nos. 18-42-520053 p_a).

References

1. D. V. Yanin, A. V. Kostrov, A. I. Smirnov, et al. Diagnostics of the Atmospheric Pressure Plasma Parameters Based on Near-Field Microwave Plasma Probing // Tech. Phys., 82, No. 4, 468 (2012).
2. Yanin D.V., Galka A.G., Smirnov A.I., Kostrov A.V., Strikovskii A.V. Resonant near-field microwave diagnostics of inhomogeneous mediums // Advances in Applied Physics. 2014, 2, No. 6. P. 555-570.
3. Martusevich A.K., Galka A.G., Krasnova S.Yu., Yanin D.V., Bogomolova E.B. Study of dielectric properties of human skin: the first experience of microwave probing. International Journal Of Applied And Fundamental Research. – 2017. – № 3

Investigation of electrodynamics parameters of biological tissues

D. V. Yanin[1], A. G. Galka[1], A. V. Kostrov[1], V.E.Zagainov[2], S.A.Vasenin[2],

[1]Federal Research Center Institute of Applied Physics of the Russian Academy of Sciences (IAP RAS), Nizhny Novgorod, Russian Federation, dyanin@appl.sci-nnov.ru
[2]Privolzhsky Federal Medical Centre Roszdrava, Nizhny Novgorod, Russian Federation.

Pathological and physiological processes in living tissues are usually accompanied by variation of their electrodynamic parameters; hence study of dielectric permittivity and conductivity of bio-objects is of considerable interest for various medical applications.

Diagnostics of skin pathologies without using the hystomorphologic method is required in dermatology. Tissue sampling (biopsy) refers to minor operations and is often undesirable for patients with abnormality in carbohydrate metabolism, vascular pathology, and eruption on unclothed parts of the body (face, neck, and hands).

The estimation of viability of organs in vivo in the case of acute pathology and organs conserved for their further transplantation is necessary in surgery. The processes of tissue ischemia and reperfusion complications are the result of violation of the fine cellular mechanisms, diagnostics of which cannot be carried out by the known diagnostic techniques (X-ray, ultrasound). These processes can be recognized only by means of biopsy with further optical or electron microscopic analysis; this procedure is time-consuming, whereas a clinician does not have much time at his disposal.

The aim of the study is to consider the opportunities of resonance near-field microwave sounding for estimation of viability of parenchymal organs in critical states, determination of pathologic processes, differential diagnostics of various dermatoses, and control of medical maintenance.

The method of resonance near-field microwave probing can be explained as follows. The area of a medium located in the near field of a probing electrically small antenna affects its impedance. This feature enables one to provide high spatial resolution. If the antenna is connected to the resonance system as a load, the resonance frequency shift and the Q-factor variation can be used to estimate the electromagnetic parameters of the medium and then the state of the examined object.

Diagnostic probes for passive measurements of the electrodynamic features of parenchymal organs and sensors for investigation of skin of dermatologic patients have been developed. A high-Q microwave resonator placed on a segment of the coaxial line is employed as a resonance system. The eigen frequencies of the sensors are $\omega_0 \sim 2\pi \cdot 800$ MHz and the Q-factor is Q_0 ~150. The spatial resolution and the sensitivity are determined by the design and sizes of the electrically small antenna.

If Z_0 is the internal impedance of the antenna, Z_{medium} is the impedance of the antenna contacting with the medium, and $Z_{medium} < \rho$ (ρ is the wave resistance of the coaxial resonator), according to [1] one can obtain the equation of the resonance curve $U_{res}(\omega)$ of the sensor

$$U_{res}(\omega) = U_0\left[16Q_0^2\left(\frac{\omega-\omega_0}{\omega_0}+\frac{1}{\pi\rho}\mathrm{Im}(\delta Z_x)\right)^2 + \left(1+\frac{4}{\pi}\frac{Q_0}{\rho}\mathrm{Re}(\delta Z_x)\right)^2\right]^{-\frac{1}{2}}$$

$\delta Z_x = Z_0 - Z_{medium}$, U_0 is the signal amplitude in the resonance curve maximum.

From the equation $U_{res}(\omega)$ one can easily obtain the relation between the resonance characteristics of the sensor and the impedance features of the electrically small antenna

$$\omega_{res} - \omega_0 = -\frac{1}{\pi}\frac{\omega_0}{\rho}\mathrm{Im}(\delta Z_x)$$

$$\max(U_{res}) = U_0\left(1+\frac{4}{\pi}\frac{Q_0}{\rho}\mathrm{Re}(\delta Z_x)\right)^{-1}$$

Electrodynamic characteristics of skin of 32 cases of psoriasis, 10 cases of atopic dermatitis, and 13 cases of lichen acuminatus (LA) were studied at the Research Institute of Dermatology and Venereology (the city of Nizhny Novgorod).

It is stated that the dielectric permittivity and conductivity of skin of dermatologic patients (cases of psoriasis, atopic dermatosis, and lichen acuminatus) are lower than those of healthy skin. The patients were examined before treatment, in the course of treatment, and after it. As the patients recovered, the dielectric permittivity and conductivity of tissues in the area of the focus of disease in all three groups of patients approximated the values of healthy skin.

In the exacerbation stage, the difference between healthy and damaged skin were more distinct in the case of psoriasis. In the regress stage, the dielectric permittivity and conductivity of tissues in the area of psoriatic focuses of disease were analogous to ε and σ of tissues in the case of atopic dermatitis. Hence in the cases of psoriasis and atopic dermatitis, the method is diagnostically significant only when a disease is active.

When studying the electrodynamic characteristics of skin in the case of lichen acuminatus, it was found out that if the dielectric permittivities of tissues in the case of psoriasis and in the case of LA coincided, the conductivities in these cases differed by factor of 2.

This permits drawing a conclusion on the possibility of diagnostics in the cases of psoriasis and LA at arbitrary stages of disease.

Differential diagnostics of pathologic processes in parenchymal organs is carried out. It is shown that the measuring systems are sensitive to physiological and pathological properties of tissues. The possibility to determine tumor focuses of disease in an organ and the limits of their growth is demonstrated. Study of a remote material (fig. 1) confirmed high accuracy and sensitivity of the measuring complex. The difference in sensor indications in measurement of various types of tissues is well seen in fig. 2.

It should be noted that near-field systems are sensitive to arbitrary, even slight, variations of blood flow in tissues abounding with blood vessels.

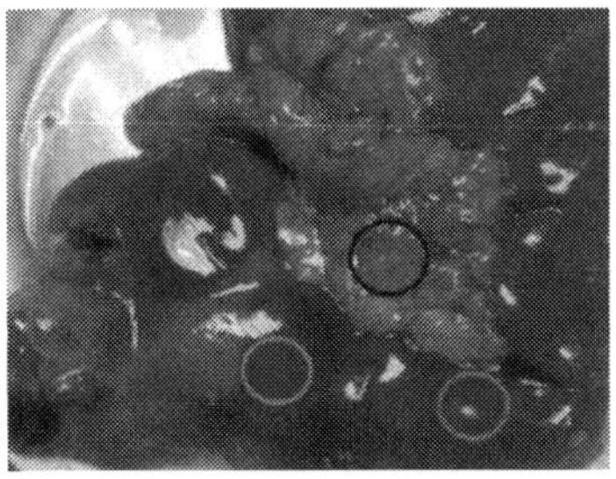

Fig. 1. Examined object (kidney). Circles show the measuring areas

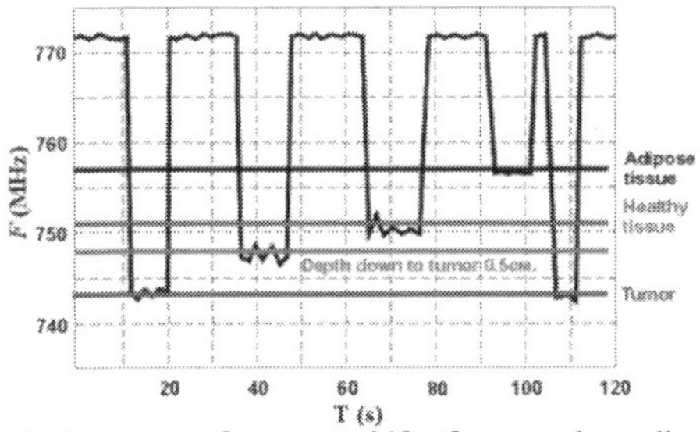

Fig. 2. Resonance frequency shift of sensor depending on the features of an examined tissue.

Electrodynamic characteristics of parenchyma of kidneys under the conditions of thermal and cold ischemia are measured in time dynamics. Laboratory animals (rabbits) were used in the studies. The process of multiorgan sampling for transplantation was simulated completely. Kidneys were irrigated with a cooled solution of kustodiol (additive) through the aorta and appropriate arteries until blood was fully eliminated from the organ. Measurement results of the resonance frequency of sensor (dielectric permittivity) on time are given in fig. 3. The results show a difference between cold and thermal ischemia of organs; the signal frequency variation depends directly on the rate of ischemic damage.

Besides the probe contacting with the examined tissue, a sensor based on the resonance contact sensor was fabricated and tested; it responded to small additives occurring as a result of diffluence in the conserving liquid irrigating the organ prepared to transplantation and being in the critical state. Additive sampling was made in definite periods. The result obtained unambiguously demonstrates the relation between the electrodynamic characteristics of kustodiol varying in the experiment, as diffluence products accumulate in it (fig. 4). Having a set of gage curves, one can carry out express diagnostics of viability of an organ during several seconds.

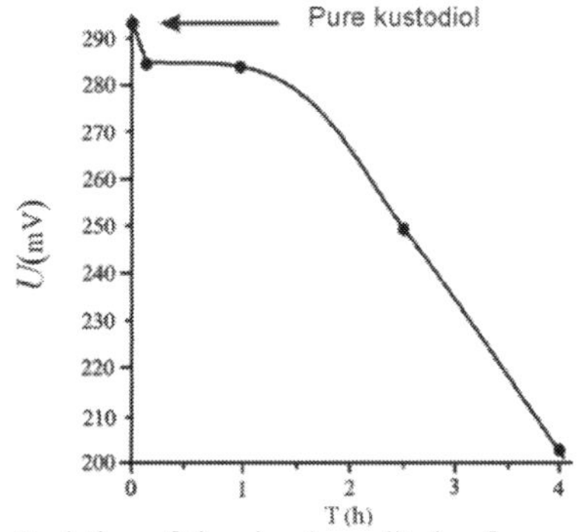

Fig. 3. Variation of the signal amplitude of sensor at the resonance frequency on pump time. The examined liquid is "kustodiol". (Thermal ischemia)

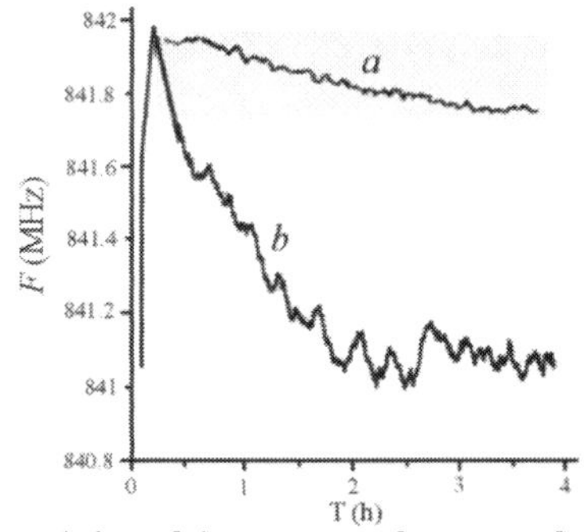

Fig. 4. Variation of the resonance frequency of sensor in time; the sensor are in contact with parenchyma of a kidney (a) cold ischemia, (b) thermal ischemia.

When comparing measurements of the electrodynamic parameters of the parenchimatous tissue and the additive, it is seen that the occurrence of diffluence products in kustodiol slightly "delays", which agrees with recent publications on results of marginal (cadaveric) transplantations of kidneys [2].

This work was supported by the Russian Foundation for Basic Research (project Nos. 18-42-520053 p_a).

References

1. *A.V. Kostrov, V.A. Kostrov, A.I. Smirnov, D.V. Yanin,A.V. Strikovskiy, G.A. Panteleeva.* Diagnostics of inhomogeneous and time varying media by means of a resonance microwave probe based on a section of the tow wire line. Preprint of IAP RAS №707, Nizhny Novgorod, 2006.

2. *M.M.Kaabak, V.A.Sandikov end others.* The remote effects of transplantations of a cadaveric kidney. Materials of conference: clinical transplantation of bodies. Moscow, 26-27 September, 2007.

 https://doi.org/10.1051/epjconf/201819509001

Influence of pollution and extraneous inclusions on the scattering of THz radiation by fabric

A. A. Angeluts[1], V. N. Aksenov[1], A. V. Balakin[1], I. A. Ozheredov[1], A. P. Shkurinov[1]

[1]Faculty of Physics M.V.Lomonosov Moscow State University, Moscow, Russia, angeluts@physics.msu.ru

One of the challenges of modern civilization now is the various manifestations of terrorist activity. It is therefore necessary to create different methods of remote diagnostics of dangerous items and substances. One of directions of development of diagnostic methods introduced in recent decades is methods based on the use of Terahertz (THz) radiation.

THz radiation is defined as electromagnetic radiation in the range from 300 GHz to 10 THz (or 30 μm to 1 mm), is of great interest to create diagnostic, measuring and inspection systems. This interest determines firstly by the fact that in recent years, started an intensive development of components for the generation and detection of THz radiation, and secondly, the specific features of the radiation. Among the important properties of this radiation should be noted his non-invasive and non-ionizing nature that allows the use of THz radiation in tasks related to human Diagnostics. In addition, this radiation weakly attenuated by most materials, so such objects, such as cloth, dry wood, paper, plastics are transparent to THz range.

Most often, the diagnostic is to irradiate the THz object of interest by probe beam and then analyze the transmitted, reflected or scattered radiation.

In the present work describes a study, based on analysis of the scattered radiation.

Experimental setup and samples

As the object of the study, tissue samples were selected with TNT traces on their surfaces or that contain a thin metal wire. Such objects simulate a clothes whose surface is contaminated with a hazardous substance or contains elements of electronic devices controlling the explosive device (current leads, antenna elements and so on).

Studies performed with the setup, the simplified diagram is shown in Fig.1. Our experimental setup detailed description can be found in [1]. Radiation with a frequency of 3 THz focused on the surface of the object along normal to its surface. The scattered radiation was collected by a system of parabolic mirrors and detected. Recorded radiation scattered at an angle of approximately 60 degrees from the normal, or 120 degrees from the direction of incidence of the probe beam. This angle allows the detector to avoid specular component of the radiation.

Samples were flaps of cotton fabric. "Contamination" of the sample was carried out by applying TNT solution in acetone to the tissue. Then the sample was dried and weighed.

The sample *Ct14* had a surface contamination density of about 3 mg/cm^2.

The sample *Ct50* had a surface contamination density of 10 mg/cm^2.

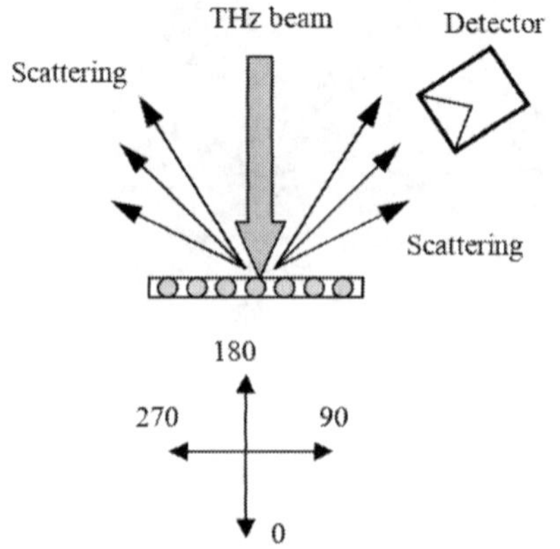

Fig. 1. Scattering experiment schematic

Since no uniform contamination was achieved in the preparation of samples, the above figures only approximately characterize the degree of contamination in the analyzed sample area.

Cm78 sample consisted of a piece of cloth of the same, in which a copper wire of 78 mkm was introduced.

In addition to the samples described above, samples were studied that were copper wires 240 and 78 mkm in diameter. With their help it was conducted an adjustment of the measuring circuit.

Experimental results

Table shows some of the results of the measurements represent the values of ambient signal from different samples. For ease of comparison values of the scattering signal, it was normalized to the value of the scattering signal of pure textiles.

Table . Relative magnitude of scattering for different samples

Sample	Copper wire *ϕ78* mkm	Copper wire *ϕ*240 mkm	Ct14	Ct50	Textile	Cm78
Scattering signal value	5.4	18.3	0.8	0.17-0.58	1	1.75

The signal value for sample *Ct50* shows a spread due to strong heterogeneity of pollution, as when drying solution came the transfer of a substance to the periphery of the sample. So most likely score pollution concentration overstated

As can be seen from the table, tissue contamination leads to suppression of scattering signal in our choice of angle observation from 20% up to 5 times. The magnitude of the suppression is determined by the concentration of the substance in the area under study

The presence of thin conductor reliably detected in tissue. The copper wire with a diameter of 78 mkm sewn into the fabric leads to an increase in the scattered signal by a factor of 1.8.

The scattered signal from a single conductor exceeds the signal from the surface of the cotton fabric from 5 to 20 times, depending on the size of the conductor

The figure 2 shows the result of yet another experiment. Here, the sample *Cm78* moved along the surface plane, simulating the scanning process of the sample. The width of the curve in Fig. 2 is one and a half times smaller than the THz size of the spot on the surface. The "pedestal" of the chart is determined by the scattering from the tissue itself. A slight decrease in the magnitude of the signal at the positions 5 and 6 mm is caused by the destructive interference of the radiation scattered by the tissue and the conductor.

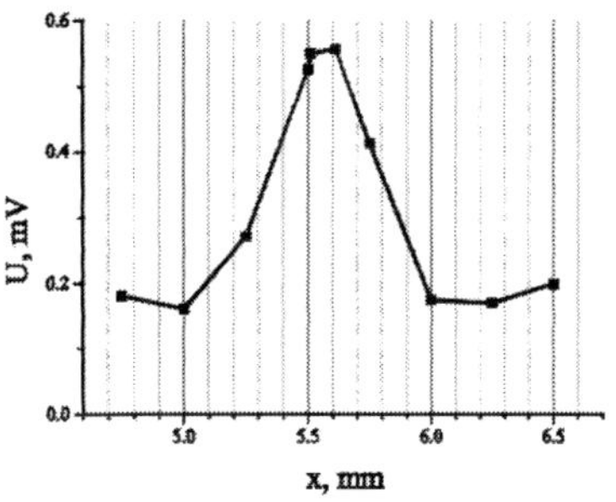

Fig. 2. The result of the movement of Cm78 sample across the probing beam. Detection of hidden wire.

Scattering numerical estimation

A numerical estimate of the scattering was carried out on the simplest model object-an infinite cylinder covered by a layer of an absorbing dielectric [2].. This model most closely describes the textile, which is a set of threads.

In our model for base cylinder, we used parameters of PET (also known as Lavsan) and cotton (see Fig. 3 and 4). The diameter of the threads equals to 100 microns. As a coating, TNT layers of different thicknesses were considered: 0.1, 1 and 10 μm.

Figures 3 and 4 show fragments of the scattering indicatrix for model objects. The angle of 180 degrees corresponds to the direction of backscattering of the incident radiation. On the graphs given, the value of the scattering signal is normalized to the value for the uncoated strand.

Given charts show that the nature of the scattering can vary significantly with increasing thickness of a polluted layer. Strong pollution (layer 10 mkm) leads to the Suppression of scattering that qualitatively corresponds to experiment. Under certain conditions (1 mkm) scattering may even increase. A small contamination (layer 100 nm) changes the magnitude of the signal is less than 10%.

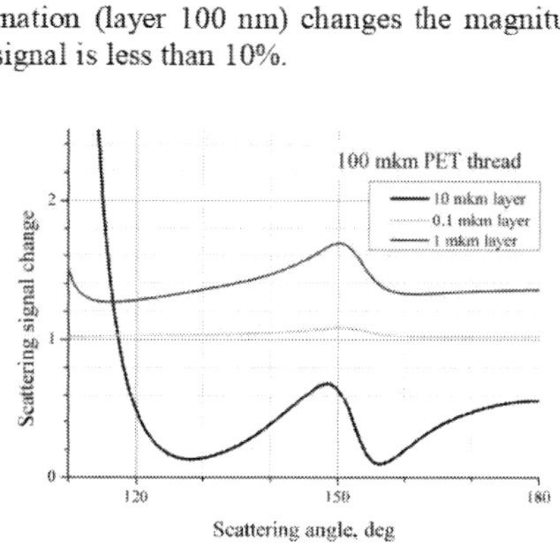

Fig. 3. The change of the scattering indicatrix with the variation in the thickness of the polluting layer on PET thread

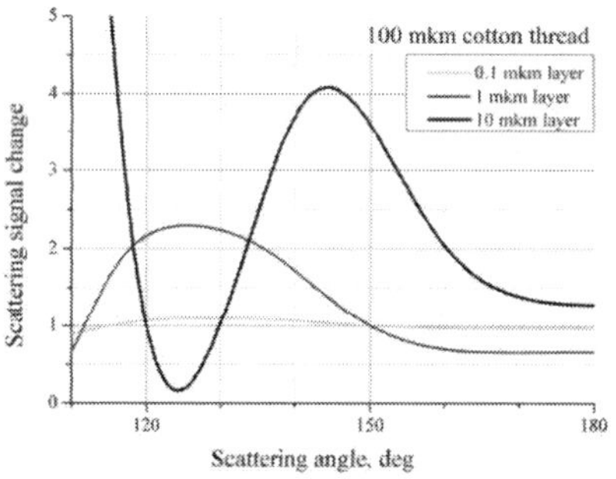

Fig. 4. The change of the scattering indicatrix with the variation in the thickness of the polluting layer on cotton thread

Conclusion

The possibility of detecting current-carrying elements or traces of explosives on the surface of the object by analyzing scattered THz radiation is investigated.

Experimental results and preliminary calculations have shown that the backscattering patterns for textile fibers differ significantly, depending on their material and the presence of contamination on them. This opens up the possibility of developing terahertz-imaging systems in the field of security.

Acknowledgements

This work was supported by the Russian Foundation for Basic Research (project No. 16–29–09593)

References

1. *Aksenov, V. N., Angeluts, A. A., Balakin., A. V. et. al.* Application of a terahertz multi-frequency radiation source based on quantum-cascade lasers for identification of substances basing on the amplitude-spectral analysis of the scattered field // Radiophysics and Quantum Electronics. 2018. V. 60, No. 11. P. 877-888

2. *Kotlar, V. V., Lichmanov, M. A.* Analysis of the diffraction of an electromagnetic wave on an infinite circular cylinder with several homogeneous layers. Computer Optics (*in rus.*). 2002. V. 24. P. 26-23

Svalbard astroclimate research: expedition and first results

Grigoriy Bubnov[1,2], V.F. Vdovin[1,2], P.M. Zemlyanukha[1], V.S. Okunev[1], V.F. Grigor'yev[3]

[1] Institute of Applied Physics RAS, N.Novgorod, Russia, bubnov@ieee.org
[2] Nizhny Novgorod State Technical University n.a. R.E. Alekseev, N.Novgorod, Russia
[3] Polar Geophysical Institute RAS, Murmansk, Russia

Expedition and equipment

This paper deals with the actual results of astroclimate investigations in mm-waveband on Svalbard archipelago. The investigations have begun in June 2018 and are still ongoing. We use the dual-wave radiometric system operating by atmospheric-dip method to estimate an optical depth in 2mm and 3mm atmospheric windows [1]. It has been installed on the polar station of Polar Geophysical Institute near Barentsburg town, Svalbard. (78° 5'42.22"N; 14°12'35.98"E) Remote control system allows to collect data via Internet.

It is also equipped with a makeshift meteostation and anti-icing heater, which is applied in autonomous field experiment for the first time. The meteostation provides a brightness temperature measurement of the sky by IR sensor, apart from typical meteo-data (eg. temperature, pressure, humidity). It allows to define a cloudiness by empiric methodic. Cloudiness calculation methodic is based on work [2,3]. The essence of methodic is to define the difference between real air temperature and brightness temperature by IR-sensor MLX90614. The more this difference is, the less clouds are in air and vice versa; the difference more than 20K corresponds a clear sky conditions, while the difference less than 5K means overcast weather.

As is known, the atmospheric-dip method can't be used in overcast conditions. The new data processing method, which is firstly applied in 2018 [4], allows to save some part of data, measured in overcast conditions. Resulting optical depth includes absorption in stratus clouds as well. Picture 1 shows 5-day record of optical depth τ (Nep) vs temperature difference dT (K). Increasing difference dT between physical air temperature and IR brightness temperature means more clear sky conditions in July 16. This is also accompanied by decreasing mm-wave Optical depth values.

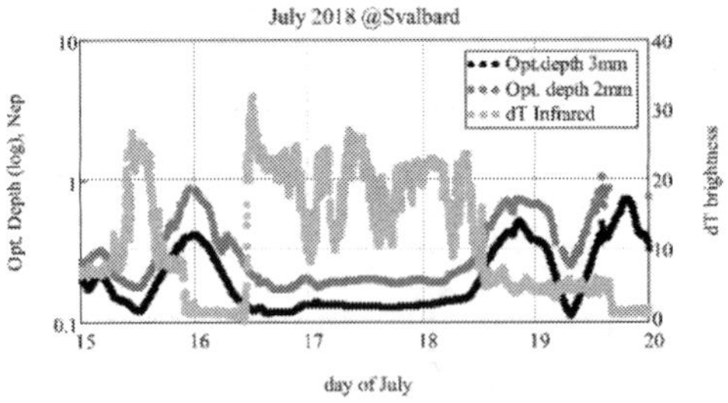

Fig. 1. MM-wave optical depth (log scale) vs temperature difference by IR sensor.

Optical depth and PWV

One of the most important result is an optical depth statistic, which is measured by our "MIAP-2" setup. A cumulative distribution shown below are representing the month-averaged optical depth values. We can characterize the astroclimate on Svalbard as poor for mm-wave observations in Summer period.

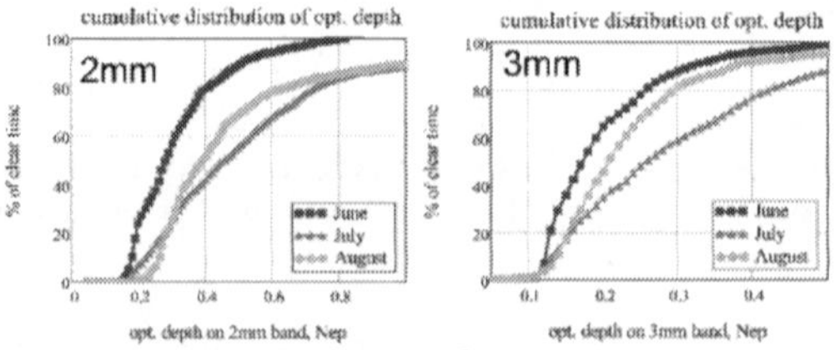

Fig. 2. Cumulative distribution of Optical depth in 2mm and 3mm atmospheric windows in Summer.

Such high values of optical depth are mostly due to high air humidity values and cloudiness. It is usually correlated in time (see pic.3).

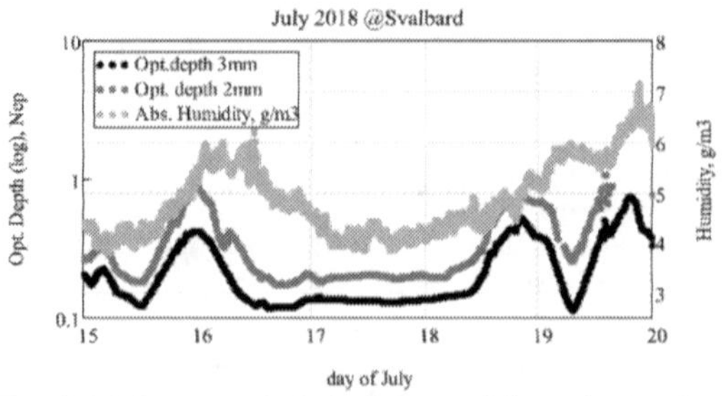

Fig. 3. MM-wave optical depth (log scale) vs Absolute humidity.

Since 2012 we have provided s several expeditions on astroclimate research, so now we can compare these places by month-median values (see pic. 4). Current Svalbard measurements are not promising, but the full-year result remains to be seen.

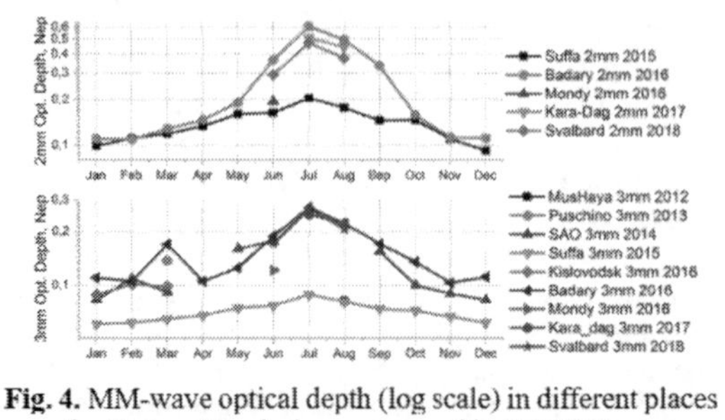

Fig. 4. MM-wave optical depth (log scale) in different places over the Eastern Hemisphere measured by MIAP-2 setup.

Knowing the Optical depth, it is easy to calculate the PWV (mm) [1,4]. The PWV values obtained below (Fig.5) includes a condensed water as well as water vapor by MIAP-2, but not for measurements in [5]. The data given in [5] obtained by Bouger method. It means that precipitable water is measured only in sunny days, and thus, it has a few statistics. It is the reason why these values are about 80% higher than presented in [5].

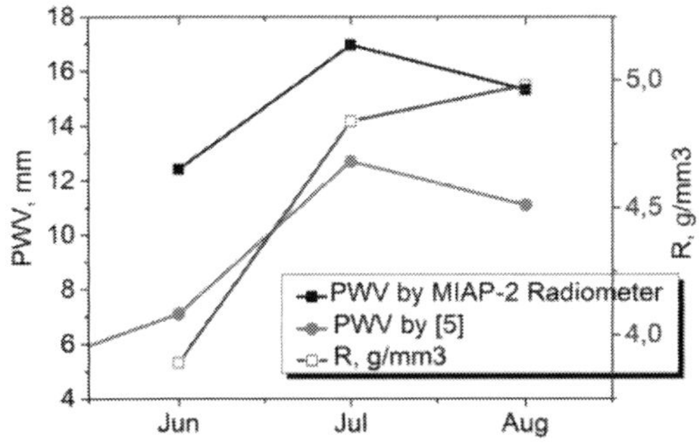

Fig. 5. Precipitable water vapor and relative humidity on Svalbard in Summer.

Using the data obtained by MLX90614 IR sensor in meteostation, it is possible to calculate the cloudiness statistics. The reference points of temperature difference were chosen as: ≤18K for Overcast, from 18 to 27K – Cloudy and upper than 27 - for Clear sky conditions. The obtained results are generally consistent with meteorological observations [6].

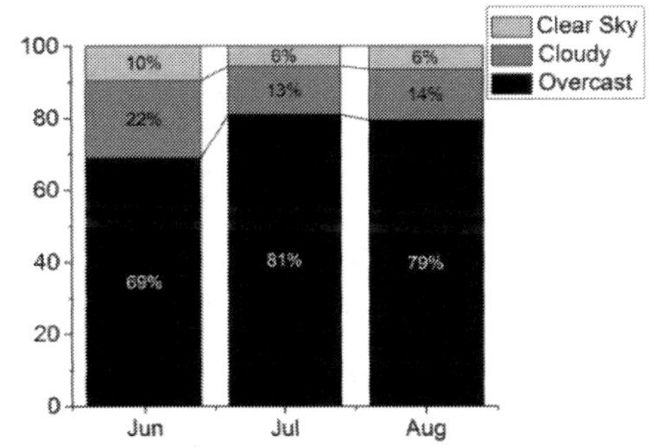

Fig. 6. Cloud statistics by IR sensor.

In **conclusion** we have to characterize Svalbard as non-suitable for mm-wave radioastronomy at least in summer period. The winter period approaching soon and inspires a hope to be much better due to low temperatures. Another important result is a good reliability of equipment: it routinely works for 3 months without any assistance. The makeshift Arduino-based meteostation gives a satisfying result. We will continue measurements in Barentsburg till June 2019 to close the year-cycle.

Acknowledgements

This work has been made with particularly financial support from IEEE APS Eugene F. Knott Memorial Doctoral Research Grant, as well as the scholarship issued by RF President [СП-5103.2018.3.]. The authors would like to grateful Oleg Kutkov and Sergey Nazarov from Krimean Astrophysical Observatory for helpful advice [3] and S.M. Sakerin, D.M. Kabanov from Institute of Atmospheric Optics of RAS for provided data. Special thanks for Barentsburg staff Yu.A. Kvashnin, V.G. Kuleshov and J.D. Mishakova for help in expedition.

References

1. *V. I. Nosov, O. S. Bolshakov et al.* A dual-wave atmosphere transparency radiometer of the millimeter wave range. // Instruments and Experimental Techniques. May 2016, Vol 59, Issue 3, pp 374-380. DOI: 10.1134/S0020441216020111

2. *Nazarov, S. V., Kutkov, O. E.* Infrared cloud sensor at CrAO // Proc. of 47th International student scientific conference, Yekaterinburg. 2018. p.261. ISBN:978-5-7996-2283-1

3. [online] http://olegkutkov.me/2017/08/10/mlx90614-raspberry

4. *G.M. Bubnov, V.F. Vdovin, P.M. Zemlyanukha and I.I. Zinchenko.* A refined method of the data processing for astroclimate measurements in mm-waves. // Proc. of 29th IEEE International Symposium on Space Terahertz Technology, Pasadena, California, USA, March 26-28, 2018.

5. *S.M. Sakerin, D.M. Kabanov et al.* Generalization of Results of Atmospheric Aerosol Optical Depth Measurements on Spitsbergen Archipelago in 2011–2016 // Atmospheric and Oceanic Optics, 2018, Vol. 31, No. 2, pp. 163–170.

6. [online] www.meteoblue.com

Microstructured sapphire shaped crystals for anitiresonant and bandgap terahertz waveguiding

G.M. Katyba[1,2,*], N.V. Chernomyrdin[2,3] K.I. Zaytsev[2,3], and V.N. Kurlov[1,2]

[1] Institute of Solid State Physics of the Russian Academy of Sciences, Chernogolovka 142432, Russia
[2] Bauman Moscow State Technical University, Moscow 105005, Russia
[3] Prokhorov General Physics Institute of the Russian Academy of Sciences, Moscow 119991, Russia
*E-mail: katyba_gm@issp.ac.ru

Terahertz (THz) radiation offers significant scientific and technological applications in many fields [1]; among them: security systems [2,3], medical diagnosis [4–7], non-destructive sensing of materials [8], etc. Nevertheless, the use of THz technologies in these demanding applications is limited by the absence of commercially-available THz waveguides and endoscopic systems, which are capable for the THz wave delivery to hardly acceptable objects with small dispersion and loss.

Recently, several types of THz waveguide have been developed relying on various physical principles. The oldest THz waveguides are based on the hollow-core metal or metal-dielectric tubes [9]. Such waveguides are characterized by relatively low propagation losses, but also feature significant dispersions. In turn, the THz waveguides based on plasmonic structures [10], such as single or dual metal wires, metal plates and ribbons, are convenient in manufacturing and characterized with small dispersion, low propagation and bending losses; however, they are plagued by low coupling efficiency and handling problem, which prevent a practical implementation of these waveguides. Significant progress in the THz waveguiding is associated with numerous developments of flexible polymer microstructured waveguide with different cross-section geometry. These waveguides realize either the anti-resonant reflecting optical waveguiding (ARROW) principle [11], or the Bragg guidance by a photonic crystal (PC) cross-section [12,13]. In such waveguides, the dispersion could be managed and the loses could be minimized by optimizing the geometry of a waveguide cross-section; however, such waveguide remain very sensitive to external perturbations and bending and, thus, are not capable for operation at high temperatures and pressures, since most of the polymers possess relatively low melting temperature and radiation resistance.

An alternative approach would be to use the THz waveguide and fibers, based on the crystalline material, which physical and mechanical characteristics are mainly independent from temperatures/pressures environments and from measurement conditions. In order to answer the challenging problem posed by the THz waveguiding, in our research work, we developed several types of THz waveguides based on the sapphire shaped crystal. These waveguide combines unique physical properties of sapphire (it is transparent in a broad range of electromagnetic spectrum, spanning the ultraviolet, visible and infrared bands; it has high hardness, melting point, thermal conductivity, chemical resistance, tensile strength and thermal shock resistance) and advantages of the edge-defined film-fed growth (EFG) technique (it yields fabrication of the sapphire shaped crystals with a pre-determined cross-section geometry, and an impressive volumetric and surface quality of the shaped crystal) [14-16]. We designed and fabricated two microstructured hollow-core sapphire THz waveguides with different cross-section geometries see Fig. 1. [17,18]. We have combined numerical analysis, using *Lumerical mode solution,* and experimental studies, using the transmission-mode THz time-domain spectroscopy, in order to demonstrate that both of the developed waveguides demonstrate advanced optical performance.

The first sapphire THz waveguide (see Fig. 1 (a)) possesses 7 hollow channels in its cross-section – i.e. 1 core channel and 6 cladding channels [17]. The diameter of each channel is 2.5 mm, while the outer diameter of the waveguide is 12.0 mm. These hollow channels form a hexagonal lattice in the waveguide cross-section with the period of 3.0 mm. Owing to rather low number of the cladding channels in the waveguide cross-section, it realizes mainly the ARROW principle of electromagnetic waveguidance. It yields guidance of THz radiation in a multimode regime with a small dispersion in the frequency range of 1.00 to 1.55 THz and the lowest propagation loss of 2.0 dB/m at 1.45 THz.

The second sapphire THz waveguide (see Fig 1 (b)) possesses the outer diameter of 24.0 mm and much more delicate PC cross-section [18]. This waveguide contains a large hollow core with the diameter of 7.15 mm, as well as 30 channels of the PC cladding, featuring the diameters of 1.6 mm and forming a hexagonal PC lattice with the period of 2.8 mm. For this shaped crystal, we observed an efficient THz waveguidance with a small dispersion in the frequency range of 0.65–1.2 THz and the lowest propagation loss of 0.01–0.03 dB/cm around 1.2 THz. We should mention that the THz radiation propagates throw this waveguide in an effectively two-mode regime; thus, leading to the intermodal interference phenomenon. In Ref. [18], we proposed an analytic model describing this two-mode guidance based on a coherent superposition of the amplitude of the two interfering waves featuring distinct amplitudes and phases. This two-mode guidance can be useful in the intrawaveguide interferometric sensing.

Finally, In Ref. [18], we demonstrated an opportunity for using the two-mode sapphire waveguide from Fig. 1 (b) in a high-temperature THz intrawaveguide interferometry of sodium nitrite ($NaNO_2$) film melts. In our experiment, the sapphire waveguide was used simultaneously as a cuvette for analyte and as a THz waveguide. During the THz wave propagation throw this waveguide, the effect of intermodal beating occurs. The interference pattern behind the output end of the waveguide forms both in the frequency domain and the spatial domain. This interference pattern strongly depends on the refractive index of the analyte placed in the waveguide hollow core. In this way, we can register the phase transitions of the $NaNO_2$ by measuring the interference pattern at different temperatures in the range of 170 to 300°C. The sharpest change in the interference pattern corresponds to the melting temperature of $NaNO_2$ film. The detailed description of this experiment is presented in Ref. [18].

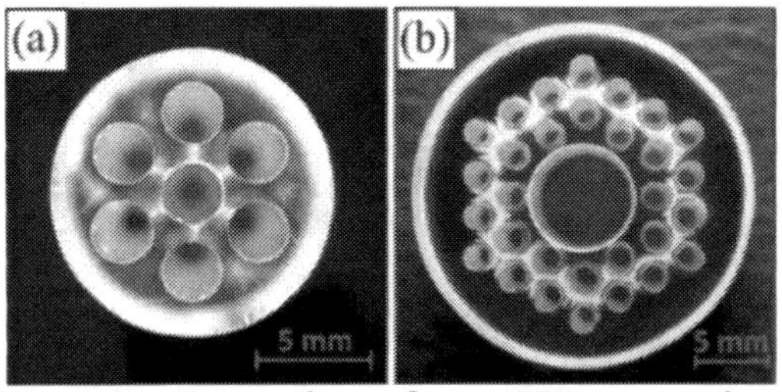

Figure 1. Cross-sections of two EFG-grown microstructured sapphire THz waveguides: (a) an ARROW multimode waveguide with 7 hollow channels [13]; (b) an effectively-two-mode PC waveguide with 31 hollow channels [14].

This work was supported by the Russian Foundation for Basic Research (RFBR), Projects # 17-08-00803 and 17-38-80057.

References

1. *Lee, Y.-S. Principles of Terahertz Science and Technology* // Springer, New York, NY, USA, 2009.

2. *Dolganova, I.N., Zaytsev, K.I., Metelkina, A.A. Karasik, V.E., Yurchenko, S.O.*, A hybrid continuous-wave tehahertz imaging system // Review of Scientific Instruments, 2015, V. 86, No. 11, P. 113704.

3. *Dolganova, I.N., Zaytsev, K.I., Yurchenko, S.O.,Karasik, V.E., Tuchin V.V.*, The Role of Scattering in Quasi-Ordered Structures for Terahertz Imaging: Local Order Can Increase an Image Quality // IEEE Transactions on Terahertz science and Technology, 2018, V. 8, No.4, PP. 403–409.

4. *Zaytsev, K.I., Kudrin, K.G., Karasik, V.E, Reshetov, I.V., Yurchenko, S.O.*, In vivo terahertz spectroscopy of pigmentary skin nevi: Pilot study of non-invasive early diagnosis of dysplasia // Applied Physics Letters, 2015, V. 106, P. 053702.

5. *Zaytsev, K.I., Gavdush, A.A., Chernomyrdin, N.V., Yurchenko, S.O.*, Highly Accurate in Vivo Terahertz Spectroscopy of Healthy Skin: Variation of Refractive Index and Absorption Coefficient Along the Human Body // IEEE Transactions on Terahertz science and Technology, 2015, V. 5, No.5, P. 817–827.

6. *Chernomyrdin, N.V. Kucheryavenko, A.S. Kolontaeva, G.S., Katyba, G.M., Dolganova, I.N., Karalkin, P.A., Ponomarev, D.S., Kurlov, V.N., Reshetov, I.V., Skorobogatiy, M. , Tuchin, V.V., Zaytsev, K.I.,* Reflection-mode continuous-wave 0.15lambda-resolution terahertz solid immersion microscopy of soft biological tissues // Applied Physics Letters, 2018, V.113, No. 11, P. 111102.

7. *Chernomyrdin, V.N., Gavdush; A.A., Beshplav, S.-I.T.,Malakhov, K.M., Kucheryavenko, A.S., Katyba, G.M., Dolganova, I.N., Goryaynov, S.A., Karasik, V.E., Spektor, I.E., Kurlov, V.N., Yurchenko,S.O., Komandin,G.A., Potapov,A.A., Tuchin,V.V., Zaytsev,K.I.,* In vitro terahertz spectroscopy of gelatin-embedded human brain tumors: a pilot study // Proceedings of SPIE, 2018, V. 10716, P. 107160S

8. *Yakovlev, E.V., Zaytsev, K.I., Dolganova, I.N., Yurchenko, S.O.* Non-Destructive Evaluation of Polymer Composite Materials at the Manufacturing Stage Using Terahertz Pulsed Spectroscopy // IEEE Transactions on Terahertz science and Technology, 2015, V.5, No.5, P. 810–816.

9. *Bowden, B., Harrington, J, A ,Mitrohanov O.,* Fabrication of terahertz hollow-glass metallic waveguides with inner dielectric coatings // Journal of Applied Physics, 2008. V. 93, P. 093110.

10. *Wang, K., Mittleman D.M.,* Metal wires for terahertz wave guiding // Nature, 2004, V. 432, P. 376.

11. *Chen, L.-J., Chen, H-.W., Kao, T.-F., Lu, J.-Y., Sun, C.-K.,* Low-loss subwavelength plastic fiber for terahertz waveguiding // Optics Express, 2006, V. 31, No. 3, P. 308–310.

11. *Hassani, A., Dupuis, A. and Skorobogatiy, M.,* Low loss porous terahertz fibers containing multiple subwavelength holes // Applied Physics Letters, 2008, V. 92, P. 071101.

13. *Ma, T., Guerboukha, H., Girard, M., Squires, A., Lewis, R., Skorobogatiy, M.,* 3D Printed Hollow-Core Terahertz Optical Waveguides with Hyperuniform Disordered Dielectric Reflectors // Advanced Optical Materials, 2016, V. 4, No.12, P. 2085.

14. *Antonov, P.I., Kurlov, V.N.,* A review of developments in shaped crystal growth of sapphire by the Stepanov and related techniques // Progress in Crystal Growth and Characterization of Materials, 2002, V. 44, No.2, P. 63–122.

15. *Abrosimov, N.V., Kurlov, V.N., Rossolenko, S.N.,* Automated control of czochralski and shaped crystal growth processes using weighing techniques // Progress in Crystal Growth and Characterization of Materials, 2003, V. 46, No. 1–2, P. 1–57.

16. Katyba, G.M., Zaytsev, K.I., Dolganova, I.N., Shikunova, I.A., Chernomyrdin, N.V., Yurchenko, S.O., Komandin, G.A., Reshotov, I.V., Nesvizhevsky, V.V. and Kurlov, V.N., Sapphire shaped crystal for waveguiding, sensing and exposure applications // Progress in Crystal Growth and Characterization of Materials (accepted, 2018)

17. *Zaytsev K.I., Katyba G.M., Kurlov V.N., Shikunova I.A., Karasik V.E., Yurchenko S.O.,* Terahertz Photonic Crystal Waveguides Based on Sapphire Shaped Crystals // IEEE Transactions on Terahertz science and Technology, 2016. V.6, No. 4, P. 576–582.

18. *Katyba G.M., Zaytsev K.I., Chernomyrdin N.V. Shikunova I.A., Komandin G.A., Anzin V.B., Lebedev S.P., Spektor I.E., Karasik V.E., Yurchenko S.O., Reshetov I.V., Skorobogatiy M., Kurlov V.N. and Skorobogatiy M.,* Sapphire photonic crystal waveguide for terahertz sensing in aggressive environments // Advanced Optics Materials, 2018, DOI: 10.1002/adom.201800573.

Investigation of the influence of the location on the rate of Sub THz space communications channels

I.V. Lesnov[1,2], G. M. Bubnov[1,2], V. F. Vdovin[1,2],

[1]Federal Research Center Institute of Applied Physics of the Russian Academy of Sciences, Nizhniy Novgorod, Russia, ss18f@mail.ru
[2]Nizhny Novgorod State Technical University n.a. R.E. Alekseev, Nizhniy Novgorod, Russia

The need to increase a rate of long-distance space links, in particular, supposed to be developed for communication with the Martian and longer space missions, as well as the space observatory Millimetron, leads to the development of a subterahertz frequency band for this purpose [1]. However, it is known that the terahertz waveband has a noticeable level of absorption in the atmosphere, which appreciably reduces the effect of increasing the channel capacity due to the expansion of the transmission band and the working frequency. Now the main long-distance space communication stations in Russia working in the centimeter range are located in Ussuriysk, Evpatoria, Kalyazin and Bear Lake, while the altitude above the sea level is very small and the absorption of mm and sub-mm waves is significant. Astronomical observations on sub THz waves are usually carried out in high-altitude conditions (over 2500 m), which significantly reduces the effect of the atmosphere on the propagation of these waves. The purpose of this work was to investigate the dependence of the capacity of long-range space communication channels on sub-THz waves from the location of the ground-based communication antenna. To compare the existing sites used for space communications now and prospective ones located in the mountains. Our purpose is to estimate a THz-data link with the best cooled receiver in the "Earth-Space" configuration.

The main parameter influencing the estimation is the data link availability in a specific region and amount of transmitted-received information throughout a year. The location of some stations is shown in Table 1.

Table 1

Object name	Geographic coordinates	Altitude above sea level, m
Suffa Radio Observatory on the Suffa plateau, Uzbekistan.	39°37'N 68°27'E	2335
East Center for Deep Space Communications, Galenki (Ussuriysk), Russia.	44°00'58"N 131°45'25"E	93
Center for Distant Space Communication (Yevpatoria, Crimea)	45°11'20.7"N 33°11'14.5"E	15

When calculating the capacity of the channel, they relied on the method described in [2] with the only difference that the attenuation of the atmosphere decreased with altitude.

At the initial stage, the program reads from the file the altitude profiles of pressure, temperature and humidity obtained by meteorological observations. These data are available at all major weather stations and are posted on the Internet two or four times a day [3]. Tabular data from the file is subjected to piecewise linear interpolation, are given in mutual correspondence of dimensions, and the measurement limits are limited by the altitude of the weather station above sea level h_min and the altitude of the probe h_max (usually 20-30km). The main stage of the program is integration of attenuation coefficients in oxygen and water vapor in height at a given frequency:

$$\tau_{oxygen}(f) = \int_{h\_min}^{h\_max} A_{oxygen}(f, P(h), W(h), T(h))dh$$

$$\tau_{water}(f) = \int_{h\_min}^{h\_max} A_{water}(f, P(h), W(h), T(h))dh$$

where: τ_{oxygen} integral attenuation in oxygen at the zenith, Np τ_{water} – integral attenuation in water vapor at the zenith, Np *h_min u h_max* – range of probe height, km A_{oxygen} – stationary absorption in oxygen, Np/km A_{water} – absorption in stationary conditions in water vapor, Np/km *f* – frequency, GHz *P(h)* – altitude pressure profile, mbar *W(h)* – altitude humidity profile, g/m^3 *T(h)* – altitude profile of temperature, K *h* – height, km.

The result is a spectrum of integral attenuation in the atmosphere. Substituting the center frequencies, the integral weakening of the corresponding transparency windows is calculated. The functions *Aoxygen* (f, P, W, T) and *Awater* (f, P, W, T) are internal functions of the MPM Liebe basic program and determine the linear absorption (He / km) under steady-state conditions, i.e. on a horizontal track. By integrating them with the corresponding profiles, we obtain an integral attenuation on the vertical path (He). The total absorption is the arithmetic sum of the absorptions in oxygen and in water vapor, and the conversion of the quantities in neperes into decibels is carried out by multiplying by a constant of 8.68.

Changing the height h_min in the integrals, we calculate the integral attenuation from the corresponding height, i.e. if the ground station were located at an altitude h_min taking into account the current atmospheric profile.

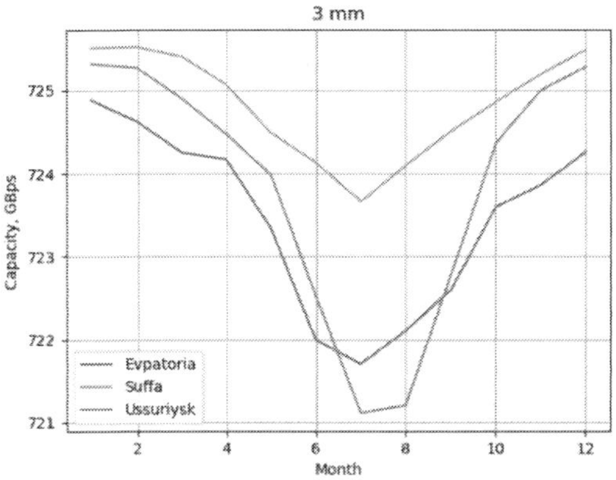

Fig. 1. Estimated THz-link budget for 3mm wave length during a year with cryogenically cooled receiver

Figure 1 showing the variation of the channel capacity for 3mm wavelengths. The calculation was carried out for the following parameters: the width of the window-band is 50GHz, the physical temperature of the receiver is 4K, the equivalent receiver noise temperature is 100K [4], the direction of transmission is zenith.

The total amount of information available for transmission for the year will be: Evpatoria $22502*10^{15}$bytes, Ussuriysk $22514.670*10^{15}$bytes, Suffa 22545 10^{15}bytes. Expectedly there is a failure in the summer months due to increased humidity, but it is not dramatic in comparison to the three locations. In the annual cycle, the difference is also less than 1%. Certainly when the space port of the channel is located far from the zenith affect of the atmosphere increase. For shorter SubTHz atmospheric windows of transparency (2, 1.3 and 0.8 mm) too. The work was supported by the IAP RAS state program (project No. 0035-2014-0021)

References

1. *V. F. Vdovin, V. G. Grachev, S. Yu. Dryagin, A. I. Eliseev, R. K. Kamaletdinov, D. V. Korotaev, I. V. Lesnov, M. A. Mansfeld, E. L. Pevzner, V. G. Perminov, A. M. Pilipenko, B. D. Sapozhnikov†, and V. P. Saurin* Cryogenically Cooled Low-Noise Amplifier for Radio-Astronomical Observations and Centimeter-Wave Deep-Space Communications Systems// Astrophysical Bulletin Vol. 71, No. 1, 2016

2. *G.M.* Bubnov, *I.V. Lesnov and V.F. Vdovin* Data rates of SubTHz wireless telecommunication channels // EPJ Web of Conferences 2017. 149, №02012 p. 2

3. *H. Liebe, P. Rosenkranz and G. Hufford* Atmospheric 60-GHz oxygen spectrum: New laboratory measurments and line parameters // J. Quant. Spectrosc. Radiat. Transfer 1992, vol. 48, no. 5, pp. 629-43,

4. *Koshelets, V., Birk, M., Boersma, D., et al.* Integrated Submm Wave Receiver: Development and Applications in the book "Nanoscience Frontiers - Funda-mentals of Superconducting Electronics", Springer Se-ries: Nanoscience and Technology 35372, pp. 263-296, Editor: Anatolie Sidorenko, 2011

Detection and Identification of a Substance with an Inhomogeneous Surface Using the Effective Time-Dependent THz Spectroscopy Method and Emission Frequency Up-Conversion

Vyacheslav A. Trofimov[1,2], Svetlana A. Varentsova[2], Yongqiang Yang[1]

[1]South China University of Technology, Guangzhou, China, vatro@cs.msu.ru
[2]M.V. Lomonosov Moscow State University, Moscow, Russian Federation

Currently, the THz radiation is actively applied for solving security and anti-terrorism problems [1], [2]. In most THz TDS setups, the substance identification occurs based on comparison of the absorption frequencies of a substance under investigation with a set of known absorption frequencies from a database (we call it as the standard THz TDS method). However, this technology has certain limitations.

It is well-known that surface roughness can lead to scattering and modulating of the spectrum, which complicates the identification [3], [4]. To overcome this problem, in [3] it was shown that by summing and averaging multiple measurements over a sample area the scattering effect can be effectively decreased. In [4] the authors proposed for the same purpose to increase the number of viewing angles. However, they can reliably identify only one absorption frequency with maximal spectral amplitude.

Unlike the methods mentioned above, in the discussed method we use only one THz signal reflected from the substance with inhomogeneous surface but measured in the long time interval duration 180-200 ps. This duration allows registering not only the main reflected THz pulse, but also several sub-pulses following it, which are due to the reflectance from the inner surfaces of the sample. These sub-pulses also contain information about spectral characteristics of the substance and can be used for the identification. In addition, we do not apply averaging of measurements over viewing angles and the sample area in order to reduce the scattering effects.

As an example of identification, we use the THz signal duration 180 ps reflected from the PWM C4 explosive (90 % RDX, 10 % plasticizer) with a rough surface grit 40 (signal PWM_40 for brevity). The signal contains the pronounced main pulse reflected from the outer surface of the sample, the first sub-pulse reflected from the inner surface, and the sub-pulses with significantly less amplitude due to multiple reflections from the inner surfaces. The Fourier spectrum of the PWM_40 main pulse is shown in Fig. 1 (a) together with that of the smooth PWM for comparison.

The spectral properties of the PWM_40 main pulse are distorted by the influence of the inhomogeneous surface (a) so that the standard THz TDS method is inefficient for identification. In the PWM_40 spectrum (a) there is a single minimum at the frequency ν=0.88 THz in the frequency range ν=[0, 1.8] THz, which is close to RDX absorption frequency ν=0.82 THz. In the frequency range ν=[1.8, 4.0] THz there are minima at the frequencies ν= 1.92, 2.24, 3.0 THz , which are close or equal to the RDX absorption frequencies ν= 1.95, 2.19, 3.0 THz [2].

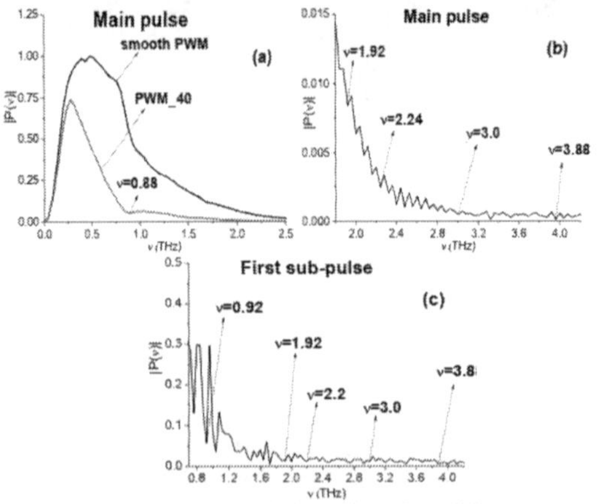

Fig. 1. Fourier spectrum of the main pulse of the PWM_40 and smooth PWM signals in the frequency ranges ν=[0.0, 2.5] THz (a), [1.8, 4.2] THz (b), spectrum of the PWM_40 first sub-pulse in the frequency range ν=[0.8, 4.2] THz (c).

The spectrum of the PWM_40 first sub-pulse contains minima at the frequencies ν= 0.92, 1.92, 2.2. 3.0 THz (c), that are close to the absorption frequencies of explosive RDX ν=0.82, 1.95, 2.19, 3.0 THz ([2]). And the spectrum of the remote part of the PWM_40 signal corresponding to the time interval t=[70, 170] ps, contains minima at the frequencies ν = 0.9, 1.96, 2.2, 3.01 THz, which are close to RDX absorption frequencies mentioned above (not shown). Therefore, the sub-pulses also contain information about the substance spectral properties and can be used for identification.

In the current report, the identification is based on the method of spectral dynamics analysis (SDA-method) together with several integral correlation criteria (ICC) [5]. We compare the dynamics of spectral intensity of a substance under analysis with the corresponding dynamics of a standard substance from database at chosen frequencies. To increase the reliability and effectiveness of the identification, we propose to use several ICC's simultaneously in different time intervals, which contain both the main pulse of the reflected THz signal and the following sub-pulses.

In Fig. 2 the evolution of the ICC $CW1_{p,P}$ [5] is calculated for the frequencies ν=0.82 THz (a), (c) and

ν =3.0 THz (b), (d) in the time intervals, which contain the main pulse (a), (b) and the first sub-pulse (c), (d) of the PWM_40 signal. In all cases (a)-(d) the frequencies ν=0.82, 3.0 THz are detected as RDX absorption frequencies. The same result is valid for the time interval t=[70, 170] ps, containing two other sub-pulses with less amplitude. The ICC $C_{p,P}$ [5] confirms these results.

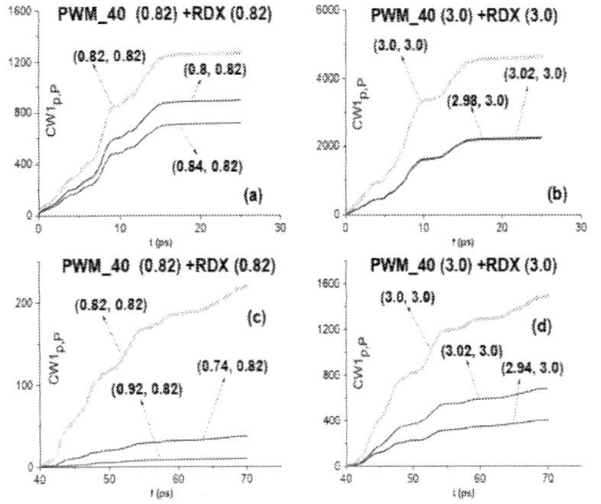

Fig. 2. ICC $CW1_{p,P}(t_n)$ detecting frequencies ν=0.82 THz (a), (c), 3.0 THz (b), (d) in the time intervals t=[0, 30] ps (a), (b), [40, 70] ps (c), (d).

Another promising approach for the substance detection and identification is related to the frequency up-conversion based on a broadband THz pulse. The THz pulse induces high energy level excitation of molecules due to essential non-stationary medium response. This leads to the substance emission at frequencies, which are absent in the incident pulse spectrum. Earlier the frequency up-conversion and the appearance of the doubled current frequency in the THz signals transmitted through or reflected from the medium without and with covering were demonstrated using a mathematical modelling in [6]. The mathematical model is based on the 1D Maxwell-Bloch equations and density matrix formalism. In the current report, we investigate the possible manifestation of the frequency-up conversion in the THz signal PWM_40 reflected from the substance with a rough surface.

Note that the spectra of the PWM_40 main pulse Fig. 1 (b) and the first sub-pulse Fig. 1 (c) contain minima at the frequency ν= 3.88 THz, which is close to the doubled frequency ν= 1.92·2 THz. The same is observed in the PWM_40 spectrum corresponding to the time interval t=[70, 170] ps.

In Fig. 3 the evolution of the ICC $CW1_{p,P}$ calculated for the frequency pair ν=(3.88, 1.92) THz is shown in the time intervals containing the PWM_40 main pulse (a) and the first sub-pulse (b). In both cases, the frequency ν = 3.88 THz is detected as the doubled RDX frequency in the reflected signal PWM_40. The ICC $C_{p,P}$ [5] also confirms these results.

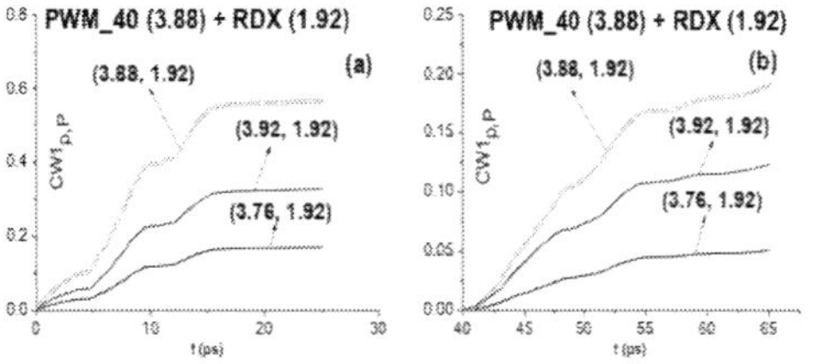

Fig. 3. ICC $CW1_{p,P}$ calculated for the PWM_40 signal at the doubled frequency ν=3.88 THz and the standard signal RDX_Air at the frequency ν=1.92 THz in the time intervals t=[0,25] ps (a), [40, 65] ps (b).

We propose an effective time-dependent THz spectroscopy method, which allows detecting and identifying a substance with inhomogeneous surface using only one long-duration reflected THz signal without averaging of the measured THz signals over the viewing angles and scanning over the surface area. For successful analysis and identification, the registered THz signal must contain both the main reflected THz pulse and several sub-pulses. This feature of the method significantly increases the signal processing speed and allows us to use it in real time.

The possibility for the substance detection and identification using the doubled current frequency is shown with the help of the SDA-method and integral correlation criteria. This result is important for practice, because the frequencies belonging to the high frequency range can be used for the substance detection and identification even in the case when the lower substance absorption frequencies are suppressed by covering or noise influence.

The THz signals reflected from the PWM C4 sample were measured in the Military University of Technology (Warsaw, Poland); THz signal transmitted through the RDX sample - in the Center for Terahertz Research, Rensselaer Polytechnic Institute (NY, USA).

References

1. *Leahy-Hoppa, M. R, Fitch, M. J., Zheng, X., et all.* Wideband terahertz spectroscopy of explosives // Chem. Phys. Lett. 2007. V. 434. P. 227–230.

2. *Liu, H. B., Zhong, H., Karpowicz, N., et all.* THz spectroscopy and imaging for defense and security applications // *Proc. IEEE*. 2007. V. 95, No. 8, P. 1514-1527.

3. *Shen, Y. C., Taday, P. F., Pepper, M.* Elimination of scattering effects in spectral measurement of granulated materials using terahertz pulsed spectroscopy // Appl. Phys. Lett. 2008. V. 92, No. 5. Paper 051103.

4. *Henry, S., Kniffin, G., Schecklman, S., et all.* Measurement and modeling of rough surface effects on terahertz spectroscopy and imaging // Proc. SPIE. 2010. V. 7601. Paper 760108.

5. *Trofimov, V. A., Varentsova, S. A.* Detection and identification of drugs under real conditions by using noisy terahertz broadband pulse // Appl. Opt. 2016. V. 55, No. 33. P. 9605–9618.

6. *Trofimov, V. A., Zakharova, I. G., Zagursky, D. Yu., Varentsova, S. A.* New approach for detection and identification of substances using THz TDS // Proc. SPIE. 2017. V. 10441. Paper 1044107.

EPJ Web of Conferences **195**, 09006 (2018)
TERA-2018 https://doi.org/10.1051/epjconf/201819509006

Using a Gyrotron as a Source of Modulated Radiation for Data Transmission Systems in the Terahertz Range

A. I. Tsvetkov, A. P. Fokin, A. S. Sedov, M. Yu. Glyavin

Institute of Applied Physics of the Russian Academy of Sciences, Nizhny Novgorod, Russia, tsvetkov@ipfran.ru

The report describes results of preliminary experiments on using low THz range gyrotron (263 GHz) with medium output power (up to 1 kW in CW mode) as a modulated radiation source for data transmission. It was demonstrated up to 1.5 Mbit/s data rate. Estimates show that at relatively low costs it is possible to build a gyrotron-based system with a speed of about 1 Gbit/s.

The scientific community is actively discussing the possibility of promotion of data transmission systems in the subterahertz and terahertz frequency range [1−5]. Data transmission in these bands is attractive for a number of reasons including noise immunity and the complexity of signal interception. It should be taken into account that communication links at such high frequencies suffer from attenuation due to polar molecules (mainly H_2O and O_2) in the atmosphere in addition to the large free-space loss. A suitable transmission window of atmospheric transparency can be found at 200 – 300 GHz [6]. Another complication in the implementation of low THz communication systems is the lack of sufficiently powerful radiation sources with flexible capabilities to control the output radiation parameters in these ranges. The experiments described below demonstrate the possibility of using a subterahertz gyrotron for the data transmission.

There are various methods of control of the output power and frequency of electron devices. One of the most used approaches for the gyrotrons is the variation of the potential of one of the electrodes of the magnetron-injection gun (MIG). The most economical way to quickly control the output parameters is to change the voltage at the isolated anode. This fact is caused by small capacity of anode relative to other electrodes and low anode current, so there is no need in complex and expensive power supplies. The change of anode voltage does not alter the electron energy, but only its pitch-factor (ratio of it orbital to longitudinal velocity). Such change of pitch-factor leads to change of the electron beam complex susceptibility and, hence the change of the interaction condition and, finally, frequency and amplitude of the oscillations. To control the gyrotrons output parameters the fast anode voltage control unit was developed that allows gyrotron anode voltage to be modulated in range of 1 kV with slopes better than 1 kV/μs. The active element of the control unit is the tetrode; the voltage drop is proportional to the external control signal. It was demonstrated up to 200 kHz modulation frequency bandwidth in recent experiments.

In the latest experiments performed in the Institute of Applied Physics of the Russian Academy of Sciences (IAP RAS), the gyrotron with an output frequency $f \approx 0.26$ THz was equipped with the PLL system for its frequency stabilization, which was implemented on the basis of the fast variation of the anode potential. Such an approach made it possible to demonstrate the width Δf of a gyrotron radiation spectrum of about 1 Hz, which corresponds to the relative value of $\Delta f / f \approx 3 \cdot 10^{-12}$[7]. The created control system was then used for the experiments described below demonstrating the principle possibility of using a lower terahertz range gyrotron for the information transmission.

This paper reports on experiments in which it was first realized the transmission of an AM modulated radio signal for several meters using a 263 GHz gyrotron [8] radiation with up to 1 kW power as a carrier.

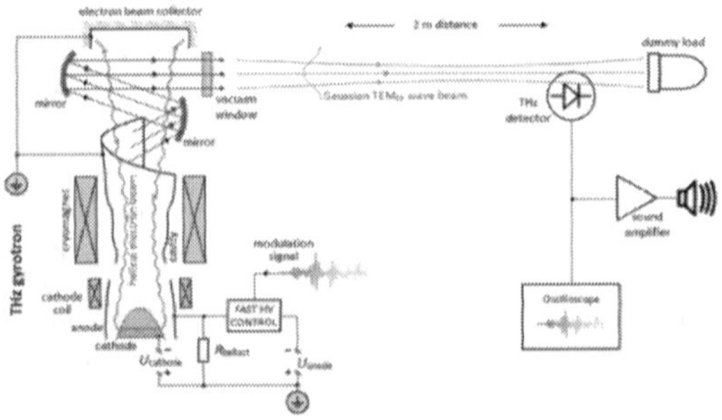

Fig. 1. Diagram of the experimental setup

An ultrafast terahertz (50 GHz – 0.7 THz) detector developed by TeraSense company [9] was used as a radiation receiver and for detecting the AM signal. The detector was installed almost on the path of the gyrotron radiation beam propagation at a distance of several meters from the output vacuum window. The voltage at the anode of the gyrotron was recorded with a NorthStar PVM-5 [10] fast high-voltage probe. Demonstration experiments were carried out at the gyrotron output power level of about 50 W, which in no way limits their generality. Similarly, the parameters of this gyrotron can be controlled at any output power level up to 1 kW.

An audio recorder that provides a level of a modulation signal in the 0 – 10 V range was used in the first experiment. At the same time, the terahertz detector signal was applied to both an oscilloscope and an audio frequency amplifier, followed by output to a loudspeaker. Successful analog transmission of several audio recordings was demonstrated. Fig. 2 shows examples of oscillograms of the corresponding sound signals.

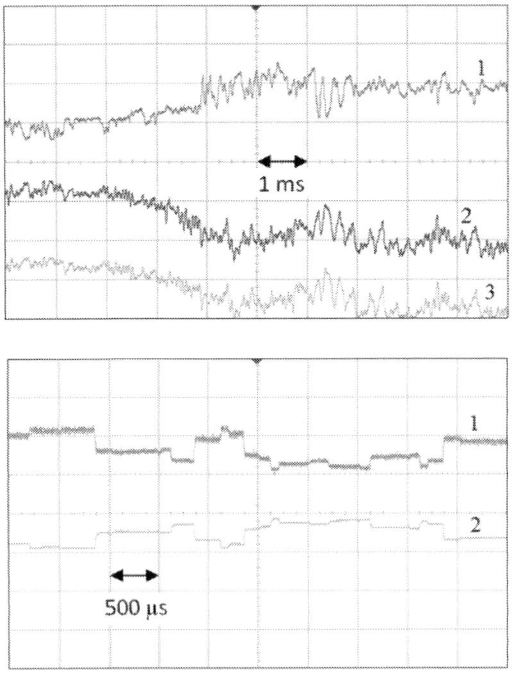

Fig. 2. Top – oscillogramms of the detector signal (1), anode voltage (2), modulation signal of the audio recorder (3); bottom – pseudo-random bit sequence: control (modulating) signal of 33521B generator (1), terahertz detector signal (2).

One can see in Fig. 2 (top) that for some values of the amplitude of the carrier signal the oscilloscope trace obtained from the terahertz detector is distorted with respect to the modulating signal. This is due to the nonlinearity of the slope of the regulating element used in the anode voltage control unit (this effect is especially noticeable near the 0 V control signal) and can in principle be compensated by the choice of a suitable control signal bias.

In the second experiment, the arbitrary waveform generator Agilent 33521B has been used as a source of the control (modulating) signal. It was demonstrated the transmission of a digital signal - a pseudo-random bit sequence (an example of an oscillogram is shown in Fig. 2). In this case, just as in the previous experiment, simple amplitude modulation of the signal was used. Up to a data rate of about 1.5 Mbit/s, the logical levels of the received (detected) signal were resolvable.

The authors believe that the achieved data transfer rate is far from the principal limit and can be increased by decreasing the time constant of the anode power circuit, applying more complex methods for gyrotron output parameters control, using advanced digital signal modulation schemes, increasing the power in the data transmission channel. Simple estimates using the Shannon-Hartley theorem show that it is possible to achieve at least 1 Gbps of channel capacity (under practically the same experimental conditions) by increasing bandwidth of the channel with an improved anode power supply. It is clear that one of the next steps should be a study of the transmission system with respect to various modulation schemes and different symbol rates.

The work was partially supported by the Russian Federation President Grant No. MK-3452.2017.8.

References

1. J. Federici and L. Moeller, "Review of terahertz and subterahertz wireless communications," Journal of Applied Physics, 107, 111101 (2010).

2. T. Kleine-Ostmann and T. Nagatsuma, "A review on terahertz communications research," J. Infrared, Millimeter, and Terahertz Waves, 32,143-171 (2011).

3. H.-J. Song and T. Nagatsuma, "Present and future of terahertz communications," IEEE Trans. Terahertz Science and Technology, 1, 256-263 (2011).

4. T. Nagatsuma et al., "Terahertz wireless communications based on photonics technologies," Opt. Express, 21, 23736-23747 (2013).

5. T. Kürner and S. Priebe, "Towards THz communications-status in research, standardization and regulation," J. Infrared, Millimeter, and Terahertz Waves, 35, 53-62

6. V. Petrov, A. Pyattaev, D. Moltchanov and Y. Koucheryavy, "Terahertz band communications: Applications, research challenges, and standardization activities," in 8th International Congress on Ultra Modern Telecommunications and Control Systems and Workshops (ICUMT), Lisbon, pp. 183-190, Oct. 2016.

7. A. Fokin et al. High-power sub-terahertz source with a record frequency stability at up to 1 Hz // Scientific Reports, 8, 4317 (2018)

8. *M. Glyavin et al.* Experimental tests of 263 GHz gyrotron for spectroscopy applications and diagnostic of various media // Review of Scientific Instruments, 86, No. 5, 054705 (2015)

9. TeraSense: «Ultrafast Terahertz Detectors». URL: http://terasense.com/products/detectors/.

10. North Star High Voltage: «High Voltage Probes». URL: http://www.highvoltageprobes.com/high-voltage-probes

EPJ Web of Conferences **195**, 10001 (2018) https://doi.org/10.1051/epjconf/201819510001
TERA-2018

Application of high resolution subTHz spectroscopy methods for analysing the content of grain odors

V.A. Anfertev[1], V. L. Vaks[1], M.B. Chernyaeva[1], E.G. Domracheva[1], A.A. Gavrilova[2], E.V. Dabakhova[2]

[1]IPM RAS, Nizhny Novgorod, Russia, anfertev@ipmras.ru
[2]NNSAA, Nizhny Novgorod, Russia

Terahertz high resolution transient spectroscopy on nonstationary effects (free damping polarization realizing in two modes – phase switching of radiation influenced on gas and fast frequency sweeping) and spectrometers, based on these methods (a wideband spectrometer, a marker detector for registering certain gases, a two-channel spectrometer with two independent radiation sources and one detector) are very interesting for analysis of multicomponent gas mixtures.[1] The possibility of use of terahertz high resolution transient spectroscopy in such agricultural (diagnostics of degree of corn mycosis based on analysis of corn odors), medical applications (veterinary diagnostics of diseases and pathology states of animals based on analysis of exhaled breath and vapors of biological liquids (urine, saliva)); monitoring the quality of foodstuff (meat, fish, poultry) based on analysis of its odors shows considerable promise.

In this work possibility of use of terahertz high resolution transient spectroscopy in such agricultural applications as analysis of corn odors for developing the methods of quality control and diagnostics is considered. The method was used to study "odors" of grain of new unique varieties: spring wheat "Sitara", winter wheat "Nemchinivskaya-17", barley "Binom" and oats "Arkhat". Researches of "odors" of different variety of grain were provided with subTHz transient backward wave oscillator based spectrometer in frequency range of 115-160 GHz which developed in IPM RAS. THz gas transient spectrometers based on transient gas absorption effects under radiation, which comes into resonance with molecular two-level quantum system in a shorter time than in the times of relaxation processes and transition radiation, when the radiation leaves resonance in a shorter time than in the relaxation times. The spectrometer has 1kHz frequency resolution and high frequency stability of radiation (1e-8). Quartz cell length of 1 meter was used as measurement volume. Grain (whole or ground) was placed in a flask, which was connected to a measuring gas cell. The cell was pumped to operating pressure with a vacuum pumping station «Pfeiffer Hi-Cube Eco 80» to avoid unwanted gases and gas collisional broadening. For dehydration of the grain sample, vacuum drying with heating of the sample was carried out for some time. According to the developed measurement procedure, it is considered that to exclude the effect of water vapor on detecting trace concentrations of the mixture components, the pressure in the cell should be $1\times10e\text{-}3 \div 5\times10e\text{-}3$ mbar. After that, the heating of the flask containing the sample of grain continues.

The preliminary results on studying the grain odors composition of wheats, barley (Binom) and oats (Arkhat) are presented in the Table 1.

Grain	Substances
Winter wheat "Nemchinovskaya-17"	Glycolaldehyde (HCOCH2OH)
	Formic acid (HCOOH)
Summer wheat (Sitara)	Formamide (NH2CHO)
barley (Binom)	Acetaldehyde (CH3CHO)
oats (Arkhat)	HCOCH2OH
	Ga-n-C3H7OH
	Acetaldehyde (CH3CHO)
	CH2(OH)-CH2-CH2(OH) Propane-1,3-diol
	CH2(OH)-CH(OH)-CH3 Propane-1,2-diol

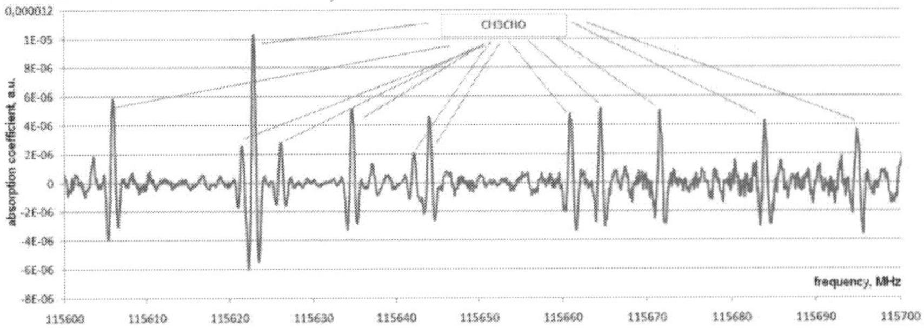

Fig. 1. The record of subTHz absorption spectrum in odor of barley ("Binom") sample after heating

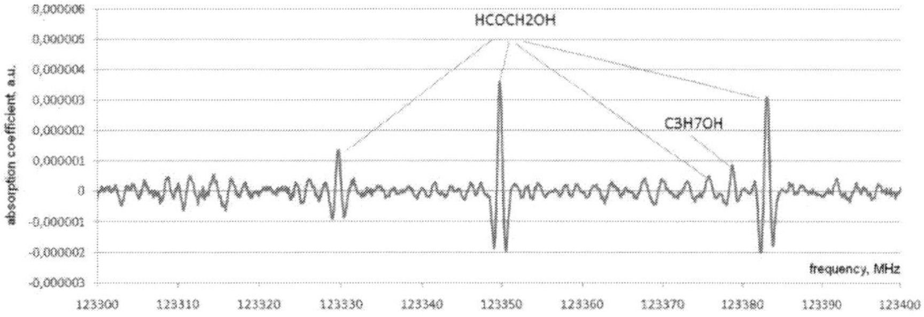

Fig. 2. The record of subTHz absorption spectrum in odor of oats ("Arkhat") sample after heating

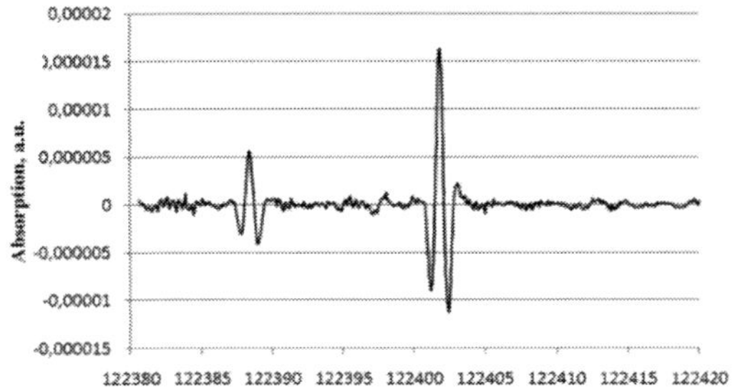

Fig. 3 The record of subTHz absorption spectrum of formamid in "odor" of summer wheat "Sitara"

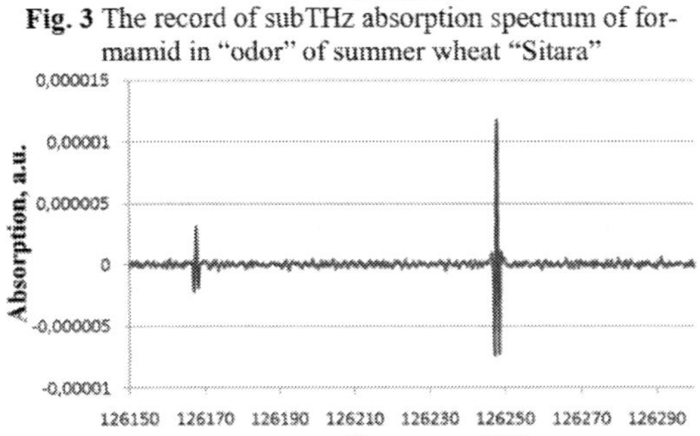

Fig. 4 The record of subTHz absorption spectrum of formamid in "odor" of summer wheat "Sitara"

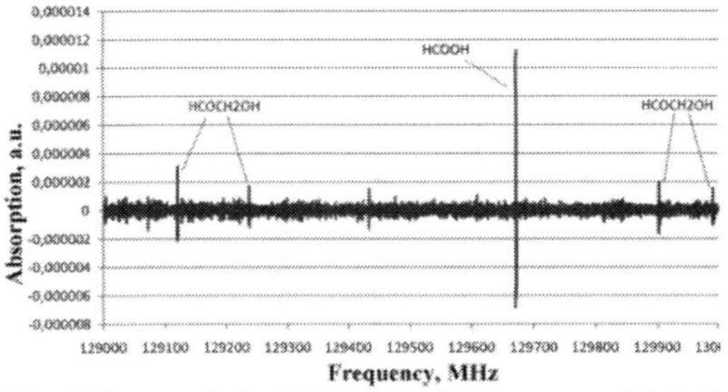

Fig. 5 The record of subTHz absorption spectrum of "odor" of winter wheat "Nemchinovskaya-17"

Fig. 3,4 shows the spectrum of formamide (NH2CHO) at 122.388 GHz 122.402 GHz, 126.168 GHz and 126.248 GHz [2,3], recorded when heated, when the grain sample begins to turn black, in the absorption spectra of the "odor" of "Sitara" wheat. The presence of formamide may be a consequence of the defeat of the grain by necrotrophic parasites. During evolution process, they developed a system of protection against toxic substances produced by the dying of infected cells, while one of the products of the transformation of toxic compounds is formamide.

The recording of the spectral region in the "odor" of winter wheat seeds of the variety "Nemchinovskaya-17" is shown in Fig. 5. A series of glycolaldehyde absorption lines (HCOCH2OH) at frequencies of 129120.22 MHz, 129.236 GHz, 129.902 GHz, 129.990 GHz [2,3] and formic acid (HCOOH) at the central frequency of 129.671 GHz have been detected. The appearance of glycolaldehyde, apparently, is characteristic of a multicomponent gas mixture that appears during the heat treatment of plants or the self-heating of grain during storage. Thus, the appearance of glycolaldehyde was detected by the pyrolysis of cellulose [4], which is one of the main components of the cell walls of plants. The appearance of formic acid in the gas mixture of a sample of wheat during heat treatment requires additional studies.

The applicability of the method of subTHz transient spectroscopy for studying of grain of various cereal crops for the diagnosis of their condition and the detection of fungal diseases was shown.

This work was carried out in the framework of state targets N 0035-2014-0206 and financially supported by the Russian Foundation for Basic Research (grant N 18-42-520050 r_a_povoljie, N 17-00-00184 KOMFI, N18-52-16017)

References

1. Vaks, V.L., Domracheva, E.G., Chernyaeva, M.B., Pripolzin, S.I., Revin, L.S., Tretyakov. I.V., Anfertev, V.A., Yablokov, A.A., Methods and approaches of high resolution spectroscopy for analytical applications //Opt Quant Electron (2017) 49: 239

2.Pickett H. M. Submillimeter, Millimeter, and Microwave Spectral Line Catalog. JPL Molecular Spectroscopy// H. M. Pickett, E. A. Cohen, B. J. Drouin, J. C. Pearson California Institute of Technology. http://spec.jpl.nasa.gov/ftp/pub/catalog/catform.html, 2003

3. C. P. Endres, S. Schlemmer, P. Schilke, J. Stutzki, and H. S. P. Müller, The Cologne Database for Molecular Spectroscopy, CDMS, in the Virtual Atomic and Molecular Data Centre, VAMDC//J. Mol. Spectrosc. 327, 95–104 (2016)

4. Cole, Daniel Paul, "High resolution mass spectrometry for molecular characterization of pyrolysis products and kinetics" (2015). Graduate Theses and Dissertations. 14342. http://lib.dr.iastate.edu/etd/14342

Definition of thresholds of the heating effects of THz radiation on cancer cells

A.A. Avseenko[1], M. K. Khodzitsky[2]

[1]ITMO University, Saint-Petersburg, Russia, arishkaavseenko@niuitmo.ru
[2]ITMO University, Saint-Petersburg, Russia

The main objective of the investigation is to determine the threshold parameters of THz radiation for its use in non-invasive therapy and diagnosis of cancer. This paper presents a model that allows identifying and calculating the thermal effects caused by THz radiation. The possibility of using these effects for the diagnosis and treatment of cancer is also being considered.

Cancer is a tumor formed by epithelial tissues that covers almost all human organs. Modern statistics show that every year cancer is registered in 10 million people around the world. In Russia, cancer mortality ranks second after heart and vascular diseases. (see [1]) Nowadays, medicine has achieved good success in the field of oncology, but the problem is that there is still no effective cure for cancer cells, and timely diagnosis in the early stages is a huge range of different tests. Therefore, it is important to create an effective technique aimed at diagnosing and combating cancer.

The scientific novelty of the study is determined by the fact that it examines the currently little-known thermal effects (see [2,3]) that occur when exposed to THz radiation (see [4]).

This model performs calculations (by solving the equation of thermal conductivity using the Debye model) for a cylindrical sample of water (see [5]), with the radiation frequency from 0 THz to 25 and temperatures from 0 C° up to 100 C°. The program interface includes several input parameters such as incident radiation parameters, sample parameters, coordinates and environmental conditions. The experimental values are presented in the table below:

Input parameter	Value
Radial coordinate	[-2:2]
Input power, mWt	1
Beam radius, mm	5
Initial temperature, K	295
Frequency, THz	1
Sample radius, mm	50
Sample thickness, mm	5

Figures 1 and 2 illustrate the results. It is obvious that with these values of the input parameters, the temperature increase reaches from 0.035 degrees in the center of the sample to 0.02 degrees at the periphery.

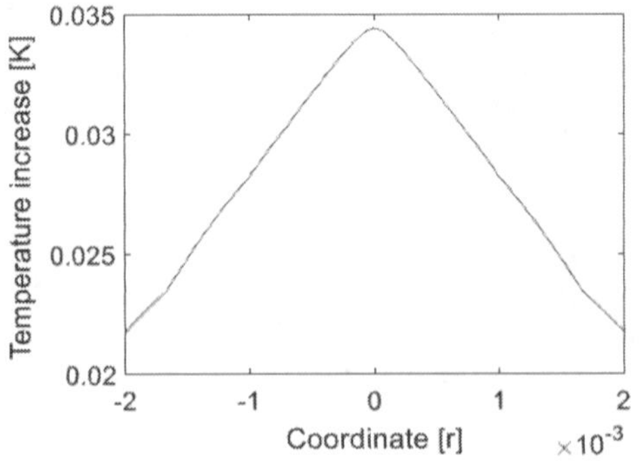

Fig. 1. The temperature increase in the sample along the radial coordinate

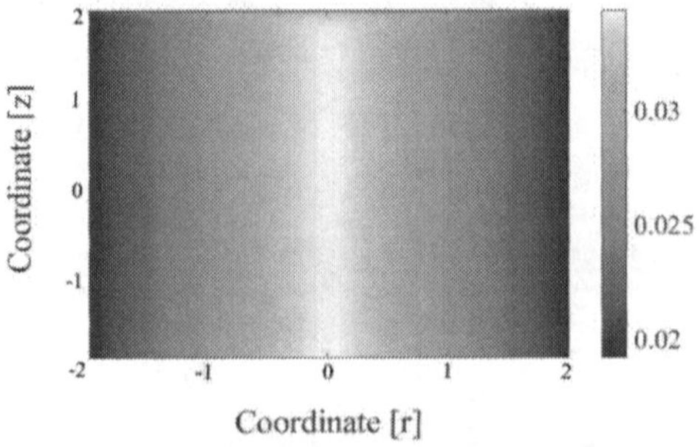

Fig. 2. The temperature increase in the sample caused by exposure to THz radiation

In addition, this model assumes the possibility of using the results of real experiments. Firstly, in the experiment, the cultivation of a line of cancer cells is carried out. Then their optical parameters are determined on the pulse THz spectrometer, such us the dispersion of the complex permittivity and the refractive index. The obtained parameters are loaded into the program. The results may lead to the conclusions about the value of the input parameters of radiation that can cause apoptosis of cancer cells and apply THz radiation for the diagnosis of diseases.

References

1. *Maksimova T. M., Belov V. B.* The morbidity and mortality of malignant neoplasms in Russia and some foreign countries // Problems of social hygiene, healthcare and history of medicine. 2012. No. 1. P.10.
2. *Borovkova, M., Serebriakova, M., Fedorov, V., Sedykh, E., Vaks, V., Lichutin, A., Salnikova, A., (...),*

Khodzitsky, M Investigation of terahertz radiation influence on rat glial cells // Biomedical Optics Express. 2017. Vol. 8. Issue 1. 2017. No. 276461. P. 273-280.

3. *Geyko, I.A., Smolyanskaya, O.A., Sulatsky, M.I., Parakhuda, S.E., Sedykh, E.A., Odlyanitskiy, E.L., Khodzitsky, M.K., Zabolotniy, A.G. Impact* of terahertz radiation on the epithelialization rate of scarified cornea // Progress in Biomedical Optics and Imaging - Proceedings of SPIE. 2015. Vol. 9542. No. 95420E.

4. *Borovkova, M., Khodzitsky, M., Demchenko, P., Cherkasova, O., Popov, A., Meglinski, I.* Terahertz time-domain spectroscopy for non-invasive assessment of water content in biological samples // Biomedical Optics Express. 2018. Vol. 9. Issue 5. No. 309306. P. 2266-2276.

5. *Torben T. L. Kristensen, Withawat Withayachumnankul, Peter Uhd Jepsen, and Derek Abbott.* Modeling terahertz heating effects on water // Optics Express. 2010. Vol. 18, No. 5. P. 4727-4739.

Study of blood and its components by terahertz pulsed spectroscopy

O.P.Cherkasova[1, 2], M. M. Nazarov[3], A. P. Shkurinov[4, 5]

[1]Institute of Laser Physics of SB RAS, Novosibirsk, Russia, o.p.cherkasova@gmail.com
[2]Tomsk State University, Tomsk, Russia
[3]Kurchatov Institute National Research Center, Moscow, Russia
[4]Crystallography and Photonics Federal Research Center, Russian Academy of Sciences, Moscow, Russia
[5]Lomonosov Moscow State University, Moscow, Russia

Introduction. The prospect of widely using THz radiation in medical diagnosis and therapy raises the question of interaction of human tissues with THz radiation. Blood is a fluid connective tissue. Blood consists of plasma, and cellular elements. There are three main types of cellular blood cells: red blood cells (RBC), white blood cells (leukocytes) and blood platelets (platelets). Blood plasma contains 90% of water, about 6.6 - 8.5% of proteins and other organic and mineral compounds, which are intermediate or final products of metabolism, transferred from some organs in others. Capillaries with blood are in the skin and can be exposed to THz radiation. Therefore it is important to know the dielectric properties of blood and its components in THz range. The report will provide a review of the known data and our own studies on THz time domain spectroscopy (THz-TDS) of blood and its components at normal states and some pathology.

Experimental. The study of blood and its components is performed both in transmission and reflection configurations. Blood is poured into a special cuvette with a thickness (d) of 50-500 μm. For d<100 μm thickness variation can introduce noticeable errors, especially in refraction index. In reflection spectroscopy, data variation is related to the time delay between the solution signal and the reference signal producing different slopes of the reflection spectrum phase. The results obtained by transmission spectroscopy are more reliable at lower frequencies (0.05–0.8 THz), whereas reflection spectroscopy is preferable for higher THz frequencies (1.0–2.5 THz) [1].

Results. THz absorption and reflection spectra of blood and its components have not any sharp spectral features (fig.1.) [1-3].

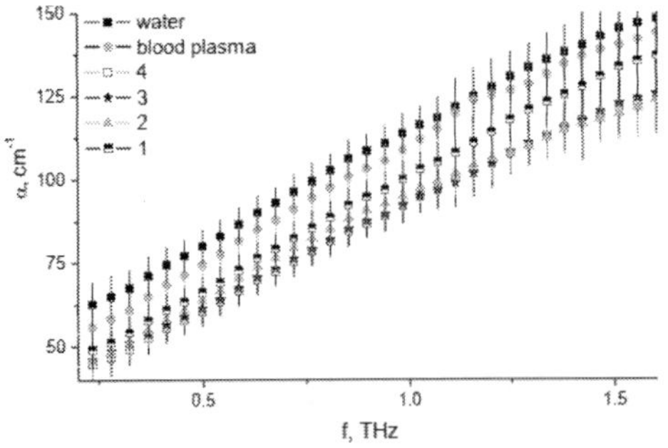

Fig. 1. Absorption coefficient of water, human blood plasma and RBC at different concentrations from $5.7 \cdot 10^6$ (1) to $7.32 \cdot 10^6$ (4) (cells/ml).

The shape of the spectra corresponds to the form of water spectrum, and for analysis, it is possible to use the two-component Debye model [1-3]. It was shown that there are some distinction in the Debye relaxation coefficients for the whole blood and its components [2]. A significant difference can observe in both the absorption and refractive index spectra of the whole blood and its thrombus: the coagulation and the formation of a thrombus have affected the absorption coefficient as well as the Debye coefficients [2, 4]. It was demonstrated the capability to distinguish thrombus formation at its early stage from uncoagulated blood by the change in the absorption coefficient at 200–270 GHz [4].

It has been shown that the absorption coefficient decrease linearly with increasing RBC concentration in whole blood [3, 5, 6]. The excellent linearity between the THz signal and the RBC concentration was also confirmed in a polyurethane resin tube using a THz imaging method. These results demonstrate that THz-TDS imaging can facilitate the quantitative analysis of blood [5].

It has been demonstrated that various biological substances of blood have a significant effect on its optical properties. A clinical study of ex-vivo fresh human whole blood showed that triglycerides and the number of RBC were two dominant factors to have significant negative correlation to the sub-THz absorption coefficients. This circumstance should be taken into account for future THz investigation in human whole blood [6]. The effect of blood glucose concentration on the optical properties of whole human blood was shown [7]. By investigating the THz spectra of different human blood, it was found out THz absorption coefficients reflect a high sensitivity to the glucose level in blood. With a quantitative analysis of 70 patients, it was demonstrated that the THz absorption coefficients and the blood glucose levels have a linear relationship [8].

The reduction of the absorption of blood plasma from mice with experimental oncology (Ehrlich carcinoma) relatively that of healthy mice have been shown in paper [9]. The authors concluded that such components of whole blood as blood red cells and platelets make the major contribution to the blood absorption in the THz range and the changes in the composition of other components cannot be reliably detected using the methods of THz spectroscopy.

It has been shown that the absorption coefficient and the refractive index of diabetic rat blood plasma are small compared to those of water and plasma of healthy rats [1, 10]. The diabetic rats have a higher level of glucose in blood and low protein concentration in comparison with healthy rats [10]. Assuming that the observed spectral changes are due to changes in the state of water in blood plasma, we have selected one of the parameters of the Debye model, $\Delta\varepsilon_1/\tau_1$, leading to the spectral features observed in the experiment. This change in the response of bound water can be the reason of the observed changes at increasing glucose concentration in blood plasma. We have demonstrated that when the concentration of glucose in blood rises to 24 mM (in rats with diabetes), $\Delta\varepsilon_1/\tau_1$ ratio decreases in 1.2 times [1].

We carried out a detailed study of the influence of glucose and protein concentrations on THz response of water solutions at 0.05-3.2 THz [11-13]. Different sensitivity to protein concentration in three spectral subregions, around 0.1, 1 and 3 THz was shown, which can influence the quantitative analysis of blood components. We proposed a simple method for analyzing experimental data [1, 11-13]

Conclusions. Whole blood and its components were studied by the THz-TDS method by a number of authors. The change of composition caused by pathological processes in the organism can considerably affect the optical properties of blood components in the THz frequency range. This can be used to create new rapid diagnostic methods. The cause of the detected differences was discussed with respect to variation in the terahertz response of water.

Acknowledgements. This work has been supported by Russian Foundation for Basic Research (project № 17-00-00275 (17-00-00270)).

References

1. *Cherkasova, O.P., Nazarov, M. M., Angeluts, A.A., Shkurinov, A.P.* Analysis of blood plasma at terahertz frequencies // Optics and Spectroscopy. 2016. V. 120(1). P. 50-57.

2. *Reid, C.B., Reese, G, Gibson, A.P., Wallace, V.P.* Terahertz time domain spectroscopy of human blood // IEEE J. Biomed. Health Inform. 2013. V. 3, No. 4. P. 363–367.

3. *Angeluts, A.A., Balakin, A.V., Evdokimov, M.G. et. al.* Characteristic responses of biological and nanoscale systems in the terahertz frequency range // Quantum Electronics. 2014. V. 44, No. 7. P. 614 – 632.

4. *Sun, C. K., Chen, H. Y., Tseng, T.F. et. al.*, High Sensitivity of T-Ray for Thrombus Sensing // Scientific reports. 2018. V. 8:3948, DOI:10.1038/s41598-018-22060-y

5. *Jeong, K., Huh, Y.-M., Kim, S.-H., et.al.* Characterization of blood using terahertz waves // J. Biomed. Opt. 2013. V. 18 (10), 107008.

6. *Tseng T.-F., You, B, Gao H.-C., et. al.* Pilot clinical study to investigate the human whole blood spectrum characteristics in the sub-THz region // Opt. Express. 2015. V. 23, No 7. P. 9440–9451.

7. *Gusev,S.I., Demchenko,P.S., Cherkasova, O.P., Fedorov, V.I., Khodzitsky, M.K.*, Influence of glucose concentration on blood optical properties in THz frequency range // Chinese Optics. 2018. V. 11(2). P. 182-189.

8. *Chen, H., Chen, X., Ma, S. et al.* Quantify Glucose Level in Freshly Diabetic's Blood by Terahertz Time-Domain Spectroscopy // Journal of Infrared Millimeter and Terahertz Waves. 2018. V. 39. P. 399.

9. *Smolianskaya, O.A., Kravtsenyuk, O.V., Panchenko, A.V. et al.* Study of blood plasma optical properties in mice grafted with Ehrlich carcinoma in the frequency range 0.1 –1.0 THz // Quantum Electronics. 2017. V. 47, No. 11. P. 1031 – 1040.

10. *Cherkasova, O.P., Nazarov, M.M., Smirnova, I.N., Angeluts, A.A., Shkurinov, A.P.* Application of Time-Domain THz Spectroscopy for Studying Blood Plasma of Rats with Experimental Diabetes // Physics of Wave Phenomena. 2014. V. 22, No. 3. P. 185-188.

11. *Nazarov, M.M., Cherkasova, O.P., Shkurinov, A.P.* Study of the dielectric function of aqueous solutions of glucose and albumin by THz time-domain spectroscopy // Quantum Electronics. 2016. V. 46, No 6. P. 488 – 495.

12. *Nazarov, M.M., Cherkasova, O.P., Shkurinov, A.P.* Properties of aqueous solutions in THz frequency range // Journal of Physics: Conference Series. 2017. V. 793. Doi:10.1088/1742-6596/793/1/012005.

13. *Nazarov, M. M., Cherkasova, O. P., Shkurinov, A. P.* Comprehensive Study of Albumin Solutions in the Extended Terahertz Frequency Range // Journal of Infrared Millimeter and Terahertz Waves. 2018. V. 39 (9). P. 840-853.

Morphological analysis of microglia in early postischemic period in the mouse local cerebral ischemia

Glyavina M.M.[1,2], Loginov P.A.[1], Dudenkova V.V.[1,2], Shirokova O.M. [1], Reunov D.G.[1,2], Karpova A.O. [1,2], Prodanets N.N.[1], Korobkov N.A. [3], Zhuchenko M.A.[4], Mukhina I.V.[1,2]

[1]Privolzhsky Research Medical University, Nizhny Novgorod, Russian Federation, mariyannov@gmail.com
[2]Lobachevsky State University, Nizhny Novgorod, Russian Federation,
[3]Saint Petersburg State University, Saint Petersburg, Russian Federation,
[4]Pharmapark LLC, Moscow, Russian Federation.

Ischemic stroke is one of the most frequent causes of death and disability of patients. Microglia is a unique example of immunocompetent cell in central nervous system that is capable protecting the brain from ischemic injury. Microglia are constantly surveying their microenvironment. Under pathological conditions microglia rapidly change their morphology and adopt activation states in order adequately react to the activation-causing stimuli. However, now there is no understanding of the specific physiological role of each of the morphological microglia phenotypes [1, 3, 6].

At present, drugs mainly aimed at minimizing the consequences of a stroke are intensely developed. One of the new lines in the development of neuroprotection methods in a stroke is to use a β- common receptor subunit (β-cR) with EPOR as a therapeutic target. One of microglia membrane's receptors is common receptor subunit (β-cR), which in the presence of a ligand can form a dimeric complex with erythropoietin. Carbamylated darbopoietin (CdEPO) (PHARMAPARK LLC, Moscow) was used as an agonist for this receptor. When the receptor is connected to erythropoietin, this triggers two types of reactions, namely, a positive one, which activates the protective mechanisms of damaged neurons, and a negative one, which is caused by the increasing number of erythrocytes, resulting in reformation of a thrombus and a rise in blood pressure [2,4]. Therefore, Carbamylated darbopoietin (CdEPO) (Pharmapark LLC, Moscow) is able protect neurons from ischemic damage, ignoring hemopoiesis, which prevents the increase in hematocrit. Approximately 80% of the thrombotic or thromboembolic strokes occur in the middle cerebral artery. Thus, a model of transient occlusion of the middle cerebral artery (tOCMA) was used to study the stroke pathogenesis mechanism [5].

The somatosensory cortex and the hippocampus play an important role in learning and memory processes. Therefore, were investigated the morphological characteristics of microglia in these areas of brain.

Laser radiation is used to image the dynamic of changing microglia morphological phenotype. An objective of this work is to study the proportion of microglia phenotype and their morphological characteristic after of the middle cerebral artery occlusion (MCAO) and CdEPO treatment, because this is important for assessing the development of ischemic damage. The report presents the results of microglia phenotype distribution in the somatosensory cortex and hippocampus of the mouse brain in the early post-ischemic period in the MCAO model.

For this experiment, we use C57BL/6 mice. Surgery has been made using a binocular microscope and under isoflurane anesthesia. One-hour exposure was proceeded by a 7-0 microfilament from Doccol Corporation. The animal behavior was tested on the first day after the injection. Preservation of basic activity and memory was studied with Open Field LE800S test, Shuttle Box LE916 (PanLab/Harvard Apparatus), "novel object recognition" test. CdEPO was injected after intravenously in 1, 3, 6, 12, 24, 48 hours after model of transient occlusion of the middle cerebral artery (tMCAO) (one-hour exposure) at 50 mkg/kg dose. Behavior was studied on the 4-th postoperative day. Brain is fixed in 4% paraformaldehyde (PFA). Microglia was marked with primary Iba-1 antibodies (Rabbit polyclonal) and secondary antibody with fluorescent agant AlexaFluor 555. Visualization was carried out by a ZEISS LSM 880 laser scanning confocal fluorescents microscope with an 40x immersion objective and a yellow-green spectrum filter. Paraffin coronal sections of the brain (5 μm thick) were studied. The images were processed using ImageJ and Python 3 software.

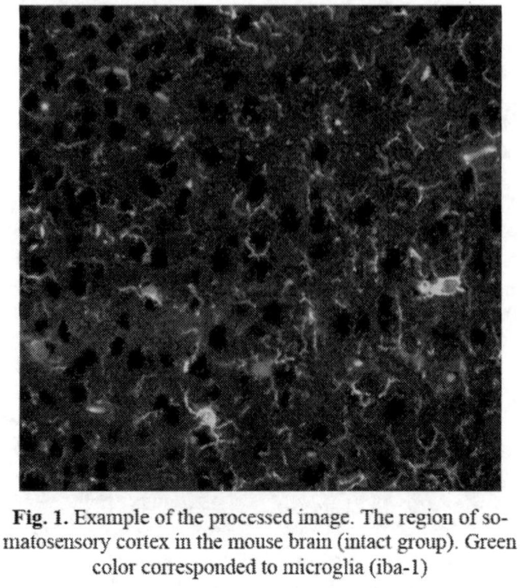

Fig. 1. Example of the processed image. The region of somatosensory cortex in the mouse brain (intact group). Green color corresponded to microglia (iba-1)

The morphological characteristics of microglia in the hippocampus is researched in the hemisphere with an ischemic injury. The number of microglial cells in the hippocampus were decreased in control group. In the group with CdEPO injection, the number of microglia cells had not changed. The area of microglial cells decreased in the control group and rise in the group with injection of CdEPO on the 4th day after the MCAO in the hippocampus. On the 4th day after MCAO, there was a decline in the number of ramified microglia in the control group and in group with injection of CdEPO. However, generation of active microglia was rise in in the control group and in group with injection of CdEPO. Moreover, generation of amoeboid microglia was doubled in control group and had not changed in group with injection of CdEPO.

Microglia in the somatosensory cortex were studied in the hemisphere with an ischemic injury and in the contralateral zone. The number of microglial cells in the somatosensory cortex in the zone with ischemic damage decreased in control group. In the group with CdEPO injection there is an opposite situation: number of microglial cells approximated the intact values. The area of microglial cells decreased on the 4th day after the MCAO in the control group and in the group with injection of CdEPO. On the 4th day after MCAO, there was a decline in the number of ramified microglia in the control group as well as in generation of active and amoeboid microglia in both hemispheres. In the group with CdEPO injection, the number of resting microglia had not changed but the amount of activated microglia increased.

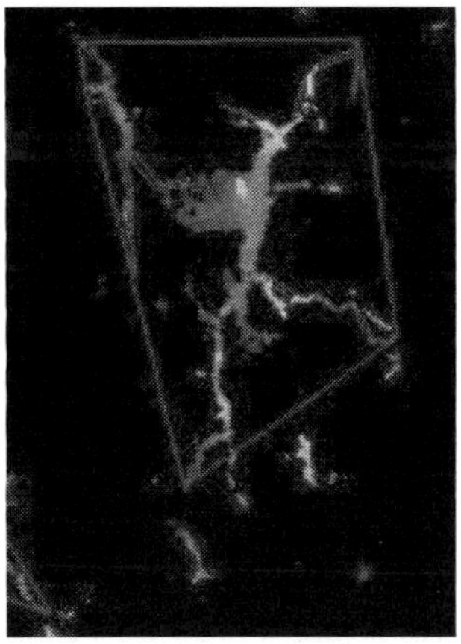

Fig. 2. Dendritic tree area (inside the solid - red in color online - line area). Area of contact microglia cell and microenvironment.

Change in microglia phenotypes and their morphology characteristic proportion in the group with CdEPO injection correlated with improved functional state of the animals after one-hour exposure: their motor activity restored, stress level was decreased and their memory was preserved.

The program of evaluating the morphological characteristics of microglia will help to understand the contribution of various microglia phenotypes to development of the ischemic damage and restoring the organism in the post-ischemic period. The ongoing research on regulation of microglia phenotypes proportion with CdEPO allows considering it as a promising neuroprotector giving a positive impact on the organism recovery in early post-ischemic period.

The reported study was funded by RFBR according to the research project № 18-34-00877

References

1. *Bachstetter AD et all*, Disease-related microglia heterogeneity in the hippocampus of Alzheimer's disease, dementia with Lewy bodies, and hippocampal sclerosis of aging// Acta Neuropathol Commun. , 2015 No. 23. P.3- 32.
2. *James M., Ian G. Young*, IL-3, IL-5, and GM-CSF signaling: crystal structure of the human beta-common receptor // Vitam Horm. 2006. No.74. P.1-30
3. *Karperien A. L et all.*, Box-counting analysis of microglia form in analysis of microglia form in schizophrenia, alzheimer's disease and affective disorder// Fractals, 2008. V. 16. No. 2. P. 103-107.
4. *Nataša Debeljak et all*, Erythropoietin and cancer: the unintended consequences of anemia correction // Front. Immunol., 2014. No. 5. P.563
5. *Rousseletl E., Kriz J., Seidah N.G.*, Mouse Model of Intraluminal MCAO: Cerebral Infarct Evaluation by Cresyl Violet Staining // J. Vis. Exp., 2012. No.69. pii: 4038.
6. *Youngjeon Lee et all*, Therapeutically Targeting Neuroinflammation and Microglia after Acute Ischemic Stroke// BioMed Research International, 2014. Article ID 297241, P.1-9.

EPJ Web of Conferences **195**, 10005 (2018) https://doi.org/10.1051/epjconf/201819510005
TERA-2018

Investigation of interaction of THz radiation with blood components

S. I. Gusev[1], T. Zhang[1], V. Yu. Soboleva[1], Yu. A. Kononova[2], V. A. Guseva[1], P. S. Demchenko[1], E. A. Segykh[1], M. K. Khodzitsky[1]

[1]ITMO University, St. Petersburg, Russia, mail@gusev-spb.ru
[2]Federal Almazov North-West Medical Research Centre, St. Petersburg, Russia

Last decades THz radiation becomes more popular for spectroscopy and imaging tasks. One of the most important directions is biomedical diagnostics [1]. In comparing with visible, IR and microwave frequency ranges, THz radiation has more unexplored topics [2]. In context of wide spreading of diabetes mellitus disease, the blood glucose sensing in THz frequency range has especial importance.

This work represents list of our results, obtained step-by-step to develop non-invasive technique of blood glucose measurement. There is list of performed approaches described below: some show dependence between blood glucose concentration and optical properties, some experiments devoted to increasement of sensitivity to glucose and some to make measurement by non-invasive way.

All physical experiments had been performed with THz time-domain spectroscopy (TDS) in Terahertz Biomedicine lab of ITMO University.

First experiment objects were cotton clothes impregnated by fresh blood and measured in transmission way. [3] Each sample had been extracted from the same patient during short time period and had different glucose concentrations.

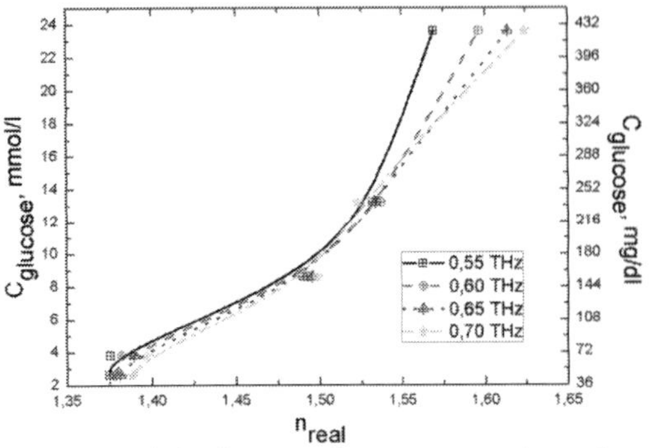

Fig. 1. Correlation between glucose concentration and refractive index had been observed at some frequencies

Second group of experiments was performed to reassure in results and increase accuracy. To prevent blood coagulation, water evaporation and provide stable thickness of the blood layer, special container had been produced. [4] It consisted from bottom and top parts, produced from polymethyl methacrylate. There were 2 experiments where blood layer was located inside the container. Blood samples had been from the same patient up to 8 times per experiment.

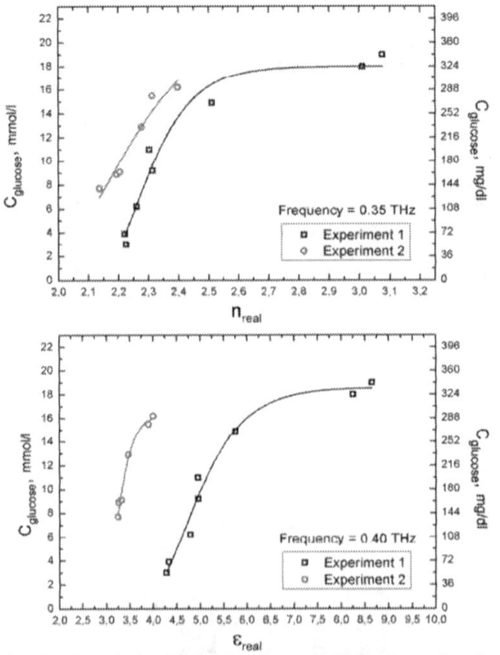

Fig. 2. Correlation behavior repeatability: for refractive index (top graph) as well as for permittivity (bottom graph)

Experimental results represented above showed method sensitivity to glucose level on condition of stable concentrations of other blood components (there were single patient during experiments). To estimate variability there were about 50 biosamples extracted from different patients, where concentration of components varies at each patient. [5] Each blood sample was analyzed for concentration of glucose, bilirubin, triglycerides, creatinine and uric acid by standard clinic laboratory tests. To observe influence of each component concentration on blood optical properties, observed parameters had been grouped by following principle: one parameter is varied, other are fixed.

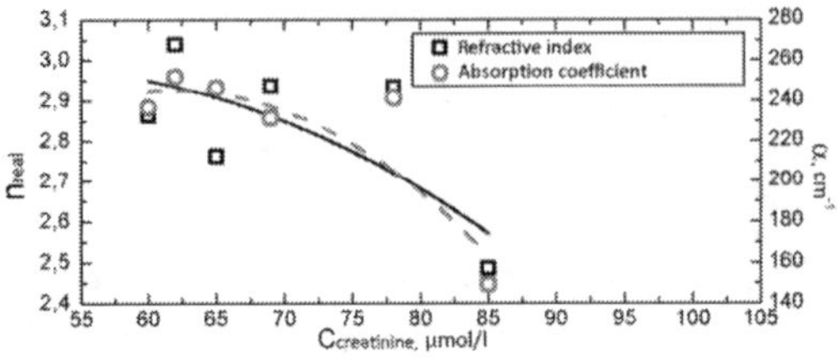

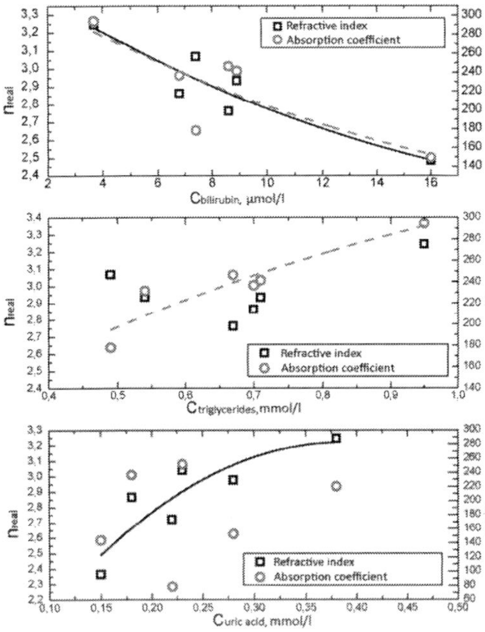

Fig. 3. In a number of cases there is possible to observe as blood optical properties in THz frequency range (such as refractive index and absorption coefficient) depend on concentrations of blood components.

To enhance sensitivity to glucose level there was performed simulation with band-pass aluminum metasurface with cross-shaped resonators sprayed on polyethylene terephthalate substrate. [6] Model consisted from this structure covered by thin layer of blood samples with the constant thickness. Each blood sample had different glucose concentration, their optical properties calculated before had been used during simuations.

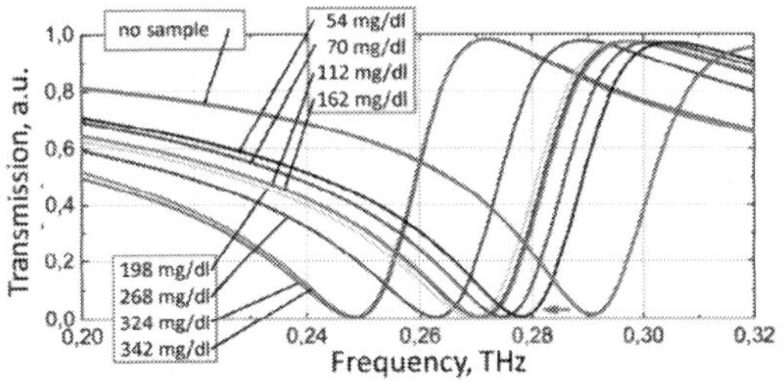

Fig. 4. With increasing of blood glucose concentration, we observe resonance frequency shift, which value directly correlates with blood glucose concentration.

For non-invasive technology development reasons, simulation of reflective experiment had been performed. [7] Described object consisted from nail plate and nail bed. Nail bed contains significant part of capillary filled by blood. Nail plate has stable thickness, flat surface and low absorption at THz frequencies. [8] As a result, modeled structure included listed objects, which were described by frequency dispersions of permittivity, calculated before with THz TDS in transmission mode.

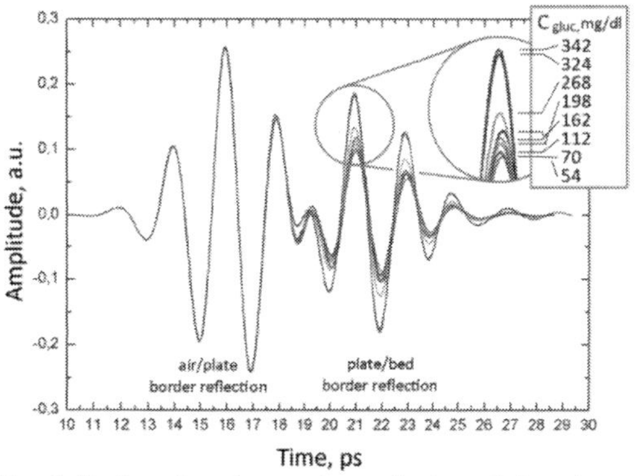

Fig. 5. Peak and peak-to-peak amplitude variations in reflection from plate/bed border shows tendency to correlation with blood glucose concentration.

Described series of studies shows dependence of blood glucose level and blood optical properties such as refractive index and permittivity. Influence of bilirubin, triglycerides, creatinine, uric acid concentration on whole blood optical properties had been investigated. There is metasurface-based method of increasing selectivity to glucose had been modeled and proposed. Non-invasive experiment using THz TDS had been simulated. Correlation of reflection amplitude and blood glucose had been observed.

There results showed THz radiation applicability and optimistic opportunities in the task of non-invasive glucose measurement.

References

1. *Pickwell, E., Wallace, V. P.* Biomedical applications of terahertz technology // Journal of Physics D: Applied Physics. 2006. V. 39, No. 17. P. R301.

2. *Tonouchi, M.* Cutting-edge terahertz technology // Nature photonics. 2007. V. 1, No. 2. P. 97.

3. *Gusev, S. I., et al.* Blood optical properties at various glucose level values in THz frequency range // Clinical and Biomedical Spectroscopy and Imaging IV. 2015. V. 9537. P. 95372A.

4. *Gusev, S. I., et. al.* Influence of Glucose Concentration on Blood Optical Properties in THz Frequency Range // Chinese Optics. 2018. V. 11, No. 2. P. 182–189.

5. *Zhang, T., et. al.* Qualitative analysis of blood components in terahertz frequency range // Scientific and Technical Journal of Information Technologies, Mechanics and Optics. 2018. V. 18, No. 5. (in Russian).

6. *Soboleva, V. Yu., Gusev, S. I., Khodzitsky, M. K.* Physical properties of macroscopically inhomogeneous media // Scientific and Technical Journal of Information Technologies, Mechanics and Optics. 2018. V. 18, No. 3. P. 377–383. (in Russian).

7. *Gusev, S. I., et al.* Study of glucose concentration influence on blood optical properties in THz frequency range // Nanosystems: Physics, Chemistry, Mathematics. 2018. V. 9, No. 3. P. 389–400

8. *Guseva, V. A., et al.* Optical properties of human nails in THz frequency range // Journal of Biomedical Photonics & Engineering. 2016. V. 2, No. 4.

https://doi.org/10.1051/epjconf/201819510006

Applications of THz laser spectroscopy and machine learning for medical diagnostics

Yu. V. Kistenev[1,2], A. V. Borisov[1,2], A.I.Knyazkova[1], E. A. Sandykova[1,2], V. V. Nikolaev[1], D. A. Vrazhnov[1]
[1]Tomsk State University, Tomsk, Russia, yuk@iao.ru
[2]Siberian State Medical University, Tomsk, Russia

THz spectroscopy allows to analyze molecular rotations associated with hydrogen bond breaking. But, the identification of pure compounds using molecular signatures with THz spectroscopy is still not straightforward because of the inherently broad spectral signatures in biotissue. A smooth shape THz of spectra causes a necessity to use the machine learning for tissue diagnosis using THz spectroscopy.

Typical machine learning pipeline includes the following steps (Fig.1) [1]:
- preprocessing of data;
- selection of informative features;
- development of predictive models for new data classification.

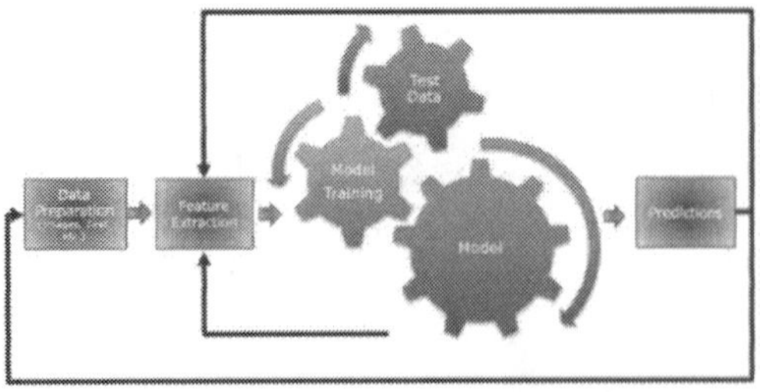

Fig.1. Typical machine learning pipeline.

In this report we present examples of peculiarities of machine learning applications for 2D THz image analysis. The results have been obtained using THz time-domain spectrometer (TDS) T-Spec by EXPLA in the range 0.2-3 THz.

At the preprocessing stage, in addition to filtering, various approaches to allocating areas of interest also should be considered. Automatic search of characteristic structures in the image is based on their formalized mathematical description of the image textures.

The example of preprocessing stage image transform connected with 2D THz TDS absorption spectra of formalin-fixed paraffin-embedded prostate cancer biopsy tissues is presented below. The goal is to remove artifacts of plastic substrate and paraffin from the image.

The Fig. 2 shows the spatial distribution of 2D THz image for a paraffin block without a sample and for a plastic substrate at frequencies 0.90 THz (Fig. 1a) and 1.05 THz (Fig. 1b).

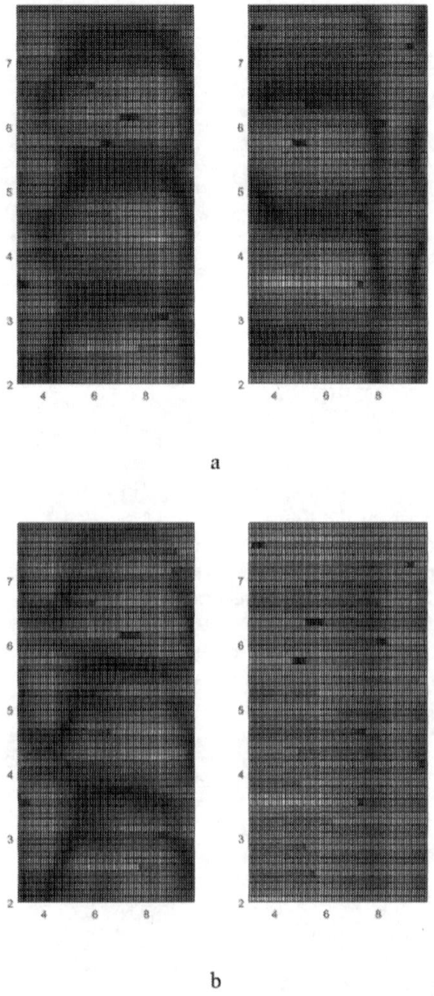

Fig. 2. Spatial distributions of the THz signal intensity for a paraffin block without a sample (on the left in each figure) and for a plastic substrate (on the right in each figure) at the frequencies: a - 0.90 THz and b - 1.05 THz.

The difference of absorption spectra allows to remove similar artifacts from the image. To realize it, an optimization algorithm was developed and implemented [3]. This algorithm allows to select pixels on the THz image with minimal influence of the paraffin and plastic substrates. The results of selection of a biopsy tissue on the 2D THz image are shown in the Fig.3.

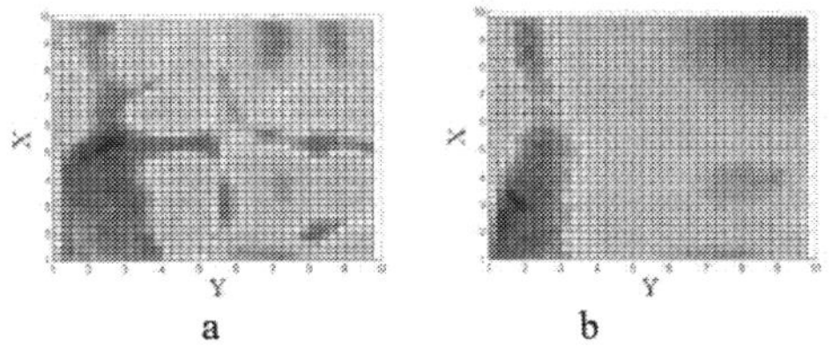

Fig.3. Selection of a biopsy tissue on the 2D THz image: initial image (a), image after artifacts removing (b).

A key step in image analysis is the informative features selection, because the quality of the created predictive model is defined by the ability of spatial separation of the various groups under study in the feature space. One of the most effective method of this task solution is the principal component analysis (PCA). The basic idea of PCA is to find the reduced number of new variables, termed the principal components, which are enough for recovery of the initial variables, possibly with insignificant errors [2].

The PCA applications for 2D THz image analysis was done on animal model (rats) lymphedema tissue.

Lymphedema is a chronic progressive disease of the lymphatic system caused by abnormal accumulation of tissue fluid with a high protein content. Early diagnosis of this disease helps to choose the right treatment and prevent its further development. The existing methods of lymphedema diagnosing at early stages are not strict and consistent. The invention of THz microscopy opens up new possibilities for lymphedema tissue analysis in vivo.

Using the optimization algorithm, mentioned above, we carried out the classification of THz spectra of the most informative areas obtained in-vivo from the lymphedema affected leg tissue (result of surgery) and obtained from healthy leg tissue. The results show good enough the separation of lymphedema tissues from and healthy tissues in the space of the principal components (see Fig. 4).

The principal components are built using the THz spectra in the 0.8-1.0 THz spectral range. Note that the separation of the groups using THz imaging became possible after three weeks from the lymphedema surgery initiation.

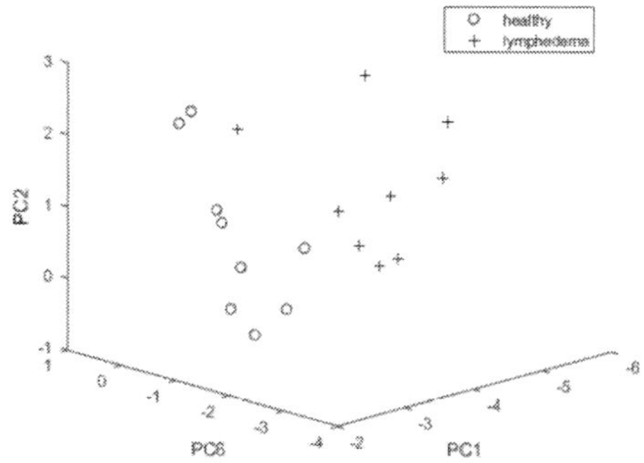

Fig. 4. Spatial distribution of the THZ spectra from lymphedema affected leg tissues and healthy leg tissues in the principal component space.

The work was carried out under partial financial support of the Russian Fund of Basic Research (grant No.17-00-00186, grant No. 18-42-703012 р_мол_а).

References

1. https://blog.westerndigital.com/machine-learning-pipeline-object-storage/machine-learning-pipeline/
2. *L. Pomerantsev and O. Ye. Rodionova,* Concept and role of extreme objects in PCA/SIMCA//*Journal of Chemometrics*. 2014. **28**(5), P.429-438.
3. *Kistenev, Yu.V., Borisov, A.V., Knyazkova, A.I., Ilyasova, E.E., Sandykova, E.A., Gorbunov, A.K., Spirina, L.V.* Possibilities of cytospectrophotometry of oncological prostate cancer tissue analysis in the TGz spectral range // Proc. of SPIE – XIII International Conference on Atomic and Molecular Pulsed Lasers (AMP17). - AMP100-94, 2018. No.10614-94.

The research method of a qualitative analysis of the composition of the blood in the terahertz frequency range

T. Zhang[1], Y.A. Kononova[1,2], M.K. Khodzitsky[1], P.S. Demchenko[1], S.I. Gusev[1], A.Y. Babenko[1,2], E.N. Grineva[1,2], Kublanova I.L.[1]

[1]ITMO University, Saint Petersburg, Russia

[2]Almazov National Medical Research Centre, Saint Petersburg, Russia

Introduction

The study of biochemical parameters of blood has an important role in the diagnosis of diseases, in honesty for patients diabetes mellitus. Of particular interest in this area is the study of the interaction of terahertz (THz) radiation with components of human blood. THz frequency range is remarkable in biomedicine in that it contains resonance frequencies of specific vibrational and rotational modes of biomolecules, therefore, using THz radiation, we can evaluate and determine the state of biological molecular bonds. In addition, THz radiation is very sensitive to different types of conformation of H_2O molecules with the rest of molecules contained in bio-samples. In the course of this work, the effect of bilirubin, triglyceride, uric acid, creatinine concentrations on the spectra of the refractive index and absorption of human blood in the THz frequency range was investigated, the results of which can be useful in the development of a method for analyzing blood composition.

Subject of study

The paper presents the study of blood biochemical composition (bilirubin, creatinine, triglycer-ides, uric acid) effect on its optical properties, i.e. refractive index and absorption coefficient, in terahertz fre-quency range was investigated.

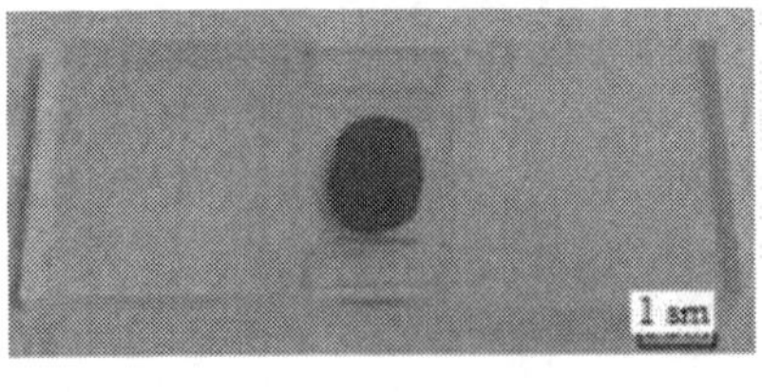

Fig. 1. Measuring cell with a blood sample

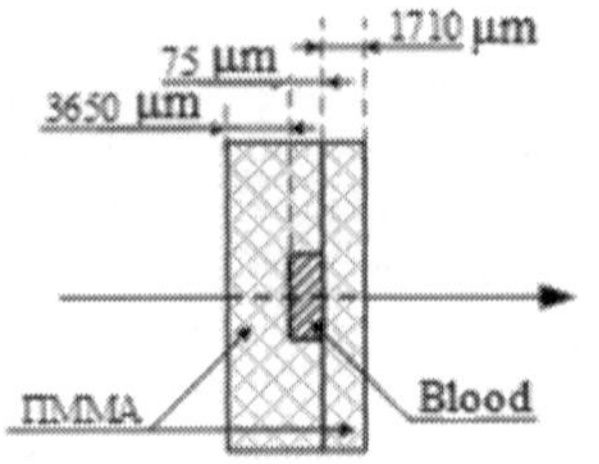

Fig. 2. Scheme of the measuring cell

Method

To obtain the values of the refractive index and the absorption coefficient, the method of terahertz time-domain spectroscopy was used in the transmission mode. Concentrations of total bilirubin, creatinine and triglycerides were measured in blood serum by the colorimetric method, the pseudokinetic method and the enzymatic method, respectively. The glucose level was determined in blood plasma by the enzyme method. All measurements of blood component concentrations were carried out at the Almazov National Medical Research Centre.

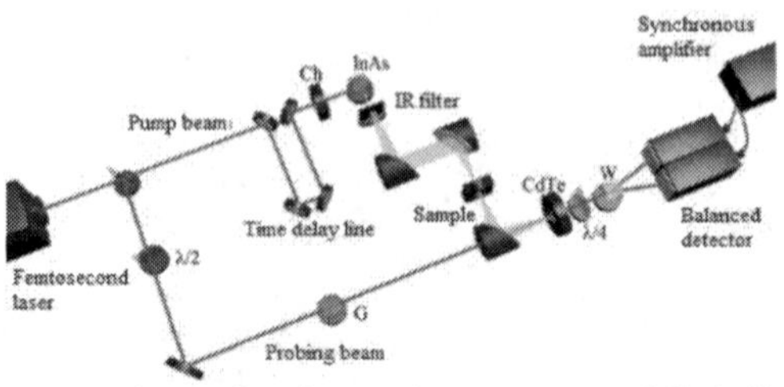

Fig. 3. Scheme of terahertz pulse spectrometer: λ/2 - half-wave plate; G - the Glan prism; Ch - breaker; λ/4 - the quarter-wave plate; W - the Wollaston prism

Main results

The optical properties of blood with various biochemical composition were ob-tained using terahertz time-domain spectroscopy at the frequency of 0.4THz. It is shown that the refractive indices and the absorption coefficients of blood decrease with an increase in the concentration of bilirubin and creatinine. It has also been found that with an increase in the concentration of uric acid and triglycerides, the refractive index of blood increases and the absorption coefficient of blood decreases, respectively. The correlation between the refractive index and the concentration of triglycerides and the correlation between the blood absorption coefficient and the concentration of uric acid were not revealed.

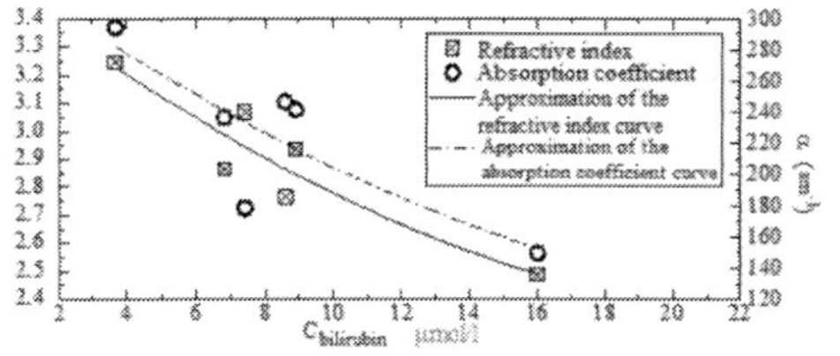

Fig. 4. Dependence of optical properties of blood on bilirubin concentration

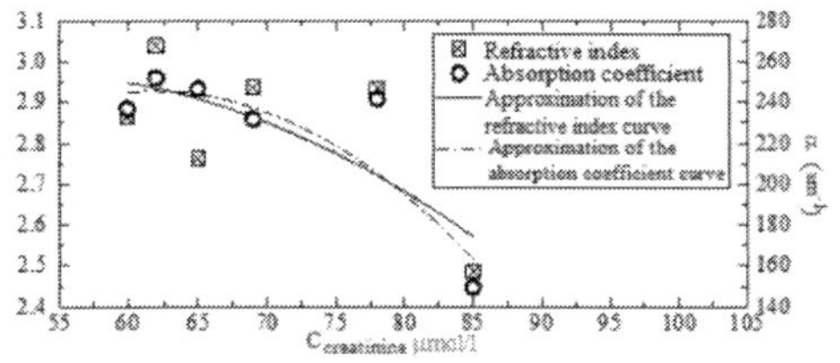

Fig. 5. Dependence of the optical properties of blood on the concentration of creatinine

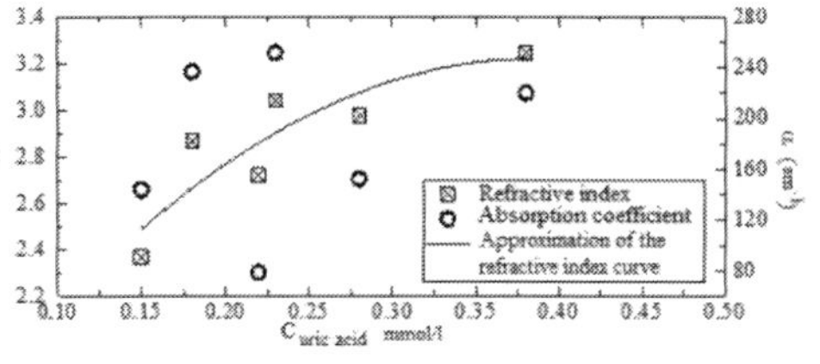

Fig. 6. Dependence of the optical properties of blood on uric acid

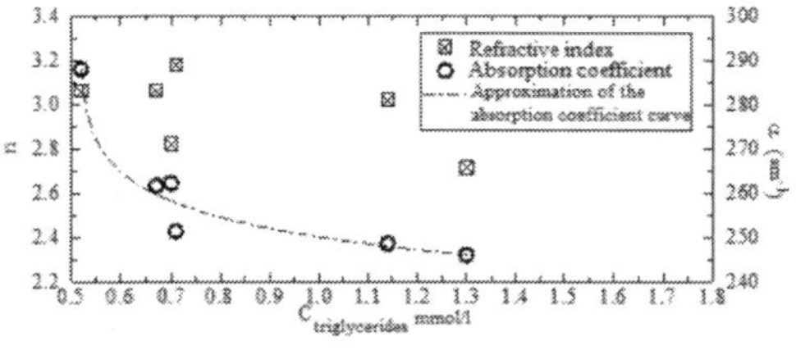

Fig. 7. Dependence of the optical properties of blood on the concentration of triglycerides

Practical significance

The observed correlations between the concentrations of blood components and the optical properties are useful in the development of a new technique for blood analysis.

References

1. Juvenile Diabetes Research Foundation Continuous Glucose Monitoring Study Group, Fox LA, Beck RW, Xing D. Variation of interstitial glucose measurements assessed by continuous glucose monitors in healthy, nondiabetic individuals // Diabetes Care. 2010. V. 33. N 6. P.1297–1299. doi: 10.2337/dc09-1971
2. Zhang X.C., Xu J. Introduction to THz wave photonics. Springer, 2010. 246 p.
3. Son J.H. Terahertz biomedical science and technology. CRC Press, 2014. 355 p.
4. Gusev S.I., Borovkova M.A., Strepitov M.A., Khodzitsky M.K. Blood optical properties at various glucose level values in THz frequency range // European Conference on Biomedical Optics. Munich, Germany, 2015. P. 95372A. doi: 10.1364/ECBO.2015.95372A
5. Gusev S.I., Demchenko P.S., Cherkasova O.P., Fedorov V.I., Khodzitsky M.K. Influence of glucose concentration on blood optical properties in THz frequency range // Chinese Optics. 2018. V. 11. N 2. P. 182–189. doi: 10.3788/CO.20181102.0182

Spectroscopy of solutions in the low frequency extended THz frequency range

Nazarov Maxim[1], O. P. Cherkasova[2, 3], A. P. Shkurinov[4, 5]

[1]Kurchatov Institute National Research Center, Moscow, Russia, nazarovmax@mail.ru
[2]Institute of Laser Physics of SB RAS, Novosibirsk, Russia
[3]Tomsk State University, Tomsk, Russia
[4]Crystallography and Photonics Federal Research Center, RAS, Moscow, Russia
[5]Department of Physics and International Laser Center, M.V.Lomonosov Moscow State University, Russia

One of the most important biological components of the living systems is water. It is the main informative feature and simultaneously the main obstacle for tissue and bio-solution spectroscopy in THz range. Its intensive absorption in this frequency range, on the one hand, limits its penetration into the samples, but, on the other hand, allows one to consider it as the subject of a separate fruitful study [1-4]. Water molecules form hydrogen bonds (HB) with their neighbors to construct the water network. Currently, water can be divided on bulk water (it does not form bonds with biomolecules) and hydration or bounded water (it surrounds biomolecules and interacts with them). Thus, in the solution, a considerable part of water molecules are in the form of a hydration shell – Fig 1.

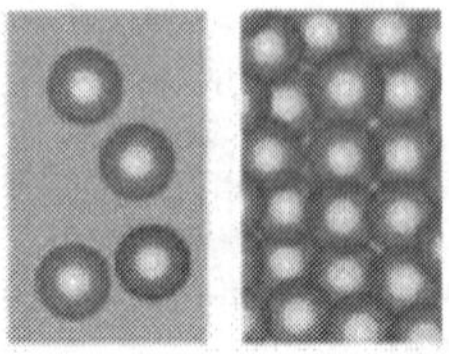

Fig. 1. Components of the solution: free water (blue background), bounded water (gray circles), solute molecules (yellow circles). Dilute solution (a) and saturated solution (b). After [4]

Bound or free water makes a valuable contribution to the THz response of biological objects. This allows us to use THz spectroscopy as a powerful tool to study different forms of water in a wide class of biological media, including the macromolecules in aqueous solutions. Actually, we do not measure the solute, its contribution is negligible. We measure how the solute modifies the solvent, the appearance and the structure of hydration shell around the solute molecules. To detect small changes in the solution smooth spectra, we should precisely describe solvent properties in a broad spectral range.

We describe a number of models of a water solution dielectric permittivity, which are applicable in the THz frequency range. We give a detailed description of biological solutions (protein and sugar solutions, blood components), analyze modern measuring techniques. All known processes (relaxation and dumped oscillation) in polar solutions below the tens of THz have very broadband responses, so it is essential to combine several experimental techniques, each for its own spectral range, to obtain spectra over many octaves. From the low frequency side, it is well established dielectric spectroscopy; from the high frequency side, it is FTIR and it is THz-time-domain spectroscopy (TDS) between them.

The main relaxation process (at 0.02 THz) reflects the cooperative reorientational dynamics of the dipole moment. It is assigned to the cooperative reorientation time of hydrogen-bonded bulk water molecules involving HB switching events. The oscillation processes indicates the overdamped modes, which correspond to several known vibration modes in the THz region: the bending mode between two water molecules forming the hydrogen-bond (at 1.5-2 THz); the intermolecular stretching of water is assigned to a hindered O...O translation (at 5.4 THz) *etc*.

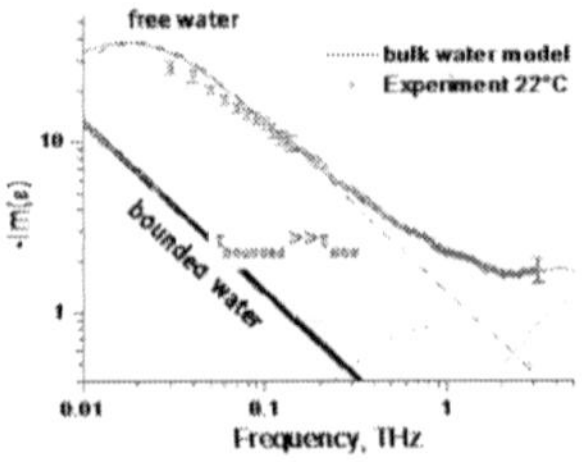

Fig. 2. Slow relaxation shift of bounded water spectra.

To fit dielectric function of water a well-known model with Debye-type relaxation and over-dumped oscillator components is usually applied [1-4].

$$\varepsilon(\omega,C)=\varepsilon_\infty+\frac{\Delta\varepsilon_1(C)}{1+i\omega\tau_1}+\frac{\Delta\varepsilon_2}{1+i\omega\tau_2}+\frac{A_1}{\omega_{01}^2-\omega^2+i\gamma_{01}\omega}+\ldots \quad (1)$$

where τ_1 and τ_2 are relaxation times for the first (main) relaxation process and the second ('fast') relaxation term; $\Delta\varepsilon_i$ are the contributions into permittivity from corresponding relaxation; A_1 is amplitude; w_{01} is frequency; γ_{01} is the linewidth of the oscillator term. Slow relaxation spectra of bounded water are shifted to much lower frequencies – Fig.2, thus, in GHz and THz range bounded water has smaller absorption and dispersion. That is the main reason of solution spectra changes. In general, we should use the effective medium approximation for the total spectrum; in a more simplified case, we should take into account the volume fractions of the three solution components with different spectral responses.

The most widely-exploited method of solution measurements using THz spectroscopy implies studying the THz-wave transmission through a thin (50–100 μm) cuvette. In such measurements, even a small uncertainty of the analyte layer thickness might cause a noticeable error in the reconstructed THz dielectric response. Thus thick cuvette is preferable. It gives considerable advantages in low frequency part of the THz spectra, where water transmission is not so small. For the high THz frequencies, the attenuated total reflection (ATR) measurement approach is preferable, since it does not suffer from strong water absorption and implies measurements of a bulk sample; thus, eliminating an error of layer thickness measurements. Actually, these methods have considerable overlap of the spectral operation range; thus, we combine the results obtained by transmission and reflection in order to achieve a broader spectral operation range and more reliable data for water solution spectroscopy [1].

Once the water spectrum is precisely measured and described, one can proceed to water solution analysis. The spectra of aqueous protein or saccharides solutions with not extremely high concentration are almost completely determined by the spectral properties of the solvent water. Therefore, there are reasons for application of Eq. (1) to describe permittivity of such biological media as aqueous solutions. To detect small-scale changes in solutions we have normalized the transmittance (and reflectance) of the solution T_{BSA} to that of water – Fig 3.

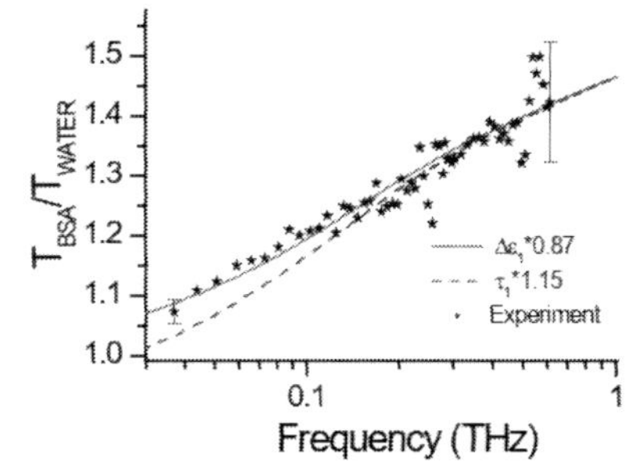

Fig 3. Transmission spectra of BSA 100 mg/ml water solution in a 500 μm cell, normalized to the case of water in the same cell, experimental data – open circles. Solid lines – model transmission with the same thickness and normalization. Red line with modified $\Delta\varepsilon_1$*0.87 ($\Delta\varepsilon_1/\tau_1$=0.87) and blue line with modified τ_1*1.15 ($\Delta\varepsilon_1/\tau_1$=0.87).

To identify the reasons for the THz transmission changes of BSA solutions we have compared the experimental spectra to the relative transmission of modified model of water [1]. Such a simplified approach, implying variation of a single parameter $\Delta\varepsilon_1(C)$, describes well the experimental spectra of aqueous solutions (particularly, of sugars and proteins) in the frequency range of at least 0.03 to 3.2 THz, which is important for practical applications of THz-TDS systems. Such a difference is best fitted by a decrease in the amplitude of the slow relaxation amplitude $\Delta\varepsilon_1$. An alternative approach is to use $\tau_1(C)$ as the only varied parameter instead of $\Delta\varepsilon_1$. At frequencies above f=0.3 THz, it is mathematically almost equal to vary τ_1 or $\Delta\varepsilon_1$ because $\omega\tau_1>>1$. In any case, in this THz band, 80 % of the spectral response is determined by the slow relaxation and its parameters, while contributions of the remaining processes in Eq. (1) are relatively small and weakly depend on the concentration.

Since there are practically no standards and calibrated instruments in the THz range, before studying a new system, it is necessary to achieve an agreement of the measured data with the well-known literature results on the THz dielectric response of some "reference" solute, for example, glucose.

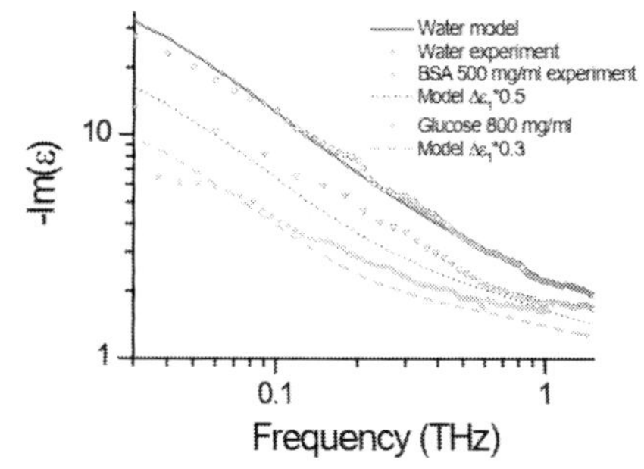

Fig 4. Dielectric function spectra of almost saturated solution of BSA (500 mg/ml) and of glucose (800 mg/ml) and of distilled water. Solid lines – model transmission with modified $\Delta\varepsilon_1$*0.5 for BSA and $\Delta\varepsilon_1$*0.3 for glucose. Symbols – experimental data.

Remarkable that different models lead to almost equal values of complex dielectric function in the frequency range where the models overlap.

Other important questions to be discussed in the talk are: Can we distinguish proteins from saccharides in water environment from the THz spectra shape? Are there narrow/resolvable spectral features of biomolecules in water environment below 5 THz? How can we extract the pure spectra of bounded water? What is useful in saturated solution spectra? When it is necessary to apply effective model for the solution?

Acknowledgements

The reported study was funded by RFBR project № 17-00-00275 (17-00-00270), 18-52-00040.

References

1. *Nazarov, M.M., Cherkasova, O.P., Shkurinov, A.P.* Study of the dielectric function of aqueous solutions of glucose and albumin by THz time-domain spectroscopy // Quantum Electronics. 2016. V. 46, No 6. P. 488 – 495.

2. *Shiraga, K., Adachi, A., Nakamura, M., Tajima, T., Ajito, K., Ogawa, Y.* Characterization of the hydrogen-bond network of water around sucrose and trehalose: Microwave and terahertz spectroscopic study // The Journal of Chemical Physics. 2017. V. 146 (10). P. 105102.

3. *Ellison, W. J.* Permittivity of pure water, at standard atmospheric pressure, over the frequency range 0–25 thz and the temperature range 0–100°C // Journal of Physical and Chemical Reference Data. 2007. V. 36 (1). P. 1–18.

4. *B. Born, B., Havenith, M.* Terahertz Dance of Proteins and Sugars with Water // J Infrared Milli Terahz Waves. 2009. V. 30. P. 1245–1254.

EPJ Web of Conferences **195**, 10009 (2018) https://doi.org/10.1051/epjconf/201819510009
TERA-2018

Application of THz radiation for *in situ* control of eye cornea hydration level

I.Ozheredov[1], M.Prokopchuk[1], T.Safonova[2], E.Sikach[2], P.Solyankin[3], A.Angeluts[1], A.Balakin[1,3] and A.Shkurinov[1,3]

[1]Lomonosov Moscow State University, Moscow, Russian Federation, ozheredov@physics.msu.ru
[2]Eye Diseases Institute, Moscow, Russian Federation
[3]Institute on Laser and Information Technologies of RAS, Branch of the FSRC "Crystallography and Photonics" RAS, Shatura, Moscow Region, Russian Federation

The epithelium of the cornea and conjunctiva needs continuous moistening. With closed eyelids, the tear fluid fills the entire space, and when open, it spreads over the anterior surface of the eye forming a thin film. Tear film has a thickness of 4 - 7 μm and performs protective, metabolic and optical functions. The direct involvement of the corneal and conjunctival epithelium into formation of tear film was demonstrated in [1]. The tear film covers the whole corneal and conjunctival surface, from the area of transition from the keratinous multilayered epithelium of the eyelid skin edges to the non-corroborating multilayer epithelium of the conjunctiva and up to the corneal center. Thus it allows considering the tear film as one of the corneal layers.

Numerous causes leading to the loss of normal properties and structure of the tear film, lead to changes in the water-electrolyte balance. They also cause a decrease in the volume of tear fluid, a decrease in sensitivity and a development of corneal epitheliopathy. A complex of signs of the corneal and conjunctival epithelium lesion due to a decrease in the quality and/or amount of tear fluid defines a concept of 'dry eye syndrome' [2].

The assessment of pre-corneal tear film is a leading direction in the diagnosis of dry eye syndrome. In clinical practice the most common method for determining the stability of the tear film is Norn test. Non-invasive methods for studying the stability of a tear film include thioscopy. Confocal microscopy expands the possibilities of studying the anatomy of the cornea at the level of its microstructure. It has been shown recently [3,4] for *in vivo* measurements of corneal tissue hydration and tear film dynamic control the terahertz (THz) reflectometry might be efficiently used.

In the present study we apply frequency-domain THz reflectometry technique for sensing of the ocular surface system. Our study allows the dynamic control of pre-corneal tear film thinning and sensing of corneal tissue hydration level in clinical applications.

The concept of the proposed technique was based on THz photomixing approach. Typical photomixer consists of two independent tunable laser sources yielding the difference frequency in a desired THz region by heterodyning. Two laser beams with different frequencies light a photoconductive antenna (PCA), where running fringes with the THz difference frequency excite carries in the semiconductor material. The solid-state lasers are used as elements of a laser heterodyne. The single-mode selection of the laser is done using distributed feedback (DFB) - the diffraction grating technologically etched close to the p-n junction of the diode structure. PCA consists of an electrical dipole and semiconductor quick enough to produce carriers in time with the beat frequency. When the heterodyned laser beams light the surface of the semiconductor, than carriers appear in the material. If the metal antenna is polarized with some voltage then carriers give a periodical short circuit for the antenna and PCA converts the photocurrent into a THz wave.

The proposed apparatus was based on two DFB diode lasers with precise resonator temperature control. The lasers generated narrow 10 kHz line in the range of 1530 – 1608 nm with average power of 22 mW each. X-type fiber optical beam splitter with the splitting ratio 50/50 combined both lasers output. The outputs of the splitter were connected to the THz emitter and receiver - low-temperature InGaAs bow-tie PCA (BATOP GmbH). The bias voltage for the THz emitter (0/6 V, 40 kHz) was provided by the function generator. The difference frequency range of DFB lasers allowed tuning of the output THz radiation within a range of 0.04-0.1 THz. High-density polyethylene lens (HDPE) with focal distance of 30 mm focused THz signal to the ocular surface system. The specularly reflected THz signal was guided through second similar HDPE-lens into the receiver PCA module. The digital lock-in amplifier SR 810 (Stanford Research Systems) was used to measure the resulting photocurrent. Both THz emitter and receiver were were embeded into the clinical ophthalmic apparatus (Fig.1) allowed measuring the eye reflection in THz range simultaneously with using of the other standard ophthalmological methodic.

For the *in vivo* study the special experimental protocol was developed. For each human eye cornea the reflected signal versus time was measured and reflection coefficient was calculated. Each measurement started at the time of eye opening. During the measurement series the testee human was requested to keep the eye open as long as possible. The same time he was able to close the eye as soon as he willing doing that. The measurements continued until the next eye closing.

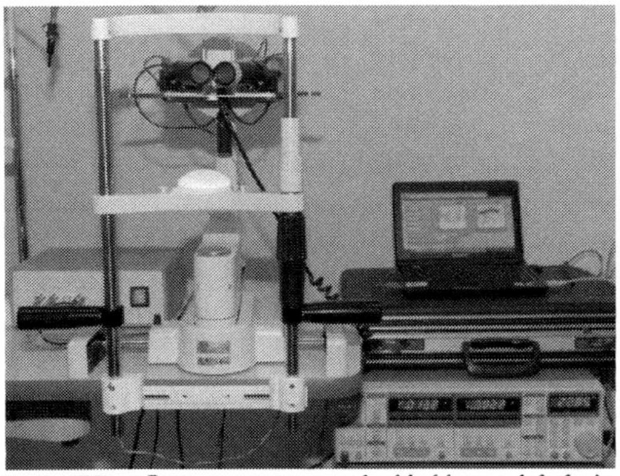

Fig. 1. THz reflectometer setup embedded into ophthalmic apparatus

The THz reflectivity of the cornea demonstrates the dehydration dynamics (Fig.2). Such dependency may be fitted by the special function and decreases with the time. At a time of eye closing the reflectivity decreases and remains constant while the eye is closed. After the next eye opening the reflectivity restores to the value similar one of the previous cycle. Based on the results of preliminary studies, it can be possible to measure of pre-corneal tear film dynamics.

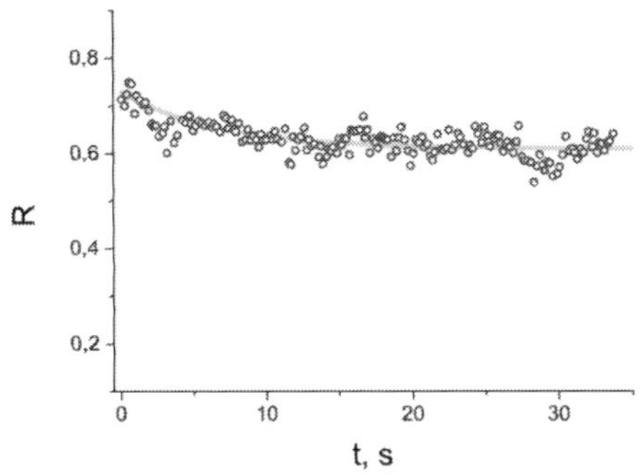

Fig. 2. Reflection coefficient of the ocular surface system as a function of time

Fig. 2 illustrates an example of the ocular surface system THz reflection coefficient versus time. This graph was obtained for the cornea of testee human without any claims. In case of complains on dryness in the eyes and suspicion of dry eye syndrome the dynamics of reflection coefficient change would be more changeable. Norn tests for each testee human were performed and the behavior of the reflection coefficient dynamics for these patients possesses good correlation with the Norn test results.

The value of THz reflection coefficient at the right edge of the graph in Fig. 2 corresponds to the case of critical thinning of the tear film and very close to the value of tear film break up case. This value could be taken for estimation the reflection coefficient of the corneal tissue and calculation of the hydration level of this tissue.

To confirm the detected correlations, similar studies were carried out for three groups of subjects. In the first group of 10 people, the eyes of people who did not express subjective complaints of discomfort in the eyes or excessive lacrimation were considered.

The second group represented a sample of people with instability of tear film established by Norn test and subjective complaints of dryness. The third group included eyes with hypersecretion of tear glands confirmed by Shirmer test. These groups of tested humans with eye abnormalities included 13 and 7 subjects respectively.

The results of these studies, presented in the form of a histogram in Fig. 3, show a correlation between the value of the linear function slope and the established diagnosis.

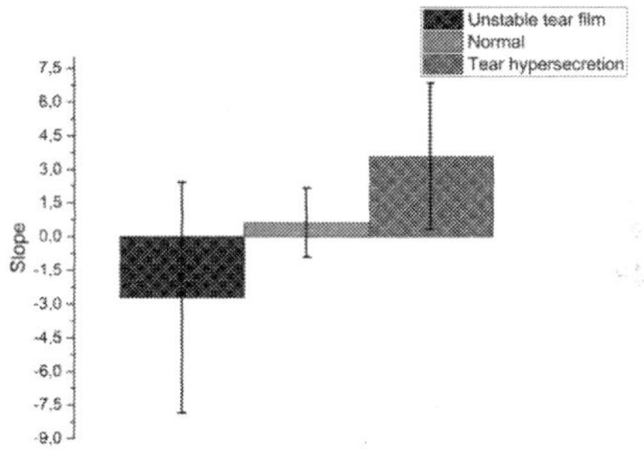

Fig. 3. Histogram of function slope coefficient for different eye type diseases

We showed the results of applications of the THz reflectometry to sensing of the ocular surface system and pre-corneal tear film. Our study allows the dynamic control of pre-corneal tear film thinning and sensing of corneal tissue hydration level in clinical applications *in vivo*.

This work was supported by RFBR (projects No.15-29-03900 and 17-00-00270).

References

1. *K.M.Meek*. The Cornea and Sclera. Boston, MA:Springer, 2008.
2. *V.V.Brzheskij, G.B.Egorova and E.A.Egorov*. Syndrome of 'dry eyes' and diseases of the eye surface: clinic, diagnosis and treatment. Moscow: GEOTAR-Media, 2016.
3. *Z.D.Taylor, J.Garritano, S.Sung, et al*. THz and mm-Wave Sensing of Corneal Tissue Water Content: In Vivo Sensing and Imaging Results. IEEE Transactions on Terahertz Science and Technology, vol.5, pp.184-196, 2015.
4. *I.Ozheredov, M.Prokopchuk, M.Mischenko, et al*. In vivo THz Sensing of Eye Cornea. Laser Physics Letters, vol.15, 055601, 2018.

https://doi.org/10.1051/epjconf/201819510010

A device to inspect a skin cancer tumour in the terahertz range, transferring the image into the infrared

A.V. Postnikov[1], K.A. Moldosanov[2], N.J. Kairyev[2], V.M. Lelevkin[2]

[1]University of Lorraine, Metz, France, andrei.postnikov@univ-lorraine.fr
[2]Kyrgyz-Russian Slavic University, Bishkek, Kyrgyzstan

The sensitivity of the THz radiation to water, which is one of the most important components of the biological tissue, can be used in practical schemes of visualizing a tumour. Namely, the water molecules absorb throughout the entire THz band (0.1–10 THz) [1], and the water content in cancerous tumours is higher than in normal tissues. One can think of a setup using the transmission geometry (see Fig. 1), which would normally involve *in vitro* study of a thin, clinically prepared tissue sample. In this approach, the necessary contrast for imaging the tumour and discerning it from the normal tissue is provided by enhanced water content in cancer cells. Otherwise, the fact that the THz waves cannot penetrate moist tissue motivated for the development of imaging in the reflection geometry, better suited for investigations *in vivo*, the most straightforwardly in relation with the skin cancer (Fig. 2). The reflectance of THz radiation, and hence the contrast in imaging the area with cancer cells, is enhanced as the water temperature in them increases [2,3,4,5]. To heat water in cancer cells, the gold nanoparticles (GNPs) can be premediately delivered therein; corresponding techniques are known [6,7]. Then the tumour is non-invasively treated by irradiating with near-infrared (NIR) laser beam at ~650–1350 nm wavelength (this is the so-called "therapeutic window" where light has its maximum depth of penetration into the tissues). Under irradiation, the surface plasmons are excited in the GNPs; on dumping out the plasmons, the water is heated around the nanoparticles in cancer cells. In consequence, the cancer cells start to reflect the incident THz radiation even more efficiently. The feasibility of the THz imaging of the body with skin cancer in reflection geometry have been demonstrated in [8].

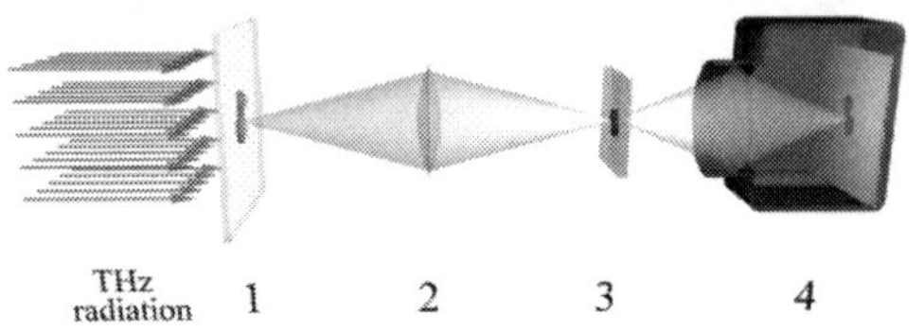

Fig. 1. Transmission mode for *in vitro* studies: 1 – a tissue sample, 2 – the THz objective (high resistivity float zone silicon, high density polyethylene or Teflon®), 3 – THz-to-IR-converter, 4 – highly sensitive IR camera

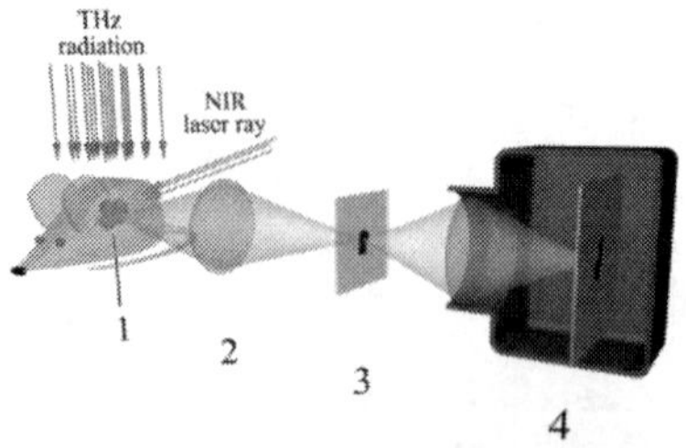

Fig. 2. Reflection mode for *in vivo* imaging with a usage of the near-infrared (NIR) laser for excitation of surface plasmons in GNPs inside a tumour (for heating water in cancer cells): 1 – tumour, 2, 3, 4 – the same as in Fig. 1

The nature of the THz radiation source to be used is not crucial for the present contribution; we note, nevertheless, that a possible design for such source was suggested in [9], having, as its working element, gold nanobars or nanorings which ought to be irradiated by microwaves in order to emit THz photons with energies within the full width at half maximum of the longitudinal acoustic phononic density of states of gold (16 –19 meV, i.e., 3.9 – 4.6 THz), with a maximum at about 4.2 THz (17.4 meV). In Ref. [10] it was shown that gold nanorhombicuboctahedra could be used as emitters of radiation at 0.54 and at 8.7 THz, important for the THz imaging of human skin cancer, in the context of findings of Ref. [11, 12] and possible impact on the improvement of spatial resolution.

The essential element for the hence suggested THz inspection device is the THz-to-infrared converter consisting of a layer (a matrix) of GNPs deposited onto (or, embedded into) a substrate, transparent in both THz and IR ranges. In both setups outlined by Figs. 1 and 2, the image (in the THz rays) originating from the tissue sample (position 1) is projected by the objective (position 2) onto the two-dimensional THz-to-IR-converter (position 3). The GNPs, on irradiation with THz rays, convert the energies of THz photons into heat, being so the bright spots for subsequent detection by a highly sensitive IR-camera (position 4). In this quality, commerciably available devices can be considered, which allow nowadays the temperature sensitivity of 14 to 50 mK [13,14].

A preliminary description of the scheme has been outlined by us in Ref. [15]. The present analysis reiterates the physical mechanism, which includes an excitation of the Fermi electron via an adsorption of a THz photon, and the subsequent relaxation with releasing a longitudinal phonon. The constraints imposed by size and shape of the nano-objects are discussed, along with the manifestation of the momentum and energy conservation laws in the preference for different channels of relaxation. Technically, calculations are done of the THz irradiation power needed to be delivered in order to enable detection of a thermal image developed on the converter by IR cameras. The static distribution of temperature around a

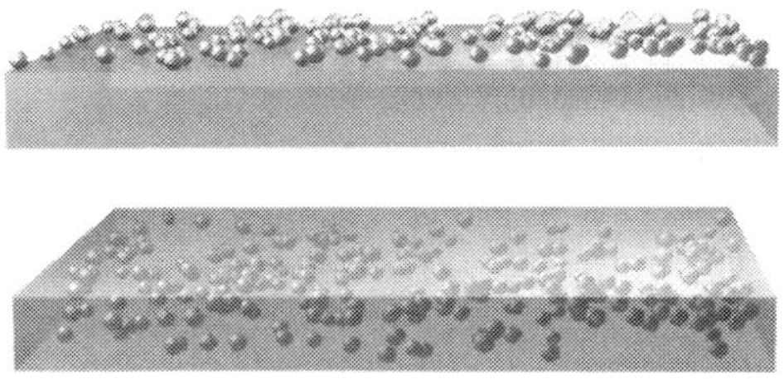

Fig. 3. Schemes of the THz-to-IR converter: above – in the form of a substrate transparent in THz wavelength range with GNPs; below – in the form of a matrix transparent in THz wavelength range with embedded GNPs

GNPs of different sizes embedded in a substrate (specifically, Teflon® and silicon have been considered) was calculated, along with dynamical onset of the temperature, in order to estimate the size and reaction time of bright spots being created in the converter, and thus to judge about the latter's spatial resolution and reaction time.

Concerning the realization of the THz-to-IR converter, two schemes shown in Fig. 3 may come into discussion. That in the form of a thick film with embedded GNPs (Fig. 3, lower scheme) seems preferable over single-layer deposition (Fig. 3, upper scheme), because it allows to achieve larger "projected" density of GNPs per surface unit, avoiding at the same time to place them too closely.

The obtained theoretical results demonstrate that the suggested approach can be realized with the THz-to-IR converter made of Teflon® film of ~ 0.1 nm thickness, containing GNPs of ~10 nm diameter. With this, the suggested design is only waiting for practical tests.

Considering a possible impact in medicine, one can recall that, in contrast to X-rays, the THz radiation is not ionizing and not harmful to living organisms. The development of practical THz imaging setups is one of the "hottest" areas in nanotechnology-supported modalities for the cancer diagnostics. The THz diagnostics in reflection geometry allows for non-invasive *(in vivo)* THz imaging of skin cancer's surface features. It could be performed *in situ*, without time losses on standard *in vitro* histological tests. Other areas of dermatology, where the THz imaging in reflection geometry might be advantageous, are (*i*) skin burns and wound inspection through bandages, and (*ii*) monitoring the treatment of skin conditions (like psoriasis), since this allows to avoid the direct contact with the skin as *e.g.* under ultrasound investigation. Even if the THz radiation is strongly absorbed by water and does not penetrate tissue to any significant depth, the use of the transmission geometry and clinically prepared tissue samples (both of inner organs and the skin) may offer an interesting extension to established techniques. One can mention in this relation that the THz imaging is less costly than, for instance, the magnetic resonance study.

References

1. *Berry, E., Walker, G. C., Fitzgerald, A.J., Zinov'ev, N.N., Chamberlain, M., Smye, S. W., Miles, R.E., Smith, M.A.* Do *in vivo* terahertz imaging systems comply with safety guidelines? // J. Laser Appl., 2003. V. 15, No. 3. P. 192–198.

2. *Rønne, C., Thrane, L., Åstrand, P.O., Wallqvist, A., Mikkelsen, K.V., Keiding, S.R.* Investigation of the temperature dependence of dielectric relaxation in liquid water by THz reflection spectroscopy and molecular dynamics simulation. // J. Chem. Phys., 1997, V. 107, No. 14, P.5319–5331.

3. *Son, J.H.* Terahertz electromagnetic interactions with biological matter and their applications. // J. Appl. Phys., 2009, V. 105, No. 10, P. 102033-1–102033-10.

4. *Oh, S.J., Maeng, I., Shin, H.J., Lee, J.,.Kang, J., Haam, S., Huh, Y.M.,.suckSuh, J., HiukSon, J.* Nanoparticle contrast agents for terahertz medical imaging. // 33rd International Conference on Infrared, Millimeter and Terahertz Waves (IRMMW-THz 2008, Pasadena, USA). P. 1–2. DOI: 10.1109/ICIMW.2008.4665813.

5. *Oh, S.J., Kang J., Maeng, I., Suh, J.S.,. Huh, Y.M., Haam, S., Son, J.H.* Nanoparticle-enabled terahertz imaging for cancer diagnosis. // Opt. Express, 2009, V. 17, No. 5, P. 3469–3475.

6. *Loo, C., Lowery, A., Halas, N., West, J., Drezek, R.* Immunotargeted nanoshells for integrated cancer imaging and therapy. // Nano Lett., 2005. V. 5, No. 4, P. 709–711.

7. *El-Sayed, I.H., Huang, X., El-Sayed, M.A.* Selective laser photo-thermal therapy of epithelial carcinoma using anti-EGFR antibody conjugated gold nanoparticles. // Cancer Lett., 2006. V. 239, No. 1, P. 129–135.

8. *Woodward, R.M., Cole, B.E., Wallace, V.P., Pye, R.J., Arnone, D.D., Linfield, E.H., Pepper, M.* Terahertz pulse imaging in reflection geometry of skin tissue using time-domain analysis techniques. // Physics in Medicine and Biology, 2002, V. 47, No. 21, P. 3853-3863.

9. *Moldosanov, K.,. Postnikov, A.* A terahertz-vibration to terahertz-radiation converter based on gold nanoobjects: a feasibility study. // Beilstein J. Nanotechnol., 2016, V. 7, P. 983–989.

10. *Postnikov, A.V., Moldosanov, K.A.* Suggested design of gold-nanoobjects-based terahertz radiation source for biomedical research. // Nanotechnology, 2018. V. 29, No. 28, P. 285704 (8 pp).

11. *Wallace, V.P., Fitzgerald, A.J., Pickwell, E., Pye, R.J., Taday, P.F., Flanagan, N., Ha, T.* Terahertz pulsed spectroscopy of human basal cell carcinoma. // Appl. Spectrosc. 2006, **60**(10), 1127–1133

12. *Pickwell, E., Wallace, V.P.* Biomedical applications of terahertz technology. // J. Phys. D: Appl. Phys., 2006, V. 39, No. 17, P. R301– R310.

13. Infrared Cameras Inc., Mirage 640 P-series | fixed / process control calibrated thermal camera with temperature measurement. https://infraredcameras.com/thermal-infrared-products/mirage-640-p-series/.

14. FLIR A6700sc MWIR, science-grade MWIR INSB camera. https://www.flir.com/ products/a6700sc-mwir/.

15. *Moldosanov, K.A., Postnikov, A.V., Lelevkin, V.M,. Kairyev, N.J.* Terahertz imaging technique for cancer diagnostics using frequency conversion by gold nano-objects. // Ferroelectrics, 2017, V. 509, P.158–166.

EPJ Web of Conferences **195**, 10011 (2018) https://doi.org/10.1051/epjconf/201819510011
TERA-2018

THz absorption spectra of glucose and its polymers

Anna Semenova[1], Yu.S. Guseva[1,2] V.L. Vaks[1,2], A.N. Panin[1], D.A. Babarina[2], S.S. Morunova[3], A.S. Vilkov[3]

[1]Institute for Physics of Microstructures, Nizhny Novgorod, Russia, semenanna@yahoo.com
[2]Lobachevsky State University, Nizhny Novgorod, Russia,
[3]Nizhny Novgorod State Agricultural Academy, Nizhny Novgorod, Russia

Introduction

Terahertz (THz) spectroscopy has tremendous potential for biological and medical applications. First, THz spectroscopy is used for chemical analysis of multicomponent gas mixture, particularly, for exhaled breath (EB) analysis. There are narrow absorption lines in the THz range caused by quantum transitions between rotation levels of molecules in gas phase and different lines are corresponded with different chemical substances. Because the exhaled breath is a mixture of not only oxygen, nitrogen and carbon dioxide but more than 800 gases which concentrations reflect the level of homeostasis, as well as the severity of pathological conditions, EB analysis is a perspective method for diagnostic purposes. It is non-invasive and unpainful for patients. The probability of infection transmission is very low for the method because there is no contact with any biological fluids. It is rather inexpensive and very sensitive for small concentration of measured substances. Therefore, THz spectroscopy of exhaled breath is becoming a powerful method of medical diagnostics, especially, incipient stage diagnostics and prevention of socially important diseases [1, 2].

Second, THz technic could provide a powerful tool for investigating and even manipulating of biomolecules *in situ*, in water solution, and *in vivo*, in cell cultures and whole organisms. THz irradiation is no ionizing and is treated as non invasive for biomolecules [2, 3] if irradiation power and exposure times do not exceed some threshold. Worth noting that absorption of organic samples and liquid water in the THz frequency range is caused by low-frequency vibration modes of the molecules, which occur than large groups of atoms of the molecule move coherent, and hydrogen bond net oscillation in the sample. Hydrogen bonds stabilize spatial structure of biomolecules whereas which determine eigen frequencies of the molecular vibration modes. Therefore, THz spectra are unique for each spatial structure of each molecule [2 - 6]. Worth noting that biomolecular spatial structures commonly described in the terms of configuration and conformation are crucial for their functions. Configuration means the mutual bracing of chemical bonds in a molecule, it can not change without chemical reaction. For example, R and S forms of any chiral molecule differs as left and right hand corresponds to different conformations. It is well known that even R and S isomers of rather simple molecules such as amino acids represent different pharmacological effects [7]. Conformation means momental spatial displacement of atoms in a molecule, it continuously oscillating near some equilibrium state by thermal fluctuations without any chemical reactions. Conformational state and dynamics determine biomolecular activity. For example, specificity of enzyme catalysis is caused by matching of space structure and electric charge displacement between the enzyme and substrate molecules; structural changes occurring at denaturation lead to loss of catalytic activity (notable that denaturation could be accompanied by no chemical conversion) [7, wiki 8]. Some conformational transitions are lethally dangerous, for example, PrP protein abnormal forms, so-called prions, stop execute their common functions and begin catalyze only PrP molecules conformational transmission from normal to the same abnomal form causing neurodegenerative diseases [прионы, 9]. Therefore there is necessary medicine task developing conformational diagnostics of biomolecules *in situ*, for example in water solutions, and *in vivo*, for example, in yeast cells for understanding how exactly they manage to control their specific prions [elements 10]. However, commonly used at present methods such as X-ray analysis, electronic and atomic force microscopy requires molecular crystals or adsorbed molecules. On the one hand, THz spectroscopy is becoming the urgent method investigating of conformational state and dynamics of biomolecules in liquid media that minimizes the influence on the explored molecules [2 - 6] On the other hand, there are announcements about non-thermal influence of strong THz field on cell life activity, cellular differentiation and animals behaviour [4, 5]. The reported effects manifested frequency dependance so powerful and frequency stable THz sources could be used for exciting specific conformational transitions, for example, for conversion from prion to normal form of PrP protein [11, 12].

Purposes

While THz spectroscopy shows great perspectives for diagnostics of molecular spatial structure, nevertheless its practical application is obstructed by some problems mainly caused by rather short story of THz technic and, consequently, rather small experience in the field. There are not enough models pointing at what to look at in THz absorption spectra. Our work is aimed at comparative analysis of absorption spectra measured for water solutions of different carbohydrates with different concentrations. The explored samples include water solutions of monosaccharides glucose and fructose, disaccharide sucrose and polysaccharide amylose.

First investigation of dependance absorption spectra on concentration is planned for the glucose solutions with concentration varying from 10% to 40%. There is some possibility of deviation from the most common linear dependance because THz absorption of water-water and water-glucose hydrogen bonds differs and dependance of the amount of both types of bonds on the solution concentration is not linear at high concentrations.

Second investigation of molecular configuration influence on absorption spectra is planned for two isomers, glucose and fructose, prepared in form of water solutions with the same concentration.

Third there is planned absorption spectra comparison for water solutions of glucose and its polymer amylose for finding out any specific corresponds to structure of long linear molecule.

Final there is planned to compare the spectrum of solution containing 20% glucose and 20% fructose with the spectrum of 40% sucrose solution. A molecule of fructose could be hydrolyzed into two molecules (one glucose and one fructose) but in neutral solutions at room temperature both glucose-fructose mixture and sucrose are stable. Thus difference between the two spectra could be caused by difference between molecular structures with resembling chemical composition.

Experimental setup

The principal scheme of the experimental setup is shown on fig. 1, where S means source of monochromatic THz irradiation with frequency turnable in the range from 112,5 to 114,8 GHz, D means detector of THz irradiation and C means cell with liquid sample or pure water (control).

The distance between the cell and the detector was fixed while the distance between the cell and the source was variated. The detected power depends quasi periodically on the distance due to standing wave forming. There are two measurements for the each frequency and each sample with two different distances between the source and the cell corresponded to two neighbour maximums of detected power.

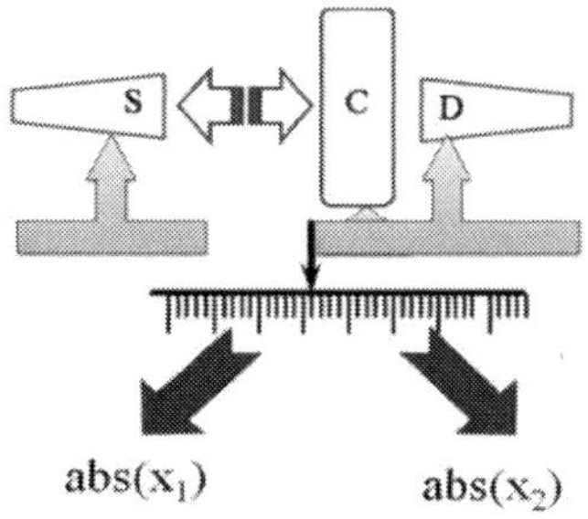

Fig. 1. Principal scheme of experimental setup. Two measurements were carried out for each sample and frequency.

Acknowledgements

The reported study was funded by RFBR according to the research project № 18-32-00328.

This work was carried out in the framework of state targets N 0035-2014-0206.

References

1. *Vaks, V.L., Domracheva, E.G., Sobakinskaya, E.A., Chernyaeva, M.B.* Exhaled breath analysis: physical methods, instruments, and medical diagnostics // Phys.-Usp. 2014. V. 57, No. 7. P. 684–701.
2. *Vaks, V.L., Domracheva, E.G., Sobakinskaya, E.A., Chernyaeva, M.B.* Methods and Instruments of High-Resolution Transient THz Spectroscopy for Diagnostics of Socially Important Diseases // Phys. Wave Phen. 2014. V. 22, No. 3. P. 177–184.
3. *Zhang, X.-C.* Terahertz wave imaging: horizons and hurdles // Phys. Med. Biol. 2002. V. 47. P. 3667–3677.
4. *Alexandrov, B.S., Phipps, M.L., Alexandrov, L.B., Booshehri, L.G. et. al.* Specificity and Heterogeneity of Terahertz Radiation Effect on Gene Expression in Mouse Mesenchymal Stem Cells // s. rep. 2013. V. 1184, No. 3. P. 1–8.
5. *Alexandrov, B.S., Gelev, V., Bishop, A.R., Ushevaet, A., et al.* DNA Breathing Dynamics in the Presence of a Terahertz Field // Phys. Lett. A 2010. V. 374, No. 10. P.1214–1217.
6. *Markelz, A., Whitmire, S., Hillebrecht, J. Birge, R.* THz time domain spectroscopy of biomolecular conformational modes // Phys. Med. Biol. - 2002 - 47; URL: stacks.iop.org/PMB/47/3797V. 57, No. 7. P. 684–701.
7. *Vol'kshtejn, M.V.* Biofizika // Nauka, Moscow 1988
8. *Koshland, D.E.* Application of a Theory of Enzyme Specificity to Protein Synthesis // Proc. Natl. Acad. Sci. U.S.A. 1958. V.44, No.2. P.98-104.
9. *Zuev, V.A.* Priony – osobyj klass vozbuditelej medlennyh infekcij cheloveka i zhivotnyh // RMZh 2013. V. 30. P. 1559–1566.
10. *Vaks, V.L., Domracheva, E.G., Sobakinskaya, E.A., Chernyaeva, M.B.* Exhaled breath analysis: physical methods, instruments, and medical diagnostics // Phys.-Usp. 2014. V. 57, No. 7. P. 684–701.
1. *Vaks, V.L., Domracheva, E.G., Sobakinskaya, E.A., Chernyaeva, M.B.* Exhaled breath analysis: physical methods, instruments, and medical diagnostics // Phys.-Usp. 2014. V. 57, No. 7. P. 684–701.

Hastings, D. E., Wang, J. K. The radiation impedance of electrodynamic tether with end connectors // Geophys. Res. Lett. 1987. V. 14, No. 6. P. 519–522.

2. *Bergman, D. J., Stroud, D.* Physical properties of macroscopically inhomogeneous media // *Solid State Physics: Advances in Research and Applications*, edited by H. Ehrenreich and D. Turnbull. New York: Academic Press. 1992. V. 46. P. 147–269.

Interaction of terahertz radiation with tissue phantoms: numerical and experimental studies

O. A. Smolyanskaya[1], Q. Cassar[2], M. S. Kulya[1], N. V. Petrov[1], K. I. Zaytsev[3,4], A. I. Lepeshkin[1], J.-P. Guillet[2], P. Mounaix[2], V. V. Tuchin[1,5,6]

[1] ITMO University, Saint-Petersburg, Russia, smolyanskaya@corp.ifmo.ru
[2] Bordeaux University, IMS Laboratory, Bordeaux, France
[3]Bauman Moscow State Technical University, Moscow, Russia
[4]Prokhorov General Physics Institute of the RAS, Moscow, Russia
[5]Saratov State University, Saratov, Russia
[6]Institute of Precision Mechanics and Control of the RAS, Saratov, Russia

Introduction

New biomedical devices require test objects to check their performance and periodic calibration to monitor the system efficiency over the time. A phantom is a bio-like object mimicking the intrinsic real-tissue properties that serves as a test object. A variety of phantoms are commercially available. Phantoms are already used in spectroscopy within various spectral range. However, currently, the THz frequency range suffers from a lack of investigations in this direction. Only the first steps have been performed [1]. In the present work, we investigate interaction of submillimeter radiation with the breast-mimicking and skin-mimicking phantoms. Results for breast-mimicking phantom were compared with numerical models based on a double-Debye model of the dielectric permittivity for different content of fat, fibrous and cancerous tissue. The wave front propagation using the angular spectrum representation was used to simulate the theoretical response of skin-mimicking phantoms.

Model of wavefront propagation

Electromagnetic fields can be represented in various ways. The angular spectrum representation is a numerical technique describing optical fields in a homogeneous media [2]. Optical fields are described as a superposition of plane waves and evanescent waves. A useful approach to describe optical field diffraction is to conduct the Fourier analysis at a given plane so that the different Fourier components of the field distribution are identified as plane waves propagating away from that plane in different directions. Thus taking less computation time for numerical reconstruction. This technique is preferable for the analysis of light diffraction by histological slides [3]. Angular spectrum representation consists of the following stages: (i) the representation of the field through the angular spectrum of 2D waves (Eq.1); (ii) multiplication by the transfer function that contains a complex refractive index of the object (Eq.2); (iii) the back transition from plane waves to the wave field in the calculated z-plane (Eq.3).

$$C(f_x,f_y,\nu)=\int_{-\infty}^{\infty}\int_{-\infty}^{\infty}G(\tilde{x},\tilde{y},\nu)_{z=0}\exp\left(-2\pi i\left(\tilde{x}f_x+\tilde{y}f_y\right)\right)d\tilde{x}d\tilde{y} \quad (1)$$

$$g_{x,y}(f_x,f_y,\nu,z)=C(f_x,f_y,\nu)\exp\left(-i\frac{2\pi\nu n(\nu)}{c}\sqrt{1-\left(\frac{f_x\cdot c}{\nu n(\nu)}\right)^2-\left(\frac{f_y\cdot c}{\nu n(\nu)}\right)^2}\cdot z\right) \quad (2)$$

$$G(x,y,\nu,z)=\int_{-\infty}^{\infty}\int_{-\infty}^{\infty}g_{x,y,z}(f_x,f_y,\nu,z)\exp\left(2\pi i\left(xf_x+yf_y\right)\right)df_xdf_y \quad (3)$$
,

where $n(\hat{x},\hat{y},\nu)=n_{re}(\tilde{x},\tilde{y},\nu)+i\cdot n_{im}(\tilde{x},\tilde{y},\nu)$ is the spatial and frequency distribution of the complex index of refraction.

Materials and methods

a. Sample preparation

We have used three-component phantoms made of fat, protein and water of different content. Vegetable oil (10 – 75%), soya (13 – 75%) and water (0 – 70%) were homogenized to generate an emulsion. Then, the phantoms were deposited into vacuum packages.

b. THz-Spectrometer

The commercially available TPS 4000 (Teraview Ltd, UK) spectrometer working in reflection mode was used. The spectral range of the system is from 0.06 to 4.50 THz. The entire system was under an air-dried dome to limit the interaction of water molecules with generated pulses. Samples were mounted on a sapphire substrate during measurement. The sapphire cut was chosen not to exhibit a birefringence that would disrupt results of measurements [4]. Each sample was measured 5 times, independently, *i.e.* the sample was removed and replaced again into system.

c. Numerical modelling

A double-Debye model of the dielectric permittivity was used to compare optical properties of the three-component phantoms and some biological tissue (adipose, fibrous and cancerous) taken from our previous paper [5].

Results and discussion

The propagation dynamics of the pulsed THz radiation through a skin sample was numerically simulated in the temporal and spectral domains. The index of refraction,

n_{re}, absorption coefficient, α, as well as real, ε_{re}, and imaginary, ε_{im}, part of dielectric permittivity was calculated from the experimentally obtained data. Figure 1a depicts the time-domain spatial evolution of the THz pulse. Agreement between measured and simulated THz waveforms is confirmed by the comparison shown in Figure 1b.

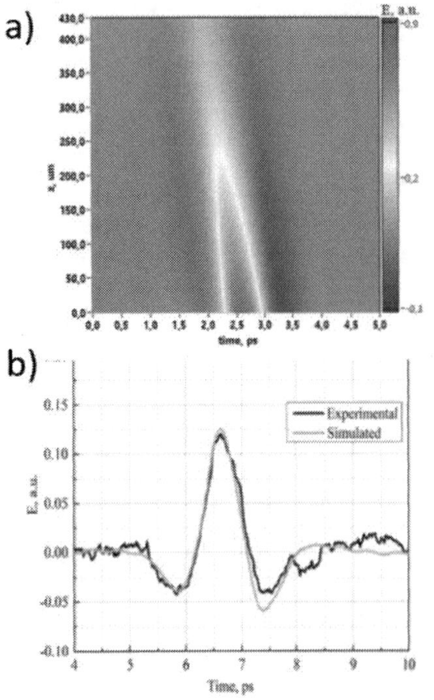

Fig. 1. Experimental and numerically simulated THz pulses propagated through a skin-mimicking sample. (a) THz spatial distribution. (b) THz waveforms.

In Figure 2 the spectral dependences of the absorption coefficient of the three-component phantoms consisting of fat, protein and water are depicted. It is shown that within the spectral window, the absorption coefficient of phantoms containing water exceeds the absorption coefficient of phantoms free of water. This is due to the fact, that water molecules, which strongly absorb THz radiations, are replaced by low absorbing components (fat and protein). Our results were compared with simulated signals of adipose, fibrous and cancerous tissues. taken from our previous paper [5].

Conclusion

To overcome the lack of knowledge on tissue-mimicking phantoms within the submillimeter spectral range, we conducted preliminary studies on skin- and breast tissue-mimicking samples. Simulations performed for skin-mimicking phantoms

have match the experimental spectral data. Thus demonstrating advantages of the technique used in this

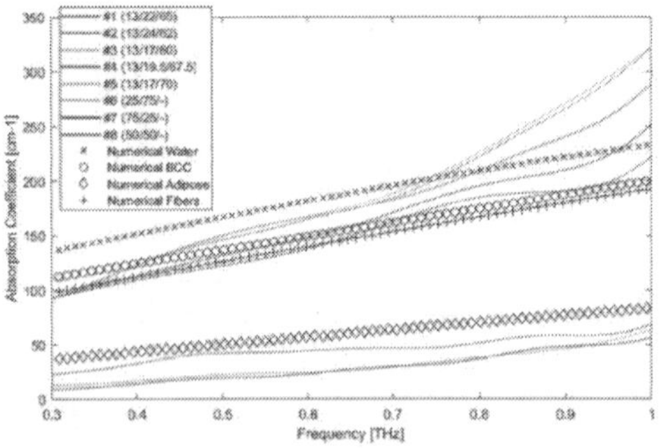

Fig. 2. Spectral dependences of the absorption coefficient of the three-component phantoms consisting of fat, protein and water in the THz frequency range (solid-colored lines) versus simulated signal of water and purely adipose, fibrous and cancerous tissue types.

study. The absorption coefficient spectral dependences extracted from measurements for breast-mimicking phantom, are clearly differentiated into two distinct groups. One for phantoms containing water and those being water-free. Such a discrimination is also shown via the double-Debye model of the dielectric permittivity. Thus, finding of a proper composition of tissue phantom is possible by an appropriate matching of model and experimental data. Deeper investigations are however still required to fulfill a set of phantoms that could be used to perform tissue measurement within the THz-range. Such a set would be profitable to design new medical THz-devices without human biopsy.

Acknowledgments

The reported study was funded by RFBR and CNRS according to the research project №18-51-16002, and by the Government of the Russian Federation (Grant 08-08).

References

1. *E Liakhov, O Smolyanskaya, A Popov, E Odlyanitskiy, N Balbekin, M Khodzitsky,* Fabrication and characterization of biotissue-mimicking phantoms in the THz frequency range // Journal of Physics: Conference Series. 2016. No. 735. P. 012080.

2. *Petrov, N. V., Kulya, M. S., Tsypkin, A. N., Bespalov, V. G., & Gorodetsky, A.* Application of terahertz pulse time-domain holography for phase imaging // IEEE Transactions on Terahertz Science and Technology, 2016, V. 6, No. 3, P. 464-472.

3. *O. A. Smolyanskaya, I. J. Schelkanova, M. S. Kulya, E. L. Odlyanitskiy, I. S. Goryachev, A. N. Tcypkin, Ya V. Grachev, Ya G. Toropova, and V. V. Tuchin* // Glycerol dehydration of native and diabetic animal tissues studied by THz-TDS and NMR methods // Biomedical Optics Express, 2018, V. 9, No. 3, P. 1198-1215.

4. *Q. Cassar, A. Al-Ibadi, la. Mavarani, Ph. Hillger, J. Grzyb, G. Macgrogan, Th. Zimmer, U.R. Pfeiffer, J.-P. Guillet, P. Mounaix* Pilot study of freshly excised breast tissue response in the 300 – 600 GHz range // Biomed Opt Express. 2018. V. 9 No. 7. P. 2930–2942.

5. *A. J. Fitzgerald, E. Pickwell-MacPherson, and V. P. Wallace,* Use of finite difference time domain simulations and Debye theory for modelling the terahertz reflection response of normal and tumour breast tissue // Plos One **9**, 1-9 (2014).

The influence of terahertz radiation on biochemical metabolism of blood in the experiment

A.G. Soloveva, P.V. Peretyagin, A.G. Polyakova, N.V. Didenko

Federal State Budgetary Educational Institution of Higher Education «Privolzhsky Research Medical University» of the Ministry of Health of the Russian Federation, Nizhny Novgorod, Russia, sannag5@mail.ru

In recent years, close attention has been paid to studies related to the impact of electromagnetic radiation (EMR) terahertz (THz) range on the living organism. Until now, the final idea of the ways and physical and chemical mechanisms of EMR THz action has not been formed, so it is important to identify all possible "points of application" of such an impact on biological systems [1]. Because the level of lipid peroxides reflecting the integrity of biological membranes changes under the action of different EMR ranges [2], and the antioxidant system controls the intensity of free radical processes, providing protection of cells and tissues [3], these systems were chosen to detect changes caused by THz radiation.

The aim of the work was to study the effect of electromagnetic radiation of the terahertz range on the functional and metabolic parameters of rat blood in the process of reparative regeneration.

Materials and methods

The experiment was carried out on 15 male Wistar rats weighing 250-300 g. in accordance with the requirements of the Geneva Convention "International Guiding Principles for Biomedical Research Involving Animals" (Geneva, 1990). The impact of EMR was carried out in the direct sound irradiation EMR THz (110-170GHz) from the experimental emitter apparatus "AMFIT" (N.Novgorod).

Animals were divided into 3 equal groups: 1– intact healthy rats, 2 – control – operated animals without any effects, 3 – operated animals with irradiation EMR THz at a dose of 0.12 MJ within the next 7 days (10 minutes 1 time per day). The impact was carried out on the base area of the displaced flap, coinciding with the localization of the skin projection of the center of vegetative regulation. Surgical intervention in 2 and 3 groups of rats and withdrawal from the experiment on day 7 after the operation was carried out under intramuscular anesthesia (Zoletil + Xyla).

Blood stabilized with sodium citrate (1:9) was used for the studies. Glucose and lactate concentrations were measured on a Super GL ambulance device (Germany) in plasma and red blood cells. The activity of lactate dehydrogenase (LDH) in direct (LDHdir) and reverse (LDHrev) reactions was determined on the spectrophotometer Power Wave XS (Bio-Tek, USA). The total content of nitric oxide metabolites was determined by the method of L.C.Green with co-authors in the modification of V. A. Metelskaya and N. G. Gumanova [4].

The intensity of lipid peroxidation (LPO) of a biological object by the level of malonic dialdehyde (MDA) in plasma and hemolysate of washed erythrocytes (1:10) [5], the activity of superoxide dismutase (SOD) [6] in hemolysate of washed erythrocytes (1:10) and catalase [7] in erythrocytes were determined by spectrophotometry.

The state of microcirculation and the degree of microcirculatory insufficiency with the analysis of the parameter of microcirculation (PM) characterizing the degree of tissue perfusion by blood, using wavelet analysis, were evaluated according to laser Doppler fluorometry on the analyzer of capillary blood flow LAKK-M (Russia).

Statistical processing was carried out using the program Statistica, version 6.0, using the Student criteria. Differences were considered statistically significant at $p<0.05$.

Results

In animals of group 2 the activation of LPO in erythrocytes was detected, MDA concentration increased by 20% compared to intact animals (Fig.1). In the erythrocytes of control animals compared with the intact group a compensatory increase of catalase specific activity on 17% ($p=0.02$) was registered, but a SOD activity decreased on 16% ($p=0.003$) (Fig.2). It indicated about an insufficient degree of compensation of the LPO by the antioxidant system in the operated animals.

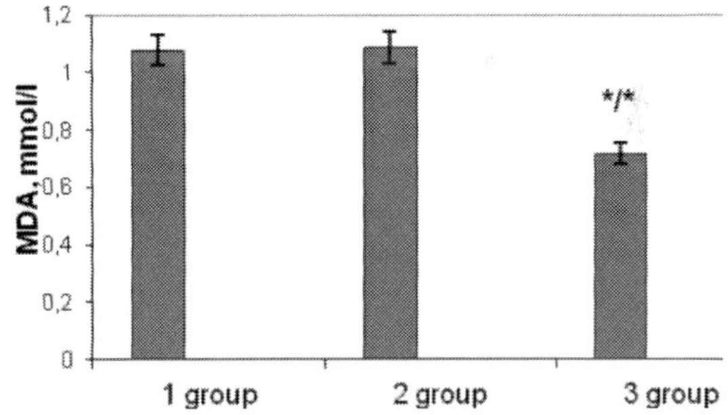

Fig.1. The concentration of malonic dialdehyde in the blood plasma of healthy and operated animals under the influence of EMR THz.
Note: * - differences are statistically significant compared to group 1; ** - differences are statistically significant compared to group 2.

The decrease of the intensity of LPO in plasma under the influence of THz-irradiation was observed. MDA concentrations in plasma decreased by 34% (relative to both comparison groups) under the influence of EMR THz.

The activity of enzymes of bioradical protection increased in the experimental group after irradiation of EMR THz. The most positive dynamics was observed for the SOD. SOD activity after surgery in the experimental group of animals increased by 30%

(p=0.018) compared to the control and reached the indicator of healthy animals. The activity of catalase also increased by 12% (p=0,025) compared to the rate of intact rats under the influence of irradiation. It reduces the free radical forms of oxygen and inactivates LPO. Consequently, THz exposure in the noise mode of radiation has an antioxidant effect through the activation of SOD and catalase, which, in turn, inhibit the release of catecholamines from the nerve endings and adrenal glands, as well as the action of these monoamines at the postsynaptic level [8]. Therefore, it is possible that one of the mechanisms that provide a decrease in the intensity of LPO under the action of EMR THz is the suppression of hyperactivity of one of the most important stress-implementing systems of the sympathoadrenal system.

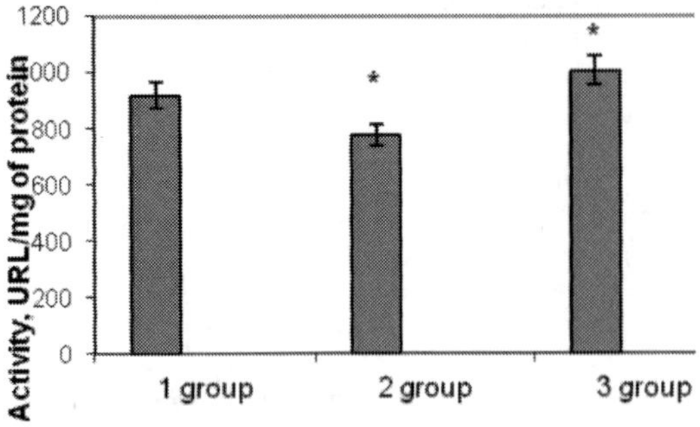

Fig.2. Activity of blood superoxide dismutase in healthy and operated animals under the influence of EMR THz

Note: * - differences are statistically significant compared to group 1; ** - differences are statistically significant compared to group 2.

Thus, in the experimental group of animals the predominance of antioxidant activity over the processes of lipoperoxidation was observed. It may indicates the inhibition of the biological oxidation system due to the suppression of reactive oxygen species under the influence of EMR THz. In response to the irradiation of the projection of the center of vegetative regulation, apparently, there is a neurohumoral activation of the antioxidant system, which blocks the processes of LPO. The revealed change in the direction of LPO processes under irradiation may be due to changes in the structure (conformation) of the cell surface of membrane components due to the weakening of hydrophobic bonds [1].

There was an increase in glucose and lactate in rats of the control group in plasma and erythrocytes. An increase in glucose concentration can contribute to a significant formation of free radicals through glycation reactions and autooxidation of glucose. The decrease of LDHdir on 33% in trauma in comparison with healthy animals.

EMR irradiation of THz in the noise range 110-170 GHz contributed to a decrease in the concentration of lactate in plasma and erythrocytes by 34% and 51%, respectively, compared with the animals of the control group. It was shown that under the influence of EMR THz activity of LDHdir increased on 47%, LDHrev – on 23% compared to the LDH activity of the rats without irradiation. In addition, it is possible that electromagnetic radiation promotes the use of lactate in carbohydrate metabolism in the formation of 2,3-diphosphoglycerate, the main source of energy in the body.

The zone of irreversible vascular changes in the flap at animals of 3 group after irradiation of EMR with a wide frequency range (110-170 GHz) containing spectra of the main metabolite molecules, was less pronounced in comparison with the control.

Since the most common among all microcirculatory disorders are local vascular spasm and decrease in blood flow rate, the registered preliminary data on animals confirm the spasmolytic mechanism of action of electromagnetic radiation of the THz range on the smooth muscle vascular wall of animals. At the same time, an increase in the content of nitrites and nitrates in the blood serum was revealed after irradiation of EMR THz.

Conclusion

The results showed that EMR of THz with a noise range of 110-170 GHz in ischemia led to an increase in the antioxidant status of the blood, as well as inhibition of peroxide states. It was shown that EMR THz of noise range 110-170 GHz increased the energy metabolism of blood in case the tissue ischemia.

References

1. *Bajinyan, S.A., Meliksetyan, A.M., Malakyan, M.G. et al.* // Millimeter waves in biology and medicine 2003. V.32, No 4. P. 12.
2. *Gapeev, A.B., Cheremis, N.K.* Mechanisms of biological action of electromagnetic radiation of extremely high frequencies at the cellular level // Biomedical technologies and Radioelectronics. 2007. No. 2-4. P. 44-61.
3. *Ulashchik, V.S.* Electromagnetic waves of the terahertz range and their therapeutic and prophylactic use // Questions of balneology, physiotherapy and physical therapy. 2007. No. 4. P. 3–7.
4. *Metel'skaya, V.A., Gumanova, N.G.* Screening method for determining the level of nitric oxide metabolites // Clinical laboratory diagnostics. 2005. No. 6. P. 15–18.
5. *Uchiyama, M., Mihara, M.* Analyt. Bbiochem. 1978. No. 86. P. 271.
6. *Sirota, T.V.* A new approach in the study of the process of adrenaline auto-oxidation and its use for measuring the activity of superoxide dismutase // Questions of medical chemistry. 1999. V. 45, No. 3. P. 109-116.
7. *Sibgatullina, G.V., Haertdinova, L.R., Gumerova, E.A. et al.* Methods for determining the redox status of cultivated plant cells: teaching aid. Kazan, Kazan (Volga) Federal University, 2011.
8. *Chuyan, E.N., Ravaeva, M.Y., Nikolskaya, V. A. et al.* Scientific notes of Taurida national University. V.I. Vernadsky. Series biology, chemistry. 2013. V. 65, No. 26. P. 223.

High resolution terahertz spectroscopy for medical, biological and agricultural applications

V. L. Vaks[1,2], E. G. Domracheva[1], S. I. Pripolzin[1], M. B. Chernyaeva[1]

[1]Institute for Physics of Microstructures RAS, Nizhny Novgorod, Russia, elena@ipmras.ru
[2]Lobachevsky University, Nizhny Novgorod, Russia

High resolution terahertz spectroscopy [1] is promising method for analysis of multicomponent gas mixtures including biological ones. Among the spectroscopic methods, the only approach to date that ensures a near-theoretical-limit sensitivity along with a good spectral resolution limited just by the Doppler effect is the nonstationary spectroscopy based on free dumping polarization effect. Other advantages of the spectrometers include easy-to-use configuration and measurement time of several microseconds that provide registration of unstable gases. Application of this method will also benefits in registration of gas-markers absorption lines at one shot without overlapping effect and performing minimal measuring time of few microseconds.

Exhaled breath analysis is a rapidly developing field of noninvasive medical diagnostics. To date, more than 800 substances have been identified among the products of physiological and biochemical processes in the human body. Their concentrations reflect the level of homeostasis, as well as the severity of pathologies and diseases [2,3]

The high-resolution THz spectroscopy method shows considerable promise for fast qualitative and quantitative analysis of various substances (metabolites) appeared in human organism including the breath at specific disease (in particular, cancer diseases, diabetes) for diagnostics at earlier stages or monitoring of patient's state. It allows to control efficiency of current treatment.

The application of high-resolution THz spectroscopy method for medicine diagnostics was used for monitoring the variation of the NO concentration at the radiotherapy course (the absence of nitric oxide in exhaled air of healthy people and its appearance in exhaled air of oncopatients with cancer of lung which had radiotherapy are determined at clinical tests); in the diagnostics and monitoring of the inflammations.

The method of detection of acetone in exhaled breath for earlier diabetes diagnostics shows its competitive ability in comparison with other methods. The experiments of detection of exhaled methanol, ethanol and H_2S and also experiments of simultaneous detection of acetone in exhaled breath and urine samples of diabetes patients show considerable promise for determination of metabolites-markers sets in exhaled breath and vapors of biological liquids unambiguously characterizing the diabetes presence. The results of simultaneous measurements of acetone concentrations at exhaled breath and urine samples of the diabetes patients at the frequency of 151646.597 MHz are presented in Fig.1.

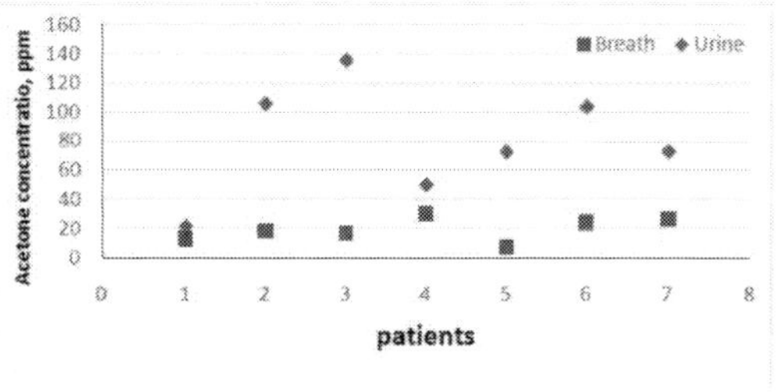

Fig. 1. Simultaneous measurements of acetone concentrations at exhaled breath and urine samples at the frequency of 151646.597 MHz

The possibility of use of terahertz high resolution nonstationary spectroscopy in such agricultural applications as veterinary diagnostics of diseases and pathology states of animals based on analysis of exhaled breath and vapors of biological liquids (urine, saliva); diagnostics of degree of corn mycosis based on analysis of corn odors; monitoring the quality of foodstuff (meat, fish, poultry) based on analysis of its odors shows considerable promise. The investigations of odors of meat and fish samples in fresh and natural decay with using the terahertz high resolution nonstationary spectroscopy methods for were carried out. The results reveal that odor of fresh meat contains ethyl formate, diethyl ether, alanine and glycine, that disappear during natural decomposition process. The latter is characterized by appearing the sulfur compound such as hydrogen sulfide and sulfur dioxide. After that the decaying process is going on with ammonia and nitrogenous organic substances appearance (formamide, ethylamine).In the spoiled fish odor propylene glycol, urethane, hydrogen sulfide, sulfur dioxide and nitriles substances have been found. The high resolution THz spectroscopy as "electronic nose" has been demonstrated to be very promising for food quality analyzing. [4]

Grain	Substances
Winter wheat (Nemchinivskaya-17)	Glycolaldehyde ($HCOCH_2OH$)
	Formic acid (HCOOH)
Summer wheat (Sitara)	Formamide (NH_2CHO)

The preliminary results on studying the grain odors composition of winter and summer wheats are presented in the Table. Some chemical substances, which can be results of defeating the grain by necrotrophic pathogens (formamide) or results of seed decomposition at heating, were identified.

Functions of biological molecules (biomolecules) such as DNA and proteins, depend not only on their chemical composition, but also on the spatial structure of the molecules – the so-called configurational and conformational state. The study of configurational and conformational structure of biological molecules is an important task of biology, medicine. Monitoring the conformational transitions occurring during biologically important reactions, such as enzymatic catalysis, will allow a better understanding of the mechanisms of these reactions. Detection of biomolecules with pathological conformations in biological products (blood or cerebrospinal fluid samples) is important for early diagnosis of some diseases, such as prion infections., there are low-frequency molecular vibrations, in which large groups of atoms in a molecule move as a unit (e.g., nucleotide in DNA) in the THz frequency region The frequency of these oscillations depends on the structure of the whole molecule. Oscillation frequencies of hydrogen bonds supporting the structure of biomolecules also lie in the THz frequency range. Therefore, THz absorption spectra contain specific information about the structure and properties of hydrogen bonds. Note that circular dichroism spectra allow one to decipher the structure of molecules and to reconstruct relative positions of atomic groups, but contain no information about hydrogen bonds. The water absorption lines also lie in the THz region. These are dependent on vibrational-rotational transitions and relaxational dielectric losses. Water absorption bands allow to estimate the amount of water in the sample.

Applying methods of THz absorption spectroscopy is difficult due to high complexity of spectra interpretation. There are no detailed spectroscopic databases for polymer molecules in the THz region to date. It is impossible to calculate polymer absorption spectra, taking into account every electron and every nucleus in the molecule. The measurements of THz spectra of DNA and modeling polymers were carried out with using the high resolution THz spectroscopy method [5].

The preliminary investigations of gas-markers in THz frequency range have demonstrated the advantages of the high resolution THz spectroscopy methods for various applications. This approach could be promising to become the compact high precise analytical tool.

This work was carried out in the framework of state targets N 0035-2014-0206 and financially supported by the Russian Foundation for Basic Research (grant N 18-42-520050 r_a_povoljie, N 17-00-00184 KOMFI, N18-52-16017)

References

1. *V. Vaks*, High-Precise Spectrometry of the Terahertz Frequency Range: The Methods, Approaches and Applications// Journal of Infrared, Millimeter and Terahertz Waves. 2012, V. 33, N. 1, P. 43-53..

2. *Stepanov E V* Diodnaya Lazernaya Spektroskopiya i Analiz Molekul-Biomarkerov (Diode Laser Spectroscopy and Analysis of Biomarker Molecules) (Moscow: Fizmatlit, 2009)

3. *V L Vaks, E G Domracheva, E A Sobakinskaya, M B Chernyaeva*. Exhaled breath analysis: physical methods, instruments, and medical diagnostics //Physics-Uspekhi 2014, V. 57, N 7, P. 684 – 701

4._*V.L.Vaks, M.B.Chernyaeva, E.G.Domracheva and S.I.Pripolzin*. Multifrequency laser spectroscopy of biological tissues odors// Proceedings of V International Symposium TOPICAL PROBLEMS OF BIOPHOTONICS, 20 – 24 July, 2015, Nizhny Novgorod, Russia pp.218-219

5. *Vaks, V.L., Semenova, A.V., Guseva, Y.S., Panin, A.N.* Phenomenological model and experimental study of DNA absorption spectra in THz range //Opt. Quant. Electron. 2017, V. 49, N 5, P. 193 (1-24).

High-power Microwaves Against Locusts and Other Harmful Animals

V.E.Zapevalov

Institute of Applied Physics RAS, Nizhny Novgorod, Russia, zapev@appl.sci-nnov.ru

Locusts, acridia - some species of insects of the present locust family (Acrididae), capable of forming large flocks (up to hundreds of millions of individuals) migrating to considerable distances [1-3]. The damage it causes to plants, incomparable. Locusts can destroy all the plants in its path, and its invasion has always been a universal scourge. Innumerable flocks of these insects destroyed the crops, which inevitably led to mass starvation. A feature of biology of the locust is the presence of two phases - single and herd. Herd and single locust phases have significant differences, both in appearance and physiology, and in the nature of behavior. Locusts of a single phase usually have a protective coloration, a well-defined sexual dimorphism, and a low-active solitary lifestyle. Insects of the gregarious phase are colored more brightly and contrastively, adults are more adapted to flight. Locusts in this phase behave much more actively are formed larvae or adults. There is practically no sexual dimorphism in the herd phase. Locust has in its life cycle such stages: eggs, then larvae, the latter – imago (fig.1). The larvae moult 5 times and after the final moult they have wings, the imago stage begins.

The hordes of locusts are able to move at a speed of 15-20 km/h and fly without interruption for up to 20 hours in a row. Huge living clouds of locusts sometimes reach 10 km in width and up to 200 km in length. Two to three days of tailwind is enough for locusts to attack even very remote areas.

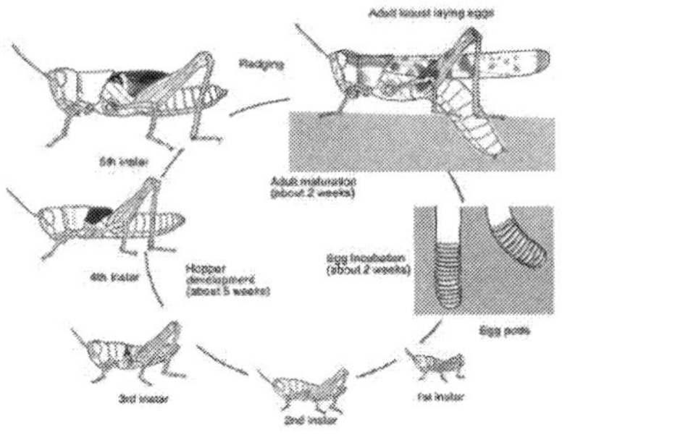

Fig. 1. Locusts: egg capsules; larvae of the 1st- 5th instar; adult insect.

The fight against this evil is conducted worldwide, but almost all used methods have a strong negative impact on the environment [1-3]. Of course, in the late XX -early XXI centuries researchers have the opportunity to obtain fundamentally new data on these insects with the help of molecular-genetic, biochemical and information methods. This especially applies to the mechanisms of the transition from a single phase to a herd and back, migrations of swarms and flocks, and so on. However, these opportunities are often not implemented. This is largely due to the fact that interest in these insects (as well as research funding) falls sharply after the suppression of another outbreak, when the danger for agriculture passes. The ultimate goal in locust control is the use of preventive and proactive methods that disrupt the environment to the least possible extent. This would make agricultural production easier and more secure in the many regions where growing crops is of vital importance to the survival of the local people.

In this publication we are going to demonstrate that for control and suppression of locust (and other harmful animals) it is appropriate to use of high-power microwave systems. Important issues are the selection of the frequency range and the required power level for creating mobile microwave complexes. These conditions are determined by technical and economic factors, availability and cost of necessary sources of microwave radiation, transportation (radiation and propagation) of electromagnetic waves, detection of insects and influence on egg capsules, kuligas (larvae), flocks (imago).

The characteristic number of individuals in the migratory flock of locusts is $N \sim 10^7$-10^9 pieces with a density of individuals n ~ 10^{-3}-10 m^{-3}. Mass of locust specimen is near ~ 3 g. The characteristic value of biomass in the flock is up to 3000 tons. The time of flock in flight is about ~ 12 hours per day. Locust locating can be remotely conducted by infrared radiation using thermal imagers since the temperature of the clusters is distinctly higher than the underlying surface. Radar and visual methods can be used also.

Important issue is the proper selection of the frequency range and required power levels for exposure to different insects. In the microwave range, the cell membranes become practically short-circuited. Therefore, the electrical properties of suspensions of cells and tissues with high water content are determined by the aqueous, salt and protein content of the intra- and intercellular environment. Fig. 2 shows dielectric properties of water which determine absorption of microwaves (pure water and with salt) as function of wavelength [4, 5].

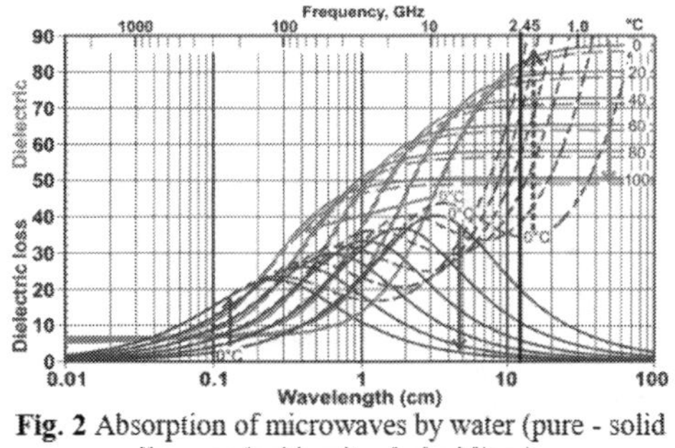

Fig. 2 Absorption of microwaves by water (pure - solid lines, and with salt - dashed lines).

Strong absorption of microwaves by water at the body of insects (pure and mainly with salt as hemolymph) effects on egg-caps, larvae and flocks of imago. In view of the specific structure of insects (fig. 3), effects of microwaves on them are mach stronger than for mammals. All vital locust organs are located close to the surface [1-3], within the skin layer, in contrast to mammals (and in general vertebrates) where 83% of the energy of this radiation is absorbed by the upper layer of the skin.

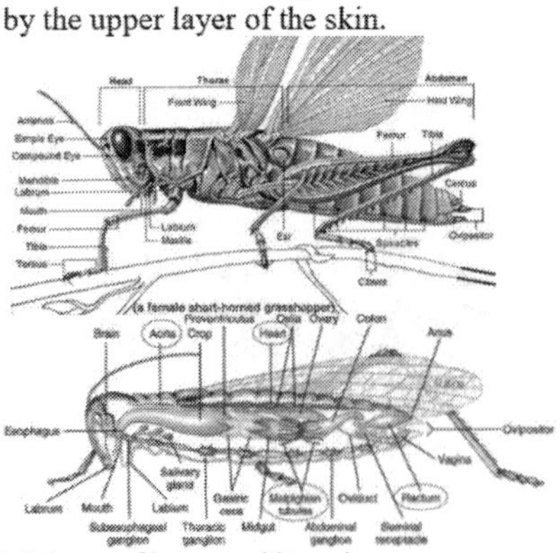

Fig. 3. Scheme of locust and its main organs

It is advisable for effects on eggs and larvae to use a high-efficient magnetrons and gyrotrons [6, 7] at the ISM frequencies (0.915, 2.45, 5.8, 24.125 GHz etc.). Requirements for such type devises for complexes against locust could be discussed. It is reasonable that due to technical and economical reason preferable power level now is 1-3 kW for magnetron (total power of complex near 100 kW) and order of 100 kW for gyrotron.

In adult stage (imago) locust clusters the clear advantages have gyrotron systems of mm range similar non-lethal weapons [8]. Fig. 4 shows vehicle-mounted Active Denial System (ADS). This advanced system uses CPI's VGB-8095 gyrotron to generate a 95 GHz, 100 kW beam which provides a safe but effective deterrent. Similar Chinese system, known as the Poly WB-1, uses millimeter-wave beams (near 30 GHz) to scald targets from up to a kilometer away.

a) b)

Fig. 4. ADS and gyrotron with cryomagnet.

It could be promising to use gyrotron operating at the harmonics of cyclotron frequency [7-10]. Fig. 5 demonstrates power and efficiency of gyrotron at the 2nd harmonics of cyclotron frequency (the TE_{03}, n=2, $\lambda_o \approx 12$ *mm*, $f_0 \approx 25$ *GHz*) vs. beam current. Some other acceptable version of gyrotrons for such system described at [10]. For those systems RF beam control by phased array antenna it will be practically useful possibility of frequency tuning of RF source [7-8].

There are a lot of other phytophagous and harmful animals of agriculture and forestry in our country and the entire world. The possibility of using similar microwave systems to combat other harmful animals, pests of agriculture and forestry could be developed also.

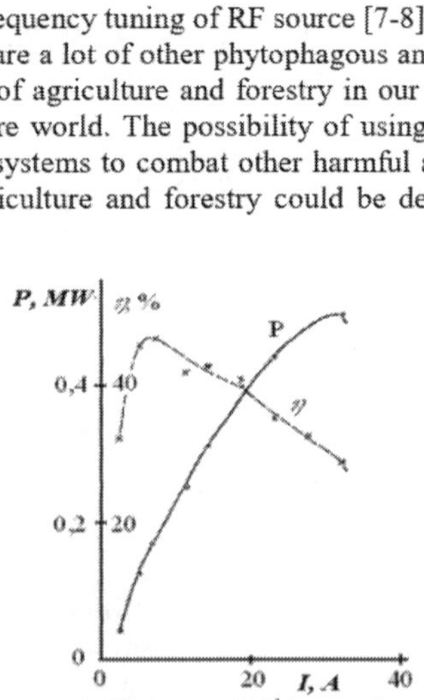

Fig. 5. Power and efficiency of 2nd harmonics gyrotron.

Summary

It is reasonable to use high-power microwaves to control locusts and other harmful animals.

Locust locating can be remotely conducted using thermal imagers because the temperature of the clusters is noticeably higher than the underlying surface.

Due to the peculiarities of the insects structure the microwaves influence for them is much stronger than for mammals.

To effect on he laying of eggs and larvae, it is advisable to use high-efficiency magnetrons and gyrotrons at ISM frequencies. At the adult stage, gyrotron systems, analogous to non-lethal weapons, have clear advantages. The most attractive versions of magnetron and gyrotrons complexes for above mentioned tasks were pointed out.

References

[1] *Resh V.H. (ed.), Carde R.T. (ed.).* Encyclopedia of Insects. Academic Press. 2003. p.666-669.

[2] *A. Steedman, (ed).* 1990. Locust Handbook (3rd edition). Chatham: Natural Resources Institute. 204 pp.

[3] *B.Uvarov,* 1966. Grasshoppers and locusts, V. I. Cambridge: University Press. 481 pp.

[4] *J. B. Hasted,* Liquid water: Dielectric properties, in Water. A comprehensive treatise, Vol 1, Ed. F. Franks (Plenum Press, New York, 1972) pp. 255-309.

[5] *Sh.E. Tsimring,* Electron beams and microwave vacuum electronics, John Wiley and Sons, Inc., Hoboken, New Jersey, 2007.

[6] *M.Thumm.* State-of-the-Art of High Power Gyro-Devices and Free Electron Masers. *KIT* Scientific Publishing, 2016.

[7] *V.E.Zapevalov,* Evolution of gyrotrons, Radiophysics and Quantum electronics, 2011, v. LIV, №8-9, pp. 559-572.

[8] *V.E.Zapevalov, S.A.Malygin, Sh.E.Tsimring,* Gyrotrons at the second harmonic of the cyclotron frequency. In the book: Gyrotrons, Gorky. IAP, 1980, pp.171-187.

[9] *A.N.Kuftin, V.A.Flyagin, V.K.Lygin, et al.* 5.8-62 GHz cw gyrotrons with warm and permanent magnets for technological application. Proc. of Int. workshop "Strong Microwaves in Plasmas", 1999, N. Novgorod, 2000. pp. 671-676.

The mechanism of the effect of microwave radiation on the parameters of homeostasis in living systems

A.G. Polyakova, A.G. Soloveva, M.V. Presnyakova, V.L. Kuznetsova, P.V. Peretyagin, I.E. Sazonova

Federal State Budgetary Educational Institution of Higher Education «Privolzhsky Research Medical University» of the Ministry of Health of the Russian Federation, Nizhny Novgorod, Russia, ag.polyakova@yandex.ru

One of the actual directions of modern Biomedicine is the study of physical and chemical mechanisms of action of electromagnetic radiation (EMR) on biological systems of various levels of organization [1,2]. An important aspect of the terahertz (THz) range is the presence of molecular spectra of radiation and absorption of the main molecules-metabolites (NO, O2, H2O), as well as large organic molecules: DNA, protein, etc., involved in the pathogenesis of various diseases, regulation of biochemical processes and physiological functions of the body [3,4]. Therefore, of particular interest is the study of the effect of EMR of the THz range on living systems [5-7].

The aim of the work was to experimentally study the effect of low-intensity electromagnetic radiation of the broad-band terahertz range on the metabolic parameters of rat blood in the process of reparative regeneration and the cellular system of hemostasis.

Materials and methods

The experiments were carried out on 15 male Wistar rats weighing 250-300 g. in accordance with the requirements of the Geneva Convention "International Guiding Principles for Biomedical Research Involving Animals" (Geneva, 1990) and blood cells of healthy volunteers (15 people) and 35 patients with burn disease. The impact of EMR was carried out in the direct sound irradiation EMR THz (110-170GHz) from the experimental emitter apparatus "AMFIT" (N.Novgorod).

Animals were divided into 3 equal groups: 1– intact healthy rats, 2 – control – operated animals without any effects, 3 – operated animals with irradiation EMR THz at a dose of 0.12 MJ within the next 7 days (10 minutes 1 time per day). The impact was carried out on the base area of the displaced flap, coinciding with the localization of the skin projection of the center of vegetative regulation. Surgical intervention in 2 and 3 groups of rats and withdrawal from the experiment on day 7 after the operation was carried out under intramuscular anesthesia (Zoletil + Xyla).

Blood stabilized with sodium citrate (1:9) was used for the studies. In plasma and erythrocytes, the activity of free radical oxidation processes was studied using the induced biochemiluminescence method on the biochemiluminometer BHL-06 (N. Novgorod). Chemiluminogenic evaluated total antioxidant activity (TAA) and the ability of a biological object to lipid peroxidation (LPO) with its intensity by the level of malondialdehyde in plasma and hemolysate of washed erythrocytes (1:10), activity of superoxide dismutase (SOD) in hemolysate of washed erythrocytes (1:10) and catalase in erythrocytes by spectrophotometry [8,9]. Glucose and lactate concentrations were measured on a Super GL ambulance device (Germany) in plasma and red blood cells. The activity of lactate dehydrogenase (LDH) was determined on the spectrophotometer Power Wave XS (Bio-Tek, USA).

In the second series of experiments coagulation and platelet units of hemostatic system were studied in 35 patients with thermal trauma and in 15 healthy people in vitro. The effect of EMR on blood cells was carried out in the mode of direct exposure to noise EMR THz (110-170GHz) ranges with exposure 1, 2, 3, 5, 10, 15, 20, 30 and 60 minutes. To assess the plasma level of hemocoagulation system, thromboelastography (TEG in citrated kaolin mode) was used on the TEG 5000 thromboelastograph ("Haemoscope Corporation", USA). Spontaneous platelet aggregation was studied using a laser aggregometer "Biola 230 LA" [6].

Statistical processing was carried out using the program Statistica, version 6.0, using the student and Wilcoxon criteria. Differences were considered statistically significant at $p<0.05$.

Result

In animals of the experimental group under the influence of EMR 110-170 GHz, in the range of which the spectra of nitric oxide and oxygen are located, a pronounced vascular effect was registered, which manifested the smallest zone of ischemia and necrosis in the flap compared to the control group of operated animals (necrosis zone was 23% of the flap area compared to 46% in the control). After the operation, the animals that did not receive radiation were found to have an increase in glucose and total cholesterol, as well as urea concentration in comparison with intact animals. When exposed to EMR 110-110GHz a significant homeostasis effect on biochemical metabolism parameters (total bilirubin, urea, total cholesterol and glucose) was registered).

Under the influence of THz irradiation in animals of the experimental group there was a significant decrease in e intensity of LPO by 14% ($p=0.003$), an increase in blood TAA (8% and 13% respectively) and activation of enzymes of bioradical protection: SOD (30%) and catalase (12%). This inactivates perekisnoe oxidation of lipids and helps to reduce free radical forms of oxygen, which inhibit the release of catecholamines from the nerve endings and adrenal glands [10]. At the same time, it is possible that one of the mechanisms of EMR THz is the suppression of hyperactivity of one of the most important stress-implementing systems of the sympathoadrenal system.

Thus, in the experimental group of animals, the prevalence of TAA over lipid peroxidation processes was observed, which may indicate the inhibition of the biological oxidation system due to the suppression of reactive oxygen species under the influence of EMR THz. In response to the irradiation of the vegetative regulation center projection, neurohumoral activation of the antioxidant system appears to occur, which blocks the processes of LPO. The revealed change in the direction of LPO processes under irradiation may be due to changes in the structure (conformation) of the cell surface of membrane components due to the weakening of hydrophobic bonds [11].

Exposure to EMR of THz in the noise range 110-170 GHz also contributed to a decrease in lactate concentration in plasma and red blood cells by 34% and 51%, respectively, compared with the control group animals. In parallel, LDH activity increased (by 47% and 23%) compared to the control group.

In addition, it is possible that electromagnetic radiation promotes the use of lactate in carbohydrate metabolism in the formation of 2,3-diphosphoglycerate, the main source of energy in the body.

Analysis of hemostasiological parameters dynamics under the influence of broadband radiation mode revealed activation of plasma hemostasis at 1 and 2 minutes of exposure ($p=0.004$ and 0.02, respectively). Dynamics of changes in coagulation activity in the period from 5 to 60 minutes recorded a gradual decrease in blood coagulation potential, reaching significant differences by 30 minutes compared to the baseline ($p=0.04$). The study of spontaneous aggregation of platelets was demonstrated in the first minutes of exposure to hypercoagulation effect after 30 minutes – hypocoagulation that corresponds to the physiological laws of development of responses of an organism to short the weak and long-strong incentives. A comparative study of the effect of broadband EMR of THz on coagulation and platelet system of hemostasis in patients with thermal injury and healthy people showed no significant differences. Study of the effect of EMR THz on spontaneous platelet aggregation.

Negative effects on the state of tissue blood flow, biochemical and hemostasiological parameters from the studied physical factors were not found.

Conclusion

Exposure to electromagnetic waves of the submillimeter range in the noise mode of radiation plays the role of control signals in the development of biological effects in the body. Based on the results obtained, it follows that low-intensity EMR of THz with a noise range of 110-170GHz in ischemic conditions leads to an increase in the antioxidant status of the blood, as well as inhibition of peroxide States. It is shown that EMR THz noise range of 110-170GHz increases the energy exchange of blood with tissue ischemia. The safety of EMR exposure to THz is evidenced by the absence of a negative side effect on the body of experimental animals.

The results are important for the development of methods of correction of poststress ischemic disorders, which can be successfully used in the process of medical rehabilitation.

Literature

1. Zotova E. A., Malinina Yu. A. et al. Biological effects of millimeter and submillimeter radiation // Izvestiya Samar. science. the centre Grew. Acad. Sciences, 2008. V. 10, No. 2. P. 636–641.
2. Betsky O. V., Kislov V. V., Kozmin A. S., Yaremenko Yu. G. et al. Terahertz waves and their application // Proceedings of the 17th international Crimean conference "Microwave equipment and telecommunication technologies", Ukraine, 2007. V. 2. P. 771–773.
3. Zhang X.C., Jingzhou Xu. Introduction to THz Wave Photonics / Springer, 2010. 248 p.
4. Vaks V. High-Precise Spectrometry of the Terahertz Frequency Range: The Methods, Approaches and Applications // Journal of Infrared, Millimeter and Terahertz Waves, 2012, V. 33, No. 1. P. 43–53.
5. Tsymbal A.A., Kirichuk V.F., Betsky O.V. Reactions of hemostatic system and fibrinolysis to electromagnetic waves of terahertz range // Millimeter waves in biology and medicine, 2011. No.2. P. 23–45.
6. Ponomarenko G.N. Innovative recovery technologies // Resort sheets. Scientific and information journal, 2010. No 5(62). P. 15–18.
7. Kirichuk V.F., Antipova V.F., Tsymbal A. A. et al. Influence of terahertz waves on complex living biological objects /edited by V.F. Kirichuk: Saratov state medical University. Saratov, 2012. 344p.
8. Sibgatullina G.V., Khaertdinova L.R., Gumerova E.A. et al. // Methods for determining the redox status of cultivated plant cells: teaching aid-Kazan: Kazan (Volga region) Federal University, 2011. 61 p.
9. Workshop on free radical oxidation. Educational and methodical manual / F.E. Putilina et al. SPb.:S.-Peterb. State. Univ.of Illinois, 2006. 108 p.
10. Chuyan E.N., Ravaeva M.Y., Nikolskaya V. A. et al.. Scientific notes of Taurida national University. V.I. Vernadsky. Series biology, chemistry 2013, No 26 (65). P. 223.
11. Bajinyan S.A., Meliksetyan A.M., Malakyan M.G. et al.. // Millimeter waves in biology and medicine 2003. No 4. (32). P. 12.

Biomedical applications of terahertz solid immersion microscopy

N. V. Chernomyrdin[1,2,3], A. S. Kucheryavenko[1,3], A. O. Schadko[1], G. A. Komandin[3], V. E. Karasik[1], V. V. Tuchin[4,5,6], and K.I. Zaytsev[1,2,3]

[1] Bauman Moscow State Technical University, Moscow 105005, Russia, chernik-a@yandex.ru
[2] Sechenov First Moscow State Medical University, Moscow 119991, Russia
[3] Prokhorov General Physics Institute of the Russian Academy of Sciences, Moscow 119991, Russia
[4] Saratov State University, Saratov 410012, Russia
[5] Institute of Precision Mechanics and Control of RAS, Saratov 410028, Russia
[6] Tomsk State University, Tomsk 634050, Russia

A number of terahertz (THz) imaging and spectroscopy techniques were proposed for biomedical purposes, i.e. for label-free differentiation of tumors and healthy tissues of different localizations: colon [1,2], breast [3,4], skin [5–7], brain [8,9] etc. Nevertheless, some problems are inherent to THz technology. We can note absence of efficient waveguides for the delivery of THz radiation and limited spatial resolution of THz imaging systems. The problem of low spatial resolution is especially important in biomedical applications of THz imaging as some tissue elements and structures have sub-wavelength dimensions. Several approaches had been already proposed for obtaining sub-wavelength resolution of THz imaging. Among them are near-field scanning-probe imaging [10], THz digital holography or synthetic aperture synthesis [11,12], which however require high-power sources, sensitive detectors or sophisticated computations. Another group of methods utilize effects of electromagnetic field localization in the shadow side of a mesoscale dielectric object, i.e. photonic jet effect [13,14] and solid immersion (SI) phenomenon [15,16]. SI microscopy implies focusing of an electromagnetic wave into the evanescent field volume behind a high-index material. In our work we propose SI microscopy assembly achieving 0,15λ resolution for handling and imaging of biological objects and soft tissues.

Developed SI lens include three optical elements: aspherical lens, forming converging beam with good aberrational correction in paraxial field [17], truncated sphere concentric to converging beam and a moveable plane window for biological tissue depositing. Truncated sphere and window were made of high-refractive index material (HRFZ-Si). Silicon window allows to deposit biological samples on the top of it during the scanning in X-Y directions. Favourable combination of aspherical singlet and HRFZ-Si truncated sphere provide 0,15λ resolution which was estimated theoretically using FDTD (finite-difference time-domain) simulations, and experimentally by means of continuous wave imaging of test object with step-like reflectivity distribution [18]. In order to demonstrate sub-wavelength resolution of SI lens for THz imaging of bioloigical objects and tissues we developed experimental setup based on backward wave oscillator as a continuous wave source of THz radiation and Golay cell coupled with mechanical chopper as a broadband detector.

We applied THz SI microscopy to study several types of biological objects and tissues, namely: leaf blades of such plants as mint [19] or poinsettia, artificially grown 300 μm tissue spheroids [20] and freshly excised human tissues of different localizations. In order to prevent tissue dehydration and i.e. minimize changes in THz properties during the measurements we embedded them into a gelatin slab. Gelatin embedding allows to conserve specimens for THz measurements for a long time after the excision. Fig. 1 demonstrate THz imaging of human breast tissue *ex vivo* obtained for the wavelength λ=500 μm. The specimen is formed by dense fibrous connective tissues with large single fat cells and their groups embedded into it. Though fat cells have sub-wavelength scales according to considered THz wavelength, they are clearly observed on the THz image 1(a).

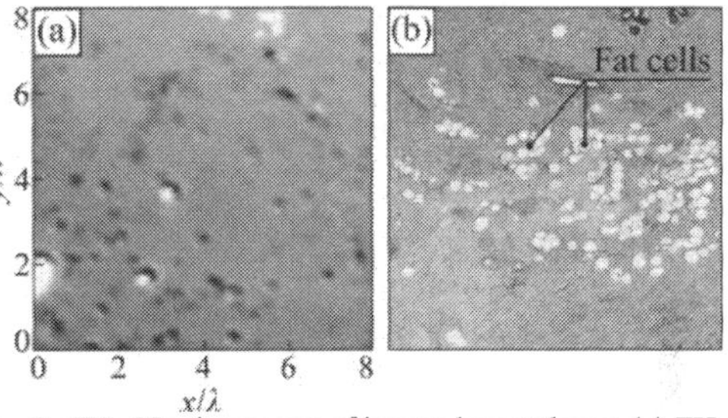

Fig. 1. THz SI microscopy of human breast tissue: (a) THz image observed on λ=500 μm; (b) histological photo.

Described configuration of THz SI microscopy provides the best of the known spatial resolution among SI imaging techniques 0,15λ. It also allows for handling of soft biological tissues during the measurements, thus making THz SI imaging technique a prospective tool for studying sub-wavelength scale inhomogeneity of biological objects.

Acknowledgements

The work was supported by Russian Science Foundation, project # 17-79-20346.

References

1. *Wahaia F. et al.* Terahertz absorption and reflection imaging of carcinoma-affected colon tissues embedded in paraffin // J. Mol. Struct. 2016. V. 1107. P. 214–219.

2. *Reid C.B. et al.* Terahertz pulsed imaging of freshly excised human colonic tissues // Phys. Med. Biol. 2011. V. 56, No. 14. P. 4333–4353.

3. *Fitzgerald A.J. et al.* Terahertz Pulsed Imaging of Human Breast Tumors // Radiology. 2006. V. 239, No. 2. P. 533–540.

4. *Ashworth P.C. et al.* Terahertz pulsed spectroscopy of freshly excised human breast cancer // Opt. Express. 2009. V. 17, No. 15. P. 12444.

5. *Woodward R.M. et al.* Terahertz Pulse Imaging of ex vivo Basal Cell Carcinoma // J. Invest. Dermatol. 2003. Vol. 120, № 1. P. 72–78.

6. *Zaytsev K.I. et al.* In vivo terahertz spectroscopy of pigmentary skin nevi: Pilot study of non-invasive early diagnosis of dysplasia // Appl. Phys. Lett. 2015. V. 106, No. 5. P. 053702.

7. *Zaytsev K.I. et al.* Highly Accurate In Vivo Terahertz Spectroscopy of Healthy Skin: Variation of Refractive Index and Absorption Coefficient along the Human Body // IEEE Trans. Terahertz Sci. Technol. 2015. V. 5, No. 5. P. 817–827.

8. *Meng K. et al.* Terahertz pulsed spectroscopy of paraffin-embedded brain glioma // J. Biomed. Opt. 2014. V. 19, No. 7. P. 077001.

9. *Chernomyrdin N.V. et al.* In vitro terahertz spectroscopy of gelatin-embedded human brain tumors: A pilot study // Progress in Biomedical Optics and Imaging - Proceedings of SPIE. 2018. V. 10716.

10. *Adam A.J.L.* Review of Near-Field Terahertz Measurement Methods and Their Applications // J. Infrared, Millimeter, Terahertz Waves. 2011. V. 32, No. 8–9. P. 976–1019.

11. *Zhang Y. et al.* Terahertz digital holography // Strain. 2008. V. 44, No. 5. P. 380–385.

12. *McClatchey K., Reiten M.T., Cheville R.A.* Time resolved synthetic aperture terahertz impulse imaging // Appl. Phys. Lett. 2001. V. 79, No. 27. P. 4485–4487.

13. *Nguyen Pham H.H. et al.* Asymmetric phase anomaly of terajet generated from dielectric cube under oblique illumination // Appl. Phys. Lett. 2017. V. 110, No. 20. P. 201105.

14. *Minin I.V., Minin O.V.* Terahertz artificial dielectric cuboid lens on substrate for super-resolution images // Opt. Quantum Electron. 2017. V. 49, No. 10.

15. *Mansfield S.M., Kino G.S.* Solid immersion microscope // Appl. Phys. Lett. 1990. V. 57, No.24. P. 2615–2616.

16. *Chernomyrdin N. V. et al.* Solid immersion terahertz imaging with sub-wavelength resolution // Appl. Phys. Lett. 2017. V. 110, No. 22. P. 221109.

17. *Chernomyrdin N.V. et al.* Wide-aperture aspherical lens for high-resolution terahertz imaging // Rev. Sci. Instrum. 2017. V. 88, No. 1. P. 014703.

18. *Chernomyrdin N. V. et al.* Reflection-mode continuous-wave 0.15 λ -resolution terahertz solid immersion microscopy of soft biological tissues // Appl. Phys. Lett. 2018. V. 113, No. 11. P. 111102.

19. *Chernomyrdin N.V. et al.* Terahertz solid immersion microscopy for sub-wavelength-resolution imaging of biological objects and tissues // Progress in Biomedical Optics and Imaging - Proceedings of SPIE. 2018. V. 10716.

20. *Chernomyrdin N. V. et al.* A potential of terahertz solid immersion microscopy for visualizing sub-wavelength-scale tissue spheroids // Proceedings of SPIE. 2018. V. 10677, No. 106771Y.

Intraoperative diagnosis of malignant brain gliomas using terahertz pulsed spectroscopy and optical coherence tomography

K.I. Zaytsev[1,2,*], I.N. Dolganova[2,3], N.V. Chernomyrdin[1,2], G.A. Komandin[1], M.A. Schcedrina[4], S.-I.T. Beshplav[5], S.A. Goryaynov[5], V.E. Karasik[2], I.V. Reshetov[4], A.A. Potapov[5], and V.V. Tuchin[6,7,8]

[1]Prokhorov General Physics Institute of the Russian Academy of Sciences, Moscow 119991, Russia;
[2]Bauman Moscow State Technical University, Moscow 105005, Russia;
[3]Institute of Solid State Physics of the Russian Academy of Sciences, Chernogolovka 142432, Russia;
[4]Sechenov University, Moscow 119991, Russia;
[5]Burdenko Neurosurgery Institute, Moscow 125047, Russia;
[6]Saratov State University, Saratov 410012, Russia, Russia;
[7]Tomsk State University, Tomsk, 634050;
[8]Institute of Precision Mechanics and Control of the Russian Academy of Sciences, Saratov 410012, Russia
*E-mail: kirzay@gmail.com

An intraoperative diagnosis of malignant brain gliomas remains a challenging problem of modern neurosurgery [1]. A complete resection of glioma is the most important factor, determining an efficiency of its treatment [2]; while an incomplete resection, caused by inaccurate detection of the tumor margins, increases a probability of its recurrence. The existing methods of the intraoperative neurodiagnosis of tumors are plagued with limited sensitivity and specificity, some of them remains laborious, time-consuming and/or rather expensive. Therefore, development of novel methods for the intraoperative diagnosis of gliomas relying on modern instruments of medical spectroscopy and imaging is a topical problem of medicine, physics, and engineering [3,4].

In our research, we studied a potential of terahertz (THz) pulsed spectroscopy (TPS) [5–9] and optical coherence tomography (OCT) [10–12] in the intraoperative diagnosis of brain tumors, with a strong emphasize on brain gliomas.

First, we performed *in vitro* THz spectroscopy of human brain gliomas of different grade. In order to fix tissues for the THz measurements, we applied the gelatin embedding, which allows to preserve tissues from hydration/dehydration; thus, sustaining the THz dielectric response of tissues unaltered for a long time after the surgery. We applied the laboratory TPS setup to study *in vitro* the refractive index and the amplitude absorption coefficient of intact tissues and gliomas. We observed significant differences between the THz characteristics of normal and pathological tissues of the brain [13]. This highlights a potential of THz technology in the label-free intraoperative diagnosis of brain tumors. Particularly, THz imaging and microscopy [14] can be applied for the intraoperative detection of the tumor margins in order to ensure its gross-total resection. Furthermore, the single-point THz spectroscopy and reflectometry could be integrated into modern neuroprobes, which are based on sapphire shaped crystals and yields combining the THz waveguiding [16,17] with diagnosis, aspiration and laser coagulation of brain tissues [18–20].

Second, we studied intact tissues and gliomas of the brain using OCT. For each tissue class, we analyzed a slope of the OCT signal in a depth. This slope reveals statistical difference between healthy and pathological tissues of the brain [10], which could be further emphasized using modern wavelet-domain denoising of OCT data [12]. Thus, the results of our study showcase a potential of OCT in intraoperative neurodiagnosis of brain gliomas.

Thereby, this work yields preliminary analysis (feasibility test), which aims to objectively uncover strengths and weaknesses of TPS and OCT from the viewpoint of their use in intraoperative diagnosis of human brain tumors before committing to a full-blown study involving measurements and analysis of a large amount of tissue sample, both *ex vivo* and *in vivo*.

This work was supported by the Russian Foundation for Basic Research (RFBR), Projects # 18-38-00504 and 18-38-00853.

References

1. *Ostrom, Q.T., Gittleman, H., Xu, J., Kromer, C., Wolinsky, Y., Kruchko, C., Barnholtz-Sloan, J.S.,* CBTRUS Statistical Report: Primary Brain and Other Central Nervous System Tumors Diagnosed in the United States in 2009-2013 // Neuro-Oncology, 2016, V.18, No.5, P.1–75.

2. *Phuphanich, S., Ferrall, S., Greenberg,* H., Long-term survival in malignant glioma. Prognostic factors // The Journal of the Florida Medical Association, 1993, V.80, P. 181–184.

3. *Pustogarov, N., Panteleev, D., Goryaynov, S.A., Ryabova, A.V., Rybalkina, E.Y., Revishchin, A., Potapov, A.A., Pavlova, G.,* Hiding in the Shadows: CPOX Expression and 5-ALA Induced Fluorescence in Human Glioma Cells // Molecular Neurobiology, 2017, V.54., No.7, P.5699–5708.

4. *Potapov, A.A., Goryaynov, S.A., Okhlopkov, V.A., Shishkina, L.V., Loschenov, V.B., Savelieva, T.A., Golbin, D.A., Chumakova, A.P., Goldberg, M.F., Spallone, A., Varyukhina, M.D.,* Laser biospectroscopy and 5-ALA fluorescence navigation as a helpful tool in the meningioma resection // Neurosurgical Review, 2016, V.39, No.3, P.437–447.

5. *Zaytsev, K.I., Kudrin, K.G., Karasik, V.E, Reshetov, I.V., Yurchenko, S.O.,* In vivo terahertz spectroscopy of pigmentary skin nevi: Pilot study of non-invasive early diagnosis of dysplasia // Applied Physics Letters, 2015, V. 106, P. 053702.

6. *Zaytsev, K.I., Gavdush, A.A., Chernomyrdin, N.V., Yurchenko, S.O.*, Highly Accurate in Vivo Terahertz Spectroscopy of Healthy Skin: Variation of Refractive Index and Absorption Coefficient Along the Human Body // IEEE Transactions on Terahertz science and Technology, 2015, V. 5, No.5, P. 817–827.

7. *Zaytsev, K.I., Chernomyrdin, N.V., Kudrin, K.G., Reshetov, I.V., Yurchenko, S.O.*, Terahertz spectroscopy of pigmentary skin nevi in vivo // Optics and Spectroscopy, 2015, V. 119, No.3, P. 404–410.

8. *Reshetov, I., Zaytsev, K., Kudrin, K., Karasik, V., Yurchenko, S., Shcherbina, V.*, Terahertz spectroscopy: Pilot study of non-invasive early diagnosis of dysplasia and melanoma // European Journal of cancer, 2015, V.51, No.3, P. S167.

9. *Zaytsev K.I., Katyba G.M., Kurlov V.N., Shikunova I.A., Karasik V.E., Yurchenko S.O.*, Terahertz Photonic Crystal Waveguides Based on Sapphire Shaped Crystals // IEEE Transactions on Terahertz science and Technology, 2016, V.6, No. 4, P. 576–582.

10. *Dolganova, I.N. Aleksandrova, P.V., Beshplav, S.-I.T, Chernomyrdin, N.V., Dubyanskaya, E.N., Goryaynov, S.A., Kurlov, V.N., Reshetov, I.V., Potapov, A.A., Tuchin, V.V. Zaytsev, K.I.*, Wavelet-domain de-noising of OCT images of human brain malignant glioma // Proceedings of SPIE, 2018, V. 10717, P. 107171X.

11. *Carrasco-Zevallos, O.M., Viehland, C., Keller, B., Draelos, M., Kuo, A.N., Toth, C.A. and Izatt, J.A.*, Review of intraoperative optical coherence tomography: technology and applications // Biomedical Optics Express, 2018, V.8, No.3., P.1607–1637.

12. *Dolganova, I.N., Chernomyrdin, N.V., Aleksandrova, P.V., Beshplav, S.-I.T, Potapov, A.A., Reshetov, I.V., Kurlov, V.N., ., Tuchin, V.V. Zaytsev, K.I.*, Nanoparticle-enabled experimentally trained wavelet-domain denoising method for optical coherence tomography // Journal of Biomedical Optics, 2018, V.23, No.9, P.091406.

13. *Chernomyrdin, V.N., Gavdush; A.A., Beshplav, S.-I.T.,Malakhov, K.M., Kucheryavenko, A.S., Katyba, G.M., Dolganova, I.N., Goryaynov, S.A., Karasik, V.E., Spektor, I.E., Kurlov, V.N., Yurchenko,S.O., Komandin,G.A., Potapov,A.A., Tuchin,V.V., Zaytsev,K.I.*, In vitro terahertz spectroscopy of gelatin-embedded human brain tumors: a pilot study // Proceedings of SPIE, 2018, V. 10716, P. 107160S.

14. *Chernomyrdin, N.V. Kucheryavenko, A.S. Kolontaeva, G.S., Katyba, G.M., Dolganova, I.N., Karalkin, P.A., Ponomarev, D.S., Kurlov, V.N., Reshetov, I.V., Skorobogatiy, M. , Tuchin, V.V., Zaytsev, K.I.*, Reflection-mode continuous-wave 0.15λ-resolution terahertz solid immersion microscopy of soft biological tissues // Applied Physics Letters, 2018, V.113, No. 11, P. 111102,

15. *Katyba, G.M., Zaytsev, K.I., Dolganova, I.N., Shikunova, I.A., Chernomyrdin, N.V., Yurchenko, S.O., Komandin, G.A., Reshotov, I.V., Nesvizhevsky, V.V. and Kurlov, V.N.*, Sapphire shaped crystal for waveguiding, sensing and exposure applications // Progress in Crystal Growth and Characterization of Materials (accepted, 2018).

16. *Zaytsev K.I., Katyba G.M., Kurlov V.N., Shikunova I.A., Karasik V.E., Yurchenko S.O.*, Terahertz photonic crystal waveguides based on sapphire shaped crystals // IEEE Transactions on Terahertz science and Technology, 2016, V.6, No. 4, P. 576–582.

17. *Katyba G.M., Zaytsev K.I., Chernomyrdin N.V. Shikunova I.A., Komandin G.A., Anzin V.B., Lebedev S.P., Spektor I.E., Karasik V.E., Yurchenko S.O., Reshetov I.V., Skorobogatiy M., Kurlov V.N. and Skorobogatiy M.*, Sapphire photonic crystal waveguide for terahertz sensing in aggressive environments // Advanced Optics Materials, 2018, DOI: 10.1002/adom.201800573.

18. *Shikunova I.A., Stryukov D.O., Rossolenko S.N., Kiselev A.M., Kurlov V.N.*, Neurosurgery contact handheld probe based on sapphire shapedcrystal // Journal of Crystal Growth, 2017, V. 457, PP. 265–269.

19. *Shikunova I.A., Zaytsev K.I., Stryukov D.O., Dubyanskaya E.N., Kurlov V.N.*, Neurosurgical sapphire handheld probe forintraoperative optical diagnostics, laser coagulationand aspiration of malignant brain tissue // Proceedings of SPIE, 2017, V. 10411, P. 104110Q

20. *Katyba G.M., Zaytsev K.I., Dolganova I.N., Shikunova I.A., Chernomyrdin N.V., Yurchenko S.O., Komandin G.A., Reshetov I.V., Nesvizhevsky V.V., Kurlov V.N.*, Sapphire shaped crystals for waveguiding, sensing and exposure applications // Progress in Crystal Growth and Characterization of Materials, 2018, accepted.

Study of PVC-based Skin Phantom with graphite particles in Terahertz Frequency Range

T. Zhang[1], M. K. Khodzitsky[1], P. S. Demchenko[1], A. V. Bykov[1,2], A. P. Popov[1,2] and I. V. Meglinski[1,2]

[1]Terahertz Biomedicine Laboratory, ITMO University, Saint Petersburg, Russia, tmzhang91@gmail.com
[2]Biophotonics Group, University of Oulu, Oulu, Finland

1. Introduction

The development of any optical imaging system requires the use of tissue-simulating objects (phantoms) to mimic human or animal tissues for device calibration. Calibration procedures need stable materials with accurately controlled optical properties. However, such a beneficial combination is hardly possible for real biological tissues and organs: due to extraction from their natural environment their properties change over time. Therefore, the phantoms being the substitutes for real biological objects are in high demand.

Many types of phantoms have been developed for THz technology in early studies. For the breast cancer, phantoms that were composed of water, oil, surfactant TX151, agar, nano-diamonds and nano-onions in varying proportions have been designed to represent the optical properties of breast tissue in THz frequency range [1], as well as the phantoms that were composed of water, lipid and collagen [2]. Using over-the-counter soft contact as a phantom of cornea has been reported in article [3]. The phantom that is made by porcine derived gelatin with water content from 83% to 95% was tested in article [4]. All these phantoms are hydrated phantoms due to their certain water content.

Since THz radiation is very sensitive to the presence of water, and water evaporation affecting the measurement, a water-free phantom needs to be developed for stable and constant usage. One of the popular and established approaches for tissue phantom manufacturing utilizes PVC as a matrix material [5]. These phantoms possess high flexibility, elasticity and serve as optical analogues of biotissue in visible and NIR spectral regions. In this article, a PVC-based, water-free phantoms to mimic human skin in THz frequency range was examined for the first time to our knowledge.

2. Sample Preparation

The proposed phantoms were originally designed for the optical and NIR spectral range and used for research in the field of optical coherence tomography (OCT) [5]. It was made using a scattering agent and a matrix material. The scattering agent was chosen to be zinc oxide (ZnO) nanoparticles with an average diameter of 0.35 μm at a concentration of 12 mg/ml to generate desired optical properties. A polyvinyl chloride-plastisol (PVCP) was used as a matrix material for fabrication. Some graphite was also added into the mixture to achieve enough absorbance. PVCP, graphite and ZnO nanoparticles were first mixed with each other. After sonication for 15 min, the mixture was then poured into a rectangular aluminum mold and put into an oven (180 °C) for 30 min. Finally, after cooling, the fabricated phantom is stored between glass slides to prevent PVC penetration into the plastic Petry dish [5]. Four phantoms with different graphite concentrations (10%, 12.5%, 16.7% 21.9%) were produced.

3. Experiment Setup and Signal Processing

To see, whether the phantom can be used as a proper replacement of human skin for medical research in the THz frequency range, the THz optical properties of the phantom should be assessed. Terahertz time-domain spectroscopy (THz-TDS) was used for this purpose. It is the most in-use THz device to measure sample's optical properties in the THz frequency domain [6]. It measures both the amplitude and phase of the THz pulse that is reflected by the sample or transmits through the sample [6]. With the help of Fourier transform, optical properties, such as refractive index, absorption coefficient and/or transmission coefficient can be finally obtained [6,7].

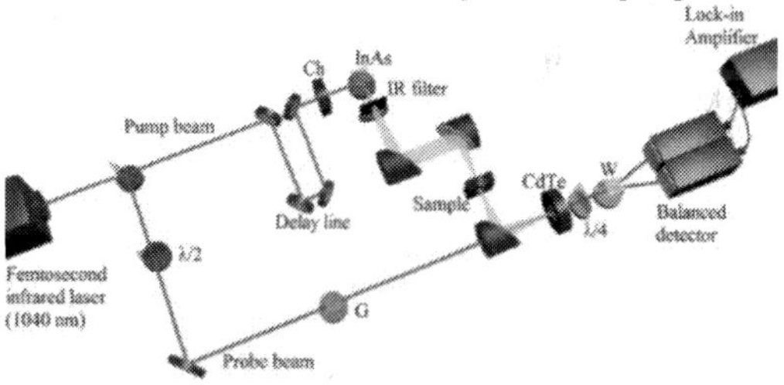

Fig. 2. The experimental setup of THz TDS system. λ/2 – half-wave plate, G - Glan prism, Ch - chopper, λ/4 - quarter-wave plate, W – Wollaston prism

The following formulas were used to calculate the refractive index and absorption coefficient basing on the spectrums and the phases we got from the fast Fourier transform (FFT) [6,7]:

$$\alpha(v) = -2\ln[\frac{T(v)E_{sample}(v)}{E_{reference}(v)}]/d \quad (1)$$

$$n(v) = 1 + c[\phi_{sample}(v) - \phi_{reference(v)}]/[2\pi v d] \quad (2)$$

$$T(v) = 1 - R = -[n(v)-1]^2/[n(v)+1]^2 \quad (3)$$

where α - the absorption coefficient, n - the refractive index of the sample, E and ϕ - the amplitude and the phase of the signal, d - the thickness of the sample, T - the Fresnel loss at the air-sample interface.

4. Results

After the calculation using formula (1) and (2), the refractive index and absorption coefficients of the phantoms were obtained and compared with the published data [8].

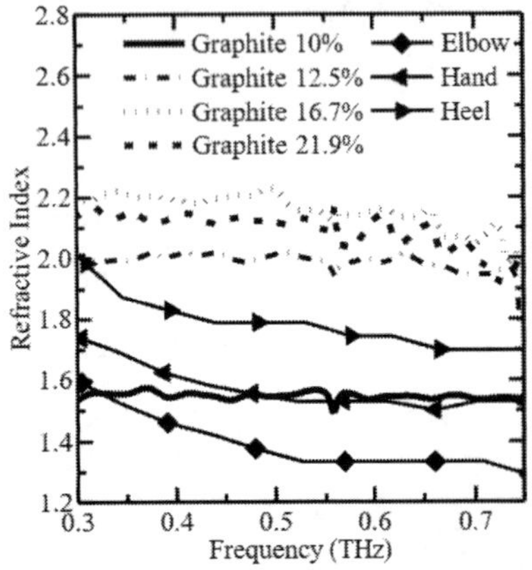

Fig. 3. Comparison of the refractive indices between phantom, elbow, hand and heel skin.

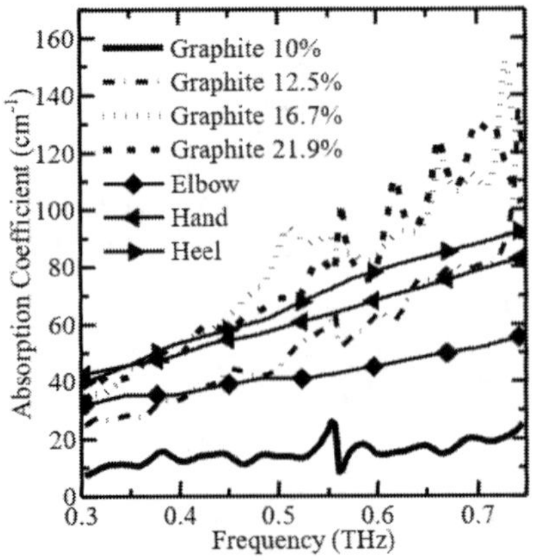

Fig. 4. Comparison of the absorption coefficients between phantom, elbow, hand and heel skin.

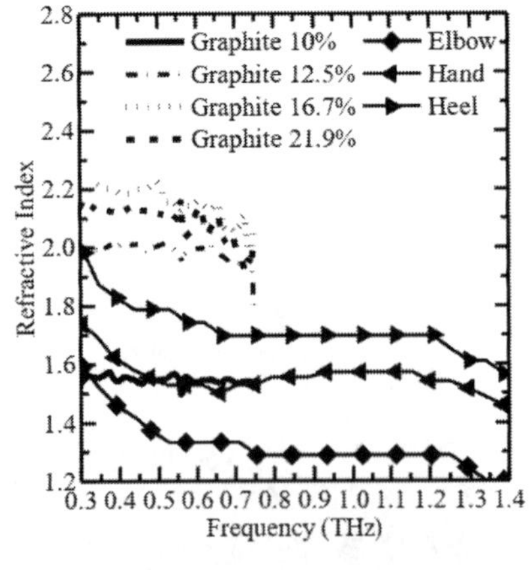

Fig. 5. Comparison of the refractive indices between phantom, elbow, hand and heel skin in a wider frequency range

Fig. 3 and Fig. 4 demonstrate the optical properties of the PVC-based phantom with graphite. The concentration of graphite influences on both refractive index and absorption coefficient of phantom. The higher the graphite concentration, the higher the value of optical properties. However, the influence of the graphite concentration also has a limit. The optical property differences between the graphite concentration of 16.7% and 21.9% are very small.

The refractive index of the phantom with 10% graphite shows good correspondence with the refractive index of hand skin in the frequency range of 0.45-0.75 THz. Fig.5 presents a good promise that in higher frequency range (>0.75THz) the phantom with 10% graphite may continue mimicking the refractive index of hand skin, since both refractive indices are nearly constant. Nonetheless, the absorption coefficient of PVC-based phantom with 10% graphite is much lower than that of any skin. The phantoms with 16.7% and 21.9% graphite possesses similar absorption coefficient as hand and heel skin in lower frequency range (0.3-0.5THz). The absorption coefficient of the phantom with 12.5% graphite is close to that of hand skin in higher frequency range (0.6-0.75THz).

5. Summary

The result proves that the presented phantoms with ZnO nanoparticles and graphite may achieve similar optical properties as human skin. The influence of different concentrations of graphites on the optical properties is also revealed. Such characteristic gives a good promise that by choosing proper concentrations of graphite, we may achieve an ideal phantom that can mimic the optical properties of human skin. Since the phantom is water-free, it can be preserved much longer than the phantom with water or lipid while the optical properties of the phantom still maintain.

Reference

1. *Bowman, T., Walter, A., et al.* phantom study of terahertz spectroscopy and imaging of micro-and nano-diamonds and nano-onions as contrast agents for breast cancer // Biomed Phys Eng Express. 2007. V. 3, No. 5. P. 055001.
2. *Truong, B.C.Q., Fitzgerald, A.J., et al.* Concentration analysis of breast tissue phantoms with terahertz spectroscopy // Biomed. Opt. Exp. 2018. V. 9. No. 3. P. 1334-1349.
3. *Sung, S., Garritano, J., et al.* Preliminary results of non-contact THz imaging of cornea // Proc. SPIE V. 9362. P. 93620C-1.
4. *Bajwa, N., Garritano, J., et al.* Reflective terahertz (THz) imaging: system calibration using hydration phantoms // Proc. SPIE. 2013. V. 8585, P. 85850W.
5. *Wróbel, M. S., Popov, A. P., et al.* Measurements of fundamental properties of human skin phantoms // J. Biomed. Opt., 2015. V. 20. No. 4. P. 045004.
6. *Dexheimer, S. L.* Terahertz Spectroscopy: Principles and Applications // CRC Press. 2007.
7. *Yasuda, H.* Measurement of terahertz refractive index of metal with terahertz time-domain spectroscopy // Jpn. J. Appl. Phys. 2008. V. 47. No. 3R. P. 1632-1634.
8. *Zaytsev, K. I., Gavdush, A. A., et al.* Highly accurate in vivo terahertz spectroscopy of healthy skin: variation of refractive index and absorption coefficient along the human body // IEEE Trans. THz Sci. Technol. 2015. V. 5. No. 5. P. 817-827

AUTHOR INDEX

AUTHOR INDEX

AUTHOR INDEX

AUTHOR INDEX

AUTHOR INDEX

AUTHOR INDEX